Tectonic Evolution of the Bering Shelf–Chukchi Sea–Arctic Margin and Adjacent Landmasses

Edited by

Elizabeth L. Miller
and
Arthur Grantz
Department of Geological and Environmental Sciences
Stanford University
Stanford, California 94305-2115
USA

and

Simon L. Klemperer
Department of Geophysics
Stanford University
Stanford, California 94305-2115
USA

THE
GEOLOGICAL
SOCIETY
OF AMERICA

Special Paper 360

3300 Penrose Place, P.O. Box 9140 ▪ Boulder, CO 80301-9140 USA

2002

Published by The Geological Society of America, Inc.
3300 Penrose Place, P.O. Box 9140, Boulder, Colorado 80301
www.geosociety.org

Printed in U.S.A.

GSA Books Science Editor Abhijit Basu
Cover design by Heather L. Sutphin

Library of Congress Cataloging-in-Publication Data

Tectonic evolution of the Bering Shelf–Chukchi Sea–Arctic Margin and adjacent
landmasses / edited by Elizabeth L. Miller, Arthur Grantz, and Simon L. Klemperer
    p. cm. — (Special paper; 360)
    Includes bibliographical references and index.
    ISBN 0-8137-2360-4
    1. Geological, Structural—Alaska. 2. Geology, Structural—Russia (Federation)—
Russian Far East. 3. Geology—Bering Sea Region. 4. Geology—Chukchi Sea
Region. I. Miller, Elizabeth L., 1951– II. Grantz, Arthur, 1927– III. Klemperer,
Simon. IV. Special papers (Geological Society of America); 360.

QE27.5.A4 T43 2002
551.8′09798—dc21

                2002024212

Cover: View looking southeast across the Kigluaik Mountains gneiss dome on Seward Peninsula, Alaska. The layering of the high-grade gneisses defines the geometry of the dome as seen in the distant ridgeline. The Kigluaik dome is cored by a 90 Ma mafic pluton capped by the granites in the foreground (see Chapter 7). The Kigluaik gneiss dome and its counterparts in Russia, the Koolen dome and Senyavin uplift, expose rocks that originated in the middle crust that were metamorphosed and exhumed during Cretaceous magmatism and extension. The reflective crust imaged in the center portion of the Bering-Chukchi deep crustal seismic reflection profile (see Chapter 1) may be similar to these gneisses, perhaps with a greater component of mafic sills and transposed mafic dikes. Photo by J. Amato.

10 9 8 7 6 5 4 3 2 1

Contents

Preface

Elizabeth L. Miller
Arthur Grantz
Department of Geological and Environmental Sciences,
Stanford University, Stanford, California 94305-2115, USA
Simon L. Klemperer
Department of Geophysics, Stanford University,
Stanford, California 94305-2115, USA

The slightly submerged Bering Shelf–Chukchi Sea–Arctic margin region comprises more than 50% of the total United States continental shelf and forms a broad isthmus of continental crust connecting the North American and Asian continents (Fig. 1). The crustal structure of the region has been shaped by Paleozoic, Mesozoic, and Cenozoic accretion related to Pacific plate margin tectonics and by Cretaceous plate motions in the Arctic that displaced crustal fragments of northwestern North America southward to form what is now the northern part of this shelf system. The many sedimentary basins across this shelf region evolved at disparate times and in various tectonic settings with respect to Pacific and Arctic plate margin tectonics. The region is currently undergoing slow localized deformation due to its position within the broad junction of the Asian and North American plates. A harsh climate and remote setting make this region and its adjacent landmasses difficult to work in; thus the overall geologic and tectonic evolution of the crust beneath this region, the continuity of geologic structures between North America and Russia, the tectonic setting of the region's basins, the plate tectonic origin of the Arctic Ocean, and the region's neotectonic history remain poorly known. This volume brings together a number of data-intensive papers that take our knowledge of this remote region a giant leap forward. This preface describes the general geologic setting of the Bering Shelf–Chukchi Sea–Arctic margin, and provides a short summary of the contributions to this volume and their relevance to the geotectonic history of this broad shelf region and surrounding landmasses.

The geologic structure of the upper crust beneath the Bering and Chukchi shelves, and the Aleutian basin of the Bering Sea, is portrayed in Plate 1. This compilation is based on seismic reflection data, exploratory drilling and stratigraphic test wells from the Bering and Chukchi shelves, and on dredge samples from the continental slope (the Beringian margin) of the Bering Shelf. Plate 1 provides the context for the chapters in this volume that discuss data acquired along the Bering-Chukchi Deep Seismic Transect and the ties of the seismic data to geologic features in adjacent parts of Alaska and northeastern Russia (Klemperer et al., this volume, Chapter 1). The tectonic setting of the Bering-Chukchi Shelf and its sedimentary basins is shown by displaying these data in the context of the tectonostratigraphic terranes and major structural features in the adjacent landmasses (Plate 1 and Klemperer et al., this volume, Chapter 19). The Bering and Chukchi shelves may be economically significant because they constitute by far the largest area of outer continental shelf in the United States, and they represent a significant portion of the Russian outer continental shelf as well. Although several exploratory wells drilled into the Alaska sectors of the Bering and Chukchi shelves have failed to encounter economic quantities of oil or gas, the petroleum potential of extensive parts of these shelves, underlain by significant thicknesses of sedimentary rocks, has not yet been fully evaluated (Grantz et al., 1987; Thurston and Theiss, 1987; Marlow et al., 1987; Craig et al., 1985; Sherwood et al., this volume).

In most areas, the offshore structural data shown for the sedimentary basins (Plate 1) were adapted from published sources cited in the plate explanation, but those from the Chukchi Shelf were interpreted by the authors from sources identified in Grantz et al. (1987, 1990). The tectonostratigraphic terranes of adjacent portions of Alaska and northeastern Russia were generalized from Silberling et al. (1994), Nokleberg et al. (1998), and other sources cited in Plate 1, including the geographic information system (GIS) database that is part of this volume (Klemperer et al., Chapter 19). Data for the stratigraphy of the offshore basins and exploratory and stratigraphic test wells of the Bering and Chukchi shelves, and of the dredge samples collected from the Beringian margin, are also summarized from sources cited in Plate 1.

The display of geologic features across an area as large and geologically diverse as the Bering and Chukchi shelves, in a manner that accurately conveys its rock units and upper-crustal structure, requires selectivity in the features displayed. Only faults and folds of significance are shown, and structural contours to basement are generalized. In particular, the disparity in age between the various sedimentary basins of the Bering and Chukchi shelves prevented a choice of a single chronostratigraphic or lithostratigraphic horizon as the structural contour

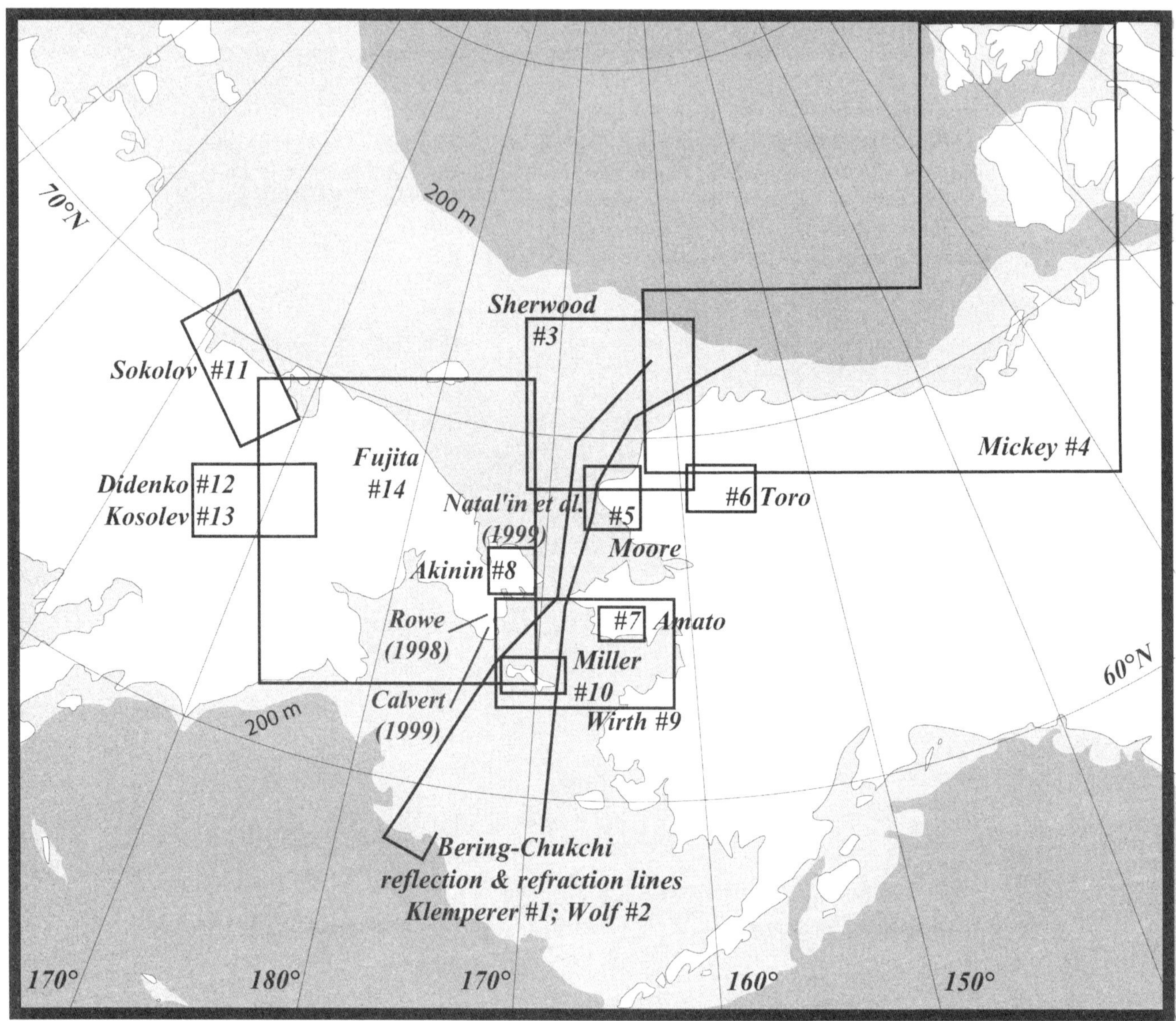

Figure 1. Index diagram for Alaska; northeastern Russia; and Bering, Chukchi, and Beaufort shelves showing approximate location of studies discussed in various chapters of this volume. Names are first authors and numbers are individual chapters. Regional studies by Blodgett et al. (Chapter 15), Dumoulin et al. (Chapter 16), Miller et al. (Chapter 17), Lawver et al. (Chapter 18), and Klemperer et al. (Chapter 19) are not shown because they cover entire region of map. The location of three additional field-based studies carried out as part of the larger seismic transect project include Rowe (1998), Calvert (1999), and Natal'in et al. (1999).

datum for the entire map area. The location and a synopsis of the stratigraphy of these basins are shown in Plate 1.

Structural contours in the Canada and Aleutian basins are drawn on the top of oceanic layer 2, which is overlain by Lower Cretaceous marine sedimentary strata in the Canada basin and by strata of unspecified Mesozoic age in the Aleutian basin. In the Hope and Norton basins of the Inner Chukchi and Bering shelves, and in the Navarin, Saint George, and Bristol Bay basins of the Outer Bering Shelf, the structural contours are also drawn on acoustic basement, but acoustic basement there is continental crust composed of several tectonostratigraphic terranes, which are overlain by marine and nonmarine sedimentary strata that are Paleocene to middle Eocene age at their base (Plate 1). In the Arctic Alaska basin of the North Slope and eastern Chukchi Shelf, acoustic basement is not a useful structural contour horizon because it is variable in its structural and stratigraphic position, and is commonly structurally complex. For this basin we chose to highlight the depth to the lowest stratigraphic horizon that displays only moderate structural complexity and that can be followed across the entire basin using public-domain seismic reflection data (cited in Grantz et al., 1987, 1990).

On the Chukchi Shelf, this horizon is the base of the Lisburne Group (Upper Mississippian) in the lower part of the Lower Mississippian to Lower Cretaceous Ellesmerian sequence of the Arctic Alaska basin (Grantz et al., 1987, 1990; Sherwood et al., Chapter 3).

The first chapter of this volume presents the main results of the Bering-Chukchi Deep Seismic Transect, the result of a multidisciplinary, multi-institutional, international collaborative effort with Russian scientists carried out with the goal of better understanding the evolution of the continental crust beneath the Bering-Chukchi-Arctic region. The project, funded by the Continental Dynamics Program of the National Science Foundation, involved the collection of two seismic reflection profiles (Fig. 1). The transect imaged the full thickness of the crust across the entire width of the continental shelf that connects North America and Asia (Klemperer et al., Chapter 1) (Plates 1 and 2). These deep seismic data provide a unique new perspective on the structure, thickness, age, and history of the entire crust beneath this broad continental shelf. The northern part of the line images intact, almost flat-lying sequences possibly as old as Precambrian and their underlying Precambrian continental crust beneath the northern Chukchi Sea. This continental crust and its sedimentary cover are part of the Arctic Alaska microplate, a fragment of rifted crust that was displaced southward from northwestern Canada during formation of the Canada Basin (Plates 1 and 2). The central part of the seismic line, from the vicinity of the Kotzebue Arch north of the Bering Strait to the Matthew Arch (Plate 1), images continental crust that exhibits prominent subhorizontal reflectivity throughout the lower crust and a sharp, well-defined Moho. This crust is believed to have been thickened during the Jurassic and Early Cretaceous Brookian orogeny and subsequently extended during Cretaceous to early Tertiary magmatic activity. The southern end of the transect images only sporadically reflective crust that thins beneath the large Cenozoic basins that developed beneath the Outer Bering Shelf. There is little or no evidence for the offshore continuation of the prominent right-lateral strike-slip faults of Alaska. The displacement on two of these faults, the Kaltag and Kobuk, appears to have dissipated in transtensional offshore fault systems that created the Norton and Hope basins (e.g., Worral, 1991).

Collection of the seismic reflection data along the transect represented an opportunity for additional geophysical experiments, including simultaneous collection of refraction data from shore-based stations to determine the crustal structure and velocity distribution beneath the Bering Shelf and Chukchi Sea (Wolf et al., Chapter 2). An important conclusion of this work, in conjunction with reflection data along the transect and gravity modeling, is that the crustal root that underlies the Brooks Range in eastern Alaska is absent along the projection of the range into the southern Chukchi Sea.

Sherwood et al. (Chapter 3) comprehensively synthesize data for the thick sedimentary cover that underlies the northern part of the transect in the Hanna Trough beneath the northern Chukchi Sea (Fig. 1; Plates 1 and 3–5). Mickey et al. (Chapter 4) summarize biostratigraphic data that support a rotational opening model for the Canada Basin, an event that created the Arctic continental margin of Alaska. This margin was imaged at the very northern end of the seismic transect (Plates 1 and 2).

The seismic reflection and refraction studies were augmented by geologic field work involving 10 separate field parties to points in northeastern Russia and westernmost Alaska (Fig. 1); these results are partially described in this volume. Moore et al. (Chapter 5) investigated the structural and thermal history of the Lisburne Hills thrust belt, which continues into the Chukchi Sea as the Herald Arch. Although post-Cenomanian thrust displacement is manifest in the Herald Arch and its foreland folds to the northeast based on offshore seismic reflection data, apatite fission-track data from onland exposures indicate that late Early Cretaceous uplift of thrust-faulted rocks on the Lisburne Peninsula is coeval with Brookian structures in the main Brooks Range to the east. This correlation is of interest because the Brooks Range is almost orthogonal in trend to the Lisburne Hills. Toro et al. (Chapter 6) detail the tectonic history of the west-central Brooks Range based on three seasons of geologic mapping and sampling. Here, the core of the Brooks Range is an elongate dome with significant structural relief that plunges westward, toward the Chukchi Sea. Early shortening (pre-112 Ma) is overprinted by tectonic exhumation ca. 90 Ma, related to the gravitational collapse of previously thickened continental crust. Late-stage north- to northwest-trending normal faults cut the Brooks Range structures and are similar in age and orientation to the Paleogene transtensional faults that created the Hope Basin, which underlies the southern Chukchi Sea. Chapters 7 and 8 describe structural and metamorphic aspects of the evolution of gneiss domes that occur on both sides of the Bering Strait. The Kigluaik dome of the Seward Peninsula, described by Amato et al. in Chapter 7, is a well-documented case of coeval but orthogonal structures formed by high-temperature flow in the crust. Data from the Kigluaik Mountains clearly indicate that the crust beneath this region was at elevated temperatures (granulite conditions in the mid-crust) and capable of flow during the Cretaceous. Geobarometry and geothermometry together with thermochronology for the Koolen dome in northeastern Russia document temperatures and pressures as high as >700°C and 4–5 kbar; uplift to shallow levels of the crust is bracketed between 104 and 94–88 Ma. These metamorphic culminations were created by the rise of mid-crustal rocks during Cretaceous magmatism and regional, mostly north-south oriented extension. Wirth et al. (Chapter 9) describe new data on Neogene basalt fields that straddle the seismic transect across a broad region of the Bering Shelf. These lavas carried mantle and crustal xenoliths from depth to the surface, thus providing hand samples of the crust and mantle imaged by the seismic profile. The abundance of gabbroic xenoliths attests that mafic intrusions are common in the middle to lower crust in the region of the transect. Miller et al. (Chapter 10) present preliminary U-Pb SHRIMP ages on individual zircons from gneissic plagioclase-pyroxene xenoliths collected by Wirth et al. (Chapter 9). These data suggest that the crust beneath the transect near Saint Lawrence Island was metamorphosed and mobilized under granulite facies conditions in the Late Cretaceous–Paleocene and that Cretaceous igneous rocks (ca. 90 Ma) are involved in this deformation and metamorphism.

Several chapters in the volume consist of regional biostratigraphic and tectonic studies that address broader aspects of the evolution of northeastern Russia and Alaska. Sokolov et al. (Chapter 11) synthesize geologic relationships along the South Anyui suture in northeastern Arctic Russia. This major suture zone is widely believed to represent part of the collision zone between the Arctic Alaska–Chukotka plate and Eurasia. The timing of events within the suture zone thus bears directly on the age of opening of the Canada Basin. Stratigraphic relations in the suture zone indicate that convergence that may be related to opening of the Canada Basin was completed by Early to middle Cretaceous time and may have been followed by post-Albian strike-slip faulting. This timing is compatible with evidence from the Canada Basin that seafloor spreading there ended by the beginning of Aptian time (e.g., Lawver et al., Chapter 18). Didenko et al. (Chapter 12) present paleomagnetic data bearing on the Jurassic-Cretaceous history of the Omolon massif, a large continental fragment involved in the greater Kolyma-Verkhoyansk deformational belt. The data are interpreted to indicate that the Omolon massif underwent translation from $60° \pm 10°$ in the Western Hemisphere via the polar region to $76° \pm 8°$ in the Eastern Hemisphere between Middle Jurassic and Early Cretaceous time, with a $30°–40°$ counterclockwise rotation relative to Siberia. Since the Early Cretaceous the Omolon massif has been a permanent part of Eurasia. Kolesov and Stone (Chapter 13) describe paleomagnetic data for Devonian strata of the Omolon massif that alternatively indicate that it could have been close to its present position during the Devonian, but rotated 90° with respect to the Siberian craton. Fujita et al. (Chapter 14) provide a summary of seismicity for the Chukotka Peninsula of northeastern Russia and surrounding regions. This seismicity defines modern plate boundaries and zones of deformation through this region and indicates that the Bering Shelf represents a fairly rigid subblock that is rotating clockwise with respect to North America about a pole in western Chukotka.

Blodgett et al. and Dumoulin et al. (Chapters 15 and 16) analyze megafossil and conodont data from Alaska, Russia, and North America. Their analyses show that during the early and middle Paleozoic, Alaska was more closely related biogeographically (and therefore probably paleogeographically) to Eurasia than to North America.

Miller et al. (Chapter 17) summarize the age of northern circum-Pacific magmatism and its tectonic setting through the Mesozoic, contrasting the tectonic history of the continental margins of northeastern Russia, Alaska, and the Cordillera. Although the boundaries and time spans of tectonic and magmatic events in the Cordillera and northeastern Russia appear to be coeval, the nature of magmatism and its tectonic setting often differ. For example, during general tectonic and magmatic quiescence in the U.S. and Canadian Cordillera between 150 Ma and 120 Ma, orogenesis, crustal shortening, and magmatism substantially modified the shelf margin of Siberia. Lawver et al. (Chapter 18) provide a summary of the plate kinematic evolution of all the individual terranes that form the present Arctic region, from the late Paleozoic to the present day. Their global reconstructions provide the framework for all the other regional and subregional studies. The final paper (Chapter 19) by Klemperer et al. is a description of the contents of the CD-ROM accompanying this volume. The CD-ROM contains a GIS-based compilation of geological and geophysical data for the Bering Shelf–Chukchi Sea and adjacent landmasses. Also included on the CD-ROM are an animation of the terrane motions on a globe at 3 m.y. intervals, and supplementary data and digital versions of figures from several other chapters.

ACKNOWLEDGMENTS

The Continental Dynamics Program, National Science Foundation, award EAR-93-17087 funded all or part of the research presented in Chapters 1, 2, 5, 6, 7, 8, 9, 10, 17, and 19. The Office of Polar Programs, National Science Foundation, grant OPP-99-05790 supported all or part of the research presented in Chapters 1, 10, and 18, and compilation of data for Plate 1, as well as the preparation of this volume. We are grateful to Exxon Exploration Company, Arco Oil and Gas Company, and Placer Dome Exploration, Inc., for support of the Russian-Alaska Tectonics Research Consortium at Stanford University, thereby helping the scientific interaction between Russian and American geologists that made this research and volume possible.

REFERENCES CITED

Bering Strait Geologic Field Party (BSGFP), 1997, Koolen metamorphic complex, NE Russia: Implications for the tectonic evolution of the Bering Strait region: Tectonics, v. 16, p. 713–729.

Calvert, A.T., 1999, Metamorphism and exhumation of mid-crustal gneiss domes in the Arctic Alaska terrane [Ph.D. thesis]: Santa Barbara, University of California, 198 p.

Craig, J.D., Sherwood, K.W., and Johnson, P.P., 1985, Geologic report for the Beaufort Sea Planning area: Regional geology, petroleum geology, environmental geology: Anchorage, Alaska, U.S. Department of the Interior, Minerals Management Service Outer Continental Shelf Report MMS85–0111, 192 p.

Grantz, A., May, S.D., and Dinter, D.A., 1987, Regional geology and petroleum potential of the United States Chukchi Shelf north of Point Hope, *in* Scholl, D.W., Grantz, A., and Vedder, J.G., eds., Geology and resource potential of the continental margin of western North America and adjacent ocean basins: Beaufort Sea to Baja California: Houston, Texas, Circum-Pacific Council for Energy and Mineral Resources, Earth Science Series, v. 6, p. 37–58.

Grantz, A., May, S.D., and Hart, P.E., 1990, Geology of the Arctic continental margin of Alaska, *in* Grantz, A., Johnson, L., and Sweeney, J.F., eds., The Arctic Ocean region: Boulder, Colorado, Geological Society of America, Geology of North America, v. L, p. 257–288.

Marlow, M.S., Cooper, A.K., and Fisher, M.A., 1987, Petroleum geology of the Beringian continental shelf, *in* Scholl, D.W., Grantz, A., Vedder, J., eds., Geology and resource potential of the continental margin of western North America and adjacent ocean basins: Beaufort Sea to Baja California: Houston, Texas, Circum-Pacific Council for Energy and Mineral Resources, Earth Science Series, v. 6, p. 103–122.

Natal'in, B.A., Amato, J.M., Toro, J., and Wright, J.E., 1999, Paleozoic rocks of northern Chukotka Peninsula, Russian Far East: Implications for the tectonics of the Arctic region: Tectonics, v. 10, p. 972–1003.

Nokleberg, W.J., Parfenov, L.M., Monger, J.W.H., Norton, I.O., Khanchuk, A.I., Stone, D.B., Scholl, D.W., and Fujita, K., 2000, Phanerozoic tectonic evolution of the circum–North Pacific: U.S. Geological Survey Professional Paper 1626, 136 p.

Rowe, H., 1998, Petrogenesis of plutons and hypabyssal rocks of the Bering Strait region, Chukotka, Russia [M.S. thesis]: Houston, Texas, Rice University, 90 p.

Silberling, N.J., Jones, D.L., Monger, J.W.H., Coney, P.J., Berg, H.C., and Plafker, G., 1994, Lithotectonic terrane map of Alaska and adjacent parts of Canada, *in* Plafker, G., and Berg, H.C., eds., Geology of Alaska: Boulder, Colorado, Geological Society of America, Geology of North America, v. G-1, plate 3, scale 1:2 500 000.

Thurston, D.K., and Theiss, L.A., 1987, Geologic report for the Chukchi Sea planning area, Alaska: Regional geology, petroleum geology, and environmental geology: Anchorage, Alaska, U.S. Department of the Interior, Minerals Management Service Outer Continental Shelf Report 87–0046, 193 p.

Worral, D.M., 1991, Tectonic history of the Bering Sea and the evolution of Tertiary strike-slip basins of the Bering Shelf: Boulder, Colorado, Geological Society of America Special Paper 257, 120 p.

Geological Society of America
Special Paper 360
2002

Crustal structure of the Bering and Chukchi shelves: Deep seismic reflection profiles across the North American continent between Alaska and Russia

Simon L. Klemperer*
Department of Geophysics, Stanford University, Stanford, California 94305-2215, USA
Elizabeth L. Miller
Arthur Grantz
*Department of Geological and Environmental Sciences, Stanford University,
Stanford, California 94305-2115, USA*
David W. Scholl
Department of Geophysics, Stanford University, Stanford, California 94305-2215, USA
Bering-Chukchi Working Group[†]

ABSTRACT

In 1994, 3750 km of crust-penetrating marine seismic-reflection profiles were acquired across the North American continent from the Aleutian basin to the Arctic Ocean. The two subparallel profiles cross the Bering and Chukchi shelves, passing through the Bering Strait between Russia and Alaska. The 40-fold, ≥15-s-penetration reflection data clearly image the major offshore sedimentary basins, lower-crustal layering, the reflection Moho, and rare sub-Moho reflectors.

The crust beneath the northern and southern segments is relatively nonreflective compared to the central part. By inference from onshore geology we associate the northern region, the Chukchi Shelf and Beaufort margin, with the Arctic Alaska–Chukotka cratonic block, and the southern region, the Outer Bering Shelf, with displaced continental and magmatic arc terranes. The central segment, extending from north of the Bering Strait to the Inner Bering Shelf south of Saint Lawrence Island, has distinctive highly reflective crust that we associate with plutonism in the middle to Late Cretaceous Okhotsk-Chukotsk magmatic belt and, in particular, with structures developed during its contemporaneous extensional history.

The reflection Moho is visible at traveltimes as great as 13.5 s (~39 km depth) beneath the Barrow Arch where sedimentary rocks are ~20 km thick. The Moho is at traveltimes as great as 13 s (~40 km?) beneath the Saint Matthew–Nunivak arch, a

---

*Corresponding author. E-mail: sklemp@stanford.edu.

[†]The Bering-Chukchi Working Group includes Jon R. Childs (U.S. Geological Survey, 345 Middlefield Road, Menlo Park, California 94025, USA), Nikita A. Bogdanov (Institute of the Lithosphere, Staromonelny Perevlok, 22 Moscow, 109180, Russia), Igor N. Belykh (Institute of Marine Geology and Geophysics, Yuzhno-Sakhalinsk, Russia; current address, Engineering Geophysics Company, Sakhalin, Russia), Helios Gnibidenko (Shirshov Institute of Oceanology, Moscow, 117851, Russia [deceased]), Brian Galloway (Department of Geophysics, Stanford University, Stanford, California 94305-2215; current address, Hewlett-Packard, 1501 Page Mill Road., MS 5LA, Palo Alto, California 94304, USA), Brian Hicks (Department of Geophysics, Stanford University, Stanford, California 94305-2215; current address, 2342 Yale Street, Palo Alto, California 94306, USA), Frances Cole (Department of Geological and Environmental Sciences, Stanford University, Stanford, California 94305-2115; current address, U.S. Department of Energy, National Energy Technology Laboratory, Morgantown, West Virginia 26507-0880, USA), and Jaime Toro (Department of Geological and Environmental Sciences, Stanford University, Stanford, California 94305-2115; current address, Department of Geology and Geography, West Virginia University, Morgantown, West Virginia 26506-6300, USA).

broad basement high underlain by Mesozoic and Paleogene arc rocks. The reflection Moho is visible at traveltimes as short as 10 s (32 km Moho depth according to re-fraction recordings) in the Bering Strait where crystalline basement is at the seafloor. It is at 11 s beneath the Navarin Basin, where the thickness of Tertiary strata is as much as ~8 km, indicating a likely crustal thickness of <25 km.

INTRODUCTION

The seismic profiles presented in this chapter (Fig. 1; for a more detailed location map see Plate 1 of the Preface) form the first continuous deep seismic reflection transect across the North American continent, and provide a first regional look at the entire crust of the Bering and Chukchi shelves between Alaska and northeastern Russia. The history of these regions is the composite result of the convergent and accretionary tectonics of the North Pacific rim and the extensional tectonics that began to form the Arctic Ocean in the Jurassic. For many years political and language barriers created artificial boundaries between Alaska and Russia. In actuality this region represents a single tectonic plate that includes parts of both North America and Russia (Mackey et al., 1997), and geologic units associated with the convergent and accretionary history of the Pacific margin and the North American Cordillera continue uninterrupted from Alaska to Russia (e.g., Nokleberg et al., 2000; Miller et al., this volume, Chapter 17). Hence, one of the most prominent puzzles presented by the geology of this region is why does the high topography of the North American Cordillera that extends from Mexico to Alaska diminish in the Bering Strait region?

The sedimentary basins in the Bering-Chukchi region have been explored for hydrocarbons by seismic surveys and exploratory wells (e.g., Craig et al., 1985; Turner et al., 1985, 1986; Thurston and Theiss, 1987; Scholl et al., 1987a; Worrall, 1991; Sherwood et al., this volume). In contrast, little was known (scattered well penetrations and a few dredge sites along the Beringian shelf edge; Plate 1, Miller et al., this volume, Preface) about the vast area of submerged continental crust that constitutes the older basement for these basins.

In this chapter a brief geological overview (organized from north to south) is followed by a review of previous crustal studies in the region and a description of our seismic data acquisition and processing. The main part of the paper presents a description and interpretation of significant reflections recorded along our seismic transect from north to south, including identification of the major sedimentary sequences, and a comparison of Moho reflections with existing refraction and potential field data. Interpretation of crustal type based on reflective character is also discussed.

GEOLOGIC OVERVIEW

The geologic history of Alaska (e.g., Plafker and Berg, 1994a) can be described by the progressive southward growth of continental crust by accretion and imbrication of continental-margin and oceanic sediments, allochthonous terranes, and magmatic arc systems and ophiolitic rocks that compose southern and central Alaska against a more rigid lithospheric block represented by the Arctic Alaska superterrane of northern Alaska. Although it is not apparent from surface geologic relations, the deep seismic data indicate that the Bering and Chukchi shelves (Fig. 1; e.g., see Scholl et al., 1987a; Plate 1, Miller et al., this volume, Preface) can be divided into three main parts with distinct crustal structure and crustal histories. These are shown schematically in Figure 1 (inset) as (1) Arctic Alaska–Chukotka plate, (2) mid-Cretaceous magmatic belt, and (3) Paleogene arc and displaced terranes.

1. The northern region consists of a displaced (rotated) fragment of the North American craton, the Arctic Alaska–Chukotka microplate, and overlying Proterozoic-to-Cenozoic sedimentary cover rocks. This area includes the offshore Chukchi and Beaufort shelves, the North Slope of Alaska, most of the Brooks Range, and the northern part of northeastern Rus-

Figure 1. Location map for Bering and Chukchi shelves. Heavy solid line from south to north (S–N): EW 94–10 deep seismic reflection tracklines shown in this chapter as line drawing (Fig. 2) and stack data (Plate 2). Numbers (0–9) are distances in hundreds of kilometers north and south of Bering Strait. Shotpoint numbers are given in Plate 1 of Miller et al., this volume, Preface. Heavy dashed line marked EW 94–10 indicates seismic reflection tracklines acquired in this experiment but not shown in this paper, including circle over Kotzebue Arch and dogleg southeast of Saint Lawrence Island, both shot in bad weather. Fine dashed lines marked EW 94–09 indicate deep reflection profiles acquired during preceding cruise EW 94–09 (McGeary et al., 1994). Dotted line marked Cooper et al. shows location of sonobuoy-refraction and gravity profile of Cooper et al. (1987). Dotted line marked Tolson shows location of industry reflection profile of Tolson (1987b, p. 131). Squares are land seismic stations that recorded onshore-offshore seismic data during cruises EW94–09 and EW94–10 (Saint Lawrence and north, Wolf et al., this volume; Cape Newenham and south, Fliedner and Klemperer, 1999, 2000). Offshore bathymetry is unshaded <200 m; shading density increases at 2000 and 4000 m. Offshore basins: sedimentary fill (Cenozoic south of Herald Arch, post-Devonian north of Herald Arch) is shaded where >3 km, and contoured at 2 km intervals (Kirschner, 1988, except Anadyr Basin: Worrall, 1991). Black circles (North Slope, Kotzebue, Norton, and Navarin Basins) are wells mentioned in text or figures, with identifying letters (north to south): B is South Barrow #1 (Collins, 1961); To is Topagoruk test well 1 (Collins, 1958); Pe is Peard #1; Tu is Tunalik #1 (Sherwood et al., this volume); PN is Point Nimiuk #1; CE is Cape Espenberg #1; No1 and No2 are Norton Sound COST#1 and COST#2; Na1 is Navarin Basin COST#1 (Plate 1). Gray ellipse (Seward Peninsula) and circle (Chukotka Peninsula) are schematic locations of gneiss domes. South Anyui suture is after Worrall (1991). Inset shows mid-Cretaceous magmatic belt after Miller et al. (this volume, Chapter 17).

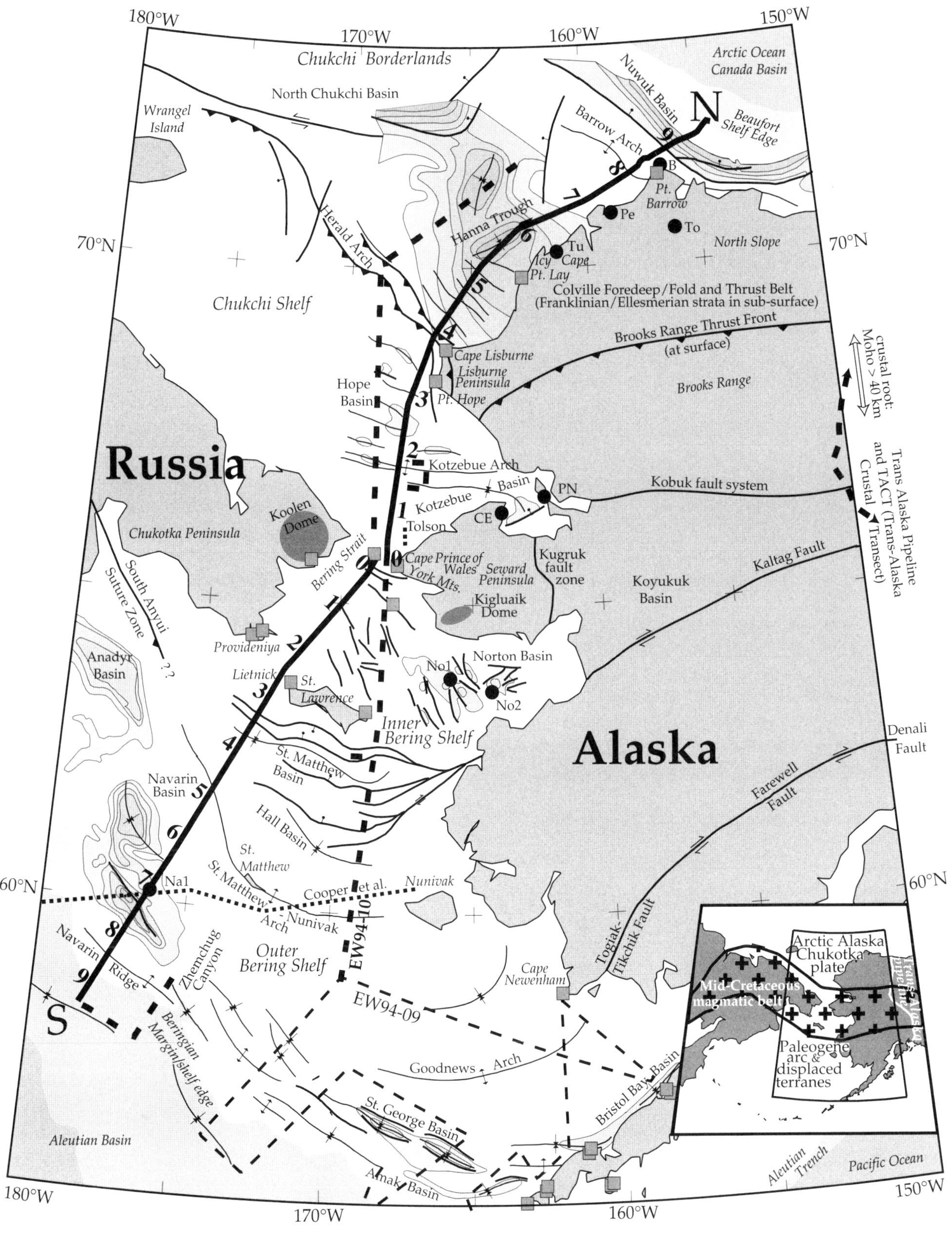

180°W
170°W
160°W
150°W
Chukchi Borderlands
Arctic Ocean
Canada Basin
North Chukchi Basin
Nuwuk Basin
Wrangel Island
N
Beaufort Shelf Edge
Barrow Arch
9
8
B
7
Pt. Barrow
Pe
70°N
Herald Arch
Hanna Trough
6
Tu
Icy Cape
To
North Slope
70°N
5
Pt. Lay
Chukchi Shelf
Colville Foredeep/Fold and Thrust Belt
(Franklinian/Ellesmerian strata in sub-surface)
Russia
4
Brooks Range Thrust Front
(at surface)
Cape Lisburne
Lisburne Peninsula
Brooks Range
Hope Basin
3
Pt. Hope
crustal root:
Moho > 40 km
2
Kotzebue Arch
Koolen Dome
Basin
PN
Kobuk fault system
Chukotka Peninsula
Kotzebue
CE
1
Tolson
Kugruk fault zone
Koyukuk Basin
Kaltag Fault
South Anvui Suture Zone
0
Cape Prince of Wales
York Mts.
Seward Peninsula
Kigluaik Dome
Trans Alaska Pipeline and TACT (Trans-Alaska Crustal Transect)
Bering Strait
1
Anadyr Basin
??
Provideniya
2
No1
Norton Basin
Alaska
Lietnick
3
St. Lawrence
No2
Inner Bering Shelf
4
St. Matthew Basin
Denali Fault
Navarin Basin
5
Farewell Fault
6
Hall Basin
St. Matthew
60°N
7
Na1
St. Matthew - Nunivak Arch
Cooper et al.
Nunivak
60°N
8
Navarin Ridge
Zhemchug Canyon
Outer Bering Shelf
EW94-10
Togiak-Tikchik Fault
Arctic Alaska Chukotka plate
9
S
EW94-09
Cape Newenham
Mid-Cretaceous magmatic belt
Beringian Margin/shelf edge
Goodnews Arch
Paleogene arc & displaced terranes
Aleutian Basin
St. George Basin
Bristol Bay Basin
Amak Basin
Aleutian Trench
Pacific Ocean
180°W
170°W
160°W
150°W

sia (e.g., Moore et al., 1994; Nokleberg et al., 2000). It is bounded on the north by oceanic crust of the Arctic Ocean.

2. The central region consists of the Inner Bering Shelf from north of the Bering Strait to south of Saint Lawrence Island, and includes the internal or southern portion of the Brooks Range orogen, the Yukon-Koyukuk region of interior Alaska, and the Seward Peninsula. Thick sequences of deformed middle to Late Cretaceous marine and paralic sedimentary rocks characterize the Yukon-Koyukuk region, which is underlain by accreted oceanic sediments and island-arc rocks and intruded by middle to Late Cretaceous (118–80 Ma) plutons (Patton et al., 1994; Nokleberg et al., 2000; Miller et al., this volume, Chapter 17). This region is inferred to have undergone variable, but locally high, amounts of extension in the Cretaceous (e.g., Miller and Hudson, 1991; Pavlis et al., 1992; however, this interpretation is not shared by all, e.g., Till et al., 1993).

3. The southern region consists of the Outer Bering Shelf, formed of continental and arc terranes assembled along transpressional fault boundaries or formed in place during Cretaceous-Paleogene convergence and strike-slip motion along the North Pacific rim. New arc material was intruded into preexisting continental material, some of Precambrian age, but largely much younger than the Arctic Alaska–Chukotka microplate to the north. Plate-boundary motion along the Beringian margin (the foot of the continental slope of the Bering Shelf) between the Kula oceanic and North American continental lithosphere ended in the early Eocene. Magmatism, subduction, and plate-boundary deformation jumped south at ca. 56 Ma from the Beringian margin to the Aleutians, trapping what is probably the former oceanic Kula plate to form the Aleutian basin of the Bering Sea (e.g., Cooper et al., 1987; Scholl et al., 1987b).

The following descriptions of the geology of these three regions or crustal provinces focus on geologic relations closest to the line of our seismic transect (line S–N in Fig. 1).

Northern region: Chukchi Shelf, Brooks Range, Herald Arch, and Hope Basin

The northern continental margin of Alaska and the Chukchi Shelf is the Beaufort passive margin, formed by Neocomian rifting and earliest Aptian (121 Ma) seafloor spreading in the Canada Basin as the Arctic Alaska–Chukotka microplate rotated counterclockwise away from the formerly contiguous North American craton (Lawver and Scotese, 1990; Embry, 1990; Grantz et al., 1990, 1998; Mickey et al., this volume; but see Lane, 1997, for a contrary model). The oldest beds at the base of the Cretaceous and Tertiary passive margin sequence of the Canada Basin on the Barrow Arch, its southern rift shoulder, are Hauterivian pelagites. This sequence progrades and thickens northward into the Canada Basin, and thins rapidly southward onto the Barrow Arch (Fig. 1). There it overlaps the northern part of an older south-facing sequence of paralic and marine stable shelf deposits of the Upper Devonian to lower Lower Cretaceous Ellesmerian sequence (Grantz et al., 1987; Plafker and Berg,

1994b) and, beneath a major Devonian unconformity, the underlying Ordovician and Silurian Franklinian sequence.

The upper Ellesmerian sequence consists of Jurassic marine clastic strata derived largely from the Barrow Arch, and the coeval Dinkum succession, deposited in the Dinkum rift-margin graben system on the north flank of the Barrow Arch (Grantz et al., 1990) during the early, prebreakup stages of rifting of the Canada Basin. The lower Ellesmerian sequence in northern Alaska consists of Lower Mississippian clastic rocks (Endicott Group) and Upper Mississippian to Triassic marine carbonates and clastic strata sourced in the Canadian Arctic Islands, from which they are now separated by the oceanic Canada Basin of the Arctic Ocean. The deepest part of this sequence could be Late Devonian. In the Topagoruk-1 test well (To in Fig. 1; Collins, 1958), the base of the Ellesmerian overlies with angular unconformity steeply dipping conglomerate and carbonaceous shale of Middle (Early?) Devonian age. Regional relations suggest that these beds unconformably overlie Ordovician and Silurian distal turbidites and graptolitic hemipelagites of the Franklinian sequence (usage after Lerand, 1973). This unconformity was attributed to the Ellesmerian orogeny by Moore et al. (1994), but Trettin (1991) considered the Ellesmerian orogeny to be latest Devonian–Early Carboniferous [Mississippian] in their type area in the Canadian Arctic Islands, so the unconformity correlates better with the Scandian phase of the Caledonian orogeny (late Early Silurian to Early Devonian [McKerrow et al., 2000]). During the Ordovician, Chukotka was separated from Arctic Alaska by a strait that connected the Iapetus Ocean and the paleo–Pacific Ocean, and was sutured to western Arctic Alaska in the Early Silurian (Nokleberg et al., 2000).

The south-facing passive continental margin of the Arctic Alaska microplate was deformed by north-vergent thrust faults during Middle or Late Jurassic to Early Cretaceous time, marking the closure of the Angayucham Ocean (Mayfield et al., 1988; Wirth and Bird, 1992; Moore et al., 1994). This compressional orogeny was succeeded by large-magnitude mid-Cretaceous extension or orogenic collapse (Miller and Hudson, 1991). The southern boundary of the Arctic Alaska–Chukotka plate is typically depicted in Russia as the South Anyui suture zone (Fujita and Newberry, 1982; Sokolov et al., this volume), and in Alaska it is placed along a large positive magnetic anomaly near the southern boundary of the modern Brooks Range (approximately along the younger Kobuk fault zone in Fig. 1). The anomaly marks the position of the updip end of a complex thrust sheet that contains the Upper Devonian to Lower Jurassic oceanic rocks of the Angayucham terrane (Grantz et al., 1991; Moore et al., 1994). On the basis of the analysis of Miller and Hudson (1991), the root zone or internides of the Brooks Range orogen (as distinct from the crustal root beneath the modern topography of the Brooks Range) must have extended an unknown distance south beneath what is now the Koyukuk basin.

The modern Brooks Range consists of a series of complexly deformed thrust sheets emplaced onto continental crust and continental-margin facies sedimentary rocks of the Arctic Alaska plate (Fuis et al., 1995). After Jurassic and Early Cretaceous thrust faulting, the Brooks Range shed large volumes of clastic debris northward into the Cretaceous-Tertiary Colville foredeep, which extends offshore onto the Chukchi Shelf (Grantz and May, 1987). Its sedimentary rocks are deformed by thrusting in the foreland of the Brooks Range fold and thrust belt in both the northern foothills of the Brooks Range and the eastern Chukchi Shelf (Moore et al., 1994). A thick Aptian-Cenomanian sedimentary section present beneath the western Colville Basin and the southern part of the Hanna Trough, referred to as the Corwin Delta, was derived from a western source area located beneath the Hope Basin and its environs (Moore et al., 1994). The Corwin Delta implies that significant topography, and hence a root to the Brooks Range, must have characterized this area in the Early Cretaceous, and that subsequent extension and collapse were required to drop the Hope Basin and its environs below sea level. Beneath the eastern Chukchi Shelf the southern margin of the Colville Basin is deformed by numerous thrust faults that are splays of the Late Cretaceous–Paleogene northeast-directed Herald thrust belt (Grantz and May, 1987). In the Lisburne Hills of westernmost Alaska, and perhaps beneath the Chukchi Shelf, this thrust belt may be superimposed on a mid-Cretaceous continuation of the Brooks Range orogen (Moore et al., this volume). Today the Herald Arch disappears southward, and the Brooks Range disappears westward, beneath the Eocene and younger deposits of the Hope Basin (Tolson, 1987a), which appears to be a transtensional basin geometrically linked to right-lateral displacement along the Kobuk fault zone (e.g., Tolson, 1987a; Worrall, 1991).

Central region: Bering Strait, Saint Lawrence Island, and Inner Bering Shelf

A broad basement high in the region around the Bering Strait and Saint Lawrence Island links the Chukotka Peninsula of Russia to the Seward Peninsula of Alaska (Fig. 1). East of the Seward Peninsula, across the Kugruk fault zone—an uncertain strike-slip or normal fault (Till and Dumoulin, 1994; Plafker and Berg, 1994b)—and south of the Brooks Range and the Kobuk fault system (Fig. 1) are mainly Cretaceous marine clastic sequences of the Koyukuk Basin. The basin is inferred to depositionally overlie the collapsed and extended root zone of the Brooks Range orogen (Miller and Hudson, 1991). The Seward Peninsula exposes mostly greenschist-facies slates, schists, marbles, and greenstone that locally preserve evidence for an older blueschist metamorphism (Till and Dumoulin, 1994). These are overprinted by biotite to granulite facies metamorphism of Cretaceous age in a series of gneiss domes such as the Kigluaik gneiss dome (Fig. 1; Amato et al., this volume; Akinin and Calvert, this volume) and are intruded by mid–Late Cretaceous plutons. Less metamorphosed lower Paleozoic carbonates underlie the York Mountains on the western tip of the Seward Peninsula (Fig. 1) (Till and Dumoulin, 1994).

Small exposures of Paleozoic and early Mesozoic rocks occur on Saint Lawrence Island and have been correlated to shelfal facies in thrust sheets of the Arctic Alaska superterrane and the overthrust Angayucham terrane in the Brooks Range of northern Alaska (Patton and Csejtey, 1980). These exposures are unconformably overlain by Cretaceous and early Paleocene volcanic rocks of intermediate composition and are intruded by Cretaceous plutons (Patton et al., 1976). The Cretaceous igneous rocks are part of a 6000-km-long mid-Cretaceous to locally Paleocene magmatic belt that is well developed in Russia and includes the Okhotsk-Chukotsk volcanic belt. Magmatism is believed to be at least partly related to subduction beneath the North Pacific margin, possibly above a southward-retreating subduction zone (Rubin et al., 1995; Amato and Wright, 1997; Bering Strait Geologic Field Party, 1997; Miller et al., this volume, Chapter 17; Amato et al., this volume). Late Cretaceous (65–77 Ma) and early Tertiary (61 Ma) arc magmatic rocks presumably linked to the Okhotsk-Chukotsk volcanic belt are found on Saint Matthew Island (Patton et al., 1976) and in dredge samples from the continental slope (Davis et al., 1989). On the Chukotka and Seward Peninsulas, Cretaceous magmatism was contemporaneous with ductile thinning and uplift of mid-crustal rocks in the Koolen and Kigluaik gneiss domes (Amato et al., this volume; Akinin and Calvert, this volume). These gneiss domes, or metamorphic culminations, involve Paleozoic and possibly older protoliths and are the result of extension within the magmatic belt and the overthickened root of the Brooks Range orogen.

Although most of the erosion and uplift of the Seward Peninsula (and by inference the Bering Strait basement high) occurred during the Cretaceous (Dumitru et al., 1995), modest onshore erosion and offshore formation of the Norton, Hope, and Saint Matthew–Hall basins starting in the Eocene testify to an extensional to transtensional environment in this region during the early Tertiary (Cooper et al., 1987; Marlow et al., 1987; Tolson, 1987a, 1987b; Worrall, 1991). Mild north-south extension and an unknown amount of strike-slip motion has continued to the present, as indicated by development of the Neogene Bering Sea basaltic province (Moll-Stalcup, 1994; Akinin et al., 1997) and normal-fault focal mechanisms from the Seward and Chukotka Peninsulas (Mackey et al., 1997; Fujita et al., this volume). The Quaternary basalt flows and cinder cones scattered across this portion of the transect (Moll-Stalcup, 1994; Wirth et al., this volume) extend to the Beringian shelf edge, as shown by exposures on the Pribilof Islands and dredge samples recovered from the continental margin near the Navarin Basin (Davis et al., 1993).

Southern region: Outer Bering Shelf

The Outer Bering Shelf is inferred to be underlain mostly by accreted terranes of the North Pacific, as suggested by rock units exposed in Alaska south of the Brooks Range and Seward Peninsula (Fig. 1), as well as Late Cretaceous and early Paleogene arc sequences. In a process beginning with the closing of the Angayucham Ocean, from Late Jurassic to early Eocene time, the ancestral circum-Pacific subduction zone migrated southward from a location beneath the Inner Bering Shelf to the modern Beringian shelf edge, and volcanism occurred in a broad subaerial magmatic belt during highly oblique subduction (Worrall, 1991). Saint Matthew Island, located between our two seismic transects (Fig. 1), is composed almost entirely of Paleogene igneous rock (Patton et al., 1976; Moll-Stalcup, 1994). Cenozoic and older terranes, developed at more southerly latitudes, were transported an unknown distance and subsequently accreted during the Late Cretaceous and Paleogene to southern and western Alaska by transpression along major dextral structures (Jones et al., 1977; Coney and Jones, 1985; Plafker and Berg, 1994b; Nokleberg et al., 2000). Dredge samples and seismic data show that the outermost Bering Shelf to at least long 174°W is composed of Peninsular terrane rocks (Naknek Formation) that may extend all the way to the Navarin Ridge (Marlow and Cooper, 1980; Jones et al., 1981; Worrall, 1991).

Major dextral features such as the Kaltag and the Farewell faults (Fig. 1) postdate the era of terrane accretion, which was mostly middle Cretaceous in the area of these faults. It remains a matter of debate whether these faults (1) ever extended across the entire Bering Shelf from northeast to southwest (e.g., Scholl and Stevenson, 1989); (2) rotate to form, or are truncated by, the postulated post-middle Eocene northwest-southeast transform trend parallel to the Beringian margin (e.g., Worrall, 1991); or (3) splay and die out under the Inner Bering Shelf beneath the Norton Basin (the Kaltag fault) or in southwestern Alaska (the Farewell fault).

The Beringian margin transform or highly oblique subduction zone was abandoned when subduction jumped southward to the modern Aleutian arc in the early Eocene (ca. 50–55 Ma), trapping most likely Kula plate oceanic crust between the Bering margin and the new arc to form the modern Aleutian basin of the Bering Sea in the process (Scholl et al., 1987b). Following this transition, transtension is thought to have continued to affect the outer edge of the Bering Shelf, forming a series of deep and elongate sedimentary basins, including the Bristol Bay, Saint George, and Navarin Basins (Cooper et al., 1987; Worrall, 1991). The largest and deepest of these, the Navarin Basin, contains to 12 km of late Eocene and younger strata (Worrall, 1991).

PREVIOUS CRUSTAL STRUCTURE STUDIES

Our only detailed knowledge of the deep-crustal structure of Alaska comes from the Trans-Alaska Crustal Transect (TACT) seismic profiles acquired intermittently along the Trans-Alaska Pipeline (Fig. 1), east of our marine transect. The TACT wide-angle reflection-refraction profile between the Kaltag fault and the Denali fault at 64°–65°N, 1000 km east of our marine profiles (Beaudoin et al., 1992) (Fig. 1), shows a thin crust, only ~30 km thick, with a low average velocity, ~6.1 km/s. Beaudoin et al. (1992) noted that this crust of the Yukon-Tanana terrane is clearly reflective from ~10 to 27 km depth (equivalent to ~3.5–

9 s). Although their wide-angle data cannot be directly compared to our reflection profiles, their data require velocity layering that is 0.2–0.5 km thick, a scale that would also be strongly reflective on our marine reflection profiles. Both the region of central Alaska sampled by the TACT transect and the central region of our marine transect are within the middle to Late Cretaceous magmatic belt defined by Miller et al. (this volume, Chapter 17). According to Pavlis et al. (1992), the Yukon-Tanana terrane of central Alaska was thinned by a factor of 2 to 4 during mid-Cretaceous extension, although as much as half of the modern crustal thickness may represent subsequent tectonic or magmatic underplating.

The TACT refraction profile across the Brooks Range (Fuis et al., 1995) extends from just south of the Kobuk fault to the North Slope at ~69°N, 800 km east of our marine transect (Fig. 1). Crustal thickness is ~37 km at the north and south ends of the profile, but Fuis et al. (1995) recognized a >100-km-wide zone of crustal thickening, the Moho depth being 50 km beneath the highest modern topography of the Brooks Range (Fig. 1). The collocated TACT low-fold near-vertical reflection profile shows that the Brooks Range is characterized by south-dipping thrust duplexes in the upper crust, overlying in turn a poorly reflective middle to lower crust, and a basal-crustal reflective layer ~5 km thick (Fuis et al., 1995). The poorly reflective middle crust of the Arctic Alaska microplate is 30 km thick beneath the North Slope, but thins south beneath the Brooks Range allochthons, apparently forming a crustal-scale wedge between the allochthons and the basal reflective layer. On the basis of this interpretation of the reflection character of the TACT profile, the southern limit of the Arctic Alaska microplate that is autochthonous with respect to Brooks Range orogenesis is at about the southern limit of the crustal root shown schematically in Figure 1. South of the modern Brooks Range, the crust appears to lack the less-reflective middle crust. The data of Fuis et al. (1995, their Fig. 2a) suggest that the crust below the Koyukuk basin is continuously reflective from 4 to 12 s, although their low-fold data acquisition makes it difficult to directly compare their data with our conventional reflection profiles.

Knowledge of the crustal structure of far-eastern Russia is rudimentary, and no active-source deep-crustal profiles have been recorded. Mackey et al. (1998) used regional earthquake recordings to model a 37 km crustal thickness at long 160°E, lat 62°–65°N. Older Russian studies of converted earthquake phases summarized by Wolf et al. (this volume) suggest crustal thicknesses of 32–39 km in the Chukotka Peninsula.

Although the sedimentary basins of the continental shelves between Russia and Alaska have been extensively explored for potential hydrocarbon reserves by seismic-reflection studies (e.g., Craig et al., 1985; Thurston and Theiss, 1987; Scholl et al., 1987a; Worrall, 1991), no crustal-penetrating profiles had been acquired, despite attempts by Cooper et al. (1987), prior to those described herein and by Wolf et al. (this volume). A regional surface-wave model of Jin and Herrin (1980) provides an average crustal thickness of 28 km for the Bering Shelf, a value that masks the considerable lateral variation shown here. Cooper et al. (1987) combined upper-crustal refraction velocities with gravity modeling to suggest crustal thicknesses in the Outer Bering Shelf ranging from 20 km beneath the Navarin Basin to 43 km beneath the Saint Matthew–Nunivak arch, but such values depend strongly on assumed densities.

ACQUISITION, PROCESSING, AND INTERPRETATION OF SEISMIC DATA

The RV *Maurice Ewing*, operated by Lamont-Doherty Earth Observatory on behalf of the University National Oceanographic Laboratory Systems, acquired 3750 km of seismic data (Fig. 1) using a 4 km, 160 channel streamer and a 20 gun, 8355 in^3 airgun source. The shot interval was either 50 m (yielding 40-fold data) or 75 m (27-fold data), the larger interval being used along line segments where strong currents prevented full recording times with 50 m shot spacing. Record lengths varied from 15 s to 23 s. Absolute shot times were recorded to enable subsequent analysis of the seismic signals recorded at offsets from 10 to >250 km at 11 land sites in Alaska and Russia around the Bering Strait and on the northwestern coast of Alaska (Fig. 1; Brocher et al., 1995; Wolf et al., this volume). Gravity and magnetic data (Fig. 2) and sonobuoy and 3.5 kHz echo-sounder data were also acquired along the tracklines.

A composite reflection profile from the Aleutian basin to the Arctic Ocean (S–N in Fig. 1) is shown as an annotated line-drawing (Fig. 2) and as an interpreted stack section (Plate 2). The seismic data were processed by CTC Pulsonic, now part of Scott-Pickford, Inc., following the processing sequence listed in Plate 2. Interpretations described in the text were made on 1:100 000 scale stack sections, then transferred to the 1:400 000 scale sections of Plate 2. Greater detail is evident on the 1:100 000 scale sections than on the reduced-scale sections of Plate 2, because of the small scale of reproduction and because the data in Plate 2 have been low-pass filtered and trace summed to match the resolution of the seismic data to that of the printing process at small scales (e.g., Klemperer and Hobbs, 1991). Our seismic data are shown as unmigrated time sections, so interpreters must be aware of the need to convert time to depth in a nonlinear way dependent on the velocity structure of the Earth, and must recognize that dipping reflections have lower dips and are downdip of the reflectors that produced them (e.g., Klemperer and Hobbs, 1991).

An important limitation on comparisons of different segments of our reflection profiles is the inevitable variability in reflection quality due to varying weather and sea state, partial equipment failure or changes in acquisition parameters, and perhaps most important, to changes in seafloor conditions. The region where our data quality suffered most from poor weather is near the Kotzebue Arch. Data from km +100 to +145 (in Plate 2 distances are given north [positive km] and south [negative km] of the Bering Strait) were acquired as a gale developed, causing noise on the recording streamer; data from km +145 to +200

were acquired after a 24 h delay as the gale abated. Fortunately, data from the parallel profile acquired ~40 km to the west 1 week later in good weather provide confirmation of the general features of the image shown in Plate 2. The strong coastal current north of Wainwright (km +750) and changes in course to avoid sea ice north of Point Barrow (km +900) led to fluctuations in streamer depth and consequent variations in the frequencies recorded. Use of a 75 m shot interval from km +0 to +145 and from km +735 to +980 provided an effective source energy in the resulting 27-fold stack trace that is about one-third less than that resulting from the 50-m-interval, 40-fold data acquired in most of the survey. Changes in seafloor geology from gassy unconsolidated sediments to exposed crystalline basement result in large changes in near-surface absorption of seismic energy, leading to abrupt changes in apparent lower-crustal reflectivity that presumably do not correspond to real changes in lower-crustal composition and structure. Examples can be seen near km –45 and km –175 in the seismic section in Plate 2.

Throughout this chapter, references are made to the (two-way) traveltime (in seconds) to reflections, and to the depth (in kilometers) of the corresponding reflector. These depths are obtained by multiplying one-half the (two-way) traveltime by the appropriate average seismic velocity. For shallow reflections (typically sedimentary reflections) we can directly measure the appropriate seismic velocity from the shape of the reflection traveltime curve (the normal-moveout curve) (e.g., Hatton et al., 1986; Yilmaz, 1987). For reflections deeper than the maximum source-receiver offset (4.2 km), such estimates of velocity become progressively more unreliable, and the appropriate seismic velocity is best estimated from nearby refraction surveys, such as those of Houtz et al. (1991), Cooper et al. (1987), and Wolf et al. (this volume). In the absence of nearby refraction profiles, averages from global compilations (e.g., Christensen and Mooney, 1995) provide a crude guide to the typical range of seismic velocity at different depths. Estimates of depth from reflection profiles unsupported by coincident refraction data have probable uncertainties ranging from a few percent within sedimentary basins to ±10% for crustal thickness.

DESCRIPTION OF DATA

All features described in this section are indicated in Plate 2 and may be located by their distance north (+km) or south (–km) of the Bering Strait. Plate 2 also shows shotpoint numbers along the profiles that may be tied to the shotpoint locations in Plate 1 of Miller et al. (this volume, Preface). Regional basins and structures are shown in Plate 1 of Grantz et al. and on the line drawing in Figure 2.

Northern region: Chukchi Shelf north of Herald Arch (Plate 2, panel A and right portion of panel B)

Beaufort rifted margin. The northern limit of our transcontinental transect is at 190 m water depth close to the modern

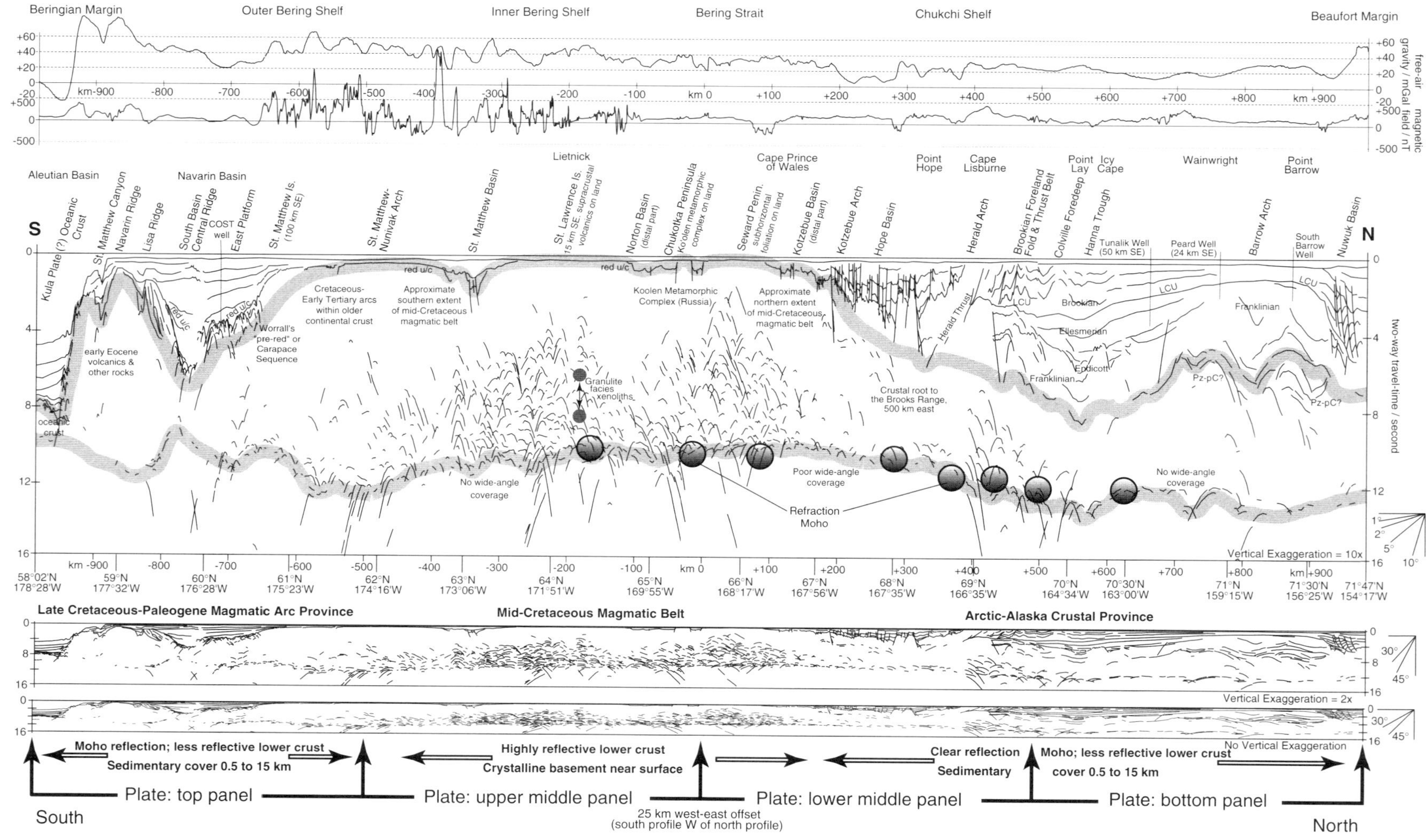

Figure 2. Summary line drawing of reflection transect from Beringian margin to Arctic margin (S–N in Fig. 1), crossing entire North American continent. Note that line drawing comprises south half of western transect merged at Bering Strait with north half of eastern transect (same data as shown in Plate 2). Bottom section is true scale for crustal velocity of 6 km/s; middle and upper sections are 2 and 10 times vertically exaggerated, respectively. Thick gray lines in upper panel: upper line is base of stratified sedimentary reflections (approximate base Franklinian below Chukchi Shelf and Beaufort margin; approximately middle Eocene [red unconformity] on Bering Shelf); lower line is reflection Moho estimated from our reflection transect (Plate 2). Small gray circles in center of upper panel indicate depth range of granulite xenoliths from Saint Lawrence Island (after Wirth et al., this volume; Miller et al., this volume, Chapter 10). Large circles in north half of upper panel are estimates of traveltime to refraction Moho (from Wolf et al., this volume). LCU is Lower Cretaceous unconformity; red u/c is Eocene red unconformity. Free-air gravity and magnetic profiles recorded along ship track, smoothed with ~1 km running filter, are shown at top of figure.

Beaufort shelf edge, and shows a progradational prism of sedimentary rocks composed of Cretaceous and Tertiary marine strata deformed by numerous listric normal faults. The listric faults have a predominantly northerly dip, and were presumably driven by gravitational body forces created by the free face of the adjacent continental slope. Tertiary strata extend to 2.5 s at the north end of our transect and overlie, in part unconformably, more-strongly rotated Cretaceous strata. The listric faults that rotate the Cretaceous strata sole into a listric normal fault of lower angle at the approximate position of the Lower Cretaceous unconformity (breakup unconformity) (see Fig. 2 and Plate 2) (Grantz et al., 1990), and extend upward through the Tertiary beds to within 0.1 s (~0.1 km) of the seafloor. The basal, low-angle, listric fault truncates synrift marine strata of Jurassic and early Early Cretaceous age that were deformed, in places intensely, by down-to-the-basin (north-dipping) listric faults (Grantz et al., 1990). These faults represent gravitationally driven failure of the northern margin of the rift shoulder during initial stages of opening of the Canada Basin.

The Lower Cretaceous unconformity becomes increasingly visible south of km +935, where it is at 3 s. A dramatic hinge line occurs at km +915 (Fig. 2), south of which Cretaceous normal faults are not present and the postrift succession thins gradually onto the Barrow Arch. The broad crest of this arch, in the vicinity of km +835, is a basement high of compound origin. Its structural morphology and relief were developed mainly during the Early or Middle Devonian, when Ordovician and Silurian strata of the Franklinian geosyncline were folded, mildly metamorphosed, and raised above sea level. Widespread erosion created a broad, southward-dipping Arctic platform (present coordinates) at the margin of the North American craton that was initially covered by nonmarine sediments at the base of the Devonian and Lower Mississippian Endicott Group. The structural elevation of the northern part of the Barrow Arch was augmented by uplift related to the Jurassic prerift extension and early Early Cretaceous seafloor spreading that created the Canada Basin. This uplift created a series of erosional unconformities along the crest and north flank of the Barrow Arch, culminating in the formation of the regionally significant Lower Cretaceous unconformity.

Franklinian strata (Ordovician and Silurian) and underlying Cambrian (?) and Precambrian bedded rocks. Although imaging is not optimal in this area (perhaps because of the strong coastal current and/or numerous course adjustments required to avoid sea ice north of the Barrow Arch), a thick reflective sequence is visible beneath the Lower Cretaceous unconformity, to 7 s and locally 8 s between the Cretaceous hinge line and at least as far south as km +630, near Icy Cape. Imaging of this sequence farther southwest, if it exists, is hampered by the overlying Brookian and Ellesmerian strata that thicken rapidly to the southwest. We follow Grantz et al. (1987) in interpreting this succession as Franklinian and underlying Cambrian (?) and Proterozoic sedimentary rocks. This may be the first time the full depth of the Cambrian and Proterozoic succession has been im-

aged here because typical industry data in this area are recorded to only 6 s. These deep reflections are not multiples of the overlying Ellesmerian and Brookian reflections (which extend to 4.5 s at km +630), because the reflections we interpret as Franklinian or older have stacking velocities of ~5 km/s and implied interval velocities (poorly determined because the far offset recorded was only 4.2 km) of >6 km/s at 7 s (consistent with Sherwood's [1992] velocities and interpretation of a deep Franklinian carbonate unit). Craig et al. (1985) interpreted our Franklinian and underlying Cambrian (?) and Proterozoic sedimentary rocks as a lower Ellesmerian clastic wedge, down-dropped by 6 km to the west across the 200-km-long coast-parallel, northwest-dipping Barrow fault; and Sherwood (1992, his Fig. 3) reinterpreted the so-called Barrow fault as a steep (70°–80°) southeast-dipping thrust fault with 10 km of vertical throw. However, our transect shows no evidence for any major fault in the position suggested by Craig et al. (1985, their Fig. 7) at ~km +750 (Plate 2), and the gravity field over the North Slope and Chukchi Sea shows no evidence of the coast-parallel gradient that would be expected across such a major fault (e.g., Barnes et al., 1994; Klemperer et al., this volume, Chapter 19, their Fig. 3).

The depth, seismic character, and low dip of reflections from 9 s at km +555 to 6.5 s at km +465 beneath the Brookian fold belt suggest that the contact between the Franklinian sequence and underlying beds may underlie the entire Arctic Alaska Basin and Colville foredeep as far south as the Lisburne Peninsula. The sporadic occurrence of similar reflectors at depths of 4–6.5 s as far south as km +195 suggests, but more speculatively, that the contact may also underlie the Hope Basin at least as far south as the north flank of the Kotzebue Arch. Paleogeographic reconstructions (Plafker and Berg, 1994b) show the Franklinian and older sedimentary rocks located on the southern margin (in the modern coordinate system) of the Arctic Alaska microplate that continues in the subsurface south of the Brooks Range orogen (Fuis et al., 1995).

The northern extension (modern coordinates) of the Franklinian basin and underlying succession imaged on our transect should now be in the conjugate rifted margin of Arctic Canada (Rickwood, 1970; Embry, 1990; Grantz et al., 1998; Mickey et al., this volume), possibly in the region of Prince Patrick Island (Sherwood, 1992). The Franklinian strata on our transect are laterally equivalent to a belt of deep basin deposits, several thousand meters thick, consisting of lutite, turbidite, radiolarian chert, carbonate debris flows, and volcaniclastic rocks of Cambrian to Early Silurian age. These strata are between the lower Paleozoic carbonate shelf strata that underlie northern Greenland and the Canadian Arctic Islands on the southeast, and the Amerasia Basin margin on the northwest (Higgins et al., 1991; Trettin et al., 1991). Restoration of counterclockwise rotation opening of the Canada Basin would place these basin deposits of the Canadian Arctic Islands against the Ordovician and Lower Silurian argillite and graywacke sequence of the North Slope of Alaska and the Franklinian sequence of the Chukchi

Shelf (Grantz et al., 1998, their Fig. 6), which we interpret to be as thick as 14 km beneath the Barrow Arch. Closure of the Canada Basin would also place the high-velocity pre-Franklinian sedimentary strata and crustal basement imaged on our seismic profiles, together as thick as 25 km, against Proterozoic and older middle and lower continental crust beneath the Canadian Arctic Islands adjacent to the Amerasia Basin (Sweeney et al., 1986), which are 30–35 km thick. The thickness (to ~10 km) of the high-velocity pre-Franklinian strata imaged on our seismic profiles suggests that they may correlate with the thick, gently folded Late Proterozoic (Hadrynian) successions of the southwestern Canadian Arctic Islands (Stockwell et al., 1970). These deposits include the shallow-water carbonates, evaporates, and sandstones (including quartzites) of the gently folded, 4-km-thick Shaler Group on Victoria Island, and have been mapped in the subsurface of Melville Island as a Proterozoic basin from 5 s to 9 s traveltime (~10 km thick) (Kanasewich and Berkes, 1990). Similar thick sequences may extend 1000 km along the strike of the Proterozoic and Franklinian margin, if the reflective strata on our transect are also laterally equivalent to the 6–8 s of layered reflections described by Coflin et al. (1990) in the Mackenzie Delta region.

Ellesmerian strata (Upper Devonian to Jurassic). Brookian strata above the Lower Cretaceous unconformity overlie the Ellesmerian sequence south of the erosional wedge-out of the Ellesmerian strata near km +880 (Fig. 2; Plate 2). These Upper Devonian to Jurassic Ellesmerian strata thicken smoothly southward from the wedge-out on the Barrow Arch to ~2 s (~4 km) at km +625, off Icy Cape. South of km +625 a series of Upper Devonian-Carboniferous grabens is present in the lowest part of the Ellesmerian sequence (the Endicott Group). The deepest of these, which is in the axial region of the Hanna Trough (Grantz et al., 1987; Sherwood et al., this volume), is ~30 km wide and centered at km +560. The Ellesmerian sequence within this graben is 1.5 s (~3.5 km) thicker (base at ~6.5 s) than adjacent Ellesmerian outside the graben. South of the graben, the base of the Ellesmerian rises from 5 s to 4 s before another 30-km-wide graben is encountered, centered at km +435 just north of Cape Lisburne. The presumed fault-plane reflections bounding this graben have true dips of ~45° (based on manual migration), and are reflective from 2.5 s at the Lower Cretaceous unconformity, which is offset ~0.2 s (~0.1 km) across at least the southern of these faults, to perhaps 7.5 s (possibly base of Franklinian). (Note that in Plate 2 interpreted faults are shown in their true structural positions [normally much steeper than the unmigrated reflections because the faults are typically steep], in contrast to the sedimentary reflections, which are marked in their unmigrated positions [normally close to the correct location because they typically dip at shallow angles].) Resolution of the Ellesmerian strata is lost immediately north of the Herald Arch (km +390), offshore Cape Lisburne, and their extent farther south is speculative.

Brookian strata (Cretaceous and Tertiary). The Colville foredeep contains clastic sedimentary rocks eroded from the area now covered by the Hope Basin, as well as the Brookian orogen to the south (Moore et al., 1994). The maximum thickness of strata in the Colville foredeep along our transect is 3.1 s (nearly 5 km), attained close to the axis of the underlying Hanna Trough. However, the general lack of faulting of the Lower Cretaceous unconformity (Fig. 2; Plate 2) suggests that the foredeep resulted from sedimentary loading rather than the reactivation of older extensional structures in the region of the Hanna Trough. The southern part of the Colville Basin is deformed in the distal Brooks Range fold and thrust belt, which is the result of deformation above the Late Cretaceous or Paleocene Herald thrust, which follows the Lower Cretaceous unconformity north of Cape Lisburne. At the deepest part of the basin (near the Hanna Trough), the Brookian foreland folds have wavelengths of ~30 km and amplitudes of ~0.3 s (~0.35 km), the amplitudes diminishing northward. At the south margin of the Colville Basin, just north of Cape Lisburne, fold wavelengths on the seismic section (Plate 2) are as short as ~15 km, fold amplitudes exceed 1 s (~1.5 km), and the folds are broken by blind thrusts. Imaging of the Brookian sedimentary rocks deteriorates to the south as fold limbs steepen, and all resolution is lost at Cape Lisburne where the thrust folds are replaced by imbricate thrust splays that bound horsts of Ellesmerian as well as Brookian strata at the Herald Arch (Plate 2) (Grantz et al., 1970).

Crystalline basement and Moho. Few intracrystalline basement reflections are visible on our transect north of the Herald Arch, yet the reflection Moho is visible almost ubiquitously from the arch at km +400 to the Cretaceous hinge line at km +915. Across this 500-km-wide region the Moho reflection is a characteristic single bright reflection separating nearly reflection-free lower crust from reflection-free upper mantle. The reflection Moho varies between 12 s and 13.5 s over this distance, more due to velocity pushdown than to laterally changing velocities in the crystalline basement. Beneath the thicker part of the Colville Basin section, from km +650 to +420, the traveltime to reflection Moho varies directly with the thickness of the basin, showing that the basin passively loaded and depressed the crust without corresponding basement deformation. For example, the Lower Cretaceous unconformity deepens from 2 s at km +645 to 3 s at km +565, an increase in depth of ~0.8 km across 80 km. Across the same distance the reflection Moho is also delayed by the same traveltime, deepening from 12 s to 13 s. We speculate that beneath the Barrow Arch changes in traveltime to reflection Moho may similarly represent changes in thickness in the overlying Franklinian and Proterozoic basins. The Moho traveltime varies from 13.5 s (km +710) to 12.5 s (km +755) to 13.5 s (km +810) across this area, seeming to correlate with the long-wavelength warping of the Franklinian and pre-Franklinian sedimentary reflections at 6–7 s.

The subsedimentary crystalline basement across the 500-km-wide region from the Cretaceous hinge line to the Herald Arch is <20 km thick, in places <15 km thick, suggesting that in Late Proterozoic and early Phanerozoic time this part of the Arctic Alaska plate must have been a very broad, highly extended, continental margin (less than half average continental crust thickness). True depth to Moho probably varies from ~39 km beneath the Barrow Arch to ~35 km beneath the axis of the Hanna Trough (depth calculated using sedimentary rock velocities from our reflection data, and crystalline-basement velocities of 6.5 km/s). Wolf et al. (this volume) use wide-angle data to determine a Moho depth of 35 km and traveltime of ~12 s beneath Point Lay, in good agreement with our reflection Moho (Fig. 2), thereby encouraging the conventional assumption that the reflection Moho and refraction Moho are equivalent (Klemperer et al., 1986; Mooney and Brocher, 1987) beneath the Chukchi Shelf.

Northern region: Herald Arch and Hope Basin (Plate 2, central portion of panel B)

Herald fault zone and arch. The Herald fault zone and arch is the northwest-southeast–trending belt of nonreflective acoustic basement at the seafloor (Grantz et al., 1987) that emerges onshore as the Tigara uplift of Ellesmerian and Franklinian strata exposed in east-vergent thrust sheets on the Lisburne Peninsula (Moore et al., 1994, and this volume) (Plate 2). These thrusts are Late Cretaceous to Paleocene in age, because they deform Albian and probably Cenomanian strata (Campbell, 1967), but do not deform or tilt Eocene and younger Hope Basin sedimentary rocks (Tolson, 1987a; see following section). Late Early Cretaceous apatite fission-track cooling ages (ca. 115 Ma) (Moore et al., this volume) date the event that caused imbricate thrusting in the Lisburnc Peninsula. Because the Brookian structures are subparallel to the Herald Arch in the offshore region, it is difficult to distinguish the two in our regional deep profile. However, we may take the north limit of the Herald Arch as the ~20° south-dipping reflections that project to the surface at km +410 and reach the prominent Lower Cretaceous unconformity reflection (Brookian detachment) at ~2 s, where the Lower Cretaceous unconformity changes from horizontal to the north to ~10° south-dipping to the south. Scattered south-dipping reflections extend at least 40 km to the south beneath the Hope Basin, as far as km +370. In addition, south-dipping reflections at least as deep as 8 s between km +340 and km +325 may represent Franklinian or older strata rotated by either north-dipping normal faults of the Hope Basin or possibly the south-dipping extension of the Herald thrust. However, clear evidence for the Herald thrust is only visible to 3 s traveltime, and it cannot be reliably traced south of the northern onlap of the Hope Basin at km +380.

Hope Basin. Our transect crosses the Tertiary Hope Basin from the Herald Arch at km +380 to the Kotzebue Arch at km +185 (Tolson, 1987a). Along our transect the dominant structural style controlling the basin is high-angle normal faulting, the two deepest points reaching 3.6 s (~6 km) in half-grabens within the basin at km +275 and at km +315. The beds in these half-grabens were downdropped by 2 s (~3 km) from a central horst that is between km +285 and +305, just southwest of Point Hope. The northern half-graben is bounded and underlain by prominent fault-plane reflections that can be traced to >6 s at km +330. The listric appearance (decrease in dip to the north) of the fault-plane reflections in this unmigrated section may be an imaging artifact (pull-up) due to the decreasing sedimentary thickness to the north that bends the reflection from a fault that is in reality planar (cf. Peddy et al., 1986). Both half-grabens are underlain by low-frequency reflections (at 5–7 s at km +265 to +275; at 4–6 s at km +315 to +335) that also dip inward toward the central horst. These may therefore be Paleozoic strata that were also rotated by the Hope Basin faults. Similar reflections are seen at 4–5 s, km +200, and at 3–4 s, km +170. North of km +315, the Hope Basin sedimentary sequence thins northward to the Herald Arch across several south-dipping normal faults (antithetic to the main fault). South of km +275 the basin gradually thins southward toward the Kotzebue Arch across many south- and north-dipping faults.

The oldest rocks in the Hope Basin are middle Eocene (pre-41 Ma) sedimentary rocks that, based on the section penetrated in two coastal wells at the eastern end of the Kotzebue Basin (Fig. 1), unconformably overlie crystalline limestone, phyllite, and dolostone of Paleozoic (?) age (Tolson, 1987a; Plate 1). The stratigraphy of the Cape Espenberg No. 1 and the Point Nimiuk No. 1 wells in the eastern part of the Hope Basin (CE and PN in Fig. 1) is summarized in Plate 1. In the Tertiary section, Tolson (1987a) defined a nonmarine and deltaic seismic-stratigraphic Unit I of Eocene and Oligocene age, a generally nonmarine lower Miocene Unit II, and a generally marine middle Miocene to Pleistocene Unit III. Tolson's (1987a) seismic stratigraphy is particularly clear at km +275, where Units I, II, and III are separated by more strongly reflective bands at 1.3 and 2.3 s. Unit I is only present in the deepest half-grabens, which appear to be bounded by growth faults initiated at the earliest stage of basin formation. Unit II (clearly visible between km +280 and the Kotzebue Arch) has a comparatively uniform thickness (~0.8 s), suggesting a broader pattern of thermal subsidence. Unit III reaches a thickness of 1.8 s (~1.8 km) at km +215, is offset by many high-angle normal faults that reach essentially to the seafloor, and thins to 0.5 s over the Kotzebue Arch.

Tolson (1987a) suggests, from indications that some of the faults bifurcate upward into flower structures, that the first phase of deformation in the Hope Basin was Eocene transtension on a dextral-slip fault system. This extension in the Hope Basin possibly accommodated dextral slip along the Kobuk fault system (e.g., Kirschner, 1994; Moore et al., 1994). The 100 km gap between onshore basement exposures that define the Kobuk fault system and Tolson's offshore seismic data precludes a more certain statement. (Crustal thinning arguably related to this transtension is discussed in the following section.) The second phase of deformation in the Hope Basin was a middle to late Miocene reinitiation of extension on both reactivated and newly formed normal faults. Extension continues to the present, indicated by the diffuse belt of seismicity with both extensional and dextral strike-slip focal mechanisms, the Bering rift zone of

Mackey et al. (1997) that extends across the southern Hope Basin. Just as the modern faulting is presumed to represent a far-field effect of shortening in southern Alaska and concomitant extrusion of western Alaska (Mackey et al., 1997), so the Eocene faulting possibly represented a far-field accommodation of changing boundary conditions imposed on the North American plate by Pacific subduction.

Crystalline basement and Moho. Deep-crustal reflectivity is fairly pronounced between 10 and 12 s below the southernmost Colville Basin and Herald Arch (km +440 to +385), and scattered reflectivity is also visible in the middle crust in this area. Beneath the southern part of the Herald Arch and the entire Hope Basin, imaging of the deep crust is poor, with only a faint reflection Moho, and few scattered reflections. One might speculate that the bright reflectivity near km +400 represents an imaging window through shallow high-velocity rocks into the lower crust that typifies a much broader region. However, it is not obvious that imaging should be dramatically clearer through the 2 s (~3 km) Brookian section at km +400 than through the 1 s (~1 km) Neogene section at the northern and southern margins of the Hope Basin at km +370 and +185, or through the <1 s Brookian section over the Barrow Arch north of km +800. Because the few scattered lower-crustal reflectors observed beneath the Hope Basin are dipping and of lower amplitude than the flat Moho reflection (in contrast to the character imaged in the central region of our transect), we argue that the lower crust is similar beneath the Barrow Arch, Colville foredeep, and Hope Basin, and that these are all parts of the autochthonous Arctic Alaska plate. The brighter reflectivity around km +400 is presumed to be a local feature.

The reflection Moho, which can be traced with only minor difficulty beneath the Hope Basin, rises from 12 s beneath the Herald Arch (km +400) to 10.5 s below the northern major half-graben of the Hope Basin (km +320), and remains fairly uniformly at this level to the Kotzebue Arch. Because Cenozoic sedimentary thickness increases from 0 to 3 s (~5 km) over this distance, the apparent shallowing of reflection Moho across the Hope Basin represents a true thinning of the pre-Tertiary crust of as much as ~10 km. This thinning implies up to ~30% extension during formation of the Hope Basin, assuming that the pre-extensional crust was of uniform thickness. However, because we infer that there was preexisting topography (to provide a western source for the mid-Cretaceous Corwin Delta; summarized in Moore et al., 1994) and hence a crustal root, more than 30% extension is locally required beneath the deepest grabens of the Hope Basin. Typical extension may have been only 20%–25% if an average sedimentary thickness of 2–3 km, and travel-time to reflection Moho of 10.5 s, is taken for the entire basin. Although the crustal refraction model of Wolf et al. (this volume) shows essentially no thinning of the crust from Cape Lisburne to the Bering Strait, their model is not inconsistent with the thinning claimed here because their recording seismographs were all located outside the Hope Basin (Fig. 1). Traveltime delays from seismic source points along the ship track above the

Hope Basin depocenter are increased because of near-surface sedimentary rocks; however, only shots with a source-receiver midpoint close to the depocenter can image the crustal thinning. Because of the onshore-offshore recording geometry, only shots more than ~200 km north of Cape Prince of Wales and more than ~200 km south of Cape Lisburne have midpoints over the Hope Basin depocenter. At these long offsets the seismic arrivals were too weak to be reliably observed and so were not used in the modeling presented by Wolf et al. (this volume). The Wolf et al. model is, however, an appropriate measure of crustal thickness, 32–34 km, north and south of the Hope Basin.

The western projection of the Kobuk fault (Patton et al., 1994) strikes toward a zone of antithetic normal faults along the north flank of the Kotzebue Arch that has far more structural relief than the south side of the Kotzebue Arch, and that forms the south side of the Hope Basin (Fig. 1; Plate 1). This geometry suggests that the westward-diverging normal faults of the Hope Basin may represent a transtensional system at the trailing edge of the Brooks Range block that created the Hope Basin as it moved eastward along the dextral Kobuk fault (Tolson, 1987a; Kirschner, 1994) during the Paleocene or early Eocene. Based on seismic mapping of the Hope Basin in United States waters (Tolson, 1987a; Kirschner, 1988) and Russian waters to 172°W (Eittreim et al., 1978), and on the Bouguer gravity anomaly of −10 to −20 mGal (Ostenso, 1968; see also Klemperer et al., this volume, Chapter 19, their Fig. 3), the Hope Basin continues for ~400 km in a west-northwest direction, and so could conceivably have accommodated 80–100 km dextral motion on the Kobuk fault system. Knowledge of the displacement history of the western end of the Kobuk fault remains poor, but W.W. Patton Jr. (2001, personal commun.) reported that Upper Cretaceous paralic deposits containing 80 Ma volcanic ash appear to have been offset dextrally a minimum of 40 km along the lower reaches of the Kobuk River near 160°–161°W. It is therefore possible that dextral displacement along the Kobuk fault created the transtensional faults that opened the Hope Basin, but specific correlation of these events is not yet established. In contrast, some terrane reconstructions suggest sinistral motion along the northeast flank of the nascent Hope Basin and the Kugruk fault zone (which separates the Seward Peninsula from the Yukon-Koyukuk Basin) in order to relocate the Seward Peninsula southward to its present location (Plafker and Berg, 1994b).

Central region: Bering Strait, Saint Lawrence Island, and Inner Bering Shelf (Plate 2, left of panel B and panel C)

Kotzebue Basin. The Kotzebue Basin (Plate 2) is little more than a shelf, locally with a broad, rather shallow sag, between the crest of the asymmetrical Kotzebue Arch and the Seward Peninsula. The Kotzebue Basin may be regarded as a southern extension of the Hope Basin; Unit II and some Unit III Neogene sedimentary rocks (Tolson, 1987a) are present for 100 km along our transect from the Kotzebue Arch at km +185 to the southern wedge out of Hope Basin fill against basement of the Seward

Peninsula at km +85. It is distinguished from the main Hope Basin by the near absence of normal faulting, and by the much smaller sediment thickness. The maximum thickness of the section along our transect is 1.4 s, ~1.5 km, from km +160 to +175, and it is never more than 0.8 s thick south of km +150.

Norton Basin. The transect of Figure 2 and Plate 2 is offset discontinuously 25 km to the west at km 0 in the Bering Strait, and misses all but the most distal part of the Norton Basin. Norton Basin has a stratigraphy that is similar to that of the Hope Basin where drilled at ~166° and 164°W (~4 km of Eocene to Pleistocene strata), but may have a latest Cretaceous-Paleocene early history represented by an additional 2 km of strata below the deepest drilled horizon (Marlow et al., 1987; Worrall, 1991). The stratigraphy of the Norton Sound COST No. 1 and No. 2 wells in the eastern part of the basin (No1 and No2 in Fig. 1) is summarized on Plate 1. Unlike in the Hope Basin, a regional unconformity is present within the Eocene section above folded early basin strata (carapace sequence of Worrall, 1991). This unconformity is widely mapped as the broadly contemporaneous, ca. 43 Ma "red" event across all basins of the Bering Shelf (Worrall, 1991), and could be equivalent to the pre-41 Ma unconformity of the oldest Hope Basin sediment onto basement. The Norton Basin has the same spatial relation to the Kaltag fault (northwest of the possible terminus of a regional dextral fault) that the Hope Basin has in relation to the Kobuk fault system, suggesting that the Norton Basin formed by distributed transtensional faulting at the trailing edge of the north block of the dextral Kaltag fault (Fisher et al., 1982; Worrall, 1991; Kirschner, 1994). The Norton Basin could thereby allow the Kaltag fault to terminate on the shelf at this system of basins (e.g., Worrall, 1991), or could alternatively transmit the Kaltag strike-slip motion to the shelf edge or plate boundary.

At km –15 (negative distances are south of Bering Strait) small normal faults allow accumulation of 0.8 s (~0.7 km) of Tertiary sediments in a basin 20 km wide where crossed by the transect. The western feather edge of the Norton Basin is visible from km –45 to –125, reaching maximum thickness on our transect of 0.7 s at km –85.

Saint Matthew Basin. The Saint Matthew–Hall Basin has attracted only scant attention in the published literature (Marlow et al., 1976), presumably because it is both remote and does not have sufficient thickness to produce hydrocarbons. For the Saint Matthew–Hall Basin, unlike other basins on the Bering Shelf, our seismic data provide a significant update of previous mapping, which suggested that it was only ~1 km deep west of Saint Lawrence Island (Kirschner, 1988; Worrall, 1991). On our seismic profile, an almost unfaulted thermal subsidence or sag phase from km –305 to km –415 reaches a maximum thickness of 1.1 s (~1 km) at km –350, and overlies two prominent tilted fault blocks in which sedimentary reflections continue to 2 s (~3 km total depth) at km –345 and km –355. By analogy with the Norton Basin and the Navarin Basin, we regard the tilted rift-phase sedimentary rocks as largely "pre-red unconformity" (Fig. 2) (Worrall, 1991; see also following Navarin Basin discussion).

Another, much smaller, trough of strata pre-red unconformity may be present at km –385, to 1.2 s.

Although our single profile precludes any firm conclusions about the areal extent of the Saint Matthew Basin, the existence of a continuous basin from the Alaskan coast to southwest of Saint Lawrence Island may suggest that the strike-slip systems believed responsible for aspects of the development of the Hope and Norton Basins may also have extended across the shelf beyond Saint Lawrence Island (Holmes and Creager, 1981; Scholl et al., 1970).

Crystalline upper crust of Bering Strait and Saint Lawrence. The Bering Strait region and the Inner Bering Shelf are characterized by broad basement arches with intervening sedimentary basins. So-called acoustic basement (labeled as such only by the absence of well-defined reflections) is at or very close to the seafloor along our transect between the Kotzebue Basin and the Norton Basin, km +85 to km –5; between the western distal onlaps of the Norton Basin from km –25 to km –45; and between the Norton Basin and the Saint Matthew Basin from km –125 to km –305. As with many, if not most, crustal-scale reflection profiles, the uppermost crystalline crust returns few laterally coherent reflections, in part due to technical reasons. These include higher-amplitude source-generated noise (refracted and reflected reverberations or multiples) and greater sensitivity to correct stacking velocity and lower stacking fold than at longer traveltimes; and coarse shot spacing and low frequencies, chosen to allow deep-crustal recording, that are inappropriate to image complex, nonstratified near-surface geology. It is therefore no surprise that we cannot attribute any specific reflections to specific rock bodies, e.g., plutons or metamorphic complexes. It is noteworthy, however, that close to the Seward Peninsula we record coherent reflections within the upper 5–10 km of crust that offer at least the prospect of mapping structural fabrics. Close to where our reflection profile passes only 10 km from Cape Prince of Wales, from km +55 to km +40, reflections from 1 s to 3 s (~3–9 km) rise to the south at ~45° (after hand migration), projecting to the surface at ~km 35; and from km +35 to km 0 reflections from 1.5 s to 3 s form an antiform with flanks dipping ~20°. As with any individual seismic line, we can only assume that our reflection line is a true dip line recording in-plane reflections. Possibly correlative reflections on profile 3B, 25 km due west, are no shallower than 3 s, suggesting that this antiform plunges 10° west. Even though these reflectors are relatively close to land, it is unclear to what they are related. Some of the reflections could represent the Cretaceous granites that underlie the region (exposed on the Diomede Islands 30 km to the west [Shumway et al., 1964] and on the Seward Peninsula [e.g., the Kigluaik pluton, Amato et al., this volume]), although these granites are typically undeformed, so reflectivity would likely represent internal compositional variability (e.g., Amato and Wright, 1997). Alternatively, and we believe more likely, the reflections might represent undulatory domal structures and shallow-dipping foliations in the low-grade metasedimentary rocks exposed 30 km to the east in the Seward

Peninsula (Amato et al., this volume). A 30-km-long industry reflection profile 30 km east of km +65 to +35, recorded to 6 s traveltime (Fig. 1; Tolson, 1987b, p. 131), shows a reflection dipping 10°N from 2.5 to 4.5 s. Tolson (1987b) suggested that this reflector corresponds to a low-angle detachment at the top of a metamorphic core complex represented by blueschist outcrops in Seward Peninsula. Our data cannot corroborate Tolson's interpretation and suggest, at the least, that the crust exhibits more complexity than a single domal structure.

The only other subsedimentary reflections on this part of the transect shallower than 3 s that are demonstrably not multiples are at km −185 and km −235 to km −260 at ~2.5 s just north of Saint Lawrence Island. These reflections may be described as the shallowest manifestation of what we loosely term lower crustal reflectivity, in that these horizontal to subhorizontal reflections continue downward to the base of the crust.

Crystalline basement, Moho, and intramantle reflectors. A gradual but significant change in the character of lower-crustal reflections and the reflection Moho occurs south of the Kotzebue Arch. Unlike the northern segment in which the reflection Moho stands out as brighter than the scattered lower-crustal reflections, across this central segment the reflection Moho is normally identified as the base of a sequence of many reflections occupying the lower one-third to two-thirds of the crust, and marking the transition to a usually nonreflective upper mantle. This layered character of the lower crust is most clearly seen in the center of our central segment, beneath the Bering Strait and near Saint Lawrence Island, and disappears southward near the Saint Matthew–Nunivak arch. Although the changes in crustal character are gradual, and consideration must be given to the effects of weather on the quality of imaging along this profile (see previous section on acquisition, processing, and interpretation of seismic data), we believe, on the basis of our two parallel profiles, that this reflective, layered lower crust distinguishes the crust beneath the central segment of our transect from less-reflective regions to the north and to the south.

Flat-lying or very gently dipping reflections, sometimes called laminated lower crust (e.g., Meissner, 1986), occupy a 500-km-long sector of the deep crust, from the southern half of the Kotzebue Basin (km +120) to the southern half of the Saint Matthew Basin (km −380). In some places the middle crust exhibits more steeply dipping reflections above the deeper crustal flat-lying reflections, as from km +50 to km 0 in the Bering Strait. In other places the middle crust is apparently nonreflective above the deepest crustal flat-lying reflections, as at km −300 to −350 north of the Saint Matthew Basin. Elsewhere, the subhorizontal laminated crustal reflectivity occupies 6–7 s (~18 to 23 km) above the Moho, e.g., around km −40, km −180, and km −250. Lower crustal reflectivity commonly occurs at shallower depths beneath basement highs, partly due to clearer imaging where attenuating sediment is absent (e.g., the vertical panel of clearer imaging from km −35 to −45 directly beneath a clearly marked change in seafloor characteristics). Although it is impossible to prove in any specific location, we believe it

likely that many of these regions of shallower lower crustal reflectivity mark real lateral changes in the rock fabric, e.g., perhaps from nonfoliated igneous bodies to gneisses. The substantial lateral changes in thickness of the laminated lower crust make it impossible to assign this reflective character to a single velocity layer in the model of Wolf et al. (this volume). Wolf's deepest crustal layer, ~8 km thick with velocity of 6.5–6.7 km/s, represents the deepest 2 s of traveltime of the crust, which in a general way corresponds to the brightest reflectivity. However, in Wolf's model no distinction occurs between the velocity of the lower crust we identify as laminated, and the lower crust farther north we identify as lacking lower-crustal reflectivity. This observation helps refine our interpretation of the reflectivity as representing extensional tectonic fabrics with only a modest proportion of intruded mafic material.

From the southern half of the Kotzebue Basin (km +140) to the north flank of the Saint Matthew Basin (km −330), the deepest Tertiary sedimentary rocks are consistently above 1 s traveltime and the reflection Moho is everywhere between 10 and 11 s traveltime. Although neither our study nor that of Wolf et al. (this volume) would be sensitive to short-distance, large-magnitude changes in crustal velocity, our observations of reflection Moho are consistent with those of Wolf et al. that the crust is a uniform 32 km thick across this large region. Here the near-vertical reflection and wide-angle reflection results are in good agreement, unlike in the Hope Basin, because the wide-angle and near-vertical reflection points are nearly coincident, and because no major lateral variations in sediment thickness occur between the source and the receivers.

Globally, intramantle reflections are the exception rather than the norm on deep reflection profiles, with some spectacular exceptions (e.g., Warner et al., 1996; Cook et al., 1999). On Plate 2 most of the steep reflected phases visible below the reflection Moho are diffractions from intracrustal discontinuities, and appropriate migration would move them into the crust. Because our profiles are only recorded to ~17 s traveltime, we cannot recognize steep reflectors from within the mantle, and on these data, only reflections that continue substantially below the Moho with a gentle dip can truly represent intramantle structures. Even these reflections could be returning from steeper crustal structures obliquely crossing the transect. Assuming the reflectors are in plane, we identify three possible intramantle structures in the central segment of our transect. A single reflector is visible offshore Saint Lawrence Island, dipping south from 11.5 s (just below the Moho) at km −230 to 17 s (maximum traveltime recorded) at km −275. The true dip after (manual) two-dimensional migration at mantle velocities is ~20°, extending from 10.5 s to 16 s. Just south and just north of the Bering Strait, reflections dip north from 10.5 s at ~km +20 to 17 s at ~km +50; and from ~11 s at km −70 to 17 s at km −35. The true dips after two-dimensional migration at mantle velocities (8 km/s) are ~40°, and the true imaged extent is only to 13.5 s, so that only a small degree of obliquity of the reflector to the profile direction would mean that these reflections originated within the crust. We cannot attribute the mantle reflectors to any specific tectonic episode, but elsewhere such mantle reflectors frequently have an association either with fossil subduction zones (e.g., Warner et al., 1996), or putative extensional shear zones in continental rifts (e.g., Klemperer and Hurich, 1990).

Southern region: Outer Bering Shelf, Saint Matthew–Nunivak Arch to Beringian margin (Plate 1, panel D)

Saint Matthew–Nunivak Arch. Below a thin veneer of Neogene sedimentary rocks (increasing from 0.4 s on the crest of the Saint Matthew–Nunivak Arch at km –450 to 0.7 s at the edge of the Navarin Basin at km –635), and the barest hints of an older stratigraphy (e.g., south-dipping reflections from 0.5 s to 1.0 s from km –540 to –550; a possible synformal sag from 0.6 s to 1.2 s between km –600 and km –615), the upper crust in this region is featureless (see Plate 2, panel D). Our upper-crustal image does not distinguish between different possibilities such as a transparent magmatic arc and an accreted terrane lacking a coherent stratigraphy.

Navarin Basin. The Navarin Basin is known from extensive industry seismic data, and so here we describe our reflection data based on the thorough interpretation of these data published by Worrall (1991) without adding significant new results. There is one industry exploration well in the basin (COST No. 1; Na1 in Fig. 1) for which depths to lithostratigraphic horizons have been tied via a sonic log and synthetic seismogram to intersecting seismic data (Turner et al., 1984, 1985; stratigraphy summarized in Plate 1). COST No. 1, crossed by our transect at km –715, passed through a minor late Miocene unconformity at 1.3 s (~1.1 km depth) visible as a basinwide bright reflection on our transect (Fig. 3) (blue reflection of Worrall, 1991), and reached the important late–middle Eocene unconformity (Figs. 2 and 3) (red unconformity of Worrall, 1991) at 3.3 s (3.9 km) (Turner et al., 1985). The angular relationships of these unconformities are better seen on data displayed with large vertical exaggeration (Fig. 3). The COST well penetrated an additional 1 km of "carapace sequence" (Worrall, 1991) nonmarine strata of Late Cretaceous age, including gas-bearing coal units that predate the red unconformity. The sequence is intruded by basaltic sills that yield Oligocene and Miocene ages (although these K-Ar whole-rock ages should be regarded as minimum ages; Turner et al., 1984; Marlow et al., 1987). The brightest reflections on our profile beneath 3.4 s (~4.2 km) are from below the top of the Cretaceous sequence and are likely from these sills or from the gas-bearing coals. Structural relations on industry seismic data, and dredge data from the Beringian margin, imply that the carapace sequence encompasses Upper Cretaceous to middle Eocene strata (Worrall, 1991), corresponding largely to the period during which the Beringian margin was an active transpressional plate boundary. The uplift and subaerial erosion that produced the red unconformity, and the immediately following rapid subsidence to deposit upper-bathyal middle Eocene strata ca. 42 Ma, are not obviously related to the tectonic abandonment of the Beringian margin as a plate boundary in the early Eocene ca. 56 Ma (Worrall, 1991).

Our profile bypasses the northern depocenter of the Navarin Basin (Fig. 1) and crosses the East Platform drilled by the COST well (the red unconformity is at 3–3.5 s, 3.5–4.5 km depth, from km –685 to –740); traverses the Central Ridge (km –750 to –755) into the South Basin (maximum traveltime to red unconformity is 6.3 s [~8.2 km] at km –770); then climbs out of the Navarin Basin across the Lisa Ridge (km –825 to –835; blue horizon unconformably overlies pre-red strata at 1.3 s, 1.1 km depth) and onto the Navarin Ridge (km –870 to –880; blue horizon directly overlies strata that predate the red unconformity at 0.7 s, 0.5 km; Fig. 3) (all terminology is after Worrall, 1991).

The sequence that postdates the red unconformity is remarkably unfaulted across the entire basin, with the notable exception of the Central Ridge (km –750 to –755), a nonreflective zone on our data. Worrall (1991) showed it as a flower structure that he inferred from prebasin structures to have been formed by a dextral strike-slip fault (his cross section A–A′ is subparallel to our section, and ~10 km to the southeast). Although we do not image faults within the Central Ridge, our profile clearly shows minor inversion of the Miocene sequence south of the Central Ridge: the pre–late Miocene sequence (1.5–2 s at km –775) thickens basinward to the north, but its upper boundary now dips south (Fig. 3). The strata that predate the red unconformity, or carapace strata, are poorly imaged on our profile, as on many industry profiles (but see Worrall, 1991, his Figs. 45–51), and their base is nowhere clearly defined. However, at least 1 s of reflections is recorded from 4 to 5 s at km –740, and at least 2 s from 2.5 to 4.5 s at km –810. Below the carapace sequence we may guess at the presence of a Lower Cretaceous and Jurassic subduction complex similar to that exposed on Cape Navarin (Worrall, 1991), but we lack any direct evidence, and the structural fabric predating the carapace strata has not been imaged on our or on industry profiles. Unlike the strata postdating the red unconformity, the sequence predating the unconformity is broadly deformed. What resemble normal faults on our data (e.g., just north of the Central Ridge where the pre-red sequence is apparently downdropped 1 s [~1 km] across a north-dipping fault at km –745), and small structures that produce a crosscutting diffractive appearance on our unmigrated profiles (e.g., on the East Platform from km –705 to –735), are interpreted on a dense grid of migrated industry data to be east-west–trending folds rather than faults (Worrall, 1991). Only on the north margin of the basin (km –670) and on the south flank of the Lisa Ridge (km –835) did Worrall (1991) interpret north-south–trending normal faults, downdropped to the east, and active contemporaneously with the orthogonal folds as expected from dextral shearing in a northwest-southeast direction.

Navarin Ridge, Beringian margin, Aleutian Basin. Late Miocene and younger strata overtop the carapace strata and underlying basement of Navarin Ridge (km –870 to –880), which separates the Navarin Basin from the sedimentary prism beneath the slope and rise of the Beringian margin. The modern shelf

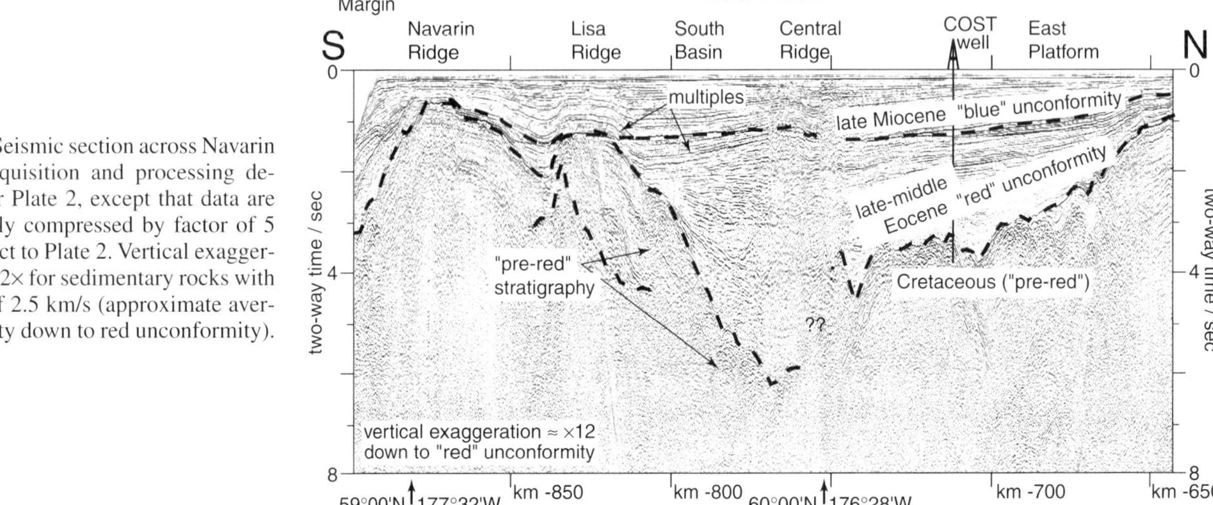

Figure 3. Seismic section across Navarin Basin. Acquisition and processing details as for Plate 2, except that data are horizontally compressed by factor of 5 with respect to Plate 2. Vertical exaggeration is ~12× for sedimentary rocks with velocity of 2.5 km/s (approximate average velocity down to red unconformity).

edge at km –895 has prograded only 15 km beyond the late Eocene shelf edge on the eroded flank of the Navarin Ridge. The slope is dissected by canyons (km –895 to –905), some of exceptional size, that produce rough seafloor topography crossed obliquely by our transect, and that cut back into the shelf eroding through the Tertiary strata into the pre-red unconformity or carapace sequence. The mean slope is about 3°, from 0.2 km water depth at the shelf edge to 3.2 km depth at km –945. The red unconformity emerges at the seafloor at ~km –935, so that pre-red strata crop out on the lower slope below ~2.8 s (2.1 km water depth). Carapace rocks including Cretaceous limestones and siltstone, and early Eocene arc volcanic rocks, have been dredged in this position (Jones et al., 1981; Davis et al., 1989). No structure is visible within the pre-red unconformity complex, here presumed to include a pre–late Eocene accretionary complex, although rock dredging has not recovered any such samples (Jones et al., 1981). At the foot of the slope the supposed accretionary complex is onlapped by oceanward-thickening terrigenous and hemipelagic sediments of the Aleutian Basin that postdate the demise of the active margin (Worrall, 1991). This presumably early Eocene to Holocene basin reaches 2 km thickness at the south end of our profile. Beneath a reflection at 6.4 s at km –980, sedimentary strata thicken toward the margin and contribute to the supposed accretionary prism, and thus presumably date from the period of active oblique subduction (pre-56 Ma), during and before the early Eocene. The diffractive horizon from 7.8 s at km –980 to 8.6 s at km –950 marks the top of oceanic crust basalts ascribed to the Kula plate by Scholl et al. (1987b) on the basis of interpreted Cretaceous magnetic anomalies. The oceanic reflection Moho is visible at 9.8 s close to the south end of the profile (Plate 2, km –990). Although oceanic crust cannot be traced inboard of the toe of the inferred accretionary wedge on this unmigrated profile, on nearby profiles a reflection can be traced to 12 s (Marlow and Cooper,

1985), a reflection we believe represents the top of oceanic crust frozen in the act of underthrusting the margin.

Crystalline basement and Moho. From the Saint Matthew–Nunivak Arch to the Beringian margin the lower crust lacks the laminated reflectivity so prominent around Saint Lawrence Island. Although changes in the lower crust are gradational, and the image of lower-crustal reflectivity constantly varies along the profile, we suggest that a change in crustal type or history is marked by the broad zone of weak reflections, dipping 15° to 20°N, that occupies the middle and lower crust from km –425 to –510. The reflection Moho is unclear over this region, but from km –500 to –600 is at 12 s to 13 s, indicating a crust that is either 10%–20% thicker, or has velocities that are 10%–20% slower, than in the central segment north of the Saint Matthew Basin. Upper-crustal refraction velocities from unreversed sonobuoys along a west-southwest–trending line south of Saint Matthew Island 200 km to the southeast (Fig. 1) reach 5.7–6.5 km/s at only 1–3 km depth, and 6.8–7.6 km/s at only 4–8 km depth (1.5–3 s) (Cooper et al., 1987). These high velocities seem to correlate with a gravity high that can be traced 200 km along strike to our profile at km –550 to –600 (Fig. 2; Klemperer et al., this volume, Chapter 19, their Fig. 3), where our reflection Moho is deepest, implying that the crust is of high density, as well as high velocity. The gravity model of Cooper et al. (1987) along their sonobuoy line shows a crust that reaches 43 km thickness. Although this estimate of 43 km is very dependent on the density contrast assumed at the crust-mantle boundary (e.g., Holliger and Klemperer, 1990), it is consistent with our observation of reflection Moho at 13 s if the average crustal velocity is 6.6 km/s.

The reflection Moho begins to shallow beneath the northern shoulder of the Navarin Basin, from 13 s at km –575 to 11 s at km –615, thereby demonstrating true crustal thinning beneath the little-subsided shoulder of the basin. Although the lower-crustal image is poor below the deep basin, bright reflections at ~11 s may be from the Moho. This traveltime implies a Moho depth of 21–27 km subsurface (13–20 km below the red unconformity) depending on local basin thickness (4–8 km), assumptions about thickness of Cretaceous and Paleogene carapace strata (5–10 km), and the velocity of crystalline basement (6.0–6.6 km/s). This estimate may be compared to the gravity model of Cooper et al. (1987) that places Moho at 20–22 km. The reflection Moho is visible at 11.5 s beneath the relatively thin sedimentary cover (<1.5 s Neogene reflectors) of the Lisa Ridge (km –830) southwest of the Navarin Basin, and even deeper beneath the Saint Matthew–Nunivak Arch, northeast of the Navarin Basin. Basement has been thinned by a factor of at least 2, if we take these traveltimes to the reflection Moho adjacent to the Navarin Basin as estimates of the crustal thickness predating the basin, and based on traveltimes to the red unconformity of as much as 6 s (km –770). Although our seismic profile demonstrates clear crustal thinning and apparent normal faulting, areal fault patterns interpreted from datasets more extensive than our single two-dimensional transect have been used to suggest that the Navarin Basin formed by transtension (Worrall, 1991) along margin-parallel strike-slip faults.

DISCUSSION

In an attempt to summarize our new data and observations of the nature of continental crust along our transect, we next note the important general characteristics of the crust and mantle beneath this region. The variable character of lower-crustal and Moho reflectivity along our transect permits us to divide the continent into three parts. The central part has highly reflective crust, whereas the northern and southern parts are comparatively nonreflective within the crustal basement. By inference from onshore geology, we associate the northern region (Beaufort margin and Chukchi Shelf) with the Arctic Alaska–Chukotka block, the central region (Bering Strait and Inner Bering Shelf) with the reworked middle to Late Cretaceous magmatic belt, and the southern region (Outer Bering Shelf and Beringian margin) with a collage of Paleozoic and Mesozoic terranes and Paleogene arcs. An absence of reflectivity is always difficult to interpret, and a lack of lower-crustal reflections beneath deep basins is particularly difficult to distinguish from poor signal penetration. However, we use the presence of Moho reflections along most of our profile as an indication that where we see no lower-crustal reflections, our data are nonetheless adequate to deduce that the lower crust is relatively nonreflective rather than obscured by noise. We summarize our observations by comparing key seismic attributes of our three continental regions with the crustal signature of the nearby active Aleutian arc (Fig. 4) (McGeary et al., 1994).

The Aleutian reflection section (Fig. 4A) is characterized by bright intramantle reflections from the subducting Pacific slab (McGeary, 1997; Holbrook et al., 1999), but only sporadic lower crustal or Moho reflections (McGeary et al., 1994). Crustal ve-

locity is high in the intraoceanic arc, especially in the lower crust that in some places may be gradational into the upper mantle (Fliedner and Klemperer, 1999). In the part of the Aleutian arc forming on older continental crust of the Alaskan Peninsula, both crustal velocity and crustal thickness are within the typical range of continental crust (Fliedner and Klemperer, 2000). The crust beneath the Alaska Peninsula formed by arc magmatism within terranes similar or identical to those believed to make up the Outer Bering Shelf, whereas the crust beneath the Aleutian Islands is that of a pristine intraoceanic arc. A comparison of the characteristics of the Aleutian arc with the characteristics of the Outer Bering Shelf helps us to interpret the nature of the southern segment of our transect.

Southern region: Paleogene arc and displaced terranes

Figure 4B cartoons the Navarin Basin above a nonreflective crust, and continentward-dipping, sub-Moho reflections as imaged beneath the inactive Beringian margin by Marlow and Cooper (1985). The lack of crustal reflectivity is consistent with reflection profiles across other intraoceanic arcs (the Aleutian intraoceanic arc; Fig. 4A; McGeary et al., 1994) and intracontinental arcs (the active Cascades arc, Keach et al., 1989; the Mesozoic Sierra Nevada, Nelson et al., 1986). The high average velocity of 6.6 km/s, which we tentatively derive by comparing the reflection and the gravity Mohos, exceeds that of mean continental crust (Christensen and Mooney, 1995), but is the same as found for the modern Aleutian oceanic arc in 30-km-thick crust (Fliedner and Klemperer, 1999). The larger thickness, ~40 km, is comparable to that of arcs such as the modern Aleutian intracontinental arc of the western Alaska Peninsula, in which part of the thickness is due to preexisting crust (Fliedner and Klemperer, 2000). Cooper et al. (1987) described the long-wavelength magnetic high that crosses our profile at ~km –550 (Fig. 2; Klemperer et al., this volume, Chapter 19, their Fig. 4) as similar to the anomaly pattern that occurs over the volcanic mass of the Aleutian Ridge, reinforcing our view that this southern region includes one or more arcs, presumably of late Mesozoic to early Tertiary age.

However, the region from the axis of the Saint Matthew–Nunivak arch to the modern shelf edge is far too broad to have formed by arc magmatism alone. It is 250–450 km wide and was ~40 km thick across the entire region before the formation of the Navarin Basin. At the crustal growth rates evidenced in the modern Aleutian arc, where a 200-km-wide arc of 25–30 km thickness grew in 55 Ma (Holbrook et al., 1999), the Outer Bering Shelf would require 100–180 Ma to be created by arc magmatism alone (and 2–4 times longer if we accept the much lower estimates of arc production rates of Reymer and Schubert, 1984), requiring arc initiation in this region at least as early as 155 Ma (Late Jurassic). The existence of the middle to Late Cretaceous Okhotsk–Chukotsk magmatic belt extending across the Bering Strait several hundred kilometers to the north further demonstrates that arc magmatism was not restricted to the re-

                                        S.L. Klemperer et al.

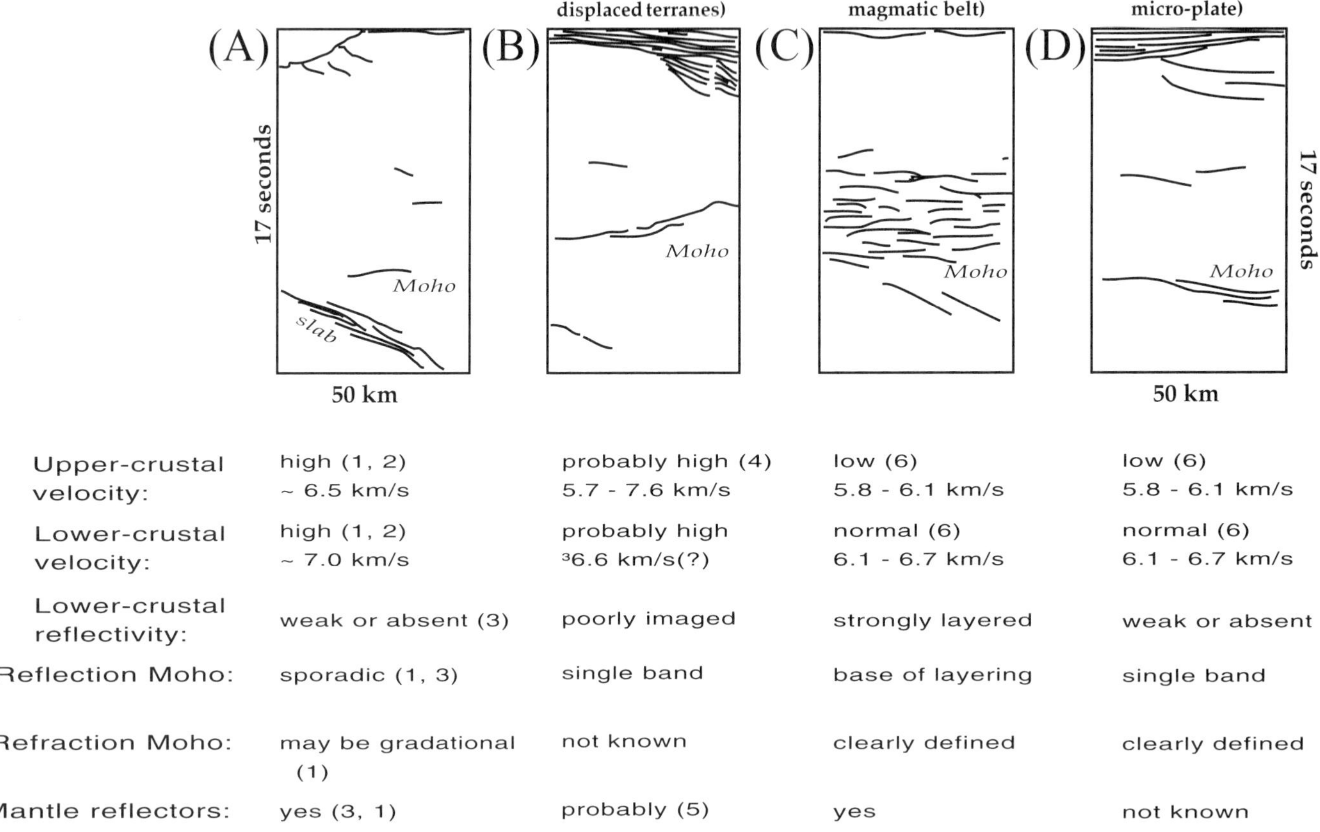

	Aleutian Arc	Outer Bering Shelf	Bering Straits	Chukchi Shelf
Upper-crustal velocity:	high (1, 2) ~ 6.5 km/s	probably high (4) 5.7 - 7.6 km/s	low (6) 5.8 - 6.1 km/s	low (6) 5.8 - 6.1 km/s
Lower-crustal velocity:	high (1, 2) ~ 7.0 km/s	probably high [3]6.6 km/s(?)	normal (6) 6.1 - 6.7 km/s	normal (6) 6.1 - 6.7 km/s
Lower-crustal reflectivity:	weak or absent (3)	poorly imaged	strongly layered	weak or absent
Reflection Moho:	sporadic (1, 3)	single band	base of layering	single band
Refraction Moho:	may be gradational (1)	not known	clearly defined	clearly defined
Mantle reflectors:	yes (3, 1)	probably (5)	yes	not known

Figure 4. Representative reflection characteristics, schematized in line drawings, from (A) active Aleutian arc (McGeary et al., 1994; McGeary, 1997); (B) Outer Bering Shelf (Navarin Basin, Plate 2); (C) Bering Strait and Inner Bering Shelf (near Saint Lawrence Island, Plate 2); (D) Arctic Alaska microplate (north margin of Colville foredeep and underlying Franklinian and older basin, Plate 2). Key velocity and reflectivity attributes of these crustal types are discussed in this chapter, or are taken from literature: (1) Fliedner and Klemperer (1999); (2) Holbrook et al. (1999); (3) McGeary et al. (1994); McGeary (1997); (4) Cooper et al. (1987); (5) Marlow and Cooper (1985); (6) Wolf et al. (this volume).

gion south of the Saint Matthew–Nunivak arch. Therefore, a large proportion of older crust must be present beneath the Outer Bering Shelf, into which the late Mesozoic to early Tertiary arc rocks were intruded. This older crust of unidentified allochthonous terranes may include the Peninsular terrane, which comprises largely Early Jurassic Talkeetna oceanic arc material (e.g., Plafker and Berg, 1994b), even though Peninsular terrane rocks have not been dredged as far west as our transect (Cooper et al., 1987; Worrall, 1991; see tabulation of dredge samples from the Beringian shelf edge in Plate 1, this volume). Paleogene arc magmatism constitutes an intrusive assemblage that has obliterated any distinctions between the different terranes beneath the Bering Shelf. Because Moho reflections are typically not clear beneath the modern Aleutian arc (McGeary et al., 1994; Fliedner and Klemperer, 1999), we speculate that the relatively sharp reflection Moho visible beneath the Navarin Basin developed after cessation of arc activity, and was created by the extension that formed that basin.

In Figure 2 we show a transitional change in lower-crustal reflectivity from the central mid-Cretaceous magmatic belt to the southern Late Cretaceous–Paleogene magmatic arc province. Cooper et al. (1987, p. 81) and Worrall (1991, p. 22) defined a broad region of "short-period, high-amplitude [magnetic] anomalies" that they suspected marks a "magmatic arc complex" and which on our transect spans the region from ~km –350 to –650 (Fig. 2). The northern and southern limits of the highest-amplitude magnetic anomalies along our transect are at the southern flank of the Saint Matthew Basin and northern flank of the Navarin Basin; however, the increasing thickness of sedimentary rocks, particularly to the south, probably suppresses the short-period magnetic anomalies. Short-period anomalies of gradually diminishing amplitude were recorded as far north as km –100 on our transect. Given that areal magnetic coverage is not complete in this region (Klemperer et al., this volume, Chapter 19, their Fig. 4), we only suggest the possibility that the gradual change in magnetic character may equate with the transitional change in crustal reflectivity that we place at km –450 ± 50 km.

Central region: Middle to Late Cretaceous extensional and magmatic belt

Figure 4C cartoons the crust of the Inner Bering Shelf and the Bering Strait region, where crustal reflectivity is significantly greater than that beneath the Outer Bering Shelf. This characteristic lower-crustal reflectivity was first observed to be widespread in the thin crust of young extensional provinces across the UK continental shelf (e.g., Matthews and Cheadle, 1986) and in the western United States (Allmendinger et al., 1987), and was assumed to be due to extensional processes. Given later discovery of similar reflectivity in many cratonic regions (e.g., BABEL Working Group, 1990; Cook et al., 1999), it now appears probable that lower-crustal reflectors typically represent tectonically imposed layering, including shear zones (e.g., Green et al., 1990) and/or primary igneous layering produced by mafic intrusions into the lower crust (e.g., Warner, 1990). In any individual location it is difficult to distinguish between these possibilities, and large-magnitude extension is likely to be associated with both shearing and new intrusions. Other possible causes of deep-crustal reflectivity (e.g., primary depositional layering, metamorphic layering, and fluid-related phenomena) seem less likely to produce sufficiently strong reflections (e.g., Warner, 1990). Modest extension in the course of intracontinental basin formation (β factor <2) seems unlikely to create significant reflectivity because it produces little magmatism, only small rotation or transposition of preexisting layering, and minor development of new foliation (Nelson, 1991). Attention has therefore focused on extension (crustal collapse) during or after crustal thickening or postorogenic delamination and uplift as likely origins for much observed highly reflective crust worldwide (Warner, 1990; Nelson, 1992; Rey, 1993). In these tectonic settings abundant magmatism and large extension can combine to generate significant new crustal reflectivity.

The limited extension (β = 1.5) beneath the Hope, Norton, and Saint Matthew Basins shown by our reflection profiles suggests that this episode of Tertiary basin formation is not the source of the observed lower-crustal reflectivity. Instead we attribute the reflectivity of the Inner Bering Shelf and the Bering Strait to the postulated collapse of Mesozoic crust overthickened by crustal shortening during formation of the Brookian orogen, and by subsequent plutonism within the middle to Late Cretaceous magmatic belt (Rubin et al., 1995; Bering Strait Geologic Field Party, 1997). The observed region of reflective crust and uniform crustal thickness coincides with the locus of Cretaceous plutonism mapped on land. Studies on land in the Chukotka and Seward Peninsulas indicate that magmatic heating, metamorphism, and deformation during the time span 108–90 Ma (Miller et al., this volume, Chapter 17) likely affected the entire crustal column. The uniform thickness of 32 km beneath the Chukchi and Inner Bering shelves (Wolf et al., this volume) matches the

global average thickness of extended continental crust of 31.5 km (Christensen and Mooney, 1995). The lower crust from Saint Lawrence Island to the Seward Peninsula has velocities in the range 6.5–6.7 km/s (Wolf et al., this volume) that are not consistent with the underplating of thick mafic units with velocities ~7 km/s. These velocities are consistent, however, with a bimodal crust consisting of ~6.5 km/s felsic to intermediate igneous and metamorphic lithologies intruded by ~7 km/s gabbroic sills. We presume such mafic sills are the source of the mafic xenoliths with cumulate textures (Wirth et al., this volume) and the 85–100 Ma ages of granulite facies crustal xenoliths (Miller et al., this volume, Chapter 10) found within younger lava flows on Saint Lawrence Island. Small-scale velocity variation between the suggested felsic to intermediate lower crust and mafic intrusions cannot be detected by our wide-angle data, which record only volumetric averages of the velocity, but we suggest that it is detected by the reflection data as the prominent reflectivity of the lower crust in this area (cf. Wenzel et al., 1987). However, the lower velocity of the middle crust, 6.1–6.4 km/s from 16 to 24 km depth (Wolf et al., this volume), precludes a large component of new mafic intrusions, and suggests that much of the reflectivity in this depth range must represent extensional fabrics formed in a ductile middle crust, such as are exposed in gneiss domes on the Chukotka and Seward Peninsulas (Fig. 1; Akinin and Calvert, this volume; Amato et al., this volume).

The low velocities of the upper crust (5.8–6.1 km/s to 16 km depth, Wolf et al., this volume) seem likely to represent Cretaceous granite, such as the granite that crops out at the surface on Saint Lawrence and the Diomede Islands and the paragneisses that crop out on Seward Peninsula (Christensen and Mooney, 1995). However, the moderate lower-crustal velocities imply that a voluminous mafic counterpart to the Cretaceous granites in the upper crust is not present today in the lower crust. Possibly a voluminous mafic counterpart never existed (which seems to require that the upper-crustal granites are crustal melts), or it formerly existed (according to Amato and Wright [1997] the granite is dominantly fractionated from mafic melt newly derived from the mantle by arc magmatism) but subsequently delaminated. Although Neogene and Quaternary basaltic volcanism (Akinin et al., 1997) presumably also left some traces within the crust, the basalts and their entrained xenoliths come directly from the mantle, and the proportion of mafic rocks intruded into the crust must be low because average crustal velocities here are close to the global continental crustal average of 6.4 km/s (Christensen and Mooney, 1995).

We conclude that the prominent lower-crustal reflectivity that characterizes the middle to Late Cretaceous magmatic belt was generated in part during magmatic thickening, but particularly by the ensuing extensional collapse and possible delamination. The magnitude of extension may have been greater in the Bering Strait, thinning the crust to the extent that it is now below sea level, than to the west or east, where gneiss domes produced by this extension rise 1 km above sea level. These gneiss domes imply ~5 km thicker crust if their topography is supported by Airy isostasy, consistent with onshore crustal thickness measurements of 38 km along the TACT profile in the Koyukuk Basin immediately south of the Brooks Range (Fuis et al., 1995). Middle Cretaceous extension and middle to Late Cretaceous magmatism certainly affected the onshore areas of Alaska (Miller and Hudson, 1991; Miller et al., this volume, Chapter 17), and we interpret the reflective crust of the Yukon-Tanana terrane (Beaudoin et al., 1992) and of the Koyukuk Basin (Fuis et al., 1995) as having formed in the same manner and at the same time, as observed along our marine transect. The more complete collapse of the orogen in the Bering Strait segment could have been a consequence of local plate geometry, with rapid slab rollback in the Bering Strait region facilitated by the presence of an open ocean south of the subduction zone west of Alaska (Rubin et al., 1995), but precluded along strike to the east by continental terranes south of the subduction zone (Amato et al., this volume).

Northern region: Arctic Alaska microplate

In Figure 4D we show the character of the northern region as poorly reflective crust beneath deep basins and above a distinct reflection Moho. Comparing D to B in Figure 4, the superficial resemblance of Arctic Alaska crust to the crust of the Outer Bering Shelf cautions against a one-to-one assignment of specific seismic characteristics to particular tectonic environments. Whereas the reflectivity character of the Chukchi Shelf resembles that of the Outer Bering Shelf, the velocity structure of the Chukchi Shelf is nearly indistinguishable from that of the Bering Strait and Inner Bering Shelf (Wolf et al., this volume). The low upper-crustal velocities of the Chukchi Shelf (5.7–6.1 km/s to 15 km depth) presumably represent the thick sedimentary succession imaged on our reflection profiles, as opposed to the granite of similar seismic velocity on the Inner Bering Shelf. Nonetheless, specific reflectivity patterns appear to characterize, and hence may be used to map, specific crustal blocks (e.g., Klemperer et al., 1990; Brown, 1991), and we suggest that the sparse crustal reflectivity of the Arctic Alaska crust on our marine profiles also characterizes the crust beneath the allochthons of the northern Brooks Range (Fuis et al., 1995, their Fig. 2A).

The northern segment of our profile, beneath the Chukchi Shelf, is the most difficult to interpret, because the basement is not exposed, and because it has the longest history, >1 Ga. The Arctic Alaska microplate has undergone at least three extensional episodes, forming the Proterozoic to Franklinian basins, the Mississippian to Early Cretaceous south-facing Ellesmerian continental margin, and the modern north-facing Beaufort margin. In addition, localized extension apparently related to dextral slip on the Kobuk fault created the Hope Basin in the continental interior during the Paleogene. If the clear image of the reflection Moho beneath Arctic Alaska crust resulted from extension, then it is likely of different age in different regions, in part Tertiary (beneath the Hope Basin) and in part Jurassic or older (beneath the Ellesmerian and Franklinian basins). Whereas high topography requires that a crustal root still underlies the Brooks Range in Alaska (Fig. 1; Fuis et al., 1995), neither topography nor a root are present in the region of the Hope and Kotzebue Basins where crossed by our profiles (see also refraction-seismic and gravity evidence; Wolf et al., this volume). Though topography and a root can be inferred to have existed during and prior to the deposition of the mid-Cretaceous Corwin Delta of the Colville foredeep, they could have been destroyed by only relatively minor extension ($\beta \leq 1.5$), without creation of lower crustal reflectivity. However, unless crustal reflectivity beneath Arctic Alaska was destroyed following its creation (e.g., through disruption by dike injection over long periods of time; Nelson, 1991), the lack of intracrustal reflectivity suggests that northern Alaska never underwent orogenic thickening and postorogenic collapse of the style evidenced farther to the south. This conclusion, drawn from the lack of crustal reflectivity, is also demonstrated by the well-known continuity of Proterozoic and younger strata across Arctic Alaska that is clearly seen on our transect.

SUMMARY AND CONCLUSIONS

This first continuous deep-crustal reflection profile acquired across the entire North American continent provides baseline information on crustal thickness and evolution for the Alaskan continental shelves from the Aleutian Basin of the Bering Sea to the Canada Basin of the Arctic Ocean. Our profiles traverse many areas that are distant from onshore outcrop, so our interpretations, although building on a history of basin exploration on the Bering and Chukchi shelves, are inevitably broad. Our transect does, however, provide new insight into various aspects of continental growth and modification in Alaska and northeastern Russia, as recorded by crustal reflectivity patterns.

Our transect (Fig. 2) shows the central Bering Strait–Inner Bering Shelf basement high, flanked to north and south by large accumulations of Tertiary sediments at the modern shelf edges (Beaufort and Beringian margins) or interior to a narrow marginal high (Navarin Basin). In contrast, neither of the deep Tertiary sedimentary basins in the continental interior (Hope Basin beneath the Chukchi Shelf and Norton Basin beneath the Inner Bering Shelf) are rift basins, but rather seem to be accommodations of strike-slip faulting.

The observed gradual thickening of crystalline basement from the Proterozoic core of the Arctic Alaska–Chukotka plate to the Bering Strait (Fig. 2) may in part reflect the accretion and growth of continental crust to the south, the crust of the Bering Shelf being dominated by Phanerozoic components. The oldest crust beneath Arctic Alaska, extending southward in the subsurface beneath Brooks Range and Herald Arch thrusts, appears to have remained a relatively stable block or microplate since the Late Proterozoic, only mildly modified by the stretching, thrusting, and loading events along its southern margin produced by the Brookian orogeny. The preservation of a great thickness of stratified sedimentary rocks beneath the Chukchi Shelf, above the thin crystalline basement of Arctic Alaska, is due to the long-term stability of Arctic Alaska, which has saved the older basins from being overprinted by younger events. It is uncertain whether this relative stability is due to crustal age or lithology, or whether it is a consequence of the distance of Arctic Alaska from the Mesozoic active-margin-related deformation.

The thicker basement of our central belt from the Seward Peninsula to south of Saint Lawrence Island was derived, at least in part, from the same Precambrian continental crust including Proterozoic to Paleozoic sedimentary successions, but also from magmatic accretion. The crust of this central belt represents a zone of former weakness in which the older crust was completely reworked at what was then the continental margin by Late Jurassic to early Late Cretaceous compression and thickening, followed by extension and magmatism, to create its present distinctive reflection character. This extension, we believe, expanded the area of central Alaska and its shelf southward and brought the topography of an overthickened crust back to near sea level. This tectonic reworking may represent one way in which the more juvenile crust of the Outer Bering Shelf could in the future acquire a more typical continental character through a similar sequence of orogenesis.

The Outer Bering Shelf is underlain by the youngest crust along our transect, and may be formed of assembled terranes of Mesozoic and older rocks that were intruded by Mesozoic and Cenozoic arc magmas. Perhaps accretionary events without substantial crustal shortening or thickening (i.e., Mediterranean-style orogens; Burchfiel and Royden, 1991), and/or translation of material along strike-slip faults into a region without major shortening, helped preserve a first-cycle crust with its original thickness and character, as seen beneath the Outer Bering Shelf. Although the major strike-slip faults that bound Alaskan terranes onshore have so far proved largely unrecognizable offshore, we can map divergent normal fault systems and crustal thinning beneath the offshore Hope and Norton Basins capable of accommodating the strain represented by two of these strike-slip faults, the Kobuk and Kaltag faults. We suggest that these faults continue west beneath our transect, albeit with progressively diminishing offset as their strike-slip displacement is transformed into extension beneath the shelf basins. Dominantly strike-slip (transpressional) accretion in the southern Bering Shelf contrasts with the collisional orogenesis and crustal thickening in the Brooks Range and the central zone of our transect. In the central zone, collisional accretion of Paleozoic and Mesozoic terrancs with the preexisting sedimentary cover and Precambrian basement led inexorably to the subsequent modification of these terranes beyond easy recognition by the processes of crustal collapse, extension, and intrusion of magmas.

ACKNOWLEDGMENTS

This cruise was funded by the National Science Foundation Continental Dynamics grant EAR-93–17087; additional scientific participation and initial seismic processing were funded by the U.S. Geological Survey Deep Continental Studies Program. We thank the shipboard scientific staff, captain, and crew of the RV *Ewing* for their efforts in acquiring high-quality seismic data, and the staff of Scott Pickford for processing of the reflection data. Tom Brocher and Walter Mooney provided helpful reviews. University of Alaska, Fairbanks, and Royal Holloway, University of London, kindly hosted Klemperer during part of this work. All the reflection seismic data are available from Stanford University for the cost of reproduction.

REFERENCES CITED

Akinin, V.V., Roden, M.F., Francis, D., Apt, J., and Moll-Stalcup, E.J., 1997, Compositional and thermal state of the upper mantle beneath the Bering Sea basalt province: Evidence from the Chukchi Peninsula of Russia: Canadian Journal of Earth Sciences, v. 34, p. 789–800.

Allmendinger, R.W., Hauge, T.A., Hauser, E.C., Potter, C.J., Klemperer, S.L., Nelson, K.D., Knuepfer, P., and Oliver, J., 1987, Overview of the COCORP 40°N Transect, western United States: The fabric of an orogenic belt: Geological Society of America Bulletin, v. 98, p. 308–319.

Amato, J.M., and Wright, J.E., 1997, Petrogenesis of the potassic Kigluaik pluton: Arc-related mafic magmatism in northern Alaska: Journal of Geophysical Research, v. 102, p. 8065–8083.

BABEL Working Group, 1990, Evidence for Early Proterozoic plate tectonics from seismic reflection profiles in the Baltic Shield: Nature, v. 348, p. 34–38.

Barnes, D.F., Marianio, J., Morin, R.L., Roberts, C.W., and Jachens, R.C., 1994, Incomplete isostatic gravity map of Alaska, *in* Plafker, G., and Berg, H.C., eds., The geology of Alaska: Boulder, Colorado, Geological Society of America, Geology of North America, v. G-1, plate 9, scale 1:2 500 000, 1 sheet.

Beaudoin, B.C., Fuis, G.S., Mooney, W.D., Nokleberg, W.J., and Christensen, N.I., 1992, Thin, low-velocity crust beneath the southern Yukon-Tanana terrane, east-central Alaska: Results from TACT refraction/wide-angle reflection data: Journal of Geophysical Research, v. 97, p. 1921–1942.

Bering Strait Geologic Field Party, 1997, Koolen metamorphic complex, NE Russia: Implications for the tectonic evolution of the Bering Strait region: Tectonics, v. 16, p. 713–729.

Brocher, T.M., Allen, R.M., Stone, D.B., Wolf, L.W., and Galloway, B.K., 1995, Data report for onshore-offshore wide-angle seismic recordings in the Bering-Chukchi Sea, western Alaska and eastern Siberia: U.S. Geological Survey Open-File Report 95–0650, 57 p.

Brown, L.D., 1991, A new map of crustal "terranes" in the United States from COCORP deep seismic reflection profiling: Geophysical Journal International, v. 105, p. 3–13.

Campbell, R.H., 1967, Areal geology in the vicinity of the Chariot Site, Lisburne Peninsula, northwestern Alaska: U.S. Geological Survey Professional Paper 395, 71 p.

Christensen, N.I., and Mooney, W.D., 1995, Seismic velocity structure and composition of the continental crust: A global view: Journal of Geophysical Research, v. 100, p. 9761–9788.

Coflin, K.C., Cook, F.A., and Geis, W.T., 1990, Evidence for Ellesmerian convergence in the subsurface east of the Mackenzie Delta: Marine Geology, v. 93, p. 289–301.

Collins, F.R., 1958, Test wells, Topagoruk area, Alaska. 5. Subsurface geology and engineering data: Exploration of Naval Petroleum Reserve No. 4 and adjacent areas, northern Alaska, 1944–1953: U.S. Geological Survey Professional Paper 305-D, p. 265–316.

Collins, F.R., 1961, Core tests and test wells, Barrow area, Alaska. 5. Subsurface geology and engineering data: Exploration of Naval Petroleum Reserve No. 4 and adjacent areas, northern Alaska, 1944–1953: U.S. Geological Survey Professional Paper 305-K, p. 569–644.

Coney, P.J., and Jones, D.L., 1985, Accretion tectonics and crustal structure in Alaska: Tectonophysics, v. 119, p. 265–282.

Cook, F.A., van der Velden, A.J., Hall, K.W., and Roberts, B.J., 1999, Frozen subduction in Canada's Northwest Territories: Lithoprobe deep lithospheric reflection profiling of the western Canadian Shield: Tectonics, v. 18, p. 1–24.

Cooper, A.K., Marlow, M.S., and Scholl, D.W., 1987, Geologic framework of the Bering Sea crust, *in* Scholl, D.W., Grantz, A., and Vedder, J.G., eds., Geology and resource potential of the continental margin of western North America and adjacent ocean basins: Beaufort Sea to Baja California: Houston, Texas, Circum-Pacific Council for Energy and Mineral Resources, Earth Science Series, v. 6, p. 73–102.

Craig, J.D., Sherwood, K.W., and Johnson, P.P., 1985, Geologic report for the Beaufort Sea planning area, Alaska: Regional geology, petroleum geology, environmental geology: Anchorage, Alaska, U.S. Department of the Interior, Minerals Management Service Outer Continental Shelf Report 85–0111, 192 p.

Davis, A.S., Pickthorn, L.-B.G., Vallier, T.L., and Marlow, M.S., 1989, Petrology and age of volcanic-arc rocks from the continental margin of the Bering Sea: Implications for early Eocene relocation of plate boundaries: Canadian Journal of Earth Sciences, v. 26, p. 1474–1490.

Davis, A.S., Gunn, S.H., Gray, L.-B., Marlow, M.S., and Wong, F.L., 1993, Petrology and isotopic composition of Quaternary basanites dredged from the Bering Sea continental margin near Navarin Basin: Canadian Journal of Earth Sciences, v. 30, p. 975–984.

Dumitru, T.A., Miller, E.L., O'Sullivan, P.B., Amato, J.M., Hannula, K.A., Calvert, A.T., and Gans, P.B., 1995, Cretaceous to recent extension in the Bering Strait region, Alaska: Tectonics, v. 14, p. 549–563.

Eittreim, S., Grantz, A., and Whitney, O.T., 1978, Isopach maps of the Tertiary sediments, Hope Basin, southern Chukchi Sea, Alaska: U.S. Geological Survey Miscellaneous Field Studies Map MF-906, scale 1:1 000 000, 1 sheet.

Embry, A.F., 1990, Geological and geophysical evidence in support of the hypothesis of anticlockwise rotation of northern Alaska: Marine Geology, v. 93, p. 317–329.

Fisher, M.A., Patton, W.W., Jr., and Holmes, M.L., 1982, Geology of Norton Basin and continental shelf beneath northwestern Bering Sea, Alaska: American Association of Petroleum Geologists Bulletin, v. 66, p. 255–285.

Fliedner, M.M., and Klemperer, S.L., 1999, Composition of an island-arc: Wide-angle studies in the eastern Aleutian islands, Alaska: Journal of Geophysical Research, v. 104, p. 10667–10694.

Fliedner, M.M., and Klemperer, S.L., 2000, Crustal structure transition from oceanic arc to continental arc, eastern Aleutian Islands and Alaska Peninsula: Earth and Planetary Science Letters, v. 179, p. 567–579.

Fuis, G.S., Levander, A.R., Lutter, W.J., Wissinger, E.S., Moore, T.E., and Christensen, N.I., 1995, Seismic images of the Brooks Range, Arctic Alaska, reveal crustal-scale duplexing: Geology, v. 23, p. 65–68.

Fujita, K., and Newberry, J.T., 1982, Tectonic evolution of northeastern Siberia and adjacent regions: Tectonophysics, v. 89, p. 337–357.

Grantz, A., and May, S.D., 1987, Regional geology and petroleum potential of the United States Chukchi Shelf north of Point Hope, *in* Scholl, D.W., Grantz, A., and Vedder, J.G., eds., Geology and resource potential of the continental margin of western North America and adjacent ocean basins: Beaufort Sea to Baja California: Houston, Texas, Circum-Pacific Council for Energy and Mineral Resources, Earth Science Series, v. 6, p. 37–58.

Grantz, A., Wolf, S.C., Breslau, L., Johnson, T.C., and Hanna, W.F., 1970, Reconnaissance geology of the Chukchi Sea as determined by acoustic and magnetic profiling, *in* Adkison, W.L., and Brosge, M.M., eds., Proceedings of the Geological Seminar on the North Slope of Alaska: American Association of Petroleum Geologists, Pacific Section, Los Angeles, p. F1–F28.

Grantz, A., May, S.D., and Dinter, D.A., 1987, Regional geology and petroleum potential of the United States Beaufort and northeasternmost Chukchi Seas, *in* Scholl, D.W., Grantz, A., and Vedder, J.G., eds., Geology and resource potential of the continental margin of western North America and adjacent ocean basins: Beaufort Sea to Baja California: Houston, Texas, Circum-Pacific Council for Energy and Mineral Resources, Earth Science Series, v. 6, p. 17–35.

Grantz, A., May, S.D., and Hart, P.E., 1990, Geology of the Arctic continental margin of Alaska, *in* Grantz, A., Johnson, G.L., and Sweeney, J.F., eds., The Arctic Ocean region: Boulder, Colorado, Geological Society of America, Geology of North America, v. L, p. 257–288.

Grantz, A., Moore, T.E., and Roeske, S., 1991, Centennial continent/ocean transect #15: Gulf of Alaska to Arctic Ocean: Boulder, Colorado, Geological Society of America, 1:500 000, 3 sheets, 72 p. text.

Grantz, A., Clark, D.L., Phillips, R.L., and Srivastava, S.P., 1998, Phanerozoic stratigraphy of Northwind Ridge, magnetic anomalies in the Canada Basin, and the geometry and timing of rifting in the Amerasia Basin, Arctic Ocean: Geological Society of America Bulletin, v. 110, p. 801–820.

Green, A.G., Milkereit, B., Percival, J.A., Davidson, A., Parrish, R.R., Cook, F.A., Geis, W.T., Cannon, W.F., Hutchinson, D.R., West, G.F., and Clowes, R.M., 1990, Origin of deep crustal reflections: Seismic profiling across high-grade metamorphic terranes in Canada: Tectonophysics, v. 173, p. 627–638.

Hatton, L., Worthington, M.H., and Makin, J.E., 1986, Seismic data processing: Theory and practice: Oxford, UK, Blackwell Science Publishers, 177 p.

Higgins, A.K., Ineson, J.R., Peel, J.S., Surlyk, F., and Sønderholm, M., 1991, Cambrian to Silurian basin development and sedimentation, North Greenland, *in* Trettin, H.P., ed., Geology of the Innuitian Orogen and Arctic Platform of Canada and Greenland: Ottawa, Canada, Geological Survey of Canada, Geology of Canada, no. 3, p. 109–161.

Holbrook, W.S., Lizarralde, D., McGeary, S., Bangs, N., and Diebold, J., 1999, Structure and composition of the Aleutian island arc and implications for continental crustal growth: Geology, v. 27, p. 31–34.

Holliger K., and Klemperer, S.L., 1990, Gravity and deep seismic reflection profiles across the North Sea rifts, *in* Blundell, D.J., and Gibbs, A.D., Tectonic evolution of the North Sea rifts: Oxford, UK, Oxford University Press, p. 82–100.

Holmes, M.L., and Creager, J.S., 1981, The role of the Kaltag and Kobuk faults in the tectonic evolution of the Bering Strait region, *in* Hood, D.W., and Calder, J.A., eds., The eastern Bering Sea Shelf: Oceanography and resources: Boulder, Colorado, National Oceanic and Atmospheric Administration, Department of Commerce, v. 1, p. 293–302.

Houtz, R.E., Eittreim, S., and Grantz, A., 1981, Acoustic properties of northern Alaska shelves in relation to the regional geology: Journal of Geophysical Research, v. 86, p. 3935–3943.

Jin, D.J., and Herrin, E., 1980, Surface wave studies of the Bering Sea and Alaska: Bulletin of the Seismological Society of America, v. 70, p. 2117–2144.

Jones, D.L., Silberling, N.J., and Hillhouse, J.W., 1977, Wrangellia: A displaced terrane in northwestern North America: Canadian Journal of Earth Sciences, v. 14, p. 2565–2577.

Jones, D.M., Kingston, M.J., Marlow, M.S., Cooper, A.K., Barron, J.A., Wingate, F.H., and Arnal, R.E., 1981, Age, mineralogy, physical properties, and geochemistry of dredge samples from the Bering Sea continental margin: U.S. Geological Survey Open-File Report 81–1297, 68 p.

Kanasewich, E.R., and Berkes, Z., 1990, Seismic structure of the Proterozoic on Melville Island, Canadian Arctic archipelago: Marine Geology, v. 93, p. 421–448.

Keach, R.W., II., Oliver, J.E., Brown, L.D., and Kaufman, S., 1989, Cenozoic active margin and shallow Cascades structure: COCORP results from western Oregon: Geological Society of America Bulletin, v. 101, p. 783–794.

Kirschner, C.E., 1988, Map showing sedimentary basins of onshore and continental shelf areas, Alaska: U.S. Geological Survey Map I-1873, scale 1:5 000 000, 1 sheet.

Kirschner, C.E., 1994, Interior basins of Alaska, *in* Plafker, G., and Berg, H.C., eds., The geology of Alaska: Boulder, Colorado, Geological Society of America, Geology of North America, v. G-1, p. 469–493.

Klemperer, S.L., 1989, Processing BIRPS deep seismic reflection data: A tutorial review, *in* Cassinis, R., Nolet, G., and Panza, G.F., eds., Digital seismology and fine modeling of the lithosphere: New York, Plenum Press, Ettore Majorana International Science Series, Physical Sciences, v. 42, p. 229–257.

Klemperer, S.L., and Hobbs, R.W., 1991, The BIRPS atlas: Deep seismic reflection profiles from around the British Isles: Cambridge, Cambridge University Press, 124 p., 99 sections.

Klemperer, S.L., and Hurich, C.A., 1990, Lithospheric structure of the North Sea from deep seismic reflection profiling, *in* Blundell, D.J., and Gibbs, A.D., eds., Tectonic evolution of the North Sea rifts: Oxford, UK, Oxford University Press, p. 37–63.

Klemperer, S.L., Hauge, T.A., Hauser, E.C., Oliver, J.E., and Potter, C.J., 1986, The Moho in the northern Basin and Range province, Nevada, along the COCORP 40°N seismic-reflection transect: Geological Society of America Bulletin, v. 97, p. 603–618.

Klemperer, S.L., Hobbs, R.W., and Freeman, B., 1990, Dating the source of lower crustal reflectivity using BIRPS deep seismic profiles across the Iapetus suture: Tectonophysics, v. 173, p. 445–454.

Lane, L.S., 1997, Canada Basin, Arctic Ocean: Evidence against a rotational origin: Tectonics, v. 16, p. 363–387.

Lawver, L.A., and Scotese, C.R., 1990, A review of tectonic models for the evolution of the Canada Basin, *in* Grantz, A., Johnson, G.L., and Sweeney, J.F., eds., The Arctic Ocean region: Boulder, Colorado, Geological Society of America, Geology of North America, v. L, p. 593–618.

Lerand, M., 1973, Beaufort Sea, *in* McCrossan, R.G., ed., The future petroleum provinces of Canada: Their geology and potential: Canadian Society of Petroleum Geologists Memoir 1, p. 315–386.

Mackey, K.G., Fujita, K., Gunbina, L.V., Kovalev, V.N., Imaev, V.S., Koz'min, B.M., and Imaeva, L.P., 1997, Seismicity of the Bering Strait region: Evidence for a Bering Block: Geology, v. 25, p. 979–982.

Mackey, K.G., Fujita, K., and Ruff, L.J., 1998, Crustal thickness of northeast Russia: Tectonophysics, v. 284, p. 283–297.

Marlow, M.S., and Cooper, A.K., 1980, Mesozoic and Cenozoic structural trends under southern Bering Sea Shelf: American Association of Petroleum Geologists Bulletin, v. 64, p. 2139–2155.

Marlow, M.S., and Cooper, A.K., 1985, Regional geology of the Beringian continental margin, *in* Nasu, N., ed., Formation of active ocean margins: Boston, Massachusetts, Terra Scientific Publishing, p. 497–515.

Marlow, M.S., Scholl, D.W., Cooper, A.K., and Buffington, E.C., 1976, Structure and evolution of Bering Sea Shelf south of St. Lawrence Island: American Association of Petroleum Geologists Bulletin, v. 60, p. 161–183.

Marlow, M.S., Cooper, A.K., and Fisher, M.A., 1987, Petroleum geology of the Beringian continental shelf, *in* Scholl, D.W., Grantz, A., and Vedder, J.G., eds., Geology and resource potential of the continental margin of western North America and adjacent ocean basins: Beaufort Sea to Baja California: Houston, Texas, Circum-Pacific Council for Energy and Mineral Resources, Earth Science Series, v. 6, p. 103–122.

Matthews, D.H., and Cheadle, M.J., 1986, Deep reflections from the Caledonides and Variscides west of Britain and comparison with the Himalayas, *in* Barazangi, M., and Brown, L.D., eds., Reflection seismology: A global perspective: American Geophysical Union Geodynamics Series, v. 13, p. 5–19.

Mayfield, C.F., Tailleur, I.L., and Ellersieck, I., 1988, Stratigraphy, structure, and palinspastic synthesis of the western Brooks Range, northwestern Alaska, *in* Gryc, G., ed., Geology and exploration of the National Petroleum Reserve in Alaska, 1974 to 1982: U.S. Geological Survey Professional Paper 1399, p. 143–186.

McGeary, S., 1997, The deep seismic reflection image of the Aleutian subducting plate as an analog for other mantle reflectors: Geological Society of America Abstracts with Programs, v. 29, no. 6, p. 233.

McGeary, S.E., and Ben-Avraham, Z., 1981, Allochthonous terranes in Alaska: Implications for the structure and evolution of the Bering Shelf: Geology, v. 9., p. 608–613.

McGeary, S., Diebold, J.B., Bangs, N.L., Bond, G., and Buhl, P., 1994, Preliminary results of the Pacific to Bering Shelf deep seismic experiment: Eos (Transactions, American Geophysical Union), v. 75, no. 44, supplement, p. 643.

McKerrow, W.S., MacNiocaill, C., and Dewey, J.F., 2000, The Caledonian orogeny redefined: Journal of the Geological Society, London, v. 157, p. 1149–1154.

Meissner, R., 1986, The continental crust: A geophysical approach: Orlando, Florida, Academic Press, International Geophysics Series, v. 34, 426 p.

Miller, E.L., and Hudson, T.L., 1991, Mid-Cretaceous extensional fragmentation of a Jurassic–Early Cretaceous compressional orogen, Alaska: Tectonics, v. 10, p. 781–796.

Moll-Stalcup, E.J., 1994, Latest Cretaceous and Cenozoic magmatic rocks of Alaska, *in* Plafker, G., and Berg, H.C., eds., The geology of Alaska: Boulder, Colorado, Geological Society of America, Geology of North America, v. G-1, plate 5, 1:2 500 000, 1 sheet.

Moll-Stalcup, E.J., Brew, D.A., and Vallier, T.L., 1994, Latest Cretaceous and Cenozoic magmatism in mainland Alaska, *in* Plafker, G., and Berg, H.C., eds., The geology of Alaska: Boulder, Colorado, Geology of North America, v. G-1, p. 589–619.

Mooney, W.D., and Brocher, T.M., 1987, Coincident seismic reflection/refraction studies of the continental lithosphere: A global review: Geophysical Journal of the Royal Astronomical Society, v. 89, p. 1–6.

Moore, T.E., Wallace, W.K., Bird, K.J., Karl, S.M., Mull, C.G., and Dillon, J.T., 1994, Geology of northern Alaska, *in* Plafker, G., and Berg, H.C., eds., The geology of Alaska: Boulder, Colorado, Geological Society of America, Geology of North America, v. G-1, p. 49–140.

Nelson, K.D., 1991, A unified view of craton evolution motivated by recent deep seismic reflection and refraction results: Geophysical Journal International, v. 105, p. 25–35.

Nelson, K.D., 1992, Are crustal thickness variations in old mountain belts like the Appalachians a consequence of lithospheric delamination?: Geology, v. 20, p. 498–502.

Nelson, K.D., Zhu, T.F., Gibbs, A., Harris, R., Oliver, J.E., Kaufman, S., Brown, L., and Schweickert, R.A., 1986, COCORP deep seismic reflection profiling in the northern Sierra Nevada, California: Tectonics, v. 5, p. 321–333.

Nokleberg, W.J., Parfenov, L.M., Monger, J.W.H., Norton, I.O., Khanchuk, A.I., Stone, D.B., Scotese, C.R., Scholl, D.W., and Fujita, K., 2000, Phanerozoic tectonic evolution of the circum-North Pacific: U.S. Geological Survey Professional Paper 1626, 122 p.

Ostenso, N.A., 1968, A gravity survey of the Chukchi Sea region, and its bearing on westward extension of structures in northern Alaska: Geological Society of America Bulletin, v. 79, p. 241–254.

Patton, W.W., Jr., and Csejtey, B., Jr., 1980: Geologic map of St. Lawrence Island, Alaska: U.S. Geological Survey, Miscellaneous Investigations Series I-1203, scale 1:250 000, 1 sheet.

Patton, W.W., Jr., Lanphere, M.A., Miller, T.P., and Scott, R.A., 1976, Age and tectonic significance of volcanic rocks on St. Matthew Island, Bering Sea, Alaska: Journal of Research of the U.S. Geological Survey, v. 4, p. 67–73.

Patton, W.W., Jr., Box, S.E., Moll-Stalcup, E.J., and Miller, T.P., 1994, Geology of west-central Alaska, *in* Plafker, G., and Berg, H.C., eds., The geology of Alaska: Boulder, Colorado, Geological Society of America, Geology of North America, v. G-1, p. 241–269.

Pavlis, T.L., Sisson, V.B., Foster, H.L., Nokleberg, W.J., and Plafker, G., 1992, Mid-Cretaceous extensional tectonics of the Yukon-Tanana terrane, Trans-Alaska Crustal Transect (TACT), east-central Alaska: Tectonics, v. 12, p. 103–122.

Peddy, C., Brown, L.D., and Klemperer, S.L., 1986, Interpreting the deep structure of rifts with synthetic seismic sections: American Geophysical Union Geodynamics Series, v. 13, p. 301–311.

Plafker, G., and Berg, H.C., editors, 1994a, The geology of Alaska: Boulder, Colorado, Geological Society of America, Geology of North America, v. G-1, 1055 p.

Plafker, G., and Berg, H.C., 1994b, Overview of the geology and tectonic evolution of Alaska, *in* Plafker, G., and Berg, H.C., eds., The geology of Alaska: Boulder, Colorado, Geological Society of America, Geology of North America, v. G-1, p. 989–1021.

Rey, P., 1993, Seismic and tectonometamorphic characters of the lower continental crust in Phanerozoic areas: A consequence of post-thickening extension: Tectonics, v. 12, p. 580–590.

Reymer, A., and Schubert, G., 1984, Phanerozoic addition rates to the continental crust and growth: Tectonics, v. 3, p. 63–77.

Rickwood, F.K., 1970, The Prudhoe Bay field, *in* Adkison, W.L., and Brosge, M.M., eds., Proceedings of the Geological Seminar on the North Slope of Alaska: American Association of Petroleum Geologists, Pacific Section, Los Angeles, p. L1–L11.

Rubin, C.M., Miller, E.L., and Toro, J., 1995, Deformation of the northern circum-Pacific margin: Variations in tectonic style and plate-tectonic implications: Geology, v. 23, p. 897–900.

Scholl, D.W., and Stevenson, A.J., 1989, The Aleutian-Bowers-Shirshov arc system, response to deformation of Alaska ("orocline") by subduction-driven impact and escape of crustal masses: Exploration of an idea: Eos (Transactions, American Geophysical Union), v. 70, no. 43, p. 1307.

Scholl, D.W., Marlow, M.S., Creager, J.S., Holmes, M.L., Wolf, S.C., and Cooper, A.K., 1970, A search for the seaward extension of the Kaltag Fault beneath the Bering Sea: Geological Society of America Abstracts with Programs, v. 2, no. 2, p. 141–142.

Scholl, D.W., Grantz, A., and Vedder, J.G., editors, 1987a, Geology and resource potential of the continental margin of western North America and adjacent ocean basins: Beaufort Sea to Baja California: Houston, Texas, Circum-Pacific Council for Energy and Mineral Resources, Earth Science Series, v. 6, 799 p.

Scholl, D.W., Vallier, T.L., and Stevenson, A.J., 1987b, Geologic evolution and petroleum geology of the Aleutian Ridge, *in* Scholl, D.W., Grantz, A., and Vedder, J.G., eds., Geology and resource potential of the continental margin of Western North America and adjacent ocean basins: Beaufort Sea to Baja California: Houston, Texas, Circum-Pacific Council for Energy and Mineral Resources, Earth Science Series, v. 6, p. 123–155.

Sherwood, K.W., 1992, Stratigraphy, structure and origin of the Franklinian, Northeast Chukchi Basin, Arctic Alaska Plate, *in* Thurston, D.K., and Fujita, K., eds., Proceedings of the 1992 International Conference on Arctic Margins: Anchorage, Alaska, Minerals Management Service, p. 245–250.

Shumway, G., Moore, D.G., and Dowling, G.B., 1964, Fairway Rock in Bering Strait, *in* Miller, R.L., Papers in marine geology, Shepard Commemorative Volume: New York, Macmillan, p. 401–407.

Stockwell, C.H., McGlynn, J.C., Emslie, R.F., Sanford, B.V., Norris, A.W., Donaldson, J.A., Fahrig, W.F., and Currie, K.L., 1970, Geology of the Canadian Shield, *in* Douglas, R.J.W., ed., Geology and economic minerals of Canada: Geological Survey of Canada, Economic Geology Report, v. 1, p. 44–150.

Sweeney, J.F., Mayr, U., Sobczak, L.W., and Balkwill, H.R., 1986, North American Continent-Ocean Transects Program, Transect G: Canadian Arctic: Somerset Island to Canada Basin: Boulder, Colorado, Geological Society of America, Centennial Continent/Ocean Transect no. 11, scale 1:500 000, 2 sheets.

Thurston, D.K., and Theiss, L.A., 1987, Geologic report for the Chukchi Sea planning area, Alaska: Regional geology, petroleum geology, and environmental geology: Anchorage, Alaska, U.S. Department of the Interior, Minerals Management Service Outer Continental Shelf Report 87–0046, 193 p.

Till, A.B., and Dumoulin, J.A., 1994, Geology of Seward Peninsula and St. Lawrence Island, *in* Plafker, G., and Berg, H.C., eds., The geology of Alaska: Boulder, Colorado, Geological Society of America, Geology of North America, v. G-1, p. 141–152.

Till, A.B., Box, S.E., Roeske, S.M., and Patton, W.W., Jr., 1993, Comment on "Mid-Cretaceous extensional fragmentation of a Jurassic-Early Cretaceous compressional orogen, Alaska": Tectonics, v. 12, p. 1076–1086.

Tolson, R.B., 1987a, Structure and stratigraphy of the Hope Basin, southern Chukchi Seas, Alaska, *in* Scholl, D.W., Grantz, A., and Vedder, J.G., eds., Geology and resource potential of the continental margin of western North America and adjacent ocean basins: Beaufort Sea to Baja California: Houston, Texas, Circum-Pacific Council for Energy and Mineral Resources, Earth Science Series, v. 6, p. 59–71.

Tolson, R.B., 1987b, Structure, stratigraphy, tectonic evolution and petroleum source potential of the Hope Basin, southern Chukchi Sea, Alaska [Ph.D. thesis]: Palo Alto, California, Stanford University, 261 p.

Trettin, H.P., 1991, Summary: Silurian–Early Carboniferous deformational phases and associated metamorphism and plutonism, Arctic Islands, *in* Trettin, H.P., ed., Geology of the Innuitian Orogen and Arctic Platform of Canada and Greenland: Boulder, Colorado, Geological Society of America, Geology of North America, v. E, p. 337–341.

Trettin, H.P., Mayr, U., Long, G.D.F., and Packard, J.J., 1991, Cambrian to Early Devonian basin development, sedimentation, and volcanism, Arctic Islands, *in* Trettin, H.P., ed., Geology of the Innuitian Orogen and Arctic Platform of Canada and Greenland: Ottawa, Canada, Geological Survey of Canada, Geology of Canada, no. 3, p. 163–238.

Turner, R.F., McCarthy, C.M., Steffy, D.A., Lynch, M.B., Martin, G.C., Sherwood, K.W., Flett, T.O., and Adams, A.J., 1984, Geological and operational summary, Navarin Basin COST No. 1 well, Bering Sea, Alaska: Anchorage, Alaska, U.S. Department of the Interior, Minerals Management Service Outer Continental Shelf Report 84–0031, 245 p.

Turner, R.F., Martin, G.C., Flett, T.O., and Steffy, D.A., 1985, Geologic report for the Navarin Basin planning area, Bering Sea, Alaska: Anchorage, Alaska, U.S. Department of the Interior, Minerals Management Service Outer Continental Shelf Report 85–0045, 140 p.

Turner, R.F., Martin, G.C., Flett, T.O., and Steffy, D.A., 1986, Geologic report for the Norton Basin planning area, Bering Sea, Alaska: Anchorage, Alaska, U.S. Department of the Interior, Minerals Management Service Outer Continental Shelf Report 86–0033, 179 p.

Warner, M.R., 1990, Basalts, water, or shear zones in the lower continental crust?: Tectonophysics, v. 173, p. 163–174.

Warner, M., Morgan, J., Barton, P., Morgan, P., Price, C., and Jones, K., 1996, Seismic reflections from the mantle represent relict subduction zones within the continental lithosphere: Geology, v. 24, p. 39–42.

Wenzel, F., Sandmeier, K.-J., and Wälde, W., 1987, Properties of the lower crust from modeling refraction and reflection data: Journal of Geophysical Research, v. 92, p. 11575–11583.

Wirth, K.R., and Bird, J.M., 1992, Chronology of ophiolite crystallization, detachment, and emplacement: Evidence from the Brooks Range, Alaska: Geology, v. 20, p. 75–78.

Worrall, D.M., 1991, Tectonic history of the Bering Sea and the evolution of Tertiary strike-slip basins of the Bering Shelf: Geological Society of America Special Paper 257, 120 p.

Yilmaz, O., 1987, Seismic data processing: Tulsa, Oklahoma, Society of Exploration Geophysicists, Investigations in Geophysics, v. 2, 526 p.

MANUSCRIPT ACCEPTED BY THE SOCIETY MAY 15, 2001.

Geological Society of America
Special Paper 360
2002

Crustal structure across the Bering Strait, Alaska: Onshore recordings of a marine seismic survey

Lorraine W. Wolf
Department of Geology, Auburn University, Auburn, Alabama 36849, USA
Robert C. McCaleb
Department of Geological Sciences, Michigan State University, East Lansing, Michigan 48824-1115, USA
David B. Stone
Geophysical Institute, University of Alaska, Fairbanks, Alaska 99775, USA,
and Departments of Physics and Geology, Rhodes University, Grahamstown 6140, South Africa
Thomas M. Brocher
U.S. Geological Survey, 345 Middlefield Road, MS 977, Menlo Park, California 94025, USA
Kazuya Fujita
Department of Geological Sciences, Michigan State University, East Lansing, Michigan 48824-1115, USA
Simon L. Klemperer
Department of Geophysics, Stanford University, Stanford, California 94305-2215, USA

ABSTRACT

As part of an experiment to study crustal structure in Bering Strait–Chukchi Sea, we collected wide-angle seismic refraction data at onshore stations on the coasts of western Alaska and eastern Russia. Receiver gathers made from the onshore recordings of air-gun shots from two deep-crustal marine seismic profiles were used to construct seismic velocity models for the region. Results of the seismic analysis, combined with analysis of gravity data, indicate that crustal thickness ranges from 32 to 35 km along the profiles, the thinner crust occurring beneath a middle Cretaceous magmatic belt that includes the Okhotsk-Chukotsk volcanic belt and plutonic rocks on both the Chukotka and Seward Peninsulas. These results are consistent with models supporting widespread crustal extension in the Bering Strait region. In addition, there is no indication in the seismic or gravity data that a crustal root associated with a westward continuation of the Brooks Range currently exists to the west of the Lisburne Peninsula.

INTRODUCTION

The Bering Strait region links the Chukchi Shelf to the north and the Bering Shelf to the south, which together form an area of shallowly submerged continental crust (Figs. 1 and 2). Although on strike with the trend of the Brooks Range, the Bering Strait and adjacent parts of the Chukchi Sea appear to have a tectonic history distinct from that of the range in northern central Alaska, where a thick crust and elevated topography currently exist (Barnes, 1977; Fuis et al., 1997; Wissinger et al.,

1997). A complex history of contraction, extension, and transtension in the Bering Strait region has produced the structural features that characterize this bridge between the Asian and North American continents (Miller and Hudson, 1991; Dumitru et al., 1995). In 1994, two deep-crustal marine seismic profiles and accompanying wide-angle refraction data, constituting the Bering-Chukchi Crustal Transect, were acquired in the Bering Strait region (Fig. 2) (Klemperer et al., 1995; Brocher et al., 1995; Klemperer et al., this volume, Chapter 1). The combination of multichannel and wide-angle coverage provides detailed

Wolf, L.W., McCaleb, R.C., Stone, D.B., Brocher, T.M., Fujita, K., and Klemperer, S.L., 2002, Crustal structure across the Bering Strait, Alaska: Onshore recordings of a marine seismic survey, *in* Miller, E.L., Grantz, A., and Klemperer, S.L., eds., Tectonic Evolution of the Bering Shelf–Chukchi Sea–Arctic Margin and Adjacent Landmasses: Boulder, Colorado, Geological Society of America Special Paper 360, p. 25–37.

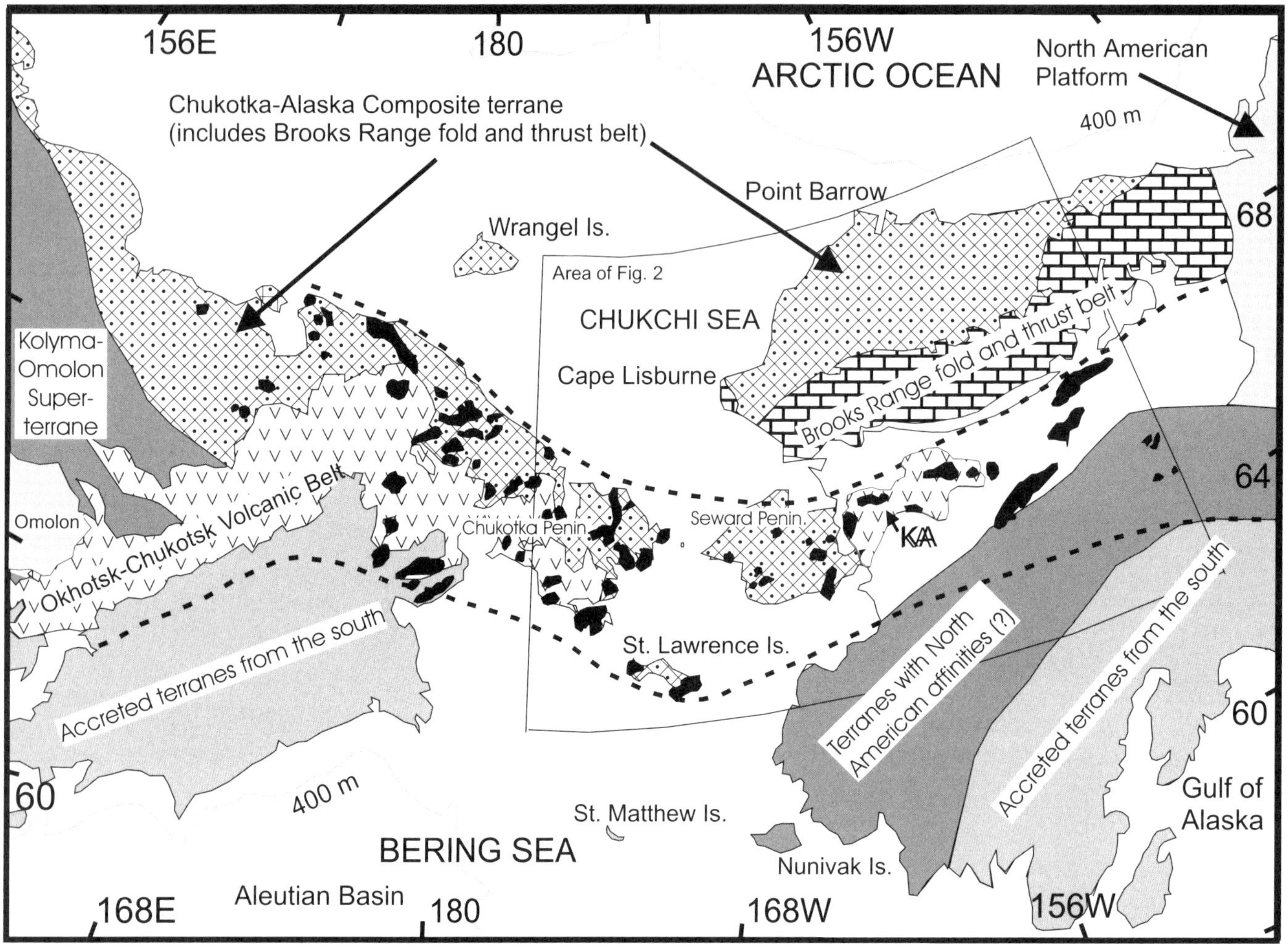

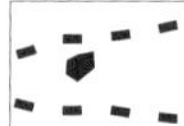

Okhotsk-Chukotsk Volcanic Belt (OCVB), dominantly Upper Cretaceous in age, and the Koyukuk arc (KA), Lower Cretaceous.

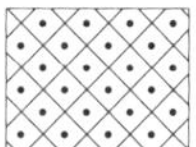

North and south boundaries of the Cretaceous magmatic belt (dashed lines) and intrusive rocks, (black pattern)(from Bering Strait Field Party, 1

Chukotka-Alaska composite terrane. (Includes Brooks Range fold and thrust belt). Parts of the southern boundary are obscured by the OCVB and the terranes east of Seward Peninsula.

The Brooks Range fold and thrust belt.

Area east of Seward peninsula consisting of accreted terranes, volcanic and plutonic complexes and overlap deposits .

Terranes with North American affinities that have moved northwards along the North American margin.

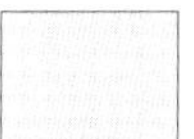

Accreted terranes that have been transported northwards on the plates of the Pacific.

Omolon Massif containing Precambrian crystalline basement and considered to be a continental fragment transported to its present location.

The Kolyma-Omolon Superterrane (includes Omolon terrane),a complex assemblage of terranes accreted to the Siberian platform.

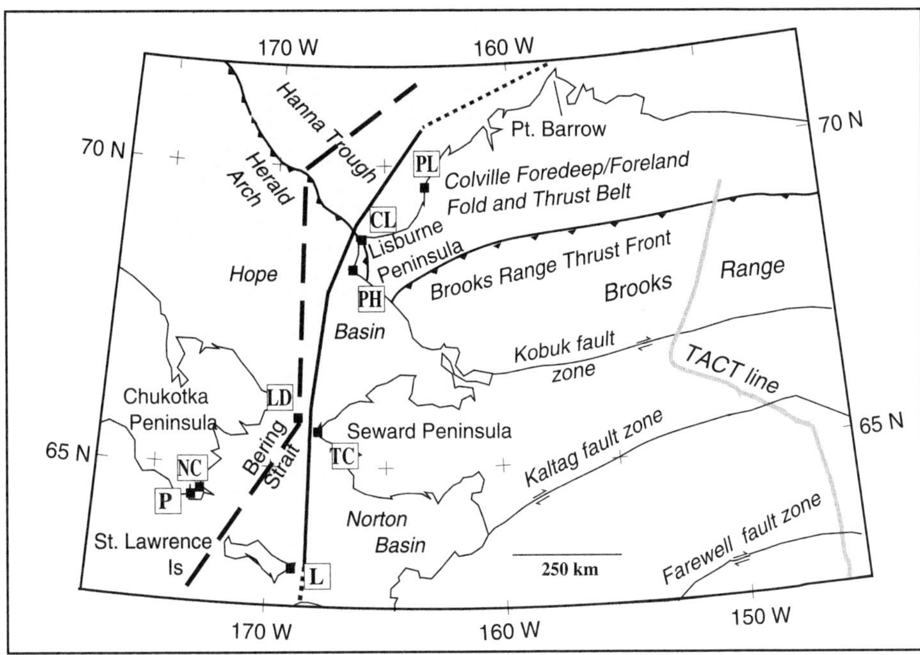

Figure 2. Tectonic map of Bering Strait region showing ship track (solid and dotted are northbound lines 1 and 2; dashed line is southbound line 3) and locations (solid squares) of onshore seismic stations used in refraction-wide-angle reflection analysis (modified from Brocher et al., 1995). CL is Cape Lisburne, L is Leitnik, LD is Little Diomede, NC is New Chaplino, PH is Point Hope, PL is Point Lay, P is Provideniya, TC is Tin City. Also shown is location of Trans-Alaska Crustal Transect (TACT) through northern central Alaska. For complete description of Bering Strait seismic experiment, operating parameters, and collected data, see Brocher et al. (1995).

images of the crust and its general attributes, such as reflectivity, crustal thickness, possible rock composition, and the deep subsurface expression of structural boundaries. Raypaths from the ship to onshore receivers crossed several geologically significant features, including the Herald Arch, the Hanna Trough, and a region of crustal extension and orogenic collapse in the northern Bering Shelf (Miller and Hudson, 1991; Dumitru et al., 1995). Of particular interest in the refraction experiment were (1) the deep structure of Cape Lisburne, the Hope Basin, and the Herald Arch; (2) the crustal thickness and seismic velocity structure between Chukotka and Alaska; (3) the deep structure beneath the Cretaceous magmatic belt that crosses Bering Strait; and (4) the location of the western extension of the northern Brooks Range, if it exists beyond the Lisburne Peninsula. Each of these has important implications for reconstructing the structural evolution of northwestern Alaska.

Figure 1. Location map showing tectonic setting of Bering-Chukchi region. Accreted terranes from south have moved northward on plates of Pacific, arriving in late Mesozoic and Tertiary time. Terranes with North American affinities have moved northward along North American margin, but may have had earlier, more exotic affinities. Chukotka-Alaska composite terrane includes Brooks Range fold and thrust belt, northern Alaska, and northeastern Russia, together with their continental shelves. Okhotsk-Chukotsk volcanic belt (OCVB) overlaps terranes from south, edges of Kolyma-Omolon superterrane, and Chukotka-Alaska composite terrane, showing that these tectonic provinces were amalgamated by latest Cretaceous–earliest Tertiary time.

GEOLOGIC BACKGROUND

Precambrian, Paleozoic, and lower Mesozoic rocks of the eastern Chukotka Peninsula and western Seward Peninsula are part of the continental crust of the Chukotka-Alaska composite terrane (Fig. 1). This terrane contrasts with a variety of accreted terranes to the south that include island arc, continental margin, oceanic sediment, and continental fragments (McGeary and Ben-Avraham, 1981; Fisher et al., 1982; Howell et al., 1987; Stavsky et al., 1990). An extensive and complex magmatic belt of Cretaceous age that extends from China to the Canadian Cordillera (Rubin et al., 1995; Bering Strait Geologic Field Party, 1997) includes extrusive rocks of the Okhotsk-Chukotsk volcanic belt and plutonic rocks on both the Chukotka and Seward Peninsulas. In the vicinity of the Bering Strait, the plutonic complexes show evidence of extension (Rubin et al., 1995; Amato and Wright, 1997; Bering Strait Geologic Field Party, 1997). In general, the Okhotsk-Chukotsk volcanic belt rocks are slightly younger (<90 Ma, P. Layer, 2000, personal commun.) than the cooling ages of the plutons in the Bering Strait area (ca. 90 Ma, Amato and Wright, 1997; Bering Strait Geologic Field Party, 1997). The lower parts of the Okhotsk-Chukotsk volcanic belt show evidence of being produced by Andean-style subduction; however, this convergent regime was followed later by extension ca. 86 Ma (Newberry et al., 1997). A curious feature of the Okhotsk-Chukotsk volcanic belt is that it ends abruptly at the eastern edge of Chukotka. The recognition of caldera structures within the Okhotsk-Chukotsk volcanic belt (Belyi, 1994) indicates minimal erosion, in stark contrast to the eastern side of the

Bering Strait, where uplift has exposed the roots of plutons of similar age.

Between the Hanna Trough on the north and the Hope Basin on the south, the Herald Arch is a prominent fold-thrust system that contrasts with the less disturbed basins to the north and south (Grantz and May, 1987). The structural relationship of the Herald Arch to the extensive east-west–trending Brooks Range fold and thrust belt has been the focus of considerable debate. Some models link development of the Herald Arch to the formation of the Brooks Range (Middle Jurassic to middle Cretaceous), but the northwest-southeast orientation of the arch has led to speculation that it postdates the orogenic events of the western Brooks Range (Late Cretaceous to Tertiary). In addition, indications are that the Herald Arch predates the Hope Basin, which contains Eocene and younger sediments, thus further defining the age of formation (e.g., Patton and Tailleur, 1977; Grantz et al., 1970, 1981; Tolson, 1987). Moore et al. (1994) proposed two orogenic events in the Brooks Range, one Jurassic and/or Early Cretaceous age and the other Late Cretaceous age. The younger of these events is associated with west-trending folds that extend smoothly across northern Alaska and continue northwesterly along the Herald Arch (Moore et al., 1994).

Structural, petrologic, geochronologic, and seismological data suggest that episodes of generally north-south extension have affected the crust of the northern Bering Sea since middle to Late Cretaceous time and appear to be continuing today (c.g., Miller and Hudson, 1991; Dumitru et al., 1995; Page et al., 1991; Turner and Swanson, 1981; Mackey et al., 1997). Exposures of metamorphic rocks in the Seward Peninsula reveal deformational histories at several crustal levels and have led to a tectonic model involving regional denudation, tilting, and normal faulting related to these extensional events (Hannula and McWilliams, 1995; Dumitru et al., 1995). Blueschist facies metamorphic rocks exposed on the peninsula are believed to have originated at mid-crustal depths and provide evidence of postmetamorphic uplift and cooling in Cretaceous time (Hannula and McWilliams, 1995; Bering Strait Geologic Field Party, 1997). Today the Bering Strait region is seismically active. Earthquake focal mechanisms indicate that north-south–directed extension is occurring from eastern Chukotka through the Seward Peninsula to the Yukon Delta (Mackey et al., 1997; Mackey, 1999).

PREVIOUS WORK

Although there have been extensive multichannel seismic reflection investigations (many related to petroleum exploration) and a few sonobuoy refraction studies of the upper crust, there has only been limited previous work to determine the deep crustal structure of the Bering Strait region. In addition, the International Boundary between the United States and the former Soviet Union has precluded joint experiments or even the routine exchange of data. Thus, combined with the logistical difficulties inherent in the region, very few studies have examined the deep structure and continuity of structure along both sides of the Bering Strait.

Prior Russian studies of crustal structure were based on the analysis of converted waves (Mishin and Dareshkina, 1966), usually *Ps,* recorded predominantly at temporary stations deployed in the 1960s (see Fujita et al., this volume). Because these analyses were based on a small number of events and unsubstantiated assumptions about intracrustal seismic velocities (Belyaevsky and Borisov, 1974; Vashchilov, 1979; Bulin, 1989), they vary somewhat in their estimates of crustal thickness (usually ~±5 km) and their reliability is probably poor. These studies indicate crustal thicknesses of 37–39 km throughout Chukotka, with slightly thinner crust at Provideniya (32–37 km), and thicker crust in the vicinity of the Omolon massif to the west (40–41 km). The only other published Russian work is a profile along 173°E meridian based on magnetic data that indicate a crustal thickness of 30 km (Demenitskaya et al., 1973).

Prior published American studies of crustal thickness in the Chukchi Sea and Bering Strait region are also lacking. Ostenso (1968) used gravity measurements to suggest that the thick crust associated with the north-central Brooks Range (e.g., Barnes, 1977; Fuis et al., 1997) thins considerably into the south Chukchi Sea; he also referred to an unreversed refraction line on the continental shelf edge that yielded a Moho depth of 30 km. A refraction line between Point Barrow and ice station Arlis II, then at ~74°N, 165°W, yielded a 32 km crustal thickness at Barrow (Hunkins, 1966) at the northern end of the present transect, and surface-wave dispersion studies of the Bering Shelf yielded a Moho depth of 27.6 km (Jin and Herrin, 1980) at the southern end.

SEISMIC DATA ACQUISITION AND ANALYSIS

The seismic refraction–wide-angle reflection data analyzed in this study are of a reconnaissance nature and were collected at accessible onshore sites on the eastern and western sides of the Bering Strait during both the northbound and southbound legs of the Bering-Chukchi Crustal Transect (Fig. 2). On the eastern side of the strait, the ship track provided approximately in-line raypaths for offsets beyond ~40 km for most of the Alaskan stations. On the western side of the strait, the ship was restricted to U.S. waters, and the resulting greater distance between the ship track and recording stations led to a fan recording geometry for stations in Russia.

Receiver gathers for each station were made from the recordings of air-gun shots fired every ~50 m and analyzed using a combination of one- and two-dimensional forward-modeling techniques (Luetgert, 1992). For receivers that lacked a significant degree of raypath reversal, only one-dimensional modeling of data from each station was performed. These one-dimensional velocity-depth functions were organized by distance into a two-dimensional pseudosection. The oblique recording geometry for all onshore stations at close range meant that no arrivals occur in the receiver gathers at offsets <~20 km, thus introducing uncertainty into estimates of the upper-crustal velocities. In many cases, estimates for near-surface velocities were obtained from the literature (e.g., Houtz et al., 1981) or inferred from published geologic maps. Although none of the receiver gathers represents completely reversed ray coverage, the section of the eastern transect between Lietnik on Saint Lawrence Island and Point Lay, Alaska, provides enough common subsurface sampling to justify a two-dimensional modeling technique for determining thickness and average velocities of the crust. Higher-order models (e.g., 2.5 dimensional or 3 dimensional) were not attempted because ray coverage was too sparse and the quality of some records too low.

Wide-angle data collected in the study range in signal-to-noise ratio from excellent to poor. For a complete discussion of all receiver gathers and processing parameters, see Brocher et al. (1995). In general, upper-crustal refracted phases (*Pg*) and prominent upper mantle reflections (*PmP*) can be traced in most gathers to offsets beyond 100 km (e.g., Fig. 3). Upper mantle refracted phases (*Pn*) were observed in only five gathers.

VELOCITY MODELS AND INTERPRETATION

Figures 4 and 5 contain crustal compressional-wave velocity models based on receiver gathers from the multichannel experiment. Figure 4 is a composite of one-dimensional velocity-depth functions from Provideniya to Tin City. Figure 5 shows a crustal velocity model for Lietnik to Point Lay based on the two-dimensional ray-tracing analysis. To derive the models, we used traveltimes of *Pg, PmP,* and, where observed, *PiP* and *Pn* phases, from 16 receiver gathers. Picks were made of first arrivals and high-amplitude secondary arrivals, only some of which are attributable to coherent phases. Observed and calculated traveltime curves for the receiver gathers used in the two-dimensional model, along with ray diagrams showing the area of subsurface sampling, are shown in Figures 6 and 7. Calculated traveltimes were generally fit to within 0.15 s of observed times of identifiable phases. Errors associated with this level of agreement, assuming that the phases were correctly identified, are ~2%–3% in velocity and 10% in depth. (For a more complete discussion of uncertainties and error estimates in forward modeling, see Ansorge et al., 1982.) Occasional outliers to 0.4 s are seen; however, these arrivals are attributed to (1) scattered or diffracted energy because they either could not be traced over significant distances and/or they exhibit phase velocities that are unrealistic, or (2) structural complexities that could not be represented effectively in the model. Due to the large number of traces in each gather, far fewer observed traveltimes for each phase are shown in the figures than were picked from the

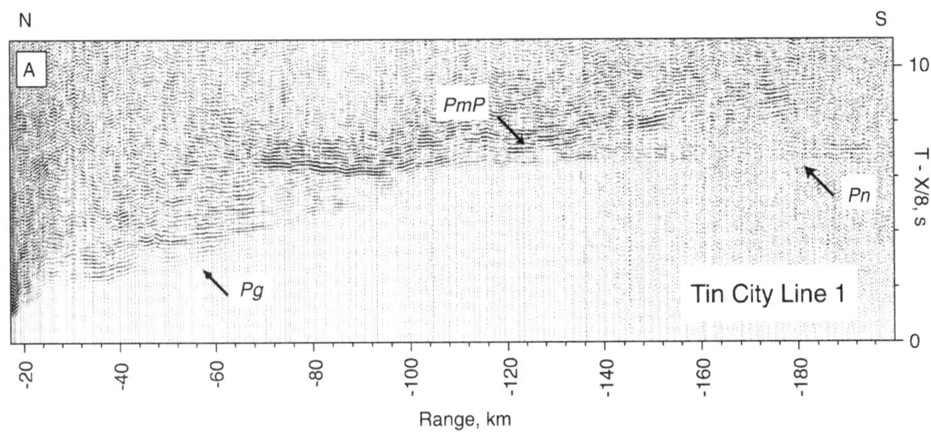

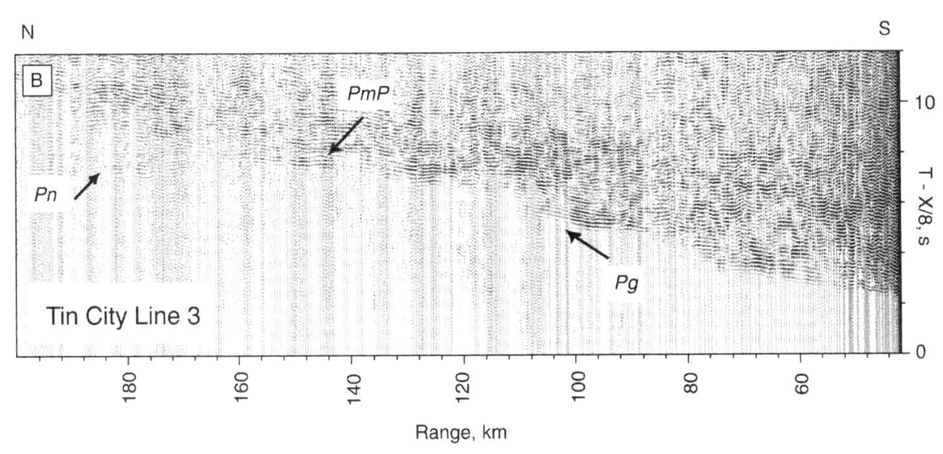

Figure 3. Wide-angle refraction receiver gathers from Tin City line 1 (A) and Tin City line 3 (B) (plotted with reduction velocity of 8 km/s, where *T* is traveltime and *X* is offset distance). Data have been bandpass filtered (6–13 Hz) and mixed over five traces. Offsets from recording station are given as positive for shots fired north of receiver and negative for shots fired south of receiver. In each gather, *Pg* (crustal phase), *PmP* (upper mantle reflection), and *Pn* (upper mantle refracted phase) are clearly apparent. Crossover of *Pn* phase with *Pg* crustal phase in each gather occurs at ~140 km offset and 7 s, indicating crustal thickness of ~32 km in this area.

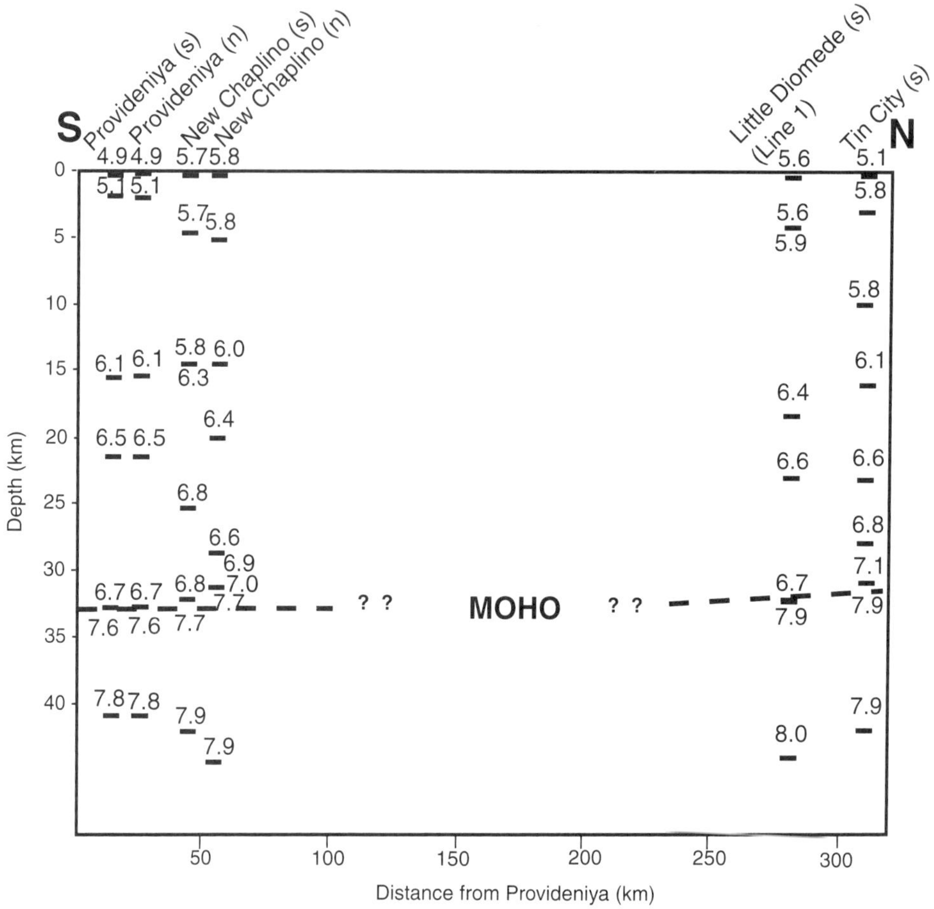

L.W. Wolf et al.

Figure 4. Pseudo two-dimensional crustal velocity model constructed from one-dimensional velocity-depth functions between Provideniya and Tin City (see Fig. 2 for transect route). Because these data lack reversed raypath coverage, two-dimensional analysis was not warranted. One-dimensional analysis suggests that crust is ~32–33 km thick along western side of Bering Strait, similar to estimates for eastern side (see Fig. 5).

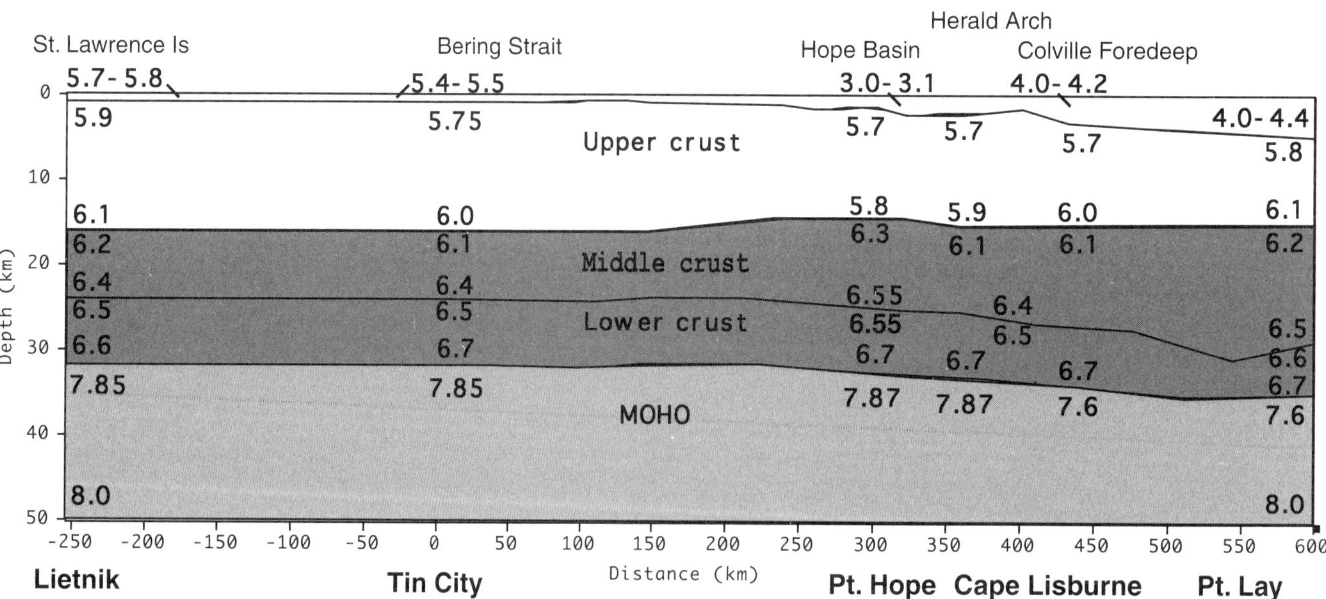

Figure 5. Crustal velocity model from Lietnik to Point Lay, Alaska, based on two-dimensional ray tracing. Variations in upper crustal rocks and sediment thickness along transect were accommodated by first model layer and based on results of reflection data along Bering-Chukchi Crustal Transect (Klemperer et al., Chapter 1) and on published geologic information for onshore station locations. Crustal thickness along profile ranges from ~32 to 35 km; thicker crust occurs to north of Cape Lisburne. See text for discussion.

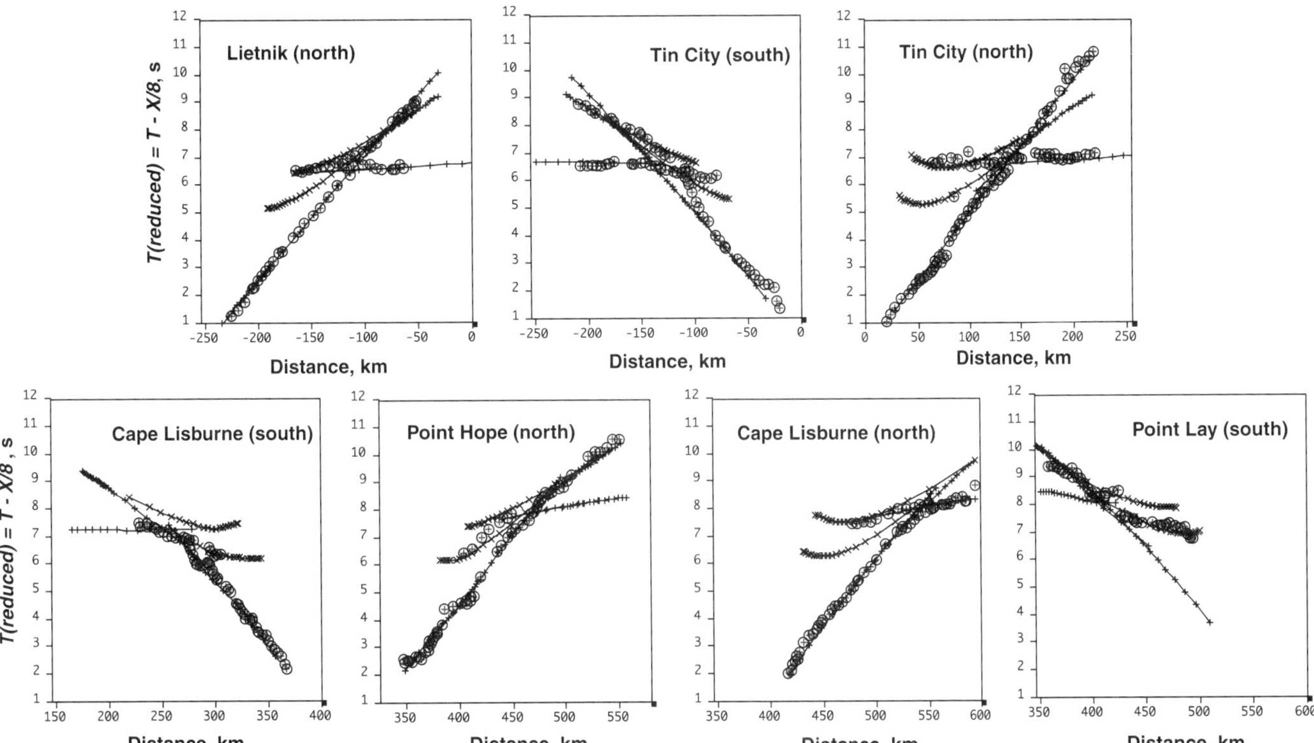

Figure 6. Observed (circles) versus calculated (pluses connected by solid line) traveltimes for stations used in two-dimensional model shown in Figure 5. Traveltimes, *T,* shown are reduced at 8 km/s with respect to offset distance, *X.* Offsets are given in model coordinates. North and south refer to raypaths arriving from air-gun shots fired to north or south of each onshore recording station. Observed traveltimes consist of first arrivals and high-amplitude secondary arrivals, but only some of latter are attributable to coherent phases. Due to large number of traces, far fewer observed traveltimes are shown than were picked from records. Generally, calculated times are fit to within 0.15 s of observed times of coherent phases (e.g., *Pg* and *PmP* phases, and where observed, *PiP* and *Pn* phases; see text). Occasional outliers, e.g., Tin City (south) at −100 km and 5 s, were not modeled and are mostly attributed to scattered or diffracted energy or to structural complexities that could not be effectively represented in model. Picks from three receiver gathers—Leitnik (south), Point Hope (south), and Point Lay (north)—are not shown but were considered in modeling.

records. In developing both models (Figs. 4 and 5), more weight was subjectively given to traveltime picks from those record sections in which all phases were clearly observed and the signal-to-noise ratio was high.

The two-dimensional modeling between Lietnik and Point Lay (Fig. 5) allowed us to include detail in the upper crust that correlates with near-surface geology, as described in published literature, and with interpretations of the reflection data (see Klemperer et al., this volume, Chapter 1). Such features as the Hope Basin and the Herald Arch fold and thrust belt significantly influence the arrival times of phases at the onshore stations and are thus considered in the model. Variable near-surface velocities representing laterally juxtaposed rocks are shown in the model's first layer. Compressional wave velocities for this layer in the region between Lietnik and Tin City are high (>5.4 km/s) relative to those at Point Hope (~3.0 km/s). Farther to the north, in the fold and thrust belt off Cape Lisburne, velocities are ~4.2 km/s.

Average velocities of the upper to middle crust in both the one- and two-dimensional models (Figs. 4 and 5) are ~5.8 km/s (excluding near-surface variations in the first model layer) and

~6.4 km/s in the middle to lower crust. These velocities are consistent with those of average continental crust (Christensen and Mooney, 1995; Rudnick and Fountain, 1995). At the base of the crust, velocities are ~6.7 km/s. Inclusion of a separate lower-crustal layer in the model was based on gathers having *PiP* intracrustal phases. Upper mantle *Pn* phases, where observed, had crossover distances with *Pg* crustal phases of ~140–160 km and estimated velocities of 7.6–7.9 km/s (Figs. 3 and 6), indicating crustal thicknesses ranging from 32 to 35 km. Our results indicate that the crustal velocity structure varies little from the eastern to the western shores of the Bering Strait between the Seward and Chukotka Peninsulas.

A combination of a late-arriving *PmP* upper mantle reflection and the lack of a clearly defined *Pn* crossover in the Point Hope record raises the possibility that the crustal structure beneath Cape Lisburne and the Herald Arch differs from that inferred for the rest of the transect (Fig. 8). Furthermore, the *Pg* phase in this receiver gather represents the first-arriving energy well beyond offsets observed in other gathers. These record attributes might be interpreted to indicate that this area has a lower

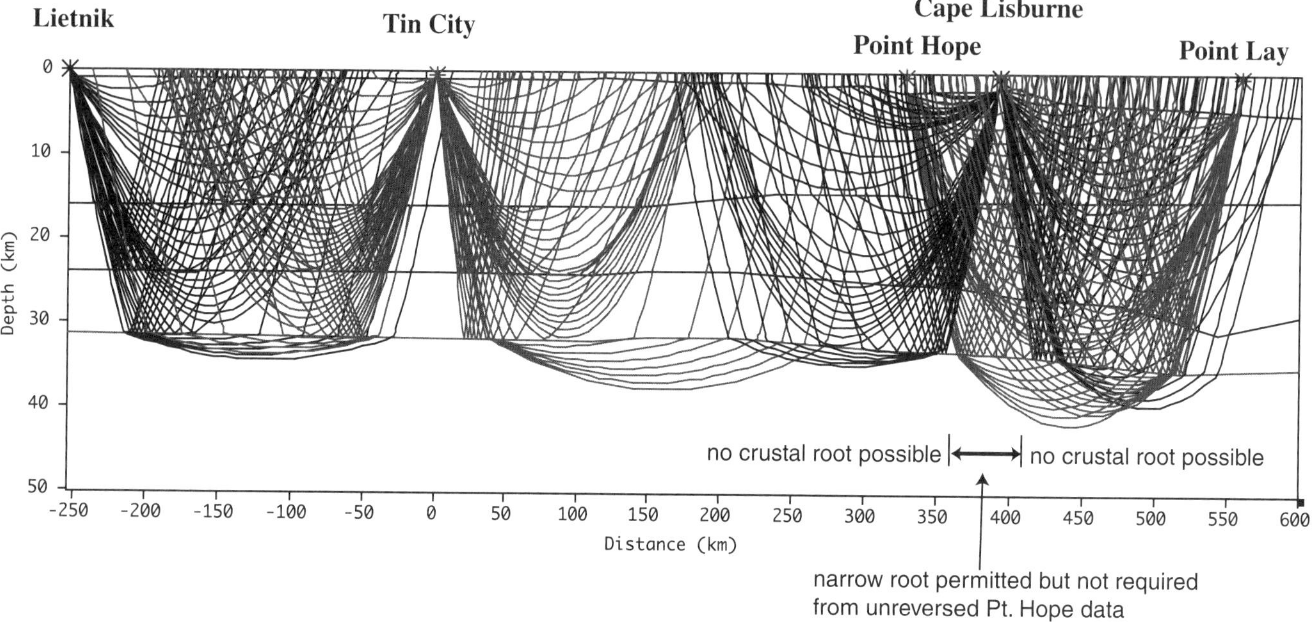

Figure 7. Ray diagrams corresponding to calculated traveltimes shown in Figure 6 and overlain on two-dimensional model shown in Figure 5. Note that raypath reversals are only partial and that at close offsets they represent oblique recording geometries. Two-headed arrow shows region of possible crustal root. However, preferred seismic and gravity models successfully match observed data without root (see text for discussion).

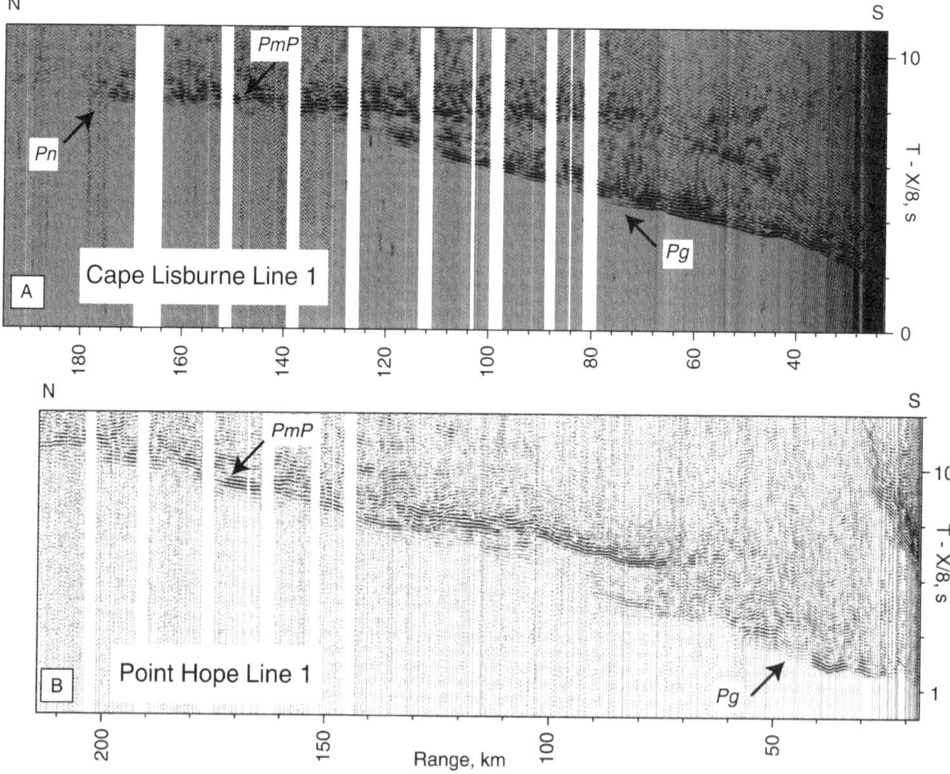

Figure 8. Wide-angle receiver gathers from Cape Lisburne line 1 (A) and Point Hope line 1 (B). Data have been band-pass filtered (6–13 Hz) and mixed over five traces. Positive offsets indicate data from shots fired north of receiver. Crossover of *Pn* upper mantle phase with *Pg* crustal phase appears at ~150 km offset in record from Cape Lisburne. No *Pn* phase is observed, however, in record from Point Hope. See text for discussion.

average crustal velocity and possibly a thick crustal root, which would be on trend with the westward extension of the northern Brooks Range. As an alternative to the model shown in Figure 5, a crustal model with a thickened root (~40 km crustal thickness) was tested against the data observations. Although the calculated traveltimes from the model could be made to match the data, the root could not extend northward of km +425 in the model (Figs. 5 and 6). The reason for this is that raypaths from the Point Hope gather have common sampling points in the lower crust and upper mantle with raypaths from the Cape Lisburne gather (Fig. 7), and the latter contains a clearly defined upper mantle *Pn* phase that indicates a crustal thickness of ~35 km near Cape Lisburne (Fig. 8). We argue that the data do not require the presence of the root in the vicinity of Point Hope to Cape Lisburne, and the lack of a clear upper mantle *Pn* phase in the Point Hope receiver gather can be attributed to a low signal-to-noise ratio or to a low or negative velocity gradient in the upper mantle.

GRAVITY MODELS

To examine independently the presence of a crustal root beneath Cape Lisburne, we compared satellite-derived gravity measurements (Sandwell and Smith, 1997) with values calculated using models without and with a thickened root (Figs. 9 and 10). Profiles for the models were extracted from gridded data from along five longitude lines between 168° and 170°W. The initial gravity model was based on a simplified version of the refraction models, in which an average bulk density for the crust was assumed to be 3.0 g/cm^3. The sedimentary basin depth for the Aleutian Basin and Bering Shelf was obtained from Worrall (1991). Water depths (20–200 m) do not vary significantly throughout the region encompassed by the model. Sediment thickness for the Chukchi Sea was obtained from Grantz et al. (1994) and supplemented from Shipilov et al. (1989). The density assigned to sedimentary basins was 2.6 g/cm^3, and was based on measured densities of clastic and carbonate rocks in test wells (Collins, 1958, 1961). The density used for mantle rocks was 3.3 g/cm^3. Results of the analysis indicate that the presence of a crustal root of the maximum thickness and extent that the suite of tested seismic models could allow would result in a gravity anomaly three times larger than that observed (Fig. 10). Instead, the gravity models indicate that the observed anomalies can be sufficiently accounted for by a slight northward thickening of the crust and the presence of thick sediments in the basins in the Point Hope–Cape Lisburne area, in keeping with the preferred and more plausible seismic model (Fig. 5).

DISCUSSION

Of significant interest is the question of whether the Brooks Range orogen continues westward from the Lisburne Peninsula and merges with the Herald Arch or if the Herald Arch thrust zone crosscuts the Brooks Range. The absence of strong evidence for a crustal root or greatly thickened crust in the seismic and gravity data supports earlier interpretations that the crust of

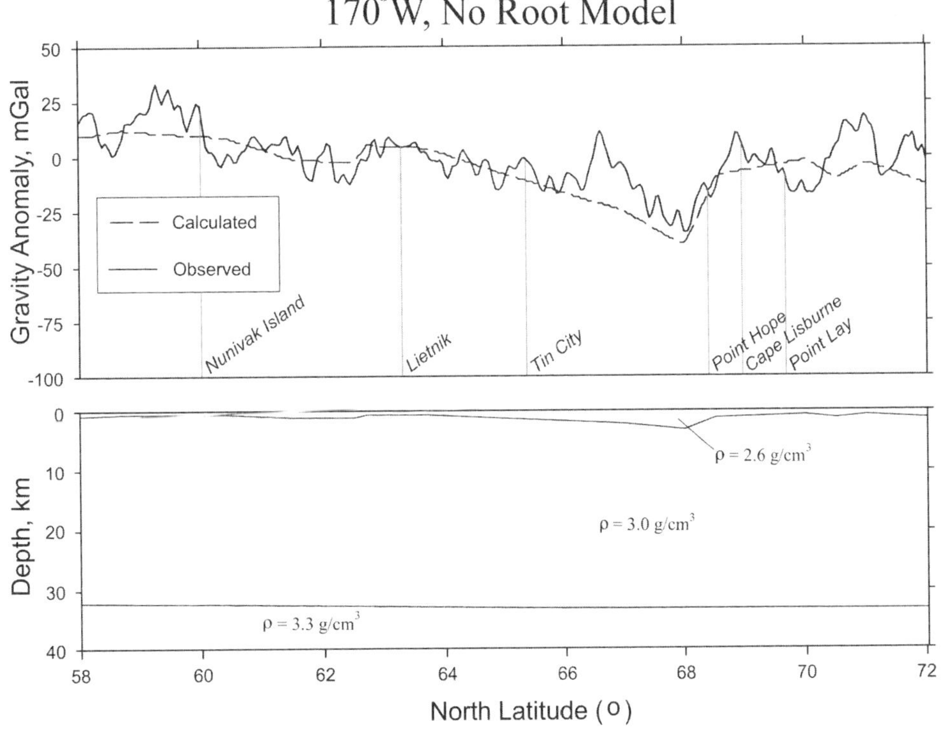

Figure 9. Gravity model of profile along 170°W meridian from 58°N to 72°N. Density, ρ, in g/cm^3 for sedimentary basins from Collins (1958, 1961). Gravity data were derived from gridded satellite altimetry data (Sandwell and Smith, 1997). Model indicates that gravity anomalies can be accounted for by configuration of near-surface sediments and gradual northward thickening of crust. Crustal root is not required.

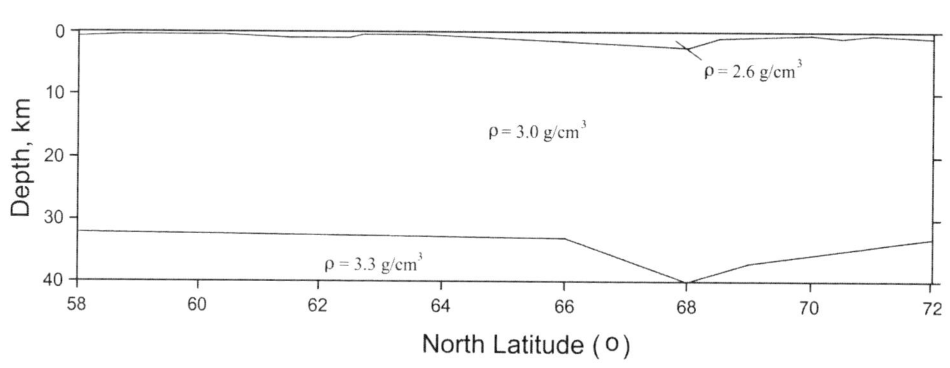

Figure 10. Gravity model (as in Fig. 9) assuming crustal root to 40 km beneath Herald Arch (~68°N). Density, ρ, in g/cm³. Addition of root produces gravity low approximately three times that seen in gridded data. These results support no-root seismic model shown in Figure 5.

the Bering Strait and areas to the north near Point Hope and Cape Lisburne are thin relative to the thickened crust (~49 km) observed in the central Brooks Range (Fuis et al., 1997; Wissinger et al., 1997). Thus our results suggest that if the Brooks Range orogen formerly extended west of Lisburne Peninsula, its topographically high elevations and corresponding thickened crustal root have been removed by subsequent extension. This interpretation is corroborated by the multichannel seismic data, which lack any indication of a thick crustal root west of the Lisburne Peninsula (Fig. 11).

Although the seismic and gravity data provide no compelling evidence for continuity of the Herald Arch with the Brooks Range, the refraction data indicate a slight change in upper-crustal velocity from 5.7 km/s south of the Herald Arch to 5.8–6.1 km/s north of the Herald Arch. The slightly higher (<0.9–1.4 km/s) upper-crustal velocity observed in this area most likely reflects a change in petrology or depth to basement. In addition, a prominent northward dip to the middle-lower crustal interface (Fig. 5), accompanied by a drop in the reflectivity of the lower crust (Fig. 11) (Klemperer et al., this volume, Chapter 1) and a slight increase in crustal thickness north of Point Hope, are consistent with interpretations that the widespread north-south extension proposed for the area beneath the magmatic belt did not affect the crust north of the Herald Arch.

Results of the refraction analysis from the Bering Strait (e.g., crustal thickness of 32–35 km and average velocities of

6.4 km/s in the middle to lower crust) are consistent with geologic interpretations of extended continental crust in this area (Miller and Hudson, 1991; Dumitru et al., 1995; Bering Strait Geologic Field Party, 1997). In addition, these findings are in agreement with interpretations of multichannel seismic data (Fig. 11), which show a highly reflective lower crust (6–12 s two-way traveltime) underlying the Cretaceous magmatic belt (Bering Strait Geologic Field Party, 1997; Klemperer et al., this volume, Chapter 1). Similar patterns of reflectivity are observed in other regions that have undergone extension or in areas that have undergone magmatism (Mooney and Brocher, 1987; Beaudoin et al., 1992). The lack of a significant change in lower-crustal velocities throughout the region covered by the two-dimensional refraction model (Fig. 5) suggests that the reflectivity of the lower crust observed between Saint Lawrence Island and the Bering Strait (Fig. 11) is unlikely to be the result of a large volume of mafic to ultramafic intrusions into the lower crust; rather, the average seismic velocity of 6.6 km/s along the profile is more consistent with a composite of felsic to intermediate igneous and metamorphic rocks with gabbroic intrusions (Klemperer et al., this volume, Chapter 1). On the Seward Peninsula, Jurassic and/or Early Cretaceous blueschist facies metamorphic rocks that formed at high pressures at mid-crustal depths were exhumed and subsequently uplifted. In view of the seismically derived crustal thickness of ~32 km west of the Seward Peninsula and assuming these now-exposed metamorphic rocks were formed at ~15 km depth, we estimate that

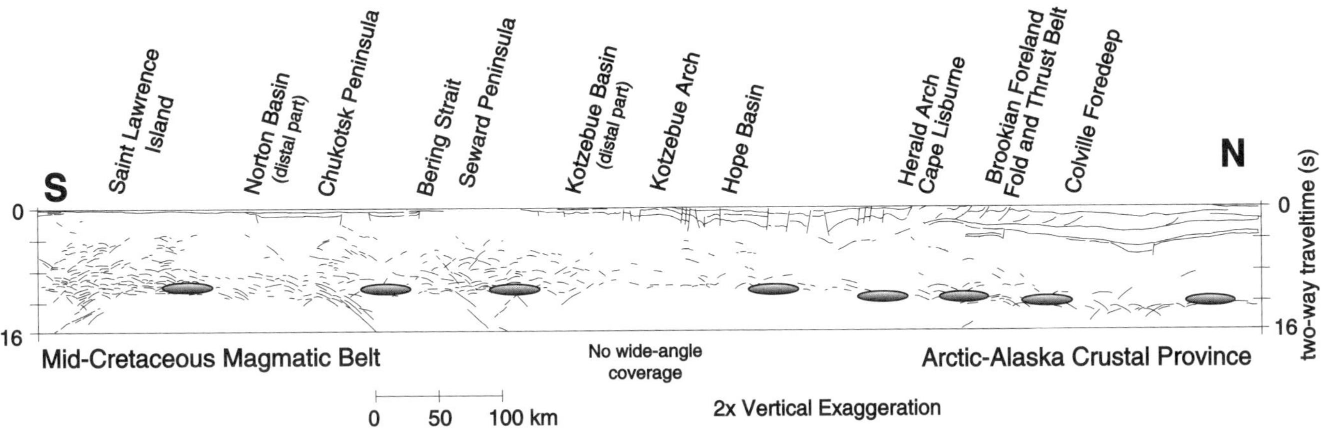

Figure 11. Line drawing of seismic reflection data interpretation (Klemperer et al., Chapter 1) along portion of transect corresponding to two-dimensional refraction model (Fig. 5). Ellipses indicate positions of wide-angle reflections that were used to define refraction Moho. In general, there is good agreement between locations of refraction and reflection Mohos. Note that reflection data lack indication of crustal root in area between Point Hope and Cape Lisburne, thus supporting no-root refraction model.

a thicker crust (>47 km) once existed in this area and at least 15 km of crust has been removed through the processes of extension and erosion.

CONCLUSIONS

Onshore refraction and wide-angle reflection data collected as part of the Bering-Chukchi Crustal Transect, in combination with gravity modeling of satellite data, have provided information on the general attributes of the crust in the Bering Strait region. Results yield present-day crustal thicknesses of 32–35 km, average velocities of 6.4 km/s in the middle to lower crust, and upper mantle velocities of 7.6–7.9 km/s. These observations are consistent with global averages for continental crust, although upper mantle velocities may be slightly low. Observations from the wide-angle data are consistent with models of widespread extension in the portion of the Bering Sea underlain by the Cretaceous magmatic belt and, when viewed in light of studies of blueschist facies rocks on the Seward Peninsula, suggest that the crust in this area may have been at least 47 km thick prior to extension. Crustal thickness and velocity estimates for the crust between the Seward and Chukotka Peninsulas support regional geologic, petrologic, geochronologic, and geochemical studies that indicate continuity of crustal structure from western Alaska to eastern Russia. The lack of compelling evidence in the seismic and gravity data for a crustal root west of the Lisburne Peninsula suggests that if the Brooks Range once continued beyond the peninsula, its root has been subsequently removed by extensional processes.

ACKNOWLEDGMENTS

We thank K. Mackey, P. Minyuk, and V. Kovalev for their assistance with the data collection on the Chukotka Peninsula and the many Russian colleagues who provided logistical support. We also thank W. Koperwhats, T. Burdette, R. Allen, S. Crumley, B. Kennedy, C. Rowe, and the PASSCAL instrument center for help with data reduction, and M. McLean for help with the gravity figures. Discussions with A. Grantz, T. Moore, and E. Miller greatly aided the interpretation of the study results, and formal reviews by A. Grantz, T. Miller, and G. Fuis greatly improved the manuscript. The deployment and operation of the land stations on Chukotka were supported by National Science Foundation (NSF) grant 92–24029, the Incorporated Research Institutions for Seismology (IRIS), and the Northeast Interdisciplinary Scientific Research Institute, Magadan. The deployment and operation of land stations in Alaska were supported by the Deep Continental Studies program of the U.S. Geological Survey. The IRIS PASSCAL program provided Refraction Technology recorders used in Alaska and the W.M. Keck Foundation provided the Portable Data Acquisition Systems–based seismic instruments deployed in Chukotka. Acknowledgment is made to the donors of the Petroleum Research Fund, administered by the American Chemical Society, for support of this research through a grant to L. Wolf and to the Michigan Spacegrant Consortium for support of gravity analyses by R. McCaleb. Additional support was provided to D. Stone by NATO grant SA.12–2–02 (ENVIR.LG 930919), NSF grant OPP-95–00241, Shell Oil Company, and Exxon Exploration Company.

REFERENCES CITED

Amato, J.M., and Wright, J.E., 1997, Potassic mafic magmatism in the Kigluaik gneiss dome, northern Alaska: A geochemical study of arc magmatism in an extensional setting: Journal of Geophysical Research, v. 102, p. 8065–8084.

Ansorge, J., Prodehl, C., and Bamford, D., 1982, Comparative interpretation of explosion seismic data: Journal of Geophysics, v. 51, p. 69–84.

Barnes, D.F., 1977, Bouguer gravity map of Alaska: U.S. Geological Survey Map GP-913, scale 1:2 500 000, 1 sheet.

Beaudoin, B.C., Fuis, G.S., Mooney, W.D., Nokleberg, W.J., and Christensen, N.I., 1992, Thin, low-velocity crust beneath the southern Yukon-Tanana terrane, east central Alaska: Results from Trans-Alaska crustal transect refraction/wide-angle reflection data: Journal of Geophysical Research, v. 97, p. 1921–1942.

Belyaevsky, N.A., and Borisov, A.A., 1974, Structure and thickness of the earth's crust of the USSR, *in* Dedeev, V.A., ed., Structure of the basement of platformal regions of the USSR: Leningrad, Nauka, p. 381–393 (in Russian).

Belyi, V.F., 1994, Geology of the Okhosk-Chukchi volcanogenic belt: Magadan, Russia, Northeast Interdisciplinary Scientific Research Institute, 76 p. (in Russian).

Bering Strait Geologic Field Party, 1997, Koolen metamorphic complex, N.E. Russia: Implications for the tectonic evolution of the Bering Strait region: Tectonics, v. 16, p. 713–729.

Brocher, T.M., Allen, R.M., Stone, D.B., Wolf, L.W., and Galloway, B.K., 1995, Data report for onshore-offshore wide-angle seismic recordings in the Bering-Chukchi Sea, western Alaska and eastern Siberia: U.S. Geological Survey Open-File Report 95–650, 57 p.

Bulin, N.K., 1989, Deep structure of the Verkhoyansk-Chukotka fold belt according to seismic data: Tikhookeanskaya Geologiya, no. 1, p. 77–85 (in Russian).

Christensen, N.I., and Mooney, W.D., 1995, Seismic velocity structure and composition of the continental crust: A global view: Journal of Geophysical Research, v. 100, p. 9761–9788.

Collins, F.R., 1958, Test wells, Umiat area, Alaska, *in* Exploration of Naval Petroleum Reserve No. 4, and adjacent areas, northern Alaska, 1944–53. 5. Subsurface geology and engineering data: U.S. Geological Survey Professional Paper 305-B, p. 71–206.

Collins, F.R., 1961, Core tests and test wells, Barrow area, Alaska, *in* Exploration of Naval Petroleum Reserve No. 4, and adjacent areas, northern Alaska, 1944–53. 5. Subsurface geology and engineering data: U.S. Geological Survey Professional Paper 305-K, p. 569–644.

Demenistakaya, R.M., Ivanov, S.S., and Volk, V.E., 1973, Crust of the Arctic seas of Eurasia: Tectonophysics, v. 20, p. 97–104.

Dumitru, T.A., Miller, E.L., O'Sullivan, P.B., Amato, J.M., Hannula, K.A., Calvert, A.T., and Gans, P.B., 1995, Cretaceous to recent extension in the Bering Strait region, Alaska: Tectonics, v. 14, p. 549–563.

Fisher, M.A., Patton, W.W., and Holmes, M.L., 1982, Geology of Norton basin and continental shelf beneath northwestern Bering Sea, Alaska: American Association of Petroleum Geologists Bulletin, v. 66, p. 255–285.

Fuis, G.S., Murphy, J.M., Lutter, W.J., Moore, T.E., and Bird, K.J., 1997, Deep seismic structure and tectonics of northern Alaska: Crustal-scale duplexing with deformation extending into the upper mantle: Journal of Geophysical Research, v. 102, p. 20873–20896.

Grantz, A., and May, S.D., 1987, Regional geology and petroleum potential of the United States Chukchi Shelf north of Point Hope, *in* Scholl, D.W., Grantz, A., and Vedder, J.G., eds., Geology and resource potential of the continental margin of western North America and adjacent ocean basins: Beaufort Sea to Baja California: Houston, Texas, Circum-Pacific Council for Energy and Mineral Resources, Earth Science Series, v. 6, p. 37–58.

Grantz, A., Wolf, S.C., Breslau, L., Johnson, T.C., and Hanna, W.F., 1970, Reconnaissance geology of the Chukchi Sea as determined by acoustic and magnetic profiling, *in* Adkinson, W.L., and Brosge, M.M., eds., Proceedings of the Geological Seminar on the North Slope of Alaska: American Association of Petroleum Geologists, Pacific Section, Los Angeles, p. F1–F28.

Grantz, A., Eittreim, S., and Whitney, O.T., 1981, Geology and physiography of the continental margin north of Alaska and implications for the origin of the Canada basin, *in* Nairn, A.E.M., Churkin, M., Jr., and Stehli, F.G., eds., The Arctic Ocean: New York, Plenum Press, The Ocean Basins and Margins, v. 5, p. 439–492.

Grantz, A., May, S.D., and Hart, P.E., 1994, Geology of the Arctic continental margin, *in* Plafker, G., and Berg, H.C., eds., The geology of Alaska: Boulder, Colorado, Geological Society of America, Geology of North America, v. G-1, p. 17–48.

Hannula, K.A., and McWilliams, M.O., 1995, Reconsideration of the age of blueschist facies metamorphism on the Seward Peninsula, Alaska, based on phengite $^{40}Ar/^{39}Ar$ results: Journal of Metamorphic Geology, v. 13, p. 125–139.

Houtz, R.E., Eittreim, S., and Grantz, A., 1981, Acoustic properties of northern Alaska shelves in relation to the regional geology: Journal of Geophysical Research, v. 86, p. 3935–3943.

Howell, D.G., Moore, G.W., and Wiley, T.J., 1987, Tectonics and basin evolution of western North America: An overview, *in* Scholl, D.W., Grantz, A., and Vedder, J.G., eds., Geology and resource potential of the continental margin of western North America and adjacent ocean basins: Beaufort Sea to Baja California: Houston, Texas, Circum-Pacific Council for Energy and Mineral Resources, Earth Science Series, v. 6, p. 1–15.

Hunkins, K., 1966, The Arctic continental shelf north of Alaska: Geological Survey of Canada, Paper 66–15, p. 197–205.

Jin, D.J., and Herrin, E., 1980, Surface wave studies of the Bering Sea and Alaska area: Bulletin of the Seismological Society of America, v. 70, p. 2117–2144.

Kirschner, C.E., 1998, Map showing sedimentary basins of onshore and continental shelf areas, Alaska: U.S. Geological Survey Miscellaneous Investigations Series, I-1873.

Klemperer, S.L., Galloway, B.K., and Childs, J.R., 1995, Preliminary results of deep seismic profiling offshore Alaska from the Aleutian Basin to the Arctic Ocean: Eos (Transactions, American Geophysical Union), v. 76, p. 590.

Luetgert, J.H., 1992, MacRay: Interactive two-dimensional seismic raytracing for the Macintosh™: U.S. Geological Survey Open-File Report 92–356, 43 p.

Mackey, K.G., 1999, Seismological studies in northeast Russia [Ph.D. thesis]: East Lansing, Michigan State University, 346 p.

Mackey, K.G., Fujita, K., Gunbia, L.V., Kovalev, V.N., Imaev, V.S., Koz'min, B.M., and Imaeva, L.P., 1997, Seismicity of the Bering Strait region: Evidence for a Bering block: Geology, v. 25, p. 979–982.

McGeary, S.E., and Ben-Avraham, Z., 1981, Allocthonous terranes in Alaska: Implications for the structure and evolution of the Bering Sea: Geology, v. 9, p. 608–613.

Mishin, S.V., and Dareshkina, N.M., 1966, Isolation of converted waves on records of distant earthquakes: Academy of Sciences of the USSR, Izvestiya Earth Physics, v. 2, p. 598–601.

Miller, E.L., and Hudson, T., 1991, Mid-Cretaceous extensional fragmentation of a Jurassic–Early Cretaceous compressional orogen: Tectonics, v. 10, p. 781–796.

Mooney, W.D., and Brocher, T.M., 1987, Coincident seismic reflection/refraction studies of the continental lithosphere: A global review: Reviews of Geophysics, v. 25, p. 723–742.

Moore, T.E., Wallace, W.K., Bird, K.J., Karl, S.M., Mull, C.G., and Dillion, J.T., 1994, Geology of northern Alaska, *in* Plafker, G., and Berg, H. C., eds., The geology of Alaska: Boulder, Colorado, Geological Society of America, Geology of North America, v. G-1, p. 49–140.

Newberry, R.J., Allegro, G.L., Cutler, S.E., Hagen-Lavelle, J.H., Adams, D.D., Nicholson, L.C., Weglarz, T.B., Bakke, A.A., Clautice, K.H., Coulter, G.A., Ford, M.A., Myers, G.L., and Szumigala, D.J., 1997, Skarn deposits of Alaska: Economic Geology Monograph 9, p. 355–395.

Ostenso, N.A., 1968, A gravity survey of the Chukchi Sea region, and its bearing on westward extension of structures in northern Alaska: Geological Society of America Bulletin, v. 79, p. 241–254.

Page, R.A., Biswas, N.H., Lahr, J.C., and Pulpan, H., 1991, Seismicity of continental Alaska, *in* Slemmons, D.B., et al., eds., Neotectonics of North America: Boulder, Colorado, Geological Society of America, Geology of North America, Decade Map Volume, p. 47–68.

Patton, W.W., Jr., and Tailleur, I.L., 1977, Evidence in the Bering Strait region for differential movements between North America and Eurasia: Geological Society of America Bulletin, v. 88, p. 1298–1304.

Rubin, C.M., Miller, E.L., and Toro, J., 1995, Deformation of the northern circum-Pacific margin: Variations in tectonic style and plate-tectonic implications: Geology, v. 23, p. 897–900.

Rudnick, R.L., and Fountain, D.M., 1995, Nature and composition of the continental crust: A lower crust perspective: Reviews of Geophysics, v. 33, p. 267–309.

Sandwell, D.T., and Smith, W.H.F., 1997, Marine gravity anomaly from GEOSAT and ERS-1 satellite altimetry: Journal of Geophysical Research, v. 102, p. 10039–10054.

Shipilov, E.V., Senin, B.V., and Yunov, A.Y., 1989, Sedimentary cover and basement of the Chukchi Sea from seismic data: Geotectonics, v. 23, p. 456–463.

Stavsky, A.P., Checkhovitch, V.D., Kononov, M.V., and Zonenshain, L.P., 1990, Plate tectonics and palinspastic reconstructions of the Anadyr-Koryak region, northeast USSR: Tectonics, v. 9, p. 81–101.

Tolson, R.B., 1987, Structure and stratigraphy of the Hope Basin, southern Chukchi Sea, Alaska, *in* Scholl, D.W., Grantz, A., and Vedder, J.G., eds., Geology and resource potential of the continental margin of western North America and adjacent ocean basins: Beaufort Sea to Baja California: Houston, Texas, Circum-Pacific Council for Energy and Mineral Resources, Earth Science Series, v. 6, p. 59–71.

Turner, D.L., and Swanson, S.E., 1981, Continental rifting—A new tectonic model for central Seward Peninsula, *in* Wescott, E., and Turner, D., eds., Geothermal reconnaissance survey of the central Seward Peninsula, Alaska: Fairbanks, University of Alaska Geophysical Institute Report UAG R-284, p. 7–36.

Vashchilov, Y.Y., 1979, Seismicity and the problem of deep structure of the northeast USSR, *in* Lin'kova, T.I., ed., Geophysical investigations of structure and geodynamics of the earth's crust and upper mantle of the northeast USSR: Magadan, Russia, Northeast Interdisciplinary Scientific Research Institute, p. 138–157 (in Russian).

Wissinger, E.S., Levander, A., and Christensen, N.I., 1997, Seismic images of crustal duplexing and continental subduction in the Brooks Range: Journal of Geophysical Research, v. 102, p. 20847–20871.

Worrall, D.M., 1991, Tectonic history of the Bering Sea and the evolution of Tertiary strike-slip basins of the Bering Shelf: Boulder, Colorado, Geological Society of America Special Paper 257, 120 p.

MANUSCRIPT ACCEPTED BY THE SOCIETY MAY 15, 2001

Geological Society of America
Special Paper 360
2002

Structure and stratigraphy of the Hanna Trough, U.S. Chukchi Shelf, Alaska

Kirk W. Sherwood, Peter P. Johnson, James D. Craig
Minerals Management Service, 949 East 36th Avenue, Anchorage, Alaska 99508-4362, USA
Susan A. Zerwick
P.O. Box 123, Seldovia, Alaska 99663, USA
Richard T. Lothamer
*Snohomish County Public Utility District, 2310 California Street, Everett,
Washington 98206, USA*
Dennis K. Thurston
Minerals Management Service, 949 East 36th Avenue, Anchorage, Alaska 99508-4362, USA
Sally B. Hurlbert
*National Park Service, 3655 U.S. Highway 211 East, Shenandoah National Park,
Luray, Virginia 22835, USA*

ABSTRACT

The Hanna Trough is a basin of probable Devonian to Late Jurassic age that trends north beneath the U.S. Chukchi Shelf off northwestern Alaska and that was filled with over 11 600 m of strata of the Devonian to Jurassic Ellesmerian sequence. The Hanna Trough overlies acoustic basement that appears to consist of an assemblage of terranes that amalgamated into a crustal unit with a northerly structural grain in pre-Middle or Late Devonian time. The subsequent Hanna Trough rift structures also developed along northerly lines. On the north, the Hanna Trough appears to have been segmented by a transform fault that links the southern segment to an independent, northern rift segment that is offset to the west. However, the hypothesized rift segment north of the transform fault is now mostly lost to observation beneath the younger North Chukchi Basin. On the south, the Hanna Trough rift structures appear to merge structurally and stratigraphically with the coeval Arctic Alaska Basin that trends east beneath northern Alaska.

Fault-related subsidence and sedimentation occurred mostly from Devonian to Permian time and controlled deposition of over 11 000 m of strata of the Lower Ellesmerian sequence. From Late Permian to Late Jurassic time, a broader "sag" phase of subsidence accommodated up to 3660 m of strata of the Upper Ellesmerian sequence. Stratigraphic control within Hanna Trough is provided by five offshore petroleum exploration wells that reached rocks as old as Late Mississippian (equivalent to the Lisburne Group).

PURPOSE

The Hanna Trough trends south beneath the U.S. Chukchi Shelf and joins the west-trending Arctic Alaska Basin that is beneath northern Alaska (Fig. 1). The Hanna Trough has been implicated in Arctic reconstructions as a seaway linking the Arctic Alaska Basin of Alaska and the Sverdrup Basin of the Canadian Arctic Islands, the latter now separated from Chukchi Shelf by

Data Repository item 2002073 contains additional material related to this article.

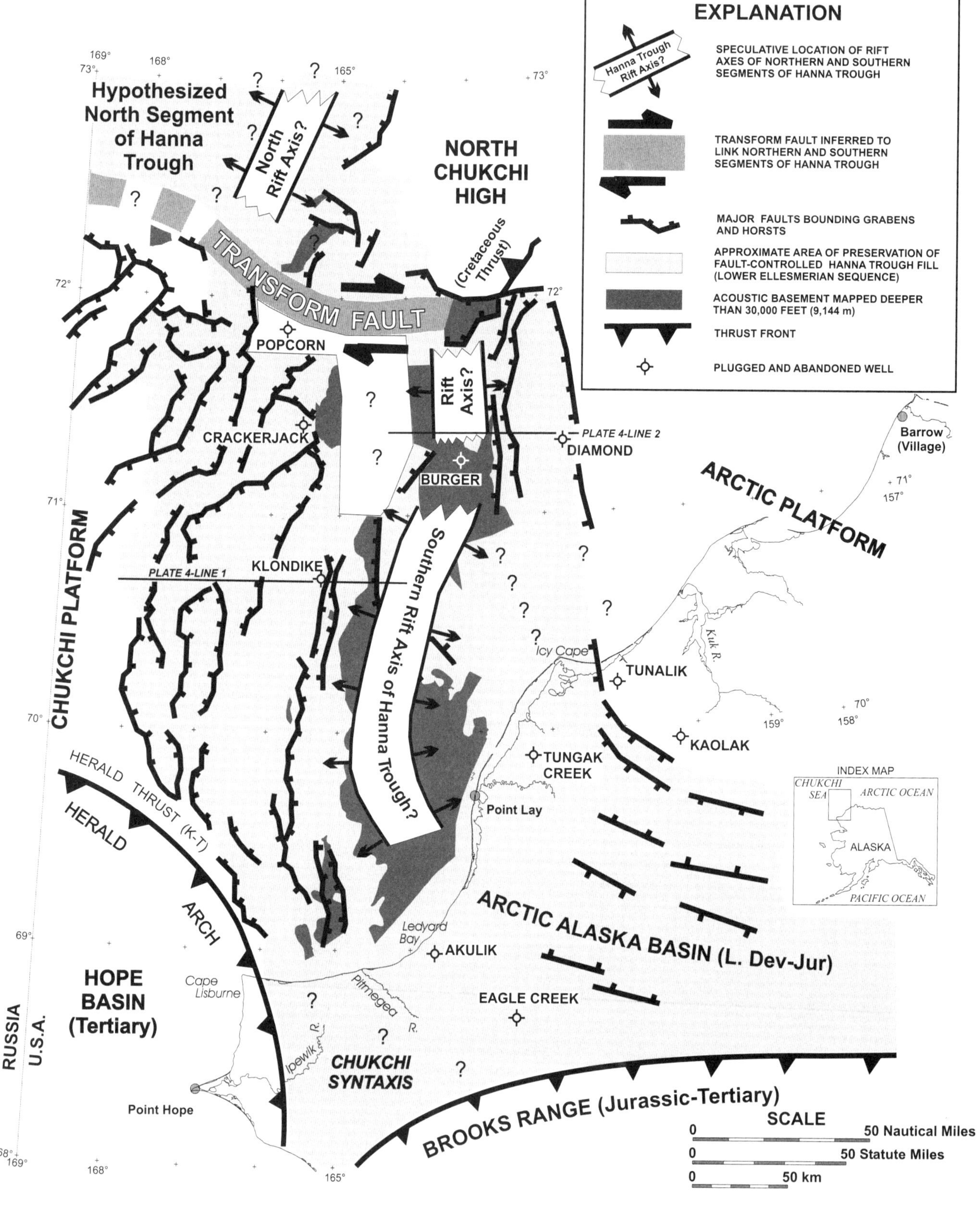

EXPLANATION
Hanna Trough Rift Axis?
SPECULATIVE LOCATION OF RIFT AXES OF NORTHERN AND SOUTHERN SEGMENTS OF HANNA TROUGH
TRANSFORM FAULT INFERRED TO LINK NORTHERN AND SOUTHERN SEGMENTS OF HANNA TROUGH
MAJOR FAULTS BOUNDING GRABENS AND HORSTS
APPROXIMATE AREA OF PRESERVATION OF FAULT-CONTROLLED HANNA TROUGH FILL (LOWER ELLESMERIAN SEQUENCE)
ACOUSTIC BASEMENT MAPPED DEEPER THAN 30,000 FEET (9,144 m)
THRUST FRONT
PLUGGED AND ABANDONED WELL
Hypothesized North Segment of Hanna Trough
North Rift Axis?
NORTH CHUKCHI HIGH
(Cretaceous Thrust)
TRANSFORM FAULT
POPCORN
CRACKERJACK
Rift Axis?
DIAMOND
PLATE 4-LINE 2
Barrow (Village)
BURGER
ARCTIC PLATFORM
CHUKCHI PLATFORM
PLATE 4-LINE 1
KLONDIKE
Southern Rift Axis of Hanna Trough?
Icy Cape
TUNALIK
Kuk R.
HERALD THRUST (K-T)
HERALD
ARCH
Tungak Creek
KAOLAK
INDEX MAP
CHUKCHI SEA
ARCTIC OCEAN
ALASKA
PACIFIC OCEAN
Point Lay
ARCTIC ALASKA BASIN (L. Dev-Jur)
HOPE BASIN (Tertiary)
RUSSIA
U.S.A.
Cape Lisburne
Ledyard Bay
AKULIK
Pitmegea R.
EAGLE CREEK
Ipewik R.
CHUKCHI SYNTAXIS
BROOKS RANGE (Jurassic-Tertiary)
Point Hope
SCALE
0   50 Nautical Miles
0   50 Statute Miles
0   50 km
169° 168° 165° 73°
73°
72°   72°
71°   71°
70°   70°
69°   69°
68°
169° 168° 165°
157°
159° 158°

seafloor spreading in the Canada Basin of the Arctic Ocean (Embry, 1990). The Hanna Trough structures may have helped shape the western end of the Brooks Range. The Hanna Trough projects south into the Chukchi syntaxis, where the westerly trends of the Brooks Range are intersected by the northerly trends of the Herald thrust system (Fig. 1). Faults and fault-bounded rift structures established in the Hanna Trough during Late Devonian to Early Mississippian time subsequently influenced Mesozoic and Cenozoic faulting and sedimentation on the Chukchi Shelf. Despite its intriguing relationships to some major geological features of the Arctic, the Hanna Trough has been described in only the most general terms by a few regional studies. The purpose of this chapter is to document in detail the structures and stratigraphy of the Hanna Trough as revealed by seismic mapping by the Minerals Management Service and by the analysis of petroleum industry wells on the Chukchi Shelf.

LOCATION AND INFORMATION BASE

The U.S. Chukchi Shelf extends 110–400 km west from Alaska to Russian waters (west of long 169°W) and 530 km north from Point Hope to the shelf edge (~90 m isobath) near lat 73°N (Fig. 1). Water depths are typically <50 m over this broad shelf, except where incised by the Barrow submarine canyon that parallels and is within 60 km of the northwest coast of Alaska. The U.S. Chukchi Shelf is flanked on the north by the oceanic Canada Basin and the Chukchi borderland and on the west by the Russian Chukchi and East Siberian shelves off northeast Siberia.

Petroleum industry investigations of the U.S. Chukchi Shelf prompted by oil and gas lease sales in 1988 and 1991 resulted in the collection of 161 000 line km of high-quality seismic-reflection data, nearly all of which are located south of lat 73°N. In addition, comprehensive gravimetric, magnetic, thermal, and geochemical surveys were conducted on the U.S. Chukchi Shelf. These data remain proprietary to petroleum industry owners and cannot be freely shared with the scientific community. This study is based upon the proprietary geophysical data and the data from petroleum exploration wells drilled following the lease sales. Previous investigations of the Chukchi Shelf and the contiguous Arctic continental shelves were carried out primarily by Arthur Grantz and colleagues of the U.S. Geological Survey. These pioneer studies, specifically those of Grantz et al. (1975, 1979, 1981, 1982a, 1982b, 1987, 1990, 1994), Grantz and Eittreim (1979), Grantz and May (1983, 1987), and Eittreim and Grantz (1979), established the basic geologic framework for the

Chukchi Shelf. Hubbard et al. (1987) and Haimila et al. (1990) published regional studies of the Chukchi and Beaufort Seas based on proprietary seismic data. Regional studies, also based on proprietary seismic data, were published as Minerals Management Service agency reports by Thurston and Theiss (1987), Craig et al. (1985), and Sherwood et al. (1998).

REGIONAL TECTONIC HISTORY OF CHUKCHI SHELF

The rocks that underlie the Chukchi Shelf can be divided into four main sequences for purposes of introduction to their place in the regional stratigraphic framework of northern Alaska. These sequences are shown in a stratigraphic column in Figure 2. In northern Alaska and probably most of the U.S. Chukchi Shelf, acoustic (or seismic) and economic basement is composed of deformed and metamorphosed rocks of Late Devonian and older age that are generally assigned to the Franklinian sequence (Lerand, 1973). The Franklinian rocks were deformed in a regional event, probably an amalgamation of crustal terranes, termed the Ellesmerian orogeny, that is widely recognized in many parts of Arctic North America (Moore, 1986; Grantz et al., 1990; Moore et al., 1994).

The post-Franklinian basin-forming events on the Chukchi Shelf are summarized in Figure 3. Deposition of the Ellesmerian sequences—the principal sedimentary fill for the Hanna Trough—ranged from Late Devonian (?) to Late Jurassic time. In the Hanna Trough, the early phase of subsidence (Late Devonian to Permian time) was rift driven or fault driven (Fig. 3A), and corresponds approximately to the deposition of the Lower Ellesmerian sequence. A second phase of regional subsidence (Permian to Late Jurassic time), largely unaccompanied by faulting (Fig. 3B), governed deposition of the Upper Ellesmerian sequence. Subsidence of the Hanna Trough and the Ellesmerian cycle of sedimentation were concluded with the development of an Upper Jurassic regional unconformity that marks the base of the overlying Rift sequence (JU, Fig. 2).

Rifting along the Beaufort continental margin extended into the northern Chukchi Shelf in Jurassic time and opened a new rift that ultimately became the North Chukchi Basin (Fig. 3C; Grantz et al., 1994, their Fig. 3). Grabens and flexural downwarps within and near the rift were filled with thick sequences of clastic sediments (Fig. 3C). These strata represent a distinct tectonic event and have been variously distinguished along the Beaufort margin as the Rift sequence (Craig et al., 1985), the Beaufortian sequence (Hubbard et al., 1987), or the Barrovian sequence (Carman and Hardwick, 1983). Because it is more general, we prefer the term Rift sequence for rocks deposited during the Mesozoic rifting episode on the northern Chukchi Shelf. The Rift sequence beneath the Chukchi Shelf ranges in age from Late Jurassic to Early Cretaceous (Aptian to Albian) and as a convenience here we extend this informal term to include time-equivalent rocks deposited to the south and possibly beyond the influence of the rift zone.

Figure 1. Major elements of the Hanna Trough rift system, U.S. Chukchi Shelf, Alaska featuring a southern segment, an hypothesized northern segment (axis location unknown—obscured beneath North Chukchi Basin), and a transform fault or rift jump that links the two segments. The axis of the southern segment of Hanna Trough south of Burger well is drawn along a gravity high mapped by Herman and Zerwick (1994) and shown in Figure 4.

U.S. CHUKCHI SHELF
STRATIGRAPHIC COLUMN

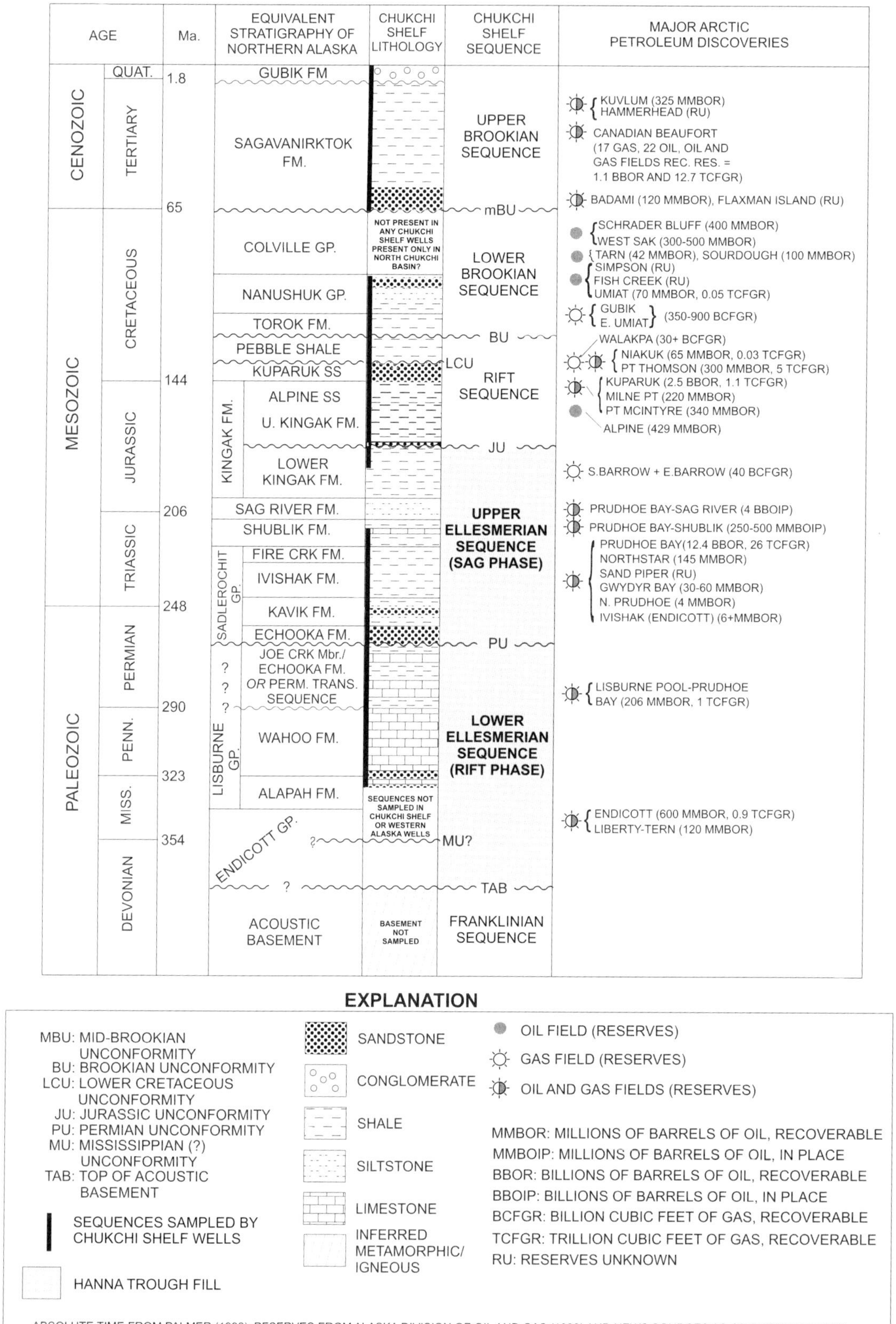

AGE		Ma.	EQUIVALENT STRATIGRAPHY OF NORTHERN ALASKA	CHUKCHI SHELF LITHOLOGY	CHUKCHI SHELF SEQUENCE	MAJOR ARCTIC PETROLEUM DISCOVERIES
CENOZOIC	QUAT.	1.8	GUBIK FM			
	TERTIARY		SAGAVANIRKTOK FM.		UPPER BROOKIAN SEQUENCE	KUVLUM (325 MMBOR) / HAMMERHEAD (RU); CANADIAN BEAUFORT (17 GAS, 22 OIL, OIL AND GAS FIELDS REC. RES. = 1.1 BBOR AND 12.7 TCFGR)
		65			mBU	BADAMI (120 MMBOR), FLAXMAN ISLAND (RU)
MESOZOIC	CRETACEOUS		COLVILLE GP. (NOT PRESENT IN ANY CHUKCHI SHELF WELLS PRESENT ONLY IN NORTH CHUKCHI BASIN?)		LOWER BROOKIAN SEQUENCE	SCHRADER BLUFF (400 MMBOR) / WEST SAK (300-500 MMBOR); TARN (42 MMBOR), SOURDOUGH (100 MMBOR); SIMPSON (RU); FISH CREEK (RU); UMIAT (70 MMBOR, 0.05 TCFGR)
			NANUSHUK GP.			GUBIK / E. UMIAT (350-900 BCFGR)
			TOROK FM.		BU	WALAKPA (30+ BCFGR)
			PEBBLE SHALE		LCU	NIAKUK (65 MMBOR, 0.03 TCFGR) / PT THOMSON (300 MMBOR, 5 TCFGR)
		144	KUPARUK SS		RIFT SEQUENCE	KUPARUK (2.5 BBOR, 1.1 TCFGR) / MILNE PT (220 MMBOR) / PT MCINTYRE (340 MMBOR)
	JURASSIC		ALPINE SS (KINGAK FM.)			ALPINE (429 MMBOR)
			U. KINGAK FM.			
			LOWER KINGAK FM.		JU	S.BARROW + E.BARROW (40 BCFGR)
		206	SAG RIVER FM.		UPPER ELLESMERIAN SEQUENCE (SAG PHASE)	PRUDHOE BAY-SAG RIVER (4 BBOIP)
	TRIASSIC		SHUBLIK FM.			PRUDHOE BAY-SHUBLIK (250-500 MMBOIP)
			FIRE CRK FM. (SADLEROCHIT GP.)			PRUDHOE BAY (12.4 BBOR, 26 TCFGR); NORTHSTAR (145 MMBOR); SAND PIPER (RU); GWYDYR BAY (30-60 MMBOR); N. PRUDHOE (4 MMBOR); IVISHAK (ENDICOTT) (6+MMBOR)
			IVISHAK FM.			
		248	KAVIK FM.			
PALEOZOIC	PERMIAN		ECHOOKA FM.		PU	
			JOE CRK Mbr./ ECHOOKA FM. OR PERM. TRANS. SEQUENCE			LISBURNE POOL-PRUDHOE BAY (206 MMBOR, 1 TCFGR)
		290			LOWER ELLESMERIAN SEQUENCE (RIFT PHASE)	
	PENN.		WAHOO FM. (LISBURNE GP.)			
		323	ALAPAH FM.			
	MISS.			(SEQUENCES NOT SAMPLED IN CHUKCHI SHELF OR WESTERN ALASKA WELLS)		ENDICOTT (600 MMBOR, 0.9 TCFGR) / LIBERTY-TERN (120 MMBOR)
		354	ENDICOTT GP. ?		MU?	
	DEVONIAN		?		TAB	
			ACOUSTIC BASEMENT	BASEMENT NOT SAMPLED	FRANKLINIAN SEQUENCE	

EXPLANATION

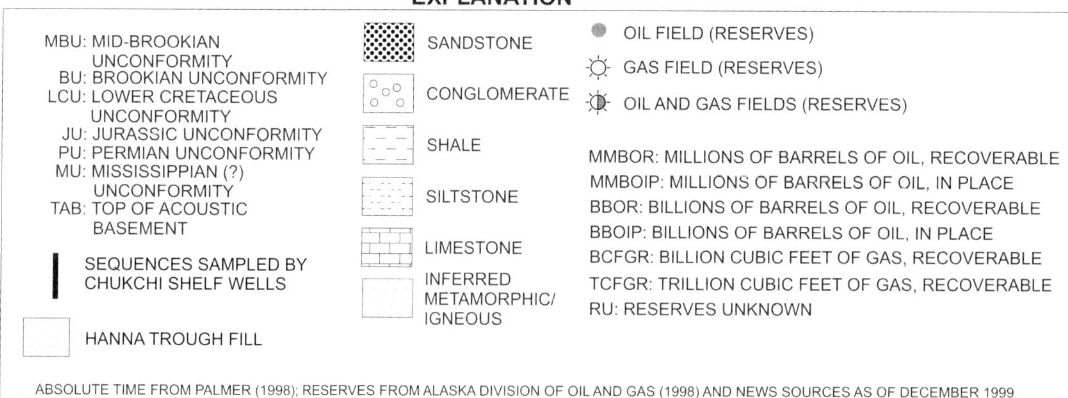

ABSOLUTE TIME FROM PALMER (1998); RESERVES FROM ALASKA DIVISION OF OIL AND GAS (1998) AND NEWS SOURCES AS OF DECEMBER 1999

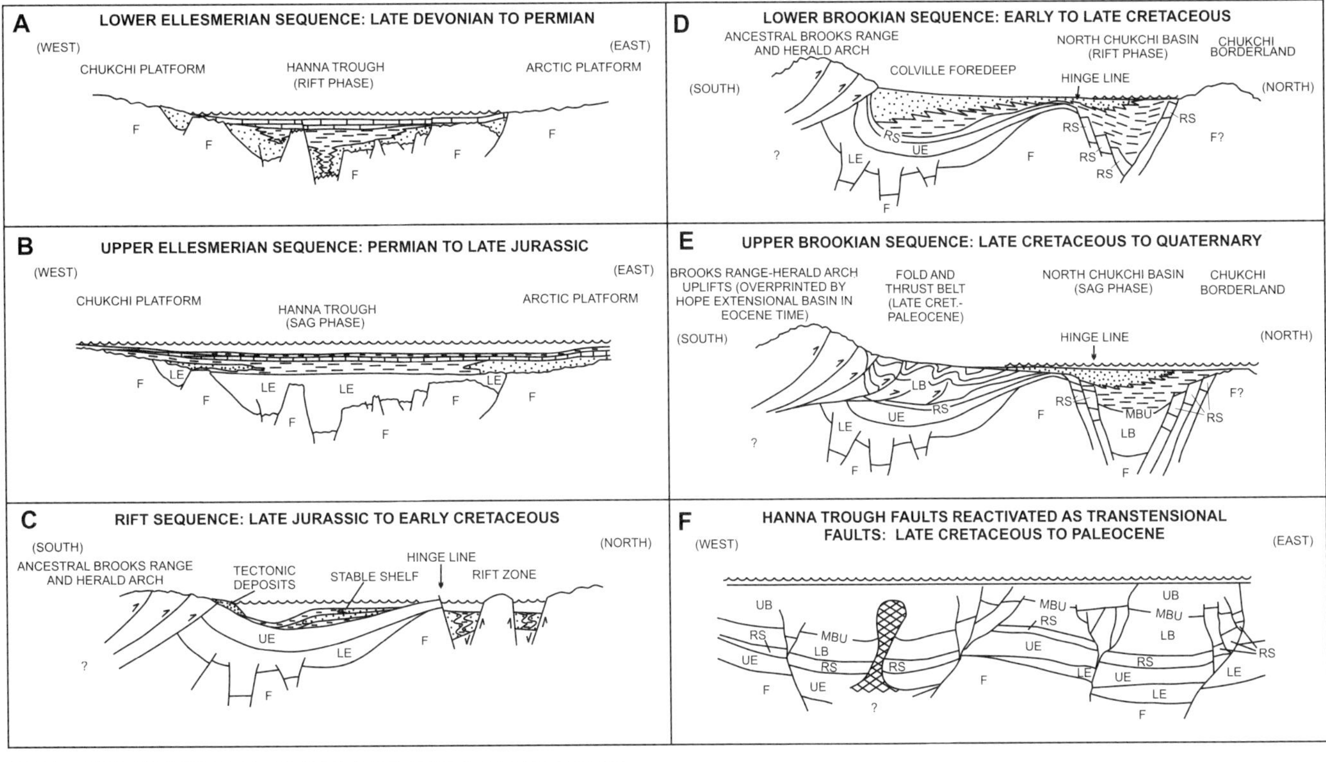

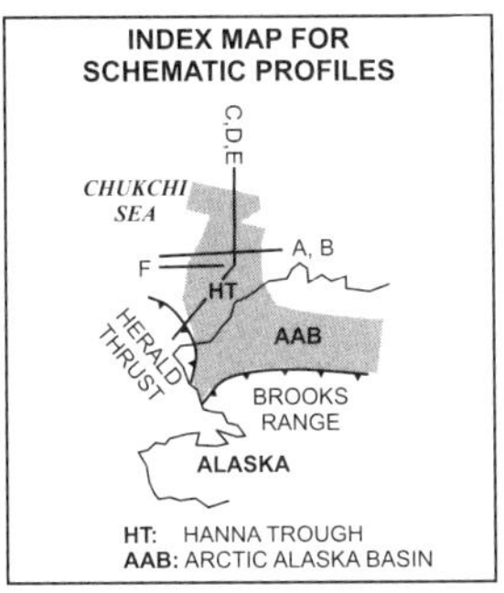

Figure 3. Geological evolution of Chukchi Shelf. Schematic cross sections illustrating the six main basin-forming events beneath the U.S. Chukchi Shelf.

The Brookian orogeny, ranging in possible age from Middle Jurassic to Early Cretaceous (ca. 175–115 Ma), completely reorganized the tectonic framework of northern Alaska and the Chukchi Shelf. Cretaceous and Tertiary rocks of the Brookian sequence, consisting mostly of sediments shed from mountain belts created during the Brookian orogeny (Moore et al., 1994, p. 54), fill several (Colville, Hope, North Chukchi) basins beneath the Chukchi Shelf (Fig. 3D). Late Cretaceous to early Tertiary deformation folded the rocks in the southern Colville basin and reactivated the north-trending faults within the Hanna Trough as the roots of upward-branching transtensional zones that complexly structured Brookian strata on the Chukchi platform (Fig. 3, E and F).

Figure 2. Stratigraphic column for U.S. Chukchi Shelf. Upper and Lower Ellesmerian sequences (shaded) represent the Hanna Trough sedimentary fill.

BASEMENT ROCKS BENEATH NORTHERN ALASKA AND THE CHUKCHI SHELF

Beneath the sedimentary rocks filling the Hanna Trough and blanketing the Arctic and Chukchi platforms is an assemblage of highly diverse rocks probably representing collapsed basins and accreted terranes that were joined together as a single crustal block in Devonian time. The basement rocks are only known in a fragmentary way, but they are generally Late to Middle Devonian or older in age and, as a convenience, are typically grouped as the Franklinian sequence. Elements of Franklinian basement are mapped in Figure 4.

Most wells that penetrated basement rocks beneath the Arctic platform in northwestern Alaska have encountered steeply inclined argillite, slate, phyllite, and minor quartzites, here termed *tectonized siliciclastics*. The tectonized siliciclastics have yielded a wide range of ages. Phyllites penetrated in wells along the north Alaskan coast yielded K-Ar radiometric ages ranging from 547–564 Ma (mica separates) to 592 Ma (whole rock) (Late Proterozoic to Cambrian; Brosgé and Dutro, 1973, Fig. 3; Drummond, 1974). Ordovician and Silurian fossils have been recovered from the tectonized siliciclastics at two wells in the Barrow area and at a third well near Prudhoe Bay (Carter and Laufield, 1975). Steeply dipping coal-bearing shale and sand-

NORTH CHUKCHI HIGH
OFFSET EXTENSION OF CHUKCHI PLATFORM MAGNETIC HIGH ?
DENSE MAFIC BODY EMPLACED BELOW 20 KM WITHIN ACOUSTIC BASEMENT DURING HANNA TROUGH RIFTING ?
NORTHEAST CHUKCHI BASIN--MODERATELY DEFORMED FRANKLINIAN ROCKS ?
POPCORN
CRACKERJACK
BURGER
DIAMOND
KLONDIKE
HANNA
TROUGH
CHUKCHI PLATFORM
RUSSIA
U.S.A.
Barrow
BRONTOSAURUS
SKULL CLIFF
PEARD
KUGRUA
Wainwright
ARCTIC PLATFORM
TUNALIK
BASEMENT NOT PENETRATED BY WELLS
KAOLAK
TUNGAK CREEK
Point Lay
MAP LOCATION
ALASKA
POSSIBLE MAGMATIC ARC TERRANE WITHIN ACOUSTIC BASEMENT ?
HERALD
HERALD
ARCH
THRUST
HOPE BASIN
Lisburne Peninsula
AKULIK
SCALE
50 Nautical Miles
50 Statute Miles
50 km
EAGLE CREEK
MISSISSIPPIAN ROCKS UNCONFORMABLE UPON TECTONIZED ORDOVICIAN-SILURIAN SILICICLASTICS[3]
CHUKCHI SYNTAXIS
Point Hope
BROOKS RANGE
EXPLANATION
PRE-LATE DEVONIAN TECTONIZED SILICICLASTICS BENEATH ARCTIC PLATFORM AND ON LISBURNE PEN.
MAGNETIC HIGHS (>100 nT) ON CHUKCHI PLATFORM AND NORTH CHUKCHI HIGH [1]
GRAVITY HIGH (> -5 mgal) IN SOUTHERN HANNA TROUGH [2]
EXPLORATORY WELL, PLUGGED AND ABANDONED
AXIS OF HANNA TROUGH
BOUNDARY OF GEOLOGIC AREA
THRUST FAULT SYSTEM (BROOKIAN)
THRUST FAULT (FRANKLINIAN)
[1] Outlines of magnetic highs on Chukchi platform and North Chukchi high from Cramer et al. (1986).
[2] Outline of Hanna trough gravity high from Herman and Zerwick (1994). [3] Localities described by Moore et al. (1984), and mapped by Martin (1970).

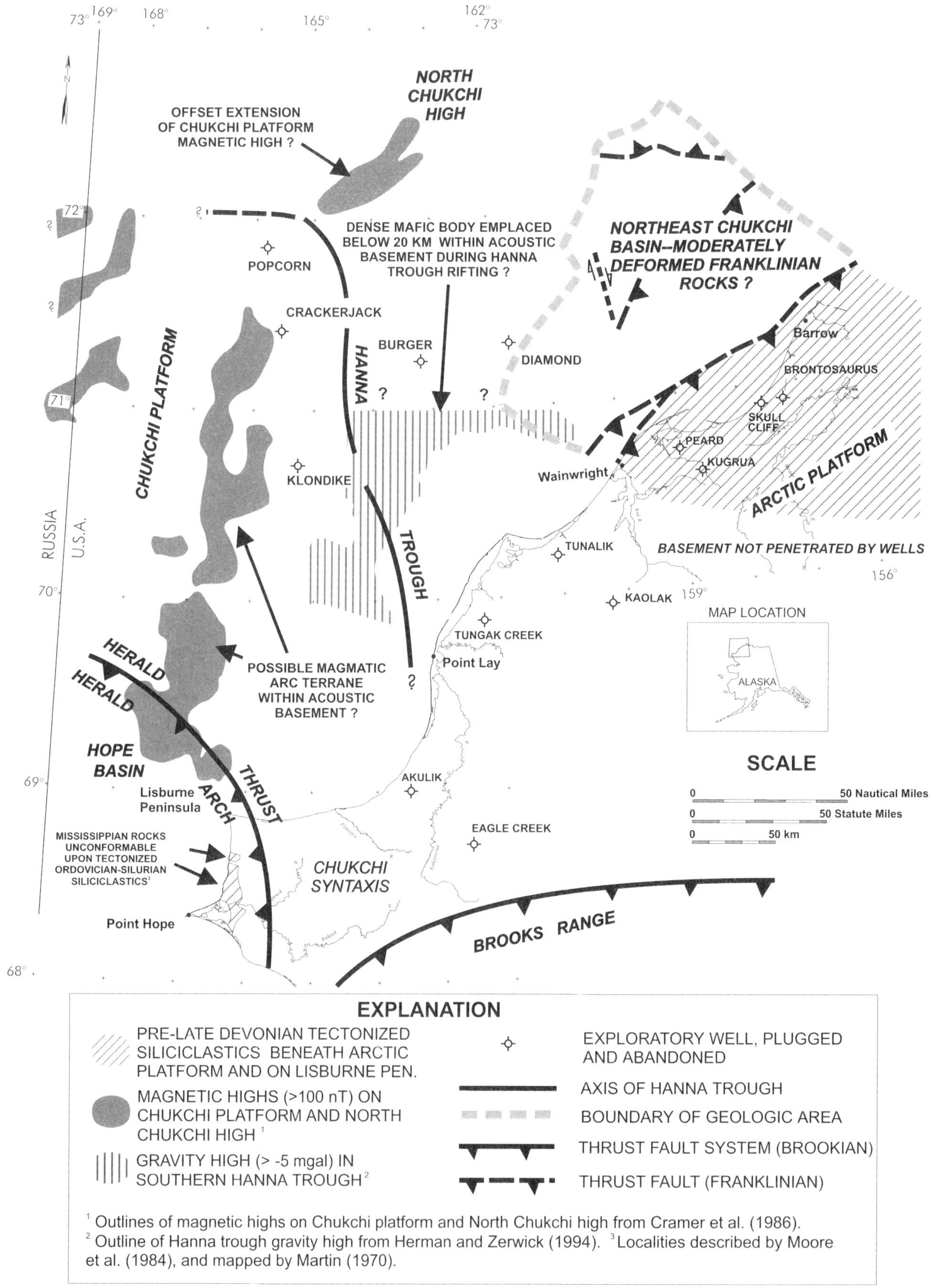

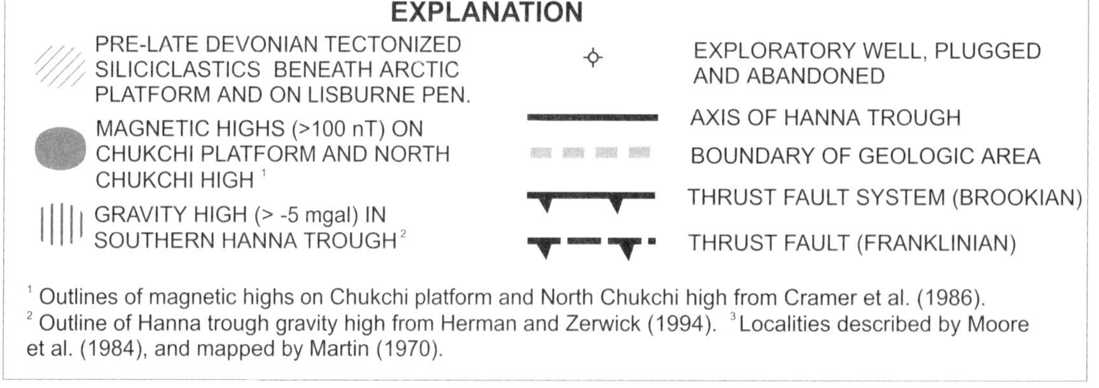

stone encountered in the Topogoruk and South Meade wells have, at the Topogoruk well, yielded fossils of Early or Middle Devonian age (Collins, 1958; Husky Oil NPR-A Operations, Inc., 1982a; Kirschner and Rycerski, 1988, p. 192). A potassium feldspar separate from a granite that intrudes the basement complex in the East Teshekpuk No. 1 well (Fig. 5) yielded a minimum K-Ar radiometric age of 332 Ma (Visean or Early Mississippian; but the sample probably underwent some thermal resetting [altered biotite yielded a date of 243 Ma] and the granite is more likely Devonian in age; Bird et al., 1978).

No wells have penetrated acoustic basement beneath the Chukchi Shelf and interpretations of basement there rely upon seismic, gravity, or magnetic data, which suggest a number of interesting features. In the northeast, seismic data reveal that the Chukchi Shelf is underlain by moderately deformed to undeformed rocks (interpreted as carbonate and clastics) that range up to 9140 m in stratigraphic thickness and are in fault contact with surrounding acoustically transparent, presumably tectonized rocks (Sherwood, 1994). This feature, called the Northeast Chukchi Basin (Fig. 4) by Craig et al. (1985), was interpreted by Sherwood (1994) as an isolated fragment of the Precambrian to Upper Devonian Franklinian basin that underlies the Canadian Arctic Islands. The base of the stratified rocks (Precambrian?) in the Northeast Chukchi Basin, although generally well defined in reflection seismic data, has not been mapped and is not shown in Plate 3. Since Late Devonian or Mississippian time, the Northeast Chukchi Basin and the acoustically transparent basement rocks beneath the Arctic platform have been united as a single structural unit.

A north-trending magnetic high and coinciding gravity high mapped (Bassinger, 1968; Cramer et al., 1986; May, 1985) on the Chukchi platform (magnetic high located in Fig. 4) have been interpreted as marking a possible magmatic complex within the acoustic basement (Herman and Zerwick, 1994). This feature is not observed as a discrete entity within basement in seismic data (crossed by line 1, Plate 4 [note that the proprietary seismic data shown in Plate 4 are presented with the express consent of the data owner]), although it is bounded on the west by a discontinuous west-dipping seismic reflection within basement that may mark a rift-reactivated basement thrust fault.

Herman and Zerwick (1994) speculated that a gravity high over the southern Hanna Trough (located in Fig. 4) may identify

a dense mafic mass below 20 km (below the Curie isotherm and magnetically neutral) within the crust. This mafic body may have been emplaced during the rifting of the Hanna Trough. This feature is not distinguished within basement in available seismic reflection data (e.g., Plate 4, line 1). Similarly, Saltus et al. (1999, p. 5) identified magnetic anomalies corresponding to the deepest parts of the Arctic Alaska Basin that they attributed to mafic rocks erupted or emplaced during Late Devonian to Mississippian rifting.

In exposures on the Lisburne Peninsula (Fig. 4), Mississippian rocks unconformably overlie well-indurated to metamorphosed clastic rocks of Ordovician and Silurian ages that much resemble the tectonized siliciclastics penetrated within acoustic basement by wells on the Arctic platform (Moore et al., 1984).

RELATIONSHIP OF THE HANNA TROUGH TO THE ARCTIC ALASKA BASIN

The Ellesmerian rocks in northern Alaska were deposited on a south-facing continental shelf that was created by breakup of a larger landmass during Late Devonian time (Moore et al., 1994). This shelf hosted the deposition of sediments derived from exposed lands to the north until Late Jurassic to Early Cretaceous time, when the northern exposed lands withdrew across a rift and the former shelf was inundated by sediment derived from mountains elevated on the south related to the Brookian orogeny. The former shelf and its sedimentary wedge thus became the Arctic Alaska Basin.

In northern Alaska at the beginning of Ellesmerian sedimentation (Endicott Group), the floor of the Arctic Alaska Basin was not a smooth, south-dipping shelf, but was highly faulted in the south and embayed by the north- and northwest-trending Meade and Ikpikpuk-Umiat subbasins (Fig. 5). Although the northern (and best-known) parts of the former shelf are only sparsely faulted, the southern part is highly dissected by west-trending faults that mostly displace Pennsylvanian and older rocks (Barnes et al., 1981; Tetra Tech, 1981, maps PDN, PML, and TPS). Through time the Ellesmerian rocks reached an aggregate thickness of 6100 m in parts of the Arctic Alaska Basin (Hubbard et al., 1987, Fig. 8; Bird, 1988b, Fig. 16.21).

The Arctic Alaska Basin passes west and offshore beneath Chukchi Shelf, where it joins the north-trending Hanna Trough (Fig. 1; Hubbard et al., 1987, Fig. 8). Unlike the Arctic Alaska Basin, which was a shelf that faced an open ocean to the south, the Hanna Trough was flanked on both sides by shelf areas and uplifted sediment sources.

STRUCTURAL FEATURES OF THE HANNA TROUGH

Structure on the top of the acoustic basement, illustrated in Plate 3, is dominated by the northerly trends of faults and fault-bounded structures that record the Devonian-Permian rift phase of subsidence and deposition in the Hanna Trough. The map

---

Figure 4. Elements of the acoustic basement or Franklinian rocks beneath the U.S. Chukchi Shelf. The moderately deformed rocks in the Northeast Chukchi Basin are surrounded by acoustically transparent "basement" (Sherwood, 1994; Grantz et al., 1990). Tectonized siliciclastics of Devonian and older ages are exposed on the Lisburne Peninsula and were encountered in wells on the Arctic platform. Herman and Zerwick (1994) suggest from gravity and magnetic models that a magnetic high on Chukchi platform may mark a magmatic arc within acoustic basement. Herman and Zerwick (1994) speculate that a gravity high corresponding to southern Hanna trough may mark a dense mafic mass below 20 km within acoustic basement.

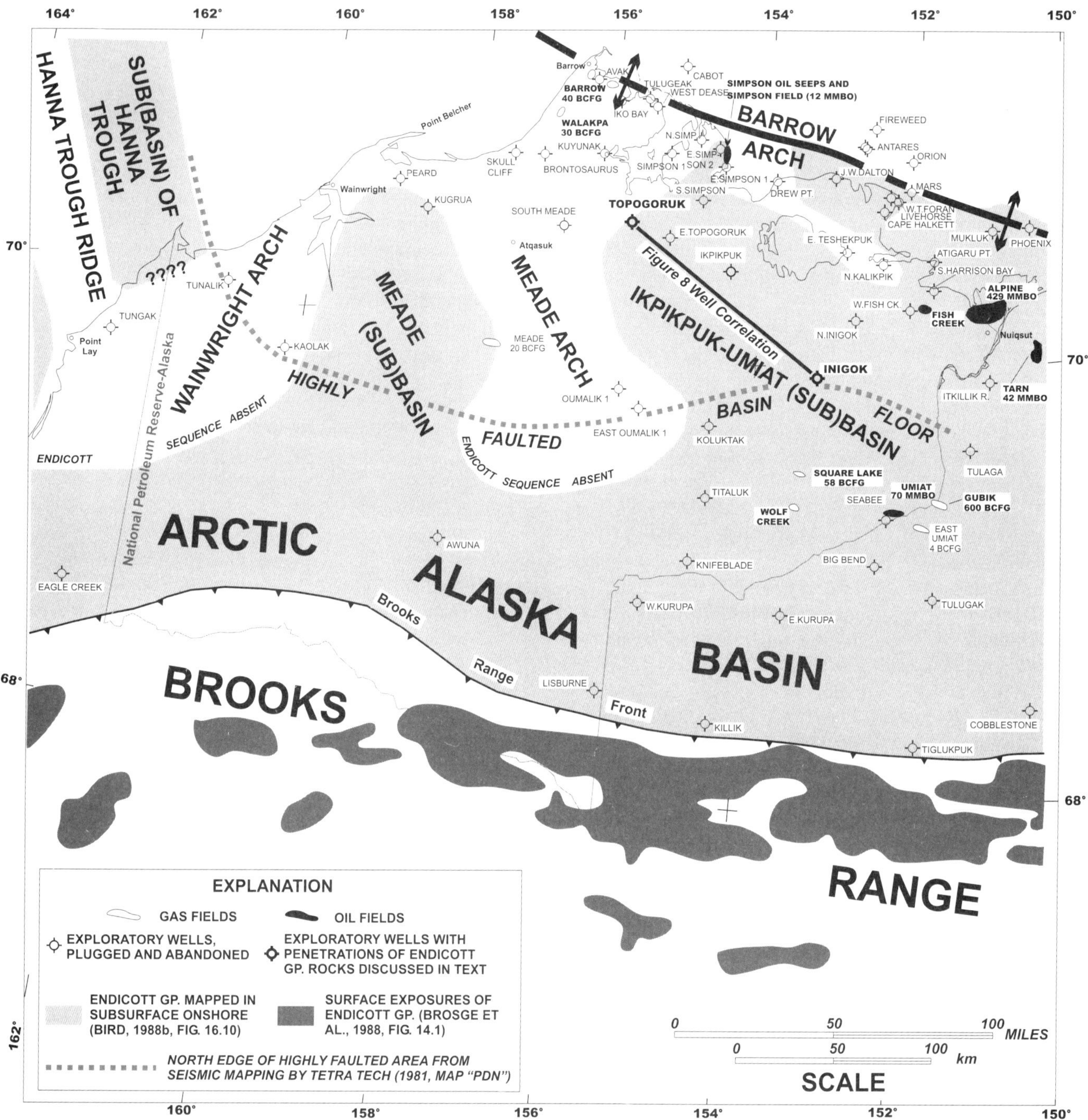

Figure 5. Distribution of the Endicott Group (Upper Devonian to Mississippian age) in Brooks Range exposures and the subsurface basin(s) of the North Slope. The westernmost penetration of Mississippian rocks of the Endicott Group is at the Topogoruk No. 1 well, where Mississippian coal-bearing clastics overlie steeply dipping Lower to Middle Devonian coal-bearing clastics, possibly also part of the Endicott Group.

datum for Plate 3 is simply the base of acoustically stratified (in seismic-reflection data) rocks and is highly diachronous. Above the map datum are found strata that range from Late Devonian to Tertiary age and overlie a number of unconformities that incise the acoustic basement. The youngest unconformities erode the basement on persistent structural uplifts such as the North Chukchi high and the Chukchi platform. The oldest unconformities floor the stratified sequence in subbasins within the Hanna Trough, but have been eroded and removed by younger unconformities over uplifts.

Generally, the faults in the Hanna Trough and on the Chukchi platform originated at the onset of subsidence of Hanna Trough (Late Devonian time or earlier). Following Permian to Early Cretaceous quiescence, they were reactivated with minor displacements during Cretaceous and Tertiary events. Because the later displacements were small, these faults still mainly record Paleozoic displacements. In some areas, the pattern of Paleozoic deformation is less clear. On North Chukchi high, faults that probably originated at the inception of the Hanna Trough clearly sustained significant displacements during Cretaceous and Tertiary uplifts.

Overall, the Hanna Trough is asymmetric in cross section, with a gently inclined west flank and a steeply inclined east flank. The west margin of the Hanna Trough is illustrated by line 1 of Plate 4. The Hanna Trough is bounded on the west by the Chukchi platform, which was elevated during early phases of Ellesmerian sedimentation, as recorded by regional stratigraphic wedging and internal angular unconformities along the west flank of the Hanna Trough (Thurston and Theiss, 1987, p. 40). There is no clear boundary that separates the Hanna Trough and the Chukchi platform, although the Hanna Trough deepens abruptly east of the longitude of the Klondike well.

The east margin of the Hanna Trough is illustrated by line 2 of Plate 4. The Hanna Trough is bounded on the east by the Arctic platform, to which the Hanna Trough rises abruptly across a narrow (80 km) zone of large-throw listric faults (approximately the Northeast Chukchi fault zone of Thurston and Theiss, 1987, Fig. 23). The Arctic platform, like the Chukchi platform, was elevated during early phases of Ellesmerian sedimentation. The east boundary of the Hanna Trough could be drawn at a fault just east of the Diamond well where the Lower Ellesmerian rocks are truncated (line 2, Plate 4; Fig. 1).

A northern segment of the Hanna Trough is conjectured to have been centered on a separate rift axis located at least 100 km west of the axis of the southern segment of the Hanna Trough. The two rift segments were mechanically linked by a west- to northwest-trending transform fault. This segmented configuration of the rift system, shown in Figure 1, presumably was created during Hanna Trough rifting. The presence of a northern segment of the Hanna Trough is mostly suggested by a westward-thickening wedge of Lower Ellesmerian rocks on the west flank of the North Chukchi high (and north of the transform fault), as shown in the isopach map of Figure 6 (and seismic records published by Grantz et al. [1990, Fig. 9] and Lothamer [1994, Fig. 4]). West of this wedge, the hypothesized northern segment of the Hanna Trough passes beneath the thick North Chukchi Basin fill (Cretaceous-Tertiary) and is not recognized in conventional (6–8 s) seismic-reflection data.

The transform fault that segments the Hanna Trough is mechanically like the rift jump fault systems—some >1000 km in length—that segment the East African rift systems (Nelson et al., 1992, Fig. 6). Outwardly, in terms of scale and associated structures, the Hanna Trough transform fault also resembles both the accommodation zone of Reynolds and Rosendahl (1984) and Coffield and Schamel (1989) and the transfer fault of Lister et al. (1986, Fig. 3). Both of the latter terms describe cross-rift structures that separate domains of oppositely dipping detachments within the deep crust that are centered on a linear, through-going rift axis. In the shallow crust above the detachments, and separated by shallow faults corresponding to the accommodation zones, are found listric fault terranes with opposing tilt-block asymmetry. Similar dispositions of shallow structures are found north and south of the Hanna Trough transform fault. For example, south of the transform fault, the overall dip of the Hanna Trough across a complex horst-graben terrane is eastward. In contrast, north of the transform fault and on the west flank of the North Chukchi high, regional dips (strata and faults) are westward. However, transform fault is probably a superior term for the feature that segments the Hanna Trough. The Hanna Trough transform fault appears to be a first-order structure that segments and offsets the actual rift axes, rather than just shallow fault terranes above a through-going, linear rift.

The transform fault follows the south margin of the North Chukchi high along a zone of east-trending faults that have up to 12 000 m of down-to-the-south throw, most of which took place during Ellesmerian sedimentation. The transform fault east of long 166°W is located along faults that are near vertical to north dipping and that pass into younger north-dipping thrusts on the southeast margin of North Chukchi high (Plate 3). During rifting of the Hanna Trough, the transform fault probably terminated on the east at the northern terminus of the graben-forming faults that form the east margin of the Hanna Trough, a configuration suggested in Figure 1. The transform fault was reactivated as the south margin of the North Chukchi high during Mesozoic and Cenozoic uplifts (Johnson, 1992; Lothamer, 1994). Deformations associated with these uplifts probably created both the currently observed north dips and the reverse senses of offset along the older transform-related faults, as well as creating the thrust faults on the southeast margin of the high.

West of long 166°W, older faults associated with the transform fault are difficult to identify because Ellesmerian rocks are absent from large areas of the northern Chukchi platform (Figs. 6 and 7) and the younger Rift and Brookian sequences directly overlie basement. Because they seem to coincide, the transform fault between long 166° and 169°W is conjectured to have later localized the easternmost 110 km of the 600 km Mesozoic-Cenozoic hinge line that controls the south margin of the North Chukchi Basin. Along this part of the hinge line, the top of basement flexes and faults down to the north and the overlying Brookian-age strata abruptly descend beneath 6 s seismic records (illustrated by Lothamer, 1994, Fig. 4A).

The fundamental structures of the Chukchi platform and the Hanna Trough consist of half-grabens that commonly link along strike to half-grabens of opposing or flipped asymmetry (Plate 3). As such, the Hanna Trough structures resemble the sinusoidal alternation of opposing fields of half-grabens described for Lake Tanganyika by Rosendahl et al. (1986) or the larger

LOWER ELLESMERIAN SEQUENCE, CHUKCHI SHELF
(Upper Devonian to Upper Permian)

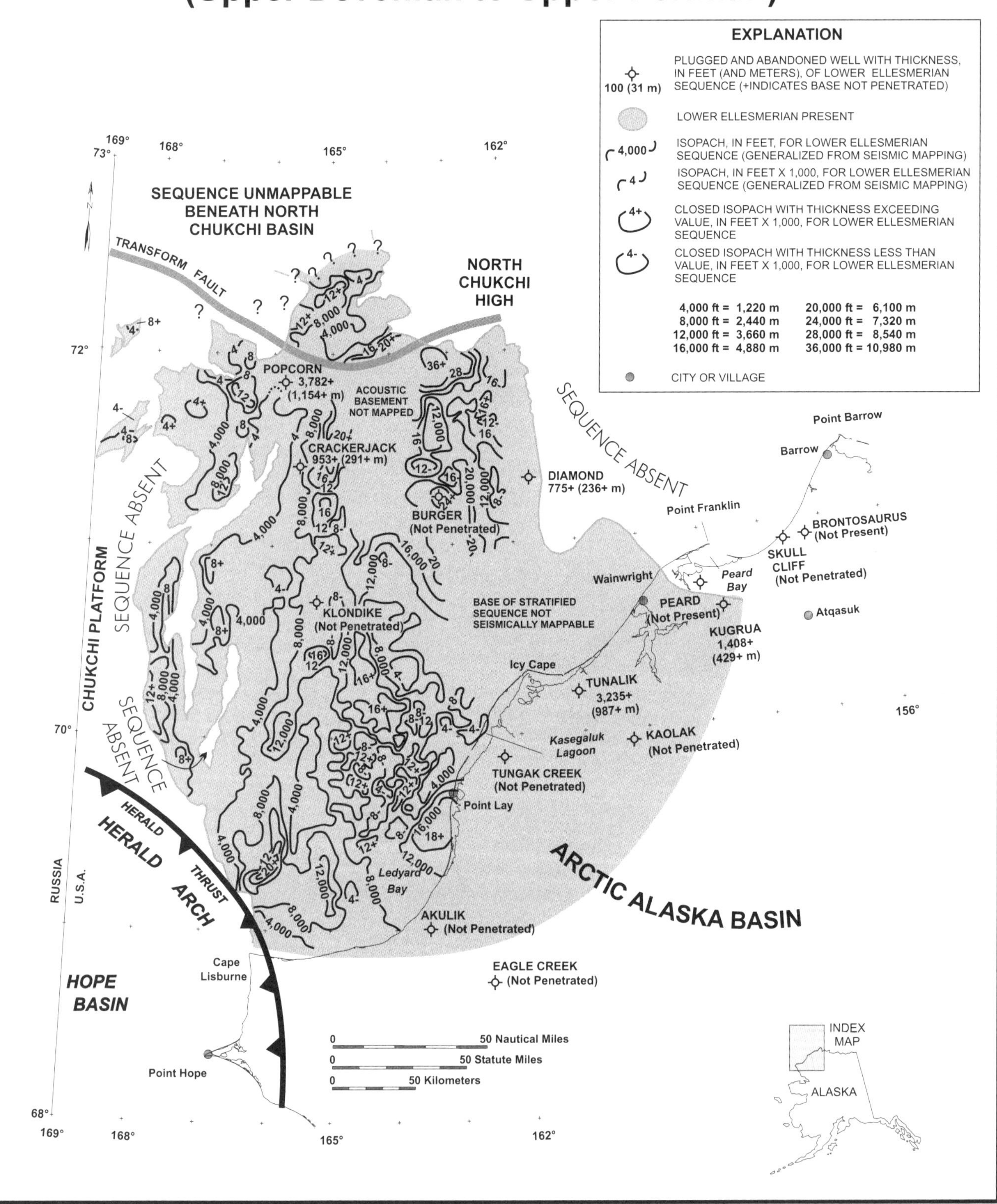

East African rift system described by Morley et al. (1990). Typical half-grabens are illustrated in seismic line 1 of Plate 4. At the latitude of this seismic line, the Chukchi platform is dominated by half-grabens bounded on the east by west-dipping normal or listric faults. Viewed on the structure map of Plate 3, we observe that the westernmost of these grabens passes southward across a broad ridge or transfer zone into a graben of opposite asymmetry bounded on the west by east-dipping normal faults. Northward, the array of half-grabens west of the Klondike well merges into a 145-km-long northeast-striking horst that terminates at a large northwest-striking fault northeast of the Crackerjack well. This horst, a major uplift in western Hanna Trough, appears to represent a larger-scale transfer zone that separates the sinuous half-graben terrane on the south and southeast from the orthogonally faulted terrane near the transform fault to the north.

Within 120 km south of the transform fault, the horsts and grabens on the Chukchi platform clearly bend northeasterly and a set of northwest-trending faults also becomes prominent (Plate 3). The northwest-trending faults may have formed as extensional faults in the mechanical environment of right-lateral shear across the transform fault linking the northern and southern segments of the Hanna Trough.

Pelka et al. (1992) suggested that the half-graben faults in the western Hanna Trough are old basement thrust faults reactivated as listric faults during Hanna Trough rifting. We have possibly observed this relationship at one locality in the seismic reflection data (west flank of postulated magmatic arc within basement; line 1, Plate 4), but not as a general case. The northerly trends of half-grabens may follow a template of old structural trends within basement. However, all elements of the regional structures of the Hanna Trough are observed in other rift systems (Rosendahl et al., 1986). The alternating asymmetry of arrays of half-grabens and bounding faults that dip both east and west would require opposing vergences of the old basement thrust faults. The last observation seems to argue against universal origination of (all) graben-bounding listric faults as reactivated basement thrusts.

The structure along the east flank of the Hanna Trough has been mapped only in the vicinity of the Diamond well (Plate 3). The area of the Diamond well is dominated by half-grabens bounded by west-dipping listric normal faults, some with throws of more than 6000 m. The half-graben structures near the Diamond well are illustrated in seismic line 2 of Plate 4. The easternmost half-graben, partially penetrated by the Diamond well, is bounded by a fault with perhaps 6100 m of down-to-the-west

throw. The half-graben faults in the area of the Diamond well pass northward and merge into the large-throw, east-trending faults along the south edge of the North Chukchi high. Southward, the half-graben faults near the Diamond well pass into an area where no coherent reflections can be identified or mapped at the top of the acoustic basement (Plate 3). Most seismic-reflection data sets in the latter area, both onshore and offshore, are dominated at the basement level by reverberations or multiple events (echoes between two horizons) that originate at shallow depths and effectively mask acoustic basement. The Hanna Trough faults emerge south of this area of poor seismic imaging as north-trending faults that were mapped onshore just west and south of the Tunalik well by Tetra Tech (1981, map PDN).

Very great depths were recorded for the base of stratified rocks in several areas of the Hanna Trough. For example, in two small areas in Plate 3—24 km northeast of the Crackerjack well and along the south margin of the North Chukchi high—basement is mapped at depths exceeding 13 700 m. Somewhat shallower depths (10 670–13 700 m) to the acoustic basement were recorded in two en echelon subbasins (near Point Lay and Icy Cape) that are separated by a north-northwest–trending ridge in the southern Hanna Trough. Similar depths are also achieved in the deep subbasin partly mapped just south of the North Chukchi high (Plate 3).

SEISMIC STRATIGRAPHY OF HALF-GRABENS IN THE HANNA TROUGH

The deepest strata observed in the seismic-reflection data in the Hanna Trough have not been penetrated by any wells. The oldest strata reached by the offshore wells are Upper Mississippian sandstones and carbonates that are probably age equivalent to the Alapah Formation of the Lisburne Group (Micropaleo Consultants, 1989b). Onshore in northwestern Alaska, the oldest rocks penetrated were Pennsylvanian carbonates of the Lisburne Group (Wahoo Formation at the Tunalik and Kugrua wells; Bird, 1982; Molenaar et al., 1986).

In the Hanna Trough, as much as 6700 m of strata beneath the deepest-drilled stratigraphic datum remain untested by any well. However, in most parts of the Hanna Trough outside the very deep subbasins, we estimate that ~3000 m of the lowermost strata of the Hanna Trough remain unprobed by any wells.

Because the oldest rocks in the Hanna Trough were not reached by any offshore wells, interpretations of the age of onset of subsidence must rely upon seismic data and analogy to drilled areas onshore >300 km to the east. We first describe the seismic stratigraphy of sequences flooring the Hanna Trough and then compare these sequences to basal sequences in the areas onshore that have been drilled.

The basal sequences filling the half-grabens that floor the Hanna Trough are typically represented in seismic data as wedge-shaped bodies of relatively high amplitude, moderately continuous (0.3–1.9 km) reflections. Typical examples are the sequences labeled Endicott Group in Figure 8 or sequence EG

Figure 6. Generalized isopach map (from seismic mapping) of the Lower Ellesmerian sequence (rocks equivalent to Endicott and Lisburne Groups, Upper Devonian (?) through Lower Permian ages) beneath the U.S. Chukchi Shelf. The isopach contouring ignores faults. The thickness of the Lower Ellesmerian rocks at wells is shown in feet (and meters). The addition symbol (+) at the thickness values indicates that the base of the sequence was not penetrated.

UPPER ELLESMERIAN SEQUENCE, CHUKCHI SHELF
(Upper Permian to Upper Jurassic)

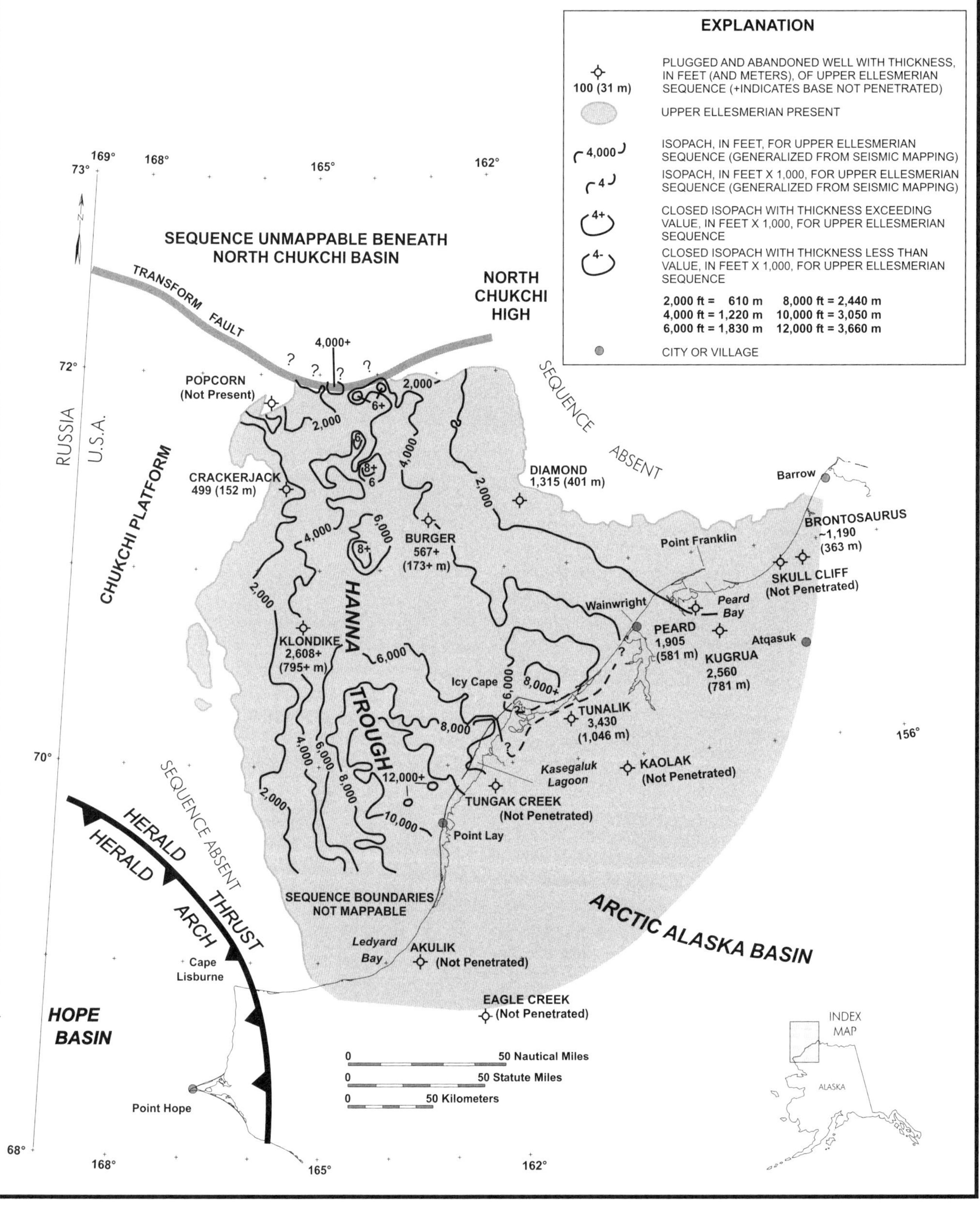

in Plate 4. The seismic reflections within the wedge-shaped bodies diverge toward the fault-bounded margins of the half-grabens and distinctly onlap basal unconformities at the top of acoustic basement (TAB). The wedge-shaped bodies are overlain with angular unconformity by seismic sequences that feature parallel, relatively moderate amplitude, highly continuous (0.6–5.5 km) reflections. The upper parts of the latter seismic sequence were penetrated at the Popcorn, Crackerjack, Tunalik, and Diamond wells and shown to be time equivalent to the Lisburne Group.

The unconformity MU in Figure 8 and Plate 4 incises the basement across the crests of some intergraben horsts, and within half-grabens it truncates tilted beds of the underlying early graben fill. The unconformity MU, which is not penetrated by any Chukchi Shelf wells, probably is within the upper part of the Endicott Group. The unconformities (MU and TAB, Fig. 8; Plate 4) bounding the wedge-shaped bodies flooring half-grabens correspond in character to those described as enclosing the "Eo-Ellesmerian" sequence identified by Grantz and May (1987, p. 41). The wedge-shaped bodies that form the earliest fill in half-grabens are widely observed and are not confined to the "Eo-Ellesmerian" depocenters mapped by Grantz and May (1987, Fig. 6).

ANALOGY TO DRILLED AREAS OF THE ARCTIC ALASKA BASIN

Drilling has shown that the Arctic Alaska Basin is floored by the Endicott Group, consisting of the Kekiktuk (conglomerate) Formation and the overlying Kayak shale, both Mississippian in age. The Endicott Group rocks overlie tectonized basement in much of the Arctic Alaska Basin, both in the western drilled areas of the basin and in exposures in eastern parts of the basin across the northeastern Brooks Range.

At distant localities south and west of the Brooks Range, the Kekiktuk Formation also unconformably overlies tectonized and metamorphosed basement rocks of Devonian and older ages. Noteworthy localities include Doonerak anticline on the south side of the Brooks Range (Kelley and Brosgé, 1995) and the Lisburne Peninsula (located in Fig. 4; Moore et al., 1984).

Seismic records from the National Petroleum Reserve-Alaska (NPR-A) in the Arctic Alaska Basin document a seismic sequence at the base of the stratified column that shares important similarities with the sequences flooring the half-grabens of the Hanna Trough. Three seismic lines published by Kirschner and Rycerski (1988, Plate 9.2, lines 6, 8, and 14) cross the Meade and Ikpikpuk-Umiat subbasins and show graben structures floored by onlapping wedge-shaped bodies containing relatively high amplitude, moderately continuous reflections. A more extensive set of regional seismic lines in NPR-A (Barnes et al., 1981) shows that similar features are widespread within the Arctic Alaska Basin, particularly in the highly faulted area of the basin floor south of lat 69°30′N (outlined in Fig. 5) and passing south beneath Brooks Range overthrust structures.

The inset well correlation in Figure 8 shows two well penetrations of the basal sequence in the Ikpikpuk-Umiat subbasin, scaled for direct comparison to the illustrated half-graben on the Chukchi platform. The upper 680 m of the basal clastic fill—the Endicott Group—was penetrated at 5447–6127 m at the Inigok No. 1 well. The Inigok well did not reach the acoustic basement, which was mapped by Tetra Tech (1981) to be ~305 m below the base of the well. At total depth, the Inigok well reached coal-bearing clastics of the Endicott Group. Mississippian fossils were recovered from the upper part of this sequence. The lower 610 m of the sequence yielded sparse, age-indeterminate foraminifera recoveries and sparse palynomorphs of probable Mississippian (Visean) age, but may include pre-Mississippian rocks (Anderson et al., 1979a; Husky Oil NPR-A Operations, Inc., 1983a). The dipmeter data for the Inigok well suggest the presence of an angular unconformity within the upper part of the Endicott Group at 5739 m, where westward dips of 5°–10° in the overlying sequence abruptly increase to 10°–25° below. Two cores below 5901 m in the Inigok well noted dips of 15°–30° (Husky Oil NPR-A Operations, Inc., 1983a) and further affirm the presence of angular unconformities within the Endicott Group.

The oldest fill in the Ikpikpuk-Umiat subbasin may have been penetrated at the Topogoruk No. 1 well. Sandstone, shale, and coal lithologically indistinguishable from the Endicott Group but bearing Early or Middle Devonian fossils (Collins, 1958) at the base of the Topogoruk well are so steeply inclined (35°–50° in cores; Collins, 1958, Plate 17) as to be seismically indistinguishable from acoustic basement (Kirschner and Rycerski, 1988, Plate 9.2, line 6; Barnes et al., 1981, line 6). From a structural standpoint, these dips are not much greater than those observed within the Endicott Group at the base of the Inigok well, although the latter return seismic reflections. The steeply dipping Lower to Middle Devonian rocks at the bottom of the Topogoruk well are unconformably overlain by near-horizontal, coal-bearing, Mississippian clastics of the Endicott Group. The angular unconformity separating these two sequences may correlate with the unconformity at 5739 m in the Inigok well, and possibly to the 355 Ma (approximate Devonian-Mississippian boundary) regional unconformity that Hubbard et al. (1987, Fig. 2) recognized between their Lower and Middle Ellesmerian sequences.

The Devonian rocks in the Topogoruk well and the lowermost 388 m of tilted strata in the Inigok well therefore have structural links to the basement assemblage (steep dips) but

Figure 7. Generalized isopach map (from seismic mapping) of the Upper Ellesmerian sequence (rocks equivalent to Sadlerochit Group, Shublik Formation, Sag River Formation, and lower Kingak Formation, Upper Permian through Upper Jurassic ages) beneath the U.S. Chukchi Shelf. The isopach contouring ignores faults. The thickness of the Upper Ellesmerian rocks at wells is shown in feet (and meters). The addition symbol (+) at the thickness values indicates that the base of the sequence was not penetrated.

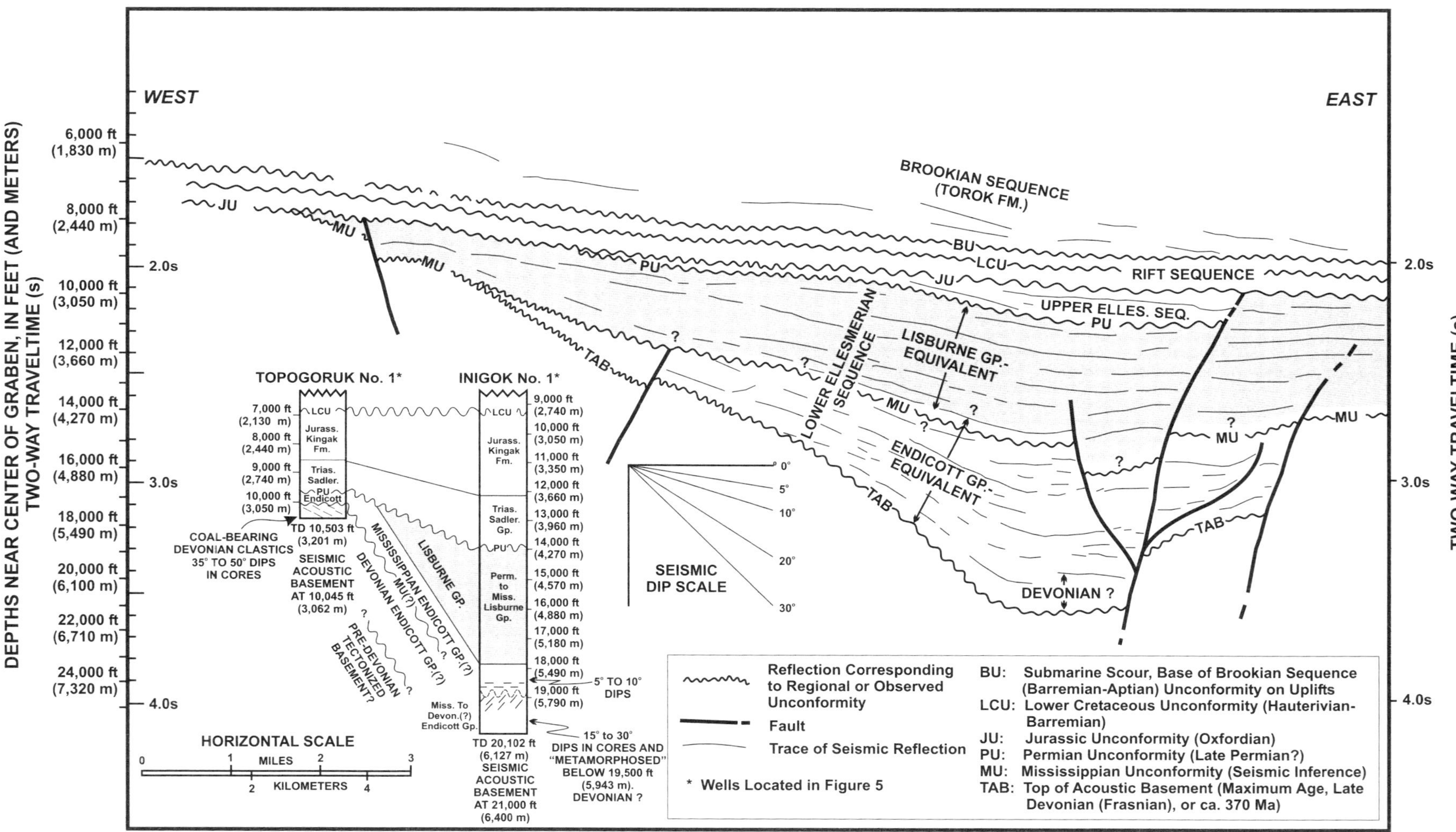

Figure 8. Stratigraphic interpretation of a half-graben on the Chukchi platform from seismic line 1, Plate 4. The inset well correlation shows the stratigraphy of probable time-equivalent units in the Ikpikpuk-Umiat Basin of the central North Slope, Alaska. See Figure 5 for the location of the inset well correlation section.

lithologic ties (coal-bearing clastics) to the post-basement Mississippian Endicott Group. The Devonian and Mississippian rocks are together transitional across the tilting event. The tilting event may represent the waning phase of the Ellesmerian orogeny, as suggested by Grantz et al. (1994, p. 25). Alternatively, the tilting and folding of Devonian and Mississippian coal-bearing strata may have occurred primarily at restraining bends along some transcurrent faults in the rift system that formed the Arctic Alaska Basin. For example, mapping by Tetra Tech (1981, map PDN) shows that the anticline tested by the Inigok well is along a northwest-striking fault that reverses throw along strike, suggesting a transcurrent displacement history. The anticline at the Inigok well probably formed in a restraining bend along this fault.

ANALOGOUS ROCKS ON WRANGEL ISLAND, RUSSIAN CHUKCHI SHELF

Rocks that are potentially analogous to the strata flooring the Hanna Trough grabens are exposed on Wrangel Island, 450 km to the west of the Hanna Trough (Kos'ko et al., 1993). Lower to Upper Devonian conglomerate, sandstone, and shale (the D_{1-3} unit) up to 1200 m thick unconformably overlie highly deformed and intruded rocks of the Wrangel complex (Kos'ko et al., 1993, p. 23). These Lower Devonian rocks may thus form the base of the Ellesmerian-equivalent sequence in this area. The Devonian rocks are overlain by as much as 350 m of Carboniferous conglomerate, sandstone, and shale that grade upward into carbonate, perhaps 700 m thick and probably equivalent to the Endicott and Lisburne Groups of the Hanna Trough and the Arctic Alaska Basin.

ANALOGOUS ROCKS IN DISRUPTED PALEOZOIC BASINS OF THE BROOKS RANGE

Throughout the central and western Brooks Range, the Endicott Group is exposed as an Upper Devonian to Mississippian deltaic assemblage comprising, from base to top, the Hunt Fork Shale, Noatak Sandstone, Kanayut Conglomerate, and Kayak Shale (Brosgé et al., 1988). The Endicott Group ranges to Late Devonian (Frasnian) in age and stratigraphically overlies the Upper Devonian (Frasnian) Beaucoup Formation (Brosgé et al., 1988, p.301). In all areas, the assemblage is observed or inferred to be floored by a regional décollement. The Endicott Group in the central and western Brooks Range, unlike the Endicott Group of the Arctic Alaska Basin, is nowhere observed to overlie a tectonized or metamorphosed basement.

The origin of the Endicott Group rocks of the central and western Brooks Range is controversial. Clearly, they are carried by thrust faults and have been displaced some distance northward. Many workers suggest that the Endicott Group rocks of the central and western Brooks Range represent an allochthon—a thrust-emplaced, rootless fragment of a basin that originated some distance south of the present Brooks Range (Mull, 1982;

Oldow et al., 1987; Mayfield et al., 1988; Grantz et al., 1991; Moore et al., 1994). Alternatively, Kelley and Brosgé (1995) suggested that thrust faulting in varying degrees merely disrupted and elevated a direct southern extension of the Arctic Alaska Basin of the North Slope subsurface. In either case, the Upper Devonian Endicott Group of the central and western Brooks Range may provide a regional analog for the very oldest rocks filling the half-grabens of the Hanna Trough and the deepest, untested parts of the Arctic Alaska Basin.

The Ellesmerian orogeny in northern Alaska may be dated by the Endicott Group rocks of the central and western Brooks Range. These rocks are inferred to have filled synorogenic basins that developed within or flanking uplifts created by the Ellesmerian orogeny (Hubbard et al., 1987, p. 7). The base of this sequence was reported by Brosgé et al. (1988, p. 301) as Frasnian or early Late Devonian in age (no older than 370 Ma in the Palmer [1998] time scale). The Ellesmerian orogeny sensu stricto in its reference area, the Canadian Arctic Islands, is somewhat younger, bracketed as ranging from Famennian to Tournaisian (late Late Devonian to Early Mississippian, ca. 364–342 Ma) (Trettin, 1991). The Ellesmerian orogeny sensu stricto corresponds in age to the younger parts of the synorogenic deposits (Noatak, Kanayut, and basal Kayak Formations, Famennian to late Kinderhookian [i.e., early Tournaisian]) in the Brooks Range. It also corresponds to the oldest postorogenic rift deposits—the Kekiktuk Formation, of early Osage age, approximately mid-Tournaisian, or post-348 Ma—in the Arctic Alaska Basin in the northeast Brooks Range (Dutro, 1987, p. 362; Brosgé et al., 1988, p. 301). The Devonian synorogenic clastics of the Brooks Range have no known corollary in the drilled parts of the Arctic Alaska Basin, save perhaps the steeply dipping Lower to Middle Devonian strata encountered at the Topogoruk well. Devonian rocks could conceivably floor the deep parts of the Arctic Alaska Basin, but if present are buried more than 7600 m deep and remain untested by any wells (Tetra Tech, 1981, map PDN).

PROBABLE AGE OF ONSET OF HANNA TROUGH SUBSIDENCE

We infer from the regional ages of the oldest synorogenic and postorogenic (Ellesmerian orogeny) sedimentary rocks in the Arctic Alaska Basin, the Brooks Range, and Wrangel Island that the very oldest fill in the Hanna Trough is at least probably Late Devonian in age, and possibly as old as Early Devonian. These rocks are probably succeeded upward by coal-bearing Mississippian clastics correlative to the typical Endicott Group (Kekiktuk Formation) in the Arctic Alaska Basin. The widely observed yet discontinuous MU unconformity in the Hanna Trough seems to mark a pause in rift extension that was associated with uplift and erosion, and was followed by renewed, but less vigorous, faulting and graben subsidence. The MU unconformity probably occurs within the upper part of the Endicott Group and may correspond to the 355 Ma (approximate Devonian-Mississippian boundary) regional unconformity recognized by Hubbard et al. (1987, Fig. 2) and the Lower Mississippian unconformity at the top of the "Eo-Ellesmerian" sequence of Grantz et al. (1990, p. 263).

TRANSITION FROM FAULT-DRIVEN TO THERMAL SUBSIDENCE IN THE HANNA TROUGH

Many faults in the Hanna Trough ceased movements prior to the MU unconformity, although faulting continued in some areas up to the time of development of the PU unconformity (Plate 4). The fault-driven or rift phase of subsidence, bracketed as occurring from Devonian to Late Permian time, accommodated up to 11 000 m of Lower Ellesmerian strata in some parts of the Hanna Trough (Fig. 6). By Late Permian time, the trough had entered a sag phase, where subsidence was largely unaccompanied by faulting, but possibly driven by thermal contraction of the extended, heated crust beneath the Hanna Trough. The sag phase of subsidence, which lasted from Late Permian to Late Jurassic (Oxfordian-Kimmeridgian) time, accommodated up to 3660 m of Upper Ellesmerian strata in southern parts of the Hanna Trough (Fig. 7).

WELL-BORE LITHOSTRATIGRAPHY OF THE HANNA TROUGH FILL

The five wells drilled on the Chukchi Shelf from 1988 to 1991 tested prospects that were leased to petroleum companies in offshore sales in 1988. Four of the wells targeted Permian-Triassic sandstones, equivalent to the Sadlerochit Group of the Upper Ellesmerian sequence—the prolific reservoir at the Prudhoe Bay oil field. A fifth well (Burger) targeted sandstones within the Rift sequence and approximately equivalent to the Kuparuk Formation—the reservoir at the Kuparuk oil field. The five Chukchi Shelf wells penetrated the east and west flanks of the Hanna Trough and collectively sampled rocks that are time equivalent to the Lisburne Group, Sadlerochit Group, Shublik Formation, and lower Kingak Formation. Post–Hanna Trough sequences were also penetrated and sampled, but are not described here. The parts of the Hanna Trough fill that were not sampled offshore include (1) the lowermost parts of the lower Kingak Formation and the Sag River Formation (estimated aggregate thickness, 250–300 m) and (2) the Lisburne (lower part) and Endicott Groups (estimated aggregate thickness, 3000–6700 m). The heavy vertical bars in the Chukchi Shelf lithology column in Figure 2 indicate the intervals of the Ellesmerian, Rift, and Brookian sequences that were sampled by the five Chukchi Shelf wells.

A regional stratigraphic correlation section for all Chukchi Shelf wells and two onshore wells is presented in Plate 5, the correlations in which are based on a combination of paleontological data[1] (Micropaleo Consultants, 1989a, 1989b, 1990a,

1990b, 1991), geophysical log correlations, and seismic correlations. Figure 9 shows a basin-to-margin stratigraphic panel that illustrates the basic facies relationships among the Hanna Trough stratigraphic units penetrated by the exploratory wells.

Porosity and permeability data are summarized where appropriate in the text and Plate 5. Averages of porosity measurements are calculated as arithmetic means. Averages of permeability measurements are calculated as geometric means, consistent with permeability characterizations of formations as practiced in the petroleum industry.

Where appropriate, summaries of geochemical data are presented. TOC is total organic carbon content, in weight%. Potential oil sources generally must contain at least 0.5% organic carbon, and rocks with >2.0% are considered excellent oil sources. HI is the hydrogen index, in 10^{-3} g of hydrocarbons generated (during pyrolysis) per gram of organic carbon. Potential oil sources generally have HI >300.

Lower Ellesmerian sequence

Alapah Formation of the Lisburne Group (Mississippian) and time-equivalent strata. The Popcorn well represents the only penetration of Mississippian rocks of any kind on the Chukchi Shelf or in northwestern Alaska. The nearest sampling of Mississippian-age carbonate sequences is 450 km to the east at the Ikpikpuk well (Fig. 5). The Popcorn well drilled and sampled 103 m of Mississippian sandstone and carbonate that are age equivalent to the uppermost part of the Alapah Formation (Lisburne Group) of the Arctic Alaska Basin and Brooks Range (Fig. 10A). It did not reach the base of the Mississippian carbonate-bearing sequence or the underlying Endicott Group.

At the Popcorn well, the Alapah-equivalent sequence consists of sandstone (aggregating 53 m or 52% of sequence) in beds to 8 m thick, carbonate (aggregating 33 m or 32% of sequence) in beds to 10 m thick, and siltstone and shale (16% of sequence). The sandstone is very fine to coarse grained and composed of angular framework grains of quartz, feldspar, plutonic igneous rock fragments, volcanic rock fragments, chert, and fossil fragments (echinoderms, bryozoa). Carbonate intraclasts, glauconite, and mica are also present. The carbonate consists of calcite-cemented sandy grainstones and packstones containing crinoids, bryozoans, and brachiopods. Noncarbonate framework grains include quartz, feldspar, granitoid rock fragments, felted porphyritic volcanic rock fragments, and chert (Reservoirs, Inc., 1991a). Minor glauconite is present. Dolomite, which is common in the Alapah Formation in the Arctic Alaska Basin (Bird and Jordan, 1977), is not present in the Mississippian carbonate penetrated at the Popcorn well. The shale and siltstone are bioturbated, micaceous, and contain broken fragments of brachiopods and bryozoa.

---

[1]GSA Data Repository item 2002073, Summaries of paleontological data, is available on request from Documents Secretary, GSA, P.O. Box 9140, Boulder, CO 80301-9140, USA, editing@geosociety.org, or at www.geosociety.org/pubs /ft2002.htm, or on the CD-ROM accompanying this volume.

Figure 9. Composite stratigraphic panel for basin-to-margin transition within the Hanna Trough, showing the lithology, facies relationships, and cumulative penetrated stratigraphic thickness.

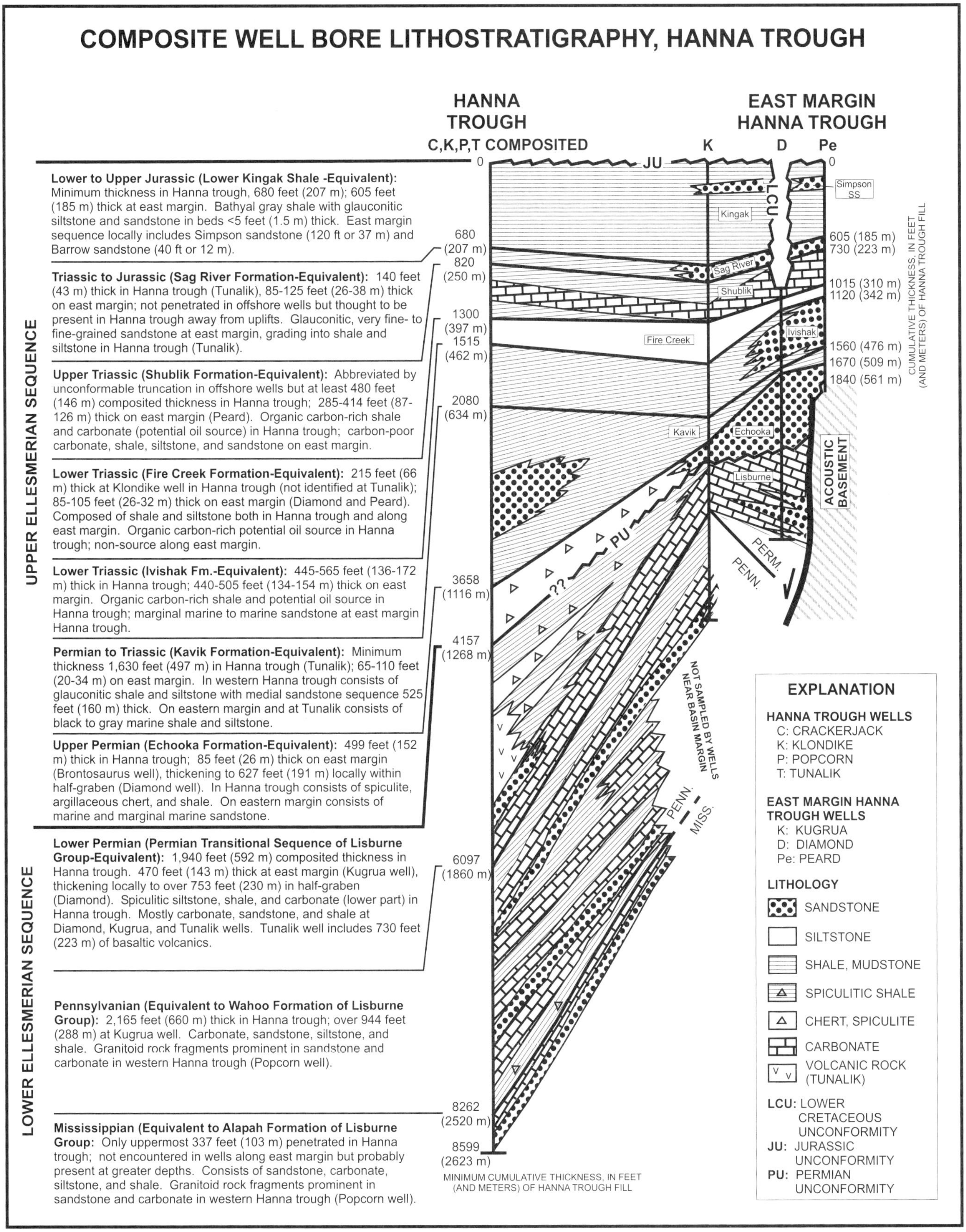

COMPOSITE WELL BORE LITHOSTRATIGRAPHY, HANNA TROUGH

UPPER ELLESMERIAN SEQUENCE

Lower to Upper Jurassic (Lower Kingak Shale -Equivalent): Minimum thickness in Hanna trough, 680 feet (207 m); 605 feet (185 m) thick at east margin. Bathyal gray shale with glauconitic siltstone and sandstone in beds <5 feet (1.5 m) thick. East margin sequence locally includes Simpson sandstone (120 ft or 37 m) and Barrow sandstone (40 ft or 12 m).

Triassic to Jurassic (Sag River Formation-Equivalent): 140 feet (43 m) thick in Hanna trough (Tunalik), 85-125 feet (26-38 m) thick on east margin; not penetrated in offshore wells but thought to be present in Hanna trough away from uplifts. Glauconitic, very fine- to fine-grained sandstone at east margin, grading into shale and siltstone in Hanna trough (Tunalik).

Upper Triassic (Shublik Formation-Equivalent): Abbreviated by unconformable truncation in offshore wells but at least 480 feet (146 m) composited thickness in Hanna trough; 285-414 feet (87-126 m) thick on east margin (Peard). Organic carbon-rich shale and carbonate (potential oil source) in Hanna trough; carbon-poor carbonate, shale, siltstone, and sandstone on east margin.

Lower Triassic (Fire Creek Formation-Equivalent): 215 feet (66 m) thick at Klondike well in Hanna trough (not identified at Tunalik); 85-105 feet (26-32 m) thick on east margin (Diamond and Peard). Composed of shale and siltstone both in Hanna trough and along east margin. Organic carbon-rich potential oil source in Hanna trough; non-source along east margin.

Lower Triassic (Ivishak Fm.-Equivalent): 445-565 feet (136-172 m) thick in Hanna trough; 440-505 feet (134-154 m) thick on east margin. Organic carbon-rich shale and potential oil source in Hanna trough; marginal marine to marine sandstone at east margin Hanna trough.

Permian to Triassic (Kavik Formation-Equivalent): Minimum thickness 1,630 feet (497 m) in Hanna trough (Tunalik); 65-110 feet (20-34 m) on east margin. In western Hanna trough consists of glauconitic shale and siltstone with medial sandstone sequence 525 feet (160 m) thick. On eastern margin and at Tunalik consists of black to gray marine shale and siltstone.

Upper Permian (Echooka Formation-Equivalent): 499 feet (152 m) thick in Hanna trough; 85 feet (26 m) thick on east margin (Brontosaurus well), thickening to 627 feet (191 m) locally within half-graben (Diamond well). In Hanna trough consists of spiculite, argillaceous chert, and shale. On eastern margin consists of marine and marginal marine sandstone.

LOWER ELLESMERIAN SEQUENCE

Lower Permian (Permian Transitional Sequence of Lisburne Group-Equivalent): 1,940 feet (592 m) composited thickness in Hanna trough. 470 feet (143 m) thick at east margin (Kugrua well), thickening locally to over 753 feet (230 m) in half-graben (Diamond). Spiculitic siltstone, shale, and carbonate (lower part) in Hanna trough. Mostly carbonate, sandstone, and shale at Diamond, Kugrua, and Tunalik wells. Tunalik well includes 730 feet (223 m) of basaltic volcanics.

Pennsylvanian (Equivalent to Wahoo Formation of Lisburne Group): 2,165 feet (660 m) thick in Hanna trough; over 944 feet (288 m) at Kugrua well. Carbonate, sandstone, siltstone, and shale. Granitoid rock fragments prominent in sandstone and carbonate in western Hanna trough (Popcorn well).

Mississippian (Equivalent to Alapah Formation of Lisburne Group: Only uppermost 337 feet (103 m) penetrated in Hanna trough; not encountered in wells along east margin but probably present at greater depths. Consists of sandstone, carbonate, siltstone, and shale. Granitoid rock fragments prominent in sandstone and carbonate in western Hanna trough (Popcorn well).

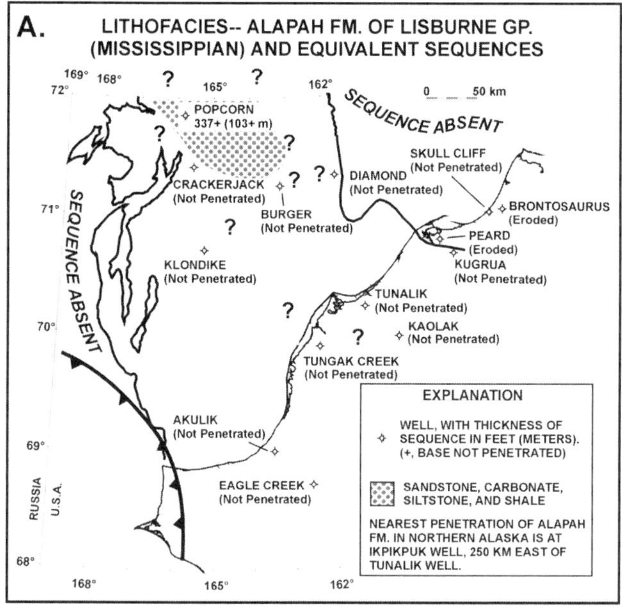

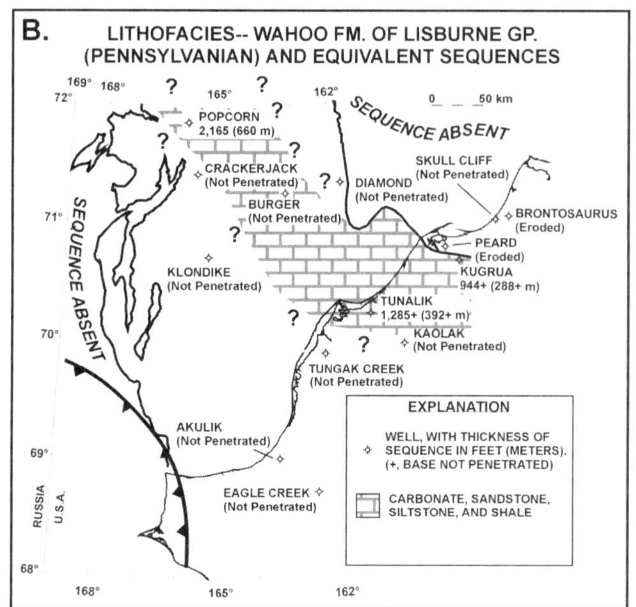

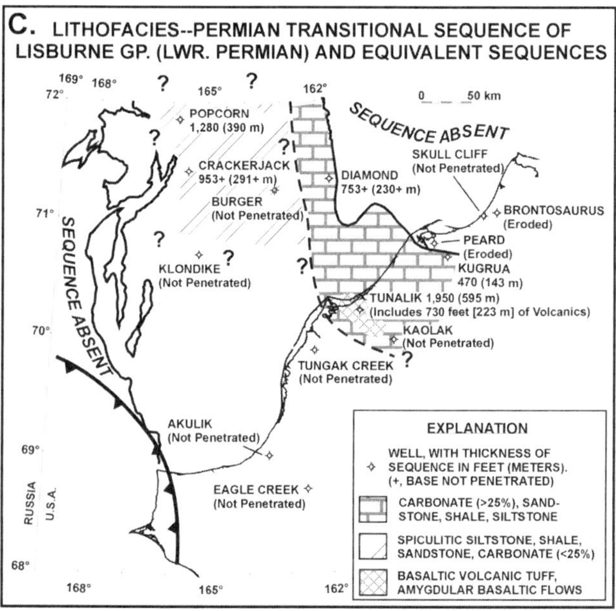

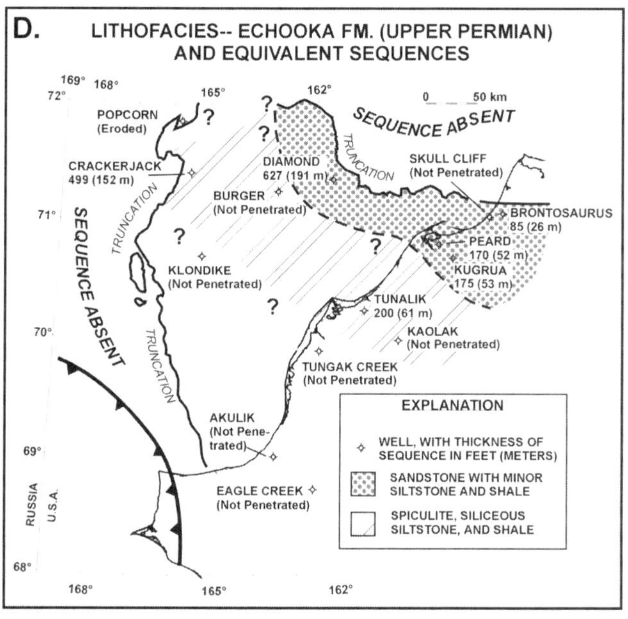

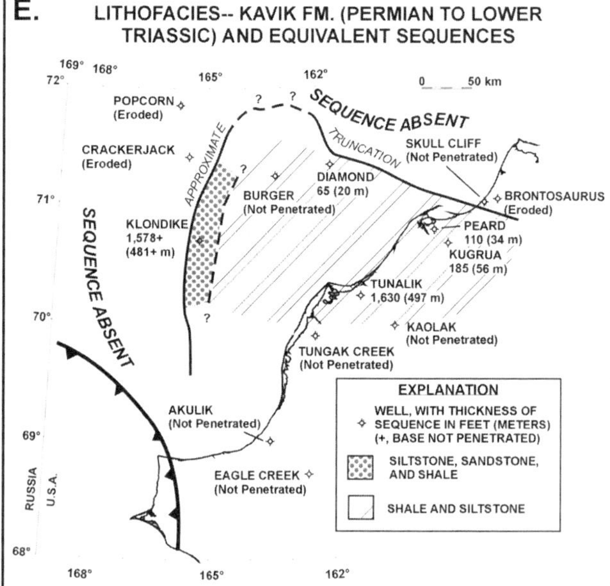

Figure 10. Lithofacies maps for the Hanna Trough and western Arctic Alaska Basin with thickness at wells for (A) Alapah Formation (Lisburne Group), (B) Wahoo Formation (Lisburne Group), (C) Permian transitional sequence (Lisburne Group), (D) Echooka Formation (Sadlerochit Group), and (E) Kavik Formation (Sadlerochit Group).

The sandstone and carbonate appear to have very little porosity in thin sections; some porosity is associated with secondary leaching of carbonate framework grains. Analyses of eight rotary sidewall cores of carbonate yielded an average porosity of 2.9% (maximum 4.8%) and permeability values of 0.01 md or less.

Foraminiferal data indicate a Late Mississippian age for the interval 3007–3109 m at the Popcorn well, which is time equivalent to the Alapah Formation of the Lisburne Group. Palynological studies of the same interval failed to recognize Mississippian rocks and indicate a Pennsylvanian age (Micropaleo Consultants, 1989b).

Wahoo Formation of Lisburne Group (Pennsylvanian) and time-equivalent strata. The Popcorn well penetrated an apparently complete sequence of Pennsylvanian strata 660 m thick that are time equivalent to the Wahoo Formation (Lisburne Group) of the Arctic Alaska Basin and Brooks Range. Onshore, 392 m of Wahoo Formation strata (base not penetrated) were drilled and sampled in the Tunalik well (Plate 5). The Kugrua well penetrated 288 m of Wahoo Formation strata but, like the Tunalik well, failed to reach the base of the Pennsylvanian sequence. Lithofacies and thickness variations for Pennsylvanian sequences are shown in Figure 10B.

The Pennsylvanian sequence at the Popcorn well consists of intercalated carbonate (aggregating 335 m or 50% of the sequence) in beds to 18 m thick, sandstone (aggregating 152 m or 23% of sequence) in beds to 9 m thick, and siltstone and shale (together 27% of sequence). High-energy grainstone and sandstone are more prevalent in the upper part of the Pennsylvanian sequence at the Popcorn well; spicular siltstone, spicular wackestones, peloidal packstones, and fossiliferous (bryozoa, brachiopods) shale are typical of the lower parts. The Pennsylvanian carbonate at the Popcorn well consists of calcite-cemented skeletal packstone and grainstone containing crinoids, bryozoans, brachiopods, miliolid foraminifera, algae, and peloids (Reservoirs Inc., 1991a). Oolites, described as common to the Wahoo Formation of the Arctic Alaska Basin (Bird and Jordan, 1977), were not observed in the Pennsylvanian carbonate at the Popcorn well. Some carbonate at the Popcorn well contains trace amounts of clasts derived from granitoid plutons (e.g., microperthite, myrmekite, quartz, feldspar) and volcanic rocks (Reservoirs Inc., 1991a). Glauconite is also present. Evaporite minerals or casts of such crystals are present locally. The Pennsylvanian sandstone in the Popcorn well is very fine to coarse grained and well compacted. The most abundant framework grains are quartz, feldspar, plutonic lithic grains (generally quite angular), and mica, with metamorphic grains, glauconite, fossil debris (mostly crinoid fragments), and micritic carbonate clasts present as accessories. Angular fragments of metamorphic and plutonic rocks dominate coarse-grained sandstone.

Carbonate and sandstone are thoroughly cemented with sparry calcite and offer little porosity. Analyses of 61 rotary sidewall cores yielded an average porosity of 3.3% (maximum 8.0%) and no permeability values greater than 0.05 md (most <0.01 md).

Foraminifera indicate a Pennsylvanian age for the Popcorn

well interval at 2383–3007 m. Palynology indicates that the strata below 2,053 m range in possible age from Pennsylvanian to Early Permian. A compromise paleontological-lithological top for the Pennsylvanian sequence at the Popcorn well was chosen at 2347 m (Micropaleo Consultants, 1989b).

Permian transitional sequence of Lisburne Group (Lower Permian). The Permian transitional sequence is an informal unit defined here to include all of the Permian rocks between the unconformity at the base of the Echooka Formation and the Pennsylvanian carbonate of the Lisburne Group (Wahoo Formation). Incomplete sections of the Permian transitional sequence were sampled at the Diamond, Popcorn, and Crackerjack wells. Correlation and compositing of the Permian transitional sequence in these wells yield an aggregate stratigraphic thickness of 517 m. Stratigraphically complete sequences 143 m and 594 m thick were penetrated onshore at the Kugrua and Tunalik wells, respectively. The Permian transitional sequence at Tunalik includes 223 m of volcanic rocks that are unknown at any other locale. Lithofacies and thickness variations for the Permian transitional sequence are shown in Figure 10C.

Our organization of stratigraphic units differs somewhat from the stratigraphic scheme established by Detterman et al. (1975) for the Echooka Formation and Lisburne Group in northeastern Alaska. There, Detterman et al. (1975, p. 10) defined a Lower and Upper Permian (Sakmarian to early Kazanian) sequence of shale, siltstone, and carbonate as the Joe Creek Member of the Echooka Formation. The Joe Creek Member is overlain by an Upper Permian (Kazanian) sandstone unit that forms the formal upper (Ikiakpaurak) member of the Echooka Formation (Detterman et al., 1975, p. 11; Bird and Molenaar, 1992, Fig. 4; Crowder, 1990). The Joe Creek Member unconformably overlies an erosionally dissected surface developed on Lisburne carbonate of Pennsylvanian age (Crowder, 1990; Robinson et al., 1992).

Mapping by Tetra Tech (1982, Fig. 45) extended the Joe Creek Member of the Echooka Formation westward from central Arctic Alaska to the Chukchi Sea coast, but the relationship of this map unit to the type or reference Joe Creek Member in northeastern Alaska is unclear. In northwestern Alaska and the Chukchi Shelf the Permian transitional sequence, which is time equivalent and lithologically similar to the Joe Creek Member, appears to simply grade downward through an increasing content of carbonate beds with no observed hiatus into the Pennsylvanian carbonate sequence. Because the important regional unconformity at the base of the type Joe Creek Member does not seem to be present in the Popcorn, Tunalik, and Kugrua wells, we are reluctant to extend the Joe Creek nomenclature to the time-equivalent rocks in the Hanna Trough or northwestern Alaska.

The Permian transitional sequence in the western Hanna Trough consists mostly of shale and other fine-grained clastic rocks interbedded with carbonate rocks that are abundant in the lower part of the sequence but sparse in the upper part. These rocks appear to mark progressive drowning of a shelf, as recorded by the transition from a shallow-water carbonate at the base to a turbid, deeper-water environment dominated by spiculitic mudstone at the top. Along the eastern margin of the Hanna Trough, carbonate and sandstone form a larger fraction of the Permian transitional sequence, presumably reflecting shallower water and proximity to a sediment source on the Arctic platform.

At the Popcorn and Crackerjack wells in the western Hanna Trough, the Permian transitional sequence consists mostly (60%–80% of upper part of sequence) of bioturbated, varicolored, spiculitic siltstone and fossiliferous, glauconitic, gray shale. Very fine grained glauconitic sandstone in beds 1.5–9 m thick are also present in the upper part of the sequence. At the Popcorn well, low gamma-ray intervals on logs mimic sandstone, but sidewall cores indicate that these intervals are actually highly spiculitic siltstone. The siltstone offers the highest porosity values (exceeding 25%) found in the sequence at the Popcorn well. The siltstone porosity is mainly associated with secondary dissolution of sponge spicules (Reservoirs Inc., 1991a). Analyses of 20 rotary sidewall cores of siltstone yielded an average porosity of 16.6% (range 3.2%–25.8%), but permeability values are generally <1 md (range 0–22 md).

Carbonate comprises 10%–20% of the Permian transitional sequence at the Popcorn and Crackerjack wells, occurring mostly in the lower parts of the sequence in beds from 1.2 to 9 m thick. At the Popcorn well, sparse carbonate high in the Permian transitional unit is generally skeletal wackestone composed mostly of calcareous mud or micrite. In contrast, carbonate in the lower part of the sequence is generally packstone or grainstone with peloids, sponge spicules, and broken fragments of gastropods, trilobites, crinoids, brachiopods, bryozoans, and various foraminifera, including fusilinids (Reservoirs Inc., 1991a). Quartz, feldspars, and granitoid rock fragments are present as minor siliciclastic constituents. The Permian carbonate in the Popcorn and Crackerjack wells offers little porosity. The average porosity of 15 rotary sidewall cores of carbonate at the Popcorn well is 3.0%, with a maximum value of 8.9%; nearly all of these samples have permeability values <0.06 md.

The Permian transitional sequence along the east margin of the Hanna Trough was penetrated at the Diamond, Kugrua, and Tunalik wells. At the Diamond well, argillaceous rocks compose ~50% of the Permian transitional sequence; subequal proportions of carbonate grainstone and sandstone form the remainder. The carbonate is generally grainstone with abundant accessory siliciclastics consisting mostly of quartz and chert, and trace amounts of glauconite and feldspars. Sandstone is composed mostly of quartz and chert clasts mixed with carbonate fragments and fossil debris (Reservoirs Inc., 1991b). Four rotary sidewall cores of Permian transitional sequence sandstone in the Diamond well yielded porosity values from 3.9% to 4.9% and permeability values all <0.02 md. Two rotary sidewall cores of carbonate yielded porosity values of 6.7% and 9.2%, and a maximum permeability of 1.90 md.

The Permian transitional sequence penetrated at the Tunalik well is unique among tests of Lower Permian rocks because it contains a sequence of basaltic volcanic tuffs and amygdular flows that is 223 m thick (Plate 5; Husky Oil NPR-A Operations, Inc., 1983b). The Tunalik well is located along the northern edge of the highly faulted southern part of the Arctic Alaska Basin, in the area where west-trending faults join the north-trending faults of the Hanna Trough (Figs. 1 and 5).

The incomplete Permian transitional sequence at the Crackerjack well (base not penetrated) is dated as probable Early Permian and Early to middle Permian on the basis of foraminiferal data (Micropaleo Consultants, 1990a). The incomplete Permian transitional sequence at the Popcorn well (top eroded at Jurassic unconformity) is dated as Early Permian on the basis of foraminifera, or Pennsylvanian to Early Permian on the basis of the palynology of the lower 305 m (Micropaleo Consultants, 1989b). The incomplete Permian transitional sequence at Diamond well (base not penetrated) is dated as probable Early Permian on foraminifera or undifferentiated Pennsylvanian to Permian on the basis of palynology (Micropaleo Consultants, 1991).

Upper Ellesmerian sequence

The Upper Ellesmerian sequence extends from the Permian unconformity at the base of the Echooka Formation to the Jurassic unconformity (Fig. 2) and includes, from base to top, the Sadlerochit Group and the Shublik, Sag River, and lower Kingak Formations. The uppermost and lowermost parts of the Upper Ellesmerian sequence were collectively sampled by the five wells on Chukchi Shelf. However, the middle parts of the sequence, including the lower Kingak and Sag River Formations, are absent at four of the five offshore well sites because of erosional truncation on the uplifts that were drilled. Part of the lower Kingak Formation was penetrated at Burger well. Strata equivalent to these absent units are surely present in the Hanna Trough away from uplifts. Onshore, the Tunalik, Kugrua, and Peard wells sampled a complete, stratigraphically continuous Upper Ellesmerian sequence.

Sadlerochit Group (Permian to Triassic). The Sadlerochit Group of northern Alaska comprises the Echooka, Kavik, Ivishak, and Fire Creek Formations (nomenclature of Jones and Speers, 1976, Fig. 9). Sadlerochit-equivalent rocks were the primary exploration target for four of the five wells on Chukchi Shelf and were actually encountered in the Klondike, Crackerjack, and Diamond wells (Plate 5). No Sadlerochit-equivalent rocks were present at the Popcorn well because the sequence there was completely truncated at the Jurassic unconformity.

The aggregate thickness of the Sadlerochit Group is 364 m at the Diamond well, which is consistent with regional thickness trends (Bird, 1988b, Fig. 16.12). Klondike well drilled a minimum (base not penetrated) of 719 m of rocks equivalent to the Sadlerochit Group. Onshore, the Tunalik well encountered a stratigraphically complete Sadlerochit Group 693 m in thickness. The unusually thick Sadlerochit sequences encountered at Tunalik and Klondike wells fit a regional pattern of stratigraphic expansion of Permian-Triassic (sag phase) sequences in the southern Hanna Trough (Fig. 7).

Echooka Formation of Sadlerochit Group (Upper Permian) and time-equivalent strata. The Echooka Formation, the basal unit of the Sadlerochit Group, was penetrated at the Diamond well on Chukchi Shelf and at the Peard, Kugrua, and Tunalik wells onshore in northwestern Alaska. Time-equivalent rocks were also penetrated at the Crackerjack well in the western Hanna Trough. The Echooka Formation (or equivalent sequence) in these wells is primarily a clastic unit and is most appropriately correlated with the Ikiakpaurak Member of the Echooka Formation in northeastern Alaska (nomenclature of Detterman et al., 1975). Along the eastern margin of the Hanna Trough, the Echooka Formation is sandstone dominated, but to the southwest grades into basinal lutites. At the Tunalik and Crackerjack wells in more central parts of the Hanna Trough, the Echooka Formation (or equivalent sequence) consists of siltstone, siliceous shale, and chert. The lithologic and thickness variations for the Echooka Formation and equivalent strata are shown in Figure 10D.

The base of the Echooka Formation corresponds to our Permian unconformity, which separates Echooka Formation sandstone from basement at the Peard well (Plate 5). At the Diamond, Crackerjack, Kugrua, and Tunalik wells, the Permian unconformity separates rocks of Late Permian age from the Permian transitional sequence (Lisburne Group) of Early Permian age. Even within the Hanna Trough, the Permian unconformity appears to be present as a subtle unconformity separating rock sequences of Late and Early Permian ages. Seismic-reflection data show that the Permian unconformity truncates underlying strata with low (<1°–3°) structural discordance on both margins of the Hanna Trough and well into the basin interior (Plate 4).

At the Diamond well, the Echooka Formation is 191 m thick, considerably thicker than the 0–76 m typically observed in northern Alaska (Jones and Speers, 1976, Fig. 11c; Tetra Tech, 1982, Fig. 46). The Echooka Formation at the Diamond well consists of glauconitic (2%–10%), very fine to coarse-grained, well-sorted sandstone with an average porosity of 19.8% and an average permeability of 6.7 md (maximum 174.0 md). The presence of glauconite, interbedded carbonate grainstones, and shallow-water fossil debris in the sandstone suggests a marginal marine environment of deposition for the Echooka Formation at the Diamond well (Micropaleo Consultants, 1991).

The Echooka-equivalent sequence at the Crackerjack well is partly truncated at the Jurassic unconformity, but a minimum thickness of 152 m is preserved. The Echooka-equivalent strata consist of silty glauconitic mudstone, bioturbated and glauconitic siltstone, argillaceous spiculitic chert, spiculite, and sparse very fine to fine-grained glauconitic sandstone (Reservoirs Inc., 1991c). Sponge spicules are ubiquitous and locally form the main framework component of shale and chert. Porosity values from 47 rotary sidewall cores average 25%; permeability values average only 0.40 md (maximum 30.0 md in a fractured sample). The highest porosity values are associated with moldic secondary porosity (dissolution of spicules) in spiculites and spiculitic cherts (Reservoirs Inc., 1991c). The spi-

culitic rocks at Crackerjack well, assigned to foraminiferal zone F-20, or Late Permian (Micropaleo Consultants, 1990a), are time equivalent to the Echooka Formation sandstone on the east margin of the Hanna Trough. Foraminiferal data and the presence of glauconite suggest a marginal marine to inner neritic depositional environment.

The Echooka Formation at Tunalik well is 61 m thick and consists of dark-gray to black, glauconitic and pyritic silicified siltstone and shale (Husky Oil NPR-A Operations, Inc., 1983b). The lithology of the Echooka Formation at Tunalik seems to indicate a basinal facies, but foraminiferal data reportedly indicate nonmarine (?) to inner neritic environments of deposition (Anderson et al., 1980). They also reported a Permian-Triassic age for the foraminifera.

Kavik Formation of Sadlerochit Group (Permian to Lower Triassic) and time-equivalent strata. Lithofacies and thickness variations within the Kavik Formation and equivalent rocks are shown in Figure 10E. At the Diamond well, the Kavik Formation is 20 m thick and consists of marine silty shale that is varicolored but mostly gray. The Kavik Formation is 34 m thick in the Peard well and 56 m thick in the Kugrua well. The Kavik-equivalent sequence thickens abruptly westward into the Hanna Trough. At Tunalik well, the Kavik Formation is 497 m thick and consists of black shale and gray siltstone with a few sandstone beds to 6 m in thickness. The Klondike well penetrated 481 m of Kavik-equivalent rocks, but did not reach the base of the sequence (Plate 5).

At the Klondike well, a 160 m sandstone-dominated sequence was encountered in the middle part of a Kavik-equivalent sequence consisting primarily of shale and siltstone (Plate 5). The shaly intervals above and below the sandstone sequence consist of gray shale and brown siltstone with sparse beds of white, very fine grained sandstone 0.3–1.2 m thick. The 160 m medial sandstone within this Kavik-equivalent sequence is the only observed manifestation of a western shoaling or littoral facies west of the Hanna Trough for the Upper Ellesmerian sequence (Fig. 10E). The sandstone sequence consists predominantly of interbedded light-gray, glauconitic, very fine to medium-grained sandstone and brown siltstone. Glauconite composes to 16% of some sandstone; lithic grains (often as pseudomatrix and difficult to separate from clay), quartz (17%–25%), feldspar and plutonic liths (0%–5%), and chert (1%–6%) are the principal framework grains (Reservoirs Inc., 1989). The sandstone is bioturbated and contains abundant pseudomatrix and detrital clay matrix. The 26 rotary sidewall cores yielded an average porosity of 5.9% (maximum 13.6%) and an average permeability of 0.01 md (maximum 0.025 md). Foraminifera and palynomorphs indicate that the Kavik-equivalent sandstone sequence at Klondike well is marine. The entire Kavik-equivalent sequence is also characterized paleontologically as marine and is assigned a Late Permian to Early Triassic age (Micropaleo Consultants, 1989a).

Ivishak Formation of Sadlerochit Group (Lower Triassic) and time-equivalent strata. The Ivishak Formation—the major sandstone reservoir at Prudhoe Bay oil field—is present as a sandstone facies only at the Diamond and Peard wells along the east margin of the Hanna Trough. Within the trough, Ivishak-equivalent sequences are predominantly shale and siltstone. Lithofacies and thickness variations within the Ivishak Formation and equivalent sequences are shown in Figure 11A.

At the Diamond well the Ivishak Formation is 134 m thick and consists of very fine to medium-grained, glauconitic, quartz-chert sandstone in beds 0.6–6 m thick that aggregate 94 m (net sandstone). Sandstone in seven rotary sidewall cores averaged 13.2% porosity and 1.8 md permeability. The maximum permeability observed among samples of Ivishak sandstone at the Diamond well was 21.0 md. The depositional environment of the Ivishak Formation at Diamond well is characterized by foraminiferal data as nonmarine to marine (Micropaleo Consultants, 1991).

At the Tunalik and Kugrua wells, the Ivishak-equivalent sequence is 136 m and 145 m thick, respectively, and consists mostly of black shale and gray siltstone with sparse beds of very fine to coarse-grained glauconitic sandstone to 3 m in thickness (Husky Oil NPR-A Operations, Inc., 1983b). Foraminiferal data characterize the Ivishak-equivalent sequence in these wells as Permian-Triassic in age and nonmarine to inner neritic in depositional setting; palynological data characterize the sequence as Triassic and marine to marginal marine (Anderson et al., 1978, 1980).

At the Klondike well, the Ivishak-equivalent sequence is 172 m thick and consists of organic carbon-rich (TOC, 0.7%–2.4%) siltstone and shale, the latter with hydrogen indices ranging from 150 to 437 and forming good to excellent potential oil source rock (Shell Development Co., 1989; Sherwood et al., 1998, Fig. 13.15c). The depositional environment at Klondike well is characterized by foraminiferal data as marine (Micropaleo Consultants, 1989a). The occurrence of organic-carbon-rich shale of Early Triassic age is atypical for the Arctic Alaska Basin and more like certain elements of the Lower Triassic to Jurassic Otuk Formation of the Brooks Range (Blome et al., 1988; Bodnar, 1989).

Fire Creek Formation of Sadlerochit Group (Lower Triassic) and time-equivalent strata. Lithofacies and thickness variations for the Fire Creek Formation and equivalent strata are shown in Figure 11B. Rocks correlative to the Fire Creek Formation were penetrated in a 26 m interval in the Diamond well consisting of light-gray, very fine grained, glauconitic sandstone and siltstone. Elsewhere, the Fire Creek-equivalent rocks are mostly siltstone. At the Peard well in northwestern Alaska, the Fire Creek Formation is 32 m thick (depth interval 2667–2699 m; correlation reinterpreted to differ slightly from the 2664–2699 m interval of Bird [1988a, Table 15.3]) and consists of gray to black siltstone and shale with minor glauconitic, very fine-grained sandstone. At Kugrua well, the Fire Creek Formation is 102 m thick (Bird, 1988a, p. 324) and consists of dark-gray to black siltstone and shale with gray sandstone in beds to 3 m thick (Husky Oil NPR-A Operations, Inc., 1983c). At

Klondike well, the Fire Creek-equivalent section is 66 m thick and consists of brown calcareous siltstone and varicolored to gray mudstone.

Paleontological studies do not recognize the Fire Creek Formation at the Tunalik well (Mickey and Haga, 1987, Plate 5A; Micropaleo Consultants, 1989a, p. 36). It may be absent, so thin that it escaped sampling, or unrecognized because it is paleontologically atypical. In an alternative, lithology-based interpretation of the Tunalik well, Bird (1988a, p. 324) assigned the entire 224 m sequence between the Kavik and Shublik Formations to the Fire Creek Formation. We have provisionally adopted the paleontological interpretation of Mickey and Haga (1987) for this report.

At the Diamond well, the TOC of the Fire Creek Formation ranges from 0.54% to 1.05% and hydrogen indices do not exceed 140, indicating that it lacks oil source potential (Dow Geochemical Service, Inc., 1991). At the Peard and Kugrua wells, the Fire Creek Formation generally has <1.0% TOC and likewise does not appear to be a potential oil source (Magoon et al., 1988, Plates 19.3, 19.4). At Klondike well the Fire Creek-equivalent rocks are rich in TOC (2.3%–4.7%) and have HI values ranging from 310 to 494, indicating excellent potential oil sources (Sherwood et al., 1998, Fig. 13.15).

The Fire Creek-equivalent rocks at Klondike are distinguished paleontologically by their Early Triassic fossils (Micropaleo Consultants, 1989a), but they are lithologically and geochemically indistinguishable from the immediately overlying shale of the Upper Triassic Shublik Formation. The Fire Creek Formation at the Diamond well was identified by log correlation with the Peard well.

Shublik Formation (Upper Triassic) and time-equivalent strata. The Shublik Formation (or equivalent rocks) was encountered at the Diamond and Klondike wells, in both cases as partial sequences abbreviated by erosion at overlying unconformities. At the Diamond well the preserved Shublik Formation is 37 m thick and at the Klondike well the preserved Shublik-equivalent sequence is 76 m thick. The Shublik Formation is preserved in its entirety in onshore wells where it ranges in thickness from 87 m at the Peard well to 126 m at the Brontosaurus well. Lithofacies and thickness variations within the Shublik Formation and equivalent strata are shown in Figure 11C.

The Shublik Formation at the Diamond well and the Shublik-equivalent sequence at the Klondike well include brown and gray carbonate-bearing filamentous bivalves (Reservoirs Inc., 1989)—presumably the *Monotis* sp. and *Halobia* sp. mud pectins commonly found elsewhere within the Shublik Formation (Jones and Speers, 1976, p. 39; Kupecz, 1995). The carbonate is interbedded with dark, carbonaceous, fossiliferous shale. Glauconitic, fine-grained sandstone is present in the Diamond well (projected to be within the glauconitic facies belt of Parrish, 1987, Fig. 1), but not at the Klondike well. The Shublik Formation at the Diamond well is generally low in TOC (0.5%–2.8%) and forms a poor to fair potential source for oil (Plate 5). In contrast, the Shublik Formation at the Klondike well is very

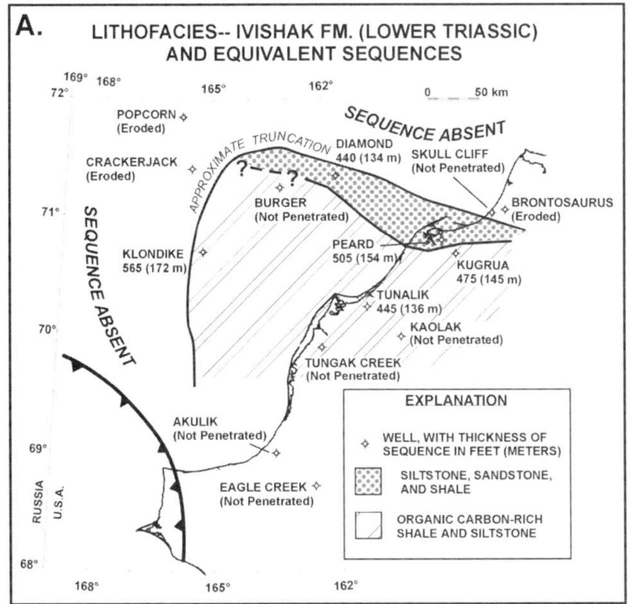

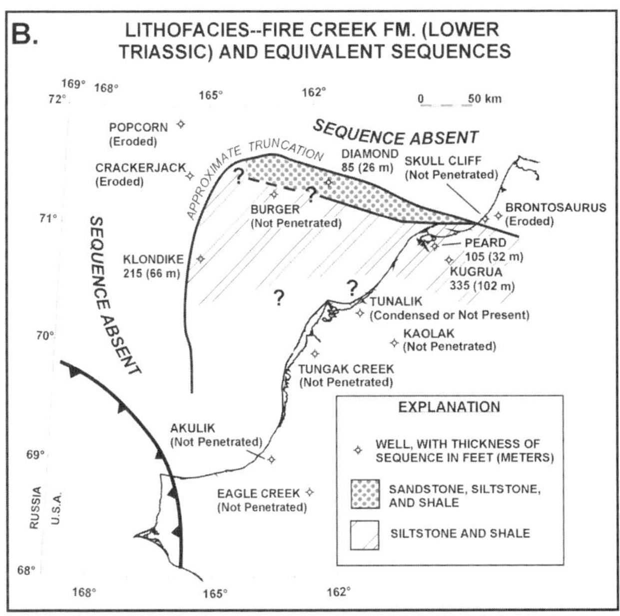

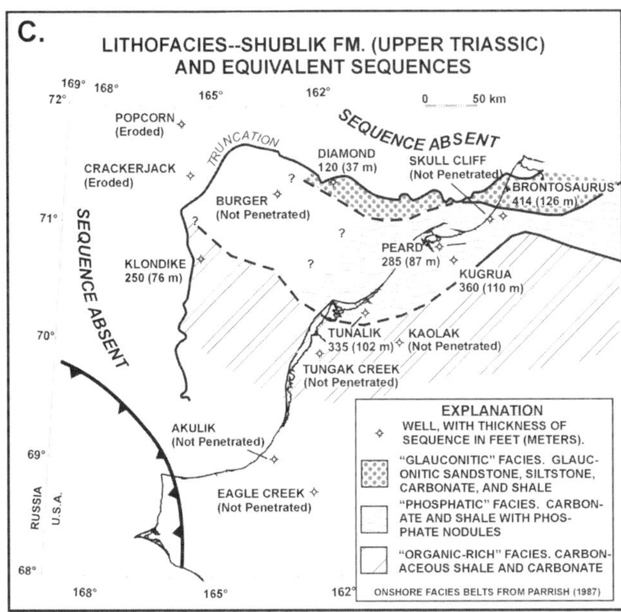

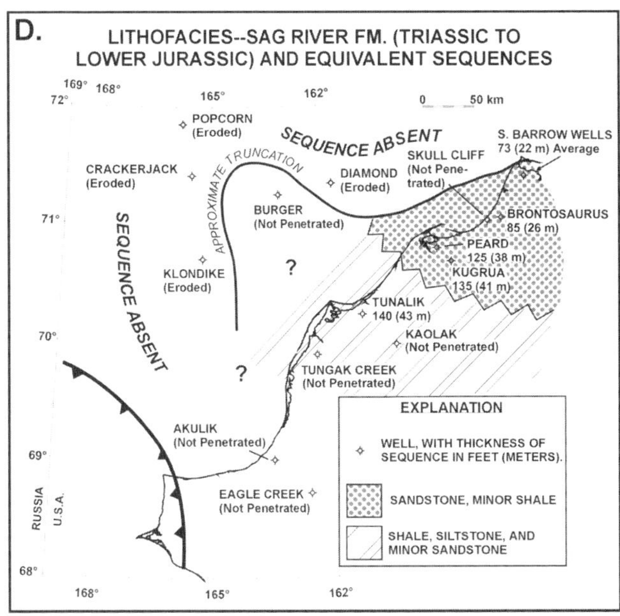

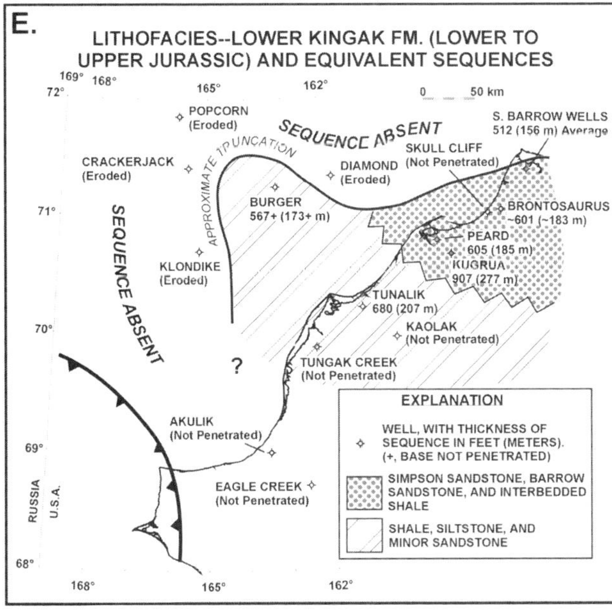

Figure 11. Lithofacies maps for the Hanna Trough and western Arctic Alaska Basin with thickness at wells for (A) Ivishak Formation (Sadlerochit Group), (B) Fire Creek Formation (Sadlerochit Group), (C) Shublik Formation, (D) Sag River Formation, and (E) Lower Kingak Formation.

rich in TOC (2.0%–8.0%) and forms an excellent potential oil source. As shown in Figure 11C, the Klondike site probably is within the offshore extension of the organic-rich facies belt of the Shublik Formation identified by Parrish (1987). Neither of the offshore penetrations of the Shublik Formation appear to contain the phosphate nodules commonly found elsewhere (Jones and Speers, 1976) in the formation and that are thought to mark an oceanic upwelling zone that created the phosphatic facies belt of the Shublik depositional system (Fig. 11C; Parrish, 1987).

Foraminiferal data indicate a probable Middle to Late Triassic age and a probable neritic environment for the Shublik Formation in the Klondike well (Micropaleo Consultants, 1989a). Paleontological studies at the Diamond well failed to identify any Late Triassic microfossils, but noted possible *Monotis* sp. (mud pectin) fragments in the Shublik Formation (Micropaleo Consultants, 1991).

Sag River Formation (Triassic to Lower Jurassic). Strata time equivalent to the Sag River Formation were not penetrated by any of the Chukchi Shelf wells (Fig. 2) but correlative rocks are presumed to exist in central areas of the Hanna Trough away from eroded uplifts. Onshore, the Sag River Formation was encountered at the Tunalik, Kugrua, Brontosaurus, and Peard wells. Lithofacies and thickness variations for the Sag River Formation are shown in Figure 11D.

The Sag River Formation is 43 m thick and consists of glauconitic shale and siltstone (Husky Oil NPR-A Operations, Inc., 1983b) in the Tunalik well. To the north, the Sag River Formation enters a sandstone facies. In the area between the South Barrow and Kugrua wells the Sag River Formation ranges from 22 to 41 m thick and consists of very fine grained, micaceous and glauconitic sandstone, siltstone, and shale (Husky Oil NPR-A Operations, Inc., 1982b).

At the Tunalik well the Sag River Formation is dated as Triassic in age and foraminiferal data indicate a middle neritic environment (Anderson et al., 1980; Husky Oil NPR-A Operations, Inc., 1983b). At the Peard and Kugrua wells the Sag River Formation is considered to be Triassic to Early Jurassic in age. Foraminifers indicate an inner neritic environment at the Kugrua well and a middle to outer neritic environment at the Peard well (Anderson et al., 1978, 1979b; Husky Oil NPR-A Operations, Inc., 1982b).

Lower Kingak Formation (Lower Jurassic to Upper Jurassic [Oxfordian]) and time-equivalent strata. The lower Kingak Formation, as informally defined here, extends from the top of the Sag River Formation to the Jurassic unconformity at the base of the upper Kingak Formation. The lower Kingak Formation consists mostly of shale in the Burger well in the Hanna Trough, but contains the Lower Jurassic Barrow and Simpson marginal marine sandstones along the east margin of the Hanna Trough. Lithofacies and thickness variations for the lower Kingak Formation and equivalent sequences are shown in Figure 11E.

The Jurassic unconformity truncates Ellesmerian strata and ultimately overlies basement across the western parts of Chukchi platform (line 1, Plate 4; Figs. 7 and 8). The Jurassic unconformity erodes pre-Jurassic rocks and is overlain by Upper Jurassic (Oxfordian to Kimmeridgian) rocks at the Klondike, Crackerjack, and Popcorn wells. Some of the Jurassic fossil material above the Jurassic unconformity in the Popcorn well was thought by Micropaleo Consultants (1989b) to be reworked and incorporated into Lower Cretaceous rocks; our correlations indicate that the strata overlying the Jurassic unconformity are probably Jurassic. At the Burger and Tunalik wells, the Jurassic unconformity correlates with depositional interfaces or conformities within Upper Jurassic (Oxfordian to Kimmeridgian) bathyal shale (Micropaleo Consultants, 1990b; Anderson et al., 1980).

The lower Kingak shale in the Burger well consists of gray to black pyritic shale interbedded with glauconitic siltstone and sparse very fine to fine-grained sandstone in beds <1.5 m thick. Dense, micritic carbonate is also present in thin beds. Paleontological data at Burger well indicate that the sequence includes both Lower to Middle Jurassic rocks and Upper Jurassic (Oxfordian-Kimmeridgian) rocks (Micropaleo Consultants, 1990b). Foraminifera data for the Burger well indicate middle to lower bathyal environments (Micropaleo Consultants, 1990b).

SUMMARY

The Hanna Trough formed as a complex north-trending rift basin, perhaps localized by an older structural grain or faults in the underlying basement, beneath the U.S. Chukchi Shelf off northwest Alaska. Rift- or fault-driven subsidence and sedimentation occurred from Devonian to Permian time and was succeeded by regional sagging, possibly driven by postrift crustal cooling, which lasted from Permian to Late Jurassic time. The Hanna Trough fill is correlative to the Ellesmerian sequence of the Arctic Alaska Basin and ranges to 11 600 m in thickness.

The Hanna Trough is asymmetric in regional cross section, with a broad, shallow-dipping west flank and a narrow, steeply dipping east flank. It is between the Chukchi platform on the west and the Arctic platform on the east, which both acted at times as sediment source highlands. On the north, a transform fault probably linked two offset rift segments within the Hanna Trough. Evidence for the segment north of the transform fault is fragmentary because it passes to great depths beneath the younger North Chukchi Basin and is inaccessible to conventional seismic-reflection data. South of the transform fault, the Hanna Trough curves east and joins the highly faulted southern part of the west-trending Arctic Alaska Basin, which shares the overall subsidence history of the Hanna Trough.

The basic structural features of the Hanna Trough are half-grabens arrayed in sinuous fields that often flip asymmetry along strike. These half-grabens were initially filled by clastic wedges derived from adjacent horsts. The basal clastic wedges within grabens were not penetrated by offshore wells, but are correlated by seismic analogy to the Devonian to Mississippian Endicott Group of onshore northern Alaska.

Some notable points of similarity and difference are found between time-equivalent sequences in the Hanna Trough and the Arctic Alaska Basin: (1) the Mississippian sequence in the Hanna Trough is time equivalent to the Alapah Formation of the Lisburne Group, but although carbonate bearing, it is mostly composed of sandstone; (2) Lisburne-equivalent carbonate and sandstone of Mississippian and Pennsylvanian ages in the western Hanna Trough contain sedimentary debris from plutonic sources—possibly eroded from a magmatic arc complex within basement (independently hypothesized from geophysical data) that was exposed at that time to the west on Chukchi platform; (3) the Permian to Jurassic sequences penetrated by wells in the Hanna Trough and northwest Alaska show basin-to-margin facies transitions that are analogous to those observed in the Arctic Alaska Basin; and (4) wells penetrating the Upper Ellesmerian sequence in the western Hanna Trough encountered mostly basinal facies; the only lithostratigraphic evidence for a Chukchi platform sediment source during Upper Ellesmerian deposition consists of sandstone that is time equivalent to the Permian to Lower Triassic Kavik Formation.

ACKNOWLEDGMENTS

The Minerals Management Service, U.S. Department of the Interior, provided support for this work. We thank Art Grantz, Elizabeth Miller, Gary J. Pelka, Chris A. Johnson, John Larson, Catherine Dunkel, and an anonymous reviewer for their thorough technical readings and constructive suggestions. The manuscript was also improved by the comments of Minerals Management Service technical editor Jody Lindemann. We thank the Minerals Management Service for financial support for the publication of this paper.

REFERENCES CITED

Alaska Division of Oil and Gas, 1998, Historical and projected oil and gas consumption: Anchorage, Alaska, Department of Natural Resources, 81 p.

Anderson, R.E., Warren, A.D., and Associates, 1978, Final Paleontological Reports (Palynology, 4 p.; Foraminifera, 10 p.) for Husky/USGS Kugrua Bay No. 1 well, northern Alaska: Report to U.S. Geological Survey, Contract Studies by Anderson, Warren and Associates (now defunct), formerly of 11526 Sorrento Valley Road, Suite G, San Diego, CA 92121.

Anderson, R.E., Warren, A.D., and Associates, 1979a, Final Paleontological Reports (Palynology, 7 p.; Foraminifera, 14 p.) for Husky/USGS Inigok No. 1 well, northern Alaska: Report to U.S. Geological Survey, Contract Studies by Anderson, Warren and Associates (now defunct), formerly of 11526 Sorrento Valley Road, Suite G, San Diego, CA 92121.

Anderson, R.E., Warren, A.D., and Associates, 1979b, Final Paleontological Reports (Palynology, 7p.; Foraminifera, 12 p.) for Husky/USGS Peard Bay No. 1 well, northern Alaska: Report to U.S. Geological Survey, Contract Studies by Anderson, Warren and Associates (now defunct), formerly of 11526 Sorrento Valley Road, Suite G, San Diego, CA 92121.

Anderson, R.E., Warren, A.D., and associates, 1980, Final Paleontological Reports (Palynology, 7 p.; Foraminifera, 10 p.) for Husky/USGS Tunalik No. 1 well, northern Alaska: Report to U.S. Geological Survey, Contract Studies by Anderson, Warren and Associates (now defunct), formerly of 11526 Sorrento Valley Road, Suite G, San Diego, CA 92121.

Barnes, D., Bowsher, A.C., Carter, R., Cobb, M., and Gutman, S., 1981, Regional Geophysics of NPRA: A series of 20 regional compressed sections prepared for Husky Oil NPR Operations, Inc., by Seiscom Delta, Inc. under contract to the U.S. Geological Survey. National Geophysical Data Center data sets TGX-0170 and TGX-0172.

Bassinger, B.G., 1968, Marine magnetic studies in the northeast Chukchi Sea: Journal of Geophysical Research, v. 73, no. 2, p. 686–687.

Bird, K.J., 1982, Rock-Unit Reports of 228 Wells Drilled on the North Slope, Alaska: U.S. Geological Survey Open-File Report 82-278, 105 p.

Bird, K.J., 1988a, Alaska North Slope stratigraphic nomenclature and data summary for government-drilled wells, *in* Gryc, G., ed., Geology and exploration of the National Petroleum Reserve in Alaska, 1974–1982: U.S. Geological Survey Professional Paper 1399, p. 317–353.

Bird, K.J., 1988b, Structure-contour and isopach maps of the National Petroleum Reserve in Alaska, *in* Gryc, G., ed., Geology and exploration of the National Petroleum Reserve in Alaska, 1974–1982: U.S. Geological Survey Professional Paper 1399, p. 355–377.

Bird, K.J., and Jordan, C.F., 1977, Lisburne Group (Mississippian and Pennsylvanian), potential major hydrocarbon objective of Arctic Slope, Alaska: American Association of Petroleum Geologists Bulletin, v. 61, p. 1493–1512.

Bird, K.J., and Molenaar, C.M., 1992, The North Slope Foreland Basin, Alaska, *in* Macqueen R., and Leckie, D., eds., Foreland basins and foldbelts: American Association of Petroleum Geologists Memoir 55, p. 363–393.

Bird, K.J., Connor, C.L., Tailleur, I.L., Silberman, M.L., and Christie, J.L., 1978, Granite on the Barrow arch, northeast National Petroleum Reserve in Alaska: U.S. Geological Survey Circular 772-B, p. B24–B25.

Blome, C.D., Reed, K.M., and Tailleur, I.L., 1988, Radiolarian biostratigraphy of the Otuk Formation in and near the National Petroleum Reserve in Alaska, *in* Gryc, G., ed., Geology and exploration of the National Petroleum Reserve in Alaska, 1974–1982: U.S. Geological Survey Professional Paper 1399, p. 725–751.

Bodnar, D.A., 1989, Stratigraphy of the Otuk Formation and a Cretaceous coquinoid limestone and shale unit, north-central Brooks Range, Alaska, *in* Mull, C.G., and Adams, K.E., eds., Dalton Highway, Yukon River to Prudhoe Bay, Alaska: Bedrock geology of the eastern Koyukuk basin, central Brooks Range, and east-central Arctic slope: Fairbanks, Alaska, State of Alaska Division of Geological and Geophysical Surveys Guidebook 7, v. 2, p. 277–284.

Brosgé, W.P., and Dutro, J.T., Jr., 1973, Paleozoic rocks of northern and central Alaska, *in* Pitcher, M.G., ed., Arctic geology: American Association of Petroleum Geologists Memoir 19, p. 361–375.

Brosgé, W.P., Nilsen, T.H., Moore, T.E., and Dutro, J.T., Jr., 1988, Geology of the Upper Devonian or Lower Mississippian(?) Kanayut Conglomerate in the central and eastern Brooks Range, *in* Gryc, G., ed., Geology and exploration of the National Petroleum Reserve in Alaska, 1974–1982: U.S. Geological Survey Professional Paper 1399, p. 299–316.

Carman, G.J., and Hardwick, P., 1983, Geology and regional setting of Kuparuk oil field, Alaska: American Association of Petroleum Geologists Bulletin, v. 67, p. 1014–1031.

Carter, C., and Laufield, S., 1975, Ordovician and Silurian fossils in well cores from the North Slope of Alaska: American Association of Petroleum Geologists Bulletin, v. 59, p. 457–464.

Coffield, D.Q., and Schamel, S., 1989, Surface expression of an accommodation zone within the Gulf of Suez rift, Egypt: Geology, v. 17, p. 76–79.

Collins, F.R., 1958, Test wells, Topogoruk area, Alaska: U.S. Geological Survey Professional Paper 305-D, p. 265–316.

Craig, J.D., Sherwood, K.W., and Johnson, P.P., 1985, Geologic report for the Beaufort Sea Planning Area, Alaska: Anchorage, Alaska, U.S. Department of the Interior, Minerals Management Service Outer Continental Shelf Report MMS-85-0111, 192 p.

Cramer, C.H., May, S.D., Hanna, W.F., Grantz, A., and Holmes, M.L., 1986, Reconnaissance magnetic anomaly map of the Chukchi Sea and adjacent northwest Alaska: U.S. Geological Survey Miscellaneous Investigations Series Map I-1182-F, scale 1:1 000 000, 1 sheet.

Crowder, R.K., 1990, Permian and Triassic sedimentation in the northeastern Brooks Range, Alaska: Deposition of the Sadlerochit Group: American Association of Petroleum Geologists Bulletin, v. 74, p. 1351–1370.

Detterman, R.L., Reiser, H.N., Brosge, W.P., and Dutro, J.T., Jr., 1975, Post-Carboniferous stratigraphy, northeastern Alaska: U.S. Geological Survey Professional Paper 886, 46 p.

Dow Geochemical Services, Inc. (DGSI), 1991, Geochemical analysis of Chevron Diamond No. 1, Alaska: Houston, Texas, DGSI, Report prepared for Chevron, USA.

Drummond, K.J., 1974, Paleozoic Arctic margins of North America, *in* Burke, C.A., and Drake, C.L., eds., The geology of continental margins: New York, Springer-Verlag, p. 802.

Dutro, J.T., Jr., 1987, Revised megafossil biostratigraphic zonation for the Carboniferous of northern Alaska, *in* Tailleur, I.L., and Weimer, P., eds., Alaska North Slope geology, Volume 1: Bakersfield, California, Pacific Section, Society of Economic Paleontologists and Mineralogists and the Alaska Geological Society, Book 50, p. 359–364.

Eittreim, S., and Grantz, A., 1979, CDP seismic sections of the western Beaufort continental margin: Tectonophysics, v. 59, p. 251–262.

Embry, A.F., 1990, Geological and geophysical evidence in support of the hypothesis of anticlockwise rotation of northern Alaska: Marine Geology, v. 93, p. 317–329.

Grantz, A., and Eittreim, S., 1979, Geology and physiography of the continental margin north of Alaska and implications for the origin of the Canada Basin: U.S. Geological Survey Open-File Report 79-288, 80 p.

Grantz, A., and May, S.D., 1983, Rifting history and structural development of the continental margin north of Alaska: American Association of Petroleum Geologists Memoir 34, p. 77–102.

Grantz, A., and May, S.D., 1987, Regional geology and petroleum potential of the United States Chukchi Shelf north of Point Hope, *in* Scholl, D.W., Grantz, A., and Vedder, J.G., eds., Geology and resource potential of the continental margin of western North America and adjacent ocean basins: Beaufort Sea to Baja California: Houston, Texas, Circum-Pacific Council for Energy and Mineral Resources, Earth Science Series, v. 6, p. 37–58.

Grantz, A., Holmes, M.L., and Kososki, B.A., 1975, Geologic framework of the Alaskan continental terrace in the Chukchi and Beaufort Seas, *in* Yorath, C.J., Parker, E.R., and Glass D.J., eds., Canada's continental margins and offshore petroleum exploration: Canadian Society of Petroleum Geologists Memoir 4, p. 669–700.

Grantz, A., Eittreim, S., and Dinter, D., 1979, Geology and tectonic development of the continental margin north of Alaska: Tectonophysics, v. 59, p. 263–291.

Grantz, A., Eittreim, S., and Whitney, O.T., 1981, Geology and tectonic development of the continental margin north of Alaska and implications for the origin of the Canada Basin, *in* Nairn, A.E.M., Churkin, M., Jr., and Stehli, F.G., eds., The Arctic Ocean: New York, Plenum Press, The Ocean Basins and Margins, v. 5, p. 439–492.

Grantz, A., Dinter, D.A., Hill, E.R., Hunter, R.E., May, S.D., McMullin, R.H., and Phillips, R.L., 1982a, Geologic framework, hydrocarbon potential, and environmental conditions for exploration and development of proposed oil and gas lease sale 85 in the central and northern Chukchi Sea: U.S. Geological Survey Open-File Report 82-1053, 84 p.

Grantz, A., Dinter, D.A., Hill, E.R., May, S.D., McMullin, R.H., Phillips, R.L., and Reimnitz, E., 1982b, Geological framework, hydrocarbon potential, and environmental conditions for exploration and development of proposed oil and gas lease sale 87 in the Beaufort and Northeast Chukchi Seas: U.S. Geological Survey Open-File Report 82-482, 71 p.

Grantz, A., May, S.D., and Dinter, D.A., 1987, Regional geology and petroleum potential of the United States Beaufort and northeasternmost Chukchi Seas, *in* Scholl, D.W., Grantz, A., and Vedder, J.G., eds., Geology and resource potential of the continental margin of western North America and adjacent ocean basins: Beaufort Sea to Baja California: Houston, Texas, Circum-Pacific Council for Energy and Mineral Resources, Earth Science Series, v. 6, p. 17–35.

Grantz, A., May, S.D., and Hart, P.E., 1990, Geology of the Arctic continental margin of Alaska, *in* Grantz, A., Johnson, L., and Sweeney, J.F., eds., The Arctic Ocean region: Boulder, Colorado, Geological Society of America, Geology of North America, v. L, p. 257–288.

Grantz, A., Moore, T.E., and Roeske, S.M., 1991, Continental-ocean transect A-3, Gulf of Alaska to Arctic Ocean: Geological Society of America DNAG Transect TRA-A3, scale 1:500 000, 3 sheets, 72 p. text.

Grantz, A., May, S.D., and Hart, P.E., 1994, Geology of the Arctic continental margin of Alaska, *in* Plafker, G., and Berg, H.C., eds., The geology of Alaska: Boulder, Colorado, Geological Society of America, Geology of North America, v. G-1, p. 17–48.

Haimila, N.E., Kirschner, C.E., Nassichuk, W.W., Ulmishek, G., and Proctor, R.M., 1990, Sedimentary basins and petroleum resource potential of the Arctic Ocean region, *in* Grantz, A., Johnson, L., and Sweeney, J.F., eds., The Arctic Ocean region: Boulder, Colorado, Geological Society of America, Geology of North America, v. L, p. 503–538.

Herman, B.M., and Zerwick, S.A., 1994, A preliminary analysis of potential field data in the southern Chukchi Sea, *in* Thurston, D.K., and Fujita, K., eds., Proceedings of the 1992 International Conference on Arctic Margins: Anchorage, Alaska, Minerals Management Service, p. 301–306.

Hubbard, R.J., Edrich, S.P., and Rattey, R.P., 1987, Geological evolution and hydrocarbon habitat of the "Arctic Alaska MicroPlate": Marine and Petroleum Geology, v. 4, p. 2–34.

Husky Oil NPR-A Operations, Inc., 1982a, Geologic Report for South Meade Test Well No. 1: Anchorage, Alaska, Report prepared by the Husky Oil Geology Department for the U.S. Geological Survey.

Husky Oil NPR-A Operations, Inc., 1982b, Geologic Report for Peard Test Well No. 1: Anchorage, Alaska, Report prepared by the Husky Oil Geology Department for the U.S. Geological Survey.

Husky Oil NPR-A Operations, Inc., 1983a, Geologic Report for Inigok Test Well No. 1: Anchorage, Alaska, Report prepared by Harry Haywood and R.G. Brockway for the U.S. Geological Survey.

Husky Oil NPR-A Operations, Inc., 1983b, Geologic Report for Tunalik Test Well No. 1: Anchorage, Alaska, Report prepared by Harry Haywood for the U.S. Geological Survey.

Husky Oil NPR-A Operations, Inc., 1983c, Geologic Report for Kugrua Test Well No. 1: Anchorage, Alaska, Report prepared by Gordon W. Legg for the U.S. Geological Survey.

Johnson, P.P., 1992, The North Chukchi high-compressional structures in a rift setting: International Conference on Arctic Margins Abstracts with Programs, Anchorage, Alaska, p. 28.

Jones, H.P., and Speers, R.G., 1976, Permo-Triassic Reservoirs of Prudhoe Bay Field, North Slope, Alaska, *in* Braunstein, J., ed., North American oil and gas fields: American Association of Petroleum Geologists Memoir 24, p. 23–50.

Kelley, J.S., and Brosgé, W.P., 1995, Geologic framework of a transect of the central Brooks Range: Regional relations and an alternative to the Endicott Mountains allochthon: American Association of Petroleum Geologists Bulletin, v. 79, p. 1087–1116.

Kirschner, C.E., and Rycerski, B., 1988, Petroleum potential of representative stratigraphic and structural elements in the National Petroleum Reserve in Alaska: *in* Gryc, G., ed., Geology and exploration of the National Petroleum Reserve in Alaska, 1974–1982: U.S. Geological Survey Professional Paper 1399, p. 191–208.

Kos'ko, M.K., Cecile, M.P., Harrison, J.C., Ganelin, V.G., Khandoshko, N.V., and Lopatin, B.G., 1993, Geology of Wrangel Island, between Chukchi and East Siberian Seas, northeastern Russia: Geological Survey of Canada Bulletin 461, 101 p.

Kupecz, J.A., 1995, Depositional setting, sequence stratigraphy, diagenesis, and reservoir potential of a mixed lithology, upwelling deposit: Upper Triassic Shublik Formation, Prudhoe Bay, Alaska: American Association of Petroleum Geologists Bulletin, v. 79, p. 1301–1319.

Lerand, M., 1973, Beaufort Sea, *in* McCrossan, R.G., ed., The future petroleum provinces of Canada: Their geology and potential: Canadian Society of Petroleum Geologists Memoir 1, p. 315–386.

Lister, G.S., Etheridge, M.A., and Symonds, P.A., 1986, Detachment faulting and the evolution of passive continental margins: Geology, v. 14, p. 246–250.

Lothamer, R.T., 1994, Early Tertiary wrench faulting in the North Chukchi Basin, Chukchi Sea, Alaska, *in* Thurston, D.K., and Fujita, K., eds., Proceedings of the 1992 International Conference on Arctic Margins: Anchorage, Alaska, Minerals Management Service Outer Continental Shelf Report 94-0040, p. 251–256.

Magoon, L.B., Bird, K.J., Claypool, G.E., Weitzman, D.E., and Thompson, R.H., 1988, Organic geochemistry, hydrocarbon occurrence, and stratigraphy of government-drilled wells, North Slope, Alaska, *in* Gryc, G., ed., Geology and exploration of the National Petroleum Reserve in Alaska, 1974–1982: U.S. Geological Survey Professional Paper 1399, p. 483–487.

Martin, A.J., 1970, Structure and tectonic history of the western Brooks Range, DeLong Mountains and Lisburne Hills, northern Alaska: Geological Society of America Bulletin, v. 81, p. 3605–3622.

May, S.D., 1985, Free-air gravity anomaly map of the Chukchi and Alaska Beaufort Seas, Arctic Ocean: U.S. Geological Survey Miscellaneous Investigations Series Map I-1182-E, scale 1:1 000 000, 1 sheet.

Mayfield, C.F., Tailleur, I.L., and Ellersieck, I., 1988, Stratigraphy, structure, and palinspastic synthesis of the western Brooks Range, northwestern Alaska, *in* Gryc, G., ed., Geology and exploration of the National Petroleum Reserve in Alaska, 1974–1982: U.S. Geological Survey Professional Paper 1399, p. 143–186.

Mickey, M.B., and Haga, H., 1987, Jurassic-Neocomian biostratigraphy, North Slope, Alaska, *in* Tailleur, I.L., and Weimer, P., eds., Alaska North Slope geology, Volume 1: Bakersfield, California, Pacific Section, Society of Economic Paleontologists and Mineralogists, Book 50, p. 397–404.

Micropaleo Consultants, 1989a, Biostratigraphy Report for Shell et al. OCS-Y-1482 No. 1 (Klondike), OCS Block #NR3-3-287, Chukchi Sea, Alaska: Technical Report M.C.I. Job No. 89-110 prepared for Shell Oil and partners in Klondike Well by Micropaleo Consultants, 329 Chapalita Drive, Encinitas, CA 92024.

Micropaleo Consultants, 1989b, Biostratigraphy Report, Shell et al. OCS-Y-1275 No. 1 (Popcorn), OCS Block #NR3-1-150, Chukchi Sea, Alaska: Technical Report M.C.I. Job No. 89-118 prepared for Shell Oil and partners in Popcorn Well by Micropaleo Consultants, 329 Chapalita Drive, Encinitas, CA 92024.

Micropaleo Consultants, 1990a, Biostratigraphy Report, Shell et al. OCS-Y-1320 No. 1 (Crackerjack), OCS Block #NR3-1-591, Chukchi Sea, Alaska: Technical Report M.C.I. Job No. 90-112 prepared for Shell Oil and partners in Crackerjack Well by Micropaleo Consultants, 329 Chapalita Drive, Encinitas, CA 92024.

Micropaleo Consultants, 1990b, Biostratigraphy Report, Shell et al. OCS-Y-1413 No. 1 (Burger), OCS Block #NR3-1-718, Chukchi Sea, Alaska: Technical Report M.C.I. Job No. 89-115 prepared for Shell Oil and partners in Burger Well by Micropaleo Consultants, 329 Chapalita Drive, Encinitas, CA 92024.

Micropaleo Consultants, 1991, Biostratigraphy Report, Chevron et al. OCS-Y-0996 No. 1 (Diamond), OCS Block #NR4-1-591, Chukchi Sea, Alaska: Technical Report M.C.I. Job No. 91-114 prepared for Chevron Oil and partners in Diamond Well by Micropaleo Consultants, 329 Chapalita Drive, Encinitas, CA 92024.

Molenaar, C.M., Bird, K.J., and Collett, T.S., 1986, Regional correlation sections across the North Slope of Alaska: U.S. Geological Survey Miscellaneous Field Studies Map MF-1907, vertical scale 1″=2000′, 1 sheet.

Moore, T.E., 1986, Stratigraphic framework and tectonic implications of pre-Mississippian rocks, northern Alaska: Geological Society of America Abstracts with Programs, v. 18, p. 159.

Moore, T.E., Nilsen, T.H., Grantz, A., and Tailleur, I.L., 1984, Parautochthonous Mississippian marine and nonmarine strata, Lisburne Peninsula, Alaska: U.S. Geological Survey Circular 939, p. 17–21.

Moore, T.E., Wallace, W.K., Bird, K.J., Karl, S.M., Mull, C.G., and Dillon, J.T., 1994, Geology of northern Alaska, *in* Plafker, G., and Berg, H.C., eds., The geology of Alaska: Boulder, Colorado, Geological Society of America, Geology of North America, v. G-1, p. 49–140.

Morley, C.K., Nelson, R.A., Patton, T.L., and Munn, S.G., 1990, Transfer zones in the East African rift system and their relevance to hydrocarbon exploration in rifts: American Association of Petroleum Geologists Bulletin, v. 74, p. 1234–1253.

Mull, C.G., 1982, Tectonic evolution and structural style of the Brooks Range, Alaska: An illustrated summary, *in* Powers, R.B., ed., Geologic studies of the Cordilleran Thrust Belt: Rocky Mountain Association of Geologists, v. 1, p. 1–45.

Nelson, R.A., Patton, T.L., and Morley, C.K., 1992, Rift-segment interaction and its relation to hydrocarbon exploration in continental rift systems: American Association of Petroleum Geologists Bulletin, v. 76, no. 8, p. 1153–1169.

Oldow, J.S., Seidensticker, C.M., Phelps, J.C., Julian, F.E., Gottschalk, R.R., Boler, K.W., Handschy, J.W., and Avé Lallement, H.G., 1987, Balanced cross sections through the central Brooks Range and North Slope, Arctic Alaska: American Association of Petroleum Geologists Map Library, 19 p., 18 plates.

Palmer, A.R., 1998, Geological Society of America geologic time scale: Boulder, Colorado, Geological Society of America, 1 p.

Parrish, J.T., 1987, Lithology, geochemistry, and depositional environment of the Triassic Shublik Formation, northern Alaska, *in* Tailleur, I.L., and Weimer, P., eds., Alaska North Slope geology, Volume 1: Bakersfield, California, Pacific Section, Society of Economic Paleontologists and Mineralogists and the Alaska Geological Society, Book 50, p. 391–396.

Pelka, G.J., Sauve, J.A., and Walsh, T.P., 1992, Franklinian tectonic framework controls on post Devonian basin development in Chukchi province, USA sector: International Conference on Arctic Margins Abstracts with Programs, Anchorage, Alaska, p. 49.

Reservoirs Inc., 1989, Geological evaluation, SWEPI et. al. Klondike OCS-Y-1482 #1 well, Chukchi Sea, Alaska, Part III (9,460–11,837 feet): Report prepared for Shell Western E&P, Inc., by Reservoirs Inc., 1151 C Brittmore Road, Houston, TX 77043, September 1989.

Reservoirs Inc., 1991a, Geological evaluation, SWEPI et. al. Popcorn OCS-Y-1275 #1 well, Chukchi Sea, Alaska, Parts III and IV (6,405–10,160 feet): Report prepared for Shell Western E&P, Inc., by Reservoirs Inc., 1151 C Brittmore Road, Houston, TX 77043, February 1991.

Reservoirs Inc., 1991b, Geological evaluation, Chevron et. al. OCS-Y-0996 #1, "Diamond" prospect, Chukchi Sea, Alaska, 4,746–6,482 feet: Report prepared for Chevron USA, Houston, TX, by Reservoirs Inc., 1151 C Brittmore Road, Houston, TX 77043, December 1991.

Reservoirs, Inc., 1991c, Geological evaluation, SWEPI et. al. Crackerjack OCS-Y-1320 #1 well, Chukchi Sea, Alaska, Part II (5,523–8,313 feet): Report prepared for Shell Western E&P, Inc., by Reservoirs Inc., 1151 C Brittmore Road, Houston, TX 77043, December 1991.

Reynolds, D.J., and Rosendahl, B.R., 1984, Tectonic expression of continental rifting [abs.]: Eos (Transactions, American Geophysical Union), v. 65, p. 1116.

Robinson, M.S., Clough, J.G., Roe, J., and Decker, J., 1992, Chronologic variations along the contact between the Echooka Formation and the Lisburne Group in the Northeastern Brooks Range, Alaska: Alaska Dept. of Natural Resources, Division of Geological and Geophysical Surveys, prepared for Bureau of Economic Geology, University of Texas at Austin, and Minerals Management Service, Anchorage, Alaska, Cooperative Agreement No. 30497-AK, 94 p.

Rosendahl, B.R., Reynolds, D.J., Lorber, P.M., Burgess, C.F., McGill, J., Scott, D., Lambiase, J.J., and Derksen, S.J., 1986, Structural expressions of rifting-lessons from Lake Tanganyika, Africa, *in* Frostick, L.E., Renaut, R.W., Reid, I., and Tiercelin, J.J., eds., Sedimentation in the African rifts: Geological Society [London] Special Publication 25, p. 29–43.

Saltus, R.W., Hudson, T.L., and Connard, G.G., 1999, A new magnetic view of Alaska: GSA Today, v. 9 no. 3, p. 1–6.

Shell Development Co., 1989, Klondike No. 1 well (Chukchi Shelf), Rock-Eval pyrolysis of sidewall cores: Project 89/1184 prepared by DGSI for Shell Development Co., 7 p.

Sherwood, K.W., 1994, Stratigraphy, structure, and origin of the Franklinian Northeast Chukchi Basin, Arctic Alaska Plate, *in* Thurston, D.K., and Fujita, K., eds., Proceedings of the 1992 International Conference on Arctic

Margins: Minerals Management Service Outer Continental Shelf Report MMS 94-0040, p. 245–250.

Sherwood, K.W., Craig, J.D., Lothamer, R.T., Johnson, P.P., and Zerwick, S.A., 1998, Chukchi Shelf Assessment Province, *in* Sherwood, K.W., ed., Undiscovered oil and gas resources, Alaska Federal Offshore (as of January 1995): Minerals Management Service Outer Continental Shelf Monograph MMS 98-0054, p. 115–196.

Tetra Tech, 1981, Structural-contour depth maps for horizons PDN (Pre-Mississippian argillite), PML (Carboniferous-"Lisburne"), and TPS (Triassic-"Shublik"), National Petroleum Reserve, Alaska: Tetra Tech Project TC 7169, by J.R. Harris, Scale 1″=32 000′, for Husky Oil NPR Operations, Inc. by Tetra Tech, Inc., Energy Management Division, 4544 Post Oak Place, Houston, Texas 77027.

Tetra Tech, 1982, Petroleum Exploration of NPR-A, 1974–1981, Final Report: Tetra Tech Reports 8200 and 8202 (3 volumes), prepared for Husky Oil NPR Operations, Inc., March 31, 1982 (Volumes 2 and 3) and July 31, 1982, (Volume 1), 132 Figs., by Tetra Tech, Inc., Energy Management Division, 4544 Post Oak Place, Houston, Texas 77027.

Thurston, D.K., and Theiss, L.A., 1987, Geologic Report for the Chukchi Sea Planning Area: Anchorage, Alaska, U.S. Department of the Interior, Minerals Management Service Outer Continental Shelf Report 87-0046, 193 p.

Trettin, H.P., 1991, Summary (Silurian–Early Carboniferous deformational phases and associated metamorphism and plutonism, Arctic Islands), *in* Trettin, H.P., ed., Geology of the Innuitian orogen and Arctic platform of Canada and Greenland: Boulder, Colorado, Geological Society of America, Geology of North America, v. E, p. 337–339.

Wheeler, W.H., and Karson, J.A., 1994, Extension and subsidence adjacent to a "weak" continental transform: An example from the Rukwa rift, East Africa: Geology, v. 22, p. 625–628.

MANUSCRIPT ACCEPTED BY THE SOCIETY MAY 15, 2001.

Geological Society of America
Special Paper 360
2002

Biostratigraphic evidence for the prerift position of the North Slope, Alaska, and Arctic Islands, Canada, and Sinemurian incipient rifting of the Canada Basin

Michael B. Mickey
Micropaleo Consultants, Inc., 329 Chapalita Drive, Encinitas, California 92024, USA
Alan P. Byrnes
Kansas Geological Survey, 1930 Constant Avenue, Lawrence, Kansas 66047, USA
Hideyo Haga
Micropaleo Consultants, Inc., 329 Chapalita Drive, Encinitas, California 92024, USA

ABSTRACT

Over the past several decades evidence has accumulated in support of a counter-clockwise rotational opening model for the Canada Basin. Reconstructed plate positions and the timing of rifting and spreading are, however, somewhat uncertain. To add to the existing data, we examined and correlated biostratigraphic tops and foraminiferal biofacies (paleoenvironments) for 69 Alaska North Slope and Chukchi Sea wells, and 59 Canadian Arctic Island onshore and offshore wells. Correlation of regional paleoenvironment distributions for 15 biostratigraphic time periods from the Middle Pennsylvanian to early Rhaetian can be interpreted to indicate that Point Barrow on the North Slope margin was juxtaposed with the southern portion of Prince Patrick Island, Canada, prior to opening of the Canada Basin. Abrupt facies changes on the North Slope from middle neritic environments in the Rhaetian to a starved basin environment in the Hettangian (?)-Sinemurian, contrasted with continued inner-to-middle neritic environments in the Arctic Islands, are interpreted to have resulted from incipient rifting and the formation of a chain of grabens and half-grabens that isolated the future Siberian-Alaskan margin from sediment sources on the Canadian margin. Missing Middle Jurassic (Bajocian–early Callovian) and Late Jurassic–Early Cretaceous (Tithonian-Berriasian) intervals on the North Slope, many of which are also absent in the Arctic Islands, are evidence of major erosion and/or tectonic events and can be interpreted to indicate episodic rifting and/or spreading events.

INTRODUCTION

The Amerasia and Canada basins are interpreted to have been formed by counterclockwise rotation of northeastern Siberia, the Chukchi borderland, and the Alaskan North Slope away from the Canadian Arctic Islands. Lawver and Scotese (1990) reviewed the numerous paleotectonic models, rotational and otherwise, that have been proposed. The counterclockwise rotation model, originally proposed by Carey (1958), or variants of this model have received support from (1) geologic and stratigraphic evidence (Hamilton, 1968; Tailleur and Brosge, 1970; Tailleur, 1969, 1973; Grantz et al., 1979, 1990, 1998; Grantz and May, 1983; Harland et al., 1984; Mayfield et al., 1988; Embry, 1990; Embry and Dixon, 1994; Embry et al., 1994), (2) aeromagnetic evidence (Taylor et al., 1981; Vogt et al., 1982; Srivastava et al., 1994; Grantz et al., 1998), (3) heat flow evidence

Mickey, M.B., Byrnes, A.P., Haga, H., 2002, Biostratigraphic evidence for the prerift position of the North Slope, Alaska, and Arctic Islands, Canada, and Sinemurian incipient rifting of the Canada Basin, *in* Miller, E.L., Grantz, A., and Klemperer, S.L., eds., Tectonic Evolution of the Bering Shelf–Chukchi Sea–Arctic Margin and Adjacent Landmasses: Boulder, Colorado, Geological Society of America Special Paper 360, p. 67–75.

(Lawver and Baggeroer, 1983; Langseth et al., 1990; Langseth and Lachenbruch, 1992), and (4) paleomagnetic and gravity evidence (Newman et al., 1979; Hillhouse and Grommé, 1983; Halgedahl and Jarrard, 1987; Laxon and McAdoo, 1997).

Jackson and Gunnarson (1990), Harbert et al. (1990), Embry (1990), and Embry and Dixon (1994) reviewed geologic and geophysical evidence for the rotational model and the timing of initiation of rifting, spreading, and cessation of spreading. In its simplest form the rotation model proposes that during the Paleozoic northern Alaska, northeastern Siberia, and the Chukchi borderland were contiguous with the Canadian Arctic Island region of the North American craton. During the Mesozoic these areas were separated by rifting from the Canadian Arctic Islands and rotated counterclockwise ~50°–75° about a pole of rotation located near the Mackenzie Delta. Based on this model, the Amerasia Basin is bounded by the passive margins of Siberia and the Alaskan North Slope and the Canadian Arctic Islands, and by a transform fault along the Makarov Basin slope of the Lomonosov Ridge (Fig. 1).

Although the counterclockwise rotational model for opening of the Amerasia Basin may be widely accepted, the precise geometry, prerotation margin positions, and timing of rifting and spreading are still uncertain and no single model has been proposed that comprehensively, and in detail, accounts for the diverse geologic, geophysical, seafloor magnetic, paleomagnetic, and paleontologic data. The generally triangular geometry of the basin necessitates that spreading did not occur as one simple parallel-spreading event, but probably involved either changes in spreading location or rate over time, and possibly nonuniform crustal shortening. Furthermore, interpreted Late Cretaceous and Paleocene spreading in the Makarov Basin (Taylor et al., 1981), Tertiary extension in the Chukchi borderland (Grantz et al., 1998), and tectonism in northern Alaska have altered postspreading geometries and complicate reconstruction.

To provide additional data to help address the issues of geometry and timing, biostratigraphic tops and foraminiferal biofacies (paleoenvironments) were interpreted for North Slope and Arctic Island wells. The prerift positions of the Alaskan North Slope and Canadian Arctic Islands were reconstructed on the basis of correlations of regional paleoenvironmental distributions for numerous biostratigraphic time periods. The timing of initial rifting was interpreted from changes in biofacies between the North Slope and the Arctic Islands.

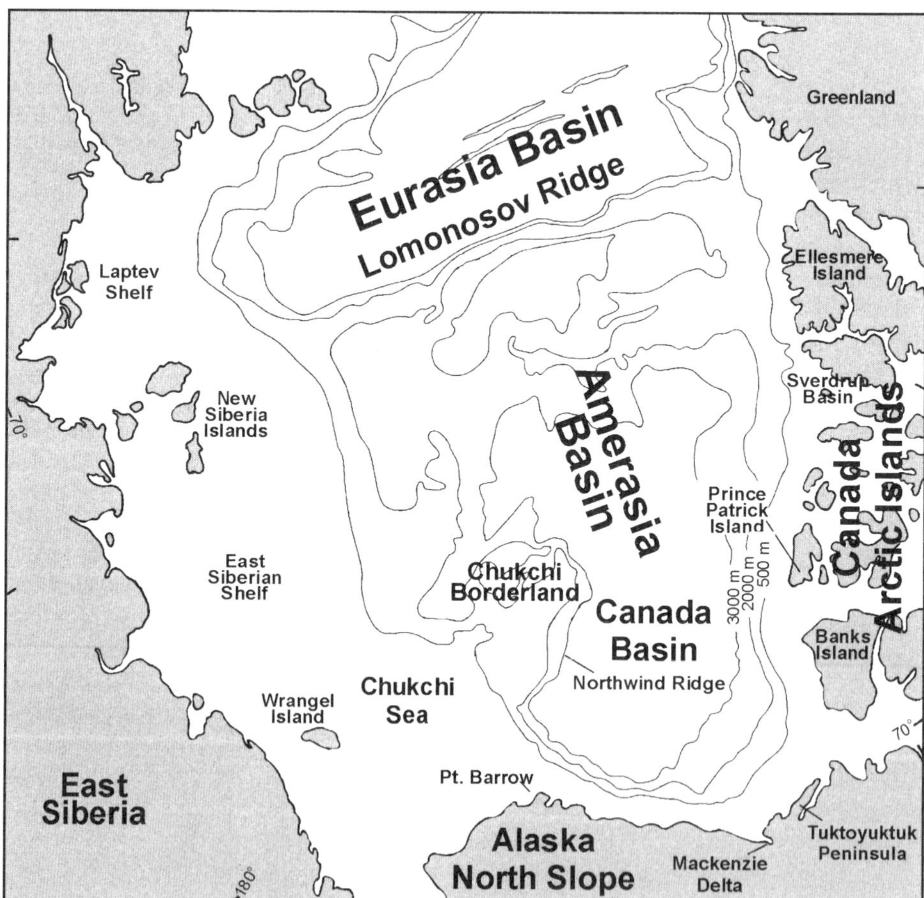

Figure 1. Map of Arctic Basin showing selected locations and important features.

BIOSTRATIGRAPHIC ZONATION AND PALEOENVIRONMENTAL ANALYSIS

Biostratigraphic ages and foraminiferal biofacies (paleoenvironments) were interpreted for 69 Alaskan North Slope and Chukchi Sea onshore and offshore wells and 59 Canadian Sverdrup Basin onshore and offshore wells (Fig. 2) for 36 biostratigraphic time periods from the Early Mississippian through the Early Cretaceous. Biostratigraphic tops were based on faunal assemblages and species distribution checklists for each well and integration of foraminiferal, palynologic, and well-log analysis data. Species checklists for wells in the National Petroleum Reserve—Alaska (NPRA) were obtained from published studies (Whitmer et al., 1981a, 1981b; Haga and Mickey, 1983a, 1983b). Checklists for the Chukchi Sea wells and the industry area of the North Slope, between the Canning and Colville Rivers, are currently proprietary and we generated them from examination of prepared slides that are available to the public at the State of Alaska Geological Material Center. Checklists for the Canadian Arctic Islands wells were obtained from open-file reports of the Geological Survey of Canada (Rice, 1972; Bujak,

1973; Clowser, 1975a, 1975b, 1975c; Fisher, 1975a, 1975b; Fisher and Sherrington, 1975; Paleo Services Ltd., 1975; Dolby and Sherrington, 1977; Robertson Research Canada, 1975a, 1975b; Bujak Davies Group, 1989) and more than 110 unpublished reports generated by commercial laboratories available through the Geological Survey of Canada (Panarctic Oils, Ltd., 1987a, 1987b, 1987c) and Riley Reproductions, Calgary.

Species checklists from the various sources were examined and ages assigned to the foraminiferal zones based on a combination of numerous published assemblage occurrences throughout the Boreal realm, comparisons with palynomorph ages, and when available, comparisons with megafossil and ostracod ages. The foraminiferal zonal system is based on assemblages and only occasionally uses index or marker species occurrences as reported previously (Tappan, 1955). In the study area most index or marker species occurrences range over two or more stages, and it is therefore only their association with other forms that allows the establishment of foraminiferal stage or substage zonal control. Palynological age determinations were based on the occurrences of both marine and nonmarine forms. These included dinocyst, spore, and pollen taxa. Age interpretations

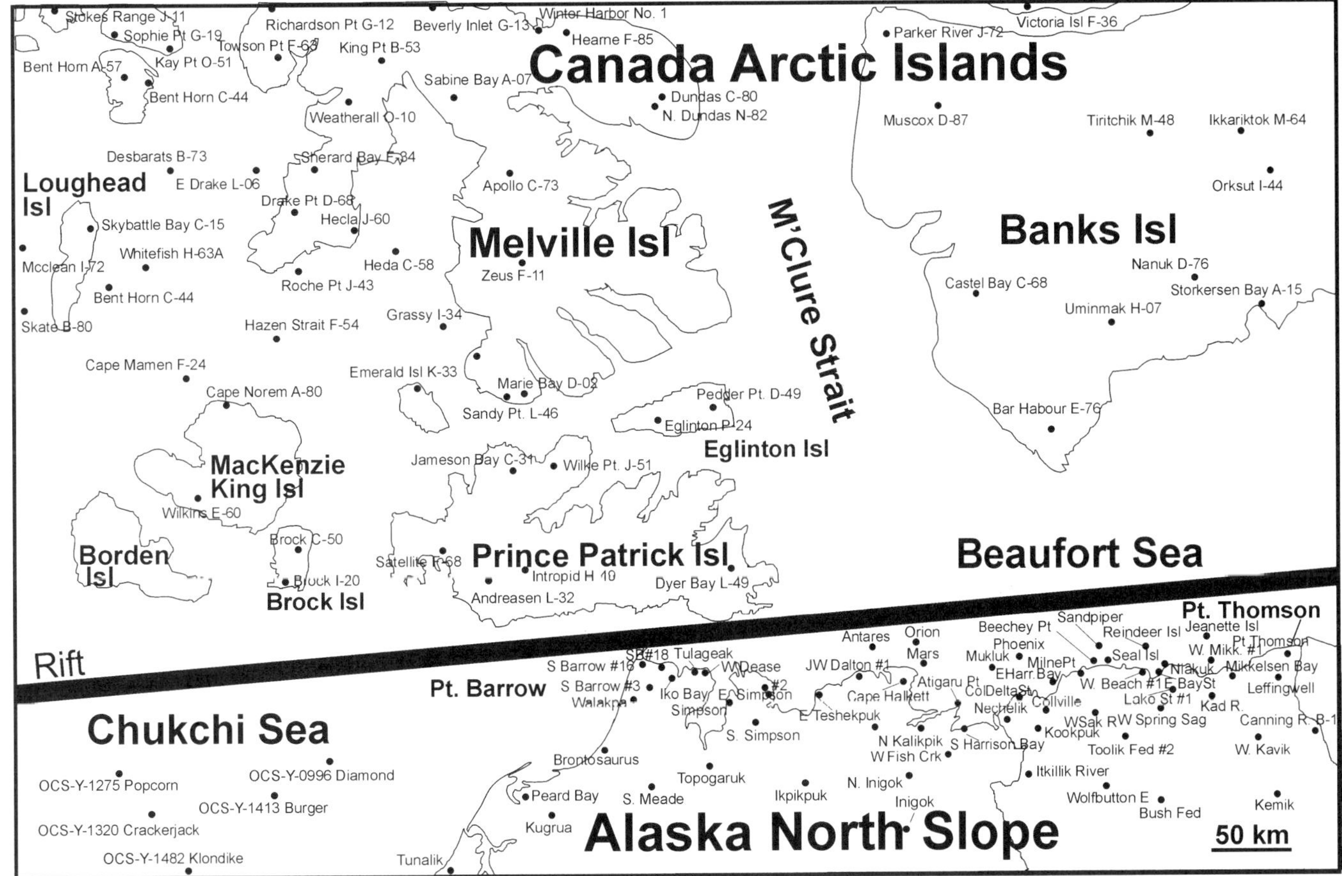

Figure 2. Location map of Canadian and Alaskan wells used in study. Relative positions of Alaska and Canadian Arctic Islands reflect their general reconstructed positions prior to opening of Canada Basin as delineated by correlation of environments shown in Figure 3. Isl—island.

were derived using age-restrictive taxa, or marker species, within assemblage zones. Age ranges of the palynomorphs have been documented in numerous published accounts from North America and Europe. The Mississippian through Middle Triassic section relies on spore and pollen species for age subdivisions. Beginning in the Late Triassic section, dinocysts make their appearance and constitute the principal basis for zonation. In the Jurassic through the mid-Cretaceous dinocysts remained the most important palynological component, although spore and pollen species are present in all assemblages. An earlier subdivision system, listing characteristic species for the Jurassic through Neocomian sections of the Alaska North Slope region, was presented by Mickey and Haga (1987). A modification of this subdivision system was used in this study. Chronostratigraphic ages were assigned to biostratigraphic intervals using the correlations presented by Gradstein and Ogg (1996).

Paleoenvironments were determined from correlations of foraminiferal assemblages with relative depths along seismically established dip sections on the Alaska North Slope that represent shelf to slope paleoenvironments. Based on palynomorph and foraminiferal assemblages, paleoenvironments were interpreted for as many as 36 biostratigraphic intervals in each well. Foraminiferal assemblages provided designation of marine paleoenvironments, and palynomorphs primarily distinguished between marine and nonmarine conditions. Paleoenvironments and their approximate associated water depths are nonmarine–marginal marine (low-high tide, 0 m), inner neritic (0–30 m), middle neritic (30–90 m), outer neritic (90–180 m), upper bathyal (180–450 m), middle bathyal (450–1450 m), and lower bathyal (1450–1850 m). Distal starved and starved basin paleoenvironments are interpreted to occur as a result of both distance from a sediment source and isolation from a source. Water depths are interpreted to range from neritic to bathyal based on the water depths of units overlying and underlying the starved intervals. Sediment transport, shelf and slope dips, and environmental conditions may all influence biofacies and therefore the interpreted paleoenvironments and water depths. Although the absolute water depths inferred by the environment names may have unspecified errors, the relative depth associations (i.e., shallow to deep, proximal to distal) of the paleoenvironments are considered correct. Therefore, regional differences in the assigned environments between localities for any given time period represent some change in paleoenvironmental conditions and provide a means for reconstructing formerly contiguous but now tectonically separated terranes.

Assignment of biofacies (paleoenvironments) to any single stage was complicated by (1) changes in environments during a single stage as a result of higher-order transgressive-regressive cycles, (2) continuous change in environments during a stage, (3) variability of thickness preserved for each environment in a section containing a range of environments, and (4) averaging associated with sample sizes representing 3–30 m sample intervals. Paleoenvironments for any single stage or zone, or even sample, can range over several paleoenvironmental niches de-

pending on the rate of sea-level rise or fall and the nature of sediment preservation associated with the deposition of the interval analyzed. The paleoenvironments we assigned to each sample represented the one or two principal assemblages exhibited in that sample. This methodology tends to average out any higher-order cycles that may be present and therefore emphasizes lower-order cycles in the environmental classification.

CORRELATION OF THE ALASKA NORTH SLOPE AND CANADIAN ARCTIC ISLANDS

Other than generalized rotational reconstructions with schematic margin positions, delineation of the prerift positions of the Chukchi borderland and Alaskan North Slope relative to the Canadian Arctic Islands has been largely based on geologic evidence. Embry (1990) and Embry et al. (1994) provided support for the general Harland et al. (1984) North Slope–Arctic Islands reconstruction based on (1) alignment of the basin axes for the Sverdrup Basin and Hanna Trough; (2) correlation of Early Triassic–Middle Jurassic sections in the Alaska South Meade No. 1 with the Canadian Sandy Point L-46 and Alaska Tunalik No. 1 with the Canadian Brock I-20 wells, and (3) reconstruction of the Alaska North Slope and Sverdrup Basin paleogeography of Early Triassic facies. This reconstruction attached Point Barrow, Alaska, to Prince Patrick Island, Canada (Fig. 2). Recent correlation of Cambrian through Triassic facies on Northwind Ridge with coeval facies in the Arctic Islands has been interpreted to indicate that the Northwind Ridge and other Chukchi borderland landmasses were attached to the Canadian Sverdrup Basin (Grantz et al., 1998).

Using the Embry et al. (1994) reconstruction as a basic premise, comparison of the foraminiferal paleoenvironmental distribution for the Alaska North Slope and the Canadian Arctic Islands, between Ellesmere Island and Mackenzie Delta for 15 time periods between the Middle Pennsylvanian and Late Triassic (Rhaetian), provided a basis for determining prerift margin positions. Comparison could not be performed for times prior to the pre-Middle Pennsylvanian because these intervals were either missing or not penetrated in the Arctic Islands wells studied, or for times after the Rhaetian because facies changes on the Alaska North Slope in the Hettangian (?)-Sinemurian are interpreted to indicate incipient rifting, as further discussed below. Foraminiferal biofacies (paleoenvironment) distributions for six Paleozoic time periods provided the best geographic constraints on the relative positions of the two margins (Fig. 3).

Assuming a simple rotation model without postrift movement, the reconstruction requires a pole of rotation near the base of the Tuktoyuktuk Peninsula in the Mackenzie Delta. The close proximity of Point Barrow, Alaska, to Prince Patrick Island, Canada, shown in Figure 3 is primarily for correlation purposes and represents ~75° clockwise rotation. A separation between the North Slope and Arctic Islands of ~100 km more than that shown in Figure 3 is probable based on hinge-line positions offshore of Point Barrow (Grantz et al., 1990) and Banks Island (Dixon and Dietrich, 1990), and the margin of the Sverdrup Basin off Prince

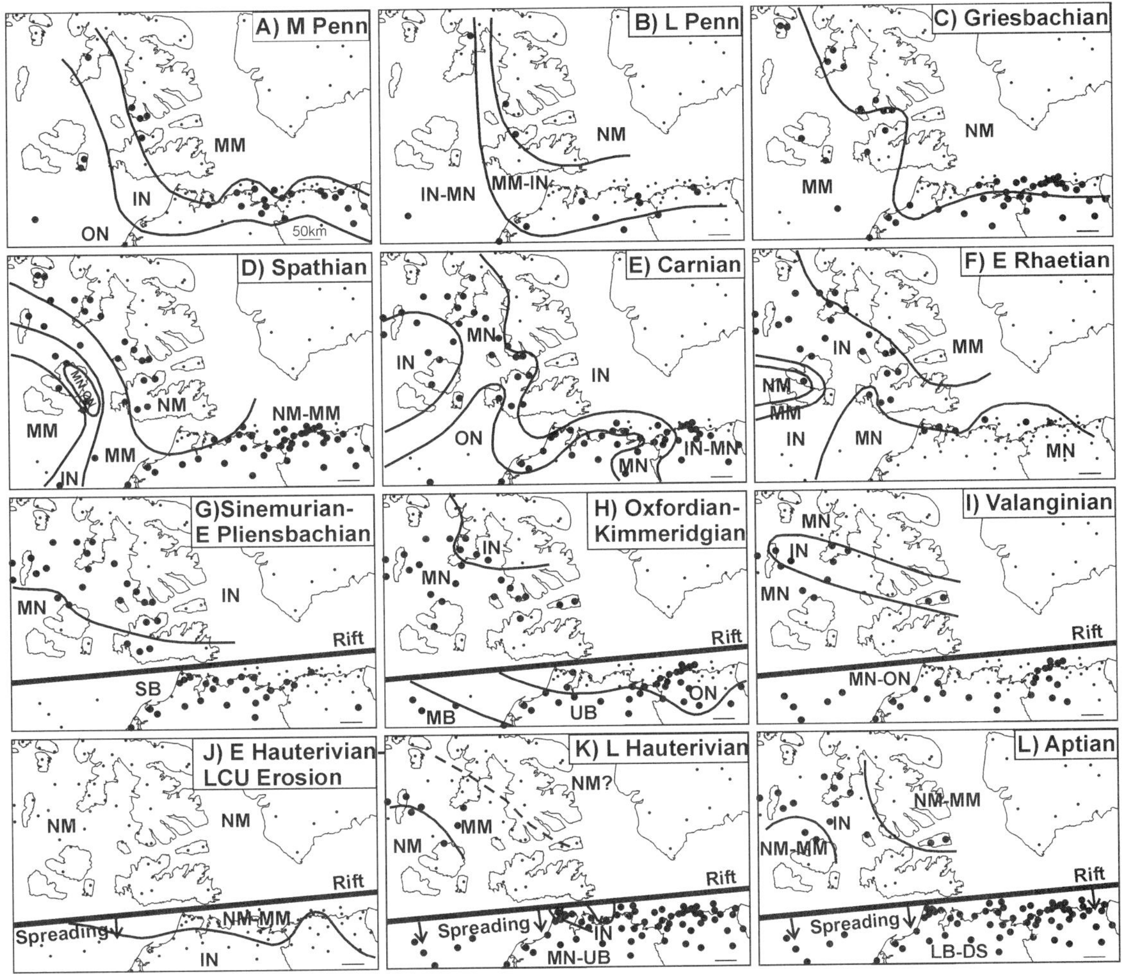

Figure 3. Interpreted distribution of foraminiferal biofacies (paleoenvironments) for selected time periods before and after rifting. Relative positions of Alaska and Canadian Arctic Islands reflect their reconstructed positions prior to opening of Canada Basin. Margins are shown in closer positions than allowed by hinge lines for graphic and comparison purposes. The following foraminifera paleoenvironmental and other abbreviations were used: NM—nonmarine, MM—marginal marine, IN—inner neritic, MN—middle neritic, ON—outer neritic, UB—upper bathyal, MB—middle bathyal, LB—lower bathyal, DS—distal starved, SB—starved basin, LCU—Lower Cretaceous unconformity.

Patrick Island (Embry, 1990). This additional separation is not interpreted to require significant change in the margin attachment positions. Uncertainty in the reconstruction due to the methodology employed results from the (1) sparse density of wells in the Chukchi Sea and Sverdrup Basin, (2) lack of data in some wells because of missing (eroded) intervals, (3) large distances covered by the map region, (4) lack of continuous correlation and the great distance from the map region to the rotation pole, and (5) imprecision in paleoenvironments assigned for some time periods. Possible lateral translation of the margins would require a change in the position of the pole of rotation.

The resulting reconstruction is similar to counterclockwise rotation reconstructions presented by Embry (1990) and others (Harland et al., 1984; Jackson and Gunnarson, 1990). Reconstruction by Grantz et al. (1998) indicates that the southern portion of Northwind Ridge was juxtaposed with the southern margin of the Sverdrup Basin. Based on the reconstruction presented here, and accounting for the greater separation between the North Slope and Arctic Islands margins than shown in Figures 2 and 3, this would suggest that prior to spreading, Northwind Ridge was immediately north-northeast of Point Barrow. Within the error range of the environmental correlations and plate positions determined here, it may also indicate that Point Barrow was not any farther north of the position shown in Figures 2 and 3 and may have been positioned slightly to the south.

TIMING OF INITIATION OF RIFTING

Estimated dates for initiation of rifting range from Early Jurassic (Hettangian; Hubbard et al., 1987) to early Middle

Jurassic (Aalenian; Embry and Dixon, 1994). Grantz and May (1983) proposed that rifting began in the Early Jurassic, on the basis of an estimated age for Dinkum graben fill; the development of a regional unconformity on Late Triassic beds; a shift in provenance from a Triassic northern source to a Jurassic southern source in the Mackenzie Delta region; and an Early Jurassic age for deposits in the Yukon, Richardson, and British Mountains that can be stratigraphically tied to the basal graben fill. Embry (1991) proposed an early Middle Jurassic age for initiation of rifting based on Middle Jurassic faulting on Prince Patrick Island. More recent detailed study of the faulting on Prince Patrick Island (Harrison et al., 1999) has been interpreted to indicate that initial rifting did not begin earlier than late Toarcian. The occurrence of a major pre–Middle Jurassic unconformity on Svalbard and the occurrence of early Middle Jurassic faults in the Mackenzie Delta region were cited as evidence for an early Middle Jurassic (Aalenian) age for initiation of rifting (Embry and Dixon, 1994).

In this study, abrupt changes in foraminiferal paleoenvironments on the Alaska North Slope that are not evident in corresponding Arctic Islands areas are interpreted to indicate the presence of rift grabens and half-grabens that trapped sediment from the Arctic Islands provenance area. The latest time period in which Arctic Island and North Slope environments exhibit continuity across the study area is early Rhaetian. Paleoenvironments in the early Rhaetian range from nonmarine to middle neritic on the Arctic Islands and from marginal marine to middle neritic on the North Slope (Fig. 3F). Beginning in the Hettangian (?)-Sinemurian, starved basin-type deposition occurred in areas of the North Slope. These deposits are characterized by black, organic-rich, high-gamma-ray, paper shales that are largely depauperate except for acid-resistant radiolaria and spore-pollen. A Hettangian (?)-Sinemurian age for these deposits is based on foraminiferal, palynologic, bivalve, and ostracod assemblages that have Sinemurian age assignments based on correlation with Siberian bivalves and ostracods (Boris Nikitenko and Boris Shurygin, Russian Academy of Sciences, Institute of Geology and Geophysics, Novosibirsk, Russia, 1999, personal commun.). The existing assemblages cannot rule out a possible Hettangian age in some wells.

The discontinuous distribution of the Hettangian (?)-Sinemurian starved basin deposits across the North Slope is interpreted to result from the formation of horsts and grabens and half-grabens. Isolated Pliensbachian inner middle neritic, coarse-grained sandstones overlying the Norian Shublik Formation are observed in several North Slope wells (e.g., Topagoruk-1, South Simpson-1, Iko Bay-1, No Name Island-1, East Simpson-1, East Simpson-2). Without provenance studies to identify the source of these sandstones, these can be interpreted to have been sourced from the Arctic Islands, from now-submerged ridges beneath the Beaufort Sea Shelf, or from upthrown portions of nearby isolated horst blocks on the North Slope Barrow arch. That these coarse-grained sandstones occur in isolated areas can be interpreted to indicate these may have resulted from

erosion of older strata on nearby ridges or wave-base exposed portions of nearby uplifted blocks. However, reestablishment of brief provenance from the Arctic Islands cannot be ruled out. Whatever the provenance for the Pliensbachian sandstones, the presence of starved basin deposits indicates the existence of isolated fault blocks and can be interpreted as an indication of incipient rifting in the Sinemurian and possibly earlier, in the Hettangian. By the late Pliensbachian–early Toarcian, deposits across the entire North Slope reflect starved basin-middle lower bathyal (?) conditions. It is interesting to note that failed incipient rifting also occurred along much of the Atlantic continental margin of the eastern United States in this same Late Triassic–Early Jurassic period and marked the beginning of the breakup of Pangea (Hutchinson and Klitgord, 1988).

TIMING OF SEAFLOOR SPREADING

As noted by Embry and Dixon (1994), the timing of the rift-drift transition and the initiation of seafloor spreading is "contentious." Tailleur and Brosge (1970) suggested that spreading occurred in the Aptian and Albian on the basis of the age of the thick Brookian sequence deposits on the western North Slope. Heat flow and lithospheric cooling models of the central Canada Basin have been interpreted to indicate that seafloor spreading occurred between the Barremian and Campanian (Lawver and Baggeroer, 1983; Langseth et al., 1990). Magnetic anomaly data in the central Canada Basin have been interpreted to support spreading from the early Kimmeridgian to Valanginian (Taylor et al., 1981; Vogt et al., 1982). Srivastava et al. (1992) and Grantz et al. (1998) reinterpreted magnetic anomalies in the central Canada Basin and concluded that these must predate the Cretaceous long normal anomaly (Aptian to Campanian). Paleomagnetic data from the North Slope Kuparuk Formation indicate that the central North Slope, near Prudhoe Bay, was still adjacent to the Arctic Islands in the Valanginian (Halgedahl and Jarrard, 1987), and data from the Brooks Range imply that motion of the North Slope relative to North America ceased in the late Albian (Harbert et al., 1990). Proposing that the best evidence for initiation of spreading is the age of the breakup unconformity, Embry and Dixon (1994) argued that the Hauterivian unconformity, present on the Alaskan North Slope and Canadian Arctic Islands, represented the breakup unconformity. Two other major unconformities, the Aptian unconformity on the North Slope and the Cenomanian unconformity present in the Mackenzie Delta region, were considered improbable as the breakup unconformity due to conflict with paleomagnetic indications of timing.

Hubbard et al. (1987) presented a detailed rifting history utilizing published and proprietary data. They recognized two rift megasequences, termed the Lower and Upper Beaufortian megasequences. The Lower Beaufortian megasequence was deposited during the Hettangian to the Tithonian and represented a failed rift episode marked by a series of extensional grabens and half-grabens, including the Dinkum graben, which they propose developed in the Aalenian. The Upper Beaufortian megasequence represented the period of seafloor spreading from Berriasian to Aptian time. This sequence of events presented by Hubbard et al. (1987) is analogous with rift events on the United States middle Atlantic continental margin (Benson and Doyle, 1988) and the Atlantic margins and the Bay of Biscay (Manspeizer, 1988), where Late Triassic–Early Jurassic failed rifting resulted in synrift deposits largely confined to rift basins. These deposits are unconformably overlain by blanketlike postrift and drift deposits. Manspeizer (1988) reported that this unconformity, similar to Falvey's (1974) breakup unconformity, represents intermittent episodes, spanning 20 m.y., of uplift, erosion, and crustal thinning along the future spreading axis. Time-transgressive eastward migration of the axis of upwelling asthenosphere and the location of failed rift grabens and half-grabens were also evident along the eastern United States continental margin (Manspeizer, 1988).

Foraminiferal paleoenvironmental distributions (Fig. 3) provide some limited information to address questions concerning the timing of seafloor spreading. The presence of starved basin deposits on the Alaskan North Slope from the Hettangian (?)-Sinemurian through the Aalenian and the thick Early Jurassic section in the Dinkum graben implies that grabens and half-grabens separating the Alaskan and Arctic Islands margins acted as effective traps for sediments sourced from the Arctic Islands during this period. The general absence of Bajocian–early Callovian age deposits in North Slope wells indicates a major erosion event in the late Callovian and/or early Oxfordian. Similar widespread late Callovian or early Oxfordian tectonism is also reported for Prince Patrick Island (Harrison et al., 1999). Late Jurassic paleoenvironments on the North Slope exhibit progressive shallowing from the early Oxfordian (Fig. 3H) through the late Kimmeridgian; no sediments of Tithonian-Berriasian age were identified in the North Slope wells studied. This implies that sediments of this age were not deposited or, more probably, that they were removed by another erosion event. Valanginian sediments that unconformably overlie Oxfordian-Kimmeridgian beds represent either similar or slightly shallower environments (Fig. 3I) than those of the shallowest Kimmeridgian.

Regional uplift and erosion along the North Slope Barrow arch and Arctic Islands in the late Valanginian (Fig. 3G) are evident from early Hauterivian sediments overlying what is referred to as the Lower Cretaceous unconformity, below which Ellesmerian sequence strata as old as Mississippian subcrop in large regions. On the North Slope, this late Valanginian unconformity is overlain by sediments exhibiting upward-deepening environments from middle to outer neritic environments in the early Hauterivian (Fig. 3K) to distal starved environments in the Barremian and lower bathyal–distal starved environments at the base of the Aptian to early Albian Brookian deposits. Distal starved environments persisted in the well study area on the North Slope Barrow arch through the Barremian (Fig. 3L), until southwesterly sourced Brookian sequence lower bathyal–distal starved basin Torok Formation and overlying neritic Nanushuk Group sediments prograded across the North Slope in the Aptian to Albian.

The preceding sequence of environments and the interpreted events are consistent with several proposed rifting models, including intermittent or episodic rifting during the Middle Jurassic to Early Cretaceous proposed in several studies, or incipient rifting that culminated in initiation of seafloor spreading after the Hauterivian (Embry and Dixon, 1994). Analysis of the age distribution of sediments in small extensional grabens and half-grabens along the Barrow arch and in the Prince Patrick Islands area, and other lines of evidence, must also be interpreted to suggest that the location of active spreading may have migrated through time (Mickey et al., 1998).

CONCLUSIONS

Correlated biostratigraphic tops and foraminiferal biofacies and/or paleoenvironments for wells of the North Slope, Alaska, and Arctic Islands, Canada, provide a basis for reconstructing the prerift configuration of the Sverdrup and North Slope margins and establishing the timing of rifting. Correlation of the distribution of biofacies and/or paleoenvironments for 15 time periods prior to rifting indicates that Point Barrow, Alaska, was originally attached to southern Prince Patrick Island, Canada. This arrangement is consistent with several previously proposed models. Between the early Rhaetian and Hettangian (?)-Sinemurian, an abrupt change from middle neritic to distal starved environments on the North Slope, contrasted with continued inner-to-middle neritic environments on the Arctic Islands, provides good evidence of incipient rifting in the Hettangian (?)-Sinemurian. Jurassic and Early Cretaceous unconformities on the North Slope and Arctic Islands suggest episodic rifting and spreading. These biostratigraphic data help to refine the timing of some of the major events involved in the formation of the Canada Basin. Continued detailed analysis and integration of the age distribution of deposits in the large- and small-scale grabens and half-grabens on the North Slope and Arctic Islands and provenance studies of the Early Jurassic sandstones should help in understanding the nature of the timing and distribution of rifting and spreading.

ACKNOWLEDGMENTS

The authors acknowledge the following peers: Art Grantz, Irv Tailleur, Gil Mull, and Ashton Embry for their helpful discussions, suggestions, and critical review of our proposed model for the opening of the Canada Basin.

REFERENCES CITED

Benson, R.N., and Doyle, R.G., 1988, Early Mesozoic rift basins and the development of the United States Middle Atlantic Continental Margin, *in* Manspeizer, W., ed., Triassic-Jurassic rifting, continental breakup and the origin of the Atlantic Ocean and passive margins: Amsterdam, Netherlands, Elsevier, p. 99–127.

Bujak, J.P., 1973, The micropaleontology, palynology and the stratigraphy of the Elf Wilkins E-60 well: Geological Survey of Canada, Institute of Sedimentary and Petroleum Geology, Paleontology Subdivision Consultant's Report #23, Robertson Research (North America) Ltd., 33 p., 3 charts.

Bujak Davies Group, 1989, Micropaleontology of Canadian Beaufort wells: Geological Survey or Canada, Institute of Sedimentary and Petroleum Geology, Paleontology Subdivision Consultant's Report #14, 14 charts.

Carey, S.W., 1958, A tectonic approach to continental drift, *in* Proceedings, Continental Drift, A Symposium, 1956: Hobart, Australia, University of Tasmania, p. 177–355.

Clowser, D.R., 1975a, The micropaleontology, palynology and stratigraphy of the Panarctic Homested Helca J-60 well: Geological Survey of Canada, Institute of Sedimentary and Petroleum Geology, Paleontology Subdivision Consultant's Report #22, Robertson Research (North America) Ltd., 38 p., 2 charts.

Clowser, D.R., 1975b, The micropaleontology, palynology and stratigraphy of the Panarctic Marie Bay D-02 well: Geological Survey of Canada, Institute of Sedimentary and Petroleum Geology, Paleontology Subdivision Consultant's Report #29, Robertson Research (North America) Ltd., 26 p., 1 biostratigraphical chart.

Clowser, D.R., 1975c, The micropaleontology, palynology and stratigraphy of the Panarctic Sandy Point L-46 well: Geological Survey of Canada, Institute of Sedimentary and Petroleum Geology, Paleontology Subdivision Consultant's Report #30, Robertson Research (North America) Ltd., 32 p., 2 plates, 1 biostratigraphical chart.

Dixon, J., and Dietrich, J.R., 1990, Canadian Beaufort Sea and adjacent land areas, *in* Grantz, A., Johnson, L., and Seeney, J.F., eds., The Arctic Ocean region: Boulder, Colorado, Geological Society of America, Geology of North America, v. L, p. 239–256.

Dolby, G., and Sherrington, P.F., 1977, Micropaleontology, palynology and stratigraphy of the Panarctic et al Drake Point D-68 well Melville Island, District of Franklin: Geological Survey of Canada Open-File Report, v. 460, 19 p.

Embry, A.F., 1990, Geologic and geophysical evidence in support of the hypothesis of anticlockwise rotation of northern Alaska: Marine Geology, v. 93, p. 317–329.

Embry, A.F., 1991, Mesozoic history of the Arctic islands, *in* Trettin, H.P., ed., Innuitian orogen and Arctic platform: Canada and Greenland: Geological Survey of Canada, Geology of Canada, v. 3, p. 369–433.

Embry, A.F., and Dixon, J., 1994, The age of the Amerasia Basin, *in* Thurston, D.K., and Fujita, K., eds., Proceedings of the 1992 International Conference on Arctic Margins: Anchorage, Alaska, Minerals Management Service Outer Continental Shelf Study MMS Report 94-0040, p. 289–294.

Embry, A.F., Mickey, M.B., Haga, H., and Wall, J.H., 1994, Correlation of the Pennsylvanian–Lower Cretaceous succession between northwest Alaska and southwest Sverdrup Basin: Implications for Hanna trough stratigraphy, *in* Thurston, D.K., and Fujita, K., eds., Proceedings of the 1992 International Conference on Arctic Margins: Anchorage, Alaska, Minerals Management Service Outer Continental Shelf Study MMS Report 94-0040, p. 105–110.

Falvey, D.A., 1974, The development of continental margins in plate tectonic theory: Australian Petroleum Exploration Association Journal, v. 14, p. 95–106.

Fisher, M.J., 1975a, The micropaleontology, palynology and stratigraphy of the Panarctic BP-Skelly-Tenneco et al Brock C-50 well: Geological Survey of Canada, Institute of Sedimentary and Petroleum Geology, Paleontology Subdivision Consultant's Report #31, Robertson Research (North America) Limited, 27 p., 2 plates, 3 biostratigraphical charts.

Fisher, M.J., 1975b, The micropaleontology, palynology and stratigraphy of the BP et al Satellite F-68 well: Geological Survey of Canada, Institute of Sedimentary and Petroleum Geology, Paleontology Subdivision Consultant's Report #32, Robertson Research (North America) Limited, 21 p., 3 plates.

Fisher, M.J., and Sherrington, P.F., 1975, The micropaleontology, palynology and stratigraphy of the Panarctic Brock I-20 well: Geological Survey of Canada, Institute of Sedimentary and Petroleum Geology, Paleontology Subdivision Consultant's Report #33, Robertson Research (North America) Limited, 25 p., 2 plates, 1 biostratigraphical chart.

Gradstein, F.M., and Ogg, J., 1996, A Phanerozoic time scale: Episodes, v. 19, nos. 1 and 2, p. 3–4.

Grantz, A., and May, S.D., 1983, Rifting history and structural development of the continental margin north of Alaska, *in* Watkins, J.S., and Drake, C.L., eds., Studies of continental margin geology: American Association of Petroleum Geologists Memoir 34, p. 77–100.

Grantz, A., Eittreim, S.L., and Dinter, D.A., 1979, Geology and tectonic development of the continental margin north of Alaska: Tectonophysics, v. 59, p. 263–291.

Grantz, A., May, S.D., and Hart, P.E., 1990, Geology of the Arctic continental margin of Alaska, *in* Grantz, A., Johnson, L., and Sweeney, J.F., eds., The Arctic Ocean region: Boulder, Colorado, Geological Society of America, Geology of North America, v. L, p. 257–289.

Grantz, A., Clark, D.L., Phillips, R.L., and Srivastava, S.P., 1998, Phanerozoic stratigraphy of Northwind Ridge, magnetic anomalies in the Canada Basin, and the geometry and timing of rifting in the Amerasia Basin, Arctic Ocean: Geological Society of America Bulletin, v. 110, no. 6, p. 801–820.

Haga, H., and Mickey, M.B., 1983a, Jurassic-Neocomian seismic stratigraphy NPRA: National Oceanic and Atmospheric Administration Report # TGY-0220BL, National Geophysical Data Center #83-TGB-08, 2 volumes, 133 p., 14 plates.

Haga, H., and Mickey, M.B., 1983b, South Barrow area Neocomian biostratigraphy: National Oceanic and Atmospheric Administration Report #TGY-0280, 2 volumes, 48 p., 20 plates.

Halgedahl, S.L., and Jarrard, R.D., 1987, Paleomagnetism of the Kuparuk River Formation from oriented drill core: Evidence for rotation of the Arctic Alaska plate, *in* Tailleur, I.L., and Weimer, P., eds., Alaska North Slope geology, Volume 2: Pacific Section, Society of Economic Paleontologists and Mineralogists and the Alaska Geological Society, Book 50, p. 581–617.

Hamilton, W.B., 1968, Continental drift in the Arctic [abs.]: Geological Society of America Special Paper 121, p. 510.

Harbert, W., Frei, L., Jarrard, R., Halgedahl, S., and Engebretson, D., 1990, Paleomagnetic and plate-tectonic constraints on the evolution of the Alaskan–eastern Siberian Arctic, *in* Grantz, A., Johnson, L., and Sweeney, J.F., eds., The Arctic Ocean region: Boulder, Colorado, Geological Society of America, Geology of North America, v. L, p. 567–593.

Harland, W.B., Gaskell, B.A., Heafford, P.A., Lind, E.K., and Perkins, P.J., 1984, Outline of Arctic post-Silurian continental displacements, *in* Spencer, A.M., ed., Petroleum geology of the north European margin: London, Graham and Trotman, Norwegian Petroleum Society, p. 137–148.

Harrison, J.C., Wall, J.H., Brent, T.A., Poulton, T.P., and Davies, E.H., 1999, Rift related structures in Jurassic and Lower Cretaceous strata near the Canadian polar margin, Yukon Territory, Northwest Territories, and Nunavit: Geological Survey of Canada, Current Research 1999–E, p. 12.

Hillhouse, J.W., and Gromme, C.S., 1983, Paleomagnetic studies and the hypothetical rotation of Arctic Alaska, *in* Mull, C.G., and Reed, K.M., eds., The origin of the Arctic Ocean (Canada Basin), Proceedings, 1981 Mini-symposium: Journal of the Alaska Geological Society, v. 2, p. 27–39.

Hubbard, R.J., Edrich, S.P., and Rattey, R.P., 1987, Geologic evolution and hydrocarbon habitat of the Arctic Alaska microplate, *in* Tailleur, I.L., and Weimer, P., eds., Alaskan North Slope geology: Bakersfield, California, Pacific Section Society of Economic Paleontologists and the Alaska Geological Society, Book 50, v. 2, p. 797–830.

Hutchinson, D.R., and Klitgord, K.D., 1988, Evolution of rift basins on the continental margin off southern New England, *in* Manspeizer, W., ed., Triassic-Jurassic rifting: Continental breakup and the origin of the Atlantic Ocean and passive margins: New York, Elsevier, Developments in Geotectonics 22, Part A, p. 81–98.

Jackson, H.R., and Gunnarson, K., 1990, Reconstructions of the Arctic: Mesozoic to present: Tectonophysics, v. 172, p. 303–322.

Langseth, M.G., and Lachenbruch, A.H., 1992, Review of thermal constraints on the age and evolution of the Amerasia Basin: Eos (Transactions, American Geophysical Union), v. 73, no. 14, supplement, p. 286.

Langseth, M.G., Lachenbruch, A.H., and Marshall, B.V., 1990, Geothermal observations in the Arctic region, *in* Grantz, A., Johnson, L., and Sweeney, J.F., eds., The Arctic Ocean region: Boulder, Colorado, Geological Society of America, Geology of North America, v. L, p. 133–153.

Lawver, L., and Baggeroer, A., 1983, A note on the age of the Canada Basin, *in* Mull, C.G., and Reed, K.M., eds., The origin of the Arctic Ocean (Canada Basin), Proceedings, 1981 Mini-symposium: Journal of the Alaska Geological Society, v. 2, p. 57–66.

Lawver, L., and Scotese, C.R., 1990, A review of tectonic models for the evolution of the Canada Basin, *in* Grantz, A., Johnson, L., and Sweeney, J.F., eds., The Arctic Ocean region: Boulder, Colorado, Geological Society of America, Geology of North America, v. L, p. 593–618.

Laxon, S.W., and McAdoo, D.L., 1997, Polar marine gravity fields from ERS1: Proceedings of the 3rd Earth Resources Satellite Symposium, March 17–21, Florence, Italy, p. 187–195.

Manspeizer, W., 1988, Triassic-Jurassic rifting and opening of the Atlantic: An overview, *in* Manspeizer, W., ed., Triassic-Jurassic rifting: Continental breakup and the origin of the Atlantic Ocean and passive margins: New York, Elsevier, Developments in Geotectonics 22, Part A, p. 41–79.

Mayfield, C.F., Tailleur, I.L., and Ellersieck, I., 1988, Stratigraphic, structure, and palinspastic synthesis of the western Brooks Range, northwest Alaska, *in* Gryc, G., ed., Geology and exploration of the National Petroleum Reserve in Alaska, 1974–1982: U.S. Geological Survey Professional Paper 1399, p. 143–187.

Mickey, M.B., and Haga, H., 1987, Jurassic-Neocomian biostratigraphy, North Slope, Alaska, *in* Tailleur, I.L., and Weimer, P., eds., Alaskan North Slope geology, Volume 1: Bakersfield, California, Pacific Section Society of Economic Paleontologist and Mineralogists and Alaskan Geological Society, p. 397–404, 8 plates.

Mickey, M.B., Byrnes, A.P., and Haga, H., 1998, Biostratigraphic correlation between the North Slope, Alaska and Sverdrup Basin, Canada and timing of the Canada Basin opening: Proceedings American Association of Petroleum Geologists Annual Convention, Salt Lake City, Utah, Extended Abstracts, v. 2, A462, 4 p.

Newman, G.W., Mull, C.G., and Watkins, N.D., 1979, Northern Alaska paleomagnetism, plate rotation, and tectonics, *in* Sisson, A., ed., The relationship of plate tectonics to Alaska geology and resources: Anchorage, Alaska Geological Society, p. C1–C7.

Paleo Services Limited, 1975, Examination of fossils, petrographic thin sections and analytical reports from wells drilled in northern Canada: Geological Survey of Canada Open-File Report, v. 256, 55 p.

Panarctic Oils Ltd., 1987a, Geological Series 1: Index of formation etc. tops as picked by Panarctic, publ. Panarctic Oils Ltd., Calgary, Canada, 2 pieces., available through Geological Survey of Canada.

Panarctic Oils Ltd., 1987b, Geological Series 2: Blackline prefold prints of Panarctic "well history" logs, Arctic Canada: publ. Panarctic Oils Ltd., Calgary, Canada, 110 individual well sheets available through Geological Survey of Canada.

Panarctic Oils Ltd., 1987c, Geological Series 3: Index of unpublished geological reports: publ. Panarctic Oils Ltd., Calgary, Canada, 3 folders, available through Geological Survey of Canada.

Rice, M.J., 1972, List of fossils, petrographic thin-sections and analytical reports from wells drilled in northern Canada available for examination at the Institute of Sedimentary and Petroleum Geology: Geological Survey of Canada Open-File Report, v. 107, 146 p.

Robertson Research Canada, 1975a, Biostratigraphy of Elf Cape Norem A-80, Panarctic Hoodoo dome H-37, and Elf Jameson Bay C-31 wells, District of Franklin: Geological Survey of Canada Open-File Report, v. 245. 65 p.

Robertson Research Canada, 1975b, Paleontological reports with biostratigraphical zonation on seven wells drilled in Arctic Canada: Geological Survey of Canada Open-File Report, v. 297, 18 maps and texts.

Srivastava, S.P., Kovacs, L.C., Verba, V.V., Maschenkov, S., Pogrebitsky, Y., Levesque, S., Stark, A., Oakey, G., Verhoef, J., and McNab, R., 1994, Opening of the Canada Basin and its relation to the opening of the rest of the Arctic Basin and the North Atlantic, *in* Thurston, D.K., and Fujita, K., eds., Proceedings of the 1992 International Conference on Arctic Margins: Anchorage, Alaska, Minerals Management Service, Outer Continental Shelf Report 94-0040, p. 57.

Tailleur, I.L., 1969, Rifting speculation on the geology of Alaska's North Slope: Oil and Gas Journal, v. 67, no. 39, p. 128–130.

Tailleur, I.L., 1973, Probable rift origin of the Canada Basin, Arctic Ocean, *in* Pitcher, M.G., ed., Arctic Geology: American Association of Petroleum Geologists Memoir 19, p. 526–535.

Tailleur, I.L., and Brosgé, W.P., 1970, The tectonic history of northern Alaska, *in* Adkison, W.L., and Brosgé, M.M., eds., Proceedings of the Geological Seminar on the North Slope of Alaska: Palo Alto, California, Pacific Section, American Association of Petroleum Geologists, p. E1–E20.

Tappan, H., 1955, Foraminifera from the Arctic Slope of Alaska. 2. Jurassic Foraminifera: U. S. Geological Survey Professional Paper 236B, p. 21–90.

Taylor, P.T., Kovacs, L.C., Vogt, P.R., and Johnson, G.L., 1981, Detailed aeromagnetic investigation of the Arctic Basin, 2: Journal of Geophysical Research, v. 86, no. B7, p. 6323–6333.

Vogt, P.R., Taylor, P.T., Kovacs, L.C., and Johnson, G.L., 1982, The Canadian Basin: Aeromagnetic constraints on structure and evolution, *in* Johnson, G.L., and Sweeney, J.F., eds., Structure of the Arctic: Tectonophysics, v. 89, p. 295–336.

Whitmer, R.J., Mickey, M.B., and Haga, H., 1981a, Biostratigraphic correlation of selected test wells of the National Petroleum Reserve–Alaska: U.S. Geological Survey Open-File Report 81-1165, 89 p.

Whitmer, R.J., Haga, H., and Mickey, M.B., 1981b, Biostratigraphic report of 33 wells drilled from 1975–1981 in NPRA: U.S. Geological Survey Open-File Report 81-1166, 47 p.

Manuscript Accepted by the Society May 15, 2001.

Geological Society of America
Special Paper 360
2002

Origin of the Lisburne Hills–Herald Arch structural belt: Stratigraphic, structural, and fission-track evidence from the Cape Lisburne area, northwestern Alaska

Thomas E. Moore*
U.S. Geological Survey, MS-901, 345 Middlefield Road, Menlo Park, California 94025, USA
Trevor A. Dumitru
*Department of Geological and Environmental Sciences, Stanford University,
Stanford, California 94305, USA*
Karen E. Adams
U.S. Geological Survey, MS-901, 345 Middlefield Road, Menlo Park, California 94025, USA
Susan N. Witebsky
*Environmental Protection and Restoration Department, Stanford Linear Accelerator Center,
Stanford, California 94309, USA*
Anita G. Harris
U. S. Geological Survey, MS-926A, 12201 Sunrise Valley Drive, Reston, Virginia 20192, USA

ABSTRACT

The Lisburne Hills fold and thrust belt forms the on-land segment of an 800-km–long arcuate structural belt that also includes the offshore Herald and Wrangel Arches in the Chukchi Sea. The regional north-south trend of structures on the Lisburne Peninsula contrasts sharply with the east-west trend of structures in the Early Cretaceous Brooks Range orogen to the southeast. In the northern Lisburne Hills, our new mapping documents three northeast-plunging, northeast-vergent structural domains: (1) imbricated Lisburne Group carbonate rocks, (2) complexly folded fine-grained strata of the Etivluk Group and Kingak Shale, and (3) isoclinally folded Brookian sequence turbidites. A balanced cross section across the Lisburne domain indicates a minimum of ~65% shortening, generally comparable to estimates for Neocomian shortening within the Brooks Range. Thermal maturity and fission-track data indicate that all three domains experienced rapid thrust-induced burial and heating bracketed between 132 and 115 Ma, followed quickly by ~5 km of erosional exhumation at ca. 115 Ma.

We conclude from similarities in age, deformational style, and percent shortening that the Lisburne Hills represents a northward projection of the frontal zone of the Brooks Range orogen. Deformation appears to be too old for previous models that ascribed the Lisburne Hills to post–Brooks Range convergence between North America and Eurasia. The Lisburne Hills may have originated at a high angle to the Brooks Range proper along the eastern margin of a western salient of the Brooks Range that developed in response to westward thinning and onlap of the Ellesmerian sequence onto the Chukchi platform.

*Corresponding author. E-mail: tmoore@usgs.gov.
Data Repository item 2002074 contains additional material related to this article.

INTRODUCTION

Two major fold and thrust belts with intersecting trends are exposed in northwestern Alaska (Brosgé and Tailleur, 1970; Grantz et al., 1970) (Fig. 1). The most prominent of these is the Brooks Range orogen, which extends east-west for more than 900 km across Alaska from the Canadian border to the Chukchi Sea. The Brooks Range orogen was formed initially by north-vergent contractional deformation associated with the emplacement of ophiolitic allochthons onto the south-facing (present co-ordinates) continental margin of Arctic Alaska during the Late Jurassic and/or Early Cretaceous (Roeder and Mull, 1978). The second fold and thrust belt is the arcuate, generally east-vergent fold and thrust belt exposed on the Lisburne Peninsula. This belt

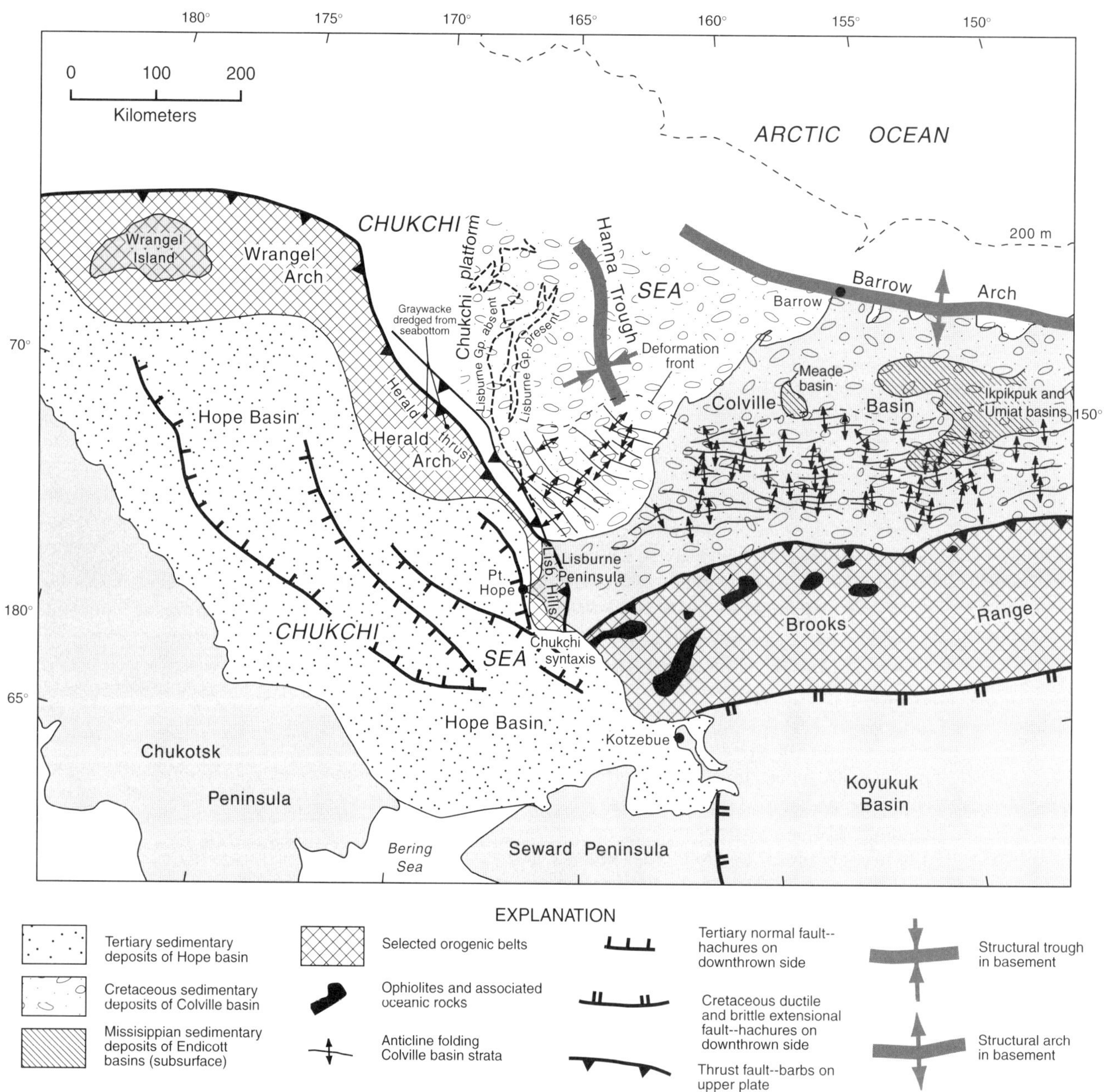

Figure 1. Map of Chukchi Sea region showing geographic and tectonic features discussed in text. Approximate northern limit of modern Arctic continental shelf is indicated by 200 m isobath. Location of onlap of Ellesmerian sequence onto Chukchi platform is approximated by westward pinch out of Lisburne Group in subsurface of Chukchi Sea (thick dashed line). Sources of data: Chapman and Sable (1960), Grantz et al. (1990a, 1990b), A. Grantz (1999, written commun.), Kirschner and Rycerski (1988), Mayfield et al. (1988), Moore et al. (1994), Sherwood et al. (1998, this volume), Tolson (1987).

extends northward for ~50 km on land and is thought to then continue offshore to the northwest as the Herald and Wrangel Arches for an additional 800 km (Grantz et al., 1975). The Lisburne Peninsula–Herald Arch–Wrangel Arch structural trend is bound on the northeast by southwest-dipping faults, including the Herald fault zone of Grantz et al. (1970, 1975, 1981), and commonly is thought to be related to crustal shortening between North America and Eurasia in the late Mesozoic or early Cenozoic (e.g., Patton and Tailleur, 1977; Grantz et al., 1991).

The nature of the intersection of the two orogenic belts, termed the *Chukchi syntaxis* (Tailleur and Brosgé, 1970), is uncertain because the intersection is offshore in the southeastern Chukchi Sea beneath younger deposits of the Hope Basin. Using single-channel seismic data from the Chukchi Sea northwest of the Lisburne Peninsula, Grantz et al. (1970) mapped foreland structures related to the Brooks Range orogen that appear to be truncated or deformed by the Herald thrust zone. Grantz et al. (1970, 1975, 1981) interpreted this relation as evidence that the Lisburne Hills–Herald Arch is a separate, younger fold and thrust belt that transects the Brooks Range orogen. On the basis of regional relations, however, Tailleur and Brosgé (1970) and Patton and Tailleur (1977) suggested that the Lisburne Peninsula-Herald Arch is the simple extension of the Brooks Range orogen that was later rotated into its present transverse trend by major oroclinal bending at the Chukchi syntaxis. If caused by oroclinal bending, the deformation would have affected the entire Bering Strait region and caused as much as 800 km of east-west shortening in the northern Alaska region. A third possibility is that the Lisburne Peninsula–Herald Arch is a continuation of the Brooks Range orogen that owes its divergent trend to Jurassic-Cretaceous paleogeographic irregularities in the region.

Key information needed to evaluate these possibilities includes stratigraphic and structural data and information on the timing of deformation in the two fold and thrust belts. Comparison of such data from the two belts would reveal similarities or differences in geologic history, style of deformation, magnitude of shortening, and time of shortening. Although there has been much recent research on these topics in the Brooks Range (e.g., Blythe et al., 1996; Christiansen and Snee, 1994; Cole et al., 1997; Fuis et al., 1997; Little et al., 1994; Moore et al., 1997; Mull et al., 1987a, 1987b, 1997; Oldow and Avé Lallemant, 1998; O'Sullivan, 1996; O'Sullivan et al., 1993, 1997; Till and Snee, 1995; Wallace, 1993; Wallace and Hanks, 1990; Wallace et al., 1997), the Lisburne Peninsula has received scant attention. Geologic information on the Lisburne Peninsula is limited to the framework studies of Collier (1906) and Martin (1970), work by Campbell (1961, 1967) in the southern Lisburne Peninsula, a few stratigraphic studies that resulted from regional coal- or petroleum-related investigations (Armstrong et al., 1971; Blome et al., 1988; Moore et al., 1984; Murchey et al., 1988; Tailleur, 1965), and very limited unpublished work by oil companies. Offshore investigations of the Herald Arch are limited to the single-channel seismic data discussed by Grantz et al. (1970), a multichannel crossing presented in Thurston and Theiss (1987),

multichannel seismic data described by Grantz and May (1988), and the two multichannel crossings of the region discussed by Klemperer et al. (this volume, Chapter 1). The geology of Wrangel Island was described by Kos'ko et al. (1993).

The purpose of this report is to present the results of a geologic investigation in the Cape Lisburne area at the northern end of the Lisburne Peninsula. This study provides a detailed view of the structural character of one location along the Lisburne Peninsula–Herald Arch–Wrangel Arch structural trend and the stratigraphic section involved in the deformation. We report thermal maturity and apatite fission-track data in an effort to determine the depth of burial and the timing of the deformation there. Although limited in scope, these data provide important information on the origin and timing of formation of the structural trend and allow development of a tectonic hypothesis for its origin. In addition, we include geologic framework observations from the Cape Lisburne area because it is perhaps the only onland exposure of source and reservoir rocks of petroliferous areas of the Chuckchi Sea to the north (e.g., Thurston and Theiss, 1987; Sherwood et al., this volume).

DEFORMATIONAL HISTORY OF THE BROOKS RANGE AND COLVILLE BASIN

The Brooks Range orogen and the Colville Basin are the dominant structural elements in northern Alaska (Fig. 1). The Colville Basin is filled with the Brookian sequence sedimentary deposits, which were shed northward from the evolving Brooks Range orogen during the Cretaceous and Tertiary. The Colville Basin deposits are underlain by the Ellesmerian sequence, a generally undeformed succession of gently south-dipping Mississippian to Lower Cretaceous rocks that record sedimentation on the inboard part of the south-facing passive margin of the Arctic Alaska terrane (Moore et al., 1994). The Ellesmerian strata are in turn underlain by deformed pre-Mississippian rocks (Moore et al., 1994).

Within the Brooks Range, Middle Jurassic ophiolitic rocks of the Angayucham terrane form the highest structural elements (Fig. 1). The ophiolites were thrust northward onto the Arctic Alaska passive-margin sequence and the thrusting resulted in detachment and northward emplacement of allochthons composed of the distal parts of the passive-margin succession onto more proximal parts of the passive margin (Mayfield et al., 1988). The thrusting resulted in a thin-skinned deformational style in the Brooks Range fold and thrust belt and shortening of as much as 400–600 km (Mull, 1982; Mayfield et al., 1988; Oldow et al., 1987). Deformation extended to deep crustal levels to the south (Till et al., 1988). High-pressure metamorphic assemblages in these rocks, coupled with thrust relations and the presence of Jurassic to Neocomian island-arc rocks south of the Brooks Range, have led to the interpretation that the Brooks Range orogen was the result of an arc-continent collision in the Late Jurassic and Early Cretaceous (e.g., Box, 1985).

Stratigraphic evidence indicates that emplacement of the

ophiolitic klippen and underlying thrust sheets of the passive-margin sequence occurred mainly in the Neocomian (Moore et al., 1994). The $^{40}Ar/^{39}Ar$ white mica ages from early metamorphic phases in the southern Brooks Range have been interpreted as indicating that contractional deformation occurred primarily 142–120 Ma and was followed by a younger episode of deformation 113–90 Ma (Christiansen and Snee, 1994; Till and Patrick, 1991; Till and Snee, 1995; Blythe et al., 1996; Gottschalk and Snee, 1998). The nature of these younger fabrics is controversial, however, and has been interpreted to reflect either (1) continued contractional deformation (Till and Patrick, 1991; Till and Snee, 1995; Gottschalk and Snee, 1998) or (2) ductile extensional deformation related to a regional episode of extensional tectonism in interior Alaska (Miller and Hudson, 1991; Little et al., 1994; Christiansen and Snee, 1994). Thus, the Albian-Cenomanian sedimentary fill in the Colville Basin may have been shed from uplift caused by either continued contractional deformation in the Brooks Range orogen or, alternatively, to uplift related to regional extensional deformation in the hinterland of the Brooks Range orogen in interior Alaska.

The Jurassic and Early Cretaceous structures of the Brooks Range orogen are deformed by younger north-vergent contractional structures, including long-wavelength folds and associated thrusts that deform middle and Late Cretaceous strata throughout the southern part of the Colville Basin (Mull, 1985) (Fig. 1). These structures have been difficult to date using stratigraphic techniques because of the absence of postdeformational strata, but insight into their age has been provided by apatite fission-track dating (O'Sullivan et al., 1993, 1997, 1998a, 1998b; Blythe et al., 1996; O'Sullivan, 1996). The fission-track data indicate that prominent, long-wavelength folds and thrusts that deform Upper Cretaceous strata in the central Colville Basin and the crustal-scale duplex at the Mount Doonerak antiform in the central Brooks Range were developed during a cooling event ca. 60 ± 4 Ma (O'Sullivan, 1996). The Mount Doonerak antiform was later reactivated ca. 24 ± 2 Ma (Blythe et al., 1996; O'Sullivan et al., 1998a). Seismic reflection and refraction data indicate that the basal detachment for the Mount Doonerak antiform is linked with thrust systems that formed the folds in the Colville Basin (Fuis et al., 1997; O'Sullivan et al., 1998a). These relations suggest that much of the modern relief of the Brooks Range formed through renewed contractional deformation ca. 60 Ma followed by renewed thrusting ca. 24 Ma (O'Sullivan et al., 1997). The Tertiary thrusting is thought to have resulted from far-field effects of north-directed subduction in southern Alaska (Grantz et al., 1991; O'Sullivan et al., 1997). The topography and structural relief of the Brooks Range decreases westward, suggesting that the amount of Tertiary uplift diminishes to the west.

DEFORMATION IN THE LISBURNE PENINSULA AND HERALD ARCH

The Lisburne Hills fold and thrust belt is delineated by uplifted pre-Cretaceous rocks ~50 km northwest of the western termination of the Brooks Range (Figs. 1 and 2). The fold and thrust belt consists of west-dipping imbricate thrust faults and associated folds that have an average easterly vergence. Folds and thrusts are generally north-northeast trending in the southernmost part of the Lisburne Peninsula, and appear to wrap around toward the northwest trend exhibited in the northernmost part of the peninsula (Campbell, 1967). In the southernmost part of the peninsula, the thrust front on the east is marked by emplacement of Ellesmerian sequence rocks onto clastic rocks that are as young as Early Cretaceous (Campbell, 1967). Progressively older parts of the Ellesmerian sequence and underlying lower Paleozoic rocks are exposed to the west, probably reflecting a progressively deeper basal detachment to the west. A structural relief of at least 2 km is evident from map relations on the peninsula.

The Herald Arch, the offshore continuation of the Lisburne Hills fold and thrust belt, is a northwest-trending belt of perched acoustic basement in seismic profiles (Grantz et al., 1975). The northeastern flank of the arch is bounded by the Herald fault zone, along which faults vary in dip on seismic profiles from near vertical to as shallow as 15° to the southwest (Grantz et al., 1970). Middle Cretaceous strata of the Brookian sequence are deformed in advance of the thrust front and have been interpreted to be overthrust more than 15 km by older rocks along the thrust front (Grantz and May, 1988). The basal detachment for the thrust system can be traced seismically from its leading edge in middle Cretaceous strata as much as 60 km downdip (Grantz and May, 1988). Rocks that, on the basis of compositional data (Platt, 1975), are possibly as old as early Paleozoic, were dredged from the crest of the Herald Arch, indicating a structural relief of more than 3 km (Sherwood et al., 1998, Plate 13.1). Strata of the Tertiary Hope Basin unconformably onlap the southwestern flank of the arch, indicating that the Hope Basin postdates formation of the arch. The Hope Basin is an extensional basin with as much as 6 km of sedimentary fill that was initially formed in the Paleogene and underwent significant subsidence in the Miocene (Tolson, 1987). This basin covers the hinterland region of the Lisburne Hills fold and thrust belt, indicating that the contractional orogen was deformed by younger extensional tectonism in the Tertiary.

The age of the thrusting in the Lisburne Hills and Herald Arch is only loosely defined and is controversial. Campbell (1967) and Grantz et al. (1970) reported that rocks as young as Albian are deformed by contractional structures, whereas Mull (1985) argued that the uplift already existed by the Albian. An Albian or older age for thrusting was inferred from paleocurrent, grain size, and facies data that indicate that sediments were shed eastward into the Colville Basin from a western source in the vicinity of the present Lisburne Peninsula and Herald Arch during the Albian (e.g., Thurston and Theiss, 1987, Fig. 20; Sherwood et al., 1998, Plate 13.2). The minimum age of thrusting is bracketed by the Paleogene strata of the Hope Basin that unconformably overlie deformed rocks of the fold belt south of Point Hope in the Chukchi Sea (Tolson, 1987).

GEOLOGY OF THE CAPE LISBURNE AREA

We undertook detailed mapping and sampling of a 30 km^2 area at Cape Lisburne, the northernmost point on the Lisburne Peninsula, for the purpose of characterizing the stratigraphy, structure, and timing of uplift of the Lisburne Hills fold and thrust belt. Our study area (Figs. 2 and 3) is located at the landward projection of the Herald thrust zone on the Lisburne Peninsula. This location afforded an opportunity to study the structures in the Ellesmerian and Brookian sequences and to determine the nature of the Herald thrust zone. The mapping was accomplished on foot over a period of 12 days from the Cape Lisburne Long-Range Radar Site.

Stratigraphy

The sedimentary strata of the Cape Lisburne area are divisible into the three regionally extensive sequences found elsewhere in northern Alaska: pre-Mississippian rocks, the Ellesmerian sequence, and the Brookian sequence (Moore et al., 1994) (Fig. 4). Pre-Mississippian rocks are not exposed in our study area, but are present in seacliff exposures in the central Lisburne Peninsula, ~20 km south of our map area (Fig. 2). Regionally, the Ellesmerian sequence consists of northerly derived quartzose and carbonate-rich strata that were deposited in a basin bounded by uplifted pre-Mississippian rocks on the north along the Barrow Arch and on the west along the Chukchi platform (Fig. 1). The Brookian sequence consists of lithic strata derived from the Brooks Range orogen to the south.

Pre-Mississippian rocks. Martin (1970) informally named pre-Mississippian strata on the Lisburne Peninsula the Iviagik Group. Grantz et al. (1983) recognized two units in the Iviagik, including (1) slaty black argillite and siliceous shale with minor siltstone and chert and (2) lithic, sand-rich turbidite deposits. The argillite and shale unit is a thin- to medium-bedded basinal succession that has an exposed thickness of >50 m. Sparse yellow-orange beds in the unit may be bentonites indicative of a tuffaceous contribution to the basin. The unit has yielded a diverse Middle Ordovician graptolite fauna (Grantz et al., 1983).

The lithic turbidite unit consists of channelized, thick-bedded, medium-grained sandstone; local pebbly sandstone; and minor siltstone. The sandstone contains abundant carbonate and clastic sedimentary lithic fragments, detrital white mica, and feldspar, indicating a dominantly sedimentary and granitic source terrain. The unit has a minimum thickness of ~300 m but is probably significantly thicker. Intercalated calcareous mudstone in the unit yielded Early Silurian (middle Llandovery to early Wenlockian) conodonts and graptolites (Grantz et al., 1983).

The previously unpublished fossil data reported by Grantz et al. (1983) from the Iviagik Group are in Tables DR1 and DR2[1] and Figure 5.

---

[1]GSA Data Repository item 2002074, Tables DR1 and DR2 and Figure DR1, is available on request from Documents Secretary, GSA, P.O. Box 9140, Boulder, CO 80301-9140, USA, editing@geosociety.org, or at www.geosociety.org/pubs/ft2002.htm, and on the CD-ROM accompanying this volume.

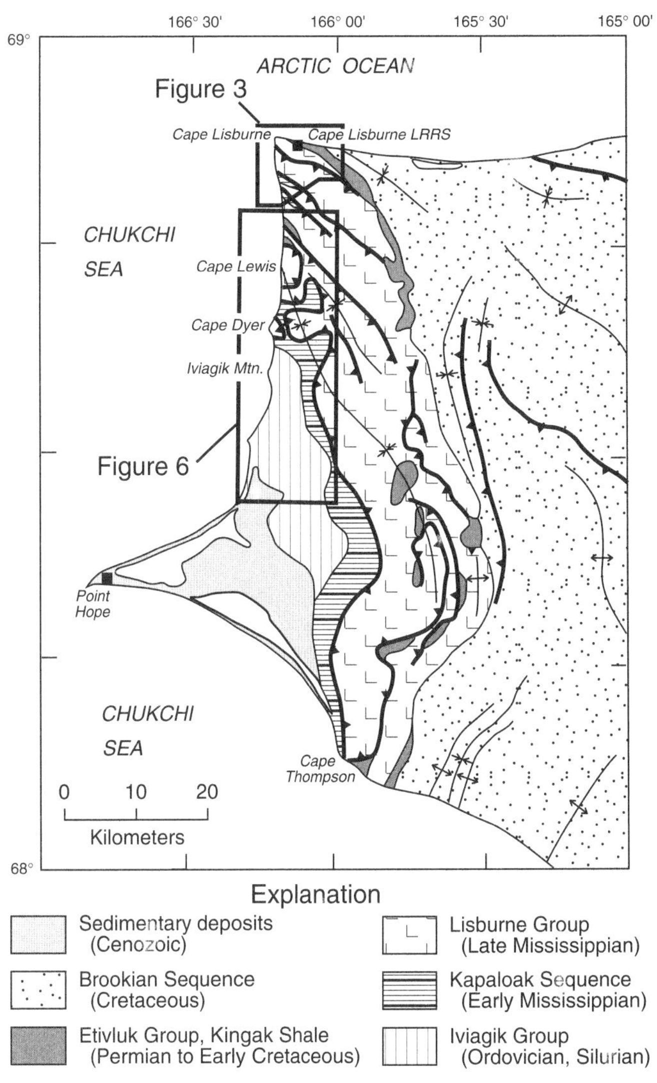

Explanation

Figure 2. Generalized geologic map of Lisburne Peninsula. Sources of data: K.J. Bird (1994, written commun.), C.G. Mull (1994, written commun.), I.L. Tailleur (1994, written commun.). LRRS is long-range radar site.

Ellesmerian sequence. The Ellesmerian sequence on the Lisburne Peninsula consists of strata assigned to the Endicott, Lisburne, and Etivluk Groups (Mayfield et al., 1988, Fig. 7.7). The pebble shale unit and/or Kingak Shale may also be present.

Endicott Group. The basal unit of the Ellesmerian sequence in the Lisburne Peninsula is a clastic unit informally named the Kapaloak sequence (Moore et al., 1984). The Kapaloak sequence, which occupies the stratigraphic position of the Kekiktuk Conglomerate elsewhere in the Arctic Alaska terrane, overlies with angular unconformity the Iviagik Group and is more than 600 m thick where measured in the central Lisburne Peninsula (Moore et al., 1984). There, it consists of interfingering thin- and medium-bedded marine and nonmarine sandstone, siltstone, and shale, and contains significant amounts of coal in its upper part (Tailleur, 1965; Moore et al., 1984). Plant fossils

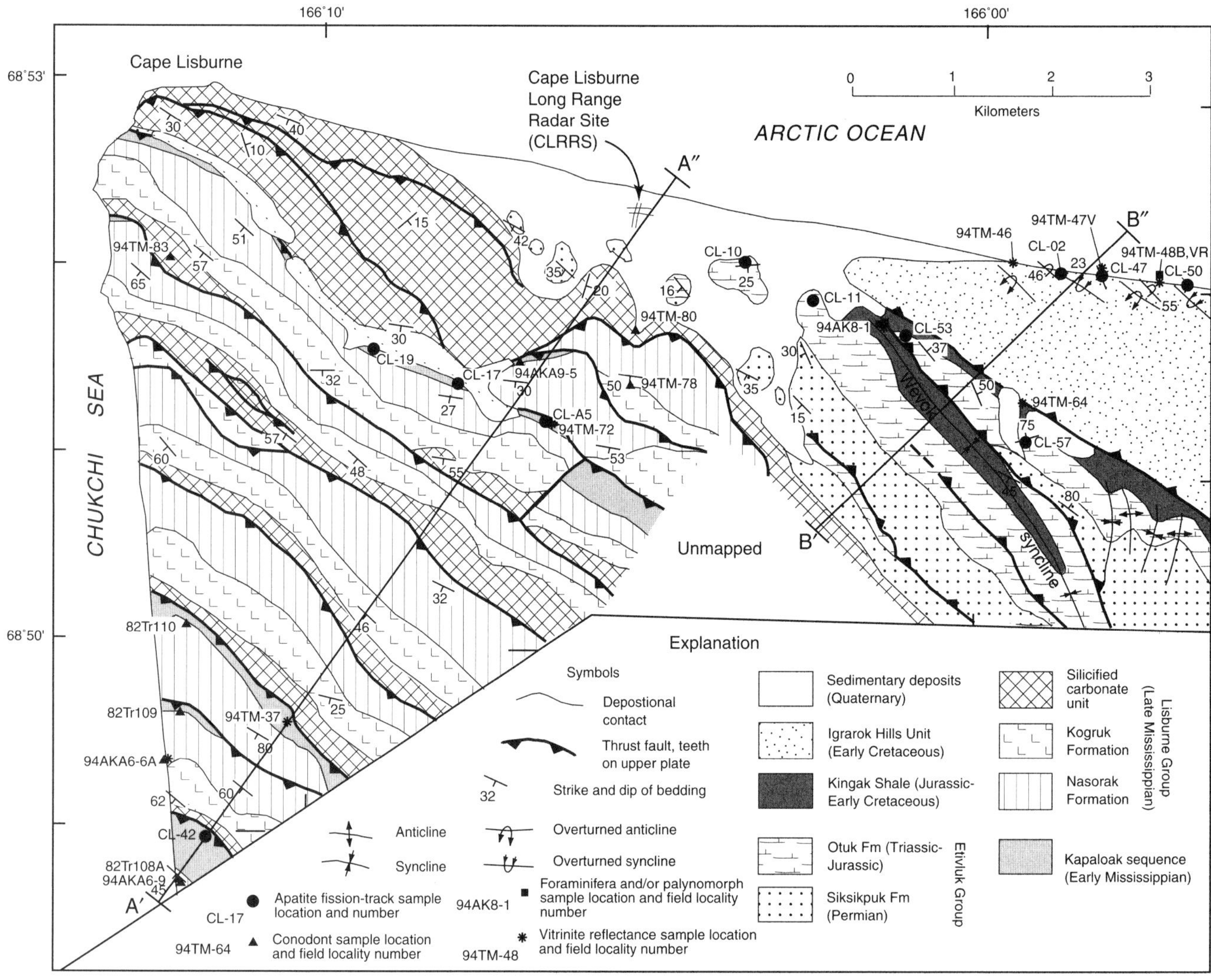

Figure 3. Geologic map of Cape Lisburne area. See Figure 12 for sections A′–A″ and B′–B″.

in the Kapaloak sequence were considered by S.H. Mamay (*in* Tailleur, 1965) to be Early Mississippian and marine fossils from near its top may be as young as Late Mississippian (Campbell, 1967). Sandstone in the Kapaloak sequence consists largely of quartz and chert and is similar to other sandstones in the Ellesmerian sequence in northern Alaska; pebble conglomerate at the base of the unit consists of chert and locally derived clasts of argillite.

In our study area, only ~50 m of the upper part of the Kapaloak sequence is exposed at the base of thrust imbricates; the complete thickness of the unit in our study area is uncertain. Where exposed, the unit consists of deformed dark-brown carbonaceous shale and siltstone and interbedded sandstone and local coal. Most of the strata contain evidence of marine deposition such as wave ripples, bioturbation, and minor carbonate, but channelized upward-thinning and fining sequences of sandstone

to 2 m thick and capped by coal indicate local nonmarine conditions (Fig. 6A). The strata were probably deposited in a deltaic region with abundant distributary bays, estuaries, lagoons, and tidal flats.

Lisburne Group. The Lisburne Group, primarily a platform carbonate succession that is widespread in northern Alaska, was named by Schrader (1902) for its widespread exposure on the Lisburne Peninsula. Campbell (1967) subdivided the Lisburne Group of the southern Lisburne Peninsula into three units: from bottom to top, these are the Nasorak, Kogruk, and Tupik Formations. Armstrong et al. (1971) extended the Nasorak and Kogruk Formations into the northern part of the peninsula. Although consisting primarily of chert and carbonate like the Tupik Formation in the southern part of the peninsula, the uppermost unit of the Lisburne Group in the Cape Lisburne area, the silicified carbonate unit in Figure 4, has significant genetic

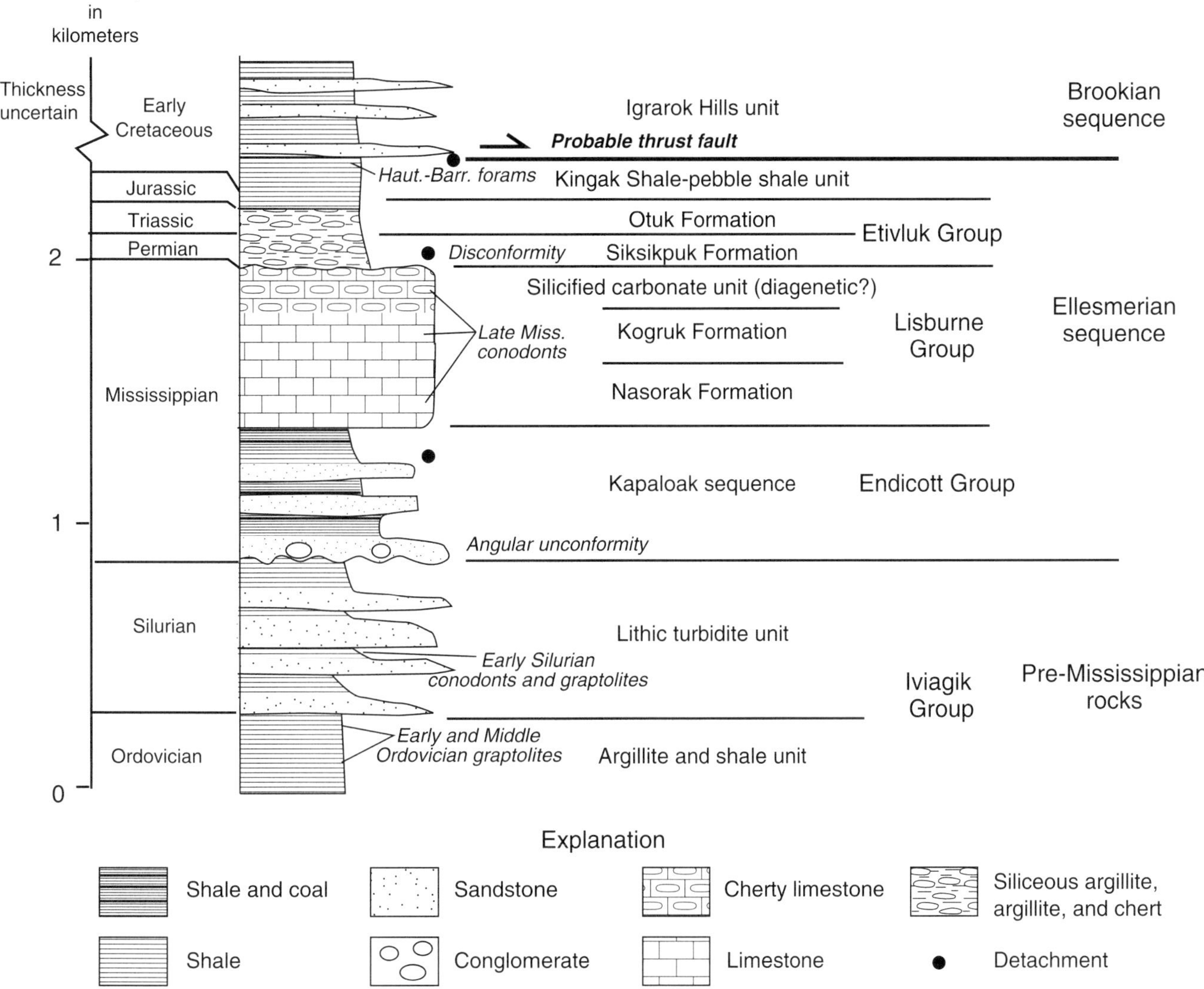

Figure 4. Stratigraphy of Cape Lisburne area, showing stratigraphic positions of fossil collections reported herein. Solid dots indicate positions of structural detachments. Thicknesses are approximate. Haut.-Barr.—Hauterivian to Barremian foraminifers.

differences from the Tupik Formation described by Campbell (1967) and may have little significance as a stratigraphic unit.

The Nasorak Formation is ~500 m thick in the southern Lisburne Peninsula (Campbell, 1967) and ~250 m thick in our study area. It overlies the Kapaloak sequence on a sharp contact and consists of thin- to medium-bedded, light yellowish-brown-weathering argillaceous limestone and minor interbedded calcareous shale. Bedding in the unit is generally laterally extensive and rhythmic. Fetid, medium- to dark-gray crinoid-brachiopod-bryozoan wackestones and packstones are the dominant lithology of the Nasorak Formation, although grainstone, including oolitic grainstone, is locally present. The Nasorak contains several intervals to 10 m thick of light-gray, thick-bedded strata similar to beds in the overlying Kogruk

Formation. Solid hydrocarbon in vugs and veins was noted locally in the Nasorak Formation.

The Kogruk Formation, the most resistant unit in the study area, gradationally overlies the Nasorak Formation, and consists of mostly medium- to thick-bedded and massive, light-gray weathering carbonate rocks that weather into rubble composed of blocks and slabs. Typically, the Kogruk Formation consists of brachiopod-echinoderm-bryozoan packstone and wackestone and minor grainstone that displays cross-bedding and channelized bedforms (Fig. 6B). Locally, beds of carbonate breccia are interstratified with these lithologies (Fig. 6C). Cryptalgal laminations and coralline limestone are locally present. These rocks are commonly variably dolomitized and contain nodular to streaky zones of silicification and chert. Although Campbell

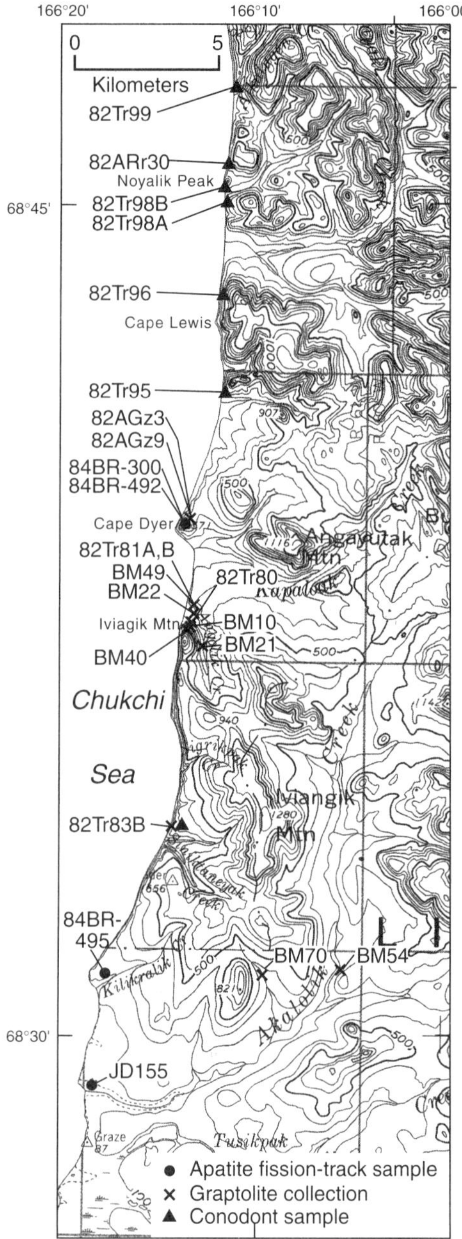

Figure 5. Fossil and fission-track localities for samples collected outside Cape Lisburne study area. Topographic base is Point Hope 1:250 000 quadrangle map.

(1967) reported a thickness of >1000 m for the unit in the southern Lisburne Peninsula, Armstrong et al. (1971) measured a thickness of only 568 m ~10 km south of our study area and an incomplete thickness of only 217 m near the mouth of Niak Creek in the southernmost part of our study area. Our mapping indicates that the Niak Creek section of Armstrong et al. (1971) represents the full thickness of the Kogruk Formation in the Cape Lisburne area. These thickness relations suggest that the Kogruk Formation thins markedly northward on the Lisburne Peninsula.

This northward thinning may reflect an oblique view of the westward onlap of Ellesmerian strata onto the Chukchi platform seen in the seismic data in the Chukchi Sea to the north (Grantz and May, 1988, Fig. 8; Sherwood et al., 1998, Fig. 13.25).

In the southern Lisburne Peninsula, the Tupik Formation gradationally overlies the Kogruk Formation, and consists of thin-bedded dark-gray chert, dolostone, and clay mudstone. Lithologies at the stratigraphic position of the Tupik Formation in the Cape Lisburne area locally include thin-bedded dark-gray to black, spiculitic chert and dark-gray fine-grained dolomite, but in most places consist of thick intervals of massive, silicified limestone. Beds of very light gray to white porcelaneous chert are commonly present in the unit and this lithology appears to increase in abundance upward in the section. Locally, medium- to thick-bedded, partly silicified limestone similar to that found in the Kogruk Formation is interbedded with the chert and silicified limestone. Spectacular coarse-grained silicified limestone breccia deposits are also present near the top of the silicified carbonate unit. In some places, the breccia grades laterally into adjacent thick-bedded, silicified limestone (Fig. 6D). The breccias do not display any deformational fabric and are not obviously associated with faults. Oncolitic clasts were observed in the breccia at one location ~1 km east of Cape Lisburne. The silicified carbonate unit is ~100 m thick.

The thin, laterally continuous bedding, argillaceous character, and abundant fauna present in the Nasorak Formation suggest that it was deposited below wave base on a carbonate platform under open-marine conditions, whereas cross-beds and channelized structures in the Kogruk Formation suggest that it was deposited in shallow-water settings. The Tupik Formation is known to have been deposited in deep-water environments (e.g., Moore et al., 1994), but most of the silicified carbonate unit that forms the uppermost part of the Lisburne Group in the Cape Lisburne area lacks evidence of deposition in deep-water environments (e.g., thin, rhythmic bedding, radiolarian chert, interbedded argillite). The top of the Lisburne Group throughout northern Alaska is marked by a widespread disconformity that separates Mississippian or Early or Middle Pennsylvanian strata from overlying Permian strata (Mull et al., 1982; Moore et al., 1994; Dumoulin et al., 1997), and a prominent unconformity beneath Upper Permian strata has been recognized in the subsurface of the north Chukchi Sea (Thurston and Theiss, 1987). Various amounts of the upper part of the Lisburne are thought to have been removed along the unconformity throughout northern Alaska. In the Cape Lisburne area, the silicified carbonate unit at the top of the Lisburne Group consists in part of heavily silicified thin- to thick-bedded carbonate rocks, some of which have prediagenetic sedimentary features (e.g., thick bedding, cryptalgal lamination) and are similar in age to strata in the underlying Kogruk Formation (see following). Coarse-grained breccias that transect bedding are common in the silicified carbonate unit and are here interpreted as karst deposits. These relations suggest that part or all of the silicified carbonate unit may represent diagenetic alteration at the top of the Lisburne plat-

Figure 6. Photographs of lower part of Ellesmerian sequence. A: Wave-rippled sandstone interval overlain by carbonaceous shale and coal, Kapaloak sequence. B: Lenticular grainstone beds in Kogruk Formation (Lisburne Group). Note channel margin (cm) eroded into underlying crossbedded grainstone bed (xbs). C: Interbedded thick-bedded carbonate breccia (br) and grainstone (gr) beds in Kogruk Formation. Stratigraphic top is to left. D: Silicified breccia (br) in silicified carbonate unit near top of Lisburne Group. Breccia grades laterally into thick-bedded silicified limestone at right.

form after a period of exposure and erosion and prior to platform drowning in the Permian. Differences in the amount of section removed by erosion along the length of the Lisburne Peninsula may have contributed to the strong northward thinning observed in the underlying Kogruk Formation and the limited age range of strata present in the Lisburne in the Cape Lisburne area (see following). It is uncertain whether this erosion removed all or part of the deep-water deposits of the Tupik Formation in the Cape Lisburne area or whether the formation was ever deposited in the Cape Lisburne area.

Armstrong et al. (1971) concluded, on the basis of corals and foraminifers, that the Nasorak and Kogruk Formations on the Lisburne Peninsula were deposited during the Meramecian and Chesterian Stages of the Late Mississippian. A sparse megafauna recovered from the Tupik Formation in the southern Lisburne Peninsula also indicates a Late Mississippian age for that unit (Campbell, 1967). Campbell (1967) and Dutro (1987)

suggested that the Pennsylvanian might be represented in beds at the top of the Tupik Formation, although they reported no Pennsylvanian fossils.

During the course of this study, we collected seven samples of the Lisburne Group for conodonts. The results are presented in Table DR1 (see footnote 1), along with nine previously unpublished samples collected by I.L. Tailleur from Lisburne strata primarily in the northern part of the Lisburne Peninsula. These data indicate that the Nasorak Formation is late Meramecian age, whereas the Kogruk Formation is late Meramecian to early Chesterian (Fig. 7). One sample collected from the middle part of the silicified carbonate unit is also Chesterian. These results suggest that the Lisburne Group as exposed on the northern Lisburne Peninsula was deposited during a time span of ~10 m.y. during the Late Mississippian (Fig. 7).

Etivluk Group. Permian, Triassic, and Jurassic rocks in the Cape Lisburne area consist of poorly exposed shale-rich strata

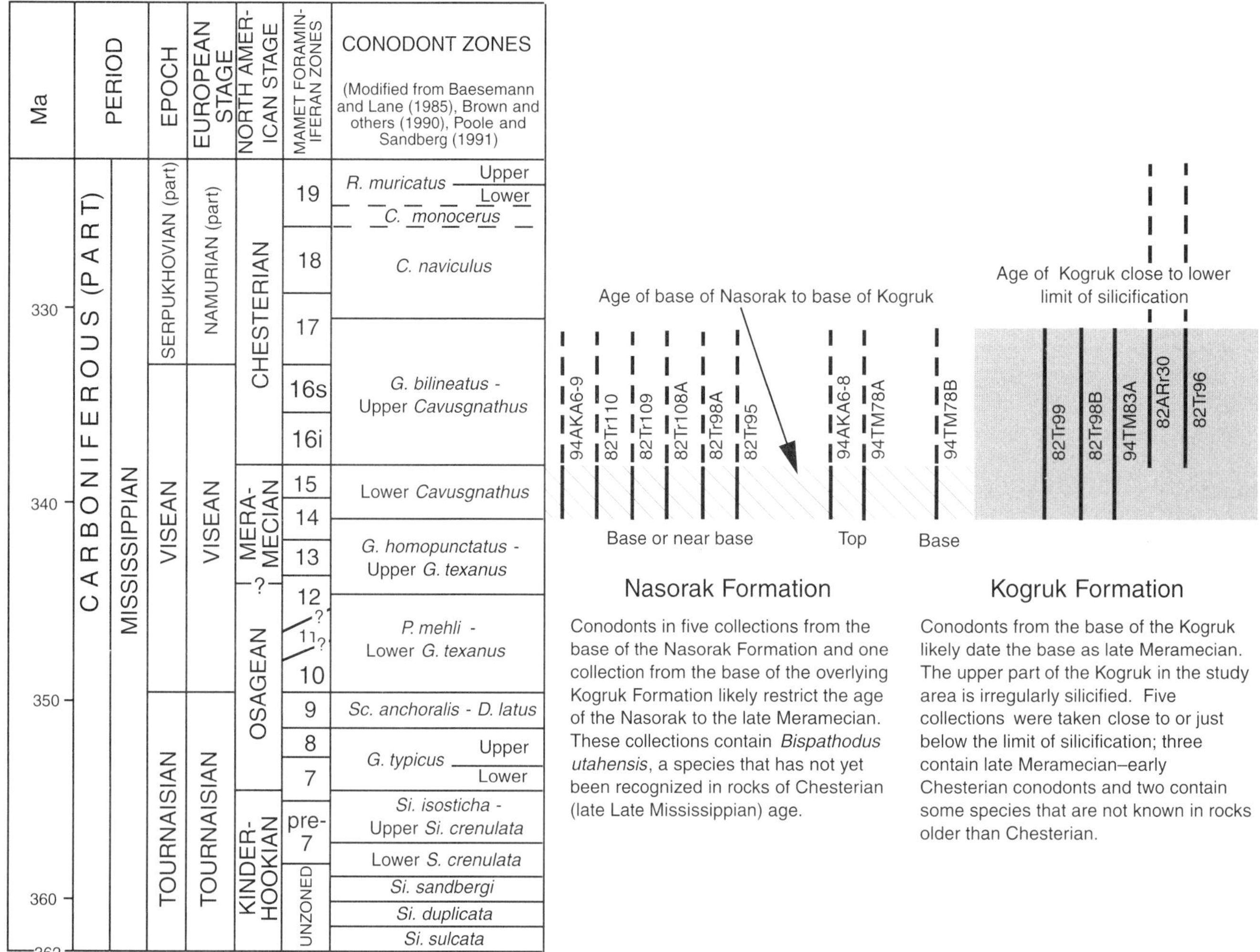

Conodonts in five collections from the base of the Nasorak Formation and one collection from the base of the overlying Kogruk Formation likely restrict the age of the Nasorak to the late Meramecian. These collections contain *Bispathodus utahensis*, a species that has not yet been recognized in rocks of Chesterian (late Late Mississippian) age.

Conodonts from the base of the Kogruk likely date the base as late Meramecian. The upper part of the Kogruk in the study area is irregularly silicified. Five collections were taken close to or just below the limit of silicification; three contain late Meramecian–early Chesterian conodonts and two contain some species that are not known in rocks older than Chesterian.

Figure 7. Chart showing age range of biostratigraphically useful conodont samples from Lisburne Group collected in Point Hope D-2 and C-2 quadrangles by T.E. Moore (TM) and K.E. Adams (KA) in 1994 and by I.L. Tailleur (Tr) and B. Reed in 1982 (see Fig. 5 for locations). Range bar for each sample is based on conodonts within that sample. Solid bar indicates age assignment is fairly certain; dashed range bar indicates age assignment or extension of age range is uncertain; diagonal pattern represents age range of base of Nasorak into base of Kogruk Formation using age data from all samples collected in that stratigraphic interval in study area; shaded pattern is age range from conodont samples collected near or just below lower limit of silicification in Kogruk Formation. Ages (Ma) from Harland et al. (1990); Devonian-Carboniferous boundary from Tucker et al. (1998).

that have been assigned to the Siksikpuk (Permian) and Otuk Formations (Triassic and Jurassic) (Mayfield et al., 1988). In the study area, the Siksikpuk consists, from bottom to top, of (1) yellowish-tan-weathering fossiliferous calcareous siltstone, (2) dark-greenish-gray bioturbated silty shale, (3) thin-bedded wispy laminated argillaceous chert and siliceous shale, and (4) dark-gray silty shale. The basal contact of the Siksikpuk Formation on the Lisburne Group was not observed in the study area. Of the four Siksikpuk members, only the chert and siliceous shale member is commonly exposed. It consists of ~8–10 m of rhythmically bedded, siliceous argillite with argillitic partings and abundant evidence of bioturbation. Because of structural disruption of the shale units and cover, the total thickness of the Siksikpuk Formation in the Cape Lisburne

area is uncertain, but probably is ~110 m. Early Permian megafossils and radiolarians were reported from the Siksikpuk Formation in the southern Lisburne Peninsula (Campbell, 1967; Murchey et al., 1988). The stratigraphic succession in the Siksikpuk Formation in the Cape Lisburne area is similar to that reported from the Endicott Mountains allochthon in the central Brooks Range (Mull et al., 1982). There, the Siksikpuk Formation is a transgressive unit that was deposited in a shelf environment. The basal member consists of storm deposits that give way upward to suspension sedimentation with sparse storm and biogenic deposits (Adams et al., 1997).

Blome et al. (1988) described the three lowest members of the Triassic and Jurassic Otuk Formation in three partial measured sections located within 7 km of our field area. From base to top, these members are (1) the shale member, consisting of black, sooty calcareous shale; (2) the chert member, consisting of thin-bedded black silicified mudstone and calcareous mudstone with an increasing abundance of limestone upward in the section; and (3) the limestone member, consisting of yellow-brown-weathering, partly silicified limestone and minor shale with coquina-like beds of *Monotis* pelecypods. Blome et al. (1988) concluded that the papery black shale of the Jurassic Blankenship Member, the highest member of the Otuk Formation, is represented instead by interbedded shale and chert.

In our study area, the Otuk Formation is found mostly in rubbly exposures that display the lithologic characteristics of the sections described by Blome et al. (1988). Throughout the map area, however, we found a previously unreported sandstone member between the shale and chert members of Blome et al. (1988). The sandstone is grayish-brown, very fine to fine-grained, massive, and ~5 m thick. *Monotis* was observed in the sandstone, suggesting a Late Triassic age. The sandstone is lithic, consisting of ~40% monocrystalline quartz, 20% altered feldspar, and 40% lithic fragments. The lithic grains, which form pseudomatrix in some thin sections possibly due in part to pervasive bioturbation, are dominated by argillitic grains, but also include chert, lithic siltstone, granitic rock fragments, and carbonate grains. Grains of opaque minerals, white mica, biotite, chlorite, and tourmaline are also present. A predominantly sedimentary provenance is inferred for this sandstone. This sandstone is notable because (1) sandstone is unusual in the Otuk Formation, a possible source rock for hydrocarbons on the North Slope and (2) sandstones that contain a high percentage of lithic grains other than polycrystalline quartz (chert) are unusual in the Ellesmerian sequence throughout northern Alaska. The Otuk Formation is thought to represent condensed sedimentation in

an outer shelf environment (Moore et al., 1994). The granitic rock fragments suggest that the lithic sandstone was shed from a source region in the Chukchi platform, which is northwest of the Cape Lisburne area (K.W. Sherwood, 2000, written commun.).

Kingak Shale–pebble shale unit, undifferentiated. A thick unit of poorly exposed, highly deformed dark-gray to black silty shale overlies the Otuk Formation east of Cape Lisburne. The shale is poorly indurated and fissile and contains rare thin beds of very fine grained quartzose sandstone. Near its contact with overlying Brookian deposits, the shale has a very soft black, sooty, organic-rich aspect, although this lithology may have resulted from milling by a fault that is inferred to occupy this position. In less deformed parts of the unit, bioturbation is apparent. A sample collected near the middle of the unit yielded Hauterivian to Barremian foraminifera, probable Neocomian palynomorphs (Table 1), and rounded, frosted quartz grains. The age and presence of floating frosted quartz grains in black shale suggest correlation with the Hauterivian and Barremian pebble shale unit, the uppermost unit of the Ellesmerian sequence, as well as the upper part of the Kingak Shale. Although the pebble shale unit has not previously been reported from the Lisburne Peninsula, it was found in the Ipewik River area, ~50 km east of our study area (Mull et al., 1999), and is a widespread unit in the North Slope subsurface. In comparison to the pebble shale unit in the western North Slope, the shale unit in our study area is thicker (~100–150 m versus 20 m in the Ipewik River area), it lacks bentonite, which is common in the pebble shale in the Ipewik River area, and it contains a total organic carbon (TOC) content of 1.18%–1.31% (P. Lillis, 1999, written commun.), generally lower than the TOC of the pebble shale unit in the western North Slope (typically >3%; C.G. Mull, 1999, personal commun.). In the northernmost parts of the North Slope, where the

TABLE 1. ANALYSIS OF FORAMINIFERA AND PALYNOMORPH SAMPLES COLLECTED FROM CAPE LISBURNE STUDY AREA

Sample number	Latitude/longitude/location details	Stratigraphic unit and lithology	Fauna	Age	Environment
94TM-48B	68.86667°/165.95722°; mouth of unnamed creek, 0.8 km east of VABM Sharpy	Igarok Hills unit; thin-bedded turbidites	Undifferentiated bisaccates *Densosporites* spp. (reworked Paleozoic) *Microdinium opacum*	Probable Early Cretaceous	Marginal marine
94KA8-1	68.86083°/166.01667°; SE tributary to unnamed creek, 3.5 km SSE of Cape Lisburne LRRS	Kingak Shale; dark-gray to black, slightly sandy paper shale	*Ammobaculites erectus* *Arenaceous* spp. (large, coarse) *Bathysiphon granulocoelia* *Glomospira subarctica* *Haplophragmoides coronis* Undifferentiated bisaccates *Classopollis classoides* *Deltoidospora* sp. *Microdinium opacum* *Oligosphaeridium complex* (thick-wall)	Probably zone F-12 to F-13; Early Cretaceous Probable Hauterivian to Barremian	Probably middle neritic to bathyal (starved basin)

Note: See Figure 3 for location of samples. Analysis by Micropaleo Consultants, Encinitas, California.

pebble shale unit was deposited in a proximal paleogeographic position to source areas in the Barrow Arch area, its thickness, grain size, and TOC values are comparable to those we observed in the Cape Lisburne area (Magoon and Bird, 1988, Fig. 17.19; Bird, 1988a, Fig. 16.16). This suggests that the Hauterivian-Barremian strata in the Cape Lisburne area may have been deposited in a proximal position to the Chukchi platform. A shelf depositional environment for the Kingak–pebble shale unit in the Cape Lisburne area is indicated from subsurface data in the Chukchi Sea; Sherwood et al. (1998, Fig. 13.6) showed the east-facing edge of the shelf as trending southward into the eastern part of the Lisburne Peninsula.

Brookian sequence. The structurally highest unit in the study area consists of tightly folded lithic sandstone of the Brookian sequence, referred to here as the Igrarok Hills unit. Despite the folding, thick intervals of well-bedded, micaceous, dark-brown to gray lithic sandstone, siltstone, and clay shale remain intact and display abundant sedimentary structures. The intact intervals include thick sections (>30 m) that consist mainly of thin-bedded turbidites. The thin-bedded sections are

Figure 8. Photographs of Igrarok Hills unit. A: Laterally continuous thick-bedded turbidites. B: Overturned Bouma Ta-e and Tb-e turbidites.

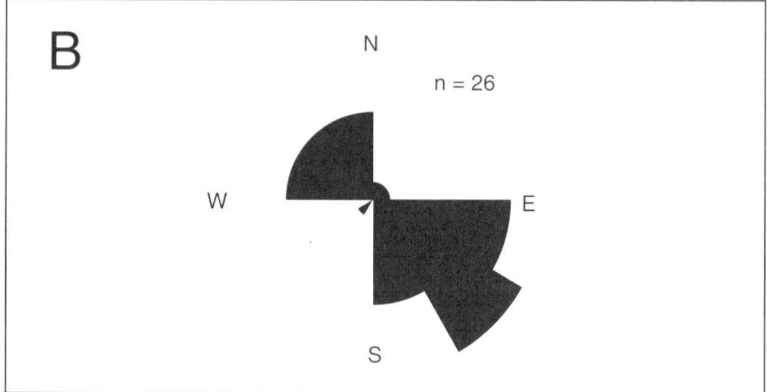

Figure 9. Paleocurrent data from Igrarok Hills unit. A: Flute casts and tool markings on basal surfaces of turbidites in Igrarok Hills unit. B: Rose diagram of paleocurrent measurements showing southeastward paleoflow. Measured features include current-ripple cross-lamination, flute casts, and tool markings.

rhythmically bedded, nonchannelized fine-grained turbidites that typically are ~10 cm thick and display partial Bouma sequences; these sections have sandstone/shale ratios of ~1:5. The thick-bedded sections consist of nonchannelized turbidites up to 2 m thick composed of medium-grained sandstone, locally with scattered granule-sized grains. Many of the thick-bedded turbidites display complete Bouma sequences, but some consist of amalgamated sandstone turbidites. The thick-bedded intervals are arranged in thickening-upward successions that are 5–8 m thick and have sandstone/shale ratios up to 10:1 (Fig. 8A). Sedimentary structures typical of turbidites, including flute casts, tool marks, convolute lamination, and climbing ripples, are very common in both the thin-bedded and thick-bedded intervals (Figs. 8B and 9A). Paleoflow, determined principally from the flute casts and tool marks, was toward the southeast (Fig. 9B). The laterally continuous bedding and thickening-upward successions suggest that deposition of the turbidites of the Igrarok Hill sequence occurred in an unconfined part of a submarine fan.

The sandstone of the Igrarok Hills unit consists of subangular grains of ~20% monocrystalline quartz, 10% feldspar (mostly K-feldspar or albite), and ~70% lithic fragments. The lithic fragments are dominantly sedimentary, consisting of sub-equal amounts of chert, argillite + siltstone, and of detrital carbonate grains. Considering that the carbonate grains (10%–35%) are the most labile of the lithic grains in the unit and that the chert might have been associated diagenetically with the carbonate strata in the source area, the composition suggests that the Igrarok Hills unit was shed from a source terrane rich in carbonate rocks.

Samples collected for microfossils from the Igrarok Hills unit were barren of foraminifers, but one sample yielded probable Early Cretaceous palynomorphs (Table 1). Because the Igrarok Hills unit is above the Hauterivian-Barremian strata of the Kingak Shale, an Early Cretaceous age may indicate that the Igrarok Hills sequence was deposited during Aptian or Albian time. Such an age might be considered likely for the Igrarok Hills sequence because its structural position and composition are similar to the Aptian or Albian Mount Kelly Tongue of the Fortress Mountain Formation, which is exposed over wide areas of the northern foothills of the Brooks Range 30–100 km east of the Lisburne Peninsula and contains abundant detrital carbonate grains (Mull, 1985; Wartes and Reifenstuhl, 1998). Alternatively, the Igrarok Hills unit may correlate with the latest Jurassic to Valanginian Okpikruak Formation of the Brooks Range. The Okpikruak Formation consists of deep-marine, locally derived lithic foredeep deposits shed from the nascent Brooks Range orogen at the time of thrusting in the Early Cretaceous (Mayfield et al., 1988; Moore et al., 1994). The Okpikruak Formation is allochthonous on underlying strata in most areas, an interpretation suggested by fission-track and vitrinite reflectance data for the Igrarok Hills unit (see following). Although sandstone in the Okpikruak Formation typically is rich in volcanic, chert, and clastic sedimentary detritus (Wilbur et al., 1987; Siok, 1989), sandstone rich in carbonate debris is also locally present (J.A. Dumoulin, 1999, oral commun.).

Structural geology

The geology in the map area is divisible into the Lisburne, Etivluk-Kingak, and Brookian structural domains, which are separated by detachments in incompetent shale units (Fig. 10). The dominant competent rock unit is different within each domain, and each domain exhibits a distinctive deformational style controlled to a large extent by the mechanical character (e.g., rigidity, thickness) of its constituent units. A fourth domain, the Iviagik, is exposed south of our study area and is inferred to underlie it.

Lisburne domain. The structural style of the Lisburne domain is characterized by moderately south dipping tabular thrust sheets that consist predominantly of Lisburne Group strata. The thrust sheets are bounded by detachments within the Kapaloak sequence and at or near the base of the Siksikpuk Formation (Fig. 11, A and B). The thrust faults cut upsection to the northeast and cause little internal deformation in the thrust sheets. Leading (hanging-wall) anticlines are mostly eroded but may be preserved at the range front along the crestal part of the Lisburne

Hills southeastward from Cape Lisburne. Although identification of anticlinal hinges is uncertain in this area because of the obscuring effects of extensive silicification and karst in the upper part of the Lisburne, local shallow dips in the highest parts of the range front and bedding attitudes in the underlying Kogruk Formation indicate the existence of a symmetrical open antiform along the northeastern margin of the domain. The large apparent thickness of the silicified carbonate unit of the Lisburne domain in this area suggests that the antiform is cut by internal faults and that an antiformal stack may be present. Trailing (footwall) synclines are either not exposed or not present in the Lisburne domain in our map area, so hanging-wall and footwall stratigraphic cutoffs cannot be matched and the amount of displacement across individual thrusts cannot be determined.

Etivluk-Kingak domain. The stratigraphic section above the detachment near the base of the Siksikpuk Formation is typically poorly exposed. Although much of this section consists of incompetent shale, the siliceous shale member of the Siksikpuk Formation and the sandstone, chert, and limestone members of the Otuk Formation form relatively competent marker units that are evident above the detachment on air photos (Fig. 10). These units define a prominent, large-scale, nearly symmetric syncline, here called the Wevok syncline, which plunges shallowly toward the northwest. Parasitic folds are abundant in the limbs of the structure and disharmonic folds are present in the Otuk Formation on the steep southwestern limb. Poor exposure prevents direct observation of the Wevok syncline and related folds in the field, but bedding attitudes suggest that it is an upright fold that trends ~N320° with a nearly vertical axial plane. An antiform, cored by a northeast-dipping thrust fault that places the Siksikpuk Formation in the hanging wall over the Otuk Formation in the footwall, forms the northeastern flank of the syncline (Figs. 3 and 12, B′-B″). On the southwestern flank of the syncline, an intact northeast-dipping section that includes the top of the Lisburne Group and the Siksikpuk and Otuk Formations is present below the detachment fault that marks the boundary of the domain (Figs. 3 and 10). Thrust imbrication within the Siksikpuk above the detachment is evident in air photos (Fig. 10).

The northeastern margin of the Etivluk-Kingak domain is defined by a prominent topographic lineament (visible in Fig. 10) at the southwestern limit of the Brookian strata. The nature of the lineament is uncertain because of a lack of exposure, but is inferred to represent a detachment in the Kingak Shale that dips to the northeast, because the structures above and below the contact are structurally discordant.

An intact section of Siksikpuk and Otuk Formations along the northeastern margin of the Etivluk-Kingak domain is deformed into a series of small-scale boxlike and chevron folds that plunge northward (Figs. 3 and 10). These folds are locally asymmetric to the west. Along strike to the southeast outside of the map area, air photos clearly show that the same section of the Otuk Formation is deformed by an apparent west-vergent series of imbricate faults (Fig. 10, location D). A western or southwestern vergence for the north-trending folds is supported by

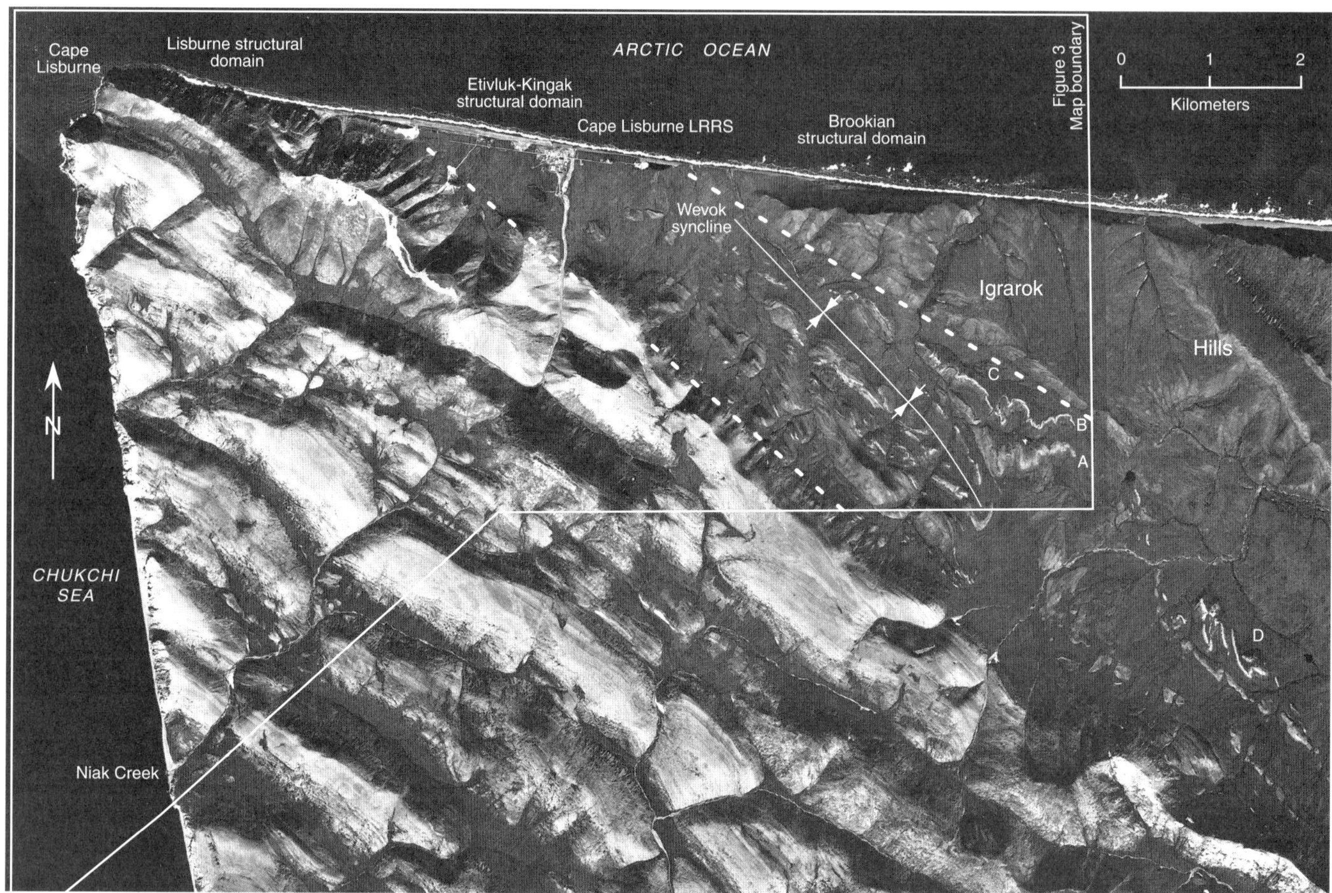

Figure 10. Infrared aerial photograph of Cape Lisburne area showing Wevok syncline and boundaries between Lisburne, Etivluk-Kingak, and Brookian structural domains (short dashed lines). LRRS is long-range radar site. Boundaries of map in Figure 3 are shown for reference. Identities of stratigraphic units in Etivluk-Kingak domain: A, Siksikpuk Formation; B, Otuk Formation (limestone member); C, Kingak Shale. Imbrication of limestone member of Otuk Formation at location D (lower right) suggests apparent hinterland-vergent structural transport.

thrusting of these rocks over the Kingak Shale on the northern limb of the Wevok syncline (Figs. 3 and 12). This relation and the difference in orientation of the north-trending structures in comparison to the dominant northwesterly structural grain in the domain suggest that the west-vergent structures may represent a relatively younger period of hinterland-vergent deformation.

Brookian domain. The Brookian structural domain consists entirely of the deformed turbidites of the Igrarok Hills unit. Although poorly exposed over much of its area, structures of the domain are well exposed in seacliffs east of the Cape Lisburne Long Range Radar Site (Fig. 3). There, the Igrarok Hills unit is deformed into a series of upright and overturned panels that extend for 200–600 m perpendicular to strike and dip moderately to the southwest (Fig. 3). Although the panels locally display discordant structures, typically they consist of intact sequences of turbidites. The boundaries between the panels are marked by faulted folds and zones as wide as 100 m with disharmonic structures. These features are interpreted to represent a wavetrain of map-scale, overturned, northwest-trending, faulted,

nearly isoclinal folds (Fig. 11C). The wavelength of the folds is ~500–1000 m. Their amplitude is difficult to estimate, but is greater than the height of the seacliffs (as much as 50 m). Subsidiary folds consistently display top-to-the-northeast structural transport, and bedding attitudes and fold axes suggest that structural transport was toward N70°E (Fig. 13). Although the amount of shortening represented by these faulted folds is difficult to estimate because of the absence of marker beds and the presence of significant zones of faulting, it seems likely that the deformation involved substantial amounts of strain.

Iviagik domain. The Iviagik domain is exposed in seacliffs south of Cape Dyer (Fig. 2). Rocks in this domain consist of largely intact sequences of the Iviagik Group and the lower part of the Kapaloak sequence, which unconformably overlies the Iviagik Group. Although not studied during this investigation, previous reconnaissance field observations suggest that these rocks may have been deformed by relatively early (Mississippian?) extensional structures prior to involvement in Mesozoic contractional deformation (Fig. 11D). The position and nature

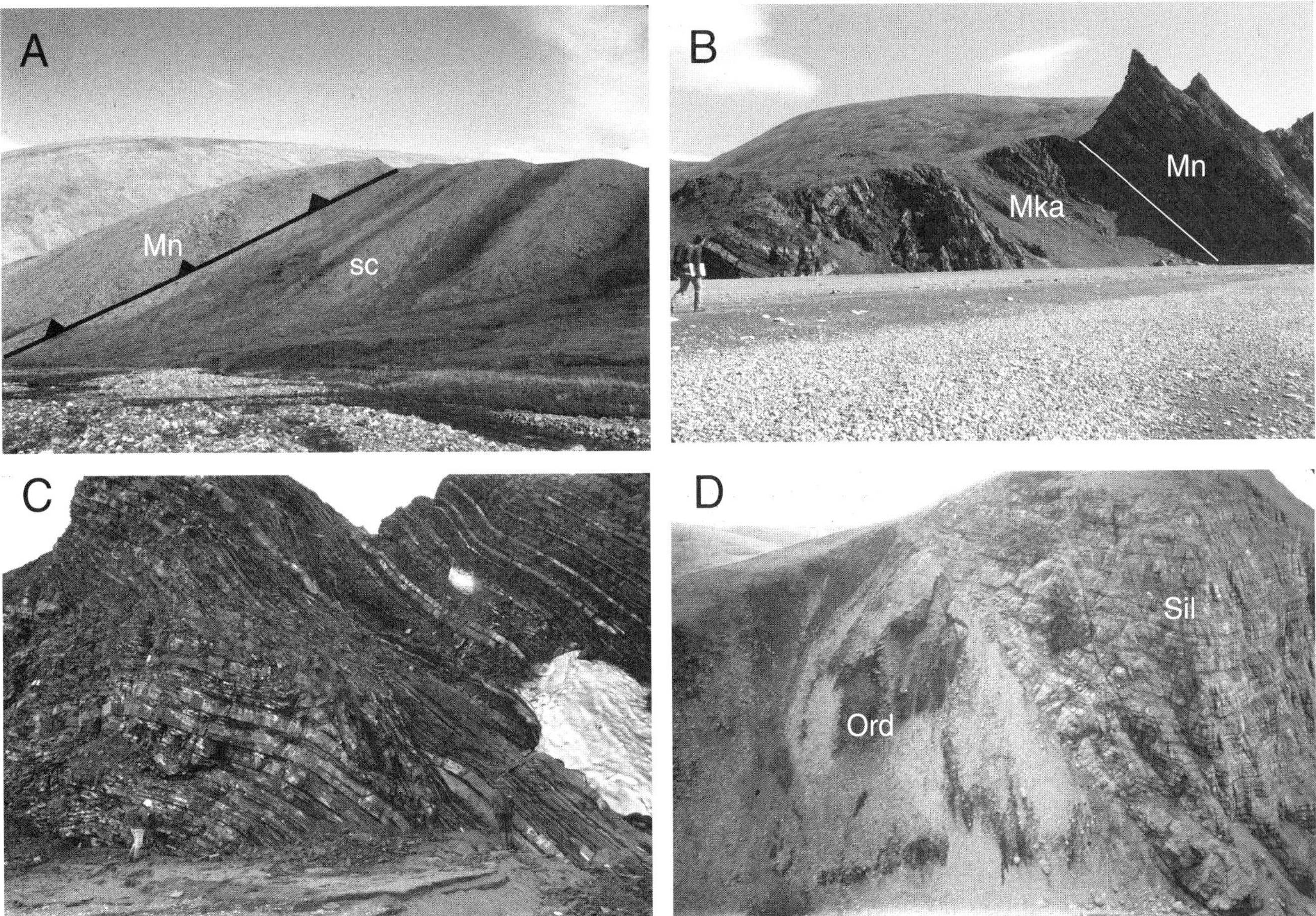

Figure 11. A: Northward view of Nasorak Formation (Mn) thrust onto silicified carbonate unit (sc). B: Southeastward view of deformed Kapaloak sequence (Mka) beneath Nasorak Formation (Mn) at base of imbricate of Lisburne Group strata. C: Nearly isoclinal syncline in Igrarok Hills unit. D: Eastward view at Cape Dyer (Figs. 3 and 5) of normal fault in Iviagik Group that drops Silurian lithic turbidites (Sil) down against Ordovician shale and argillite (Ord). Normal fault is unconformably overlain by Mississippian strata, suggesting that fault was active in Devonian or Early Mississippian.

of the contractional structures are poorly defined, but thrust panels are probably thick because of the large stratigraphic thickness of the Iviagik Group and the absence of clear repetition of its constituent units in its area of exposure.

Structural model. The interlayering of mechanically competent and incompetent stratigraphic units has played a fundamental role in determining the position of structural detachment surfaces and the size and geometry of folds in northern Alaska (Wallace, 1993; Wallace et al., 1997). Deformation of mechanically competent units (e.g., the carbonate rocks of the Lisburne Group) has resulted in shortening of the competent intervals along detachments in underlying and overlying incompetent units (generally shales). Where more than one competent interval is present in the stratigraphic section, each competent unit is shortened independently of underlying or overlying duplexes. The style of the deformation of each competent unit is dependent on its internal mechanical characteristics and may vary from that of overlying or underlying competent units. This style of de-

formation was termed a *multistoried duplex* by Wallace et al. (1997).

Duplexing at multiple structural levels is evident in the Cape Lisburne area. The carbonate rocks of the Lisburne Group form one structurally competent package that is bound above and below by shale-rich, less competent units at the base of the Siksikpuk Formation and near the top of the Kapaloak sequence, respectively. The Brookian strata and their thick-bedded sandstone (Igrarok Hills unit) form a second independent competent package, above a detachment in the incompetent Kingak Shale. A third relatively competent package consists of the upper Siksikpuk Formation and Otuk Formation, bound by detachments in the lower Siksikpuk Formation below and the Kingak Shale above. The stratigraphic positions of the detachments are shown in Figure 4.

Deformation of the competent units of the Cape Lisburne area is shown as a stacked series of stratigraphically controlled duplexes separated by detachments in shales of the Kapaloak,

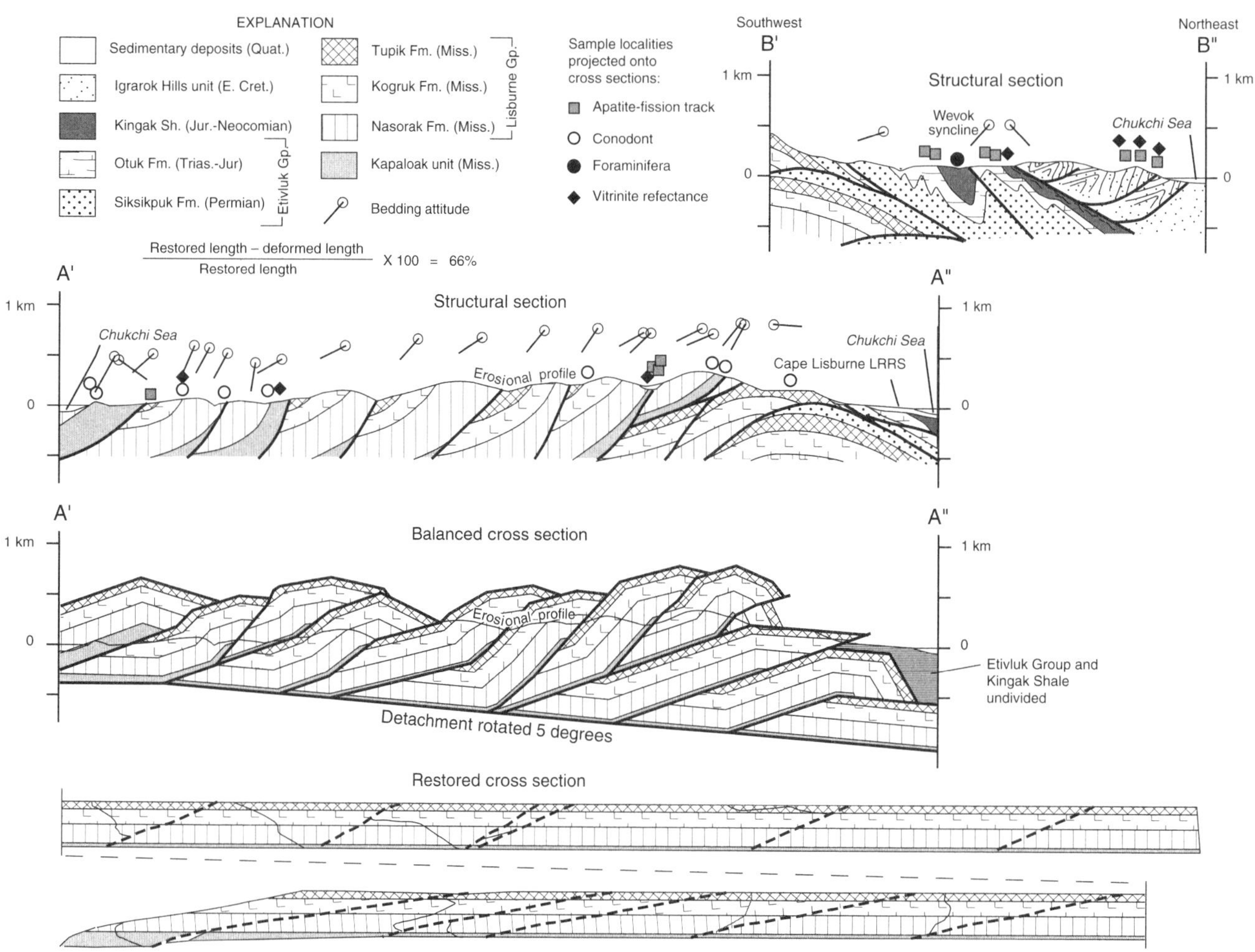

Figure 12. Cross sections of study area (location of sections shown in Fig. 3). Structural sections A′–A″ and B′–B″ show observed field relations along profile and positions of nearby sample localities projected onto profile. Homogeneous deformational style in Lisburne structural domain allowed construction of balanced cross section, A′–A″ (constructed using Geosec software program). Restoration of balanced section (bottom) indicates minimum of 13.7 km (66%) of shortening calculated at base of Lisburne Group from northeasternmost footwall ramp to trailing end of section. Model includes internal errors equivalent to 6% shortening.

Siksikpuk, and Kingak units (Fig. 12). The deformed competent intervals between the detachments compose the primary lithologic units of the three structural domains of the Cape Lisburne area. The resulting structure is tilted regionally to the northeast such that the modern erosional surface has exposed progressively older and deeper rocks toward the southwest.

Also shown in Figure 12 is a balanced cross section across the Lisburne domain. This domain was modeled for the purpose of estimating the amount of shortening in the Cape Lisburne area. For simplicity, and because the geometries of the hanging-wall anticlines are not well exposed in the field, a fault-bend fold style of deformation was assumed in the model. Imbricated Lisburne domain units are typically deformed as thrust-truncated detachment folds elsewhere in the Brooks Range, suggesting that that deformational geometry may prove to be more appropriate with further work. This assumption, however, should have little im-

pact on estimates of the minimum amount of shortening present in a regional cross section such as that shown in Figure 12.

The balanced cross section is modeled as consisting of 11 imbricates with the minimum amount of shortening required by the geologic relations. A basal detachment is shown as a single thrust in the upper part of the Kapaloak unit. Exposures of Kapaloak rocks appear in the field to be thicker in the southwestern part of the map area than to the northeast, suggesting that the detachment in the Kapaloak sequence may drop stratigraphically to the southwest, thus increasing the thickness of the allochthonous part of this unit in that direction. Recognizing these limitations, a minimum of ~13 km of shortening is calculated across the map area at the base of the Lisburne, representing ~65% shortening for the domain.

Through trial and error, the basal detachment for the model was estimated to have a regional dip of ~5° to the northeast. A

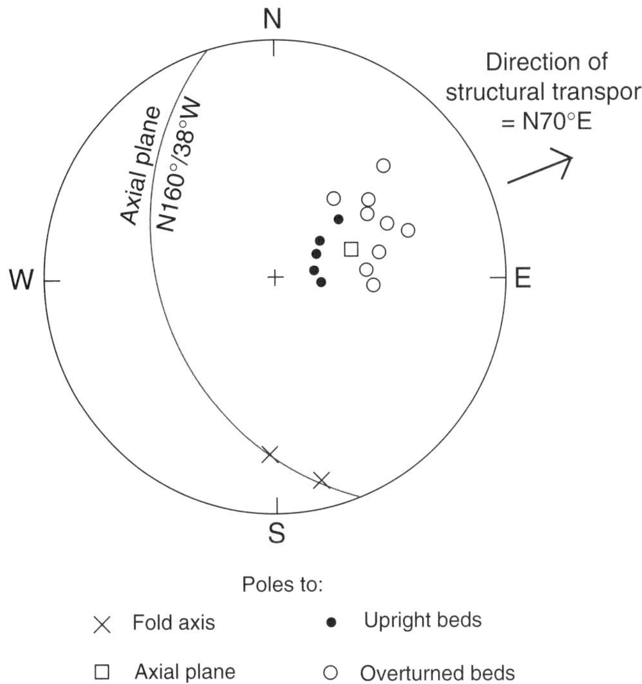

Figure 13. Lower-hemisphere plot of structural features in Brookian structural domain (Igrarok Hills unit) showing northeastward structural transport.

regional dip of this magnitude is required to explain the gradual southward downcutting into the Lisburne duplex (Fig. 2). This regional dip may have been caused by tilting due to development of an additional level of duplexing at depth beneath the Lisburne Group carbonate rocks, tilting of the orogen to the northeast after deformation, or some combination of the two. The fission-track results from the pre-Mississippian Iviagik Group from outside our study area, presented herein, indicate the same cooling history as for the Ellesmerian and Brookian rocks of the study area. This suggests that regional tilting in the Cape Lisburne area was primarily the result of development of another level of duplexing at depth in the pre-Mississippian section, rather than a younger, superimposed tilting event.

Apatite fission-track thermochronology

Apatite fission-track thermochronology is a radiogenic dating method for reconstructing the time-temperature histories of rock samples within the temperature window of 60–125 °C (e.g., Green et al., 1989a, 1989b; Miller et al., 1999; Ketcham et al., 1999, 2000; Dumitru, 2000). Assuming a nominal geothermal gradient of ~25 °C/km, this temperature window is equivalent to a depth window of ~2–5 km beneath the earth's surface. At Cape Lisburne, the method can potentially provide information on the maximum burial temperatures (T_{max}) and depths reached by various units during thrust-induced tectonic burial and the times of their subsequent exhumation back toward the earth's surface.

The ability of the fission-track system to record thermal histories relies on the interplay between two opposing processes. First, new tracks are created at an essentially constant rate by radioactive decay, so accumulated numbers of tracks may be used to calculate fission-track ages. Second, tracks are erased by thermally induced annealing at elevated subsurface temperatures. Annealing may be total, resetting the fission-track clock completely to zero age, or only partial, reducing the apparent fission-track age and shortening the lengths of individual tracks. The degree of track annealing at a given temperature has been calibrated in laboratory and borehole annealing experiments, and these data are incorporated into computer simulation models of the track formation and erasure processes (Green et al., 1989b; Gallagher, 1995; Ketcham et al., 2000). These models may be used to determine the range of time-temperature histories compatible with the observed track-length distributions and apparent ages of samples.

Track length data are especially important in defining details of the time-temperature paths. Because new fission tracks are formed continually over time, each individual track formed at a different time and has thus existed for a different fraction of the total history of the sample and has potentially been shortened a different amount. As a simple example, if a sample cools slowly and steadily, early formed tracks will be exposed to relatively high temperatures and be shortened substantially, whereas later formed tracks will be exposed to only relatively low temperatures and will remain longer. Therefore track-length distributions, generally displayed as histograms of the lengths of 50–150 individual tracks, retain valuable thermal history information that can be extracted using modeling procedures.

Fission-track samples and data. We analyzed 15 samples from the Iviagik, Lisburne, Etivluk-Kingak, and Brookian structural domains (Table 2; Figs. 3, 5, 14, 15, 16, and DR1 [see footnote 1]). All the apatite occurs as trace detrital mineral grains within samples of sandstone. Specific laboratory methods followed those used by Dumitru et al. (1995, 2001) and Ketcham et al. (1999, 2000) and are summarized in the footnote to Table 2. The observed data and modeled time-temperature histories are virtually identical among the four domains (Figs. 14, 15, 16, and DR1 [see footnote 1]). For example, Figure 14B shows sample ages plotted versus north-south location on our cross sections, and shows that ages are very similar across the entire 50 km length of the transect.

Apatite yields and data quality were good for the Iviagik, Etivluk-Kingak, and Brookian domains, which yielded apparent ages of 96 ± 5 to 104 ± 6 Ma (n = 4 samples), 105 ± 5 to 115 ± 6 Ma (n = 4), and 98 ± 6 to 103 ± 5 Ma (n = 3), respectively (Figs. 14 and 15). All of these are apparent ages that have been slightly reduced by partial annealing, as evidenced by their slightly shortened mean track lengths. To more accurately determine the true time of cooling, we used the fission-track simulation program of Ketcham et al. (2000) to model the eight samples for which at least 31 fission-track lengths could be measured. The

TABLE 2. FISSION-TRACK SAMPLE LOCALITY, COUNTING, AND AGE DATA

Sample number	Irradiation number	Latitude (°N)	Longitude (°W)	No xls	Spontaneous Rho-S	NS	Induced Rho-I	NI	$P(\chi^2)$ (%)	Dosimeter Rho-D	ND	Age ± 1σ (Ma)
Igrarok Hills Unit in Cape Lisburne Area (Neocomian)												
94CL-50	SU031-10	68°52.05′	165°56.98′	7	0.7451	113	2.4990	379	5.6	1.7710	4818	101 ± 11
94CL-47	SU031-09	68°52.11′	165°58.40′	19	0.8372	424	2.9060	1472	8.0	1.7710	4818	98 ± 6
94CL-02	SU031-01	68°52.13′	165°59.18′	25	1.1310	854	4.0600	3067	2.3	1.8580	4818	103 ± 5
Otuk Formation in Cape Lisburne Area (Triassic)												
94CL-57	SU031-12	68°51.24′	165°59.48′	16	0.5898	168	1.8190	518	37.0	1.7500	4818	109 ± 10
94CL-53	SU031-11	68°51.67′	166°00.88′	30	0.6543	666	2.0880	2125	7.1	1.7500	4818	105 ± 5
94CL-11	SU031-04	68°51.98′	166°02.54′	31	0.8844	1146	2.7780	3600	14.0	1.8360	4818	112 ± 4
94CL-10	SU031-03	68°52.12′	166°03.58′	18	1.0070	590	3.0810	1806	44.0	1.8360	4818	115 ± 5
Kapaloak Sequence in Cape Lisburne Area (Mississippian) (very poor quality samples, ages not reliable)												
94CL-A5	SU031-14	68°52.31′	166°03.39′	2	1.6760	101	6.2380	376	23.0	1.7170	4818	88 ± 10
94CL-17	SU031-05	68°51.38′	166°07.63′	1	0.5715	5	1.3716	12	—	1.8140	4818	144 ± 77
94CL-19	SU031-06	68°51.58′	166°09.18′	5	0.8728	84	5.2680	507	10.0	1.8140	4818	58 ± 7
94CL-42	SU031-08	68°48.99′	166°11.53′	2	1.5430	51	4.9920	165	21.0	1.7930	4818	106 ± 17
Basalmost Kapaloak Sequence at Cape Dyer (Mississippian)												
84BR-492	SU047-24	68°39′20″	166°13′30″	35	0.6928	852	2.3550	2896	60.0	1.7540	4390	99 ± 4
Iviagik Group at Cape Dyer (Silurian)												
84BR-300	SU046-25	68°39′20″	166°13′30″	19	0.3903	431	1.2130	1340	28.0	1.6940	4268	104 ± 6
Iviagik Group South of Kilikralik Point (Silurian)												
84BR-495	SU047-25	68°31′10″	166°17′50″	17	0.8973	437	3.1480	1533	46.0	1.7540	4390	96 ± 5
Iviagik Group Near Akalolik River (Silurian)												
94JD-155	SU031-15	68°29.12′	166°18.63′	17	1.4060	283	4.6660	939	45.0	1.7170	4818	99 ± 7

Note: Abbreviations: No xls, number of individual crystals (grains) dated; Rho-S, spontaneous track density ($\times 10^6$ tracks per cm^2); NS, number of spontaneous tracks counted; Rho-I, induced track density in external detector (muscovite) ($\times 10^6$ tracks per cm^2); NI, number of induced tracks counted; $P(\chi^2)$, χ^2 probability (Galbraith, 1981; Green, 1981); Rho-D, induced track density in external detector adjacent to dosimetry glass ($\times 10^6$ tracks per cm^2); ND, number of tracks counted in determining Rho-D. Age is the sample central fission-track age (Galbraith and Laslett, 1993), calculated using zeta calibration method (Hurford and Green, 1983). Analyst: T.A. Dumitru.

The following is a summary of key laboratory procedures. Apatites were etched for 20 s in 5N nitric acid at room temperature. Grains were dated by external detector method with muscovite detectors. Samples were irradiated in well-thermalized positions of Oregon State University reactor. CN5 dosimetry glasses with muscovite external detectors were used as neutron flux monitors. External detectors were etched in 48% HF. Tracks counted with Zeiss Axioskop microscope with 100× air objective, 1.25× tube factor, 10× eyepieces, transmitted light with supplementary reflected light as needed; external detector prints were located with Kinetek automated scanning stage (Dumitru, 1993). Only grains with c axes subparallel to slide plane were dated. Ages calculated using zeta calibration factor of 385.9. Confined track lengths were measured only in apatite grains with c axes subparallel to slide plane; only horizontal tracks measured (within ±~5°–10°), following protocols of Laslett et al. (1982). Lengths were measured with computer digitizing tablet and drawing tube, calibrated against stage micrometer (Dumitru, 1993). For 8 samples that contained more than 30 measurable confined tracks, track lengths were remeasured along with angles of tracks to the grains' c-axes and the Dpar track entrance diameter, following protocols of Donelick (1993) and Ketcham et al. (1999, 2000), except that confined tracks hosted by surface tracks and by cleavage surfaces were both measured. Age calculations were done with program by D. Coyle.

The following is a summary of thermal history modeling methods. Modeling done with the AFTSolve 1.1.2 program of Ketcham et al. (2000). Modeling parameters: (1) used raw track-length data (actual lengths, Dpar, and angle to c-axis of each track) and actual track counts (NS, NI, and Dpar of each grain); (2) used annealing model of Ketcham et al. (1999) with Dpar kinetic variable; (3) for each sample, grouped all data (lengths and track counts) into a single kinetic population and used midpoint Dpar value; (3) for length reduction in age standard, used default value of 0.893; (4) for initial track length, used default relations Lom = (0.283)(Dpar) + 15.630 μm and Loc = (0.205)(Dpar) + 16.100 μm; (5) projected track lengths to c-axis parallel; (6) calculated 10 000 model paths with Monte Carlo scheme, with each path segment monotonic and each segment halved 2 times, and without enforcing maximum heating or cooling rates; (7) output plots show best single run, good model fit envelope, and acceptable model fit envelope. No adjustments made for slight difference in etching conditions between Ketcham et al. (1999, 2000) (20 s, 21 °C, 5.5 N HNO$_3$) and our laboratory (20 s, room temperature (~22 °C), 5.0 N HNO$_3$), nor for likely minor interlaboratory differences in length reduction in age standard or initial track length relationships.

modeling results (Fig. 16) from all samples are highly consistent: all samples were at burial temperatures hotter than ~125 °C in the Early Cretaceous (other data discussed here indicate T_{max} ~193 °C), then underwent an episode of major cooling from hotter than ~125 °C (actually from ~193 °C) to ~80 °C, with a weighted mean time of cooling to below 120 °C of ca. 115 Ma (middle Aptian, Gradstein and Ogg, 1996). All samples show additional protracted slow cooling in middle Cretaceous to Holocene time. As discussed here, we interpret the rapid Early Cretaceous cooling to reflect exhumation of the samples by major erosion soon after major thrusting, and the later slow cooling to reflect later relatively minor erosion.

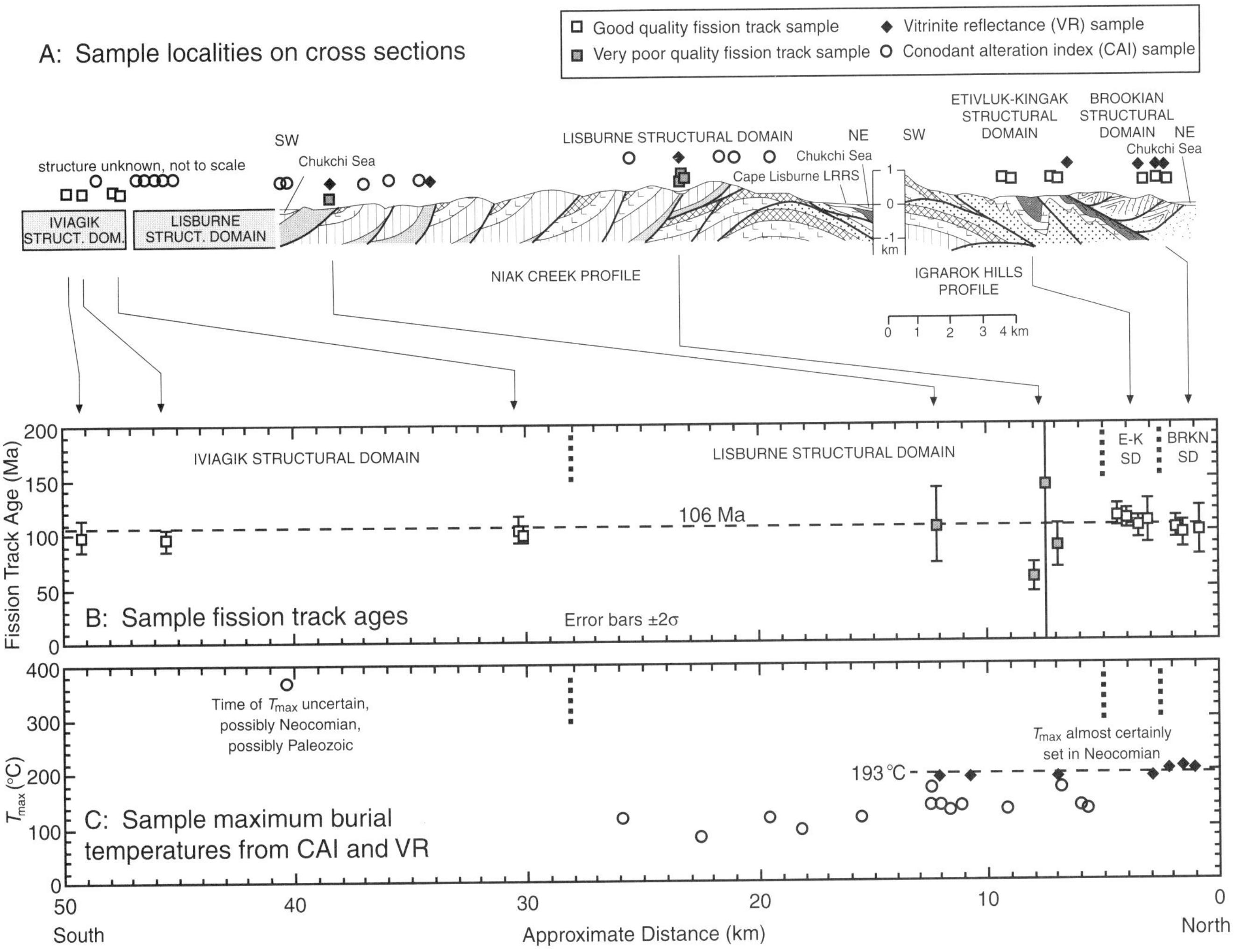

Figure 14. Schematic summary of thermal history. A: Sample localities plotted on structural sections of Cape Lisburne fold and thrust belt (Fig. 12). LRRS is long-range radar site. B: Sample apparent fission-track ages (Table 2; Fig. 15) plotted versus position along structural sections. Note consistent ages across entire thrust belt, with average apparent age of ca. 106 Ma. C: Maximum burial temperatures (T_{max}) estimated from vitrinite reflectance (VR) and conodont alteration index (CAI) data (Table DR1, see footnote 1, and Table 3). VR temperature estimates are probably more reliable; note consistent T_{max} from VR data from Lisburne, Etivluk-Kingak, and Brookian structural domains (mean of 193 °C). D: Summary time-temperature history from integration of representative fission-track model (sample 94CL-11 in Fig. 16) and geologic data. We infer from these data that major late Neocomian folding and thrusting thickened section, burying rock samples to great depth (7–8 km) and forming major mountain belt. Subsequent rapid erosion exhumed samples most of way back to present Earth's surface in Early Cretaceous time, with cooling through 120 °C isotherm ca. 115 Ma. Relatively minor cooling in middle Cretaceous to Holocene time reflects relatively minor additional erosion over that time, which is poorly defined due to poor sensitivity of fission-track system at such low temperatures.

D: Time-depth history from fission track modeling and geologic constraints

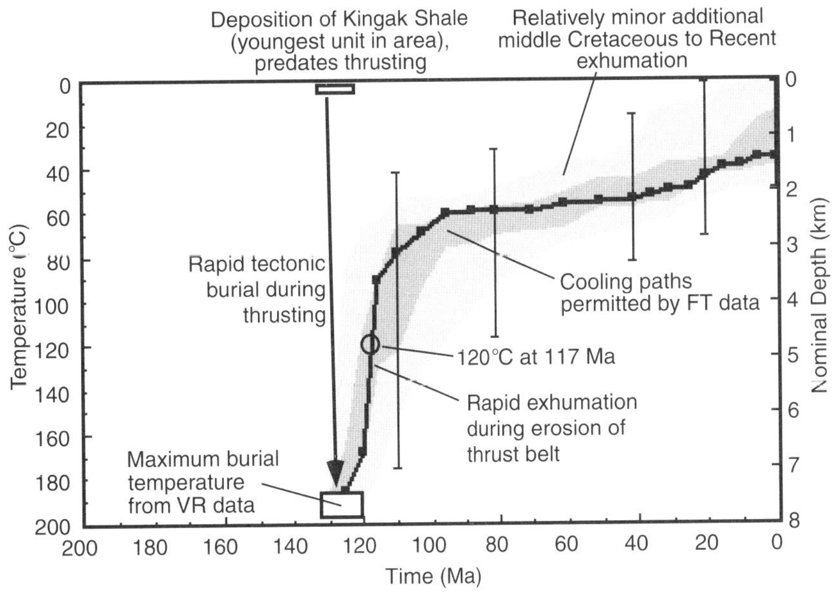

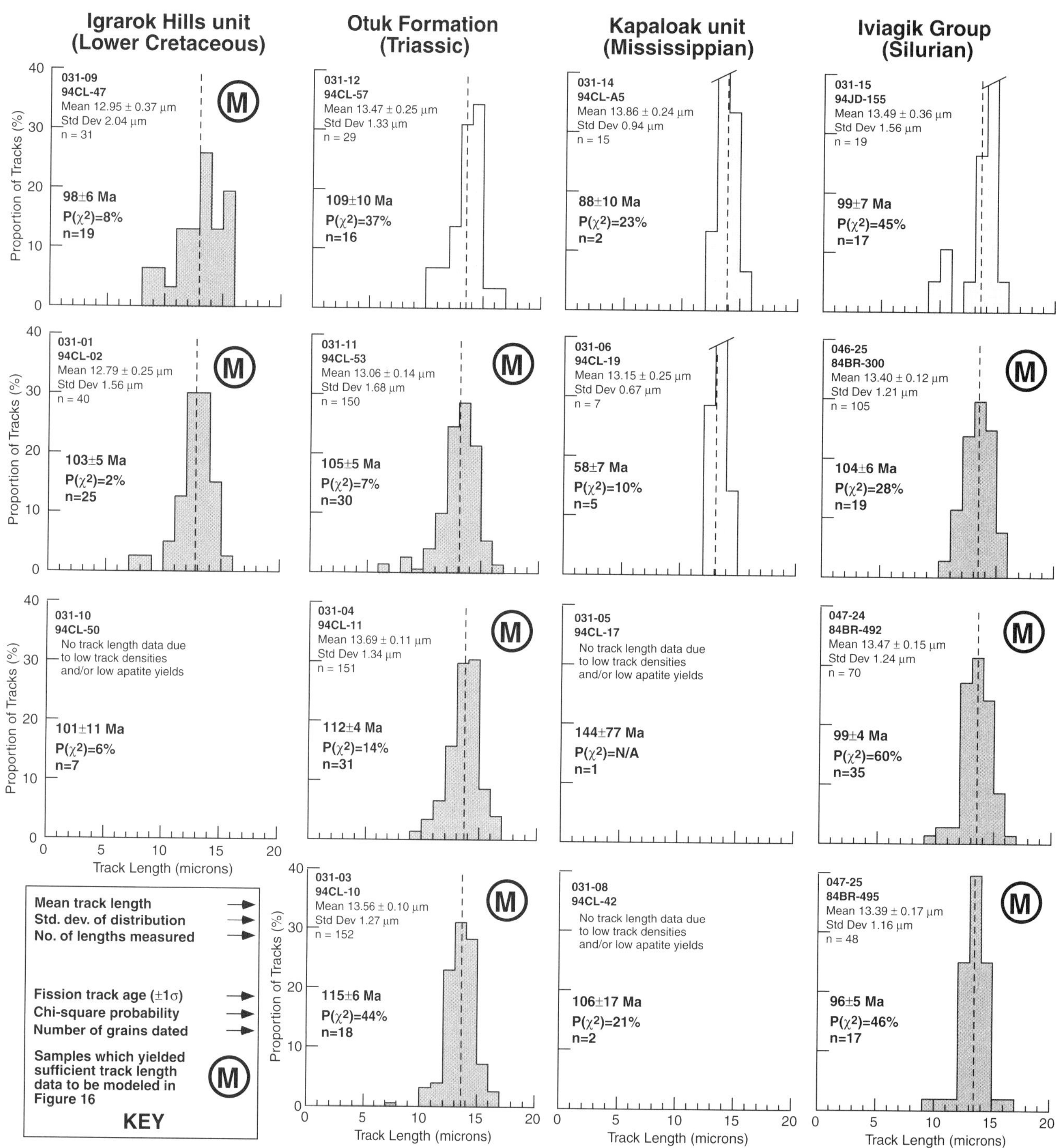

Figure 15. Apatite fission-track age and track-length data from Cape Lisburne samples. Track-length data are presented in histograms of lengths of up to 152 tracks per sample (length data could not be collected from three samples due to low track densities and sparse apatite yields). Eight samples (labeled M) permitted at least 31 track lengths to be measured, and are further analyzed in Figure DR1 (see footnote 1) and modeled in Figure 16.

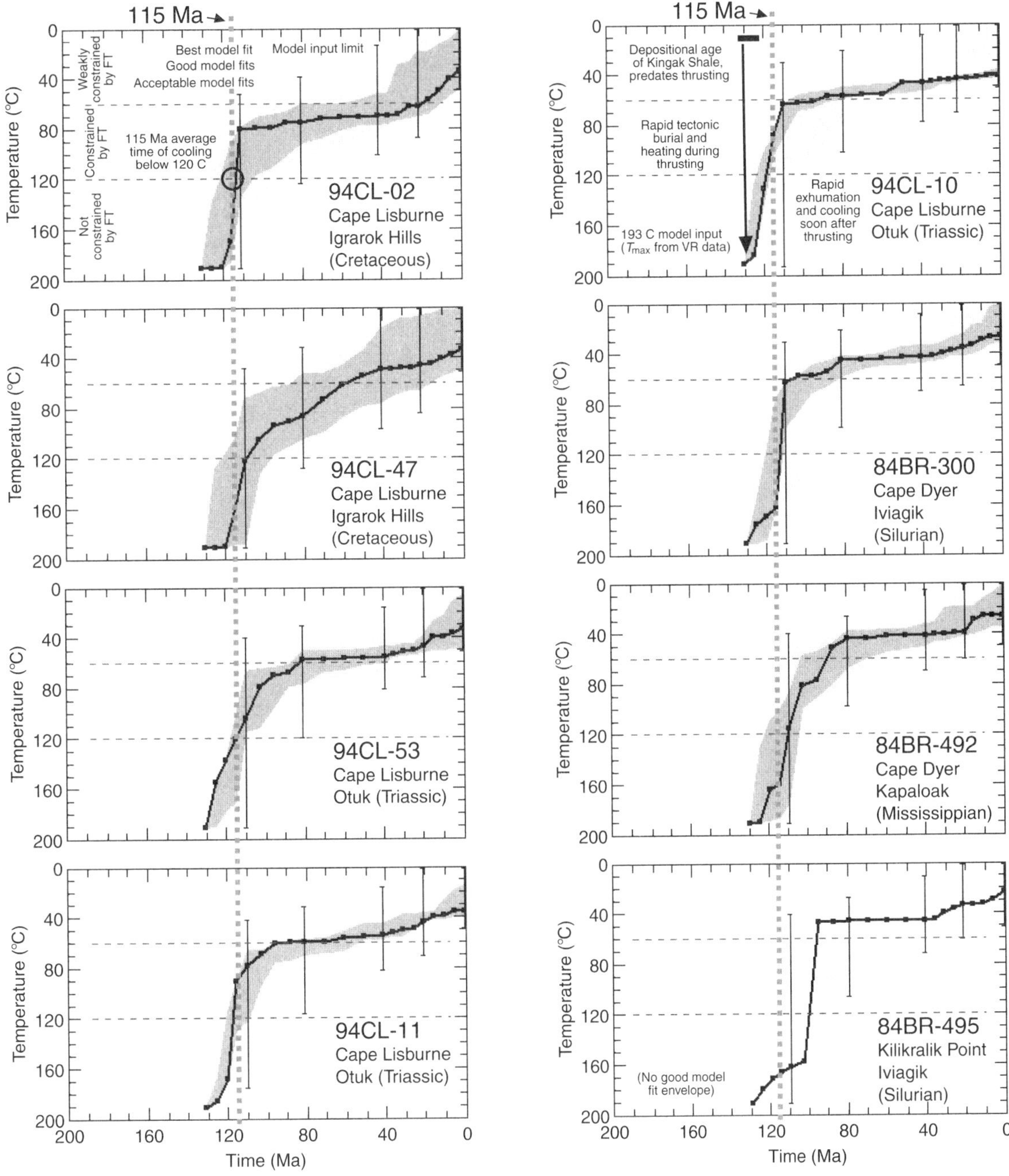

Figure 16. Modeling results for Cape Lisburne fission-track (FT) samples, computed using model of Ketcham et al. (2000; used by permission) and parameters summarized in footnote to Table 2. Modeling runs for different samples yield highly consistent results, with major cooling in Early Cretaceous time. We infer that these cooling episodes started in late Neocomian time at ~193 °C, based on maximum burial temperatures (T_{max}) from vitrinite reflectance (VR) and conodont alteration index data and involvement of Neocomian Igrarok Hills rocks and Hauterivian and/or Barremian Kingak Shale in cooling (Fig. 14D). Samples cooled through 120 °C isotherm ca. 115 Ma. Fission-track data cannot define histories at temperatures hotter than ~120 °C and are not very sensitive to temperatures cooler than ~60 °C, so exact paths hotter than 120 °C are not significant and paths cooler than ~60 °C are not very reliable. Fission-track record before Early Cretaceous cooling is totally overprinted, so time-temperature paths in pre-Cretaceous time are omitted.

Fission-track analysis proved more difficult in the Lisburne domain, which consists mainly of carbonate rocks, a lithology generally barren of apatite. Imbricated with the carbonate rocks is the Kapaloak sequence, which includes intervals of mature sandstone. We processed five Kapaloak samples; one yielded no apatite and four yielded small amounts of apatite. Ages of these samples were 58 ± 7, 88 ± 10, 106 ± 17, and 144 ± 77 Ma ($\pm 1\sigma$), with a weighted mean age of 95 Ma. These ages are based on only one to five grains of apatite per sample, primarily very poor quality grains that did not meet established criteria for reliable analysis (e.g., improper grain c-axis orientation), and we do not consider these ages reliable. However, the mean age is similar to reliable ages from the one underlying and two overlying structural domains, and this is probably sufficient to infer that the Kapaloak samples had time-temperature histories generally similar to those domains. We discount the significance of the age range 58–144 Ma, because 58, 88, and 144 Ma samples were collected from essentially the same structural position (Figs. 3, 14, and 15). One other Kapaloak sample from the Iviagik structural domain (84BR-492) yielded good-quality data. This sample is from the basal part of the Kapaloak sequence and contains abundant detritus sourced from the underlying Iviagik Group, and its apatite is probably reworked detrital apatite derived from Iviagik rocks.

Constraints from other paleotemperature indices. The time-temperature histories of the fission-track samples are also constrained by vitrinite reflectance (VR) and conodont alteration index (CAI) data, which are indicators of the maximum burial temperatures (T_{max}) of the samples, and by estimates of the depositional thicknesses of units in the section, which help control burial depths. VR and CAI data are summarized in Table DR1 (see footnote 1), Table 3, and Figure 14C.

Various calibrations are available to convert VR values to T_{max}. The calibration of Burnham and Sweeney (1989, equation 2) has perhaps the best theoretical foundation, and we use it here. In this calibration, the duration of exposure to maximum temperatures has a minor effect on the calibration. Samples from the Cape Lisburne area apparently underwent maximum burial due to thrusting and cooled rapidly thereafter (see following). Therefore we have assumed a relatively short heating duration of 3 m.y., but for comparison we have also computed T_{max} for other heating durations (Table 3; Fig. 14C). With these assumptions, the average T_{max} was ~203 °C for the Cretaceous unit, 188 °C for the Kingak Shale, and 189 °C for the Kapaloak sequence. The Kapaloak VR data, derived from coal samples, are of particularly good quality. CAI values from the Lisburne Group within our map area on the northern Lisburne Peninsula range from 2–2.5 to 3; the average value is ~2.5 (Table DR1, see footnote 1; Fig. 14C). Using the calibration of Harris et al. (1978) and assuming a heating duration of 3 m.y., this suggests T_{max} of ~150 °C, more or less consistent with the T_{max} estimate from the Kapaloak VR data, given the uncertainties in the calibrations of the CAI scale.

Our only conodont sample from the Iviagik Group yielded a CAI value of 5 (Table DR1, see footnote 1), equivalent to a much higher T_{max} of ~365 °C, assuming a 3 m.y. heating duration (Fig. 14C). However, it is not clear that the Iviagik Group

TABLE 3. VITRINITE REFLECTION DATA COLLECTED FROM THE POINT HOPE D-2 1:63 360 QUADRANGLE, LISBURNE PENINSULA, ALASKA

Sample number	Latitude/ longitude	Stratigraphic unit	Age	Lithology	Mean vitrinite reflectance	T_{max}— 1 m.y. (B&S)	T_{max}— 3 m.y. (B&S) Preferred	T_{max}— 10 m.y. (B&S)	T_{max}— (Barker)
94AKA6-6A	68.81639°/ 166.19111°	Kapaloak sequence	Mississippian	Coal	1.79	197 °C	189 °C*	180 °C	229 °C
94AKA8-1	68.86083°/ 166.01667°	Kingak Shale	Neocomian	Shale	ND	ND	ND	ND	ND
94TM-37	68.82556°/ 166.17556°	Kapaloak sequence	Mississippian	Coal	1.81	197 °C	189 °C*	180°C	230 °C
94TM-46	68.86833°/ 165.99222°	Igrarok Hills unit	Early Cretaceous	Shale	2.19	213 °C	203 °C*	194 °C	257 °C
94TM-47V	68.86722°/ 165.99278°	Igrarok Hills unit	Early Cretaceous	Shale	2.23	214 °C	205 °C*	196 °C	259 °C
94TM-48VR	68.86667°/ 165.95722°	Igrarok Hills unit	Early Cretaceous	Shale	2.17	212 °C	202 °C*	193 °C	256 °C
94TM-64	68.85556°/ 165.98889°	Kingak Shale	Neocomian	Shale	1.78	197 °C	188 °C*	179 °C	228 °C
94TM-72	68.85361°/ 166.10972°	Kapaloak sequence	Mississippian	Coaly shale	1.77	196 °C	188 °C*	179 °C	227 °C

Note: Analyst: Mark Pawlewicz. Abbreviations: ND, not determined; T_{max}, maximum burial temperature. Maximum burial temperatures calculated from mean vitrinite reflectances using the time-dependent calibration of Burnham and Sweeney (1989, eq. 2) assuming 1, 3, and 10 m.y. durations at T_{max}; 3 m.y. results are preferred and are used in the text. For comparision, T_{max} also calculated with the time-independent calibration of Barker (1988), which was used for the Thermal Maturity Map of Alaska (Johnsson and Howell, 1996a, 1996b).
*Preferred maximum burial temperature (T_{max}) used in text.

underwent its maximum burial temperatures in the Early Cretaceous. The CAI value of 5 may have instead been established earlier in pre-Mississippian time over some unknown heating duration. If the heating duration was 50 m.y. or longer, a CAI of 5 suggests a T_{max} of ~300 °C in Paleozoic time (Harris et al., 1978).

Figure 4 shows the stratigraphic section from the Cape Lisburne area, which has a preserved total thickness of ~3 km. It is clear that the fission-track, VR, and CAI data cannot be explained by simple stratigraphic burial in an undeformed section composed of the currently preserved units. For example, assuming a nominal 25 °C/km thermal gradient, the Cretaceous strata require 7–8 km of burial prior to cooling at 115 Ma. Additional, since eroded, younger Brookian strata could have contributed to this overburden. However, seismic data in the Chukchi Sea (Thurston and Theiss, 1987, Plate 8) and outcrop exposures east of our study area (Chapman and Sable, 1960) suggest that the total thickness of pre–middle Aptian Brookian strata was probably <1 km. In addition, the Cretaceous strata in our study area (Igrarok Hills unit) have higher rather than lower vitrinite reflectance values than the underlying Mississippian strata, a configuration that cannot be attained by simple stratigraphic burial and instead is better attributed to tectonic burial prior to cooling at 115 Ma.

Structural implications of thermal history data. The thermal history data lend support to the field-based structural reconstruction developed in Figure 12. The maximum burial temperature estimates derived from the various indicators imply very deep burial, 7–8 km, while the currently preserved stratigraphic section is only ~3 km thick. The strong shortening indicated by the structural model can easily explain the deep burial. During imbrication, the total thickness of the section increased, causing tectonic burial of the deeper parts of the structural stack, including the Igrarok Hills unit. This suggests the possibility that thick allochthons once covered the Lisburne Peninsula and have been subsequently eroded away.

Presumably, the emplacement of the allochthons by thrusting would have created a high mountain range subject to erosion. Such erosion, accompanied by isostatic rebound, would have exhumed the rocks within the range. It is the cooling of rocks caused by such exhumation that was recorded by the fission-track data, so the time of cooling derived from fission-track data (ca. 115 Ma for cooling below 120 °C) should postdate the time of initiation of thrusting. The Igrarok Hills unit, which yielded one probable Early Cretaceous palynomorph, and the Kingak Shale–pebble shale unit, which yielded Hauterivian to Barremian foraminifers (Table 1), were involved in the cooling at Cape Lisburne, suggesting initiation of thrusting occurred sometime after 132–121 Ma (Hauterivian and Barremian, Gradstein and Ogg, 1996). A history involving Hauterivian and Barremian marine deposition, overthrusting and deep tectonic burial, and subsequent exhumation ca. 115 Ma indicates that tectonic burial and subsequent exhumation were quite rapid, having been completed within no more than ~17 m.y. (Fig.

14D). The fission-track data indicate that all four structural domains cooled at essentially the same time, and at least the Lisburne, Brookian, and Etivluk-Kingak domains were apparently at similar depths (~7–8 km) prior to cooling. From this, we infer that all four domains were thrust, buried, and exhumed at essentially the same time. This chronology indicates that the fold and thrust belt exposed at Cape Lisburne developed principally in the Early Cretaceous (late Neocomian, possibly extending into Aptian time). Rocks currently exposed at the land surface near Cape Lisburne were at subsurface temperatures of ~200 °C before exhumation started and cooled to temperatures cooler than ~70 °C during the Early Cretaceous event, equivalent to ~5 km of erosional exhumation (Fig. 14D).

The fission-track modeling suggests relative slow additional cooling in mid-Cretaceous to Holocene time (Figs. 14D and 16). The fission-track system is not very sensitive at low temperatures (<~60 °C) and so provides only crude limitations on this additional cooling. In addition, interpretation is complicated by the fact that mean Earth surface temperatures in this region of the Arctic probably decreased from ~5 to −12 °C in Miocene time with the restriction of ocean circulation through the Bering Straits (O'Sullivan and Brown, 1998). Allowing for these climatic effects, the models crudely suggest ~2–3 km of exhumation during mid-Cretaceous to Holocene time. This exhumation may reflect protracted slow erosion of the mountain belt originally raised during Early Cretaceous time, and/or may reflect additional, relatively minor shortening sometime in mid-Cretaceous to Holocene time.

COMPARISON OF THE LISBURNE HILLS–HERALD ARCH FOLD BELT WITH THE BROOKS RANGE OROGEN

The stratigraphic, structural, and geochronologic results that we have presented allow comparisons to be made between key characteristics of the Lisburne Hills–Herald Arch fold belt and the Brooks Range orogen to the east. These comparisons provide insight into the timing and formation of the Lisburne Hills–Herald Arch fold belt and its role in the geology of northern Alaska.

Stratigraphic comparisons

The stratigraphy of the Cape Lisburne area displays many similarities to the autochthonous and parautochthonous succession of the North Slope. As noted by Grantz et al. (1983), for example, the Iviagik Group consists in part of graptolitic shale that is lithologically similar to Ordovician-Silurian argillite found in the subsurface of the North Slope (e.g., Carter and Laufeld, 1975). In addition, the Lower Mississippian Kapaloak sequence overlies these rocks on an angular unconformity that probably correlates with the regional sub-Mississippian unconformity present at the base of the Ellesmerian sequence throughout the North Slope and northeastern Brooks Range (Moore et al.,

1984). The Kapaloak appears to correlate with the transgressive coarse-grained clastic deposits of the Lower Mississippian Kekiktuk Conglomerate, a characteristic unit at the base of the Ellesmerian sequence of the autochthonous succession of the North Slope (Moore et al., 1984). In addition, the conodont results presented here (Table DR1, see footnote 1), coupled with the age data reported by Armstrong and Mamet (1978), indicate that the base of the Lisburne is within the Upper Mississippian (late Meramecian–early Chesterian), at least in the northern Lisburne Peninsula. This age of onset of Lisburne deposition is similar to that found in the northeastern Brooks Range and North Slope (Armstrong and Mamet, 1978; Dumoulin et al., 1997) and is substantially younger than the Lower Mississippian Lisburne strata found in the allochthonous rocks of the Brooks Range. A northward younging in age of the basal strata of the Lisburne Group from the Brooks Range to the Arctic margin is thought to have resulted from northward transgressive onlap of the Lisburne onto older rocks during the Mississippian (Armstrong and Mamet, 1978).

In addition to these lines of evidence indicating stratigraphic affinities between the Lisburne Peninsula and the North Slope is the possibility that the Kapaloak sequence and overlying Lisburne Group on the Lisburne Peninsula may represent an exposed, local basin succession analogous to the Endicott basins of the North Slope (Eo-Ellesmerian basins of Grantz and May, 1988). The Endicott basins (e.g., Meade, Umiat, Ikpikpuk basins, Fig. 1) are structural sags and extensionally faulted depressions that postdate development of the Mississippian unconformity. The Kekiktuk thickens into these basins to as much as 5 km (Bird, 1988b, Fig. 16.10; Grantz and May, 1988). While the Kapaloak sequence occupies the stratigraphic position of coeval Kekiktuk Conglomerate on the Lisburne Peninsula, it is significantly thicker (as thick as 600 m) than the Kekiktuk Conglomerate (which is typically <100 m thick), possibly because it represents sedimentary fill in an Endicott-type basin. Where drilled in the central North Slope (Magoon et al., 1988, Plates 19.15 and 19.21), basin-filling strata of Endicott basins consist of coal-bearing paralic strata that are lithologically very similar to the Kapaloak sequence. The thickness of strata in the overlying Lisburne sequence also increases into the Endicott basins in the North Slope subsurface, indicating that the subsidence that produced the Endicott basins was still active when the Lisburne rocks were deposited (Bird, 1988a). Although the southward increase in thickness of Lisburne Group strata in the Lisburne Peninsula (more than 1 km) may be partly the result of increased erosional downcutting to the north along the sub-Siksikpuk unconformity, as indicated by karst in the uppermost part of the Lisburne sequence, part of the thickening may be related to differences in accommodation space caused by an increased rate of subsidence over an Endicott Basin. This interpretation is supported by the apparent equivalence in age of Lisburne strata in northern and southern parts of the peninsula despite the large north-to-south increase in its stratigraphic thickness. If correct, the Kapaloak and Lisburne sequences of the Lisburne Peninsula

may represent part of an exhumed Endicott Basin subsequently inverted during Mesozoic shortening. The seismic data of Thurston and Theiss (1987, Plate 8) and Klemperer et al. (this volume, Chapter 1) indicate the presence of another Endicott Basin more than 2 km deep ~15 km north of the Herald Arch in the Chukchi Sea and several similar basins farther north. These Endicott basins are located along the east flank of the Chukchi platform, where a series of north-trending fault blocks formed in the Mississippian controlled to varying degrees the sedimentation patterns of Endicott and Lisburne strata (Thurston and Theiss, 1987; Sherwood et al., 1998). The position of the westward pinch-out of Mississippian strata in the subsurface of the Chukchi Sea, due to onlap onto the Chukchi platform, was mapped by Sherwood et al. (1998), and projects southward toward Cape Lisburne. This suggests that the western margin of the Mississippian basin exposed in the Lisburne Peninsula may also pinch out westward onto the Chukchi platform.

In contrast to the older rocks, the Permian and Triassic deposits of the Lisburne Peninsula display notable similarities to coeval allochthonous rocks of the Brooks Range orogen. Permian and Triassic deposits in the Brooks Range consist principally of shale, siliceous shale, chert, and silicified limestone deposited in middle to outer shelf conditions (Mull et al., 1982; Blome et al., 1988; Murchey et al., 1988; Mayfield et al., 1988), whereas Permian and Triassic deposits in the North Slope subsurface and northeastern Brooks Range consist of shale, siltstone, sandstone, and limestone deposited under more proximal shelfal conditions (Parrish, 1987; Crowder, 1990; Adams et al., 1997). The presence of abundant chert and siliceous argillite in the Permian deposits; chert, siliceous shale, and silicified limestone in the Triassic deposits; and the overall fine-grained character of both units indicate that Permian-Triassic deposition on the northern Lisburne Peninsula probably occurred in a middle to outer shelf environment analogous to that found in the allochthonous Brooks Range deposits. The presence of lithic sandstone in the Triassic deposits of the northern Lisburne Peninsula, however, may indicate proximity to a source area not recognized elsewhere in northern Alaska, possibly to the west of the Lisburne Peninsula on the Chukchi platform. This interpretation is supported by the presence of granitic detritus in the sandstone, a common detrital component of westerly derived sandstone in the subsurface of the Chukchi Sea (K.W. Sherwood, 2000, written commun.). Instead of being deposited on distal parts of the south-facing passive margin like the allochthonous Permian and Triassic deposits of the Brooks Range, the distal character of the Permian and Triassic deposits of the Cape Lisburne area may reflect deposition at a location that was isolated from eastern source areas on the North Slope by the presence of the Hanna Trough (Fig. 1). The Hanna Trough is the north-south–trending structural low and axial region of Ellesmerian deposition developed on pre-Mississippian rocks under the Chukchi Sea (Grantz and May, 1988). The southward extend of the structural low beneath the sedimentary fill of Hanna Trough is uncertain, but has been traced southward in seismic data to the coastal region northeast of the Lisburne Peninsula (Thurston and Theiss, 1987). These observations suggest that the stratigraphy of the Lisburne Peninsula probably reflects deposition along the eastern flank of the Chukchi platform and represents the only onland exposure of that flank of the platform.

The origin of the Brookian deposits of the Igrarok Hills unit is an important question for the geology of the Lisburne Peninsula. As discussed herein, these strata may be related to the Aptian to Albian Mount Kelly Graywacke Tongue of the Fortress Mountain Formation or the Upper Jurassic and Neocomian Okpikruak Formation. Although the paleontological age data from the Igrarok Hills unit are inconclusive, indicating only deposition sometime in the Early Cretaceous, the fission-track data from the Igrarok Hills unit suggest that the unit is most likely correlative with the Okpikruak Formation. The fission-track data indicate major exhumation and cooling ca. 115 Ma. Because 115 Ma corresponds to the middle Aptian (121–112 Ma, Gradstein and Ogg, 1996), it seems unlikely that the Igrarok Hills unit could have been deposited in the Aptian at the time of Fortress Mountain deposition, buried to depths sufficient to totally erase fission tracks in apatite and set its high vitrinite reflectance levels (nominally 7–8 km depth), and exhumed sufficiently to set its fission-track ages, all in Aptian time. This scenario would be possible only if the fission-track ages are pushed toward the youngest limits permitted by their uncertainties and the depositional age of the Igrarok Hills unit is assumed to be entirely early Aptian.

The more likely interpretation of the data suggests that deposition of the Igrarok Hills unit occurred in the Neocomian prior to the time of deposition of the Fortress Mountain Formation and during the time of deposition of the Okpikruak Formation. If deposition of the Igrarok Hills unit occurred in the early part of the Neocomian, it would require that it is allochthonous because it overlies upper Neocomian (Hauterivian-Barremian) strata of the Kingak Shale. An allochthonous origin for the unit is supported by the vitrinite reflectance data, which indicate higher thermal maturity for the Igrarok Hills unit (Ro = 2.17–2.23) than for any of the underlying units (Ro = 1.77–1.81). These data suggest that rocks that were at relatively higher paleotemperatures (Igrarok Hills unit) were emplaced onto rocks that have not been at such high paleotemperatures, a situation most easily explained by thrusting. Our preferred interpretation, therefore, is that the Igrarok Hills unit is an allochthonous unit that was thrust into its present position on the Kingak Shale–pebble shale unit shortly before 115 Ma. This interpretation suggests that the Igrarok Hills unit is an erosional remnant of the basal part of the tectonic load that provided the 7–8-km-thick overburden that caused the high paleotemperatures recorded by the vitrinite reflectance and CAI data from the Cape Lisburne area.

Structural comparisons

The structural behavior of the Lisburne Group when deformed under varying mechanical conditions and amounts of strain in the Brooks Range were discussed by W.K. Wallace and his colleagues (Wallace and Hanks, 1990; Wallace, 1993; Wallace et al., 1997; Homza and Wallace, 1997). Where underlain by a detachment horizon, shortening in the Lisburne Group at low amounts of strain results in the development of upright, symmetrical detachment folds. At increased amounts of strain, shortening is accommodated by increases in the wavelength, amplitude, and asymmetry of the detachment folds. At still higher amounts of strain, shortening is expressed by thrust breakthrough of the steep limbs of the detachment folds, leading to development of thrust-truncated detachment folds (Wallace, 1993). The truncation of the detachment folds by thrusting allows development of an imbricate thrust system under conditions of high strain and accommodation of a geometrically unlimited amount of shortening.

Where shortening in the Lisburne Group is mainly represented by detachment folding with only rare thrust duplication, overall shortening calculated from balanced sections ranges from as little as 26% to a maximum of 46% (Wallace, 1993; Hanks, 1993; Moore, 1999). Where deformation of the Lisburne Group is expressed as an imbricate thrust system of thrust-truncated detachment folds, schematic but balanced representations of the structure suggest >50% shortening (Wallace et al., 1997; Blythe et al., 1996; Wissinger et al., 1998). The larger amounts of shortening in the latter areas are supported by the widespread presence of large-scale thrusts, klippen and windows, and stratigraphically out of place units. These relations are common along the main axis of the Brooks Range and are particularly well documented west of long 149°W (e.g., Mull, 1982; Mayfield et al., 1988; Moore et al., 1997; Mull et al., 1987a, 1987b, 1994, 1997).

The balanced cross section of the Lisburne structural domain presented in this study indicates that a minimum of ~65% shortening is required in the Cape Lisburne area. If our interpretation of the Igrarok Hills unit as a correlative of the Okpikruak Formation is correct, it probably requires large-scale thrust emplacement of a stratigraphically out of place Brookian Neocomian turbidite unit onto coeval or younger fine-grained basinal deposits of the Kingak Shale–pebble shale unit, an interpretation that is consistent with the isoclinal style of folding in the Igrarok Hills unit. These features are characteristics of the higher displacement thrusts present along the main axis of the Brooks Range and suggest that the deformation of the Lisburne Hills fold and thrust belt represents a relatively comparable high amount of strain.

Comparisons of timing of deformation

The apatite fission-track and thermal maturity data presented here indicate that the Ellesmerian and Brookian section in the Cape Lisburne area was subjected to maximum burial temperatures of ~200 °C and then cooled through the 125–75 °C temperature window ca. 115 Ma. We interpret this to indicate that major thrusting in the area postdated deposition of the youngest deposits of the Kingak Shale–pebble shale (Hauterivian and/or Barremian) and Igrarok Hills units (Early Cretaceous) and predated 115 Ma (Fig. 14D). This places initiation of major thrusting between 132 Ma and ca. 115 Ma (Hauterivian to early Aptian; Gradstein and Ogg, 1996).

In the Brooks Range, stratigraphic relations and isotopic age data indicate that the major thrust faulting that emplaced the allochthonous units in the main axis of the Brooks Range occurred during the Neocomian, although thrusting probably initiated in the latest Jurassic (Mull, 1985; Moore et al., 1994). Paleontologic ages of syntectonic strata (Okpikruak Formation) are largely Berriasian and Valanginian (144–132 Ma; Gradstein and Ogg, 1996), and these strata are involved in the thrusting, indicating that thrusting continued into the late Neocomian (Hauterivian and/or Barremian) (Moore et al., 1994). The overlap in ages of thrusting in the Lisburne Peninsula and the Brooks Range strongly suggests that the two events were contemporaneous in the late Neocomian.

In the Brooks Range, however, apatite fission-track data record widespread cooling events at 100 ± 5 Ma, 60 ± 4 Ma, and 24 ± 3 Ma (e.g., O'Sullivan et al., 1997), events that are not apparent in the Cape Lisburne fission-track data. The 100 Ma event was apparently restricted to the hinterland of the Brooks Range orogen, and its absence in the Cape Lisburne area is not surprising given its position in the frontal part of the Lisburne Hills fold and thrust belt. The Tertiary cooling events from the Brooks Range, however, have been linked to widely distributed folds and thrusts that deform middle Cretaceous strata in the Colville Basin and northeastern Brooks Range (O'Sullivan et al., 1997). These structures are present within 30 km to the east of the Cape Lisburne area in the southwestern Colville Basin, where they have yielded early Tertiary fission-track cooling ages (J. Murphy, 1998, written commun. to C.G. Mull; P.B. O'Sullivan, 2000, written commun.). These structures (Chapman and Sable, 1960, Plate 9) can be traced westward well into the Chukchi Sea region to the northwest of the Cape Lisburne area, where they terminate north of the Herald Arch, as pointed out by Grantz et al. (1970) and Sherwood et al. (1998) (Fig. 1). This observation suggests that the early Tertiary contractional event deformed areas around Cape Lisburne, but did not cause significant erosion at Cape Lisburne. The reason that the Cape Lisburne area did not undergo significant exhumation in the early Tertiary is uncertain and remains to be resolved. Perhaps the amount of displacement represented by the early Tertiary structures diminishes to the west toward the Lisburne Peninsula and was insufficient to cause significant structural thickening and exhumation in that area. Alternatively, the early Tertiary shortening may have occurred on a regional flat in the basal detachment that did not result in significant uplift in the Cape Lisburne area.

TECTONIC MODEL

On the basis of the various lines of evidence described here, we have concluded the following.

1. The stratigraphic characteristics of the pre-Mississippian and Ellesmerian sedimentary section in the Cape Lisburne area appear to be similar in most respects to the sedimentary section in the North Slope subsurface, and different from the allochthonous sections in the interior of the Brooks Range.

2. The available data suggest that the Igrarok Hills unit consists of Neocomian-age flysch similar to the Okpikruak Formation, a syntectonic unit in the main axis of the Brooks Range. The Okpikruak Formation occupies an allochthonous structural position within the Brooks Range, and we tentatively interpret the Igrarok Hills to be allochthonous at Cape Lisburne.

3. The Cape Lisburne area exposes a northeast-vergent fold and thrust belt composed of a multistoried duplex. The thrusting involved relatively large amounts of displacement, with a minimum of ~65% shortening required. The thrusting emplaced allochthons over the area, resulting in tectonic burial of the currently exposed land surface to depths on the order of 7–8 km.

4. The deformation is bracketed by stratigraphic and fission-track methods to have occurred between 132 and 115 Ma and was followed by ~5 km of exhumation. The timing, style, and percent of shortening of Lisburne Group strata are similar to those documented in the main axis of the Brooks Range and suggest that the Lisburne Hills fold and thrust belt developed as part of the Brooks Range orogen in the late Neocomian or early Aptian.

5. Modeling of the fission-track data suggests that ~2–3 km of exhumation occurred in the Cape Lisburne area since the Aptian. The cause of the post-Aptian exhumation is not known.

Figure 17 portrays the tectonic evolution of the Cape Lisburne area based on these data and interpretations. Figure 17A shows inferred tectonic-stratigraphic relations in the latest Jurassic and Neocomian. Brooks Range contractional deformation was active during this time, emplacing allochthonous oceanic rocks northward onto the Arctic Alaska terrane, but had apparently not advanced into the Cape Lisburne area. Deep-marine lithic clastic strata, including the Igrarok Hills unit, may have begun to accumulate in a foredeep associated with the orogen in the early Neocomian while approximately coeval shaly basinal strata of the Kingak Shale–pebble shale unit were continuing to be deposited in the foreland region.

Sometime between the oldest possible age of footwall strata (132 Ma) and the time of exhumation indicated by apatite fission-track ages (115 Ma) (Fig. 17, B–D), Brookian deformation propagated unusually far northward into the Cape Lisburne area, initially thrusting the Igrarok Hills unit northward onto the Kingak Shale. As deformation progressed, it stepped to progressively deeper structural levels, forming first a duplex in Lisburne Group strata and later a duplex at a still deeper level in pre-Mississippian rocks. The thrusting initially buried the currently exposed rocks in the Lisburne, Etivluk-Kingak, and Brookian domains to depths of 7–8 km. Soon after the termination of thrusting, the currently exposed rocks in the Iviagik, Lisburne, Etivluk-Kingak, and Brookian domains cooled to ~60 °C, indicating that erosion of the thrust belt had totaled ~5 km. While structural thickening in the Ellesmerian and Brookian sections contributed to the uplift that permitted this erosion, much of the uplift must have been accomplished on the deeper duplex in the pre-Mississippian section.

Following the termination of thrusting, an Albian to Cenomanian sedimentary succession (Torok Shale and Nanushuk Group) accumulated north and east of the deformed Cape Lisburne section (Fig. 17E). Although the Albian to Cenomanian sedimentary section >7 km thick east of the Lisburne Peninsula (Bird, 1988a; Sherwood et al., 1998), a thickness of this magnitude would have reset fission-track ages at Cape Lisburne. For this reason, we conclude that the thickness of now-eroded middle and Upper Cretaceous sedimentary rocks in the Cape Lisburne area was significantly less than that in the Colville Basin, probably no more than 1–2 km. This conclusion is supported by the regional westward thinning of these strata in the subsurface of the Chukchi Sea (Sherwood et al., 1998, Plate 13.2).

Contractional deformation that resulted in the folding and thrusting of Colville Basin strata began ca. 60 Ma (O'Sullivan, 1996). The basal detachment for this deformation ramped upward in the section into mid-Cretaceous strata (Oldow et al., 1987), forming low-displacement, map-scale folds in Colville Basin strata. A similar style of deformation was portrayed by Thurston and Theiss (1987, Plate 8) in the Chukchi Sea north of the Lisburne Peninsula. Although this deformation resulted in regional uplift-related cooling recorded by fission-track data to the east of the Lisburne Peninsula, no structural or fission-track evidence of this deformation has been recognized in the Cape Lisburne area. To resolve this dilemma, we suggest that the basal detachment of the contractional deformation at 60 Ma was located on a footwall flat at or near the base of mid-Cretaceous strata north of the Lisburne Peninsula, and joined a reactivated Early Cretaceous thrust southwestward beneath the Igrarok Hills unit (Fig. 17F). Because there was little structural relief between these positions of the basal detachment, no appreciable uplift (and hence exhumation and cooling) of the Igrarok Hills unit or stratigraphically lower units would have occurred. This interpretation also may explain the presence of late, hinterland-vergent folds in the upper part of the Etivluk-Kingak structural domain, a common sense of folding above Tertiary duplexes that deform Colville Basin strata to the east (Moore and Potter, 2000). Figure 17G schematically depicts extensional faulting related to the formation of the Hope Basin, the sedimentary fill of

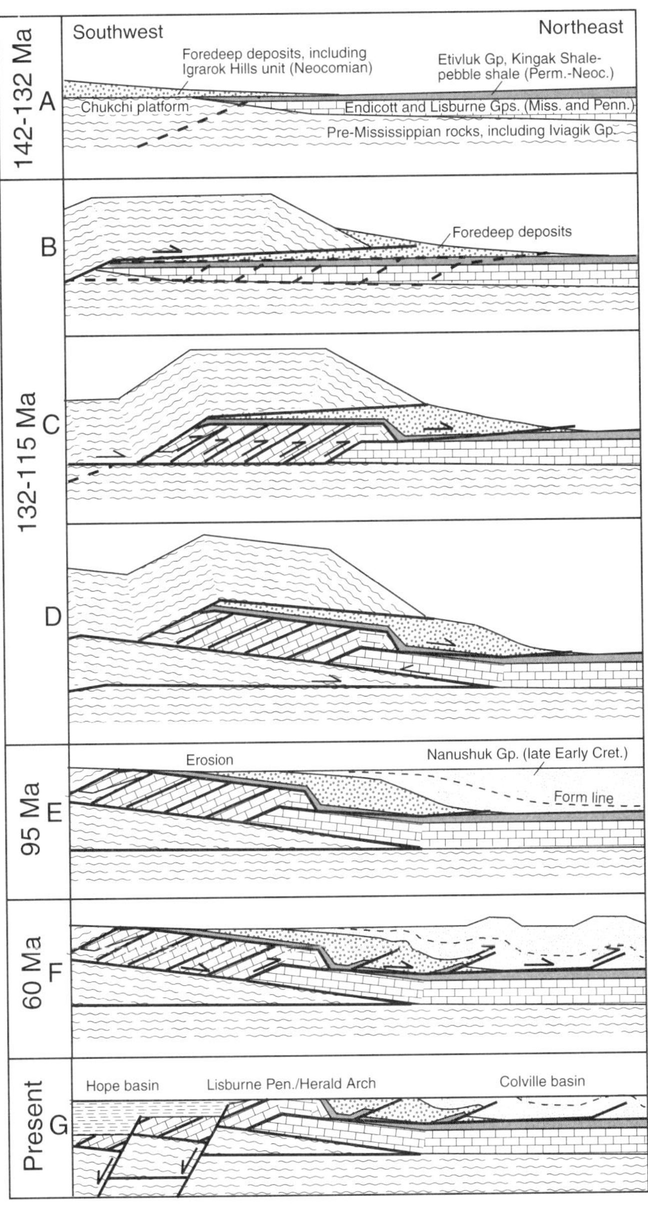

Figure 17. Schematic tectonic model for development of Lisburne Hills fold and thrust belt. A: Depositional onlap of Ellesmerian sequence onto high composed of deformed lower Paleozoic rocks (Chukchi platform) that persisted into early Neocomian. Lower Neocomian foredeep deposits shed from southwest outside of diagram deposited in basin-parallel submarine fans. Dashed line shows location of hypothesized future thrust fault. B–D: During late Neocomian, allochthons composed of lower Paleozoic rocks and earlier deposited foredeep deposits were thrust northeastward onto Ellesmerian sequence and overlying foredeep deposits (B). Thrusting was accompanied or followed shortly by imbrication of Lisburne Group, northeastward thrusting of foredeep deposits (C), and pre-Mississippian rocks at depth (D). Similarity of fission-track ages throughout structural stack suggests largest fraction of vertical uplift in study area was provided by structural thickening on thrust faults in lower Paleozoic rocks at depth, as shown schematically in D. E: About 5 km of overburden was removed by erosion at time of exhumation in Cape Lisburne area (ca. 115 Ma). Beginning at about this time, Colville Basin was filled with thick succession of clastic sediments (Nanushuk Group) shed from west. Fission-track data do not allow full depositional thickness of Nanushuk Group to have been deposited in Cape Lisburne area, suggesting that it was located near source area for Nanushuk sedimentary detritus. F: Renewed thrusting ca. 60 Ma in Brooks Range to east propagated northward into Colville Basin sedimentary fill. Absence of fission-track evidence for exhumation in Cape Lisburne area at this time indicates that amount of structural thickening was negligible, possibly because thrusting in Cape Lisburne area occurred on footwall flats. G: Hinterland of Lisburne Hills fold and thrust belt extended and buried by deposition in Hope Basin, forming south flank of Lisburne Hills and Herald Arch.

which onlaps eastward to within 10 km of our study area. The Hope Basin, as described by Tolson (1987), is a 500-km-long extensional basin located between the Wrangel Arch, Herald Arch, and Lisburne Peninsula to the north and Chukotsk Peninsula to the south (Fig. 1). Extensional faults in the basin clearly postdate older contractional structures (Grantz and May, 1988). Tolson (1987) concluded from seismic data that, although the basin initially consisted of small and narrow half-grabens formed in the early Paleocene, large-scale subsidence and sedimentary filling of the basin took place during the Miocene. This subsidence resulted in burial of the hinterland of the Lisburne-Herald fold and thrust belt.

ORIGIN OF THE LISBURNE HILLS–HERALD ARCH-WRANGEL ARCH STRUCTURAL BELT

The data presented herein indicate that the Lisburne Hills fold and thrust belt underwent a comparable percentage of shortening at about the same time in the Early Cretaceous as did the Brooks Range orogen. The key distinctions between the Lisburne Hills belt and the Early Cretaceous Brooks Range orogen are in the Lisburne Hills' position north of the Brooks Range and its generally north-south structural grain. These distinctions were the basis for the questions about the origin of the Lisburne Hills–Herald Arch–Wrangel Arch structural trend discussed by Grantz et al. (1970, 1975) and Patton and Tailleur (1977). The apparent warping or truncation of regional folds developed in middle and Upper Cretaceous rocks in the Chukchi Sea suggested to these workers that the Lisburne Hills–Herald Arch structural belt was developed in response to relative east-west convergence between North America and Eurasia that was younger than the Brooks Range orogen, either by oroclinal bending of the Brooks Range orogen or development of a fold belt that transected the Brooks Range. Our results allow for as much as 2–3 km of exhumation in the Cape Lisburne area following the main period of exhumation that occurred in the Early Cretaceous.

Because the apatite fission-track system is insensitive to cooling at temperatures appropriate for rocks at this depth, the 2–3 km of later exhumation may be ascribed partially or completely to either protracted exhumation, or to one or more episodes of contraction-related exhumation. It is possible that much of the late exhumation may have resulted from the relatively low magnitude shortening events that produced the long-wavelength folds observed in middle Cretaceous rocks in the Colville Basin and Chukchi Sea. Recent apatite fission-track results of other workers from the deformed middle Cretaceous rocks onshore suggest that an early Tertiary age for the younger period of shortening is most likely.

We believe, however, that the relatively small amount of post–middle Aptian exhumation recorded in the Cape Lisburne area is probably insufficient to accommodate a large amount of east-west tectonic shortening, whether by oroclinal bending or by development of a separate orogen, because of the large thickness of continental crust that would have been involved in the deformation. Relatively large amounts of shortening between crustal blocks of continental thicknesses would almost certainly have produced large amounts of uplift and exhumation that would be highly evident in fission-track data collected from the uplifted region. We conclude that the Lisburne Hills fold and thrust belt could not have been formed as a separate orogen in response to east-west convergence between North America and Eurasia in the early Tertiary, but instead originated as a component of the Brooks Range orogen that may have been modified by relatively minor shortening in the early Tertiary. This interpretation implies that the intersection of the Lisburne Hills and Brooks Range fold-thrust belts, buried beneath Cenozoic sediments in the southern Chukchi Sea, is probably an orogenic loop that was formed in its present position.

The reason for the change of Brookian structural trends in northwestern Alaska is unclear. Several alternatives are possible but are difficult to evaluate because of subsequent burial of the rocks that contain the change in trend beneath Hope Basin strata and the Chukchi Sea. One alternative is that the northward bend in structural trends was produced by a northward bend or bight in the paleogeography of the south-facing continental margin of the Arctic Alaska terrane prior to its collapse during the Early Cretaceous Brookian tectonic event. A second possibility is that a Mississippian Endicott Basin analogous to the Meade, Umiat, and Ikpikpuk basins may have been relatively far outboard on the continental margin near the leading edge of Early Cretaceous thrusting. The outboard location of this basin and its relatively thick sedimentary fill may have allowed the thrusts to exploit mechanical weaknesses in the sedimentary cover to a more northerly position than at other points along the orogen to the east. This may have caused inversion of the Mississippian Endicott Basin and a northward excursion or salient in the east-west structural trend of the Early Cretaceous Brooks Range orogen. These alternatives seem unlikely, however, because Ellesmerian sequence strata have not been recognized in the Herald Arch, and facies observations in the Lisburne Peninsula and Chukchi Sea suggest that a basement highland was to the west of the Lisburne Peninsula in the late Paleozoic and Mesozoic. Thus, there is no evidence of a late Paleozoic and early Mesozoic embayment in the continental margin that could have been detached and thrust eastward.

A third possibility, suggested by Ken Bird (2000, written commun.) and favored by us is that the Early Cretaceous thrusts exploited a preexisting deep, north-trending crustal structure in the Cape Lisburne area and reactivated it as an orogen-scale lateral ramp. The Hanna Trough is a north-trending basin underlain by a broad, southward-widening structural low developed on pre-Mississippian strata and bounded by the Chuckchi platform on the west and Arctic platform on the east (Grantz and May, 1988). The Hanna Trough is thought to have been formed by Late Devonian (?) or Early Mississippian extensional processes (Grantz and May, 1988), possibly by rifting as an aulacogen (Sherwood et al., 1998). The structural low beneath the basin projects southward into the region east of the Lisburne

Peninsula (Sherwood et al., 1998, Fig. 13.1) (Fig. 1), suggesting that the western flank of the Hanna Trough, i.e., the Chukchi platform, may have once existed west of the Lisburne Peninsula. This interpretation and the presence of extensional structures in the Iviagik Group (Fig. 11D) suggest that the extensional faults present along the western flank of the Hanna Trough (Thurston and Theiss, 1987) once continued southward into the Cape Lisburne area. In the Neocomian, these extensional structures may have been reactivated as eastward-vergent thrusts that ramped upward from positions in pre-Mississippian strata west of the Lisburne Peninsula into Ellesmerian strata in the Lisburne Hills. This model predicts that structures are basement involved west of the Lisburne Peninsula and become thin skinned to the east. This model is supported by the following. (1) Facies relations in the Chukchi Sea and Cape Lisburne area indicate the Ellesmerian sequence pinches out to the west against the Chukchi platform, requiring that the thrusts present in the Lisburne Peninsula must root into pre-Mississippian rocks and thus must be basement-involved to the west. (2) Fission-track data show that the Iviagik Group strata underwent the same amount of uplift-related cooling as Ellesmerian sequence strata in the Early Cretaceous. (3) The composition of westerly derived Albian strata of the Nanushuk Group on the western North Slope suggests that the now-eroded Neocomian allochthons in the Lisburne Peninsula were composed mainly of pre-Mississippian sedimentary and granitic rocks similar to those thought to underlie the Chukchi platform (C.G. Mull, Alaska Division of Geology and Geophysics, 2000, written commun.). An implication of this idea is that the Brookian orogenic belt undergoes a westward transition at the Lisburne Peninsula, from a thin-skinned orogen developed in Ellesmerian sequence strata in the Brooks Range to a thick-skinned orogen developed in pre-Mississippian rocks with little or no Ellesmerian sedimentary cover under the Chukchi Sea. A salient might have developed in the Brooks Range orogen west of the Lisburne Peninsula in order to accommodate deeper-seated thrust faults with larger amounts of vertical throw but lesser amounts of horizontal shortening than individual structures developed in Ellesmerian strata to the east. The location of the Lisburne Peninsula at the transition from thin-skinned to thick-skinned deformation would explain the presence and involvement in the deformation of pre-Mississippian rocks (i.e., Iviagik Group) uncommonly close to the thrust front compared to the central and western Brooks Range.

If the Lisburne Hills–Herald Arch structural belt originated as part of the Brooks Range orogen, why are metamorphic rocks not exposed in a hinterland to the structural belt like that of the southern Brooks Range? We suggest that the structural belt is a composite structural feature, the northern flank of which first developed as the thrust front of the Early Cretaceous Brookian orogen along the eastern flank of a salient located in the present-day southern Chukchi Sea. The thrust front was reactivated in the early Tertiary by low-magnitude thrusting that folded middle Cretaceous strata north and east of the Lisburne Peninsula, but resulted in little or no exhumation in the Cape Lisburne area. The

southern and western flanks of the Lisburne Hills–Herald Arch structural belt developed by the formation of the Hope Basin in the Cenozoic. As noted by Grantz et al. (1975), the rearward parts of the contractional belt subsided along extensional faults located at the northern and eastern margins of the Hope Basin (Tolson, 1987, Figs. 5 and 9) (Fig. 1) and were buried beneath the sedimentary fill that accumulated in the basin in the Paleogene and Neogene. This extension stranded the previously uplifted region that contained the frontal thrusts and foreland of the Early Cretaceous contractional orogen along the northern margin of the Hope Basin.

ACKNOWLEDGMENTS

We benefited greatly from the discussions and unpublished maps provided by Kenneth J. Bird, Arthur Grantz, C. Gil Mull, and Irvin L. Tailleur, who previously worked in the Cape Lisburne area. In addition, Grantz graciously contributed the previously unpublished fossil data from the Iviagik Group. We are also indebted to Julie Dumoulin, who provided petrologic information about the Okpikruak Formation in the western Brooks Range; Paul Lillis, who provided total organic carbon data from the Kingak Shale; Mark Pawlewicz, who provided vitrinite reflectance data; and Jay Namson, who aided in the construction of the balanced cross section. Foraminifers and palynomorphs were analyzed by Micropaleo Consultants, Inc. For the fission-track work, we are grateful to Tina Bush-Rester for help with sample preparation; David Coyle for providing software; Raymond Donelick, Richard Ketcham, and Donelick Analytical for permission to use their Dpar methods and software (Donelick, 1993; Ketcham et al., 2000); and the University of Oregon Radiation Center for sample irradiations. Access to fission-track samples from the Iviagik Group was provided by ARCO Alaska, Inc. Discussions with Elizabeth Miller and Jaime Toro about the tectonics and structural geology of the Cape Lisburne area are gratefully acknowledged. The manuscript was significantly improved by ideas and suggestions of reviewers Bird, Dumoulin, Grantz, Mull, Toro, and Kirk W. Sherwood. We also thank the crew at the Cape Lisburne Long-Range Radar Site, Walt Giletta, Myron Husband, Skip Linke, Calvin Lumpkin, Joe Peterson, Todd Pillars, and Gary Signs, for their hospitality and help with logistics in the field. This project was partially supported by National Science Foundation grant EAR-9317087 to Elizabeth Miller and Simon Klemperer.

REFERENCES CITED

Adams, K.E., Mull, C.G., and Crowder, R.K., 1997, Permian deposition in the north central Brooks Range, Alaska: Constraints for tectonic reconstructions: Journal of Geophysical Research, v. 102, p. 20727–20749.

Armstrong, A.K., and Mamet, B.L., 1978, Microfacies of the Carboniferous Lisburne Group, Endicott Mountains, Arctic Alaska, *in* Stelck, C.R., and Chatterton, B.D.E., eds., Western and Arctic biostratigraphy: Geological Association of Canada Special Paper 18, p. 333–394.

Armstrong, A.K., Mamet, B.L., and Dutro, J.T., Jr., 1971, Lisburne Group, Cape Lewis–Niak Creek, northwestern Alaska, Chapter B, *in* Geological Survey Research: U.S. Geological Survey Professional Paper 750–B, p. B23–B34.

Baesemann, J.F., and Lane, H.R., 1985, Taxonomy of the genus *Rhachistognathus* Dunn (Conodonta: Late Mississippian to Early Pennsylvanian), *in* Lane, R.H., and Ziegler, W., eds., Toward a boundary in the middle of the Carboniferous: Stratigraphy and paleontology: Courier Forschungsinstitut Senckenberg, no. 74, p. 93–135.

Barker, C.E., 1988, Geothermics of petroleum systems: Implications of the stabilization of kerogen maturation after a geologically brief heating duration at peak temperature, *in* Magoon, L.B., ed., Petroleum systems in the United States: U.S. Geological Survey Bulletin 1870, p. 26–29.

Bird, K.J., 1988a, Structure-contour and isopach maps of the National Petroleum Reserve in Alaska, *in* Gryc, G., ed., Geology and exploration of the National Petroleum Reserve in Alaska, 1974 to 1982: U.S. Geological Survey Professional Paper 1399, p. 355–377.

Bird, K.J., 1988b, Alaskan North Slope stratigraphic nomenclature and data summary for government-drilled wells, *in* Gryc, G., ed., Geology and exploration of the National Petroleum Reserve in Alaska, 1974 to 1982: U.S. Geological Survey Professional Paper 1399, p. 317–353.

Blome, C.D., Reed, K.M., and Tailleur, I.L., 1988, Radiolarian biostratigraphy of the Otuk Formation in and near the National Petroleum Reserve in Alaska, Chapter 33, *in* Gryc, G., ed., Geology and exploration of the National Petroleum Reserve in Alaska, 1974 to 1982: U.S. Geological Survey Professional Paper 1399, p. 725–751.

Blythe, A.E., Bird, J.M., and Omar, G.I, 1996, Deformational history of the central Brooks Range, Alaska: Results from fission-track and $^{40}Ar/^{39}Ar$ analyses: Tectonics, v. 15, p. 440–455.

Box, S.E., 1985, Early Cretaceous orogenic belt in northeastern Alaska: Internal organization, lateral extent, and tectonic interpretation, *in* Howell, D.G., ed., Tectonostratigraphic terranes of the circum-Pacific region: Houston, Texas, Circum-Pacific Council for Energy and Mineral Resources, Earth Science Series, v. 1, p. 137–145.

Brosgé, W.P., and Tailleur, I.L., 1970, Depositional history of northern Alaska, *in* Adkison, W.L., and Brosgé, M.M., eds., Proceedings of the Geological Seminar on the North Slope of Alaska: American Association of Petroleum Geologists, Pacific Section, Los Angeles, p. D1–D18.

Brown, L.M., Rexroad, C.B., Geard, J., and Williams, D., 1990, Phylogenetic and zonal implications of *Cavusgnathus tytthus* n. sp (Conodonta) from the Kinkaid Limestone (Upper Mississippian) of western Kentucky, U.S.: Geologica et Palaeontologica, v. 24, p. 77–87.

Burnham, A.K., and Sweeney, J.J., 1989, A chemical kinetic model of vitrinite maturation and reflectance: Geochimica et Cosmochimica Acta, v. 53, p. 2649–2657.

Campbell, R.H., 1961, Thrust faults in the southern Lisburne Hills, northwest Alaska: U.S. Geological Survey Professional Paper 424–D, p. 194–196.

Campbell, R.H., 1967, Areal geology in the vicinity of the Chariot site, Lisburne Peninsula, northwest Alaska: U.S. Geological Survey Professional Paper 395, 71 p.

Carter, C., and Laufeld, S., 1975, Ordovician and Silurian fossils in well cores from North Slope of Alaska: American Association of Petroleum Geologists Bulletin, v. 59, p. 457–464.

Chapman, R.C., and Sable, E.G., 1960, Geology of the Utukok-Corwin region, northwestern Alaska: U.S. Geological Survey Professional Paper 303–C, p. 47–167.

Christiansen, P.P., and Snee, L.W., 1994, Structure, metamorphism, and geochronology of the Cosmos Hills and Ruby Ridge, Brooks Range Schist belt, Alaska: Tectonics, v. 13, p. 193–213.

Cole, F., Bird, K.J., Toro, J., Roure, F., O'Sullivan, P.B., Pawlewicz, M., and Howell, D.G., 1997, An integrated model for the tectonic development of the northern Brooks Range and Colville Basin 250 km west of the Trans-Alaska Crustal Transect: Journal of Geophysical Research, v. 102, p. 20645–20684.

Collier, A.J., 1906, Geology and coal resources of the Cape Lisburne region, Alaska: U.S. Geological Survey Professional Paper 278, 54 p.

Crowder, R.K., 1990, Permian and Triassic sedimentation in the northeastern Brooks Range, Alaska: Deposition of the Sadlerochit Group: American Association of Petroleum Geologists Bulletin, v. 74, no. 9, p.1351–1370.

Donelick, R.A., 1993, A method of fission track analysis utilizing bulk chemical etching of apatite: U.S. Patent 5,267,272.

Donelick, R.A., Ketcham, R.A., and Carlson, W.D., 1999, Variability of apatite fission-track annealing kinetics. 2. Crystallographic orientation effects: American Mineralogist, v. 84, p. 1224–1234.

Dumitru, T.A., 1993, A new computer-automated microscope stage system for fission track analysis: Nuclear Tracks and Radiation Measurements, v. 21, p. 575–580.

Dumitru, T.A., 2000, Fission-track geochronology, *in* Noller, J.S., Sowers, J.M., and Lettis, W.R., eds., Quaternary geochronology: Methods and applications: American Geophysical Union Reference Shelf, v. 4, p. 131–156.

Dumitru, T.A., Miller, E.L., O'Sullivan, P.B., Amato, J.M., Hannula, K.A., Calvert, A.C., and Gans, P.B., 1995, Cretaceous to recent extension in the Bering Strait region, Alaska: Tectonics, v. 14, p. 549–563.

Dumitru, T.A., Zhou, D., Chang, E.Z., Graham, S.A., Hendrix, M.S., Sobel, E.R., and Carroll, A.R., 2001, Uplift, exhumation, and deformation in the Chinese Tian Shan, *in* Hendrix, M.S., and Davis, G.A., eds., Paleozoic and Mesozoic tectonic evolution of central Asia: From continental assembly to intracontinental deformation: Boulder, Colorado, Geological Society of America Memoir 194, p. 71–99.

Dumoulin, J.A., Watts, K.F., and Harris, A.G., 1997, Stratigraphic contrasts and tectonic relationships between Carboniferous successions in the Trans-Alaska Crustal Transect corridor and adjacent areas, northern Alaska: Journal of Geophysical Research, v. 102, p. 20709–20726.

Dutro, J.T., Jr., 1987, Revised megafossil biostratigraphic zonation for the Carboniferous of northern Alaska, *in* Tailleur, I.L., and Weimer, P., eds., Alaskan North Slope geology: Bakersfield, California, Pacific Section, Society of Economic Paleontologists and Mineralogists and the Alaska Geological Society, Book 50, p. 359–364.

Fuis, G.S., Murphy, J.M., Lutter, W.J., Moore, T.E., Bird, K.J., and Christensen, N.I., 1997, Deep seismic structure and tectonics of northern Alaska: Crustal-scale duplexing with deformation extending into the mantle: Journal of Geophysical Research, v. 102, p. 20873–20896.

Galbraith, R.F., 1981, On statistical models for fission track counts: Mathematical Geology, v. 13, p. 471–478.

Galbraith, R.F., and Laslett, G.M., 1993, Statistical models for mixed fission track ages: Nuclear Tracks and Radiation Measurements, v. 21, p. 459–470.

Gallagher, K., 1995, Evolving temperature histories from apatite fission-track data: Earth and Planetary Science Letters, v. 136, p. 421–435.

Gottschalk, R.R., and Snee, L.W., 1998, Tectonothermal evolution of metamorphic rocks in the south central Brooks Range, Alaska: Constraints from $^{40}Ar/^{39}Ar$ geochronology, *in* Oldow, J.S., and Avé Lallemant, H.G., eds., Architecture of the central Brooks Range fold and thrust belt, Arctic Alaska: Boulder, Colorado, Geological Society of America Special Paper 324, p. 225–252.

Gradstein, F.M., and Ogg, J.G., 1996, A Phanerozoic time scale: Episodes, v. 19, no. 1–2. p. 3–5.

Grantz, A., and May, S.D., 1988, Regional geology and petroleum potential of the United States Chukchi shelf north of Point Hope, *in* Gryc, G., ed., Geology and exploration of the National Petroleum Reserve in Alaska, 1974 to 1982: U.S. Geological Survey Professional Paper 1399, p. 209–230.

Grantz, A., Wolf, S.C., Breslau, L., Johnson, T.C., and Hanna, W.F., 1970, Reconnaissance geology of the Chukchi Sea as determined by acoustic and magnetic profiles, *in* Adkison, W.L., and Brosgé, M.M., eds., Proceedings of the Geological Seminar on the North Slope of Alaska: American Association of Petroleum Geologists, Pacific Section, Los Angeles, p. F1–F28.

Grantz, A., Holmes, M.L., and Kososki, B.A., 1975, Geologic framework of the Alaskan continental terrace in the Chukchi and Beaufort Seas, *in* Yorath,

C.J., Parker, E.R., and Glass, D.J., eds., Canada's continental margins and offshore petroleum exploration: Canadian Society of Petroleum Geologists Memoir 4, p. 669–700.

Grantz, A., Eittreim, S., and Whitney, O.T., 1981, Geology and physiography of the continental margin north of Alaska and implications for the origin of the Canada basin, *in* Nairn, A.E.M., Churkin, M., and Stehli, F.G., eds., The ocean basins and margins, Volume 5: New York, Plenum Publishing Corporation, p. 439–492.

Grantz, A., Tailleur, I.L., and Carter, C., 1983, Tectonic significance of Silurian and Ordovician graptolites, Lisburne Hills, northwest Alaska: Geological Society of America Abstracts with Programs, v. 15, no. 5, p. 274.

Grantz, A., May, S.D., and Hart, P.E., 1990a, Geology of the Arctic continental margin of Alaska, *in* Grantz, A., Johnson, L., and Sweeney, J.F., eds., The Arctic Ocean region: Boulder, Colorado, Geological Society of America, Geology of North America, v. L, p. 257–288.

Grantz, A., Green, A.R., Smith, D.G., Lahr, J.C., and Fujita, K., 1990b, Major Phanerozoic tectonic features of the Arctic Ocean region, *in* Grantz, A., Johnson, L., and Sweeney, J.F., eds., The Artic Ocean region: Boulder, Colorado, Geological Society of America, Geology of North America, v. L, pl. 11, scale 1:6 000 000, 1 sheet.

Grantz, A., Moore, T.E., and Roeske, S.M., 1991, Continent-Ocean Transect A-3: Gulf of Alaska to Arctic Ocean: Geological Society of America, Centennial Continent-Ocean Transect 15, scale 1:500 000, 3 sheets, 72 p. text.

Green, P.F., 1981, A new look at statistics in fission-track dating: Nuclear Tracks and Radiation Measurements, v. 5, p. 77–86.

Green, P.F., Duddy, I.R., Gleadow, A.J.W., and Lovering, J.F., 1989a, Apatite fission-track analysis as a paleotemperature indicator for hydrocarbon exploration, *in* Naeser, N.D., and McCulloh, T.H., eds., Thermal history of sedimentary basins: Methods and case histories: New York, Springer-Verlag, p. 181–195.

Green, P.F., Duddy, I.R., Laslett, G.M., Hegarty, K.A., Gleadow, A.J.W., and Lovering, J.F., 1989b, Thermal annealing of fission tracks in apatite. 4. Quantitative modelling techniques and extension to geological timescales: Chemical Geology, v. 79, p. 155–182.

Hanks, C.L., 1993, The Cenozoic structural evolution of a fold-and-thrust-belt, northeastern Brooks Range, Alaska: Geological Society of America Bulletin, v. 105, no. 3, p. 287–305.

Harland, W.B., and others, 1990, A geologic time scale 1989: Cambridge, Cambridge University Press, 263 p.

Harris, A.G., Harris, L.D., and Epstein, J.B., 1978, Oil and gas data from Paleozoic rocks in the Appalachian basin: Maps for assessing hydrocarbon potential and thermal maturity (conodont color alteration isograds and overburden isopachs): U.S. Geological Survey Miscellaneous Investigations Map I-917-E, scale 1:2 500 000, 4 sheets.

Homza, T.X., and Wallace, W.K., 1997, Detachment folds with fixed hinges and variable detachment depth, northeastern Brooks Range, Alaska: Journal of Structural Geology, v. 19, nos. 3–4, p. 337–354.

Hurford, A.J., and Green, P.F., 1983, The zeta age calibration of fission-track dating: Chemical Geology, v. 41, p. 285–317.

Johnsson, M.J., and Howell, D.G., 1996a, Generalized thermal maturity map of Alaska: U.S. Geological Survey Miscellaneous Investigations Map I- 2494, scale 1:2 500 000, 1 sheet.

Johnsson, M.J., and Howell, D.G., 1996b, Thermal maturity of sedimentary basins in Alaska: An overview, *in* Johnsson, M.J., and Howell, D.G., eds., Thermal evolution of sedimentary basins in Alaska: U.S. Geological Survey Bulletin 2142, p. 1–9.

Ketcham, R.A., Donelick, R.A., and Carlson, W.D., 1999, Variability of apatite fission-track annealing kinetics. 3. Extrapolation to geological time scales: American Mineralogist, v. 84, p. 1235–1255.

Ketcham, R.A., Donelick, R.A., and Donelick, M.B., 2000, AFTSolve: A program for multi-kinetic modeling of apatite fission-track data: Geological Materials Research, v. 2, no. 1, p. 1–32 (an entirely electronic journal at http://gmr.minsocam.org/papers/v2/v2n1/v2n1abs.html).

Kirschner, C.E., and Rycerski, B.A., 1988, Petroleum potential of representative stratigraphic and structural elements in the National Petroleum Reserve in Alaska, *in* Gyrc, G., ed., Geology and exploration of the National Petroleum Reserve in Alaska, 1974 to 1982: U.S. Geological Survey Professional Paper 1399, p. 191–208.

Kos'ko, M.K., Cecile, M.P., Harrison, J.C., Gamelin, V.G., Khandoshko, N.V., and Lopatin, B.G., 1993, Geology of Wrangel Island, between Chukchi and east Siberian Seas, northeastern Russia: Geological Survey of Canada, Bulletin 461, 73 p.

Laslett, G.M., Kendall, W.S., Gleadow, A.J.W., and Duddy, I.R., 1982, Bias in the measurement of fission track length distributions: Nuclear Tracks and Radiation Measurements, v. 6, p. 79–85.

Little, T.A., Miller, E.L., Lee, J., and Law, R.D., 1994, Extensional origin of ductile fabrics in the schist belt, central Brooks Range, Alaska: Geologic and structural studies: Journal of Structural Geology, v. 16, p. 899–918.

Magoon, L.B., and Bird, K.J., 1988, Evaluation of petroleum source rocks in the National Petroleum Reserve in Alaska, using organic-carbon content, hydrocarbon content, visual kerogen, and vitrinite reflectance, *in* Gryc, G., ed., Geology and exploration of the National Petroleum Reserve in Alaska, 1974 to 1982: U.S. Geological Survey Professional Paper 1399, p. 381–450.

Magoon, L.B., Bird, K.J., Claypool, G.E., Weitzman, D.E., and Thompson, R.H., 1988, Organic geochemistry, hydrocarbon occurrence, and stratigraphy of government-drilled wells, North Slope, Alaska, *in* Gryc, G., ed., Geology and exploration of the National Petroleum Reserve in Alaska, 1974 to 1982: U.S. Geological Survey Professional Paper 1399, p. 483–487.

Martin, A.J., 1970, Structure and tectonic history of the western Brooks Range, De Long Mountains, and Lisburne Hills, northern Alaska: Geological Society of America Bulletin, v. 81, p. 3605–3622.

Mayfield, C.F., Tailleur, I.L., and Ellersieck, I., 1988, Stratigraphy, structure, and palinspastic synthesis of the western Brooks Range, northwestern Alaska, *in* Gryc, G., ed., Geology and exploration of the National Petroleum Reserve in Alaska, 1974 to 1982: U.S. Geological Survey Professional Paper 1399, p. 143–186.

Miller, E.L., and Hudson, T.L., 1991, Mid-Cretaceous extensional fragmentation of a Jurassic–Early Cretaceous compressional orogen, Alaska: Tectonics, v. 10, p. 781–796.

Miller, E.L., Dumitru, T.A., Brown, R.W., and Gans, P.B., 1999, Rapid Miocene slip on the Snake Range–Deep Creek Range fault system, east-central Nevada: Geological Society of America Bulletin, v. 111, p. 886–905.

Moore, T.E., 1999, Balanced cross section, Bathtub syncline to Beaufort Sea through Niguanak structural high, Arctic National Wildlife Refuge (ANWR), northeastern Alaska, *in* ANWR Assessment Team, The oil and gas resource potential of the 1002 Area, Arctic National Wildlife Refuge, Alaska: U.S. Geological Survey Open-File Report 98–34, p. BC1–BC61.

Moore, T.E., and Potter, C.J., 2000, Torok–Fortress Mountain duplex, north-central Brooks Range foothills, Alaska [abs.]: Geological Society of America Abstracts with Programs, v. 32, no. 6, p. A32.

Moore, T.E., Nilsen, T.H., Grantz, A., and Tailleur, I.L., 1984, Parautochthonous Mississippian marine and non-marine strata, Lisburne Peninsula, Alaska, *in* Reed, K.M., and Bartsch-Winkler, S., eds., The United States Geological Survey in Alaska: Accomplishments during 1982: U.S. Geological Survey Circular 939, p. 17–21.

Moore, T.E., Wallace, W.K., Bird, K.J., Karl, S.M., Mull, C.G., and Dillon, J.T., 1994, Geology of northern Alaska, *in* Plafker, G., and Berg, H.C., eds., The geology of Alaska: Boulder, Colorado, Geological Society of America, Geology of North America, v. G-1, p. 49–140.

Moore, T.E., Wallace, W.K., Mull, C.G., Adams, K.E., Plafker, G., and Nokleberg, W.J., 1997, Crustal implications of bedrock geology along the Trans-Alaska Crustal Transect (TACT) in the Brooks Range, northern Alaska: Journal of Geophysical Research, v. 102, p. 20645–20684.

Mull, C.G., 1982, The tectonic evolution and structural style of the Brooks Range, Alaska: An illustrated summary, *in* Powers, R.B., ed., Geological studies of the Cordilleran thrust belt: Denver, Colorado, Rocky Mountain Association of Geologists, v. 1, p. 1–45.

Mull, C.G., 1985, Cretaceous tectonics, depositional cycles, and the Nanushuk Group, Brooks Range and Arctic Slope, Alaska, *in* Huffman, A.C., Jr., ed., Geology of the Nanushuk Group and related rocks, North Slope, Alaska: U.S. Geological Survey Bulletin 1614, p. 7–36.

Mull, C.G., Tailleur, I.L., Mayfield, D.F., Ellersieck, I.F., and Curtis, S., 1982, New Upper Paleozoic and lower Mesozoic stratigraphic units, central and western Brooks Range, Alaska: American Association of Petroleum Geologists Bulletin, v. 66, p. 348–362.

Mull, C.G., Adams, K.E., and Dillon, J.T., 1987a, Stratigraphy and structure of the Doonerak fenster and Endicott Mountains allochthon, central Brooks Range, Alaska, *in* Tailleur, I.L., and Weimer, P., eds., Alaska North Slope geology: Bakersfield, California, Pacific Section, Society of Economic Paleontologists and Mineralogists and the Alaska Geological Society, Book 50, p. 663–679.

Mull, C.G., Crowder, K., Adams, K.E., Siok, J.P., Bodnar, D.A., Harris, E.E., and Alexander, R.A., 1987b, Stratigraphy and structural setting of the Picnic Creek allochthon, Killik River quadrangle, central Brooks Range, Alaska: A summary, *in* Tailleur, I.L., and Weimer, P., eds., Alaska North Slope geology: Bakersfield, California, Pacific Section, Society of Economic Paleontologists and Mineralogists and the Alaska Geological Society, Book 50, p. 649–662.

Mull, C.G., Moore, T.E., Harris, E.E., and Tailleur, I.L., 1994, Geologic map of the Killik River quadrangle: U.S. Geological Survey Open-File Report 94–679, scale 1:125 000, 1 sheet.

Mull, C.G., Glenn, R.K., and Adams, K.E., 1997, Tectonic evolution of the central Brooks Range mountain front: Evidence from the Atigun Gorge region: Journal of Geophysical Research, v. 102, p. 20749–20772.

Mull, C.G., Reifenstuhl, R.R., Harris, E.E., Kirkham, R., and Moore, T.E., 1999, Future exploration plays in the National Petroleum Reserve, Alaska (NPRA) and adjacent western Arctic Slope [abs.]: American Association of Petroleum Geologists Annual Convention, v. 8, p. A97.

Murchey, B.L., Jones, D.L., Holdsworth, B.K., and Wardlaw, B.R., 1988, Distribution patterns of facies, radiolarians, and conodonts in the Mississippian to Jurassic siliceous rocks of the north Brooks Range, Alaska, *in* Gryc, G., ed., Geology and exploration of the National Petroleum Reserve in Alaska, 1974 to 1982: U.S. Geological Survey Professional Paper 1399, p. 697–724.

Oldow, J.S., and Avé Lallemant, H.G., editors, 1998, Architecture of the central Brooks Range fold and thrust belt, Arctic Alaska: Boulder, Colorado, Geological Society of America Special Paper 324, 317 p.

Oldow, J.S., Seidensticker, C.M., Phelps, J.C., Julian, F.E., Gottschalk, R.R., Boler, K.W., Handschy, J.W., and Avé Lallemant, H.G., 1987, Balanced cross sections through the central Brooks Range and North Slope, Arctic Alaska: American Association of Petroleum Geologists Publication, scale 1:200 000, 8 plates, 19 p. text.

O'Sullivan, P.B., 1996, Late Mesozoic and Cenozoic thermal-tectonic evolution of the North Slope foreland basin, Alaska, *in* Johnsson, M.J., and Howell, D.G., eds., Thermal evolution of sedimentary basins in Alaska: U.S. Geological Survey Bulletin 2142, p. 45–79.

O'Sullivan, P.B., and Brown, R.W., 1998, Effects of surface cooling on apatite fission-track data: Evidence for Miocene climatic change, North Slope, Alaska, *in* Van den Haute, P., and De Corte, F., eds., Advances in fission-track geochronology: Boston, Kluwer Academic Publishers, p. 255–267.

O'Sullivan, P.B., Green, P.F., Bergman, S.C., Decker, J., Duddy, I.R., Gleadow, A.J.W., and Turner, D.L., 1993, Multiple phases of Tertiary uplift and erosion in the Arctic National Wildlife Refuge, Alaska, revealed by apatite fission track analysis: American Association of Petroleum Geologists Bulletin, v. 77, p. 359–385.

O'Sullivan, P.B., Murphy, J.M., and Blythe, A.E., 1997, Late Mesozoic and Cenozoic thermotectonic evolution of the central Brooks Range and adjacent North Slope foreland basin, Alaska, including fission track results from the Trans-Alaska Crustal Transect (TACT): Journal of Geophysical Research, v. 102, p. 20821–20846.

O'Sullivan, P.B., Moore, T.E., and Murphy, J.M., 1998a, Tertiary uplift of the Mt. Doonerak antiform, central Brooks Range, Alaska: Apatite fission-track evidence from the Trans-Alaska crustal transect, *in* Oldow, J.S., and Avé Lallemant, H.G., eds., Architecture of the central Brooks Range fold and thrust belt, Arctic Alaska: Geological Society of America Special Paper 324, p. 179–194.

O'Sullivan, P.B., Wallace, W.K., and Murphy, J.M., 1998b, Fission-track evidence for apparent out-of-sequence Cenozoic deformation along the Philip Smith Mountain front, northeastern Brooks Range, Alaska: Earth and Planetary Science Letters, v. 164, p. 435–449.

Parrish, J.T., 1987, Lithology, geochemistry, and depositional environment of the Triassic Shublik Formation, northern Alaska, *in* Tailleur, I.L., and Weimer, P., eds., Alaska North Slope geology: Bakersfield, California, Pacific Section, Society of Economic Paleontologists and Mineralogists and the Alaska Geological Society, Book 50, p. 391–396.

Patton, W.W., Jr., and Tailleur, I.L., 1977, Evidence in the Bering Strait region for differential movement between North America and Eurasia: Geological Society of America Bulletin, v. 88, p. 1298–1304.

Platt, J.B., 1975, Petrography of University of Washington dredge samples from the central Chukchi Sea: U.S. Geological Survey Open-File Report 75–269, 23 p.

Poole, F.G., and Sandberg, C.A., 1991, Mississippian paleogeography and conodont biostratigraphy of the western United States, *in* Cooper, J.D., and Stevens, C.H., eds., Paleozoic paleogeography of the western United States, Volume 2: Pacific Section, SEPM (Society for Sedimentary Geology), v. 67, p. 107–136.

Roeder, D., and Mull, C.G., 1978, Tectonics of Brooks Range ophiolites, Alaska: American Association of Petroleum Geologists Bulletin, v. 62, p. 1696–1702.

Schrader, F.C., 1902, Geological section of the Rocky Mountains in northern Alaska: Geological Society of America Bulletin, v. 13, p. 233–252.

Sherwood, K.W., Craig, J.E., Lothamer, R.T., Johnson, P.P., and Zerwick, S.A., 1998, Chukchi shelf assessment province, *in* Sherwood, K.W., ed., Undiscovered oil and gas resources, Alaska federal offshore: U.S. Minerals Management Service Outer Continental Shelf Report 98–0054, p. 115–196.

Siok, J.P., 1989, Stratigraphy and petrology of the Okpikruak Formation at Cobblestone Creek, northcentral Brooks Range, *in* Mull, C.G., and Adams, K.E., eds., Dalton Highway, Yukon River to Prudhoe Bay, Alaska: Alaska Division of Geological and Geophysical Surveys, Guidebook 7, v. 2, p. 285–292.

Tailleur, I.L., 1965, Low-volatile bituminous coal of Mississippian age on the Lisburne Peninsula, northwestern Alaska: U.S. Geological Survey Professional Paper 525–B, p. 34–38.

Tailleur, I.L., and Brosgé, W.P., 1970, Tectonic history of northern Alaska, *in* Adkison, W.L., and Brosgé, M.M., eds., Proceedings of Geological Seminar on the North Slope of Alaska: American Association of Petroleum Geologists, Pacific Section, Los Angeles, p. E1–E19.

Thurston, D.K., and Theiss, L.A., 1987, Geologic report for the Chukchi Sea Planning Area, Alaska: Anchorage, Alaska, Minerals Management Service Outer Continental Shelf Report 87–0046, 193 p.

Till, A.B., and Patrick, B.E., 1991, ^{40}Ar/^{39}Ar evidence for a 110–105 Ma amphibolite-facies overprint on blueschist in the south-central Brooks Range, Alaska: Geological Society of America Abstracts with Programs, v. 23, p. A436.

Till, A.B., and Snee, L.W., 1995, ^{40}Ar/^{39}Ar isotopic evidence that blueschists formed during collision, not subduction in the Nanielik antiform, western Brooks Range, Alaska: Journal of Metamorphic Geology, v. 13, p. 41–60.

Till, A.B., Schmidt, J.M., and Nelson, S.W., 1988, Thrust involvement of metamorphic rocks, southwestern Brooks Range, Alaska: Geology, v. 10, no. 5, p. 930–933.

Tolson, R.B., 1987, Structure and stratigraphy of the Hope Basin, southern Chukchi Sea, Alaska, *in* Scholl, D.W., Grantz, A., and Vedder, J.G., eds., Geology and resource potential of the continental margin of western North

America and adjacent ocean basins: Beaufort Sea to Baja California: Houston, Texas, Circum-Pacific Council of Energy and Mineral Resources, v. 6, p. 59–72.

Tucker, R.D., Bradley, D.C., VerStraefen, C.A., Harris, A.G., Ebert, J.R., and McCutcheon, S.R., 1998, New U-Pb zircon ages and the duration and division of Devonian time: Earth and Planetary Science Letters, v. 158, p. 175–186.

Wallace, W.K., 1993, Detachment folds and a passive-roof duplex: Examples from the northeastern Brooks Range, Alaska, *in* Solie, D.N., and Tannian, F., eds., Short notes on Alaska geology: State of Alaska, Department of National Resources, Division of Geological and Geophysical Surveys Geologic Report 113, p. 81–99.

Wallace, W.K., and Hanks, C.L., 1990, Structural provinces of the northeastern Alaska Brooks Range, Arctic National Wildlife Refuge, Alaska: American Association of Petroleum Geologists Bulletin, v. 74, p. 1100–1118.

Wallace, W.K., Moore, T.E., and Plafker, G., 1997, Multistory duplexes with forward dipping roofs, north central Brooks Range, Alaska: Journal of Geophysical Research, v. 102, p. 20773–20796.

Wartes, M.A., and Reifenstuhl, R.R., 1998, Preliminary petrography and provenance of six Lower Cretaceous sandstones, northwestern Brooks Range, Alaska, *in* Clough, J.G., and Larson, F., eds., Short notes on Alaska geology, 1997: State of Alaska, Department of National Resources, Division of Geological and Geophysical Surveys Professional Report 118, p. 131–140.

Wilbur, S., Siok, J.P., and Mull, C.G., 1987, A comparison of two petrographic suites of the Okpikruak Formation: A point count analysis, *in* Tailleur, I.L., and Weimer, P., eds., Alaska North Slope geology: Bakersfield, California, Pacific Section, Society of Economic Paleontologists and Mineralogists and the Alaska Geological Society, Book 50, p. 441–447.

Wissinger, E.S., Levander, A.R., Oldow, J.S., Fuis, G.S., and Lutter, W.J., 1998, Seismic profiling constraints on the evolution of the central Brooks Range, Arctic Alaska, *in* Oldow, J.S., and Avé Lallemant, H.G., eds., Architecture of the central Brooks Range fold and thrust belt, Arctic Alaska: Boulder, Colorado, Geological Society of America Special Paper 324, p. 269–292.

MANUSCRIPT ACCEPTED BY THE SOCIETY MAY 15, 2001.

Geological Society of America
Special Paper 360
2002

Deformation and exhumation of the Mount Igikpak region, central Brooks Range, Alaska

J. Toro*
*Deaprtment of Geology and Geography, West Virginia University,
Morgantown, West Virginia 26505, USA*
P.B. Gans*
*Department of Geological Sciences, University of California, Santa Barbara,
California 93106, USA*
William C. McClelland*
*Geology and Geological Engineering Department, University of Idaho,
Moscow, Idaho 83844-3022, USA*
Trevor A. Dumitru*
*Department of Geological and Environmental Sciences, Stanford University,
Stanford, California 94305, USA*

ABSTRACT

The Mount Igikpak–Arrigetch region exposes a crustal section more than 15 km thick that includes the highest–grade metamorphic rocks in the central Brooks Range. We carried out detailed field mapping and thermochronologic analyses along a north-south transect from the deepest part of the crustal section, near Mount Igikpak, to its northern flank.

Metamorphic grade along the transect ranges from upper greenschist grade around the Mount Igikpak orthogneiss to very low grade in the Endicott Mountains allochthon to the north. Evidence of two foliations is found throughout the metasedimentary section. The S_2 fabric controls the metamorphic layering and the distribution of map units. $^{40}Ar/^{39}Ar$ analyses from white mica in the lowest–grade rocks indicate that peak metamorphism associated with the development of the S_1 foliation occurred before 112 Ma, probably owing to crustal thickening caused by collision of an island arc against the Arctic Alaska margin.

The dominant foliation (S_2) formed during rapid cooling from hornblende closure temperature (500 ± 50 °C) ca. 98 Ma to biotite closure temperature ($\sim 300 \pm 50$ °C) by ca. 90 Ma. Mineral and elongation lineations associated with S_2 plunge to the north, and associated kinematic indicators show top-to-the-north shear, reflecting north-south extension of the crust in mid-Cretaceous time. The episode of extensional deformation took place on a crustal scale and resulted in tectonic exhumation of the core of the orogen. The driving mechanism may have been gravitational collapse of crust that was overthickened during the preceding collision.

---

*E-mails: Toro—jtoro@wvu.edu; Gans—gans@geology.ucsb.edu; McClelland—wmcclell@uidaho.edu; Dumitru—Trevor@pangea.Stanford.edu.
Data Repository item 2002075 contains additional material related to this article.

Toro, J., Gans, P.B., McClelland, W.C., and Dumitru, T.A., 2002, Deformation and exhumation of the Mount Igikpak region, central Brooks Range, Alaska., *in* Miller, E.L., Grantz, A., and Klemperer, S.L., eds., Tectonic Evolution of the Bering Shelf–Chukchi Sea–Arctic Margin and Adjacent Landmasses: Boulder, Colorado, Geological Society of America Special Paper 360, p. 111–132.

INTRODUCTION

The Brooks Range orogen extends for ~1000 km across northern Alaska from east of the Canadian border to the Chukchi Sea, where it disappears abruptly (Fig. 1). The metamorphic rocks of the Brooks Range have been correlated with those of the Seward Peninsula and the Chukotka Peninsula of Russia (Natal'in et al., 1999). In Early Cretaceous time a continuous compressional orogenic belt probably extended across the Bering Strait. Extensional tectonic processes that affected the Bering-Chukchi region during the mid-Cretaceous (Dumitru et al., 1995) and early Tertiary (Tolson, 1987) contributed to the collapse of the orogen and the formation of the wide expanse of submerged shelves and basins that are imaged by the deep crustal seismic line presented in this volume (Klemperer et al., this volume, Chapter 1). Thus the tectonic history of rocks exposed today in the Brooks Range has a direct bearing on the evolution of the Bering-Chukchi crust.

We chose to study the Mount Igikpak area, located 350 km east of the Chukchi Sea, because the highest-grade metamorphic rocks of the central Brooks Range are exposed in a large-scale antiformal structure (Fig. 1). The north flank of this antiform exposes a crustal section more than 15 km thick that presumably was once above the Mount Igikpak orthogneiss (Figs. 2 and 3). Metamorphic grade ranges from lower amphibolite facies in the core of the Mount Igikpak gneiss to subchlorite grade in the sedimentary rocks of the Endicott Mountains allochthon north of the Alatna River (Nelson and Grybeck, 1981). This structural relief makes it possible to study deep-seated features that are not exposed elsewhere in the region but are critical to understanding the development of its metamorphic infrastructure.

In the Brooks Range of northern Alaska (Fig. 1) large-magnitude shortening within the thrust belt has been attributed to the Late Jurassic to Early Cretaceous collision of an island arc against the Arctic Alaska margin along the present-day southern boundary of the range (Roeder and Mull, 1978). It has been proposed that metamorphic fabrics exposed in the internal part of the range also formed during this collisional event (Gottschalk, 1990; Moore et al., 1994; Till and Snee, 1995). In this chapter we present structural and geochronologic evidence suggesting that the gently north dipping dominant foliation (S_2) in the Mount Igikpak area of the central Brooks Range is related to

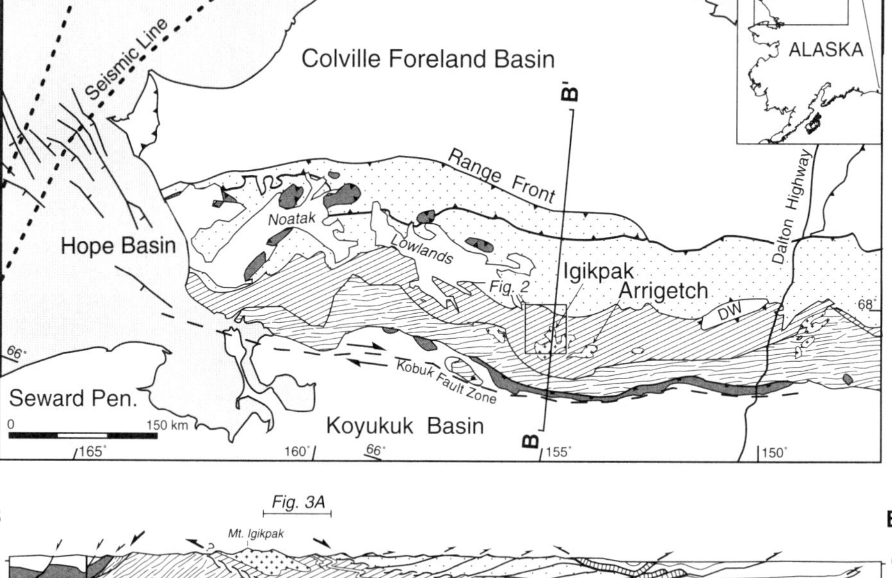

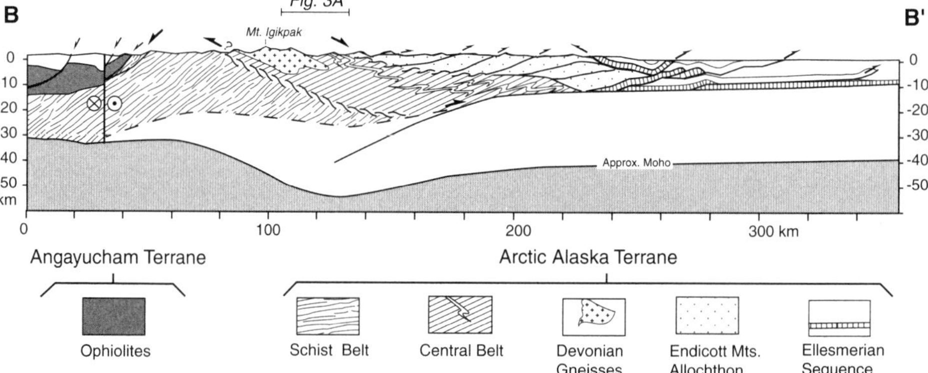

Figure 1. A: Location map of Brooks Range modified from Mull et al. (1987) and Moore et al. (1994). DW is Doonerak window. Seismic line is Bering-Chukchi deep crustal seismic profile (Klemperer et al., this volume, Chapter 1). B: Regional cross section through central Brooks Range after Mull et al. (1987) and Moore et al. (1994). Approximate depth to Moho was determined by correlating Bouguer gravity along this cross section with that of TACT transect, where Moho depths are known from seismic refraction and reflection (Fuis et al., 1997).

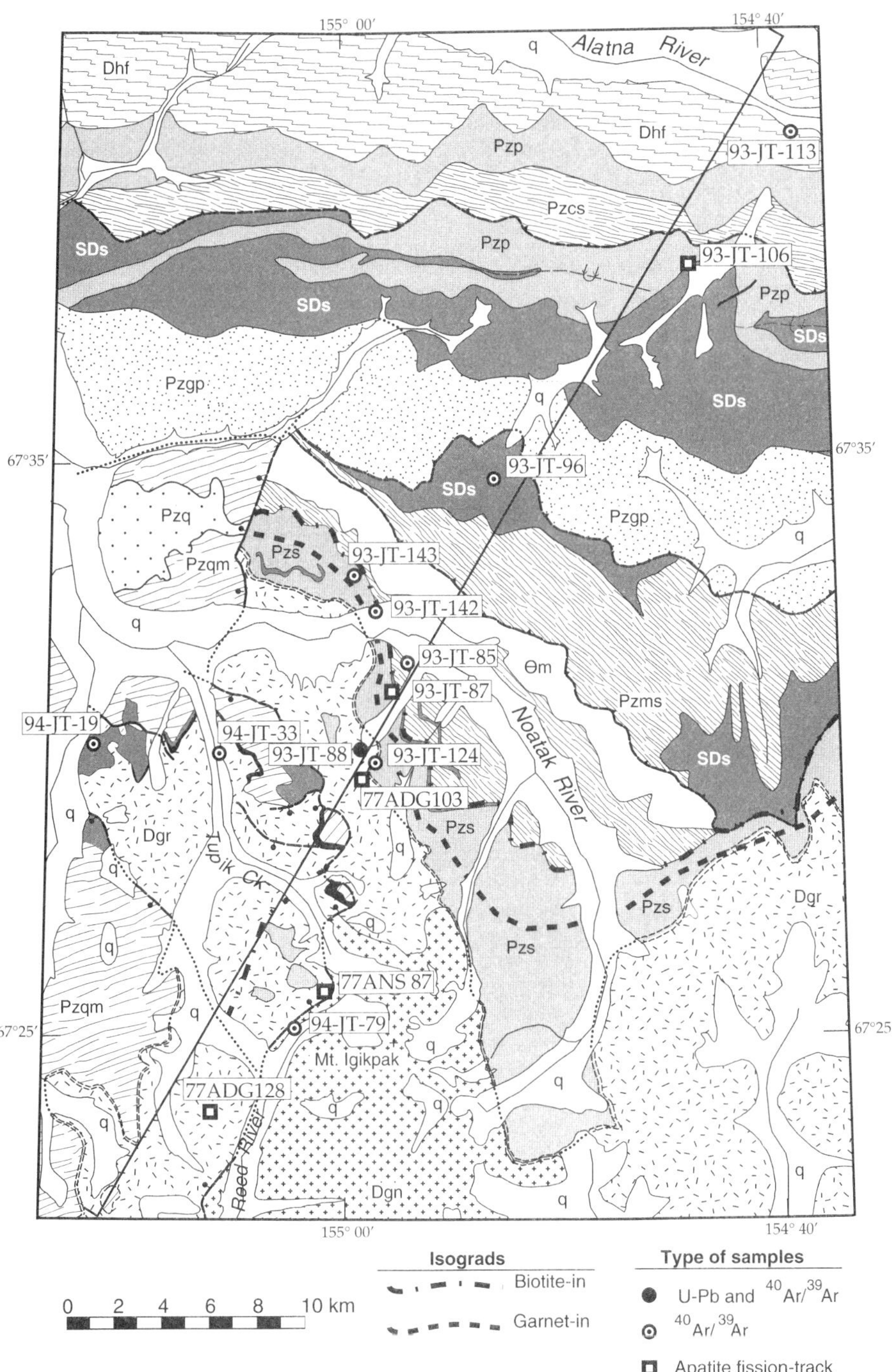

Figure 2. Geologic and sample locality map of northern flank of Mount Igikpak orthogneiss compiled from our mapping and Nelson and Grybeck (1980). Refer to Figure 1 for location and to Figure 3 for legend. Samples with prefix 77ADG are from O'Sullivan et al. (1993). A color version of this figure is on the CD-ROM accompanying this volume.

mid-Cretaceous uplift and exhumation. The S_2 foliation overprints a higher-grade fabric (S_1) presumably related to the Late Jurassic to Early Cretaceous collisional event.

Stratigraphic and seismic reflection data from the northern foothills of the Brooks Range (Cole et al., 1997) indicate that minor north-vergent thrusting in the foreland was active in the mid-Cretaceous at the same time that the metamorphic hinterland was undergoing down-to-the-north ductile deformation and exhumation. This suggests that the two modes of deformation were kinematically linked. Mid-Cretaceous exhumation of the core of the Brooks Range may have been driven by the gravitational instability of an orogenic pile thickened during the preceding collision. Contemporaneous and parallel thrusting and extension have been documented in other collisional orogens such as the Himalayas (Burchfiel et al., 1992), and have been attributed to gravity-driven topographic collapse.

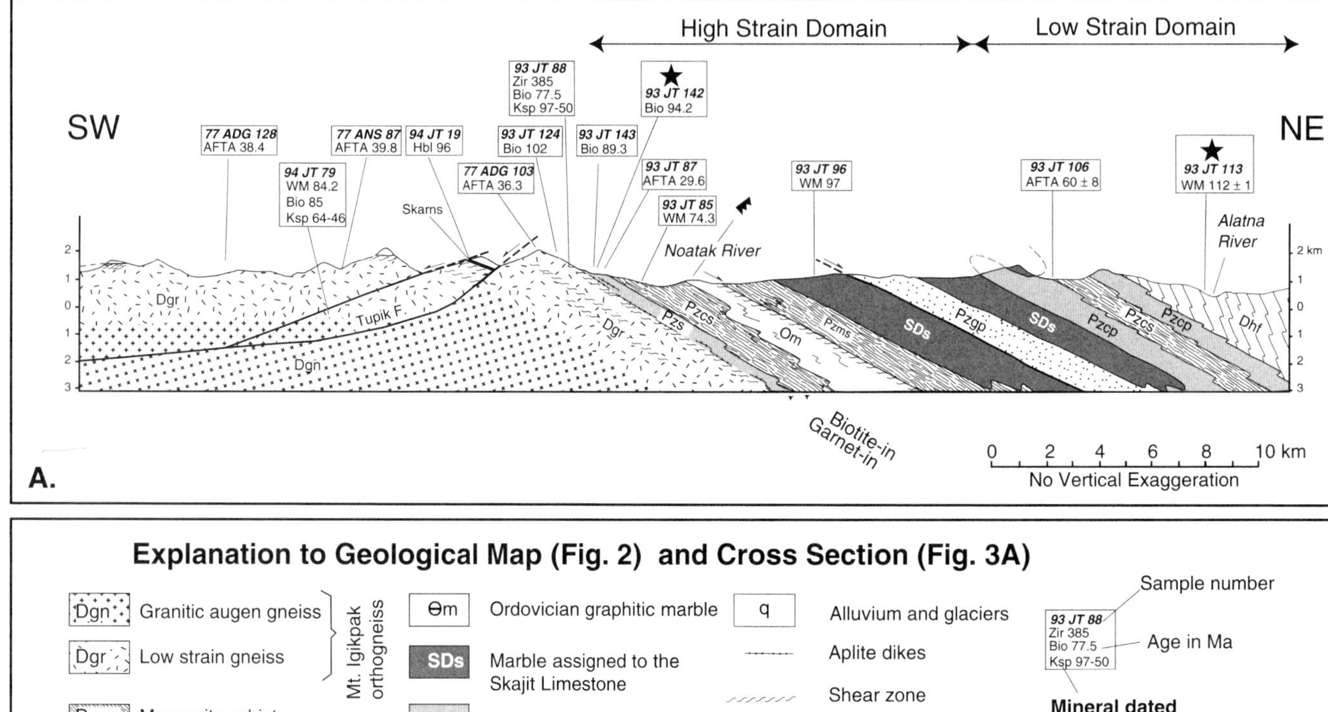

Figure 3. A: Structural cross section through study area. U-Pb, ^{40}Ar/^{39}Ar, and fission-track ages are projected onto plane of cross section. B: Legend to geologic map (Fig. 2) and cross section (Fig. 3A).

The Mount Igikpak antiform is at first sight similar to the Doonerak antiform, a much better studied structure located near the Dalton Highway to the east (Fig. 1A). The Doonerak structure has been interpreted to be underlain by a duplex of Tertiary age (Oldow et al., 1987; Moore et al., 1997; O'Sullivan et al., 1998). In this chapter we argue for a very different origin for the Mount Igikpak antiform. Our data show that doming of the area started concurrently with north-vergent ductile deformation associated with S_2. Therefore the S_2 fabric in the north flank of the dome is a primary feature and corresponds to a broad down-to-the-north extensional shear zone.

In further contrast to the Doonerak area, the youngest structures mapped on Mount Igikpak are north- to northwest-trending brittle normal faults of probable early Tertiary age. The orientation of these normal faults indicates that this part of the central Brooks Range underwent a period of orogen-parallel extension perhaps even as north-south shortening was active to the east.

GEOLOGIC SETTING

The Brooks Range orogen is believed to have originated during collision of an island arc with a portion of the North American continental margin (the Arctic Alaska terrane) during the Late Jurassic and Early Cretaceous (see review in Moore et al., 1994). The Arctic Alaska terrane consists of continental margin rocks of differing lithology, metamorphic grade, and structural style, which, in the central Brooks Range, include the Endicott Mountains allochthon, the Central Belt, and the Schist Belt (Figs. 1 and 2) (Moore et al., 1994). The Angayucham terrane, which occupies the highest structural level, represents oceanic assemblages and parts of the island arc that collided with Arctic Alaska.

The Endicott Mountains allochthon is composed of Upper Devonian through Lower Cretaceous sedimentary rocks imbricated by north-directed Brookian thrust faults. Stratigraphic relations in the higher thrust sheets indicate that they have undergone several hundred kilometers of northward displacement (Mayfield et al., 1988; Moore et al., 1994). The strata at the base of the Endicott Mountains allochthon are low greenschist grade, but at higher structural levels sedimentary rocks are not metamorphosed. Data presented in this study suggest that the Endicott Mountains allochthon and the Central Belt are separated by a broad zone of distributed ductile strain.

The Central Belt is composed mostly of greenschist facies metasedimentary and metavolcanic rocks of probable Late Proterozoic to Paleozoic age and includes marble, calcareous schist, chlorite schist, and minor pelitic schist. Granitic gneisses, surrounded by greenschist- to amphibolite-grade metasedimentary rocks, are exposed along the crest of the Brooks Range. The orthogneiss bodies are variably deformed and have yielded Proterozoic and Late Devonian ages. The Central Belt is distinguished from the Schist Belt, which is to the south, by the lack of blueschist facies minerals and the presence of mainly lower Paleozoic marbles of the Skagit Limestone unit (Moore et al., 1994, 1997).

In the Mount Igikpak area (Figs. 2, 3, and 4), rocks assigned to the Central Belt are polydeformed and the dominant foliation has completely transposed original sedimentary layering. Sedimentary structures are detectable only at the highest structural

levels of the Central Belt, near the transition into the overlying Endicott Mountains allochthon. South of the Mount Igikpak orthogneiss the nature and location of the boundary between the Central Belt and the Schist Belt are not clear. Along the Dalton Highway, 200 km to the east of our study area, the boundary is a north-dipping ductile shear zone (Moore et al., 1997). Mull (1977) and Till et al. (1988) proposed that the Schist Belt–Central Belt boundary is a backthrust, and Oldow et al. (1998) argued that it corresponds to the basal detachment of the overlying Skagit allochthon. Although we do not have data to resolve this controversy, we choose the first hypothesis because it leads to a simpler kinematic history (Fig. 1A).

Geobarometric studies have shown that peak pressures in the Schist Belt exceeded those in the Central Belt (Patrick, 1995). The Schist Belt is composed of polydeformed greenschist to epidote-amphibolite facies metasedimentary and metavolcanic rocks with relict blueschist facies minerals. Protoliths appear to be mostly Proterozoic to mid-Paleozoic continental margin- to arc-type sedimentary and volcanic rocks (Till et al., 1988). High-pressure–low-temperature metamorphism within the Schist Belt took place during collision that led to underthrusting of the Arctic Alaska continental margin beneath oceanic and arc rocks (Till, 1992a). The timing of this event has been difficult to determine due to effects of younger thermal

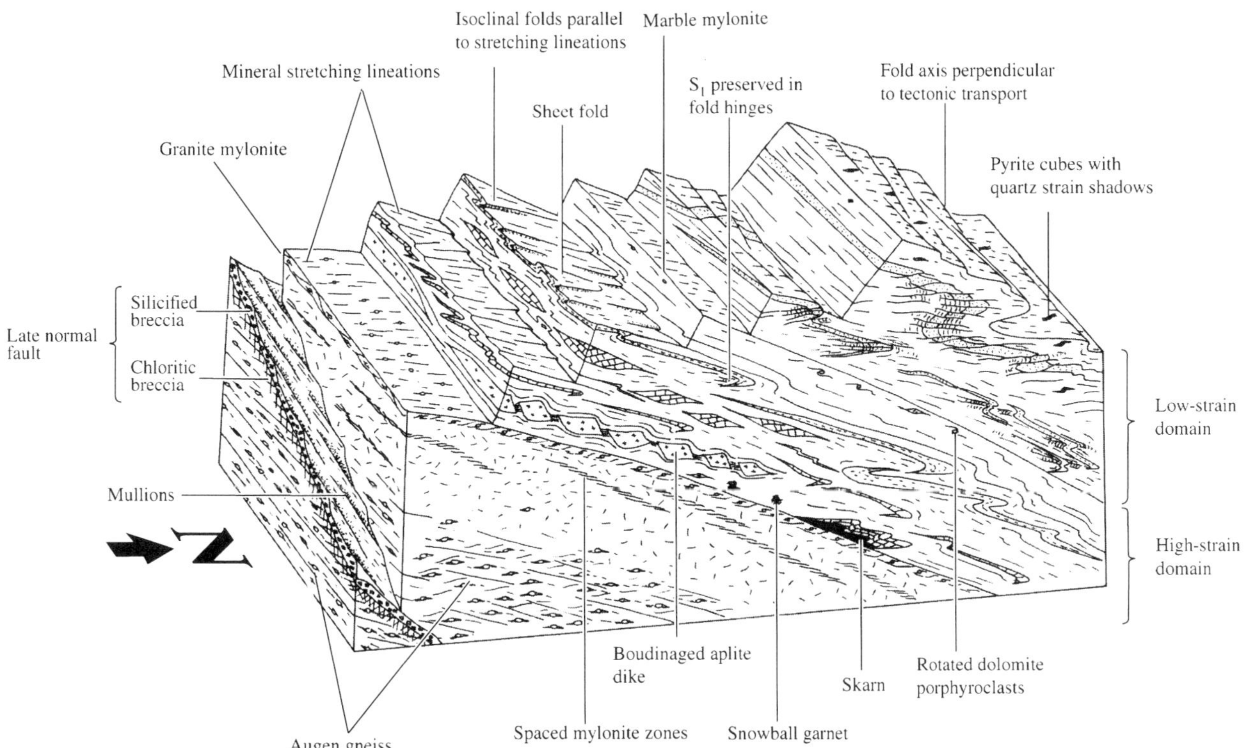

Figure 4. Synthetic block diagram showing main textural and structural features of north flank of Mount Igikpak. Inspired by a figure of Malavieille (1993).

overprints, but is thought to have taken place in Middle to Late Jurassic time, between 171 Ma (Christiansen and Snee, 1994) and 142 Ma (Gottschalk and Snee, 1998). Remnants of the overriding oceanic allochthon are represented by the Angayucham terrane, a belt of mafic and ultramafic rocks, chert, and pelagic sedimentary rocks ranging from Devonian to Jurassic age, extending along the southern flank of the Brooks Range. A narrow belt of low-grade metasedimentary rocks intervenes between the Angayucham terrane and the Schist Belt and is bounded by down-to-the-south normal faults (Box, 1987; Gottschalk and Oldow, 1988; Little et al., 1994). The presence of this system of normal faults has been cited as evidence that the Brooks Range collisional orogen was modified by extensional deformation after emplacement of the allochthons (Miller and Hudson, 1991; Little et al., 1994).

The structure of the southern boundary of the Brooks Range is further complicated by the Kobuk fault zone (Fig. 1), a complex zone of brittle deformation that operated as a right-lateral strike-slip fault zone during Tertiary to recent time (see review in Avé-Lallemant et al., 1998). This fault is thought to be part of the system of major strike-slip faults that cut through interior Alaska including the Tintina, Denali, and Kaltag faults. The Kobuk fault dies out to the west in the Chukchi Sea and has been linked with Eocene transtensional deformation in the offshore Hope Basin (Tolson, 1987).

GEOLOGY OF THE MOUNT IGIKPAK REGION

The Mount Igikpak area is mainly within the Central Belt (Figs. 1A and 2). This area was previously mapped in reconnaissance as part of the Survey Pass quadrangle (Nelson and Grybeck, 1980). The Mount Igikpak orthogneiss, exposed at the headwaters of the Noatak River, is overlain to the north by a thick section of probable Paleozoic greenschist-grade metasedimentary rocks; the rocks have a strong foliation that dips consistently to the north at ~25° (Figs. 2 and 3).

Evidence for at least two foliations is found in the surrounding metasedimentary section. S_1 is only locally preserved. The younger foliation (S_2) is the dominant fabric that controls the geometry and orientation of compositional layering in the metasedimentary rocks. S_2 is a high-strain transposition fabric and is axial planar to recumbent isoclinal folds of all scales. The degree of penetrative strain within the metasedimentary sequence decreases upsection and to the north. Likewise, metamorphic grade decreases from upper greenschist grade near the gneiss to very low grade at the north end of the transect (Fig. 3A).

Repetition of distinctive marble panels of Skajit Limestone (Fig. 3A) provides the only direct evidence of thrust imbrication in the Mount Igikpak area (e.g., Nelson and Grybeck, 1980). There is no remaining evidence for brittle fault motion along these contacts, because the entire section has been overprinted by intense ductile strain.

Mount Igikpak orthogneiss

The Igikpak orthogneiss is the largest body of metagranite exposed in the central Brooks Range. It crops out as two irregular lobes, but our mapping was confined to the western and larger one (Figs. 1 and 2). There are two texturally distinct phases within the orthogneiss: (1) megacrystic K-feldspar augen gneiss, which constitutes most of the internal portion of the metapluton (Fig. 5A) and (2) fine-grained unfoliated granite, which is found along its northern rim. K-feldspar porphyry and aplite also occur as border phases.

Throughout the augen gneiss there is a well-developed north-trending mineral and stretching lineation that parallels the lineation in the overlying metasedimentary rocks. S-C fabrics and the asymmetry of porphyroclasts consistently indicate top-to-the-north shear (Fig. 5A). Deformation at ~450–500 °C is evidenced by the dynamic recrystallization of the K-feldspars, which display core-mantle textures, and by the recrystallized rims of myrmekite (Passchier and Trouw, 1996). On the basis of these petrographic and structural data, and the field relationships and cross-sectional geometry (Figs. 2 and 3), we interpret the coarse augen gneiss to represent a deeper structural level than the finer-grained phase of the orthogneiss.

The finer-grained, marginal northern portion of the orthogneiss does not have a pervasive macroscopic metamorphic fabric. Instead, deformation is concentrated along spaced mylonitic bands. In thin section, quartz shows dynamic recrystallization, manifested by mantles of small new grains around larger internally strained quartz grains, and by convoluted grain boundaries.

We attribute the lack of a penetrative metamorphic fabric in the northern portion of the orthogneiss to the rheological contrast between the metagranite and the more readily deformed metasedimentary rocks that overlie it. During deformation, strain was concentrated within the schist and carbonate section, but only along discrete shear zones in the upper portion of the orthogneiss.

There are many contact relationships between the Mount Igikpak orthogneiss and surrounding schists. The upper contact of the gneiss is generally coplanar with the foliation in metasedimentary rocks. In many places a low-angle mylonitic shear zone is developed along the orthogneiss-metasediment boundary (Figs. 2 and 5B). Within this zone, the orthogneiss exhibits dramatic grain-size reduction and quartz ribbons are developed. S-C fabrics and porphyroclasts with asymmetric strain shadows in the shear zone consistently indicate top-to-the-north sense of shear (Fig. 5B). Brittle normal faults, oriented north-south, and with down-to-the-west displacement, also offset the margin of the gneiss (Figs. 2 and 3).

There is evidence for an original intrusive relationship along the northern and western flanks of the gneiss. Skarns are developed in calcareous lithologies (Newberry et al., 1986) and numerous aplitic dikes are found adjacent to the orthogneiss.

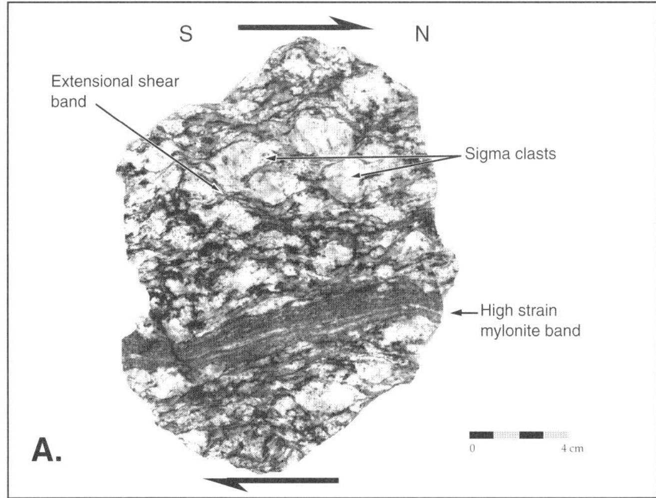

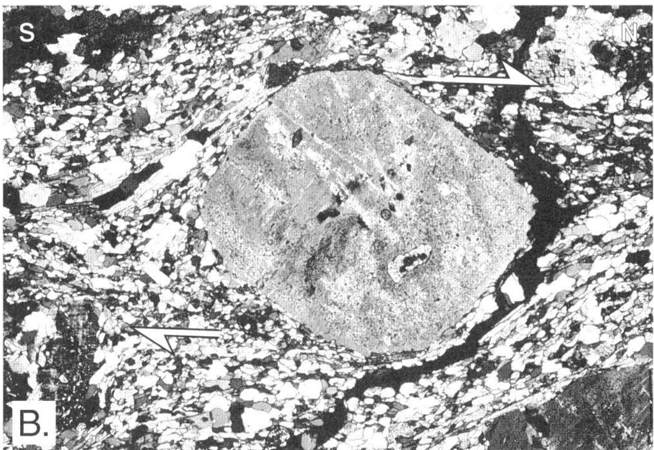

Figure 5. A: Polished slab of Mount Igikpak orthogneiss. Asymmetric strain shadows of K-feldspar augen (sigma clasts) and extensional shear bands indicate top-to-north sense of shear. There is intense grain-size reduction in high-strain mylonitic band that cuts slab. B: K-feldspar porphyroblast in mylonite zone developed along margin of Igikpak orthogneiss north of Noatak River. Field of view is 14.4 mm wide.

The aplite dikes, which are concordant with the dominant foliation, rapidly diminish in abundance upsection. Some are conspicuously boudinaged, indicating at least 100% extension in a north-south direction.

Granitic gneisses in the central Brooks Range were initially believed to be Cretaceous in age and syntectonic with Brookian metamorphism (Brosgé and Pessel, 1977). Nelson and Grybeck (1978) were the first to suggest a pre-Cretaceous age for the Arrigetch and Igikpak orthogneisses on the basis of textural and field relationships. Later Rb/Sr (Silberman et al., 1979) and U-Pb isotopic work demonstrated that many of the Brooks Range orthogneisses are Devonian, although the U-Pb zircon data are often complex (Dillon et al., 1980; Aleinikoff et al., 1993). Pat-

ton et al. (1987) carried out U-Pb analyses on samples from the foliated core and the unfoliated northern portion of the Mount Igikpak orthogneiss. Both analyses yielded discordant Devonian ages (Table 1; Fig. 6A).

In an attempt to better define the age of crystallization of the Mount Igikpak orthogneiss, we analyzed zircons from a sample of the unfoliated northern portion of the gneiss (sample 93–JT-88; location in Fig. 2). Three conventional zircon fractions yielded discordant analyses with U/Pb and Pb/Pb ages ranging from 320 to 627 Ma (Table 1; Fig. 6A). To evaluate the zircon systematics of this sample, a fraction of coarse zircon was subjected to a step-wise dissolution experiment following the technique of Mattinson (1994) and McClelland and Mattinson (1996). Sequential dissolution steps resulted in progressively increasing U/Pb and $^{207}Pb/^{206}Pb$ ages (Table 1), such that the analyses define a line that generally parallels the concordia curve (Fig. 6A). The first three steps, interpreted to reflect the effect of Pb-loss on primary magmatic components, define a chord with a lower intercept of 120 Ma. This is consistent with the age of peak metamorphism in the area as independently defined by the $^{40}Ar/^{39}Ar$ data presented in the following. Step L4 and the residue define a chord from ca. 385 to 1375 Ma. We interpret this chord to reflect a mixture between primary Devonian magmatic zircon and the average age of Proterozoic inherited components. This interpretation is consistent with earlier observations, based on bulk composition and ε_{Nd} signature, that the Mount Igikpak orthogneiss represents a highly evolved magma with a large component of contamination from Proterozoic continental crust (Nelson et al., 1993; Newberry et al., 1986).

We assign an approximate crystallization age of 375–395 Ma to this phase of the Igikpak orthogneiss, which is consistent with the ages determined for similar metaplutonic bodies near the Dalton Highway (Aleinikoff et al., 1993).

The tectonic setting of Devonian magmatism in the Brooks Range is not well understood. Both arc (Dillon et al., 1980) and extensional (Hitzman et al., 1986) environments have been proposed. A possible scenario is a continental arc undergoing extension in the backarc region to account for the inherited crustal component seen in the U-Pb analyses.

Patrick (1995) carried out a thermobarometric study of the Brooks Range orthogneisses and estimated metamorphic conditions for two samples of the Mount Igikpak gneiss to be in excess of 5.5 kbar and 373–416 °C. The moderately high pressure resulted from crustal thickening during Late Jurassic–Early Cretaceous collisional deformation in the Brooks Range.

Metasedimentary rocks

The metamorphic section overlying the Mount Igikpak orthogneiss consists of alternating layers of calcareous schist, marble, chlorite schist, carbonaceous slate, and graphitic phyllites (Fig. 3). In addition, minor pelitic schist, micaceous quartzite, metaconglomerate, and small metagabbro intrusives

TABLE 1. U-PB ISOTOPIC DATA AND APPARENT AGES

Fraction Size[#]	Wt	Concentration[†]		Isotopic Composition			Apparent Ages (Ma)[§]		
				$\frac{^{206}Pb}{^{204}Pb}$	$\frac{^{206}Pb}{^{207}Pb}$	$\frac{^{206}Pb}{^{208}Pb}$	$\frac{^{206}Pb^*}{^{238}U}$	$\frac{^{207}Pb^*}{^{235}U}$	$\frac{^{207}Pb^*}{^{206}Pb^*}$
(µm)	(mg)	U	Pb*						
Sample 93-JT-88									
Conventional Fractions									
a 63-80	0.1	532	27.9	831 ± 1	13.154	5.511	320.3 ± 0.6	349.7 ± 1.0	550 ± 5
b 100-145A	0.3	458	43.8	1167 ± 3	13.687	4.644	551.8 ± 1.1	566.6 ± 1.4	627 ± 3
c 100-350A	0.2	313	18.3	4569 ± 8	16.959	6.501	354.9 ± 0.7	366.9 ± 0.8	444 ± 1
Partial Dissolution Results									
%U									
L1	14.1%	84	2.0	106 ± 0.2	5.281	1.923	141.6 ± 0.4	145.7 ± 3.4	212 ± 53
L2	6.5%	39	1.1	803 ± 5	14.339	5.071	179.5 ± 0.5	185.5 ± 0.9	263 ± 8
L3	51.6%	308	15.5	1204 ± 3	15.062	5.619	306.5 ± 0.6	315.4 ± 0.8	382 ± 4
L4	27.4%	163	10.9	4403 ± 37	16.774	6.289	402.1 ± 0.8	411.5 ± 0.9	465 ± 2
Residue	0.4%	3	0.2	935 ± 41	12.181	4.719	526.5 ± 1.1	587.7 ± 1.5	832 ± 3
Total 145–350	1.1	543	27.1	772 ± 12	13.578	5.040	302.6 ± 0.6	314.3 ± 1.2	402 ± 10
Samples from Patton et al. (1987)									
84APa11		1183.2		274.7	10.037	4.418	322.2	326.9	359.2
84APa13		2594.3		813.0	1.382	9.344	288.5	299.6	387.5

[#]A designates fractions abraded to 30%–60% of original mass; L1, etc., designate leachate steps; Total is mathematically recombined steps and residue. Partial dissolution procedures modified from Mattinson (1994). See Toro (1998) for analytical details.
[†]Pb* is radiogenic Pb (in ppm) for conventional analyses and nanograms for step-wise dissolution steps. U is in ppm for conventional analyses and nanograms and for step-wise dissolution steps.
[§]Uncertainties reported as 2 sigma.

within the marbles are present. The protoliths are probably Late Proterozoic to middle Paleozoic continental margin sedimentary and volcaniclastic rocks (Nelson and Grybeck, 1980).

Graphitic marbles overlying the Mount Igikpak orthogneiss contain Middle Ordovician conondonts (Toro, 1998). Massive white marbles assigned to the Skajit Limestone (Silurian to Devonian) are exposed along the divide between the Noatak and Alatna Rivers (Figs. 2 and 3). Graphitic phyllites that overlie the Skajit Limestone have been correlated with the Upper Devonian Hunt Fork Shale of the Endicott Mountains allochthon (Nelson and Grybeck, 1980).

Metamorphic fabrics and structures

The study area is divided into two structural domains: (1) a southern high-strain domain adjacent to the Mount Igikpak orthogneiss and continuing northward as far as the upper marble panel of Skajit Limestone and (2) a northern lower-strain domain comprising the phyllitic and slaty rocks that extend into the Endicott Mountains allochthon (Figs. 2, 3A, and 4).

High-strain domain. The high-strain domain is characterized by a well-developed schistosity that completely transposes all sedimentary bedding. In addition, north-trending mineral and stretching lineations are present in all lithologies. Axes of isoclinal folds are parallel to these lineations and to the direction of tectonic transport. Two metamorphic fabrics can be recognized.

An S_1 foliation is locally preserved, while the dominant fabric, S_2, controls mesoscopic and map-scale layering.

Early foliation (S_1). This older fabric is preserved within the high-strain domain in the hinges of folds, in inclusion trails within garnets (Fig. 7A), and by bands of garnet porphyroblasts that have not been entirely transposed into parallelism with S_2. Garnet-biotite schists are present in pelitic lithologies along the northern and eastern margin of the Mount Igikpak gneiss. Microscopic evidence indicates that garnet is associated with this older fabric that predates the dominant foliation. The garnets exhibit spiral inclusion trails and snowball textures that are clearly syntectonic. Coarse porphyroblastic S_1 biotite is locally preserved at high angle to the dominant fabric. Early garnet and biotite are partially replaced by chlorite, muscovite, and new fine-grained biotite S_2 biotite (Fig. 7A). Relict S_1 mica survives in fold hinges and within small domains that were not fully transposed.

Previous workers noted that the garnet-in and biotite-in isograds generally follow the margin of the orthogneiss (Nelson and Grybeck, 1979; Newberry et al., 1986) (Fig. 2). This suggests the possibility that the higher-grade assemblages are related to the intrusion of the orthogneiss. However, in a thermobarometric study on garnet-biotite schists from the north side of the Arrigetch gneiss, where similar relationships exist, Vogl and Patrick (1995) obtained temperatures in the 500–600 °C range and pressures of 6–8 kbar. Because of the dynamic textures ob-

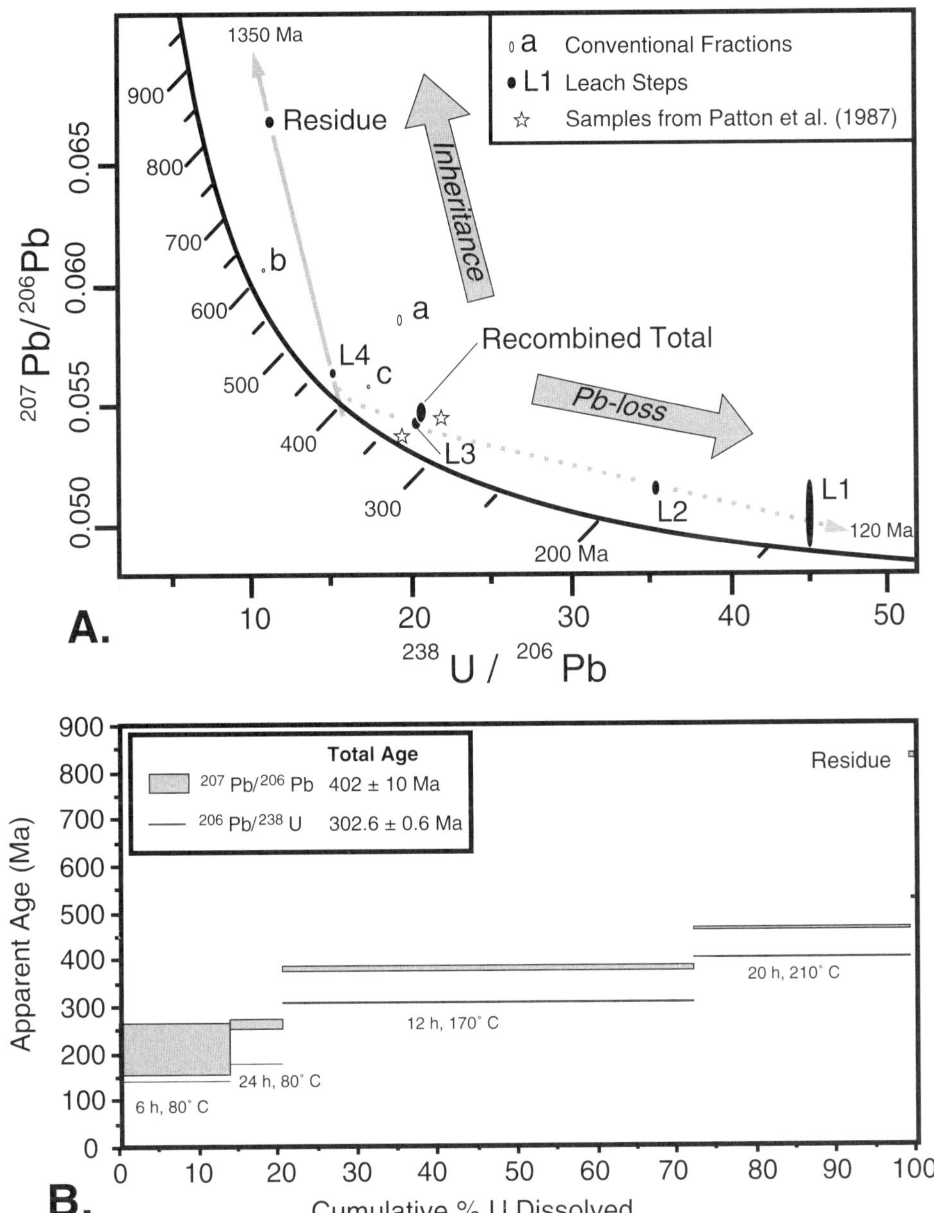

Figure 6. U-Pb zircon data from Mount Igikpak orthogneiss. A: Tera-Wasserberg plot showing both conventional and partial dissolution analyses of sample 93-JT-88. Conventional fractions b and c were abraded. Two samples from eastern and central part of orthogneiss, previously published by Patton et al. (1987), are shown for comparison. B: U release spectra for partial dissolution experiment.

served in garnet, the moderately high pressures and temperatures suggested by thermobarometry, and the evidence of an S_1 fabric throughout the structural section, we conclude that the early metamorphic and deformational event associated with S_1 took place during the initial phase of Brookian metamorphism, rather than with the intrusion of the gneiss in Devonian time. The presence of the higher-grade assemblages in the vicinity of the orthogneiss is a function of the structural level exposed.

Dominant foliation (S_2). The gently north dipping foliation (Figs. 3 and 8A) is the most obvious and laterally continuous structural feature of the metamorphic rocks on the north flank of Mount Igikpak. This fabric is axial planar to recumbent isoclinal folds that range from a few centimeters to hundreds of meters in wavelength (Figs. 3 and 7B). The fold axes range from north trending, with east-vergent asymmetry in the high-strain domain, to more east-west trending in the low-strain areas at higher structural levels (Figs. 4 and 8C). All the metamorphic assemblages associated with S_2 in the metasedimentary rocks correspond to medium and low greenschist facies.

The S_2 foliation is defined by well-aligned late biotite, muscovite, and chlorite and by elongated grain-shape fabric in matrix minerals. Chlorite is common, generally growing after early biotite and garnet in the deeper part of the section (Fig. 7A). The metamorphic event associated with S_2 produced retrograde as-

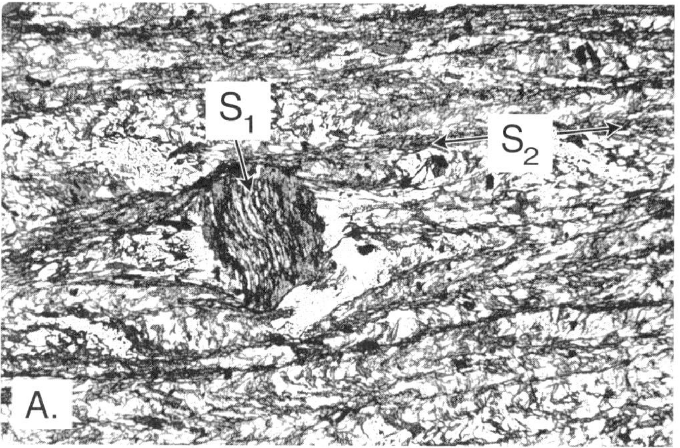

Figure 7. Metasedimentary rocks overlying Mount Igikpak orthogneiss. A: Photomicrograph showing garnet porphyroblast with spiral inclusion trails at high angle to S_2 foliation and with alteration rim of chlorite. Garnet + biotite + white mica + quartz is S_1 assemblage, white mica + chlorite + quartz is S_2 assemblage. Field of view is 7.2 mm wide. Sample was collected near 93-JT-124 locality shown in Figure 2. B: Isoclinal folds in banded marble and quartzite (map unit Pzqm).

semblages that recrystallized the earlier, upper greenschist-grade minerals.

Mineral and elongation lineations are strongly developed on S_2 planes in the high-strain domain. They consistently plunge to the north, parallel to isoclinal fold axes (Fig. 8C). Both microscopic and macroscopic kinematic indicators, such as asymmetric strain shadows of porphyroclasts in granitic mylonites, oblique grain-shape fabrics, shear bands, and asymmetric boudins, consistently indicate a top-to-the-north sense of shear. There are few good strain markers, but judging by the degree of quartz grain–size reduction, the complete transposition of all layering, and the intensity of development of the foliation and lineation, it is clear that deformation involved high strains.

Nature of the contacts. Most major lithologic contacts were interpreted as thrust faults during reconnaissance mapping of the metamorphic rocks of the central Brooks Range (Nelson and Grybeck, 1980). However, unambiguous older-over-younger relationships are difficult to find in the field because of poor age control and because a strong S_2 fabric overprints all contacts. The best evidence for structural imbrication is the repetition of the marbles of the Skajit Limestone.

The base of both panels of the Skajit Limestone is transitional, but the upper contacts are sharp. Mylonitic textures are developed near the top of the upper Skajit panel with strong oblique grain-shape fabric, reduced grain size, and dolomite porphyroclasts with asymmetric tails indicative of top-to-the-north shear. In addition, this marble is involved in large-scale recumbent isoclinal folds so that its upper contact is not a planar fault zone, although it may be a folded fault (Figs. 2 and 3). We conclude that ductile deformation and metamorphism continued after the initial juxtaposition of the units, probably by thrust faults, erasing much of the textural evidence of the original nature of the contacts, and modifying the initial geometry of the structures. Similar relationships are found in the Central Belt along the TACT transect (Moore et al., 1997).

Low-strain domain. North of the upper panel of Skajit Limestone, textures become phyllitic or even slaty. Metamorphic minerals are fine grained, sedimentary layering is only partially transposed, and penetrative strain is noticeably lower than deeper in the section (Fig. 4). The phyllites and slates of the low-strain domain have a spaced crenulation cleavage that is parallel to the dominant fabric in the high-strain domain, while S_1 constitutes the main penetrative fabric in hand specimen and thin section. We tentatively correlate S_1 in the low-strain domain with S_1 in the schists of the high-strain domain.

In the low-strain structural domain, folds are typically open or kink shaped with east-west–trending fold axes (Fig. 8C). The orientation of fold axes in the low-strain domain is in sharp contrast to the high-strain domain where fold axis are consistently north trending and parallel to the stretching lineations. The transition between structural domains appears to be abrupt: fold axes with intermediate orientations are not easily found (Fig. 8C). This suggests that different modes of deformation were at play at different structural levels of the orogenic system. Similar relationships have been documented in other mountain belts. For example, in the western Alps, fold axes and stretching lineations are east-west trending and are parallel to the dominant direction of tectonic transport in the high-grade internal crystalline massifs, while in the external zones of the Briançonnais and the Jura, the fold axes are north-south trending and perpendicular to the direction of tectonic transport (Malavieille et al., 1984).

Late-stage brittle structures

A system of normal faults trending northwest-southeast to north-south cut the Mount Igikpak massif (Figs. 2 and 3) (Nelson and Grybeck, 1980). A very well exposed example of these faults is the Tupik fault (Fig. 3). This fault places marble and quartzite in the hanging wall against hydrothermally altered and

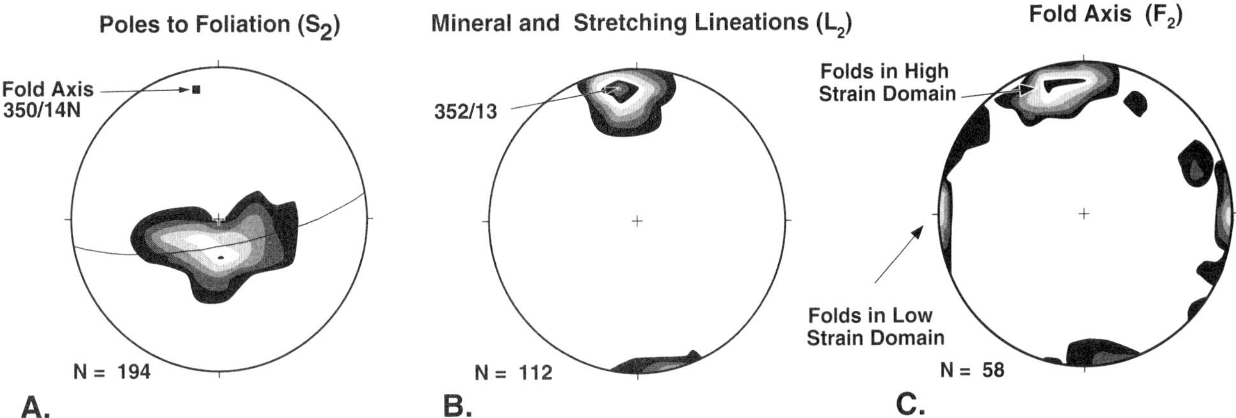

Figure 8. Lower-hemisphere stereonet plots of structural data from metasedimentary rocks overlying Mount Igikpak gneiss. A: Poles to dominant foliation (S_2) and calculated fold axis. B: Mineral and stretching lineations. C: Fold axes from two main structural domains; isoclinal folds in high-strain domain are parallel to stretching lineation and direction of tectonic transport; folds in low-strain domain are perpendicular to tectonic transport.

fractured Mount Igikpak orthogneiss in the footwall. The fault surface dips 40°SW, is marked by a zone of silicified breccia several meters thick, and has 1-m-wide mullions that plunge downdip. The metamorphic fabric in the quartzites and marbles of the hanging wall is folded by normal-sense drag. The offset on the Tupik fault is ~1.8 km, based on restoration of the cross section (Fig. 3).

The most spectacular normal fault is exposed in the headwaters of the Reed River (Fig. 2). It places high-level, hydrothermally altered Mount Igikpak orthogneiss on deep-level augen gneiss. The fault surface, which dips 40°W, is marked by a 5-m-thick zone of chloritic breccia underlain by a 30 m thickness of fractured augen gneiss. The fault has a minimum displacement of 1.2 km because the augen gneiss in the footwall extends to the summit of Mount Igikpak (2594 m), high above the metasedimentary roof pendants on the hanging-wall orthogneiss. The maximum displacement is undefined. The brittle nature of these faults, together with thermochronologic data discussed in the following, suggests that they were active during the Tertiary. Their geometry shows that the Brooks Range underwent extension parallel to the length of the orogen at that time.

K-Ar AND ^{40}Ar/^{39}Ar GEOCHRONOLOGY

Previous K-Ar and ^{40}Ar/^{39}Ar studies

Turner et al. (1979) reported 80 K/Ar ages from the southern Brooks Range Schist Belt. The ages range from 86 to 756 Ma, but most are in the 100–130 Ma range. This latter cluster of ages was interpreted as resulting from cooling after a greenschist facies metamorphic event that ended in the mid-Cretaceous and overprinted earlier blueschist facies assemblages.

A few of the ages reported by Turner et al. (1979) were obtained from schists collected near our field area. White mica

from 12 km south of the Mount Igikpak orthogneiss was 114 ± 3 Ma, and three biotite samples from the same structural level were 102–100 Ma. One muscovite from 5 km south of the orthogneiss yielded 106 ± 3 Ma. Four mica samples from the Arrigetch orthogneiss, which occupies a structural position similar to that of the Mount Igikpak gneiss, ranged between 92 ± 5 Ma (muscovite) and 86 ± 4 Ma (biotite). In addition, biotite and hornblende from a skarn at the eastern margin of the Arrigetch orthogneiss yielded 95 ± 3 Ma and 109 ± 3 Ma K/Ar ages (Silberman et al., 1979). These ages are consistent with the pattern Turner et al. (1979) observed along the southern Brooks Range: biotites are slightly younger than white micas, reflecting their lower closure temperature; and ages become younger toward the core of the Brooks Range. The biotite ages from the Arrigetch Peaks, which are some of the youngest K/Ar ages in the Brooks Range, suggest that the orthogneiss was exhumed later than the Schist Belt.

Patrick et al. (1994) mapped a metamorphic field gradient south of the Arrigetch orthogneiss that ranges from amphibolite facies in the north to low greenschist, with relict blueschist facies minerals, to the south. A hornblende sample from the high-grade rocks yielded a slightly disturbed ^{40}Ar/^{39}Ar spectrum and an isochron age of 110 ± 1 Ma, which was interpreted as the time of peak metamorphism. Coexisting muscovite yielded a plateau age of 96 Ma. White mica samples from the lower-grade rocks to the south produced hump-shaped spectra with steps ranging from 102 to 133 Ma, indicative of incorporation of excess argon. In general, these recent ^{40}Ar/^{39}Ar data are in agreement with previous K-Ar work.

New ^{40}Ar/^{39}Ar results

We present here 12 new ^{40}Ar/^{39}Ar ages obtained by step-heating experiments (Table 2; Figs. 9 and 10) of samples col-

J. Toro et al.

TABLE 2. SUMMARY OF ^{40}Ar/^{39}Ar DATA

Sample	Lithology	Mineral	Latitude (N)	Longitude (W)	Total Fusion Age (Ma)	Isochron Age (Ma)	MSWD	^{40}Ar/^{36}Ar	WM Plateau Age (Ma)	Steps Used	%^{39}Ar Used
93-JT-85	Qtz Mu Schist	White mica	67°31.5′	154°57.5′	72.9	74.5 ± 0.4	3.38	286 ± 12	74.3 ± 0.2	1/7/12	77
93-JT-88	Granite	K-feldspar	67°30.1′	154°59.4′	48.4	47	9.08	1111 ± 43	47.0 ± 0.2	2/20/21	96
93-JT-88	Granite	Biotite	67°30.1′	154°59.4′	77	78	26.95	179 ± 23	77.5 ± 0.1	2/11/12	91
93-JT-96	Mu Marble	White mica	67°34.7′	154°52.8	96.2	97.2 ± 0.3	2.14	266 ± 91	97.2 ± 0.2	4/10/12	84
93-JT-113	Graph. Phyllite	White mica	67°30.8′	154°38.7′	113	111.4 ± 1.2	7.93	311 ± 42	112 ± 1	3–8,11–12	84
93-JT-124	Bio Gt Schist	Biotite	67°29.8′	154°58.5′	102	102.5 ± 2	2.27	248 ± 7	102 ± 0.2	2/12/12	97
93-JT-142	Bio Gt Schist	Biotite	67°32.4′	154°58.4′	94.9	94.9	15.26	293 ± 15	94.2 ± 0.2	2/8/13	59
93-JT-143	Bio Gt Schist	Biotite	67°33.1′	154°59.4′	89.2	89.5 ± 0.2	4.11	240 ± 21	89.3 ± 0.1	2/10/11	95
94-JT-19	Skarn	Hornblende	67°30.1′	155°08.4′	94	95.5	185.64	317 ± 28	96 ± 1	3/5/09	72
94-JT-79	Augen Gneiss	White mica	67°25.2′	155°02.2′	84.6	84	87.09	311 ± 35	84.2 ± 0.8	2/12/15	95
94-JT-79	Augen Gneiss	Biotite	67°25.2′	155°02.2′	85.1	85	91.36	305 ± 47	85.0 ± 0.8	2/16/21	90
94-JT-79	Augen Gneiss	Kspar	67°25.2′	155°02.2′	53.9	47	147.2	806 ± 46	47.0 ± 0.1	2/9/12	69

Note: J is the irradiation parameter; *MSWD* is the mean square weighted deviation. *WM Plateau Age* is the weighted mean plateau age of the release spectrum. *Steps Used* and *%^{39}Ar Used* refer to the plateau age. Analyses with prefix 93 were preformed at P.B. Gans's laboratory at UCSB. Analyses with prefix 94 were preformed at M. McWilliam's laboratory at Stanford University. Uncertainties reported are 1 sigma. See Toro (1998) for analytical details. Full data tables can be found in GSA Data Repository (see footnote 1).

lected along a transect from the core of the Mount Igikpak gneiss to the base of the Endicott Mountains allochthon, 35 km to the north (Figs. 2 and 3). This represents a nearly complete sampling of a crustal transect through the Central Belt of the Brooks Range. We analyzed minerals with different closure temperatures (biotite, white mica, K-feldspar, and hornblende) and from different structural levels in order to characterize the cooling history of each sample and of the area as a whole.

Mica ages. These samples are discussed in order of their geographic position from north to south along the transect, i.e., toward increasing structural depth and metamorphic grade. The approximate Ar closure temperature of biotite is 300 ± 50 °C and of white mica is 350 ± 50 °C for moderate cooling rates (McDougall and Harrison, 1988).

Sample 93-JT-113 was collected from a deformed quartz vein concordant with S_1 in graphitic slates within the low-strain domain at the northern end of the transect (Figs. 2 and 3). The spectrum has a near-plateau age of 112 ± 1 Ma, calculated using 78% of the released ^{39}Ar (Fig. 9A). The ^{40}Ar/^{36}Ar intercept for this sample is within error of the atmospheric ratio, indicating that the age is reliable. Because the metamorphic grade in this part of the transect is very low, it is not likely that these rocks were ever heated above the closure temperature of white mica. We interpret the 112 Ma age as dating the growth of S_1 mica at this structural level, which probably corresponds to local peak metamorphic conditions.

Sample 93-JT-96 is white mica from a coarse-grained, well-foliated, white marble of the Skajit Limestone from the upper part of the high-strain structural domain (Figs. 2 and 3). The micas are aligned with the dominant (S_2) fabric. The sample yielded a reliable near-plateau age of 97.2 ± 0.2 Ma (Fig. 9B). The ^{40}Ar/^{39}Ar age from this sample may either represent the age of recrystallization of the marble, in which case it would be dat-

ing the S_2 deformational event, or alternatively, cooling after the metamorphic event.

Five mica samples from rocks that directly overlie the Mount Igikpak orthogneiss, and from the margin of the orthogneiss, yielded ages ranging from 102 ± 0.2 to 74.3 ± 0.2 Ma (Table 2; Figs. 2 and 3). The oldest age came from sample 93-JT-124 (Fig. 9C), a garnet-biotite–white mica schist that contains both garnet and coarse biotite porphyroblasts associated with the S_1 fabric. Samples 93-JT-142 and 93-JT-143, also garnet-biotite schists, yielded 94.2 ± 0.2 and 89.3 ± 0.1 Ma, respectively (Fig. 9, D and E). The remaining two samples from this part of the section were of metamorphic white mica from schists (93-JT-85), which had a ^{40}Ar/^{39}Ar age of 74.3 ± 0.2 Ma, and relict magmatic biotite from the unfoliated portion of the Mount Igikpak gneiss (93-JT-88), which yielded 77.5 ± 0.1 Ma (Fig. 9, F and G). These two ages are significantly younger than samples from nearby, and are the youngest ^{40}Ar/^{39}Ar mica ages reported so far from the central Brooks Range.

The wide range of ages (ca. 102–74 Ma) in micas from a small area and similar structural levels is difficult to explain.

Figure 9. ^{40}Ar/^{39}Ar and K/Ca spectra. Steps used in calculating weighted mean plateau age are shown in solid black. A, B: White mica samples from shallow structural level. C–E: Biotite samples from schists directly overlying Mount Igikpak gneiss. F, G: Anomalously young biotite and white mica from shallow-level Igikpak gneiss and overlying schist. See GSA Data Repository[1] or accompanying CD-ROM for ^{40}Ar/^{36}Ar data tables for samples.

[1]GSA Data Repository item 2002075, ^{40}Ar/^{36}Ar Data Tables for Samples, is available on request from Documents Secretary, GSA, P.O. Box 9140, Boulder, CO 80301–9140, USA, editing@geosociety.org, or at www.geosociety.org/pubs/ft2002.htm, or on the CD-ROM accompanying this volume.

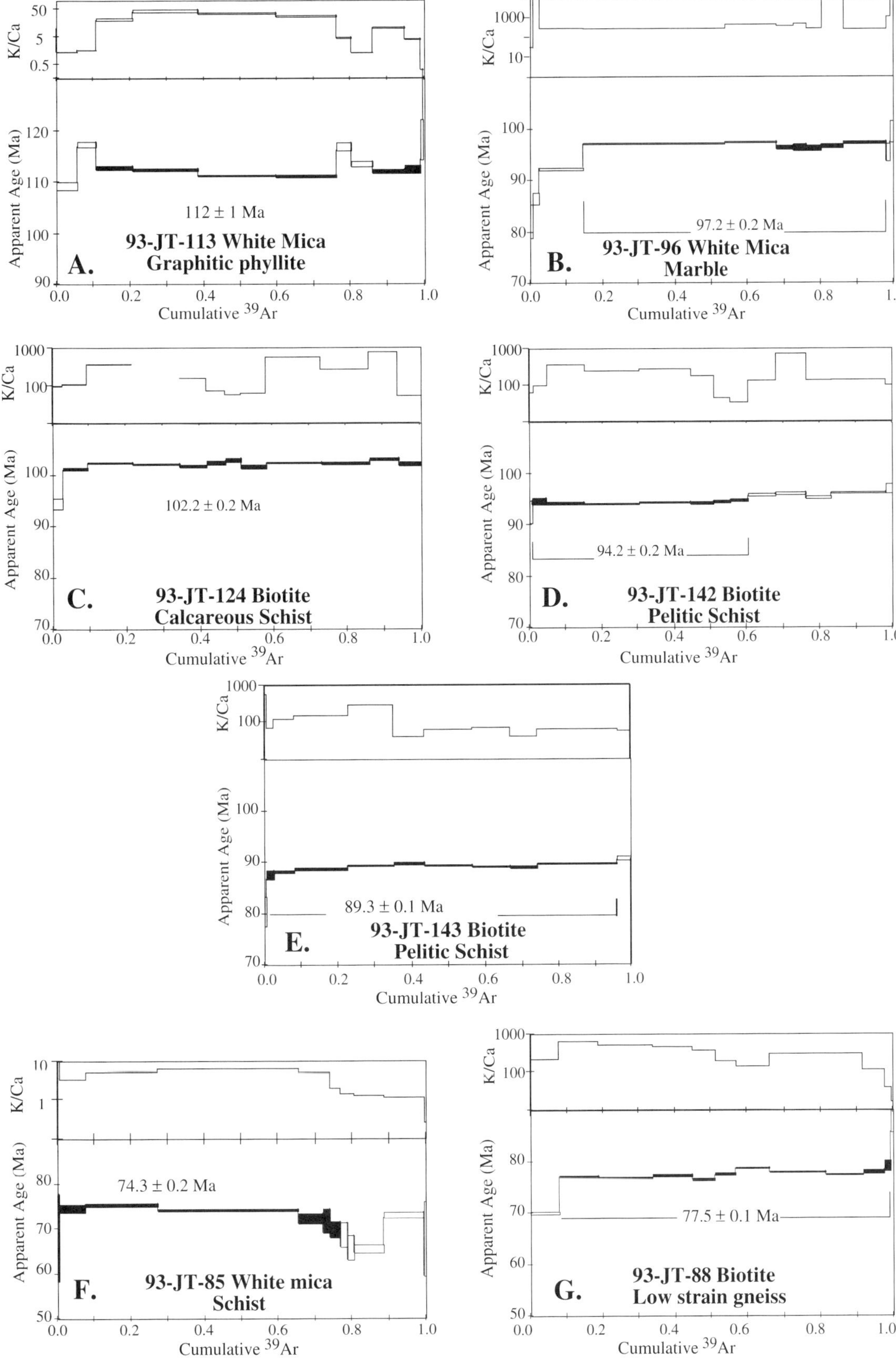

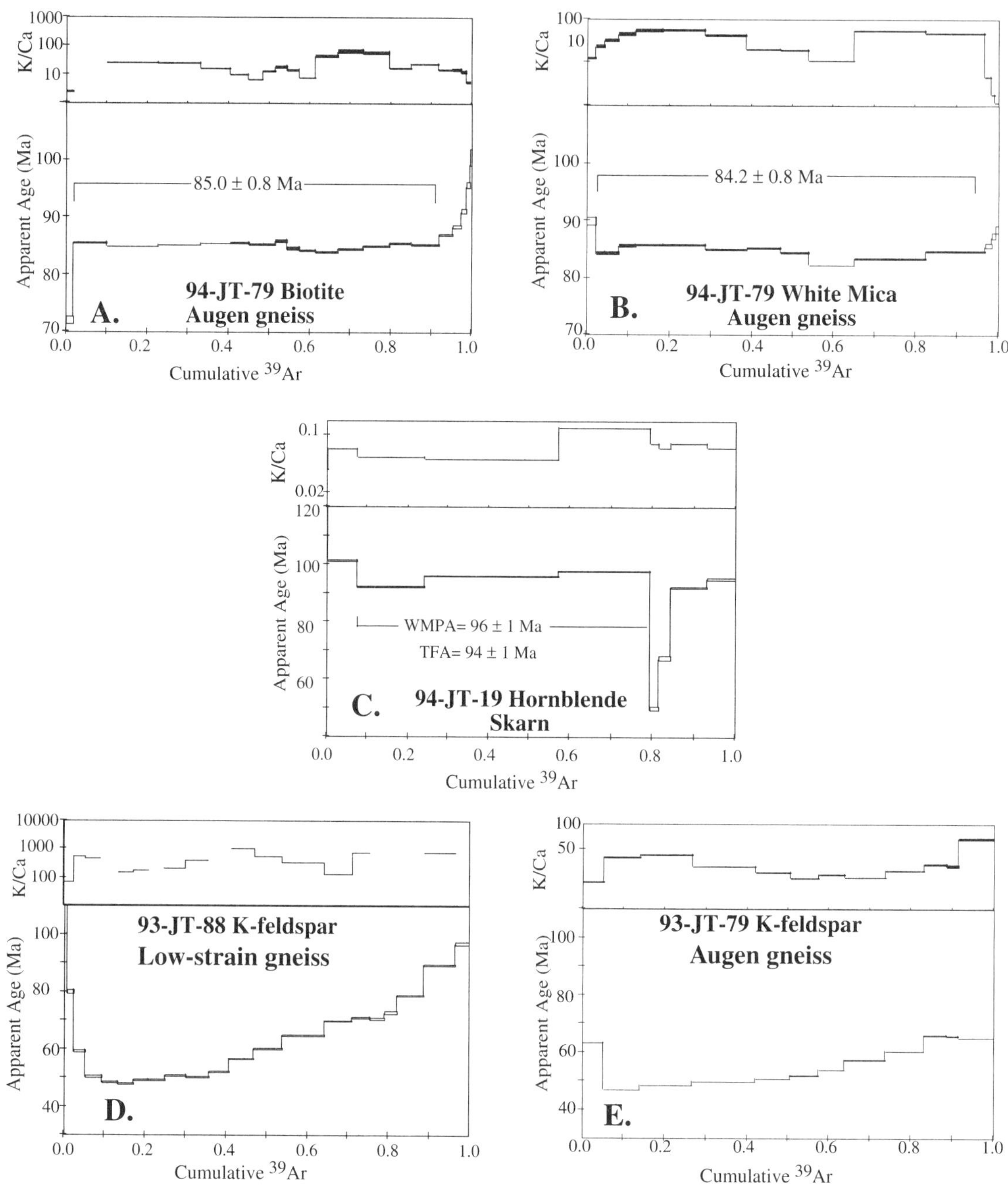

Figure 10. ^{40}Ar/^{39}Ar and K/Ca spectra. Steps used in calculating weighted mean plateau age are shown in solid black. A, B: Micas from coarse augen gneiss. C: Hornblende from Devonian skarn. D, E: K-feldspar samples from foliated and un-foliated portions of Igikpak gneiss.

There is no field evidence for late magmatic activity that could have thermally reset some of the samples, nor are there any major faults or breaks in metamorphic grade separating the samples. One possibility is that the ages reflect distinct episodes of metamorphic growth, the older ages reflecting a larger component of S_1 micas and the younger ages being dominated by S_2 micas. However, this explanation cannot be applied to all the samples. One of the youngest ages (77 Ma, sample 93-JT-88) came from what is clearly relict magmatic biotite of the Devonian orthogneiss.

We also analyzed coexisting metamorphic white mica and biotite from the coarse and well-foliated interior of the orthogneiss (sample 94-JT-79), which yielded near-plateau ages of 84.2 ± 0.8 and 85.0 ± 0.8 Ma (Fig. 10, A and B). We interpret these ages as dating postmetamorphic cooling. These two analyses behaved as expected, the biotite age being slightly younger than the white mica age and both being younger than most of the mica samples from shallower structural levels.

Hornblende age. We obtained a $^{40}Ar/^{39}Ar$ age from hornblende from a garnet-bearing skarn collected from the northwestern margin of the Mount Igikpak orthogneiss (sample 94-JT-19, Figs. 2 and 3). A plateau age calculated using the flat portions of the spectrum (Fig. 10C), representing 75% of the released ^{40}Ar, is 96 ± 1 Ma. That metasomatic hornblende, likely of Devonian age, was completely reset in the mid-Cretaceous suggests that the peak temperature during the metamorphic event associated with S_2 exceeded the 500 ± 50 °C closure temperature of hornblende.

K-feldspar ages. Two K-feldspar samples (93-JT-88 and 94-JT-79) from the Mount Igikpak gneiss help define low-temperature portions of the cooling history. Both samples yielded $^{40}Ar/^{39}Ar$ age gradients (Fig. 10, D and E), as is commonly observed in plutonic and metamorphic K-feldspars (Harrison, 1990; Lovera et al., 1989). In both samples the lowest-temperature steps are probably disturbed by excess ^{39}Ar trapped in fluid inclusions. In sample 93-JT-88 (Fig. 10D), the spectrum decreases from a high temperature at 97 Ma to a pseudoplateau of 50 Ma. This age pattern probably reflects cooling from above the closure temperature of biotite (300 ± 50 °C), which is 74 Ma for this sample, down to <200 °C ca. 50 Ma. Sample 94-JT-79 (Fig. 10E) yielded an age gradient that climbs from 46 to 64 Ma. Both samples suggest that cooling through 200 ± 50 °C took place during early Tertiary time. This is consistent with apatite fission-track data (see following).

APATITE FISSION-TRACK DATA

Previous apatite fission-track studies

A total of 22 apatite fission-track ages from the Arrigetch and Mount Igikpak orthogneisses were previously published (Blythe et al., 1997; Murphy et al., 1994; O'Sullivan et al., 1993). The samples from the Arrigetch gneiss indicate cooling through the apatite annealing zone (65–110 °C) between 22 and 25 Ma (Murphy et al., 1994). Three samples from the eastern lobe of the Mount Igikpak orthogneiss, which is geographically closest to the Arrigetch orthogneiss (Fig. 1), averaged 25 ± 2 Ma. In contrast, ages from the western lobe of the Mount Igikpak gneiss, the site of our study, were distinctly older, ranging between 42 ± 6 and 33 ± 6 Ma (samples 77ANS87, 77ADG103, 77ADG128; Fig. 2). The age difference between the western and eastern lobes of the Mount Igikpak orthogneiss was attributed to differential uplift across a north-trending vertical fault separating the two areas that was active sometime between 42 and 25 Ma (Murphy et al., 1994; Blythe et al., 1997).

New apatite fission-track results

Only two samples yielded enough apatite to obtain reliable ages (Table 3; Fig. 11). Sample 93-JT-87, from highly strained quartzite near the margin of the Igikpak orthogneiss (Fig. 2), yielded an age of 29.6 ± 4.0 Ma with a mean track length of 13.2 ± 0.3 μm. This age overlaps within the uncertainty with the previously reported ages from the Igikpak gneiss (O'Sullivan et al., 1993). All the samples from the Igikpak area are characterized by reduced track lengths and often negatively skewed length distributions. This suggests that the samples resided for a considerable time within the partial annealing zone of apatite, and therefore the ages that are reported may not reflect a time of rapid cooling.

Sample 93-JT-106, from low greenschist-grade quartz-pebble metaconglomerate located 20 km north and 9 km up structural section from the orthogneiss (Figs. 2 and 3), yielded an age of 60.0 ± 8.1 Ma. The difference between this age and the ca. 42–30 Ma ages from the core of the gneiss indicates that uplift and denudation of the core of the Brooks Range with respect to the flanks was ongoing during the early Tertiary.

Thermal and tectonic evolution

We have integrated the available stratigraphic, structural, thermochronologic, and petrological data from the Mount Igikpak area with regional tectonic considerations in order to construct a thermal history (Fig. 12A) and a pressure-temperature

TABLE 3. FISSION-TRACK DATA

Sample	Latitude (N)	Longitude (W)	Elevation (m)	No. Xls.	Spontaneous		Induced		$P(\chi^2)$ (%)	Dosimeter		Age $\pm 1\sigma$ Ma
					Rho-S	NS	Rho-I	NI		Rho-D	ND	
					Mt. Igikpak Area Samples							
93-JT-87	67°31.1′	154°57.5′	1280	15	0.206	74	2.339	839	12	1.738	5198	29.6 ± 4
93-JT-106	67°38.5′	154°43.3′	780	6	0.415	65	2.332	365	63	1.754	5198	60.0 ± 8

Note: Abbreviations: *No Xls,* number of individual grains dated; *Rho-S,* spontaneous track density ($\times 10^6$ tracks per square centimeter); *NS,* number of spontaneous tracks counted; *Rho-I,* induced track density in external detector (muscovite) ($\times 10^6$ tracks per square centimeter); *NI,* number of induced tracks counted; *P(χ^2),* χ^2 probability; *Rho-D,* induced track density in external detector adjacent to CN5 dosimetry glass ($\times 10^6$ tracks per square centimeter); *ND,* number of tracks counted in determining Rho-D. *Age* is the sample fission-track pooled age, calculated using zeta of 385.9 for CN5. See Toro (1998) for analytical details.

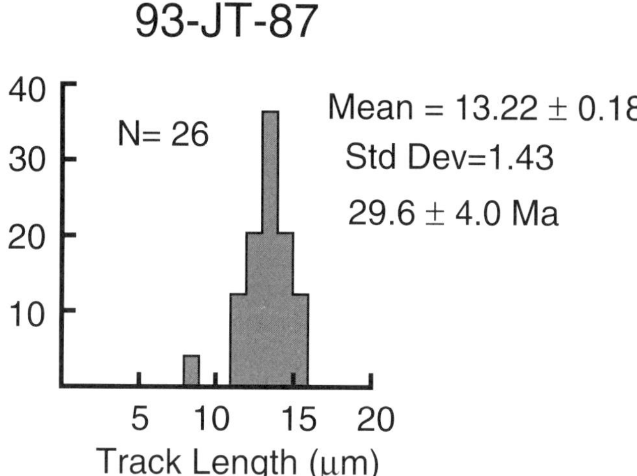

Figure 11. Apatite fission-track-length histogram.

(*P-T*) path (Fig. 12B) for rocks along our structural transect. The thermal history plot tracks the progress of two points in our structural transect (Figs. 3, 12, and 13): (1) a point at the base of the Endicott Mountains allochthon (path A, shown in white in Fig. 12a) and (2) a point in the metasedimentary rocks that di-rectly overlie the Mount Igikpak gneiss (path B, shown in gray in Fig. 12A). These two points represent the top and bottom of a crustal section through the Brooks Range Central Belt.

We use a geothermal gradient of 25 °C/km to estimate pa-leotemperature in pre-Brookian time. We assume that a lower geothermal gradient (15 °C/km) existed during the initial phase of Brookian orogeny, as suggested by the development of blueschist facies assemblages in the Schist Belt (Till and Snee, 1995), and that the gradient reequilibrated to 25 °C/km after crustal thickening was accomplished.

The following discussion is keyed to numbered stages along the time-temperature and *P-T* paths (Fig. 12). In addition, the Mesozoic and Tertiary evolution is schematically illustrated in a series of cross sections (Fig. 13) that incorporate the available geological and geochronological data.

Stage 1. Preintrusion. Our burial and thermal histories be-gin in pre-Devonian time, prior to the intrusion of the Mount Igikpak pluton and prior to deposition of the Endicott Group. A moderate depth of emplacement of ~5 km is consistent with tex-tures observed in the shallow portion of the Mount Igikpak or-thogneiss and with the mineralogy of the skarns in the contact aureole (Newberry et al., 1986).

Stage 2. Intrusion of granite. The Mount Igikpak granite was emplaced ca. 380 Ma, as determined by our U/Pb data. The

Figure 12. A: Time-temperature paths for two points of our transect. Path A (white) tracks thermal history of point at base of Endicott Mountains allochthon. Path B (gray) tracks point within schists directly above Mount Igikpak or-thogneiss. Points are located in Figures 3 and 13. Numbers along paths are keyed to geologic events discussed in text. B: Qualitative pressure-temperature dia-gram for point directly above Mount Igikpak orthogneiss. Dashed line is 25 °C/km geotherm. Boundary for end-member reaction from chloritoid + bi-otite to almandine in pelitic lithologies is shown for reference (Spear, 1993).

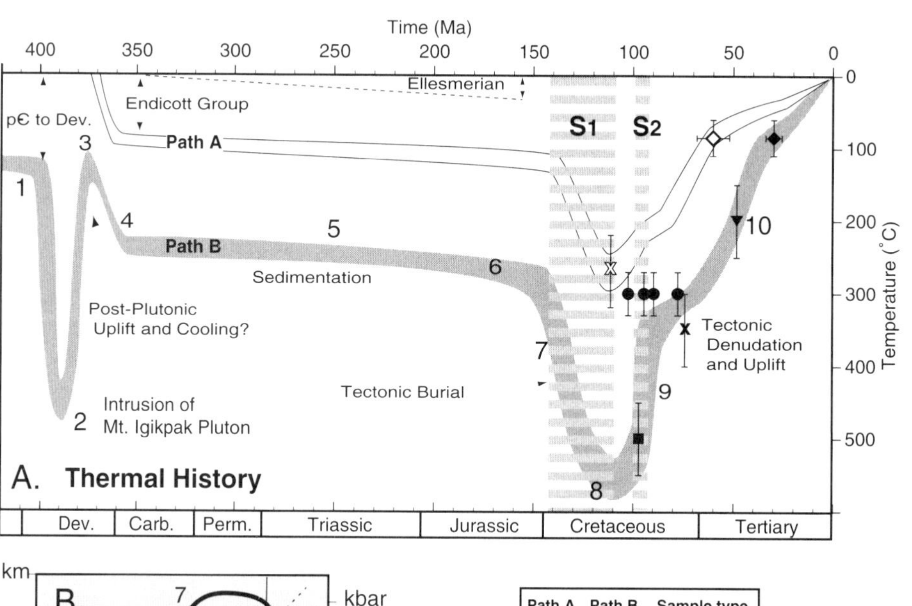

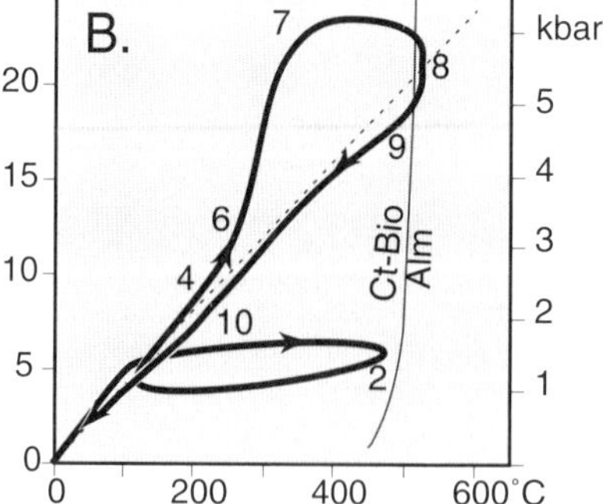

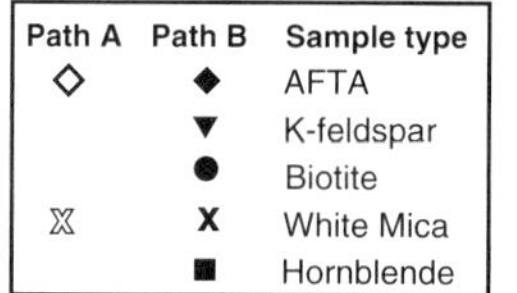

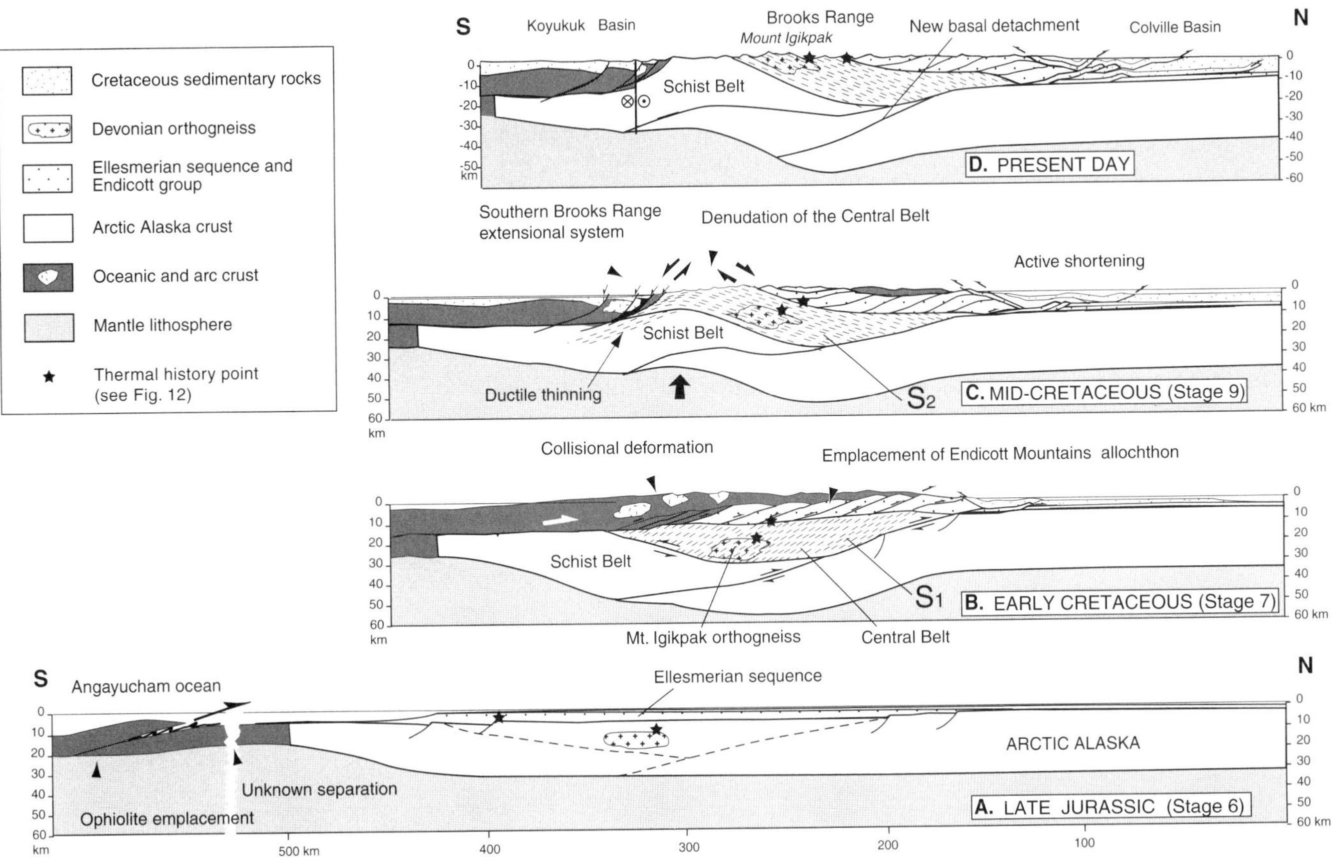

Figure 13. Schematic crustal cross sections showing evolution of Brooks Range after Patton and Box (1989), Moore et al. (1994), Cole et al. (1997), and many others. Location is shown in Figure 1. A: Late Jurassic initial emplacement of Angayucham terrane ophiolites. B: Early Cretaceous peak of collisional deformation. C: Mid-Cretaceous extensional collapse. D: Present-day structure.

temperature shown at stage 2 in Figure 12A (450 °C) was calculated using a one-dimensional conductive thermal model (e.g., Peacock, 1989). This is the temperature in the country rock 200 m away from an intrusive body comparable in size to the Mount Igikpak orthogneiss (20 km thick), with a magmatic temperature of 825 °C. This calculation is consistent with paleotemperature estimates based on phase relations of the skarn mineralogy (Newberry et al., 1986). Under these conditions, the thermal pulse associated with the pluton would decay within 20 m.y. (Fig. 12).

Stage 3. Postintrusion. This stage represents postintrusion conditions in Late Devonian time (Fig. 12). Although we know little about the tectonic setting of Brooks Range Devonian magmatism, we depict some thermally driven uplift and erosion at this time, in response to the intrusion of the granitic belt.

Stage 4. Deposition of Endicott Group. This stage spans Late Devonian to Early Mississippian time, during which a thick clastic wedge was deposited. The total stratigraphic thickness of the Endicott Group in the Endicott Mountains allochthon is ~4.5 km (Moore et al., 1994). We assume that the full thickness of the Endicott Group was deposited above the location of the Mount Igikpak orthogneiss. This assumption helps to account for the

crustal thickness attained during Brookian deformation, leading to a more plausible thermal history during metamorphism.

Stage 5. Deposition of Ellesmerian Group. This stage spans Mississippian to Late Jurassic time. During this time the Ellesmerian Group was deposited along the passive margin of Arctic Alaska. This stage corresponds to a period of slowly increasing temperature and burial.

Stage 6. Onset of Brookian orogeny. Brookian deformation is generally attributed to the collision of an island arc with the Arctic Alaska microplate (Fig. 13A), which led to the development of high-P, low-T metamorphic assemblages in the southern Schist Belt, the emplacement of ophiolitic allochthons on the continental margin, and large-scale crustal shortening within the orogen (Mayfield et al., 1988; Mull, 1982; Roeder and Mull, 1978; Till, 1992a).

The age of initiation of deformation is not directly defined by our data. The earliest indications of Brookian tectonism are the Middle Jurassic cooling ages from the metamorphic sole of the western Brooks Range ophiolites (Wirth and Bird, 1992), but it is likely that initial obduction of the ophiolite took place in an oceanic setting some distance from the continental margin, because shelf sedimentation continued uninterrupted until the Late

Jurassic along the Arctic Alaska passive margin (Moore et al., 1994; Patton and Box, 1989). Reliable $^{40}Ar/^{39}Ar$ white mica ages from the southern Schist Belt are as old as 130 Ma (Christiansen and Snee, 1994; Gottschalk and Snee, 1998), indicating that crustal thickening and deformation associated with the collision began prior to this time. However, one relict 171 Ma phengite age reported by Christiansen et al. (1994), from an area of the Schist Belt that yielded mostly Early Cretaceous ages, suggests that an earlier metamorphic event could have affected the southern Brooks Range.

Stage 7. Crustal thickening. This stage represents Early Cretaceous time. Most of the shortening and crustal thickening in the Brooks Range is inferred to have taken place at this time (Fig. 13B), based in part on the age of southerly derived synorogenic clastic rocks found in the Endicott Mountains allochthon (Moore et al., 1994). This is the most likely age of high-pressure metamorphism of the continental-margin rocks of the Schist and Central Belts (Till and Snee, 1995), and also corresponds to an episode of very rapid subsidence in the Colville foreland basin, which has been attributed to flexure of the lithosphere due to the emplacement of the Endicott Mountains and higher allochthons (Nunn et al., 1987; Cole et al., 1997). The pressures attained within the Brooks Range orthogneisses of the Central Belt during metamorphism have been estimated as 5–8 kbar (Patrick, 1995). We have chosen the lower end of this pressure range for the rocks that directly overlie the Mount Igikpak orthogneiss (gray curve in Fig. 12C). This implies a doubling of the thickness of the overburden from ~10 km to 23 km, assuming an average density of 2700 kg/m³. These values of thickening are consistent with estimates of shortening from the foreland fold and thrust belt to the north (Cole et al., 1997).

Stage 8. Peak metamorphism. This stage corresponds to peak temperatures, recorded by the garnet-biotite schists associated with relict S_1 foliation in the pelitic metasedimentary rocks that overlie the Mount Igikpak gneiss. The last stages of this metamorphic and deformational event were probably recorded by the 112 ± 1 Ma $^{40}Ar/^{39}Ar$ age of S_1 white mica in the low-grade slates from the base of the Endicott Mountains allochthon, although it is risky to correlate foliations from such different structural levels and metamorphic grades. Thermobarometry from lithologies and structural levels similar to those found adjacent to the Mount Igikpak orthogneiss yielded temperatures in the 500–600 °C range and pressures of 6–8 kbar on the north side of the Arrigetch orthogneiss (Vogl and Patrick, 1995).

Stage 9. Extensional collapse and exhumation. This stage represents initial exhumation of the metamorphic rocks of the Brooks Range Central Belt synchronous with ductile deformation that produced the S_2 fabrics (Fig. 13C). The dominant foliation (S_2) in the metasedimentary rocks that overlie the Mount Igikpak orthogneiss is mostly defined by white mica and chlorite that replaced older biotite and garnet. We interpret this texture as evidence that the S_2 fabric formed along a retrograde metamorphic path, as the crustal section was being exhumed (Fig. 12B). This deformational and metamorphic event is also recorded in the $^{40}Ar/^{39}Ar$ thermochronology from these rocks. Hornblende from the Devonian skarns along the margin of the Igikpak orthogneiss yielded an $^{40}Ar/^{39}Ar$ age of 96 ± 1 Ma, while most of the mica ages from this structural level range between 89 and 77 Ma (Figs. 9 and 10). We interpret this pattern of ages as the result of initial rapid cooling from above the hornblende Ar closure temperature (~500 °C) ca. 96 Ma to near the closure temperature of biotite (~300 °C) ca. 90 Ma, followed by a period of residence at this temperature (Fig. 12A).

Many 90–100 Ma mica $^{40}Ar/^{39}Ar$ and K-Ar ages have been reported from the Brooks Range Schist and Central Belts (Turner et al., 1979; Blythe et al., 1990; Christiansen and Snee, 1994; Patrick et al., 1994). These cooling ages probably date the exhumation of the metamorphic hinterland after earlier high-*P* metamorphism. It is widely agreed that extension played a role in the exhumation of the deep-level rocks of the Schist Belt (Gottschalk and Oldow, 1988; Miller and Hudson, 1991; Christiansen and Snee, 1994; Little et al., 1994), and our work in the Mount Igikpak region suggests that the same is true for the Central Belt. The driving mechanism for extension is still a matter of discussion (Miller and Hudson, 1993; Till et al., 1993).

Kinematic indicators in the Mount Igikpak region show that the deformation associated with the S_2 fabric took place under top-to-the-north shear. The structure exposed today is a north-dipping panel deformed by top-to-the-north shear, with metamorphic grade decreasing upsection (Fig. 4), as expected in an extensional shear zone (Fig. 13C). The variation of $^{40}Ar/^{39}Ar$ and fission-track ages along our transect shows that the north tilt of the section is not just a late-stage feature, but has persisted since the mid-Cretaceous.

We propose that rapid exhumation of the Central Belt took place as the overthickened orogenic wedge collapsed under gravitational stress both along the southern Brooks Range extensional system and along the northern boundary of the Central Belt. It is likely that collapse was facilitated by the loss of strength of the crust by conductive heating as the isotherms reequilibrated after arc-continent collision. Probably shallow-level normal faulting accompanied ductile deformation at depth. Low-angle normal faults have been identified at the base of the ophiolitic allochthons found at the highest structural level in the western Brooks Range (Harris, 1988).

The presence of small piggyback basins that contain folded early and middle Albian sedimentary rocks in the northern fold and thrust belt suggests that shortening was active in the foreland while the metamorphic core was collapsing (Cole et al., 1997). These contractional structures probably accommodated the ductile north-directed strain evidenced by S_2 in the metamorphic core. The geometry of contemporaneous thrusting and extension in the northern Brooks Range is somewhat different from that documented in the Himalayas. In the Himalayas, sense of shear on the South Tibetan detachment system was in the opposite direction from that along the Main Central thrust (Burchfiel et al., 1992). Thus the intervening rocks were extruded toward the foreland. In contrast, in the central Brooks

Range shear within the Central Belt, metamorphic rocks had the same sense as the structures in the thrust belt, producing a geometry akin to a crustal-scale gravity slide.

Stage 10. Cooling and east-west extension. This stage corresponds to the Late Cretaceous and early Tertiary portion of the thermal history as defined by the age gradients observed in the K-feldspar ^{40}Ar/^{39}Ar spectra from the orthogneiss and by the fission-track data (Fig. 12A). Both methods are compatible with a moderate cooling rate of ~5 °C/m.y. between 70 Ma and the present. However, a more complicated cooling path, which is not resolved by our data, is likely. Rapid cooling events ca. 60 and 25 Ma have been identified along the Dalton Highway, along the range front, and in the northeastern Brooks Range (O'Sullivan et al., 1997, 1998). The 60 Ma cooling event has been attributed to renewed shortening because rocks as young as latest Cretaceous are folded in the North Slope (Chapman and Sable, 1960). Along the Dalton highway, early Tertiary deformation is believed to have produced crustal-scale duplexes that underlie the Mount Doonerak antiform and that were reactivated at 25 Ma (Moore et al., 1997).

We see no direct evidence for Tertiary contraction along our transect. The youngest structural features are north- to northwest-trending normal faults that cut the orthogneiss. The brittle nature of these faults indicates that they were not active until early Tertiary time, when the orthogneiss was within ~10 km of the surface. Their timing and orientation suggest that they may be related to the Eocene-Miocene normal fault system in the Hope Basin, located directly west of the Brooks Range in the Chukchi Sea (Fig. 1) (Tolson, 1987). The origin of the Hope Basin has been attributed to transtensional deformation associated with right-lateral strike-slip motion on the Kobuk fault, which extends along the southern flank of the Brooks Range. Normal faulting in the Igikpak area may also be linked to movement on the Kobuk fault. To the west of Mount Igikpak the topographic elevation of the Brooks Range decreases progressively, and there are large low-lying areas underlain by alluvium (the Noatak Lowlands, Fig. 1). We speculate that these areas of low topography may also be controlled by normal faults and correspond to the structural transition from the high central Brooks Range to the Hope Basin. The onset of normal faulting in the Hope Basin was contemporaneous with development of several other basins on the Bering Shelf (Worral, 1991), and correlates with an important change in plate motions in the North Pacific. The rate of subduction along the paleo-Pacific margin of Alaska decreased sharply in the Eocene, and convergence changed to an oblique northwest direction (Engebretson et al., 1985). Other consequences of this plate regime were dextral movement on the major strike-slip faults that cut interior Alaska and development of continental basins in association with these faults south of the Brooks Range (Kirschner, 1994; Plafker and Berg, 1994). The presence of north- and northwest-trending normal faults in the Mount Igikpak area suggests that the central and perhaps the western Brooks Range were also affected by this regional deformational event.

CONCLUSIONS

Structural and thermochronologic studies of a 15-km-thick crustal section exposed in the Mount Igikpak area reveals that deformation in the Central Belt of the Brooks Range is characterized mostly by penetrative ductile strain. We attribute peak metamorphic conditions and the development of S_1 to crustal shortening and thickening during collision of an island arc against the Arctic Alaska margin during latest Jurassic to Early Cretaceous time. A strong north-dipping foliation (S_2) is well developed throughout the area. Metamorphic grade decreases upsection in the metasedimentary rocks from upper greenschist facies near the Mount Igikpak gneiss to very low grade at the northern end of the transect. Kinematic indicators, both macroscopic and microscopic, yield top-to-the-north sense of shear. Chlorite, white mica, and biotite associated with the dominant foliation overprint earlier, higher-grade minerals related to S_1. This demonstrates that ductile deformation within the Central Belt occurred during the retrograde path of the *P-T* history, while exhumation was taking place.

Our ^{40}Ar/^{39}Ar thermochronologic data show that the schists near the Mount Igikpak gneiss cooled from 500 to 300 °C between ca. 98 and 90 Ma, probably during development of the S_2 foliation. Many ^{40}Ar/^{39}Ar and K-Ar ages from the Schist and Central Belts of the Brooks Range demonstrate that this cooling event was regional in nature. It also coincides with the main episode of rapid sedimentation in the Colville foreland basin (Cole et al., 1997).

It is generally assumed that the large-scale antiform of the central Brooks Range is the result of Tertiary deformation, as is the case for the Doonerak antiform near the Dalton Highway (Oldow et al., 1987; Moore et al., 1997). However, our thermochronologic analysis shows that in the Mount Igikpak region the northward tilt of the section has persisted since at least mid-Cretaceous time, while ductile deformation was taking place. Therefore, the structure on the north flank of Mount Igikpak must be interpreted as a gently north dipping extensional shear zone.

We attribute mid-Cretaceous rapid cooling to extensional denudation of the metamorphic core due to gravitational collapse of the overthickened orogenic pile. Deformation by top-to-the-north shear in the Central Belt was contemporaneous with the waning stages of north-directed thrusting in the northern foothills. It is possible that extensional structures in the Central Belt were kinematically linked to coeval compressional structures in the foreland. This would make the mid-Cretaceous Brooks Range another example of contemporaneous and linked shortening and extension in a collisional orogen, as is the case of the Miocene to Pliocene Himalayas (Burchfiel et al., 1992).

The implications for the Bering-Chukchi region are that collisional deformation that took place in northern Alaska and Chukotka during Late Jurassic to Early Cretaceous time was immediately followed by extensional collapse of the orogen culminating ca. 90 Ma. This affected not only the Seward Peninsula

(Dumitru et al., 1995), Chukotka (Bering Straits Geological Field Party, 1997), and the southern Brooks Range (Miller and Hudson, 1991), but also the internal part of the Brooks Range.

The youngest structures observed along our transect are large-offset north- to northwest-trending normal faults that were probably active in early Tertiary time. Normal faults of this age and orientation are also found offshore to the west of the Brooks Range, in the Hope Basin, and may be more widespread in the western Brooks Range than has been appreciated, as suggested by the large low-lying areas that are found west of Mount Igikpak. Tertiary orogen-parallel extension may be linked to dextral slip on the Kobuk fault, related to change in the plate motions that took place in the Eocene and that affected much of interior Alaska and the Bering-Chukchi shelves.

ACKNOWLEDGMENTS

This paper is part of J. Toro's Ph.D. research at Stanford University. This project received financial support from Arco Oil and Gas Co., McGee Fund, and National Science Foundation grant EAR-9317087 to E.L. Miller and S. Klemperer. We thank Dave Howell of the U.S. Geological Survey for generous logistical support. We are grateful to the National Park Service for allowing us to work in the Gates of the Arctic National Park. Tom Seeliger, Pip Darvall, and Frances Cole bravely put up with being wet and hungry for weeks on end, chased by bears, and rolled over by large boulders. This manuscript benefited from comments by E.L. Miller, F. Cole, and T. Moore, and helpful reviews by H. Avé Lallemant, D. Foster, and S. Klemperer.

REFERENCES CITED

Aleinikoff, J.N., Moore, T.E., Walter, M., and Nokleberg, W.J., 1993, U-Pb ages of zircon, monazite, and sphene from Devonian metagranites and metafelsites, central Brooks Range, Alaska: U.S. Geological Survey Bulletin, v. B2068, p. 59–70.

Avé Lallemant, H.G., Gottschalk, R.R., Sisson, V.B., and Oldow, J.S., 1998, Structural analysis of the Kobuk fault zone, north-central Alaska, *in* Oldow, J.S., and Avé Lallemant, H.G., eds., Architecture of the Central Brooks Range Fold and Thrust Belt, Arctic Alaska: Boulder, Colorado, Geological Society of America Special Paper 324, p. 261–268.

Bering Straits Geological Field Party, 1997, Koolen metamorphic complex, NE Russia: Implications for the tectonic evolution of the Bering Strait region: Tectonics, v. 16, p. 713–729.

Blythe, A.E., Wirth, K.R., and Bird, J.M., 1990, Fission track and ^{40}Ar/^{39}Ar ages of metamorphism and uplift Brooks Range, northern Alaska: Geological Association of Canada Program and Abstracts, v. 15, p. A12.

Blythe, A.E., Murphy, J., and O'Sullivan, P.B., 1997, Tertiary cooling and deformation in the south-central Brooks Range: Evidence from zircon and apatite fission-track analyses: Journal of Geology, v. 105, p. 583–599.

Box, S.E., 1987, Late Cretaceous or younger SW-directed extensional faulting: Cosmos Hills, Brooks Range, Alaska: Geological Society of America Abstracts with Programs, v. 19, no. 6, p. 361.

Brosgé, W.P., and Pessel, G.H., 1977, Preliminary reconnaissance geologic map of the Survey Pass quadrangle: U.S. Geological Survey Open-File Report 77–27, scale 1:250 000, 1 sheet.

Burchfiel, B.C., Chen Z., Hodges, K.V., Liu Y., Royden, L.H., Deng C., and Xu, J., 1992, The South Tibetan detachment system, Himalayan Orogen: Extension contemporaneous with and parallel to shortening in a collisional mountain belt: Boulder, Colorado, Geological Society of America Special Paper 269, 41 p.

Chapman, R.M., and Sable, E.G., 1960, Geology of the Utukok-Corwin region, northwestern Alaska: U.S. Geological Survey Professional Paper 303–C, p. 47–167.

Christiansen, P.P., and Snee, L.W., 1994, Structure, metamorphism, and geochronology of the Cosmos Hills and Ruby Ridge, Brooks Range schist belt, Alaska: Tectonics, v. 13, p. 193–213.

Cole, F., Bird, K., Toro, J., Roure, F., O'Sullivan, P.B., Pawlewicz, M., and Howell, D.G., 1997, A kinematic model for the north-central Brooks Range fold- and thrust-belt, Alaska: Journal of Geophysical Research, B, Solid Earth and Planets, v. 102, p. 20685–20708.

Dillon, J.T., Pessel, G.H., Chen, J.H., and Veach, N.C., 1980, Middle Paleozoic magmatism and orogenesis in the Brooks Range, Alaska: Geology, v. 8, p. 338–343.

Dumitru, T.A., Miller, E.L., O'Sullivan, P.B., Amato, J.M., Hannula, K.A., Calvert, A.T., and Gans, P.B., 1995, Cretaceous to recent extension in the Bering Strait region, Alaska: Tectonics, v. 14, p. 549–563.

Engebretson, D.C., Cox, A., and Gordon, R.G., 1985, Relative motions between oceanic and continental plates in the Pacific Basin: Boulder, Colorado, Geological Society of America Special Paper 206, 59 p.

Fuis, G.S., Murphy, J.M., Lutter, W.L., Moore, T.E., Bird, K.J., and Christiansen, N.I., 1997, Deep seismic structure and tectonics of northern Alaska: Crustal-scale duplexing with deformation extending into the upper mantle: Journal of Geophysical Research, B, Solid Earth and Planets, v. 102, p. 20873–20896.

Gottschalk, R.R., 1990, Structural evolution of the schist belt, south-central Brooks Range fold and thrust belt, Alaska: Journal of Structural Geology, v. 12, p. 453–469.

Gottschalk, R.R., Jr., and Oldow, J.S., 1988, Low-angle normal faults in the south-central Brooks Range fold and thrust belt, Alaska: Geology, v. 16, p. 395–399.

Gottschalk, R.R., and Snee, L.W., 1998, Tectonothermal evolution of metamorphic rocks in the south-central Brooks Range, Alaska: Constraints from ^{40}Ar/^{39}Ar geochronology, *in* Oldow, J.S., and Avé Lallemant, H.G., eds., Architecture of the Central Brooks Range Fold and Thrust Belt, Arctic Alaska: Boulder, Colorado, Geological Society of America Special Paper 324, p. 225–251.

Hacker, B.R., and Wang, Q., 1995, ^{40}Ar/^{39}Ar geochronology of ultrahigh-pressure metamorphism in central China: Tectonics, v. 14, p. 994–1006.

Harris, R.A., 1988, Origin, emplacement, and attenuation of the Misheguk Mountain allochthon, western Brooks Range, Alaska: Geological Society of America Abstracts with Programs, v. 20, no. 7, p. 112.

Harrison, T.M., 1990, Some observations on the interpretation of feldspar ^{40}Ar/^{39}Ar results: Chemical Geology, Isotope Geoscience Section, v. 80, p. 219–229.

Hitzman, M.W., Smith, T.E., and Proffett, J.M., 1986, Geology and mineralization of the Ambler District, northwestern Alaska: Economic Geology, v. 81, p. 1592–1618.

Kirschner, C.E., 1994, Interior basins of Alaska, *in* Plafker, G., and Berg, H.C., eds., The geology of Alaska: Boulder, Colorado, Geological Society of America, Geology of North America, v. G-1, p. 469–493.

Little, T.A., Miller, E.L., Lee, J., Law, R.D., 1994, Extensional origin of ductile fabrics in the Schist Belt, central Brooks Range, Alaska. 1. Geologic and structural studies: Journal of Structural Geology, v. 16, p. 899–918.

Lovera, O.M., Richter, F.M., and Harrison, T.M., 1989, The ^{40}Ar/^{39}Ar thermochronometry for slowly cooled samples having a distribution of diffusion domain sizes: Journal of Geophysical Research, v. 94, p. 17917–17935.

Malavieille, J., 1993, Late orogenic extension in mountain belts: Insights from the Basin and Range and the Late Paleozoic Variscan belt: Tectonics, v. 12, p. 1115–1130.

Malavieille, J., Lacassin, and R., Mattauer, M., 1984, Signification tectonique des lineations d'allongement dans les Alpes occidentales (Tectonic significance of the lineations of elongation in the Western Alps): Bulletin de la Société Géologique de France, v. 26, p. 895–906.

Mattinson, J.M., 1994, A study of complex discordance in zircons using stepwise dissolution techniques: Contributions to Mineralogy and Petrology, v. 116, p. 117–129.

Mayfield, C.F., Tailleur, I.L., and Ellersieck, I., 1988, Stratigraphy, structure, and palinspastic synthesis of the western Brooks Range, northwestern Alaska, *in* Gryc, G., ed., Geology and exploration of the National Petroleum Reserve in Alaska, 1974 to 1982: U.S. Geological Survey Professional Paper 1399, p. 143–186.

McClelland, W.C., and Mattinson, J.M., 1996, Resolving high precision U-Pb ages from Tertiary plutons with complex zircon systematics: Geochimica et Cosmochimica Acta, v. 60, p. 3955–3965.

McDougall, I., and Harrison, T.M., 1988, Geochronology and thermochronology by the ^{40}Ar/^{39}Ar method: Oxford, UK, Oxford University Press, 212 p.

Miller, E.L., and Hudson, T.L., 1991, Mid-Cretaceous extensional fragmentation of a Jurassic–Early Cretaceous compressional orogen, Alaska: Tectonics, v. 10, p. 781–796.

Miller, E.L., and Hudson, T.L., 1993, Mid-Cretaceous extensional fragmentation of a Jurassic–Early Cretaceous compressional orogen, Alaska: Reply: Tectonics, v. 12, p. 1082–1086.

Moore, T.E., Wallace, W.K., Bird, K.J., Karl, S.M., Mull, C.G., and Dillon, J.T., 1994, Geology of northern Alaska, *in* Plafker, G., and Berg, H.C., eds., The geology of Alaska: Boulder, Colorado, Geological Society of America, Geology of North America, v. G-1, p. 49–140.

Moore, T.E., Wallace, W.K., Mull, G.M., Adams, K.E., Plafker, G., and Nokleberg, W.J., 1997, Crustal implications of bedrock geology along the Trans-Alaska Crustal Transect (TACT) in the Brooks Range, northern Alaska: Journal of Geophysical Research, B, Solid Earth and Planets, v. 102, p. 20645–20684.

Mull, C.G., 1977, Apparent south vergent folding and possible nappes in Schwatka Mountains: U.S. Geological Survey Circular, C 0751–B, p. 29–30.

Mull, C.G., 1982, The tectonic evolution and structural style of the Brooks Range: An illustrated summary, *in* Powers, R.B., ed., Geological studies of the Cordilleran thrust belt: Denver, Colorado, Rocky Mountain Association of Geologists, p. 1–45.

Mull, C.G., Roeder, D.H., Tailleur, I.L., Pessel, G.H., Grantz, A., and May, S.D., 1987, Geologic sections and maps across Brooks Range and Arctic slope to Beaufort Sea, Alaska: U.S. Geological Survey Map and Chart Series MC-28S, scale 1:2 500 000, 1 sheet.

Murphy, J.M., O'Sullivan, P.B., and Gleadow, A.J.W., 1994, Apatite fission track evidence of episodic early Cretaceous to late Tertiary cooling and uplift, central Brooks Range, Alaska, *in* Thurston, D., and Fujita, K., eds., Proceedings of the 1992 International Conference on Arctic Margins: Anchorage, Alaska, Minerals Management Service Outer Continental Shelf Report 94–0040, p. 257–262.

Natal'in, B.A., Amato, J.M., and Toro, J., 1999, Paleozoic rocks of the northern Chukotka Peninsula, Russian Far East: Implications for the tectonics of the Arctic region: Tectonics, v. 18, p. 977–1003.

Nelson, B.K., Nelson, S.W., and Till, A.B., 1993, Nd- and Sr-isotope evidence for Proterozoic and Paleozoic crustal evolution in the Brooks Range, northern Alaska: Journal of Geology, v. 101, p. 435–450.

Nelson, S.W., and Grybeck, D., 1978, The Arrigetch Peaks and Mount Igikpak plutons, Survey Pass quadrangle, Alaska, *in* Johnson, K.M., ed., The United States Geological Survey in Alaska: Accomplishments during 1977: U.S. Geological Survey Circular 772–B, p. 7–9.

Nelson, S.W., and Grybeck, D., 1979, Tectonic significance of metamorphic grade distribution, Survey Pass quadrangle, Alaska, *in* Johnson, K.M., and Williams, J.R., eds., The United States Geological Survey in Alaska: Accomplishments during 1978: U.S. Geological Survey Circular 804–B, p. 16–18.

Nelson, S.W., and Grybeck, D., 1980, Geologic map of the Survey Pass Quadrangle Map: U.S. Geological Survey Map MF-1176, scale 1:250 000, 2 sheets.

Nelson, S.W., and Grybeck, D., 1981, Map showing the distribution of metamorphic rocks in the Survey Pass Quadrangle, Brooks Range, Alaska: U.S. Geological Survey Map MF-1176–C, scale 1:250 000, 1 sheet.

Newberry, R.J., Dillon, J.T., and Adams, D.D., 1986, Regionally metamorphosed, calc-silicate-hosted deposits of the Brooks Range, northern Alaska: Economic Geology, v. 81, p. 1728–1752.

Nunn, J.A., Czerniak, M., and Pilger, R.H., Jr., 1987, Constraints on the structure of Brooks Range and Colville basin, northern Alaska, from flexure and gravity analysis: Tectonics, v. 6, p. 603–617.

Oldow, J.S., Seidensticker, C.M., Phelps, J.C., Julian, F.E., Gottschalk, R.R., Boler, K.W., Handschy, J.W., and Avé Lallemant, H.G., 1987, Balanced cross sections through the central Brooks Range and North Slope, Arctic Alaska: American Association of Petroleum Geologists, Special Publication, 19 p., 8 plates.

Oldow, J.S., Gottschalk, R.R., Avé Lallemant, H.G., Boler, K.W., Julian, F.E., and Seidensticker, C.M., 1998, Envelopment thrusting and the structure of the eastern Skajit allochthon, central Brooks Range, Arctic Alaska, *in* Oldow, J.S., and Avé Lallemant, H.G., eds., Architecture of the Central Brooks Range Fold and Thrust Belt, Arctic Alaska: Boulder, Colorado, Geological Society of America Special Paper 324, p. 127–139.

O'Sullivan, P.B., Murphy, J.M., Moore, T.E., and Howell, D.G., 1993, Results of 110 apatite fission track analyses from the Brooks Range and North Slope of northern Alaska, completed in cooperation with Trans-Alaska Crustal Transect (TACT): U.S. Geological Survey Open-File Report 93–545, p. 104.

O'Sullivan, P.B., Murphy, J.M., Blythe, A.E., 1997, Late Mesozoic and Cenozoic thermotectonic evolution of central Brooks Range and adjacent North Slope foreland basin, Alaska, including fission track results from the Trans-Alaska Crustal Transect (TACT): Journal of Geophysical Research, B, Solid Earth and Planets, v. 102, no. 9, p. 20821–20845.

O'Sullivan, P.B., Moore, T.E., and Murphy, J.M., 1998, Tertiary uplift of the Mt. Doonerak antiform, central Brooks Range, Alaska: Apatite fission-track evidence from the Trans-Alaska crustal transect, *in*, Oldow, J.S., and Avé Lallemant, H.G., eds., Architecture of the Central Brooks Range Fold and Thrust Belt, Arctic Alaska: Boulder, Colorado, Geological Society of America Special Paper 324, p. 179–193.

Passchier, C.W., and Trouw, R.A., 1996, Microtectonics: Berlin, Springer-Verlag, 289 p.

Patrick, B., 1995, High-pressure–low-temperature metamorphism of granitic orthogneiss in the Brooks Range, northern Alaska: Journal of Metamorphic Geology, v. 13, p. 111–124.

Patrick, B., Till, A.B., and Dinklage, W.S., 1994, An inverted metamorphic field gradient in the central Brooks Range, Alaska, and implications for exhumation of high-pressure/low-temperature metamorphic rocks: Lithos, v. 33, p. 67–83.

Patton, W.W., Stern, T.W., Arth, J.G., and Carlson, C., 1987, New U/Pb ages from granite and granite gneiss in the Ruby Geanticline and southern Brooks Range, Alaska: Journal of Geology, v. 95, p. 118–126.

Patton, W.W.J., and Box, S.E., 1989, Tectonic setting of the Yukon-Koyukuk basin and its borderlands, western Alaska: Journal of Geophysical Research, v. 94, no. B11, p. 15807–15820.

Peacock, S.M., 1989, Thermal modeling of metamorphic pressure-temperature-time paths: A forward approach, *in* Spear, F.S., and Peacock, S.M., eds., Metamorphic pressure-temperature-time paths: Short course in geology: American Geophysical Union Geophysical Monograph 7, p. 57–102.

Plafker, G., and Berg, H.C., 1994, Overview of the geology and tectonic evolution of Alaska, *in* Plafker, G., and Berg, H.C., eds., The geology of Alaska: Boulder, Colorado, Geological Society of America, Geology of North America, v. G-1, p. 989–1021.

Roeder, D., and Mull, C.G., 1978, Tectonics of the Brooks Range Ophiolites, Alaska: American Association of Petroleum Geologists Bulletin, v. 62, p. 1696–1713.

Silberman, M.L., Brookins, D.G., Nelson, S.W., and Grybeck, D., 1979a, Rubidium-strontium and potassium-argon dating of emplacement and metamorphism of the Arrigetch Peaks and Mount Igikpak plutons, Survey Pass quadrangle, Alaska, *in* Johnson, K.M., and Williams, J.R., eds., The United States Geological Survey in Alaska: Accomplishments during 1978: U.S. Geological Survey Circular 804–B, p. 18–19.

Spear, F.S., 1993, Metamorphic phase equilibria and pressure-temperature-time paths: Washington, D.C., Mineralogical Society of America, 799 p.

Till, A.B., 1992a, Blueschists developed during collision, not subduction, in the internal zone of the Brooks Range fold and thrust belt: Geological Society of America Abstracts with Programs, v. 24, no. 5, p. 86.

Till, A.B., 1992b, Detrital blueschist-facies mineral assemblages in Early Cretaceous sediments of the foreland basin of the Brooks Range, Alaska, and implications for orogenic evolution: Tectonics, v. 11, p. 1207–1223.

Till, A.B., and Snee, L.W., 1995, $^{40}Ar/^{39}Ar$ evidence that formation of blueschists in continental crust was synchronous with foreland fold and thrust belt formation, western Brooks Range, Alaska: Journal of Metamorphic Geology, v. 13, p. 41–60.

Till, A.B., Box, S.E., Roeske, S.M., and Patton, W.W., Jr., 1993, Comment on "Mid-Cretaceous extensional fragmentation of a Jurassic–Early Cretaceous compressional orogen, Alaska" by E.L. Miller and T.L. Hudson: Tectonics, v. 12, p. 1076–1081.

Till, A.B., Schmidt, J.M., Nelson, S.W., 1988, Thrust involvement of metamorphic rocks, southwestern Brooks Range, Alaska: Geology, v. 16, p. 930–933.

Tolson, R.B., 1987, Structure and stratigraphy of the Hope Basin, southern Chukchi Sea, Alaska, *in* Scholl, D.W., Grantz, A., and Vedder, J.G., eds., Geology and resource potential of the continental margin of western North America and adjacent ocean basins: Beaufort Sea to Baja California: Houston, Texas, Circum-Pacific Council for Energy and Mineral Resources, Earth Science Series, v. 6, p. 59–71.

Toro, J., 1998, Structure and thermochronology of the metamorphic core of the central Brooks Range, Alaska [Ph.D. thesis]: Palo Alto, California, Stanford University, 200 p.

Turner, D.L., Forbes, R.B., and Dillon, J.T., 1979, K-Ar geochronology of the southwestern Brooks Range, Alaska: Canadian Journal of Earth Sciences, v. 16, p. 1789–1804.

Vogl, J.J., and Patrick B.E., 1995, Structure and metamorphism across the northern boundary of the metamorphic core of the central Brooks Range, Alaska: Geological Society of America Abstracts with Programs, v. 27, no. 5, p. 82.

Wirth, K.R., and Bird, J.M., 1992, Chronology of ophiolite crystallization, detachment, and emplacement: Evidence from the Brooks Range, Alaska: Geology, v. 20, p. 75–78.

Worrall, D.M., 1991, Tectonic history of the Bering Sea and the evolution of Tertiary strike-slip basins of the Bering Shelf: Geological Society of America Special Paper 257, 120 p.

MANUSCRIPT ACCEPTED BY THE SOCIETY MAY 15, 2001.

Geological Society of America
Special Paper 360
2002

Orthogonal flow directions in extending continental crust: An example from the Kigluaik gneiss dome, Seward Peninsula, Alaska

Jeffrey M. Amato
*Department of Geological Sciences, New Mexico State University,
Las Cruces, New Mexico 88003, USA*
Elizabeth L. Miller
*Department of Geological and Environmental Sciences, Stanford University,
Stanford, California 94305, USA*
Kimberly A. Hannula
Department of Geology, Fort Lewis College, Durango, Colorado 81301, USA

ABSTRACT

A 15-km-thick crustal section is found within the Kigluaik gneiss dome on Seward Peninsula, Alaska. Metamorphism at 91 Ma overprints a pre–120 Ma blueschist-facies event. This overprint increases from greenschist facies in the Nome Group to granulite facies in the structurally lower Kigluaik Group. Core gneisses ceased deforming before cooling from high temperatures. Although the rocks are L-S tectonites, shear-sense indicators are rarely preserved. Rocks higher in the section cooled during deformation and preserve evidence for upward motion of core rocks relative to surrounding rocks. The rocks of the Nome Group contain two foliations. The older fabric formed during the blueschist-facies event. The younger fabric is undated but is parallel to the fabrics in the amphibolite-facies rocks of the dome and thus may be related to the same event. Shear-sense indicators adjacent to the dome show motion down and away from the core; elsewhere, top-to-the-north shear sense indicators prevail.

Lineations developed at ~90 Ma record a rotation of stretching directions with depth. Lineations in the core are east-west. An ~4-km-thick transition zone contains lineations which progressively rotate upward in the section to nearly north-south within the Nome Group. The continuous transition of lineation azimuth and the lack of overprinting relationships indicate orthogonal stretching directions at different structural levels. Mid-crustal east-west flow beneath the Seward Peninsula may reflect compensation at depth for differential crustal thinning at higher crustal levels. Thinning may have occurred during regional north-south extension associated with the vertical rise of the gneiss dome.

INTRODUCTION

Gneiss domes are common elements of orogenic belts, yet several processes, such as interference folding (Ramsay, 1967), extension (e.g., Davis and Coney, 1979), and diapirism (Ramberg, 1980), have been invoked to explain these still-controversial structural and metamorphic culminations. Gneiss domes are apparently not unique to specific tectonic environments. They

Amato, J.M., Miller, E.L., and Hannula, K.A., 2002, Orthogonal flow directions in extending continental crust: An example from the Kigluaik gneiss dome, Seward Peninsula, Alaska, *in* Miller, E.L., Grantz, A., and Klemperer, S.L., eds., Tectonic Evolution of the Bering Shelf–Chukchi Sea–Arctic Margin and Adjacent Landmasses: Boulder, Colorado, Geological Society of America Special Paper 360, p. 133–146.

are found in extensional settings, such as in the metamorphic core complexes of the Basin and Range (e.g., Davis and Coney, 1979), as well as in contractional settings, such as in the northern Himalaya (Burg et al., 1984; Chen et al., 1990) and in the European Alps (Selverstone, 1985). Regardless of their origin, gneiss domes are of great interest because they permit detailed study of processes occurring in the deeper crustal levels of orogenic belts. Metamorphic tectonites within gneiss domes may also allow qualitative estimates of the magnitude of deformation as well as indicate directions of stretching and flow during the genesis of the domes.

Rocks in the infrastructures of orogenic belts commonly contain evidence for large-scale stretching nearly orthogonal to the transport direction indicated by structures at supracrustal levels. One example of this phenomenon can be found in the Austro-alpine nappes, where nearly orogen-parallel stretching lineations (formed beneath upper crustal nappes) indicate stretching parallel to the orogenic belt and perpendicular to the transport direction of the nappes (Selverstone, 1985). Another example is from the Basin and Range Province of western North America, where lineations within core complexes like the Ruby Mountains are orthogonal to the direction of movement on a coeval supracrustal mylonitic shear zone (MacCready et al., 1997). Orthogonal orientations of structures in different parts of an orogen have also been described from transpressional and transtensional tectonic settings (e.g., Fossen et al., 1994). The phenomenon of varying stretching and transport directions developed at different structural levels of orogenic belts is not a well-documented feature in the field at either the local or the regional scale, partly because the relative timing of such structures is often unknown, and the exact geometric relationships between structures developed at different crustal levels is not always clear.

This chapter describes the structures that evolved during development of the Kigluaik gneiss dome on the Seward Peninsula of northwestern Alaska (Fig. 1). The study and description of these structures follow and build on earlier studies of the petrologic evolution and geochronology of the high-grade rocks of the gneiss dome (Amato and Wright, 1997, 1998). We determined elongation or stretching directions during deformation in the gneiss dome mainly from mineral lineations in amphibolite to granulite facies gneisses. We describe the structures developed within this gneiss dome in the context of their structural level, metamorphic history, and geochronology (Amato et al., 1994; Hannula and McWilliams, 1995; Dumitru et al., 1995; Amato and Wright, 1998; Calvert et al., 1999). These data suggest that the azimuth of the stretching lineation changes direction from nearly east-west to nearly north-south over <4 km of structural section, and that the observed orthogonal stretching directions at different crustal levels may have been formed at approximately the same time during the same metamorphic event. Although questions remain, our favored interpretation for the relationships documented in the Kigluaik Mountains gneiss dome involves the lateral, east-west, and upward flow of middle crustal material into a region undergoing localized north-south extension.

GEOLOGIC SETTING

The Kigluaik gneiss dome is one of three metamorphic and structural culminations on the Seward Peninsula (Fig. 1). Similar domes are found in Arctic Russia across the Bering Strait (Bering Strait Geological Field Party, 1997). The Kigluaik gneiss dome consists of sillimanite-grade paragneiss and orthogneiss units known as the Kigluaik Group, which are intruded by a large Late Cretaceous pluton (Fig. 2; see also Amato and Miller, 1997). The dome is cut by a late Cenozoic normal fault on its north side and is flanked by the structurally higher, lower-grade Nome Group, which consists of quartzose and pelitic schist, marble, metabasite, and rare felsic orthogneiss (Till and Dumoulin, 1994). There are no significant tectonic breaks or faults between the Kigluaik Group and the Nome Group, and fabrics and metamorphic grade vary continuously throughout the exposed structural section. Previous studies of the geology of the region include reconnaissance mapping by the U.S. Geological Survey (Brooks et al., 1901; Moffit, 1913; Hummel, 1962; Sainsbury et al., 1972; Till et al., 1986), investigations of metamorphic conditions in the core of the gneiss dome (Lieberman, 1988; Patrick and Lieberman, 1988; Todd and Evans, 1993) and in the overlying Nome Group (Patrick and Evans, 1989; Hannula et al., 1995), structural and geochronologic studies of the gneiss dome (Miller et al., 1992; Amato et al., 1994; Amato and Wright, 1998), and petrogenetic studies of the plutonic rocks (Amato and Wright, 1997).

The rocks of the Kigluaik and Nome Groups have protolith ages of Late Proterozoic to Paleozoic. The Nome Group, and by inference the Kigluaik Group, were both metamorphosed to blueschist and transitional greenschist facies in a high-pressure (P)–low-temperature (T) event that occurred before ca. 120 Ma (Hannula and McWilliams, 1995). A subsequent Late Cretaceous upper amphibolite facies to granulite facies metamorphic event associated with the development of the gneiss dome caused thermal overprinting of the blueschist facies rocks (Moffit, 1913; Throckmorton and Hummel, 1979; Lieberman, 1988). Evidence for the earlier metamorphic event is now preserved only at higher structural levels within the Nome Group, where a Cretaceous overprint exists but is much lower in metamorphic grade (Patrick and Evans, 1989; Hannula et al., 1995; Hannula and McWilliams, 1995). The region of closely spaced metamorphic isograds (referred to as the isograd zone) at the upper structural levels of the Kigluaik Group represents a high metamorphic field gradient. This steep gradient was first interpreted as the result of a static thermal overprint associated with the relaxation of isotherms following tectonic burial and blueschist facies metamorphism (Lieberman, 1988; Patrick and Evans, 1989). Miller et al. (1992), however, suggested that synmetamorphic thinning of the overlying crust during the rise of

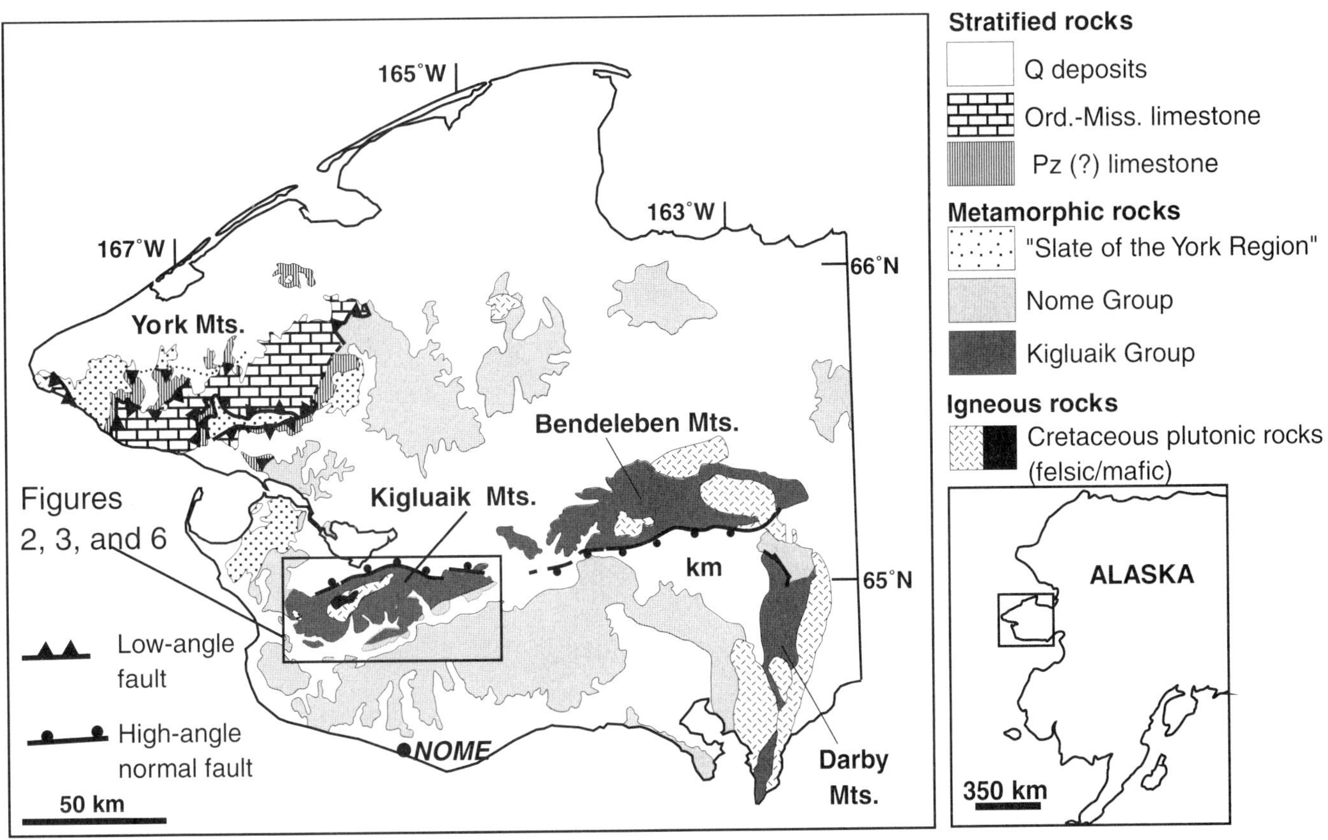

Figure 1. Simplified geologic map of Seward Peninsula, showing high-grade metamorphic culminations of Kigluaik, Bendeleben, and Darby Mountains, surrounded by lower-grade blueschist and greenschist facies metasedimentary rocks of Nome Group, and weakly metamorphosed limestones of York Mountains. Modified from Sainsbury (1972), Sainsbury et al. (1972), and Till et al. (1986).

the gneiss dome was largely responsible for telescoping the crustal section and collapsing the isograds.

The minimum age of high-temperature metamorphism is ca. 91 Ma, on the basis of U-Pb dating of metamorphic monazite from amphibolite facies metapelite and from pegmatite inferred to be derived from partial melting of metapelitic rocks during this metamorphism (Amato et al., 1994; Amato and Wright, 1998). It is possible, however, that magmatism and heating of the crust may have begun as early as ca. 110 Ma, on the basis of the oldest U-Pb zircon ages obtained from Cretaceous granitic rocks involved in deformation and metamorphism related to the genesis of the gneiss dome (Amato and Wright, 1998).

The dominant foliation and layering within the Kigluaik Group defines an east-west–trending, doubly plunging arch cut by the largely post-tectonic Kigluaik pluton and a swarm of northeast-striking diabase dikes (Fig. 2). Mafic magmatism, represented by gabbro and diorite phases of the Kigluaik pluton, was likely a heat source for Late Cretaceous metamorphism in the Kigluaik gneiss dome (Amato et al., 1994; Amato and Wright, 1997, 1998). High-precision U-Pb zircon ages of ca. 90 Ma from the pluton and U-Pb monazite ages of ca. 91 Ma from the high-

grade metamorphic rocks strongly suggest that magmatism and metamorphism were linked (Amato et al., 1994; Amato and Wright, 1998). Hornblende from one of the post-tectonic diabase dikes yields a $^{40}Ar/^{39}Ar$ age of 83 Ma (Amato and Wright, 1998).

The Kigluaik pluton and other plutons of similar age in Alaska occur within a broad belt of magmatism that crosses the Bering Strait and continues in Russia. In Russia, these plutons (and associated calc-alkaline volcanic rocks) are believed to be the product of northward subduction beneath the continent. On the basis of the geochemistry of the Alaska plutons, Amato and Wright (1997) proposed that this belt represents subduction-related magmatism that evolved during the reconfiguration of a north-dipping subduction zone beneath the region, perhaps due to slab rollback and trench retreat. These petrologic studies reinforce earlier structural and geochronologic studies suggesting that the Bering Straits region underwent large-magnitude north-south extension during this period of Cretaceous magmatism (Miller and Hudson, 1991; Dumitru et al., 1995). These relationships permit an extensional or transtensional setting for the formation of Cretaceous gneiss domes of the Seward Peninsula and adjacent Arctic Russia (Bering Strait Geological Field Party, 1997).

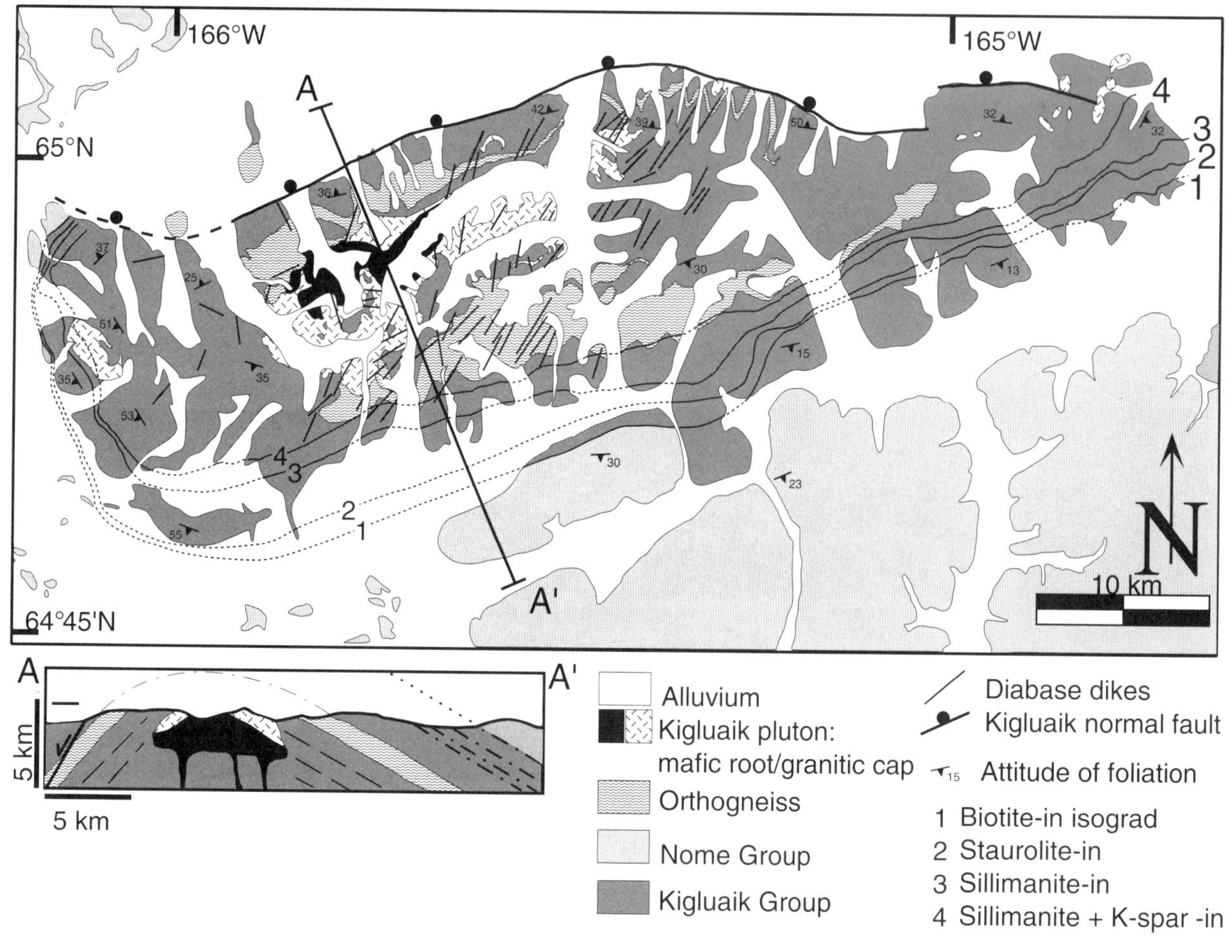

Figure 2. Geologic map of Kigluaik Mountains, showing overall geometry of dome and location of metamorphic isograds.

STRUCTURAL RELATIONSHIPS

Introduction

A pronounced metamorphic foliation of Late Cretaceous age defines the overall geometry of the Kigluaik gneiss dome (Fig. 2). The range has been divided into structural domains based on the dip direction of foliation (Fig. 3). The foliation generally dips away from the pluton in the core of the dome; dips on the flanks of the dome are shallow in domains 1–4, 9, and 10. Dips are steeper in the central domains (5 and 6), in part because the foliation is locally steeper near the intrusive contact, but also because the range-bounding fault has more offset in this region. Overall, the foliation defines a doubly plunging antiform that plunges ~3° away from the apex of the dome in the central part of the range and more steeply on both the east and west flanks.

Stretching lineations as measured in rocks in the core of the dome are mainly defined by the long axis of sillimanite, but in lower-grade rocks, quartz rods and elongate pressure shadows around porphyroblasts define the lineation. These lineations are highly variable in map view (Fig. 3), but are consistent in a given

region with respect to structural depth in the section. In order to better understand the distribution of lineations, we created 1-km-thick structural contours of the gneiss dome as measured orthogonal to the foliation, then averaged the lineation azimuths for each structural slice and plotted them with respect to structural depth. Figure 4 shows a remarkably consistent gradual transition from east-west stretching lineations in the core of the dome to nearly north-south (~160°) lineations in the region structurally higher than the biotite-in isograd. The significance of this transition is discussed with respect to structural and metamorphic features developed at various levels within the gneiss dome along with published geochronologic studies of these rocks.

Structural zones within the gneiss dome

The structural section represented by rocks of the Kigluaik gneiss dome is apparently intact and not cut by faults. Although faults have been mapped within the isograd region on the south flank of the gneiss dome (e.g., Kaufman, 1985; Sturnick, 1984), the offsets appear to be minor and normal in sense. Therefore, field relationships preclude the hypothesis that the two crustal

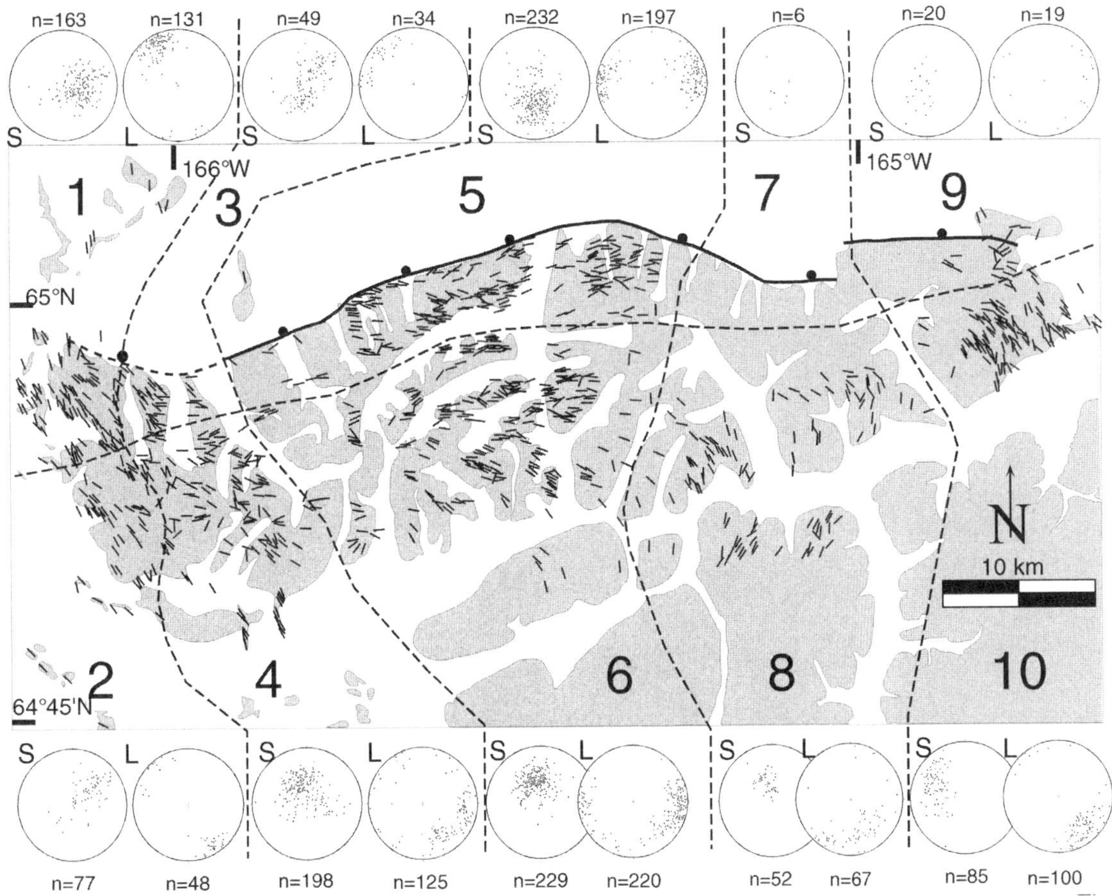

Figure 3. Map shows trend of stretching lineations throughout Kigluaik gneiss dome. Spatial distribution and orientation of foliation and lineation measurements are shown on lower-hemisphere equal-area projections. S plots show poles to dominant foliation. L plots show trend and plunge of mineral stretching lineations. Area depicted is same as that of Figure 2.

levels characterized by orthogonal mineral lineations or stretching directions were juxtaposed by subsequent deformation.

The exposed crustal section in the gneiss dome can be divided into three zones based on structural style of deformation, the textural relationship of structures with respect to peak heating and metamorphism, and the orientation of mineral elongation lineations, which rotate with structural depth. These zones are described from structurally deepest to shallowest.

Zone I (core gneisses) consists of upper amphibolite-grade to locally granulite-grade gneisses in the core of the dome. These rocks have a regional, shallowly dipping foliation that dips radially away from the pluton in the center of the range and exhibits a well-developed east-west stretching lineation (Figs. 2, 3, and 4).

Zone II (transition zone) is the structural transition zone with respect to the orientation of mineral elongation or stretching lineations. Foliation is parallel to that in the core of the dome. The lineations change from the east-west orientations characteristic of zone I to the north-south orientations characteristic of higher structural levels. The lower half of zone II consists of rocks metamorphosed at sillimanite + K-feldspar grade or higher (Figs. 2, 3, and 4). The upper half of this transition zone is the isograd zone, characterized by a series of closely spaced Barrovian isograds that indicate decreasing metamorphic grade upward in the succession. Here, foliation is parallel to that at deeper levels, but lineation azimuth is generally northwest-southeast (Figs. 2, 3, and 4).

Zone III (Nome Group) rocks are entirely above the biotite-in isograd and exhibit a polyphase structural history (Figs. 2, 3, and 4). The youngest foliation in the Nome group is a conspicuous, regionally developed high-strain fabric that can be followed continuously into the isograd region (Miller et al., 1992). Stretching lineations associated with the youngest foliation have a consistent azimuth of approximately 160° (north-northwest–south-southeast).

In the lower parts of the section (zones I and II), peak metamorphic temperatures outlasted deformation, producing coarse-grained equigranular textures and fabrics in highly strained rocks. At the higher structural levels, within the isograd region (zone II) and within the Nome Group (zone III), deformation occurred at lower metamorphic grade and continued after peak temperatures were obtained. Owing to high strain and high

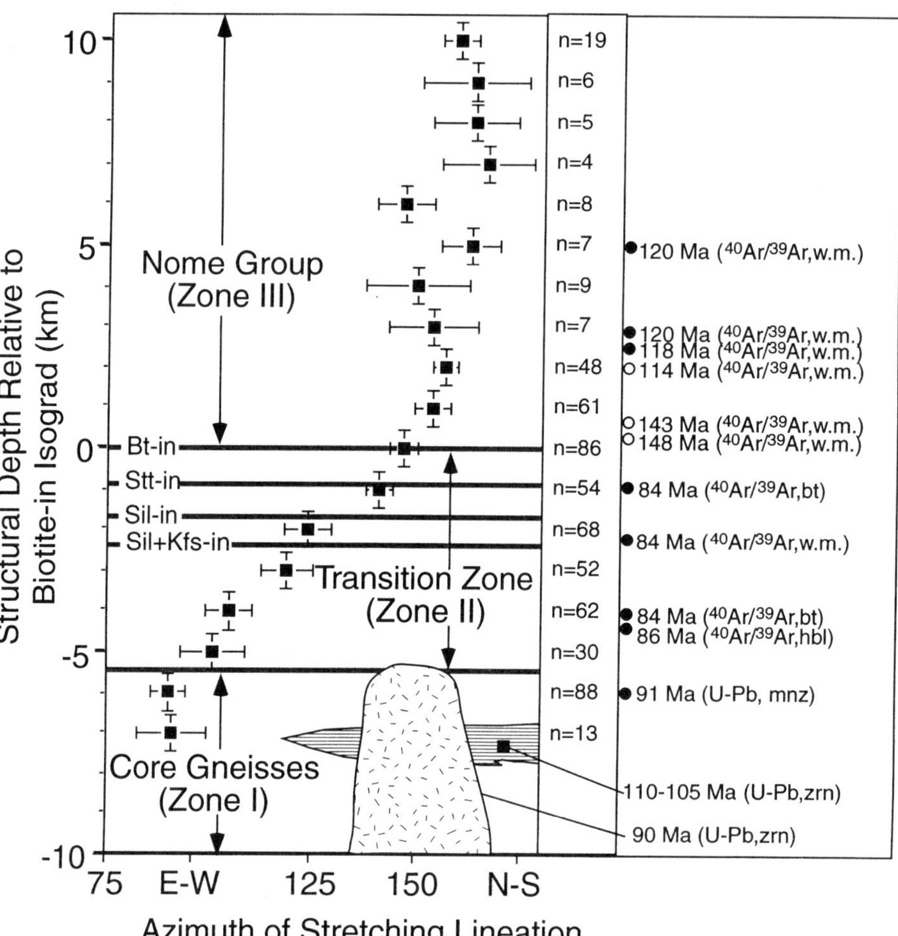

Figure 4. Graph of lineation azimuth vs. structural depth. To produce this figure, gneiss dome was contoured based on structural depth relative to biotite-in isograd (as measured orthogonal to foliation), and mean of all lineation determinations within each ~1 km slice of section was plotted with uncertainty of ±2°SE (n in second column is total number of measurements within each slice). Locations of structural zones (I, II, III) are indicated. Nome Group data are from Hannula (1993); all other data are from this study. Geochronology in third column is from Hannula and McWilliams (1995) (Nome Group), Calvert et al. (1999) (transition zone), and Amato et al. (1994) (core gneisses). Closed circles are cooling ages from metamorphic rocks. Open circles in Nome Group data indicate disturbed or hump-shaped spectra; 120 Ma is average of data that yielded plateaus in Nome Group. Squares indicate intrusive ages of igneous rocks; symbols as in Figure 2. Mineral abbreviations: w.m.—white mica; bt—biotite; hbl—hornblende; mnz—monazite; zrn—zircon.

metamorphic temperatures within zones I and II, all structural and metamorphic evidence of the high-*P*–low-*T* pre-120 Ma metamorphic event has been obliterated. These three zones are described in more detail in the following.

Zone I: Core gneisses. The gneisses in the core of the Kigluaik dome consist of coarse-grained marble, calc-silicate–bearing gneisses, metapelite, migmatite, amphibolite, and orthogneiss, together with lenses of coarse pegmatite that were derived from partial melting of metasedimentary host rocks (Amato et al., 1994). These rocks exhibit well-defined compositional and metamorphic layering at all scales. Foliation is defined by biotite and alternating biotite-rich and felsic layers within metapelite and orthogneiss and by calc-silicate–bearing layers in marble units. The majority of the units contain a strong mineral elongation lineation that is oriented east-west (Figs. 3 and 4). The lineation is defined by elongate bundles of coarse sillimanite as well as by elongate masses of fibrolite replacing biotite within metapelite and orthogneiss (Fig. 5, A and B), aligned hornblende within amphibolite layers, and by streaking and rodding within the marble and calc-silicate gneisses. Sillimanite also forms elongate masses adjacent to or in the pressure shadows of garnet porphyroblasts, and these pressure shadows are parallel to the dominant lineation direction. The fabrics in the core gneisses developed during upper amphibolite facies metamorphism, as indicated by the growth of metamorphic minerals parallel to these fabrics.

The magnitude of total strain represented by the core gneisses is difficult to quantify as in most high-grade gneisses, but it is clearly significant. Formation of the dominant foliation involved large strains, as exemplified by the pronounced flaggy nature of the rocks, the well-defined tectonic foliation, and the parallelism of metamorphic and compositional layering at all scales. Furthermore, the development of mineral elongation lineations within this foliation requires large strains, probably exceeding 10:1:0.1, supported by (1) the fact that foliation transposes layering, (2) the presence of isolated fold hinges, and (3) highly strained feldspar porphyroclasts within pegmatite bodies. Intrusive bodies involved in this deformation, such as the Precambrian–early Paleozoic Thompson Creek orthogneiss (Amato and Wright, 1998), are so highly strained and recrystallized that their fabric is nearly identical to that of the surrounding paragneiss. A fine-grained, garnet-bearing biotite granite orthogneiss dated as ca. 105–110 Ma (Amato et al., 1994) shares most of the deformation undergone by the surrounding metasedimentary

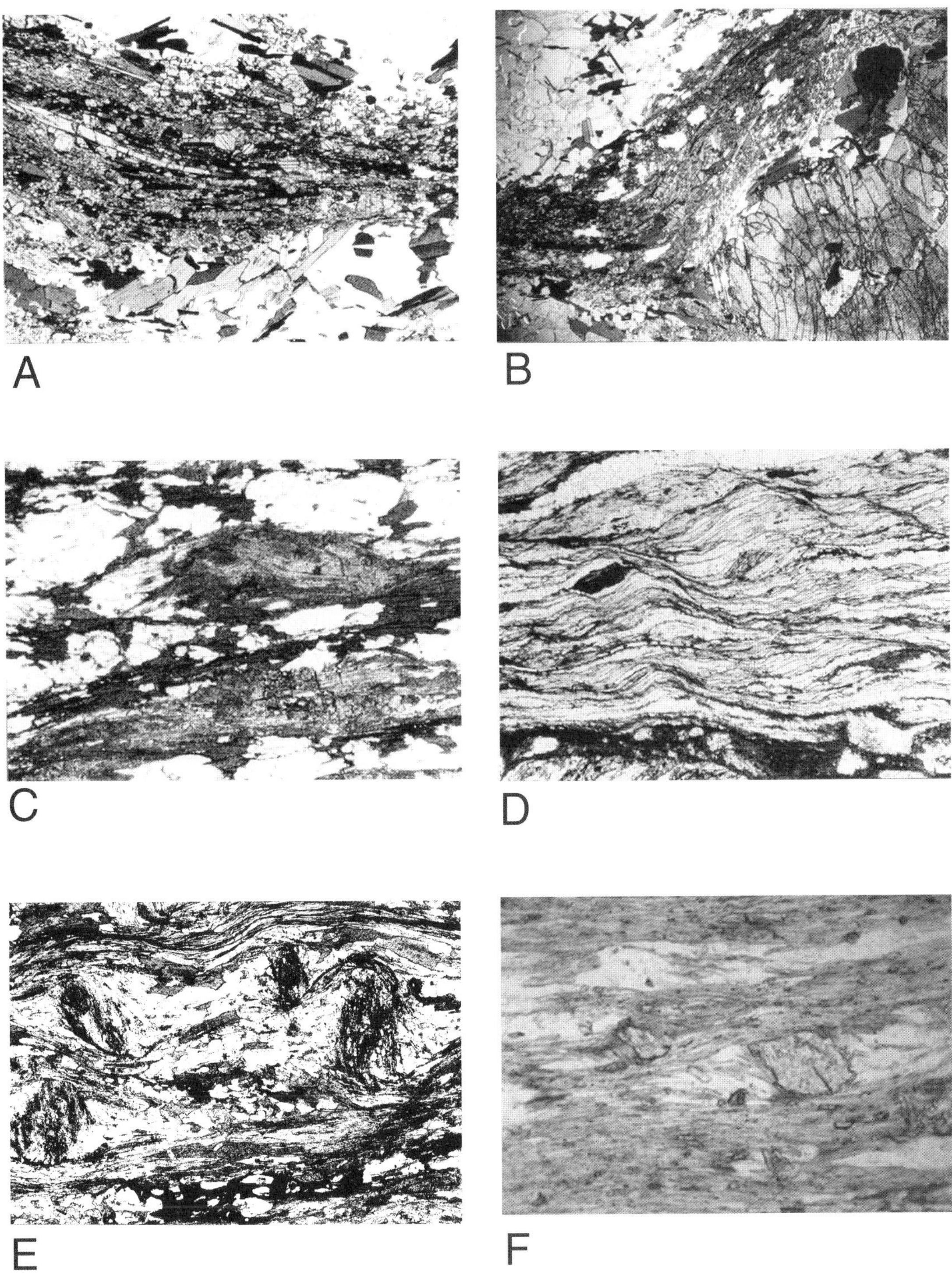

Figure 5. Photomicrographs from Kigluaik Mountains metamorphic rocks showing relationship of metamorphism to deformation. Field of view is ~7 mm across for A–E (taken with cross-polarized light). Field of view is 1.44 mm across for F (taken with plane-polarized light). A: Photomicrograph of rock containing sillimanite lineation from zone I (core gneisses). Section is cut orthogonal to foliation and parallel to lineation. Note elongate bundles of oriented sillimanite within foliation, defined mostly by alignment of biotite. B: Photomicrograph of coarse garnet porphyroblast (lower right) from zone I (core gneisses). Foliation wraps around garnet, and pressure shadow is filled with biotite and sillimanite. C: In zone II (transition zone), sillimanite forms fibrolite bundles rather than coarse crystals as in core gneisses. D: In isograd region of transition zone (zone II), deformation outlasted attainment of peak metamorphic temperatures. This biotite schist contains shear bands at angle to foliation (which is generally parallel to long axis of photo) and some evidence for grain-size reduction. E: Albite porphyroblasts in biotite zone at top of transition zone contain internal foliation at high angle to dominant foliation. This older foliation may be related to Nome Group fabric, or could be result of progressive rotation during Late Cretaceous deformation. F: In Nome Group, relict high-pressure garnets are partly replaced by chlorite ± biotite. Lineation in this rock is defined by elongate biotite pressure shadows around garnet.

rocks (Fig. 4). This relationship, described by Amato et al. (1994), supports the inference that most of the measurable strain in the core gneisses is post-105–110 Ma. We have interpreted the monazite U-Pb ages as indicating that the attainment of peak metamorphic temperatures occurred ca. 91 Ma (Fig. 4; Amato and Wright, 1998). Most of the deformation within the Kigluaik Group probably accompanied peak metamorphism, but the possibility remains that deformation in the core of the gneiss dome could have begun somewhat earlier.

Coarse pegmatite lenses and dikes derived from partial melting of the metasedimentary rocks during metamorphism are variably deformed (Amato and Wright, 1998). Metapelitic units show evidence of migmatization, and the leucosomal segregations are isoclinally folded and transposed into parallelism with the foliation, providing additional evidence for large synmetamorphic strains. The Thompson Creek orthogneiss contains late-stage tension gashes and extensional shear bands filled with partial melts, indicating that even the latest stages of deformation within the core gneisses took place at elevated temperatures.

Shear-sense indicators are only rarely preserved in the core gneisses. Of the 34 oriented thin sections studied from these rocks, only nine provided clear evidence of sense of shear. Seven of these indicate top-to-the-west sense of shear, whereas the two others, from rocks on the southern flank of the dome, suggest top-to-the-east movement (Fig. 6). More commonly, shear-sense indicators from any given thin section are conjugate, indicating only that shortening continued perpendicular to the foliation after it developed.

These observations and relationships indicate that, within the core gneisses, deformation was ongoing during peak metamorphism and partial melting and ceased prior to cooling from high temperatures. In addition, static textures, which we interpret to be almost entirely postkinematic, are observed adjacent to the Kigluaik pluton. These include spinel + cordierite assemblages that grow across the foliation (Amato et al., 1994). Although this particular assemblage indicates elevated temperatures (~650 °C), the implied pressures are <5–6 kbar (Bhattacharya, 1986), which is lower than estimates of peak pressure (~8–10 kbar) for the core gneisses during prograde metamorphism and deformation (Lieberman, 1988).

Zone II: Transition zone. The transition zone is defined as that part of the crustal section in the Kigluaik Mountains that preserves evidence for the rotation of lineations from dominantly east-west orientations at depth to dominantly north-south orientations at higher crustal levels (Fig. 4). This transition occurs over ~4 km of structural section as measured perpendicular to the foliation. The minerals that define the lineation vary with structural position, from elongate, fine-grained bundles of sillimanite in the structurally higher part of the core gneisses (Fig. 5C) to elongate quartz rods and elongate bundles of chlorite and white mica within the upper parts of the isograd zone and the lower parts of the Nome Group (Fig. 4). No overprinting relationships between the two dominant lineation directions were observed either in the field or in thin section.

The structurally higher part of the transition zone (Fig. 4) is characterized by closely spaced isograds (Miller et al., 1992). Metamorphic index minerals developed with increasing structural depth in metapelitic rocks include biotite, staurolite, sillimanite, and sillimanite + K-feldspar (Patrick and Lieberman, 1988; Miller et al., 1992). Relict staurolite is found as inclusions in garnet within sillimanite-grade rocks, preserving part of the prograde history of these rocks. The present structural distance between the biotite-in isograd (representing temperatures of ~400–450 °C) and the sillimanite + K-feldspar-in isograd (representing temperatures of ~700–750 °C; Lieberman, 1988) is ~3 km, yielding a metamorphic field gradient of ~100 °C/km (Miller et al., 1992). The pressure gradient, calculated using data from Lieberman (1988), is >0.6 kbar/km. Although condensed, the isograd zone records a continuous transition from upper amphibolite facies conditions at depth in the Kigluaik Group to the greenschist facies conditions in the overlying Nome Group.

Thin-section textural relationships described by Miller et al. (1992) indicate that within the isograd zone, significant deformation occurred in conjunction with Cretaceous high-*T* metamorphism. This deformation is documented by a pronounced foliation that becomes better developed as metamorphism increases from greenschist to amphibolite facies, and by the local development of high-strain mylonitic fabrics within biotite-zone rocks (Fig. 5D). Evidence for grain-size reduction, the development of quartz grain-shape foliations oblique to the dominant foliation, and strong preferred orientations of quartz C-axes are present in these mylonites (Miller et al., 1992; Patrick, 1988). Lineations within the isograd zone are defined by the elongate direction of mineral growth and by quartz rods and mineral streaking in lower-temperature rocks. The lineation direction developed within the isograd zone is oriented northwest-southeast at the base of the zone and rotates to nearly north-south (160°) near the contact with the Nome Group at the biotite-in isograd (Fig. 4).

In contrast to the coarse-grained core gneisses, the rocks of the isograd zone contain abundant shear-sense indicators. These include sigmoidal internal foliation trails developed within staurolite, albite porphyroblasts that are systematically rotated with respect to the attitude of foliation in the surrounding matrix (Fig. 5E), and shear bands developed perpendicular to the stretching lineation. Shear bands are commonly decorated with fresh biotite and/or white mica and thus are compatible with development at low to high greenschist grade. The majority of samples that contain extensional shear bands exhibit shear bands with a preferential dip and a unidirectional sense of shear. These indicate motion away from the core of the dome: top-to-the-southeast sense of shear in the southeast, top-to-the-south in the southwest, and top-to-the-northwest in the northwest part of the dome (Fig. 6). These motions are consistent with at least a component of this shear being related to the diapiric rise of the gneiss dome rocks through the crust.

Zone III: Nome Group. Unlike the Kigluaik Group (zones I and II), which preserves evidence for only one major deformational and metamorphic event, the rocks of the Nome Group

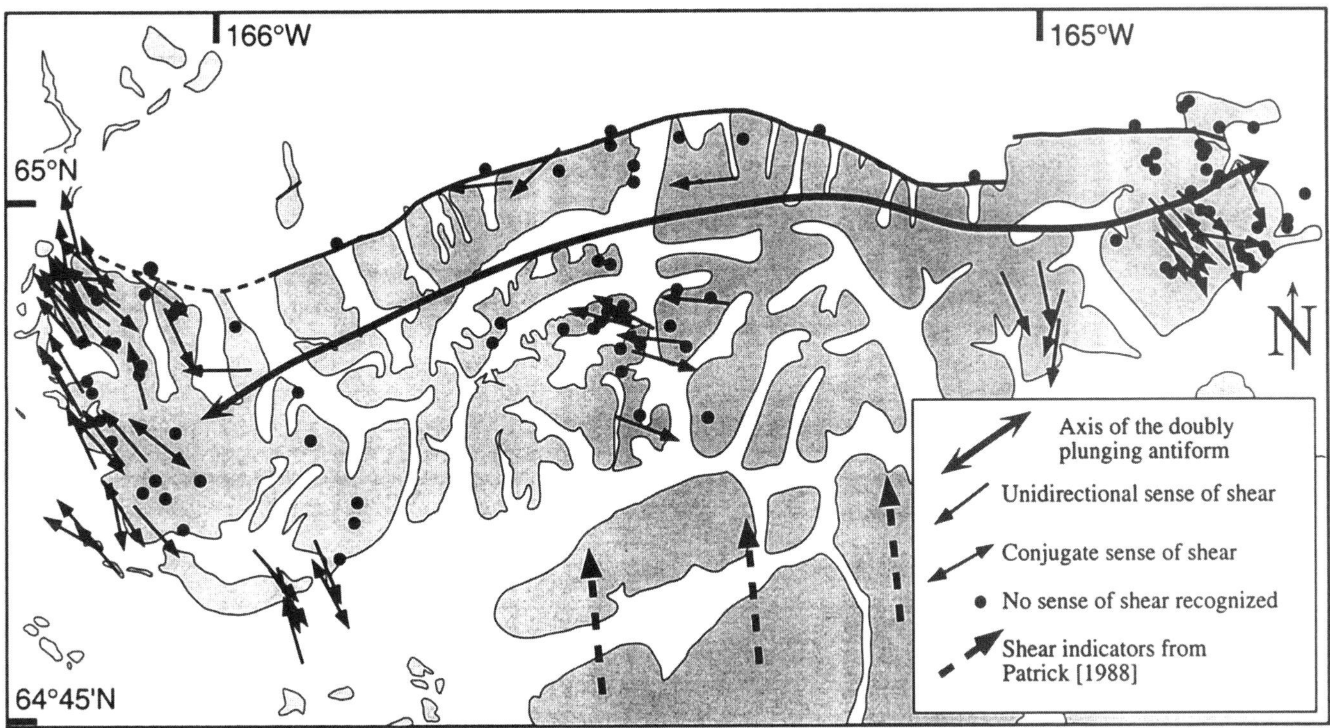

Figure 6. Map showing individual determinations of shear sense from thin sections of tectonites from core and transition zone of gneiss dome. Large arrows indicate general shear sense for various locations. Three large northward-pointing arrows in southern part of map are taken from Patrick (1988) and represent top-to-north shear. Area depicted is same as in Figure 2.

retain evidence of a more complex, polyphase metamorphic and structural history (Hannula et al., 1995). The first metamorphic event to affect the Nome Group took place under relatively high-P–low-T conditions, pumpellyite-actinolite facies assemblages being present in the upper part of the section and blueschist facies assemblages at deeper structural levels. Metamorphism progressed into the greenschist facies during the final stages of the earlier metamorphic and deformational event (Hannula et al., 1995). This event has a minimum age of ca. 120 Ma (Hannula and McWilliams, 1995). A younger greenschist facies metamorphic event accompanied by the development of a high-strain foliation overprints earlier assemblages. The thermal effects of this younger metamorphism appear to increase continuously with depth into the isograd region, providing a critical link between the metamorphism and deformation in the Nome Group and that in the transition zone and underlying core gneisses (Miller et al., 1992).

Previous workers identified two main foliations in the Nome Group: an earlier foliation subparallel to lithologic layering (S_1), and a second (S_2) foliation that is axial planar to isoclinal folds of S_1 (Thurston, 1985; Patrick, 1988; see also Fig. 5 in Hannula et al., 1995). Both of these foliations were previously interpreted to have formed during a single progressive deformation coeval with high-P metamorphism (Thurston, 1985; Patrick, 1988). Hannula et al. (1995), however, suggested that the two observed foliations are the result of separate deformational events. The dominant foliation within the Nome Group (S_2) is the younger of the two. At greater structural depth, this foliation is concordant to the foliation developed within the isograd zone and core gneisses (Miller et al., 1992). Although S_2 in the Nome Group is concordant with the dominant foliation in the Kigluaik Group, which is thought to have formed ca. 91 Ma, the precise age of S_2 in the overlying Nome Group is unknown and it could possibly be somewhat older (Fig. 4; Hannula and McWilliams, 1995; Dumitru et al., 1995; Amato and Wright, 1998).

A north-south–oriented stretching lineation is associated with the S_2 foliation in the Nome Group. This lineation is first observed in lower greenschist facies rocks northwest of the Kigluaik Mountains, where it is exclusively developed in more phyllitic units. It becomes better defined southward, closer to the gneiss dome (Hannula et al., 1995). This lineation is defined by fibrous quartz pressure shadows around pyrite in phyllitic units, mineral streaking, elongate epidote porphyroblasts, and pressure shadows around albite and rare garnet porphyroblasts. Rare biotite is found in pressure shadows around relict, partly resorbed garnet porphyroblasts that presumably grew during the earlier high-P metamorphic event (Fig. 5F). These pressure shadows provide evidence that the north-northwest–south-southeast stretching lineations formed during the later greenschist facies metamorphic event rather than the early blueschist facies metamorphism (Hannula et al., 1995). The orientation of stretching lineations in the Nome Group is somewhat variable owing to a younger superimposed warping or gentle folding event (Hannula et al., 1995). The overall general azimuth of the lineations associated with S_2 in the Nome Group, however, is north-northwest–south-southeast.

Patrick (1988) measured quartz C-axis fabrics and kinematic indicators associated with the formation of the dominant foliation (S_2) within the Nome Group south of the Kigluaik Mountains. His data show that these formed as a consequence of top-to-the-north shear. Because he inferred that these fabrics formed during blueschist facies metamorphism, he postulated that they were related to northward overthrusting of ophiolites during A-type subduction. High-P–low-T metamorphism accompanying this event has been difficult to date, but the age is believed to be Late Jurassic to Early Cretaceous in northern Alaska (for discussion, see Moore et al., 1994; Hannula et al., 1995). However, the same foliations and associated quartz deformational fabrics were reinterpreted by Miller et al. (1992) and by Hannula et al. (1995) as having formed during a younger deformational and metamorphic event, unrelated to the earlier high-P–low-T metamorphism. On the basis of both the nature of quartz fabric development and textural relationships in thin section, this younger deformational event occurred during progressive denudation and cooling of the Nome Group. The top-to-the-north shear-sense indicators in the Nome Group occur to the north and south of the gneiss dome and are mostly unidirectional (Fig. 6); thus, they do not appear to be directly related to the rise of the gneiss dome. Miller et al. (1992) and Hannula et al. (1995) suggested that these fabrics could be related to extension on a more regional scale than the gneiss dome. Data supporting the proposed extensional unroofing of the Nome Group come from the York Mountains, northwest of the Kigluaik Mountains (Fig. 1). Here, upper crustal Paleozoic carbonate sections are only weakly metamorphosed (A. Harris, 1991, written commun.) and cut by normal faults that sole into a low-angle fault that parallels foliation in the underlying Nome Group. The faulting predates ca. 70–80 Ma intrusions (Hudson and Arth, 1983). These relations are similar to those characterizing detachment faults of extensional core complexes, and the geology of the York Mountains may be representative of what once covered a larger region of the Nome Group metamorphic rocks.

Cooling history

The cooling history of the Kigluaik gneiss dome and the surrounding Nome Group rocks was investigated in detail by Calvert et al. (1999) and Hannula and McWilliams (1995) using $^{40}Ar/^{39}Ar$ techniques and by Dumitru et al. (1995) using fission-track analysis. It is apparent from this work that the Cretaceous metamorphism was the last significant thermal event to affect the region around the Kigluaik gneiss dome. Hornblende ages from structural levels near the top of the core gneisses indicate cooling through ~550 °C ca. 86 Ma, whereas those from the deepest

levels yield ca. 84 Ma ages (Fig. 4). Biotite ages, indicating cooling through ~300 °C, range from 84 Ma to 82 Ma, and decrease with increasing structural depth (Fig. 4). K-feldspar cooling ages range from about 82 Ma to 79 Ma and show the same trend as the biotite ages (Calvert et al., 1999). Fission-track data from a variety of structural levels record the final cooling (to <~120 °C) of the isograd zone rocks ca. 55 Ma, and final cooling of the core gneisses ca. 30 Ma (Dumitru et al., 1995).

Summary of structural relationships

Rock units in the Kigluaik gneiss dome together with the overlying Nome Group compose a continuous, structurally coherent crustal section that records an increase in metamorphic grade with structural depth from greenschist facies to granulite facies. The effects of this thermal event decrease upward in the section, where evidence is preserved for at least one older metamorphic and deformational event. The metamorphic isograds within this section are closely spaced, in part due to deformation during and late in the metamorphic history, resulting in development of a foliation and attenuation of rocks between isograd surfaces in the region of the gneiss dome. Compositional layering and foliation developed during Cretaceous metamorphism are subparallel and structurally concordant throughout the crustal section.

Stretching lineations developed within the gneiss dome range from east-west in the deeper levels of exposure through an intermediate transition zone, where they vary continuously in orientation to north-northwest–south-southeast in the overlying Nome Group (Fig. 4). No evidence for two metamorphic events is discernible in the core and transition zones, even though they are presumed to have undergone at least the same two events recorded in the overlying Nome Group. The observed continuous variation in lineation orientations apparently developed during a single thermal event because they are developed in rocks that exhibit a continuous metamorphic gradient. Additional support for this conclusion is based on geochronology and thermochronology. U-Pb monazite dates from the core indicate cooling through ~725 °C ca. 91 Ma (Amato and Wright, 1998) and $^{40}Ar/^{39}Ar$ dates from the upper levels of the transition zone indicate cooling through 300–400 °C by ca. 84 Ma (Calvert et al., 1999). Therefore, the field relationships as well as the geochronology suggest that the structures, regardless of their orientation, must be broadly coeval.

DISCUSSION

The stretching or mineral elongation lineations in the Kigluaik gneiss dome are interpreted as representing the dominant and final direction of stretching within the core gneisses during high-T deformation. The most intriguing and puzzling aspect of the deformational fabrics documented in the Kigluaik Mountains is the rotation of lineations with structural depth. Field relations preclude the hypothesis that the two crustal levels characterized by orthogonal flow directions were juxtaposed by subsequent deformation. Metamorphic and thermochronologic studies imply that lower and upper levels of the complex underwent heating and metamorphism during the same event. It is possible, however, that deformation became less pervasive and/or continued in discrete zones within the crustal section while ceasing at other crustal levels.

There are two possible interpretations for the observed structural relationships in the Kigluaik gneiss dome, and the critical factor is knowing the relative timing of fabric development at various structural levels. One possible explanation for the observed relationships is that an older fabric is overprinted by a younger fabric. Overprinting is commonly developed in gneiss domes characterized by lineations of varying or orthogonal orientation (e.g., Compton et al., 1977), but in this particular case, lineations representing transitional directions were not observed. In zones I and II, there is an absence of evidence at any scale for an overprinting relationship. The presence of what appears to be a continuous transition in lineation orientation would seem to preclude the possibility that two superimposed deformational events are represented. In addition, ^{40}Ar/^{39}Ar data on biotite from the transition zone give ca. 84 Ma cooling ages, and monazite data from the core yield ca. 91 Ma ages; thus, the lineations throughout the structural section, despite their different orientations, are Late Cretaceous in age.

The scenario currently preferred for the observed relations in the Kigluaik Mountains involves the concept of divergent directions of flow or stretching within the crust during high-T metamorphism. The Kigluaik gneiss dome may be an excellent example of an area that has undergone approximately coeval and orthogonal crustal flow at varying levels of the crust. We suggest that north-south extension in the Kigluaik Mountains area, as shown by north-south–directed lineations within the Nome Group, was compensated in part by the intrusion of mantle-derived magma into the crust and in part by east-west flow of the crust (on a scale of perhaps tens of kilometers) into the region undergoing north-south extension (Fig. 7).

Although a proposed extensional tectonic setting for this region during Late Cretaceous time is still somewhat controversial, there are several regional relationships that are compatible with this interpretation. (1) Trace element and isotopic studies of the Kigluaik pluton and comparison to data from other plutons of similar age in Alaska indicate that the magmatism in the Kigluaik Mountains was mantle derived (Amato and Wright, 1997), and that this magmatism likely occurred within a subduction-related tectonic setting, possibly above a north-dipping, southward-retreating subduction zone (Amato and Wright, 1997). (2) The regional development of subhorizontal foliations indicating crustal attenuation is consistent with extensional tectonism following an earlier history of compression (Miller et al., 1992). (3) There is evidence for a major pulse of exhumation during the documented crustal attenuation event, dated as occurring between 100 and 70 Ma (Dumitru et al., 1995), and this exhumation is corroborated by the cooling history described here.

The relationship of orthogonal extension directions at different crustal levels has been described for some core complexes of the Basin and Range Province, where mid-crustal temperatures are thought to exceed 600–700 °C, comparable to the peak metamorphic temperatures in the Kigluaik gneiss dome. Gans (1987) described a crustal stretching model for the Basin and Range in which the upper and middle crust are decoupled. In this model, extension at depth is more uniformly distributed than extension in the upper crust, and is accommodated by wholesale ductile flow and the intrusion of mantle-derived magmas that raise the geothermal gradient and facilitate flow. Such a process would maintain a flat Moho beneath these areas of large-scale extension, as is required by seismic reflection data from the Basin and Range (Klemperer et al., 1986). Deep seismic reflection profiling through the Bering Strait region shows that the Moho at the latitude of Seward Peninsula has little topography (Klemperer et al., this volume, Chapter 1; Wolf et al., this volume, Chapter 2). One of the consequences of this model is that middle and lower crustal rocks are capable of flowing laterally from unextended regions into more extended regions (Gans, 1987).

A good example of an area with apparently coeval and orthogonal lineation directions is the Ruby Mountains core complex in the Basin and Range Province (MacCready et al., 1997). There, a high-level mylonitic shear zone with a top-to-the-west and/or northwest shear sense structurally overlies and overprints

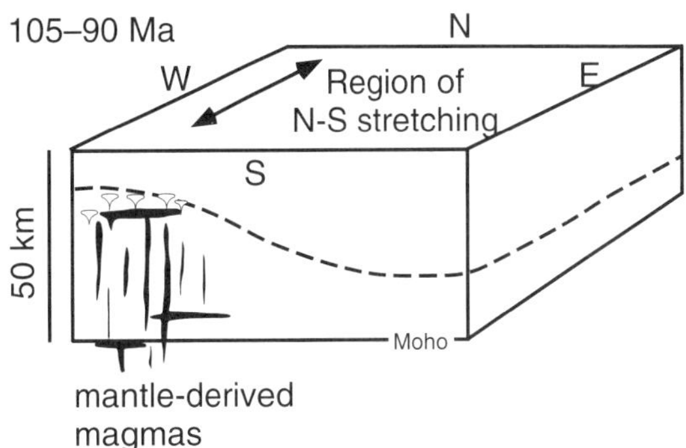

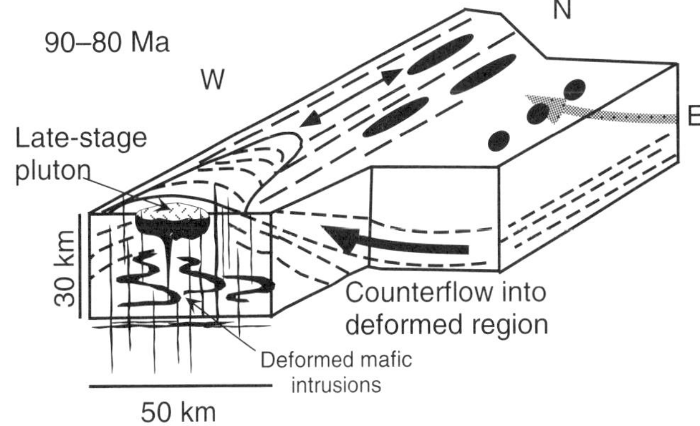

Figure 7. Block diagram showing orthogonal flow into north-south extended region. See text for details. Upper: 105–90 Ma. Lower: 90–80 Ma.

ductilely deformed subhorizontally layered gneisses with north-south stretching lineations. Movement along the shear zone is thought to be coeval with deformation at lower crustal levels based on field evidence indicating structures with a transitional orientation. Because the regional extension direction is known to be parallel to the west-northwest shear-sense indicators in the mylonite zone, the north-south lineations deeper in the lower plate do not reflect the overall strain field of the crust in the Basin and Range Province. Instead, these lineations record local north-south flow that is believed to compensate for differential crustal thinning at higher levels. Continued movement along the shear zone may have cut deeper into the crust, overprinting and thus obscuring evidence for the diffuse zone of decoupling that may have once existed between the higher-level mylonitic rocks and the deeper infrastructure (MacCready et al., 1997). In a discussion on the extensional history of the U.S. Cordillera, Wernicke (1992) noted that a fluid middle crust may locally flow in a direction orthogonal to the regional extension direction.

CONCLUSIONS AND REGIONAL IMPLICATIONS

This chapter describes an intact 15 km structural section through the crust in which lineations developed during high-strain Late Cretaceous deformation indicate a continuous and distinct rotation of stretching direction with structural depth. The most plausible explanation for the observed structural relationships is that the upper and lower structural levels were being stretched in nearly orthogonal directions during this deformational and metamorphic event. This observation, together with local and regional structural, petrological, and geochronological evidence, implies that north-south crustal extension was in part compensated by east-west stretching or flow at deeper crustal levels.

The development of orthogonal flow directions at depth beneath the Kigluaik Mountains may be attributable to counterflow, as argued for the Ruby Mountains in the Basin and Range Province (MacCready et al., 1997). This flow could be locally driven during the development of the gneiss dome. Alternatively, east-west flow into a region undergoing overall north-south extension could be explained in part by regional geologic relationships that suggest lateral westward extrusion of material from interior Alaska as far back as Cretaceous time (Dumitru et al., 1995). Building on the ideas of Scholl and Stevenson (1991) for the Cenozoic history of Alaska, Dumitru et al. (1995) suggested a possible model for extension in the Bering Strait region that integrates a history of Cretaceous through Tertiary shortening in eastern and central Alaska with extension in western Alaska during this same interval (Fig. 8). Together with right-lateral transcurrent faulting, this would result in the counterclockwise rotation of crustal blocks in Alaska and the extrusion of material westward into the region of north-south extension around the longitude of the Bering Strait (Scholl and Stevenson, 1991; Dumitru et al., 1995). This area may also have been affected by counterclockwise rotation of crustal elements during

extension. At this time, it is unclear if the relationships described for the Kigluaik Mountains are of local significance related to formation of the gneiss dome, or if they could have formed within a more regional context of westward extrusion and north-south stretching.

Further understanding of the relationships described in this chapter would require additional work in other metamorphic culminations on the Seward Peninsula, such as in the Darby and Bendeleben Mountains (Fig. 1). The relative timing of magmatism and metamorphism in these other areas should be comparable to the relationships described from the Kigluaik Mountains if a general model for regional extension is valid. Regardless of the regional interpretations, the Kigluaik gneiss dome preserves an excellent example of stretching directions that vary with structural depth. This geometry may be more common than previously thought and should be considered in cross-sectional models of flow within the infrastructures of orogenic belts.

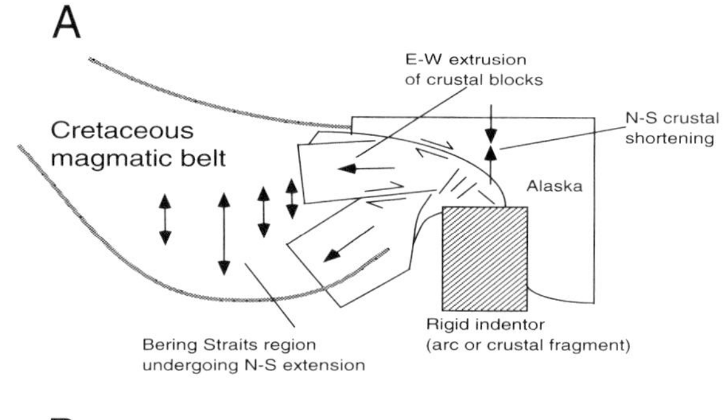

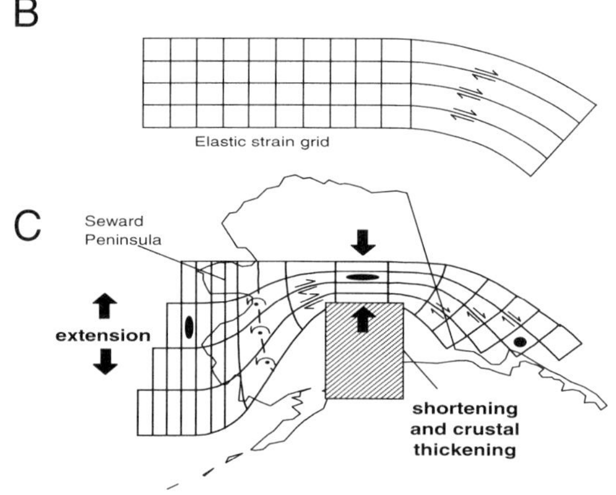

Figure 8. Proposed model for Cretaceous tectonic setting of Bering Straits region. A: Rigid indentor model of Tapponier (1982) for Asia, applied to Alaska. B and C: Schematic illustration of strain pattern resulting from deforming originally square grid by simultaneous shortening of eastern and central Alaska and extension of western Alaska. If operating in conjunction with right-lateral strike-slip faulting, this strain pattern would result in bending of preexisting faults, counterclockwise rotation of both faults and fault-bound crustal blocks, extension of western parts of fault-bound crustal fragments as evidenced by formation of sedimentary basins in western part of system, and southward translation of crustal fragments in Bering Straits region.

ACKNOWLEDGMENTS

Field work was supported by National Science Foundation (NSF) grant EAR-90–18922 to Miller; Geological Society of America (GSA) John T. Dillon awards (1990, 1991), GSA Research Grants (1991, 1992), and Stanford School of Earth Sciences McGee Fund grants to Amato and Hannula; and an NSF Graduate Fellowship to Hannula. Acquisition of geochronologic data was supported by NSF grant EAR-91–17419 to J.E. Wright. This project was based on field work carried out by J.M. Amato, P.B. Gans, A.T. Calvert, T.A. Little, K.A. Hannula, E.L. Miller, J. Lee, J. Toro, J.E. Wright, C.M. Rubin, and T.A. Dumitru, assisted by J.Y. Amory and E. King. Discussions with Gans, Calvert, and Toro helped clarify the ideas presented here. Careful reviews provided by S. Roeske, G. Axen, S. Klemperer, and T. Lawton are greatly appreciated.

REFERENCES CITED

Amato, J.M., and Miller, E.L., 1997, Bedrock geologic map of the Kigluaik Mountains, Seward Peninsula, Alaska: Alaska Division of Geological and Geophysical Surveys Public Data File 97–31, scale 1:42 240, 8 sheets, 5 p. text.

Amato, J.M., and Wright, J.E., 1997, Petrogenesis of the potassic Kigluaik pluton: Arc-related mafic magmatism in northern Alaska: Journal of Geophysical Research, v. 102, p. 8065–8083.

Amato, J.M., and Wright, J.E., 1998, Geochronologic investigations of magmatism and metamorphism within the Kigluaik Mountains gneiss dome, Seward Peninsula, Alaska, *in* Clough, J.G., and Larson, F., eds., Short notes on Alaska geology 1997: State of Alaska, Department of Natural Resources, Division of Geological and Geophysical Surveys Professional Report 118, p. 1–22.

Amato, J.M., Wright, J.E., Gans, P.B., and Miller, E.L., 1994, Magmatically induced metamorphism and deformation in the Kigluaik gneiss dome, Seward Peninsula, Alaska: Tectonics, v. 13, p. 515–527.

Bering Strait Geological Field Party, 1997, Koolen metamorphic complex, NE Russia: Implications for the tectonic evolution of the Bering Strait region: Tectonics, v. 16, p. 713–729.

Bhattacharya, A., 1986, Some geobarometers involving cordierite in the FeO-Al_2O_3-SiO_2 ($\pm$ H_2O) system: Refinements, thermodynamic calibration, and applicability in granulite facies rocks: Contributions to Mineralogy and Petrology, v. 94, p. 387–394.

Brooks, A.H., Richardson, G.B., Collier, A.J., and Mendenhall, W.C., 1901, Reconnaissances in the Cape Nome and Norton Bay Regions, Alaska, in 1900: Washington, D.C., U.S. Government Printing Office, p. 1–222.

Burg, J.P., Guiraud, M., Chen, G.M., and Li, G.C., 1984, Himalayan metamorphism and deformation in the North Himalayan Belt (southern Tibet, China): Earth and Planetary Science Letters, v. 69, p. 391–400.

Calvert, A.T., Gans, P.B., and Amato, J.M., 1999, Diapiric ascent and cooling of a sillimanite gneiss dome revealed by $^{40}Ar/^{39}Ar$ thermochronology: The Kigluaik Mountains, Seward Peninsula, Alaska, *in* Ring, U., Brandon, M.T., Lister, G.S., and Willett, S.D., eds., Exhumation processes: Normal faulting, ductile flow and erosion: Geological Society [London] Special Publication 154, p. 205–232.

Chen, Z., Liu, Y., Hodges, K.V., Burchfiel, B.C, Royden, L.H., and Deng, C., 1990, The Kangmar Dome: A metamorphic core complex in southern Xizang (Tibet): Science, v. 250, p. 1552–1556.

Compton, R.G., Todd, V.R., Zartman, R.E., and Naeser, C.W., 1977, Oligocene and Miocene metamorphism, folding, and low-angle faulting in northwestern Utah: Geological Society of America Bulletin, v. 88, p. 1237–1250.

Davis, G.H., and Coney, P.J., 1979, Geologic development of the Cordilleran metamorphic core complexes: Geology, v. 7, p. 120–124.

Dumitru, T.A., Miller, E.L., O'Sullivan, P.B., Amato, J.M., Hannula, K.A., Calvert, A.T., and Gans, P.B., 1995, Cretaceous to recent extension in the Bering Strait region, Alaska: Tectonics, v. 14, p. 549–563.

Fossen, H., Tikoff, B., and Teyssier, C., 1994, Strain modeling of transpressional and transtensional deformation: Norsk Geologisk Tidsskrift, v. 74, p. 134–145.

Galloway, B.K., Klemperer, S.L., Childs, J.R., Brocher, T.M., and Allen, R.M., 1995, Lower crustal and upper-mantle reflectivity, Bering Strait, Alaska [abs.]: Eos (Transactions, American Geophysical Union), v. 76, p. F593.

Gans, P.B., 1987, An open-system, two-layer crustal stretching model for the eastern Great Basin: Tectonics, v. 6, p. 1–12.

Hannula, K.A., 1993, Relations between deformation, metamorphism, and exhumation in the Nome Group blueschist-greenschist terrane, Seward Peninsula, Alaska [Ph.D. thesis]: Palo Alto, California, Stanford University, 171 p.

Hannula, K.A., and McWilliams, M.O., 1995, Reconsideration of the age of blueschist facies metamorphism on the Seward Peninsula, Alaska, based on phengite $^{40}Ar/^{39}Ar$ results: Journal of Metamorphic Geology, v. 13, p. 125–139.

Hannula, K.A., Miller, E.L., Dumitru, T.A., Lee, J., and Rubin, C.M., 1995, Structural and metamorphic relations in the southwest Seward Peninsula, Alaska: Crustal extension and the unroofing of blueschists: Geological Society of America Bulletin, v. 107, p. 536–553.

Hudson, T., and Arth, J.G., 1983, Tin granites of Seward Peninsula, Alaska: Geological Society of America Bulletin, v. 94, p. 768–790.

Hummel, C.H., 1962, Preliminary geologic map of the Nome D-1 quadrangle, Seward Peninsula, Alaska: U.S. Geological Survey Miscellaneous Investigations Field Studies Map MF-248, scale 1:63 360, 1 sheet.

Kaufman, D.S., 1985, Windy Creek and Crater Creek faults, Seward Peninsula, *in* Bartsch-Winkler, S., and Reed, K.M., eds., The United States Geological Survey in Alaska: Accomplishments during 1983: U.S. Geological Survey Circular 945, p. 22–24.

Klemperer, S.L., Hauge, T.A., Hauser, E.C., Oliver, J.E., and Potter, C.J., 1986, The Moho in the northern Basin and Range province, Nevada, along the COCORP 40°N seismic-reflection transect: Geological Society of America Bulletin, v. 97, p. 603–618.

Lieberman, J.E., 1988, Metamorphic and structural studies of the Kigluaik Mountains, western Alaska [Ph.D. thesis]: Seattle, University of Washington, 192 p.

MacCready, T., Snoke, A.W., Wright, J.E., and Howard, K.A., 1997, Mid-crustal flow during Tertiary extension in the Ruby Mountains core complex, Nevada: Geological Society of America Bulletin, v. 109, p. 1576–1594.

Miller, E.L., and Hudson, T., 1991, Mid-Cretaceous extensional fragmentation of Jurassic–Early Cretaceous compression orogen, Alaska: Tectonics, v. 10, p. 781–796.

Miller, E.L., Calvert, A.T., and Little, T.A., 1992, Strain-collapsed metamorphic isograds in a sillimanite gneiss dome, Seward Peninsula, Alaska: Geology, v. 20, p. 487–490.

Moffit, F.H., 1913, Geology of the Nome and Grand Central Quadrangles, Alaska: U.S. Geological Survey Bulletin 533, 140 p.

Moore, T.E., Wallace, W.K., Bird, K.J., Karl, S.M., Mull, C.G., and Dillon, J.T., 1994, Geology of northern Alaska, *in* Plafker, G., and Berg, H.C., eds., The geology of Alaska: Boulder, Colorado, Geological Society of America, Geology of North America, v. G-1, p. 49–140.

Patrick, B.E., 1988, Synmetamorphic structural evolution of the Seward Peninsula blueschist terrane: Journal of Structural Geology, v. 10, p. 555–565.

Patrick, B.E., and Evans, B.W., 1989, Metamorphic evolution of the Seward Peninsula blueschist terrane: Journal of Petrology, v. 30, p. 531–555.

Patrick, B.E., and Lieberman, J.E., 1988, Thermal overprint on blueschists of the Seward Peninsula: The Lepontine in Alaska: Geology, v. 16, p. 1100–1103.

Ramberg, H., 1980, Diapirism and gravity collapse in the Scandinavian Caledonides: Journal of the Geological Society of London, v. 137, p. 261–270.

            J.M. Amato, E.L. Miller, and K.A. Hannula

Ramsay, J.G., 1967, Folding and fracturing of rocks: New York, McGraw-Hill, 568 p.

Sainsbury, C.L., 1972, Geologic map of the Teller quadrangle, western Seward Peninsula, Alaska: U.S. Geological Survey Miscellaneous Geologic Investigations Map I-685, scale 1:63 360, 1 sheet.

Sainsbury, C.L., Hummel, C.L., and Hudson, T., 1972, Reconnaissance geologic map of the Nome Quadrangle, Seward Peninsula, Alaska: U.S. Geological Survey Open-File Report 72–326, scale 1:63 360, 1 sheet.

Scholl, D.W., and Stevenson, A.J., 1991, Exploring the idea that early Tertiary evolution of the Alaska orocline and the Aleutian-Bering Sea region is a manifestation of Kula-plate-driven Cordilleran tectonism and escape tectonics?: Geological Society of America Abstracts with Programs, v. 23, no. 5, p. A435.

Selverstone, J., 1985, Petrologic constraints on imbrication, metamorphism, and uplift in the SW Tauern Window, Eastern Alps: Tectonics, v. 4, p. 687–704.

Sturnick, M.A., 1984, Metamorphic petrology, geothermo-barometry, and geochronology of the eastern Kigluaik Mountains, Seward Peninsula, Alaska [Ph.D. thesis]: Fairbanks, University of Alaska, 175 p.

Tapponier, P., Peltzer, G., LeDain, A.Y., Armijo, R., and Cobbold, P., 1982, Propagating extrusion tectonics in Asia: New insights from simple experiments with plasticine: Geology, v. 10, p. 611–616.

Throckmorton, M.L., and Hummel, C.L., 1979, Quartzofeldspathic, mafic, and ultramafic granulites identified in the Kigluaik Mountains, Seward Peninsula, Alaska: U.S. Geological Survey Circular 804–B, p. 70–72.

Thurston, S.P., 1985, Structure, petrology, and metamorphic history of the Nome Group blueschist terrane, Salmon Lake area, Seward Peninsula, Alaska: Geological Society of America Bulletin, v. 96, p. 600–617.

Till, A.B., and Dumoulin, J.A., 1994, Geology of Seward Peninsula and Saint Lawrence Island, *in* Plafker, G., and Berg, H.C., eds., The geology of Alaska: Boulder, Colorado, Geological Society of America, Geology of North America, v. G-1, p. 141–152.

Till, A.B., Dumoulin, J.A., Gamble, B.M., Kaufman, D.S., and Carroll, P.I., 1986, Preliminary geologic map and fossil data, Solomon, Bendeleben, and Southern Kotzebue quadrangles, Seward Peninsula, Alaska: U.S. Geological Survey Open-File Report 86–276, scale 1:63 360, 1 sheet.

Todd, C.S., and Evans, B.W., 1993, Limited fluid-rock interaction at marble-gneiss contacts during Cretaceous granulite-facies metamorphism, Seward Peninsula, Alaska: Contributions to Mineralogy and Petrology, v. 114, p. 27–41.

Wernicke, B., 1992, Cenozoic extensional tectonics of the U.S. Cordillera, *in* Burchfiel, B.C., Lipman, P.W., and Zoback, M.L., eds., The Cordilleran orogen: Conterminous U.S.: Boulder, Colorado, Geological Society of America, Geology of North America, v. G-3, p. 553–581.

MANUSCRIPT ACCEPTED BY THE SOCIETY MAY 15, 2001.

Geological Society of America
Special Paper 360
2002

Cretaceous mid-crustal metamorphism and exhumation of the Koolen gneiss dome, Chukotka Peninsula, northeastern Russia

Vyacheslav V. Akinin*
Russian Academy of Science, Northeast Interdisciplinary Science Research Institute,
Portovay a Street, 16, Magadan, 685000, Russia
Andrew T. Calvert*
Department of Geological Sciences, University of California, Santa Barbara, California 93106, USA

ABSTRACT

The upper amphibolite facies Koolen gneiss dome was subjected to 600–800 MPa pressures and >700 °C temperatures at deep structural levels, and 400–500 MPa and ~600 °C temperatures at shallow levels during peak metamorphism in the Cretaceous. High strains accompanied the pre-104 Ma metamorphism and formed a penetrative L-S tectonite fabric that defines the Koolen dome. The complex was exhumed in three stages between 104 Ma and 84 Ma based on $^{40}Ar/^{39}Ar$ (hornblende, mica, and K-feldspar) thermochronology. Decompression of the complex was between 104 and 94 Ma and was followed by emplacement of epizonal plutons. Rapid cooling to <200 °C, likely associated with doming of the complex, occurred between 94 and 88 Ma. Final cooling and exhumation of the complex was coeval with motion along a low-angle fault between 88 and 84 Ma. Volcanic and subvolcanic rhyolites were emplaced across the area ca. 84 Ma. Decompression and doming of the complex was coeval with plutonism and is interpreted to be due to internal buoyancy. Final exhumation was coeval with extensional (?) low-angle faulting.

INTRODUCTION

Metamorphic culminations of shallowly dipping ductilely deformed rocks, also known as gneiss domes, are common constituents of orogenic belts (e.g., Crittenden et al., 1980). They commonly are cored by gneisses subjected to mid-crustal conditions. Previous studies have suggested that their exhumation is related to contractional (Burg et al., 1984) or extensional (Lister and Davis, 1989; MacCready et al., 1997) tectonics, or diapiric events in the crust (Ramberg, 1981; Lister and Baldwin, 1993) possibly linked to the ascent of granitoid magmas (Hill et al.,

1995). Several of these domes occur on either side of the Bering Strait in the hinterland of the Brooks Range orogenic belt (Shuldiner and Nedomolkin, 1976). These include the Kigluaik, Bendeleben, and Darby gneiss domes on the Seward Peninsula, Alaska (Till et al., 1986; Amato et al., 1994; Calvert et al., 1999); and the Koolen, Neshkan, and Senyavin domes on the Chukotka Peninsula, Russia (Gnibidenko, 1969; Gelman, 1973; Shuldiner and Nedomolkin, 1976; Bering Strait Geological Field Party, 1997) (Fig. 1; Table 1). This chapter presents analytical data on the mineral assemblages, thermobarometry, and thermochronology of the southeastern Koolen dome, northeasternmost Russia,

---

* E-mail, Akinin: petrolog@neisri.magadan.ru. Present address, Calvert: U.S. Geological Survey, 345 Middlefield Road, MS-910, Menlo Park, California, 94025, USA, acalvert@usgs.gov.
Data Repository item 2002076 contains additional material related to this article.

Akinin, V.V., and Calvert, A.T., 2002, Cretaceous mid-crustal metamorphism and exhumation of the Koolen gneiss dome, Chukotka Peninsula, northeastern Russia, *in* Miller, E.L., Grantz, A., and Klemperer, S.L., eds., Tectonic Evolution of the Bering Shelf–Chukchi Sea–Arctic Margin and Adjacent Landmasses: Boulder, Colorado, Geological Society of America Special Paper 360, p. 147–165.

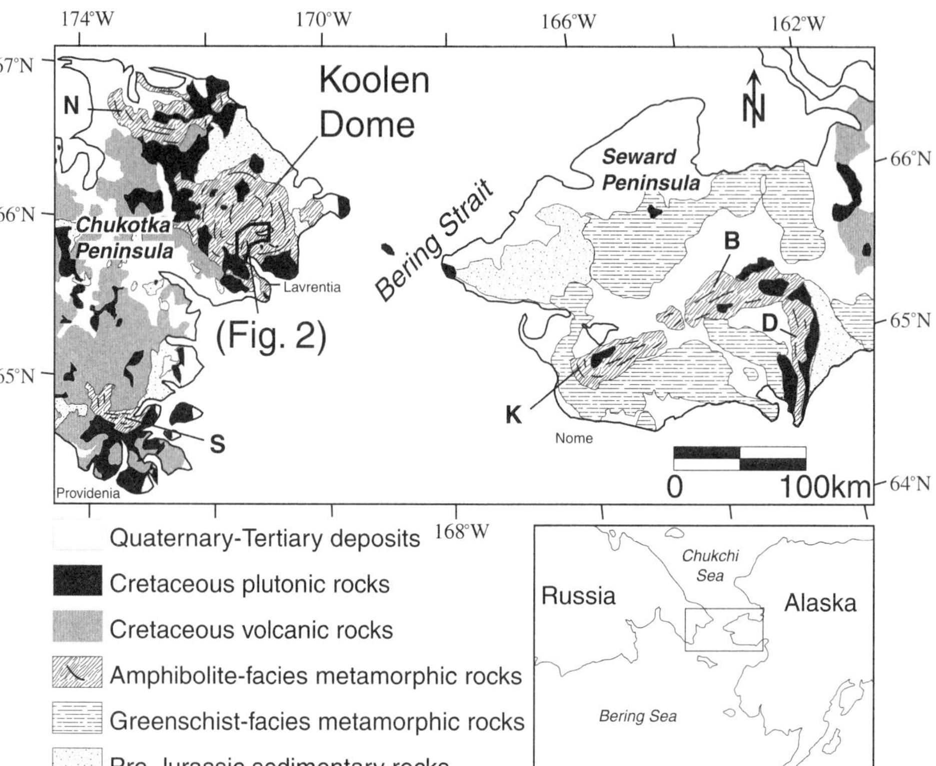

Figure 1. Index map of Bering Strait region showing metamorphic culminations and Mesozoic units (after Belyi, 1977; Beikman, 1980). Metamorphic culminations are labeled: K—Kigluaik Mountains; B—Bendeleben Mountains; D—Darby Mountains; N—Neshkan dome, S—Senyavin uplift. Lines on metamorphic culminations indicate general trends of gneissic foliations.

which provide critical constraints on peak metamorphic conditions and the timing and conditions during exhumation of the Koolen dome.

REGIONAL GEOLOGIC SETTING

Rocks exposed adjacent to the Bering Strait include Paleozoic sedimentary rocks, early Mesozoic mafic volcanic and sedimentary rocks, Cretaceous intermediate to silicic volcanic and plutonic rocks, and several culminations of greenschist to amphibolite (locally granulite) facies metamorphic rocks (Fig. 1). Where studied in detail, these metamorphic culminations (gneiss domes) on the Chukotka and Seward Peninsulas generally show limited evidence for relatively high-pressure (P)-low-to-moderate-temperature (T) Jurassic (?) to Cretaceous metamorphism overprinted by higher-T Late Cretaceous metamorphism (Table 1; Till et al., 1986; Patrick and Lieberman, 1988; Gelman, 1995).

A critical question concerning all of the gneiss domes is their relationship with lower-grade or unmetamorphosed rocks. Gnibidenko (1969) investigated many of the contacts between schists and nonmetamorphosed rocks on the Chukotka Peninsula and generally found them to be transitional. Similarly, on the Seward Peninsula, the contact on the flanks of the Kigluaik Mountains between amphibolite facies rocks and their low-grade metamorphic cover is transitional (Miller et al., 1992). These relationships are commonly obscured in the region, and are in the

mapped boundaries of the Koolen dome (Bering Strait Geological Field Party, 1997; Natal'in et al., 1999), because the key areas are in topographic lows or are displaced by younger faults.

The domes on both the Seward and Chukotka Peninsulas are spatially associated with regionally voluminous Late Cretaceous magmatism of the Okhotsk-Chukotsk volcanic belt (Belyi, 1977), but much of the metamorphism may have preceded the magmatism (Table 1). The U-Pb and $^{40}Ar/^{39}Ar$ geochronology from the Kigluaik Mountains, the most studied of these domes, reveals a Late Cretaceous age for the metamorphic and deformational event related to gneiss dome genesis (Amato and Wright, 1998; Calvert et al., 1999). The high-T–low-P metamorphism has been linked spatially and temporally to ca. 90 Ma granitoid magmatism (Amato et al., 1994), but there is no plutonism linked to the earlier higher-P metamorphism. Isotope geochemistry from plutons in the Kigluaik Mountains suggests an arc origin for the Late Cretaceous magmatism (Amato and Wright, 1997). Peak metamorphism of the Kigluaik gneiss dome on the Seward Peninsula is syndeformational and has been variably ascribed to (1) thermal equilibration following crustal thickening (Patrick and Lieberman, 1988) and (2) high regional heat flow associated with plutonism during regional crustal extension (Amato et al., 1994, and this volume), or (3) local diapiric ascent of hot, buoyant gneisses and plutonic rocks (Calvert et al., 1999). Ductile deformation in the dome was polyphase: some strain appears to be older than 120 Ma (Hannula et al., 1995), but the bulk was developed between ca. 100

TABLE 1. SUMMARY OF THERMOBAROMETRY, TIMING DATA, AND HYPOTHESIZED GEOLOGIC SETTING FOR SIX GNEISSIC CULMINATIONS BORDERING THE BERING STRAIT

	Peak pressure (age)	Inferred setting	Peak temperature (age)	Inferred setting	Cooling and/or exhumation age	Inferred setting	Reference
Kigluaik	800–1200 MPa (>120 Ma)	Shortening					Patrick and Lieberman, 1988; Hannula and McWilliams, 1995
			>750 °C (91 Ma)	Plutonism		Extension	Miller et al., 1992
						Extension, diapirism	Amato et al., 1994; Amato et al., this volume
					90–83 Ma	Diapirism	Calvert et al., 1999
Bendeleben					90–80 Ma	Plutonism	Westcott and Turner, 1981; Miller and Bunker, 1976
Darby					103–95 Ma	Plutonism	Miller and Bunker, 1976; Till and Dumoulin, 1994
Koolen	600–800 MPa (≥104 Ma)	Shortening	700–800 °C (≥104 Ma)	Plutonism	92–85 Ma	Plutonism, extension	This study; Natal'in et al., 1999
Neshkan					108–104 Ma		Natal'in et al., 1999
					87–75 Ma		Akinin and Kotlyar, 1997
Senyavin	600–800 MPa (≥139 Ma)	Shortening	600–650 °C (139 Ma)	Thermal equilibrium following thickening	132–115 Ma	Shortening/ erosion	Calvert and Gans, 1999
South-Central Brooks Range (~154°W)	600–950 MPa (≥120 Ma)	Shortening	500–600 °C (<105 Ma)	Thermal equilibrium following thickening	105–90 Ma	Extension/ ductile thinning/ erosion	Vogl et al., 1999; Vogl, Gans, and Calvert, in preparation

and 82–83 Ma (Calvert et al., 1999). The Senyavin uplift on the southern Chukotka Peninsula (Fig. 1) was subjected to an Early Cretaceous amphibolite facies metamorphism, cooling, and exhumation followed by Late Cretaceous magmatism (Calvert and Gans, 1996).

KOOLEN DOME SETTING

The Koolen gneiss dome is an ~3000 km^2 metamorphic dome exposing mostly amphibolite facies rocks near the easternmost tip of the Chukotka Peninsula (Fig. 1) (Shuldiner and Nedomolkin, 1976; Bering Strait Geological Field Party, 1997; Natal'in et al., 1999). The dome is defined by a high-*T*, gneissic foliation with a strong stretching lineation in the study area. Only the southwest quarter of the dome was mapped, but arcuate trends of foliation strikes support previous descriptions of a domal structure, and consistent north–north-northwest–trending stretching lineations suggest unidirectional flow during peak metamorphism. Metamorphic rocks are mostly metasedimentary, but deformed and undeformed intrusions with ages ranging from Devonian to Cretaceous (Natal'in et al., 1999) are an important component. The portion of the dome mapped in detail is entirely amphibolite facies, peak metamorphic grade increasing to the northeast. Ms + Sil + Qtz ± St (mineral abbreviations af-

ter Kretz, 1983) is the mineral assemblage in the southwest portion of the map area near Lavrentia Bay, but Ms is not present in the central part of the dome near Koolen Lake (Fig. 2). This Ms-out isograd is generally parallel to the map trace of the metamorphic foliation (Bering Strait Geological Field Party, 1997). Granitic and quartzofeldspathic gneiss of the Etelchvyleut Series crops out in the core of the dome, and higher structural levels consist of marble, calcareous schists, quartzofeldspathic schist, amphibolite, and scattered ultramafic rocks of the Lavrentiev Series (Shuldiner and Nedomolkin, 1976; Bering Strait Geological Field Party, 1997). The structural thickness of rock units exposed in the map area is ~15 km.

Three general orientations of metamorphic fabrics were recognized in Bering Strait Geological Field Party (1997): (1) a gneissic, high-*T* fabric that defines the dome and is plotted in Figure 2; (2) a locally developed, low-*T* mylonitic foliation beneath a low-angle fault in the western half of Figure 2; and (3) a steeply dipping, high-*T* fabric developed between the dome and a mid-Cretaceous deformed quartz monzonite intrusion on the west edge of the map area. Here we attempt to place limitations on pressures, temperatures, and timing of those fabric-development episodes.

1. The regional, gneissic, amphibolite facies fabric (D$_1$) that defines the dome involves all phases (e.g., sillimanite, horn-

                                      V.V. Akinin and A.T. Calvert

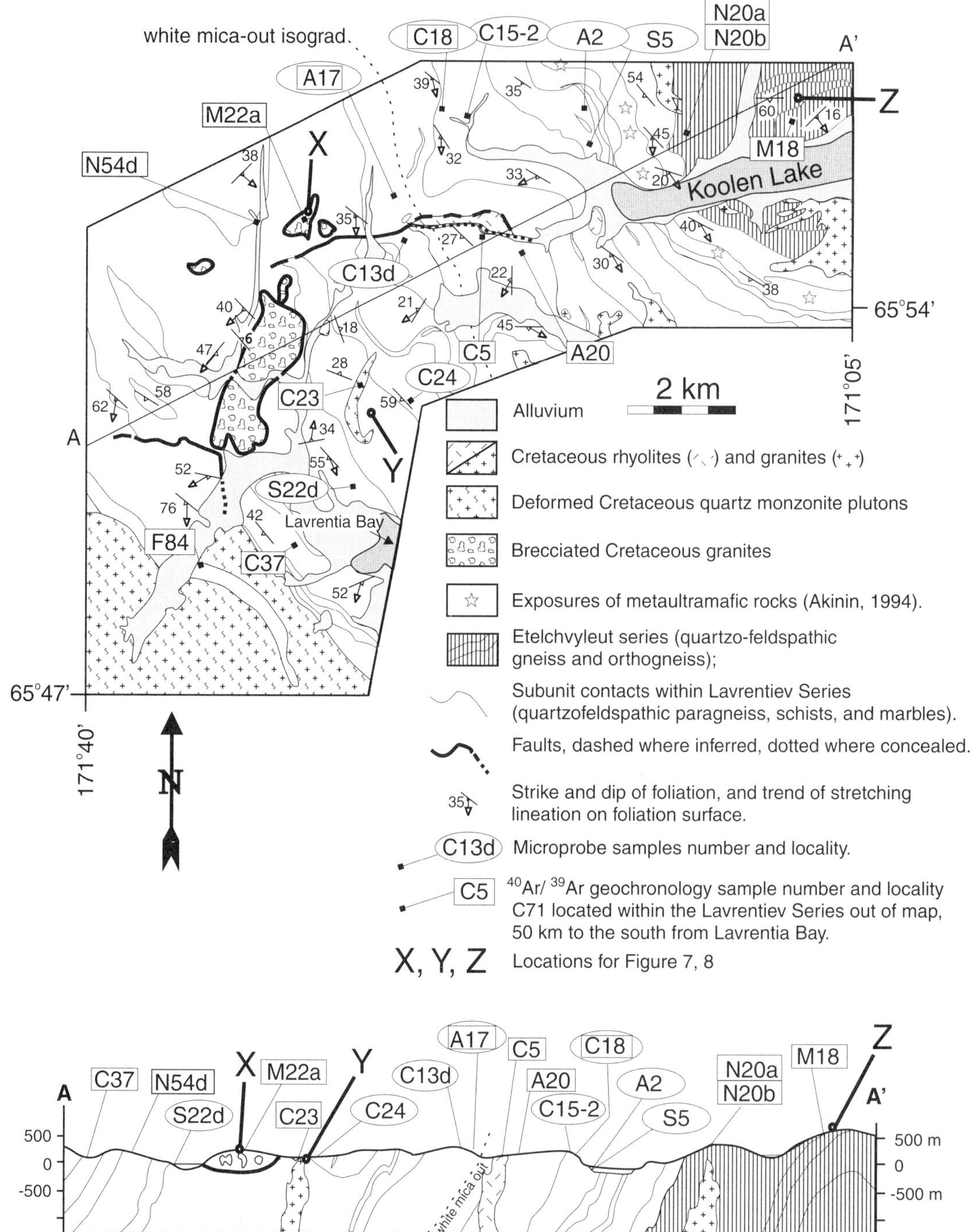

Figure 2. Geologic map and cross section of southwestern part of Koolen dome (after Bering Strait Geological Field Party, 1997).

blende, mica) and leucosomes in the high-*T* assemblage. A stretching lineation (L_1) associated with this foliation trends north-northwest–south-southeast in the gneissic core of the dome and is variable, but more north-south between Koolen Lake and Lavrentia Bay. An area on the northwest flank of the dome mapped in 1995 by Natal'in et al. (1999) has a consistent north-south stretching lineation. Similar to adjacent domes, most rocks within the Koolen dome show little or no penetrative deformation at subamphibolite facies.

2. The second fabric (D_2) is a lower-*T*, locally developed, gently dipping mylonitic foliation inferred to have formed during exhumation of the complex. We mapped an ~10-m-thick

mylonitic zone within Sil + Grt + Bt + Ms + Qtz rocks, and Zhulanova (1990) reported one at deeper structural levels. Foliations (S$_2$) within the mylonite zone have variable orientations, but lineations trend consistently north-south, with top-to-the-south shear sense. Despite having nearly the same lineation orientation as the high-*T* L$_1$, these mylonites are clearly postmetamorphic. Marble and quartzose mylonites have well-developed grain-shape foliations. This mylonite zone is below a controversial, gently dipping fault developed at high structural levels that places brecciated, hydrothermally altered granitoids on the Lavrentiev Series metasedimentary rock. This structure is important because it juxtaposes very different rock types, but its nature is debated; two possibilities were discussed in Bering Strait Geological Field Party (1997): (1) the brecciated rocks are highly altered and fractured intrusive rocks deformed during intrusion or (2) it is a gently dipping fault. The latter interpretation is supported by strongly foliated and lineated marble mylonites immediately below the fault that have oblique grain-shape foliations indicating top-to-the-south motion. As described in the following, the granitoids yield significantly older cooling histories than rocks below, ruling out the intrusive option and suggesting that the contact is a normal fault. Because these relationships are poorly and only locally exposed, it is difficult to determine their tectonic significance. If it is a normal fault, it must postdate doming, because restoring foliations below the fault to horizontal would make the fault dip north with reverse-sense offset.

3. The third foliation (S$_3$?) occurs at the edge of the dome on the westernmost part of Figure 2 adjacent to a gneissic quartz monzonite. It dips steeply westward and has a subhorizontal stretching lineation (L$_3$?), suggesting strike-slip motion. The relative age of this fabric to the previous two is poorly understood. Foliations developed in the D$_2$ mylonitic event described here appear to be folded into an open syncline, and this may be related to the strike-slip motion on the western margin of the dome, but insufficient data were collected to address this.

Metamorphic rocks are intruded by variably deformed plutons commonly with strongly developed, solid-state fabrics. These plutons range in age from Devonian (Natal'in et al., 1999) to Late Cretaceous (Bering Strait Geological Field Party, 1997). Plutons older than 104 Ma are penetratively deformed, plutons between 104 and 94 Ma are variably deformed, and younger intrusive and extrusive rocks show no deformation (Bering Strait Geological Field Party, 1997). In general, the plutons are poorly exposed.

Regional geologic, petrographic, and K/Ar and Rb/Sr geochronologic studies have argued that the polymetamorphic history of the Koolen dome spanned the Proterozoic to Cretaceous (Shilo and Zagruzina, 1965; Natal'in, 1979; Zhulanova, 1990), but the results of more recent work indicate a Cretaceous age for peak metamorphism (Bering Strait Geological Field Party, 1997). The Proterozoic ages include three Rb/Sr whole-rock isochrons: 1990 ± 150 Ma, 1770 ± 80 Ma and 782 ± 55 Ma (Zhulanova, 1990), using several samples from dispersed locations—a method requiring regional isotopic homogenization during peak metamorphism to be accurate (Faure, 1986). In addition, several K/Ar ages in the Northeast Interdisciplinary Science Research Institute (NEISRI) database (Akinin and Kotlyar, 1997) from the core of the Koolen dome yield ages of 1095 and 1615 Ma, but because nearby samples yield Cretaceous cooling ages, these are interpreted here as a result of excess argon. The importance of Cretaceous events in the Koolen dome was emphasized by joint Russian-American field studies in 1994 and 1995. U-Pb ages of 104 Ma on monazite fractions from a deformed pegmatite, derived from the partial melting of rocks of the Lavrentiev Series, are interpreted as the time of peak metamorphism (Bering Strait Geological Field Party, 1997). The metasedimentary protolith of these rocks is Paleozoic or older, because U/Pb zircon ages from orthogneisses within the Etelchvyleut Series in the core of the dome yielded ages of 370 and 375 Ma (Natal'in et al., 1999). Lithologically, the metamorphic succession of the Lavrentiev Series is similar to the lower Paleozoic units of the Chegitun Unit of Natal'in et al. (1999) and probably the Proterozoic to Paleozoic rocks of the Seward Peninsula (Till et al., 1986). All rocks in the Koolen dome have undergone amphibolite facies metamorphism; mapped isograds indicate metamorphic conditions ranging from Sil + Grt + Bt + Qtz + Kfs in the core of the dome to Sil + Grt + Bt + Qtz + Ms ± St at its periphery. We found a single granulite facies assemblage in the Lavrentiev Series (Fig. 2), supporting the claim of Ivanov and Kryukov (1973).

THERMOBAROMETRY

All thermobarometry samples analyzed and discussed in this chapter are from the Lavrentiev Series collected across a 15 km transect from the core of the dome at Koolen Lake to its periphery near Lavrentia Bay (Fig. 2). The structurally deepest Etelchvyleut Series is inferred to be the highest grade, but it does not contain assemblages conducive to thermobarometry. Metapelitic rocks are not common in this section, but are present in thin (1–50 m thick) lenses within the quartzofeldspathic gneiss and calc-silicate–bearing marble units. Metaultramafic rocks occur as boudinaged lenses 3–30 m across in one specific member of the quartzofeldspathic schists and amphibolites near the base of the Lavrentiev Series (Fig. 2). Microprobe mineral analyses were carried out on 13 garnet-biotite schist samples, 1 garnet amphibolite, 1 two-pyroxene-garnet-hornblende schist, and 2 serpentinized tremolite-bearing spinel metaultramafic rocks. Eight samples containing essential and compositionally homogeneous assemblages were chosen for quantitative thermobarometry. Mineral compositions (see Appendix 1[1] or Internet site www.geol.ucsb.edu/~calvert/koolen.html) were deter-

---

[1]GSA Data Repository item 2002076, Appendixes 1 and 2, is available on request from Documents Secretary, GSA, P.O. Box 9140, Boulder, CO 80301-9140, USA, editing@geosociety.org, or at www.geosociety.org/pubs/ft2002.htm, or on the CD-ROM accompanying this volume.

mined using the Camebax microprobe at NEISRI and the JEOL-7031 microprobe at Stanford University.

All analyzed garnets were round, xenomorphic porphyroblasts primarily of almandine with $Pyr_{9-30}Grs_{3-20}Sps_{1-21}$. Differences in the composition of garnets are insignificant within a thin section. Garnet is usually not zoned, or zoned with a decrease in pyrope content from core to rim, whereas the grossular content remains nearly constant. The composition of biotite is generally uniform within samples, and there are few samples that indicate more than one generation of biotite growth. For example, in sample C24, the larger Ti-rich grains of the biotite ($Mg/[Mg + Fe] = 0.34$; $Ti = 0.34$ per 11 oxygens) crosscut the main schistosity containing older biotite of different composition ($Mg/[Mg + Fe] = 0.43$; $Ti = 0.12-0.17$ per 11 oxygens), and are possibly the result of contact metamorphism by younger plutons. Muscovite is present in association with sillimanite (fibrolite) along the southern and southwestern edge of the area of Figure 2 and St + Sil + Ms + Qtz is present ~50 km south, near the town of Lavrentia (Fig. 1). Plagioclase compositions vary from An_{16} to An_{84}, but typically do not vary by more than An_{1-3} within individual samples.

A two-pyroxene-garnet assemblage, possible evidence of granulite facies, is reported from this area for the first time. The two-pyroxene-garnet-hornblende gneiss (sample C15–2) occurs as a small (10 cm in diameter) round boudin enclosed within a sillimanite-biotite schist ~6 km to the northwest of Koolen Lake (Fig. 2). In thin section the gneiss has a granoblastic texture with poorly oriented brown iron-bearing pargasite. Garnet is present as regularly disseminated small (<0.5 mm) oval grains. The sample contains ~35% hornblende ($Al_2O_3 = 11.5-11.9$ wt%, $TiO_2 = 1.8-2$ wt%), 20% clinopyroxene ($En_{30-31}Fs_{23-24}Wo_{44-45}$), 10% orthopyroxene ($En_{40-42}Fs_{56-58}Wo_{1.5-1.9}$), 20% plagioclase ($An_{42-50}$), 5% garnet ($Alm_{62-63}Sps_{2-3}Pyr_{13-14}Grs_{19-20}$), 3% ilmenite, 1% quartz, and 1% biotite. The lack of zoning and consistent compositions of all minerals except for plagioclase testify to a high degree of equilibration in this sample. The bulk-rock composition corresponds to an aluminous high-iron (20 wt% Fe_2O_3) basalt.

For all samples in the thermobarometric study, several areas of each thin section were analyzed and included analyses of garnet cores and rims and of areas where garnet is in direct contact with biotite and plagioclase. The core and the rim composition of the crystals were examined where possible, because they presumably represent early and late stages of mineral growth. Table 2 lists calculated P and T for various areas of studied thin sections and the type of equilibrium relations that characterize the nature of the examined point within a given crystal (i.e., core versus rim). Garnet-biotite thermometers (Ferry and Spear, 1978; Perchuk et al., 1983; Aranovich et al., 1988; Bhattacharya et al., 1992) and the GASP barometer (3An = Grs + 2Ky/Sil + Qz) (Aranovich, 1983; Perchuk, 1986) were used for metapelitic rocks. Pressures and temperatures were calculated simultaneously by resolution of two equations. Linearized calibrations of the Grt + Pl + Al_2SiO_5 + Qz and Grt + Pl + Ms + Bt geobarom-

eters for medium-grade metapelites based on the Holland and Powell (1995) data set were used also. The calibration uncertainty for those thermobarometers is estimated as 15–20 °C and 80–100 MPa (Holland and Powell, 1998). With typical mineral compositions the analytical uncertainty contributes ~20 °C and 100 MPa to the overall uncertainty. Low Ca content in garnet and plagioclase (when the product of X_{Grs} and X_{An} is <0.05) raises the error of pressure estimation by GASP and Grt + Pl + Ms + Bt barometers to as much as 200–300 MPa (Todd, 1998). Unfortunately, metapelitic rocks in the Lavrentiev Series were low Ca, so samples A17, A2, C13d, C18, C68–1, N30–5, S22d, and A5 were under Todd's (1998) threshold. Crystalline sillimanite or fibrolite is present in five of the metapelite thin sections, and only in these cases was the GASP barometer used. In addition, the values given for the two-pyroxene-garnet-amphibole schist were calculated using the garnet-clinopyroxene thermometer (Ellis and Green, 1979) and the Di + An = 2/3Grs + 1/3Py + Qz barometer (Avchenko, 1990). It is encouraging to note that thermobarometers yield similar results for the compositionally distinct samples C15–2 and C18 (Table 1) collected from adjacent outcrops.

The thermobarometric results provide a new understanding of the *P-T* history of the southwestern part of the Koolen dome. Interpretation of the combined *P-T* trend and the individual sample trends allow the following conclusions to be drawn.

1. In general, the temperature and pressure values display a wide range ($T = 780-540$ °C and $P = 780-240$ MPa), a difference of more than 500 MPa and 240 °C corresponding to granulite to epidote-amphibolite facies. Most of the *P-T* trends are regressive and reflect uplift to surface and cooling.

2. The maximum temperature and pressure values were obtained for the deepest structural level of the Lavrentiev Series, in accord with field evidence that these rocks have undergone the highest grade of metamorphism. The minimum pressure and temperature were obtained from the shallowest levels of the Lavrentiev Series, and the individual *P-T* trends of the four samples (A17, C13d, C24, S22d) examined in detail are quite similar to each other (Fig. 3). Based on field and petrographic studies, it is clear that the peak mineral assemblage is preserved in the upper part of the Lavrentiev Series. Thus, the thermobarometric data, which testify to a decrease in the metamorphic grade from the core of the dome toward its periphery, agree with the field and petrographic evidence for index-mineral zonation. Given the fact that samples are 5–8 km away in cross section (Fig. 2), but 500 MPa away in *P-T* space, suggests that equilibration occurred at different times in the structural column during exhumation and/or there was significant ductile thinning during exhumation.

3. We analyzed two samples collected near the base of the Lavrentiev Series in the core of the dome that define a local thermal maximum. One of these samples, C15–2, contains a granulite facies assemblage. The results (C15–2 garnet-two-pyroxene-hornblende schist and C18 garnet-sillimanite-biotite schist) indicate peak temperatures of 690–780 °C. Such high garnet-biotite temperatures have not previously been reported, even for

**TABLE 2. CALCULATED PRESSURE AND TEMPERATURE FOR VARIOUS AREAS
OF STUDIED THIN SECTIONS AND THE TYPE OF EQUILIBRIUM RELATIONS**

Number and type of equilibrium						T (°C)			P (MPa)			
grt	bt	pl	cpx	opx	hbl	T1	T2	T3	P1	P2	P3	P4
A17: Qtz + Pl + Bt + Grt + Sil + Ms												
1c	1c	1c				641	600	659	370	370	450	
1r	1ct	1ct				572	553	586	210	240	310	
2c	2c	2c				627	597	644	340	350	420	
2r	2r	2ct				619	597	635	320	340	400	
C13d: Qtz + Pl + Bt + Ms + Grt + Sil												
1ct	1r	1ct				613	655	587	270	410	370	410
1c	1in	1c				569	601	638	220	370	400	440
2c	1in	1c				592	626	651	250	390	460	490
2r	1r	1ct				574	618	590	220	370	370	390
C15–2: Hbl + Cpx + Opx + Grt + Pl + Qtz + Bt												
1c	1c	1c	1c			686	746	725	830*	680	1120*	
1sm	1c	1c	1c			650	725	688	740*	640	1050*	
4sm	1c	4c	4r			653	715		770*	600		
4sm	1c	4c	4ct	4ct		646	693		750*	630		
4sm	1c	4c	4c	4c		797	720		750*	660		
4sm	1c	4c	4r	4r		653	715		770*	640		
2r	1c	2ct	2r			664	744		780*	670		
2r	1c	2ct	2ct			664	754		780*	670		
2c	1c	2c	2c			673	767		800*	710		
C18: Qtz + Pl + Bt + Ms + Grt + Sil												
1c	1c	1c				731	723	787	590	650	750	820
1r	1c	1c				713	705	744	490	530	550	660
3c	3in	3in				755	778	783	680	780	670	740
3c	3in	4c				754	775	763	640	740	720	770
C24: Qtz + Pl + Bt + Grt + Sil												
1c	1in	1in				632	623	653	300	440	400	
2ct	2in	2ct				569	556	594	230	370	330	
3c	3in	3c				562	542	586	160	270	250	
3ct	3in	3r				574	554	597	180	290	270	
S22d: Qtz + Pl + Kfs + Bt + Grt + Sil + Ms												
1c	1c	1sm				605	596	621	270	380	420	
1c	1in	2sm				575	547	579	200	290	370	
1ct	1ct	3c				610	604	629	230	340	370	
S5: Qtz + Pl + Bt + Grt + Ms												
1c	1in	1c				707	729	733	600	780	670	
2c	2ct	2c				649	663	675	500	650	590	
2r	2r	2s				638	657	666	460	620	570	
A2: Qtz + Pl + Bt + Grt												
1c	1c	2c				639	651	662	480	600	600	
1r	1r	2ct				606	621	625	350	460	480	
3r	1r	3ct				610	631	634	370	500	500	

Note: Temperatures are calculated using Grt-Bt thermometers and pressures are calculated using the GASP (3An = Gros + 2Ky/Sil + Qtz) barometer. Mineral abbreviations after Kretz (1983). Sources of data set calibrations: T1—Perchuk et al. (1983); P1—Perchuk (1986); T2—Aranovich et al. (1988); P2—Aranovich (1983); T3—Bhattacharya et al. (1992); P3—Holland and Powell (1998); P4—garnet-plagioclase-muscovite-biotite barometer; Holland and Powell (1998). Muscovite was not analyzed in samples A17, S22d, and S5. For sample C15–2 we used: T2 and P2—Cpx-Grt thermometer; Ellis and Green (1979) and DAK2–barometer (Di + An = 2/3Gros + 1/3Prp + Qtz), Avchenko (1990).
*—P1 for C15–2 assuming equilibrium within sillimanite stability field. Types of equilibria are keyed to Appendix 1 (see footnote 1): c = core, r = rim, ct = contact, in = inclusion, sm = small.

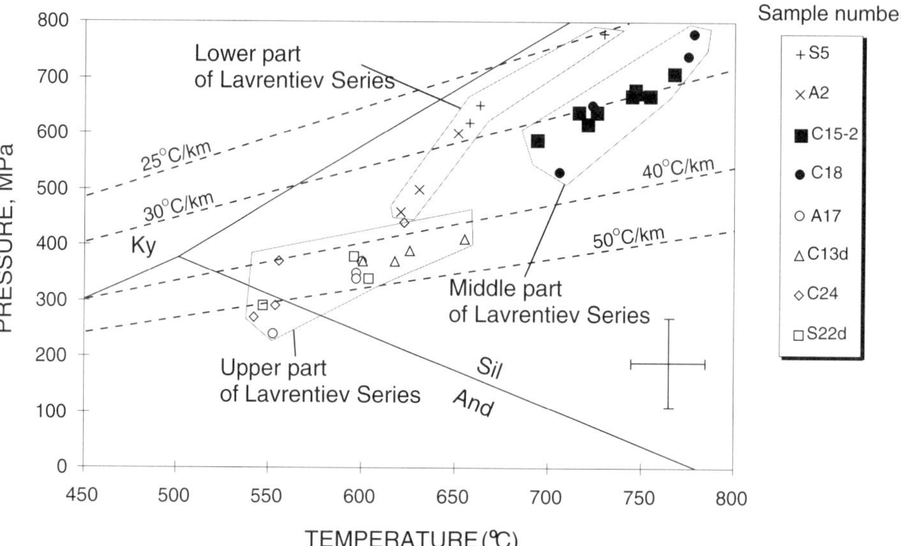

Figure 3. Pressure-temperature (*P-T*) diagram of metamorphic conditions for Lavrentiev Series of southwestern Koolen dome. Points represent *P-T* calculations of phases interpreted to be in textural equilibrium with each other. Aluminosilicate phase diagram is after Holdaway (1971). Dotted line shows geothermal gradient assuming density = 2.7 g/cm^3. Error bracket shows approximate confidence of each *P-T* determination (see text). Mineral abbreviations are after Kretz (1983).

the rocks near the core of the dome. This temperature maximum may be interpreted as an area of preserved peak mineral assemblages. The sampled bodies consist of boudinaged lenses of more resistant rocks surrounded by quartzose schist that continued to flow later in the metamorphic and deformational history.

4. Combined *P-T* trends obtained for individual samples support the conclusion that all rocks underwent a single metamorphism, and rim-to-core trends indicate substantial cooling during decompression (Fig. 3). The combined data for the upper part of the Lavrentiev Series correspond to a high geothermal gradient (40–50 °C/km) compared to normal continental crust.

THERMOCHRONOLOGY

Analytical techniques

Samples for ^{40}Ar/^{39}Ar geochronology were collected from metasedimentary rocks and orthogneisses in the high-grade core of the dome, the moderate-grade margin, and from brecciated granitic rocks near Lavrentia Bay (Fig. 2). From these samples, 17 mineral separates of hornblende, muscovite, biotite, and K-feldspar were analyzed (Table 3; Fig. 4). Complete data tables, age spectra, isochron plots, and a description of analytical procedures for all samples are available in Appendix 2 (see footnote 1) or on the Internet (www.geol.ucsb.edu/~calvert/koolen.html). All ^{40}Ar/^{39}Ar dates were obtained by conventional step-heating experiments. We used Fish Canyon sanidine as a flux monitor with an assigned age of 27.8 Ma.

Many of the samples yielded plateau ages defined as 50% of the ^{39}Ar released in consecutive steps within 2σ. One hornblende sample had a very disturbed spectrum (M18), and for this we interpret an age using an isotope correlation (isochron) plot (Fig. 4A). We interpreted ages for other samples solely from age spectra. None of the biotite spectra are true plateaus; they consistently have one or two steps with old apparent ages in the mid-

dle of the spectrum (Fig. 4C). Generally these ages are removed and we call them discontinuous plateau ages (Calvert et al., 1999). Because maximum metamorphic temperatures attained in the complex were >500 °C, all ^{40}Ar/^{39}Ar ages obtained from metamorphic rocks are cooling ages, with the possible exception of the Lavrentiev Series hornblende (F84). For these relatively rapid cooling rates (>20 °C/m.y.), we estimate closure temperatures based on published diffusion data as follows: hornblende, 535 ± 50 °C (Harrison, 1981); muscovite, 370 ± 50 °C (Lister and Baldwin, 1996); and biotite, 335 ± 50 °C (Harrison et al., 1985; Grove and Harrison, 1996).

While closure temperatures calculated with diffusion data from hydrothermal experiments show considerable variability, empirical studies (Hanson and Gast, 1967; Snee et al., 1988) yield results in reasonable agreement. K-feldspar data were modeled directly using the technique of Lovera (1992) to obtain quantitative cooling histories between ~150 and ~350 °C.

For modeling, K-feldspar samples were analyzed over 47–67 step experiments with individual step isolation times ranging from 15 min to more than 11 h. Initial steps commonly yielded very old apparent ages, but analyzing gas from replicate steps at these low temperatures allows correction of the low-*T* ages. We were not able to use the Cl-correlated, excess-argon correction technique of Harrison et al. (1994) because of low production of ^{38}Ar from chlorine in the Oregon State University TRIGA reactor. Diffusion modeling of K-feldspar data (Fig. 5) is based on the theory of Lovera et al. (1989) that alkali feldspars have a distribution of diffusion domain sizes and that conventional step-heating experiments can be modeled to extract the diffusion parameters and reconstruct quantitative temperature-time histories. This powerful technique uses Arrhenius data to construct a log (R/R$_o$) plot (a plot of the deviation from single-domain diffusion versus ^{39}Ar released, defined by Richter et al., 1991), which is then modeled for information about diffusion domain sizes independent of heating schedule. Once diffusion

TABLE 3. SUMMARY OF ^{40}Ar/^{39}Ar ANALYSES

Sample	Mineral	Total steps	Age (Ma) ± 2σ	Temperature (°C) steps used	^{39}Ar spectrum (%)	Isochron age (Ma) ± 2σ	Isochron MSWD	Total fusion age
F84	Hbl	25	**103.3 ± 0.3**	700–1250	99	103.3 ± 0.7	3.6	103.9
N20b	Hbl	22	**91.7 ± 0.2**	950–1250	98	91.9 ± 0.4	1.5	91.7
M18	Hbl	21	92.0 ± 0.3	1030–1040	22	**91.1 ± 0.6**	2.2	93.8
C71A	Ms	16	**96.2 ± 0.3**	760–1250	77	96.1 ± 0.4	1.8	96.3
C37	Ms	14	**93.0 ± 0.3**	840–970	58	92.8 ± 0.6	2.4	93.1
A17	Ms	13	**90.6 ± 0.2**	770–1250	93	90.4 ± 0.5	0.6	90.8
M22a	Bt	16	**105.6 ± 0.3**	800–850, 1000–1250	61	105.6 ± 1.2	71.0	104.0
N54d	Bt	13	**91.8 ± 0.2**	650–1250	98	91.9 ± 0.4	6.6	91.6
C23	Bt	14	**91.9 ± 0.2**	700–800, 1020–1250	58	92.4 ± 0.6	18.4	91.6
C18	Bt	13	**90.2 ± 0.2**	600–800, 990–1080	81	90.1 ± 1.0	8.8	90.1
N20a	Bt	11	**90.3 ± 0.2**	675–1250	99	90.1 ± 0.5	5.5	90.3
M18	Bt	11	**90.7 ± 0.2**	675–880, 1000–1100	85	90.7 ± 0.5	4.6	90.6
M22a	Kfs	50	**85–100**					97.4
C23	Kfs	57	**85–94**					92.6
A20	Kfs	48	**89–93**					92.9
M18	Kfs	68	**88–91(?)**					94.2
C5	Sa	17	**84.1 ± 0.2**	800–1140	71%	84.0 ± 0.3	5.7	84.3

Note: Accepted ages are in bold and consist of weighted mean plateau ages (≥50% of gas released within 2σ error), interpreted ages, age gradients (Kfs), or isochron ages. Goodness of fit index, MSWD (mean square of weighted deviates), is from Roddick (1978). C71A is from a Lavrentiev Series metapelitic schist collected 30 km south of Figure 2.

domain sizes are understood, model age spectra can be synthesized (Fig. 5) by iterative calculation of different cooling histories (Lovera, 1992) and matched to laboratory-obtained spectra. Temperature-time curves calculated from these model age spectra and a modeling example are presented in the following.

Cooling from peak metamorphic conditions

All samples collected from the Koolen dome cooled from >500 °C to surface temperatures in the middle to Late Cretaceous. The highest-grade gneisses from near Koolen Lake cooled rapidly, from hornblende closure temperatures (~535 °C) at 92 Ma to <200 °C at 86 Ma, while moderate-grade schists and gneisses cooled more slowly, from hornblende closure at 103 Ma to <200 °C at 88 Ma. At closure temperatures for mica and K-feldspar, cooling at deep structural levels lagged behind high levels by ~2 m.y. Hydrothermally altered granite samples collected from above a gently dipping mylonite zone near the head of Lavrentia Bay yielded older cooling ages at lower temperatures between 105.5 and 88 Ma. The entire map area was at near-surface temperatures (<150 °C) by the time rhyolite domes were emplaced at all structural levels ca. 84 Ma.

High-grade metamorphic core near Koolen Lake.
Two hornblende samples from the deepest portions of the complex were concordant (Fig. 4A). N20b, a biotite-quartz-plagioclase amphibolite, yielded a plateau age of 91.7 ± 0.2 Ma. Sample M18, a Devonian hornblende-biotite granodioritic orthogneiss from the Etelchvyleut Series, yielded a U-shaped spectrum with young ages in the middle of the spectrum that are concordant with N20b. We interpret the spectrum as a 92.0 Ma maximum age, and suggest that the best age for the sample is an isochron

fit to the first 75% of the gas with an age of 91.1 ± 0.6 and a mean square of weighted deviates of 2.24 (Fig. 4A). Muscovite (Fig. 4B) from near the Ms-out isograd and biotite data (Fig. 4C) from deep structural levels are all within analytical error of each other and ~1 m.y. younger than the hornblende ages. Biotite discontinuous plateau ages on M18 (90.7 ± 0.3 Ma), N20a (90.3 ± 0.3 Ma), and C18 (90.2 ± 0.2 Ma), and a muscovite plateau age on A17 (90.6 ± 0.2 Ma) show that the metamorphic complex cooled rapidly and as a coherent block.

K-feldspar samples from deep structural levels are complicated (Fig. 4D). M18 K-feldspar yielded an age gradient with old apparent ages in the first 5% of ^{39}Ar released, a climbing age spectrum from 88 to 102 Ma, the highest-temperature steps being ca. 96 Ma. There is a large amount of ^{36}Ar, and replicate isothermal steps yield younger ages, both hallmarks of excess argon. Because no ages should be older than the 91.1 ± 0.6 Ma hornblende from the same rock, the spectrum is interpreted to climb from 88 Ma to a maximum of ca. 91 Ma. Diffusion modeling indicates closure temperatures from ~200 °C to 330–350 °C for the sample (Fig. 6). Given the complexity in the age spectrum we interpret rapid cooling from hornblende closure (~535 °C) to <200 °C between 91.1 Ma and 88 Ma. A biotite leucogranite (A20) from 5 km south of Koolen Lake has evidence for excess argon throughout. Old apparent ages at low temperatures (first 10% of gas) give way to 87–88 Ma ages that climb to 92 Ma and then decrease to 90.5 Ma. We interpret the true age profile to be 87–88 to 90.5 Ma. This sample was not modeled due to thermocouple problems during the experiment.

Moderate metamorphic-grade interval near Lavrentia Bay.
Cooling ages of these amphibolite facies metamorphic rocks are somewhat older than the Koolen Lake gneisses. Sam-

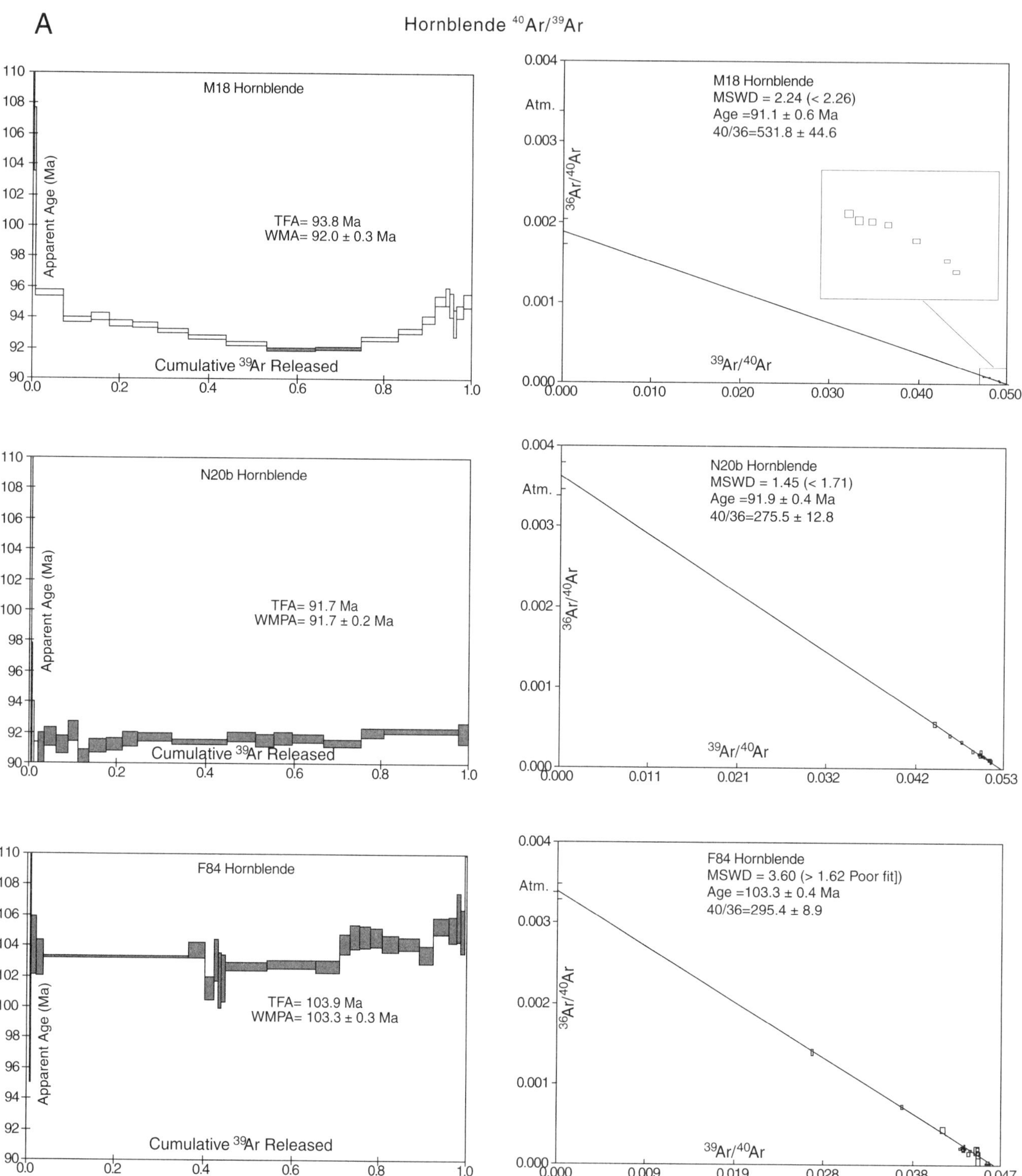

Figure 4 (on this and next three pages). A: ^{40}Ar/^{39}Ar hornblende spectra and isochron plots. Steps are shown with 1σ errors, calculated ages are shown with 2σ errors. Shaded steps are those included in weighted mean plateau age (WMPA) or weighted mean age (WMA) calculations. All gas is used to calculate total fusion ages (TFA). WMPA ages contain ≥50% of ^{39}Ar within 2σ error. Inverse isochron plots (^{36}Ar/^{40}Ar vs. ^{39}Ar/^{40}Ar with 1σ errors) are shown with best fits (lowest mean square of weighted deviates [MSWD], Roddick, 1978) and confidence interval of 40/36 ratio with respect to atmosphere. Roddick (1978) defined good fits and poor fits by whether they are lower than threshold in parentheses; this threshold varies according to number of points to be fit. B: ^{40}Ar/^{39}Ar muscovite spectra. Steps are shown with 1σ errors, calculated ages are shown with 2σ errors. Shaded steps are those included in WMPA calculations. C: ^{40}Ar/^{39}Ar biotite spectra. Steps are shown with 1σ errors, calculated ages are shown with 2σ errors. Shaded steps are those included in WMPA or weighted mean discontinuous plateau age (WMDPA) calculations. D: ^{40}Ar/^{39}Ar K-feldspar and sanidine spectra. Steps are shown with 1σ errors.

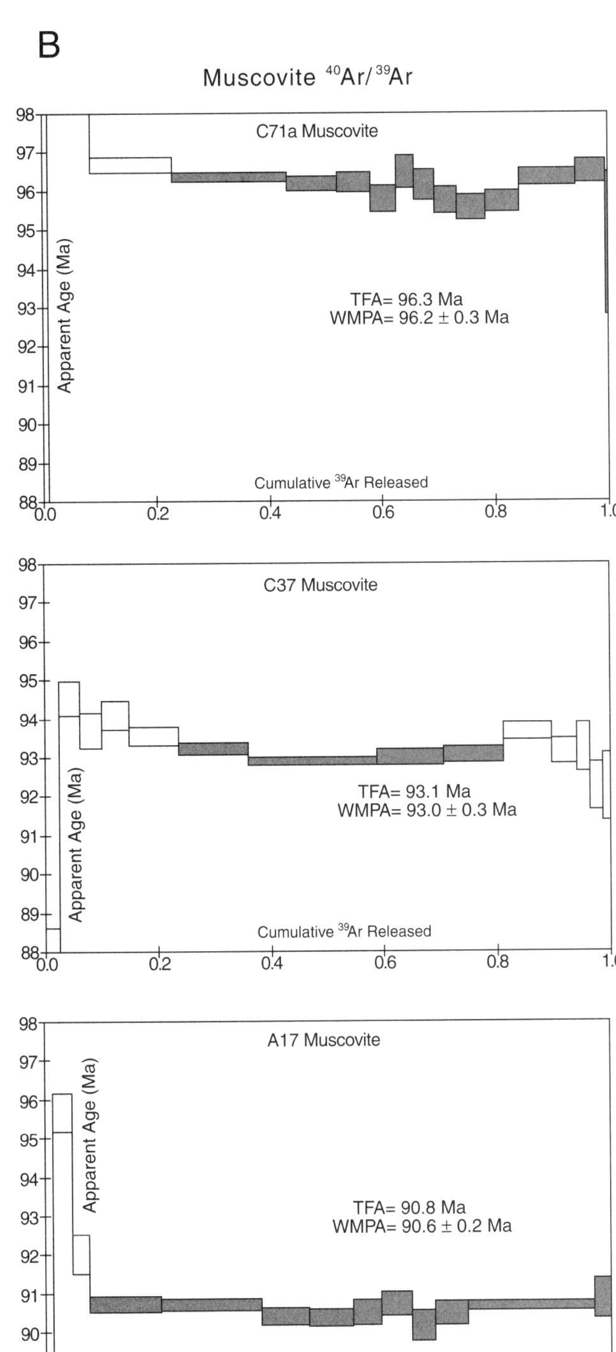

Figure 4 (*continued*)

ples from the structurally highest levels (sample F84, clinopyroxene-hornblende-plagioclase amphibolite) yielded a plateau age of 103.3 ± 0.3 Ma (Fig. 4A). Sample C37 (tourmaline-white mica leucogranite) was collected ~5 km structurally above the white mica–out isograd at the head of Lavrentia Bay and yielded a plateau age of 93.0 ± 0.3 Ma (Fig. 4B). Two biotite samples yield concordant ages of 91.8 ± 0.2 Ma (N54d) and 91.9 ± 0.2 Ma (C23) (Fig. 4C). A single sample (C71a) from high structural levels (staurolite grade) 50 km south of the map area, still within the Lavrentiev Series, yielded a plateau age of 96.2 ± 0.3 Ma (Fig. 4B).

Modeling complicated K-feldspar diffusion and age data from C23—a deformed biotite granodiorite at moderate structural levels—yields rapid cooling from 350 °C to <200 °C between 93 and 88 Ma (Fig. 4D). There is significant excess argon in the sample, and duplicate isothermal steps consistently yielded younger second steps throughout the step-heating experiment. Choosing only the second of duplicate steps gives a smooth age gradient from 88 to 93 Ma and modeling yields a linear cooling history from 350 to 200 °C over this time interval (Fig. 6).

Subvolcanic and/or volcanic rhyolites. C5 sanidine yielded an age of 84.1 ± 0.2 Ma (Fig. 4D). This series of quartz-biotite-sanidine rhyolite domes and flows crops out locally from within the Etelchvyleut Series at Koolen Lake to shallow levels near Lavrentia Bay, indicating that the complex was completely exhumed by this time.

Altered granite above low-angle fault. In the western portion of the map area, several isolated brecciated granite masses are above subhorizontal mylonite zones and yield old cooling ages. A discontinuous biotite plateau age of 105.6 ± 0.3 Ma (Fig. 4C) and a monotonically increasing K-feldspar age gradient with ages that climb from 88 to 100 Ma (Fig. 4D) are remarkably different from data from adjacent samples beneath the fault (C23 and N54d). Modeling this K-feldspar spectrum yields a good fit to the analytical data (Fig. 6), gives closure temperatures of 290–140 °C, and suggests that motion along the fault juxtaposing them is younger than the high- and moderate-temperature cooling. Modeling data for this sample are shown in Figure 5.

Summary of cooling histories

From these data we construct a temperature-time plot for three levels of the map area, each with several discrete $^{40}Ar/^{39}Ar$ cooling ages and a modeled $^{40}Ar/^{39}Ar$ K-feldspar spectrum: deepest structural levels near Koolen Lake, shallow structural levels near Koolen Bay, and brecciated granitoids above a fault on the west side of the area (Fig. 7). Data are divided just structurally above the white mica–out isograd so that white mica sample A17 is included with deep structural levels. Monazite data are plotted from Bering Strait Geological Field Party (1997) with an assigned closure temperature of 725 ± 50 °C (Parrish, 1990) and $^{40}Ar/^{39}Ar$ data are plotted with closure temperatures described previously.

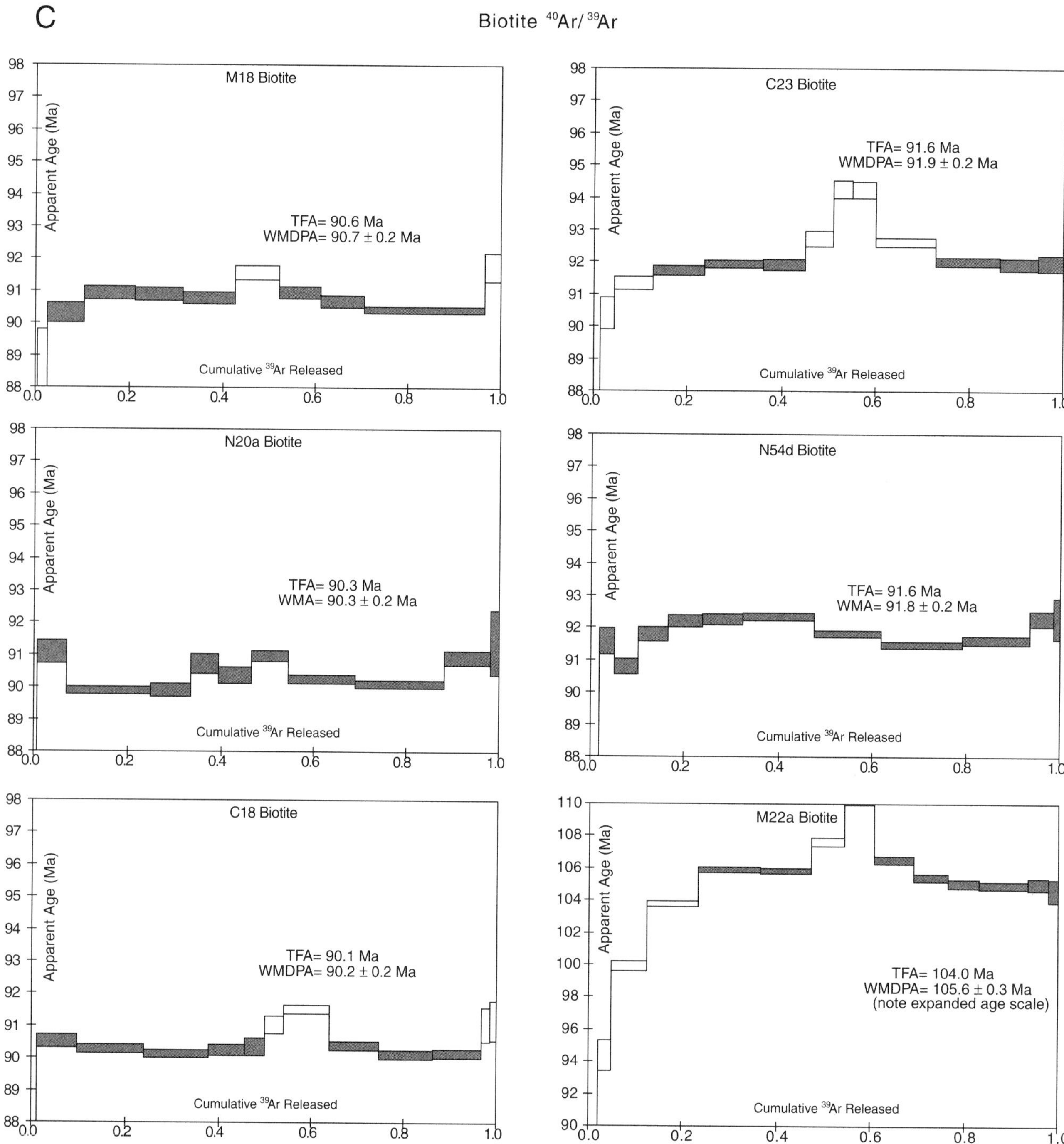

Figure 4 (*continued*)

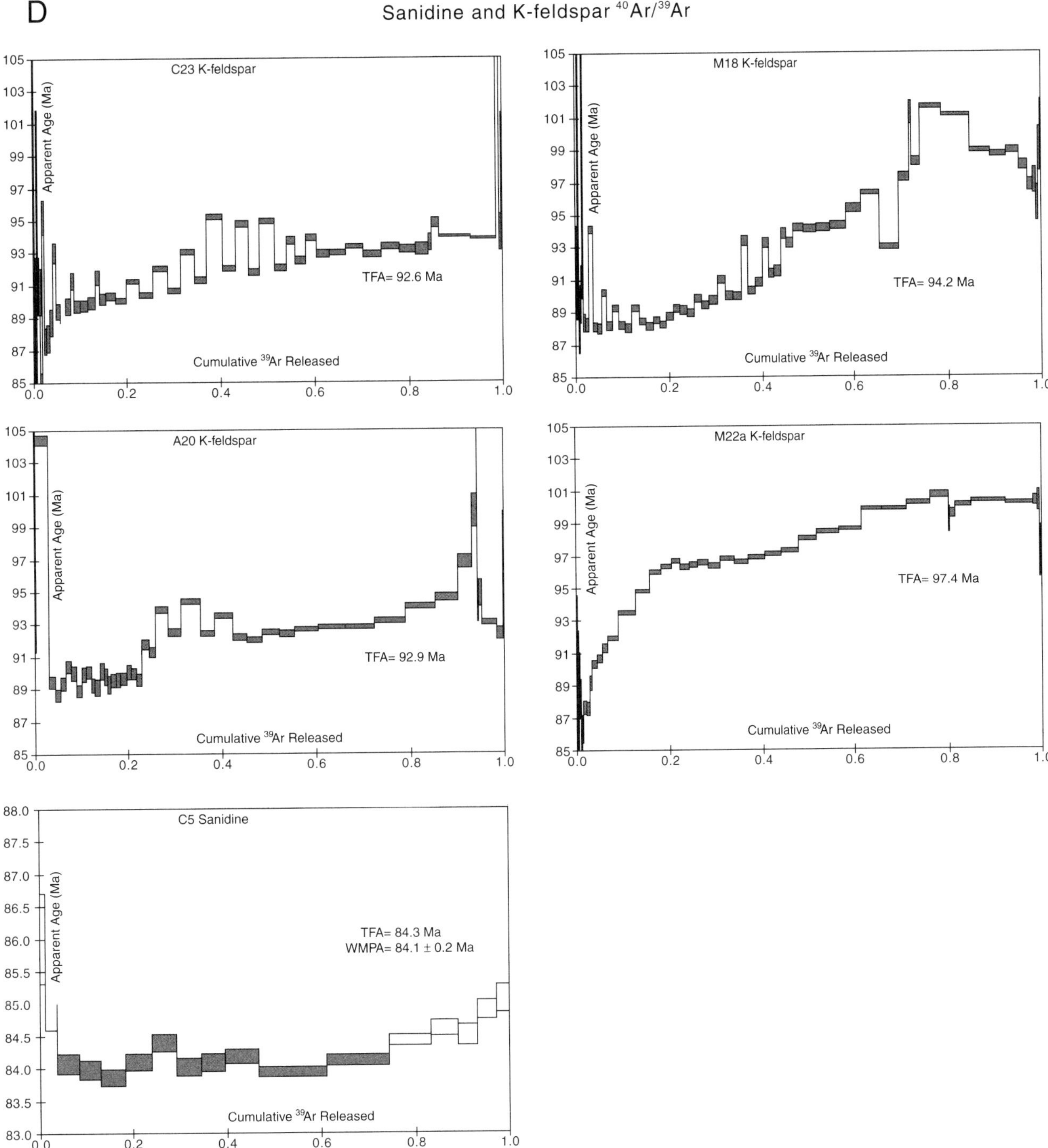

Figure 4 (*continued*)

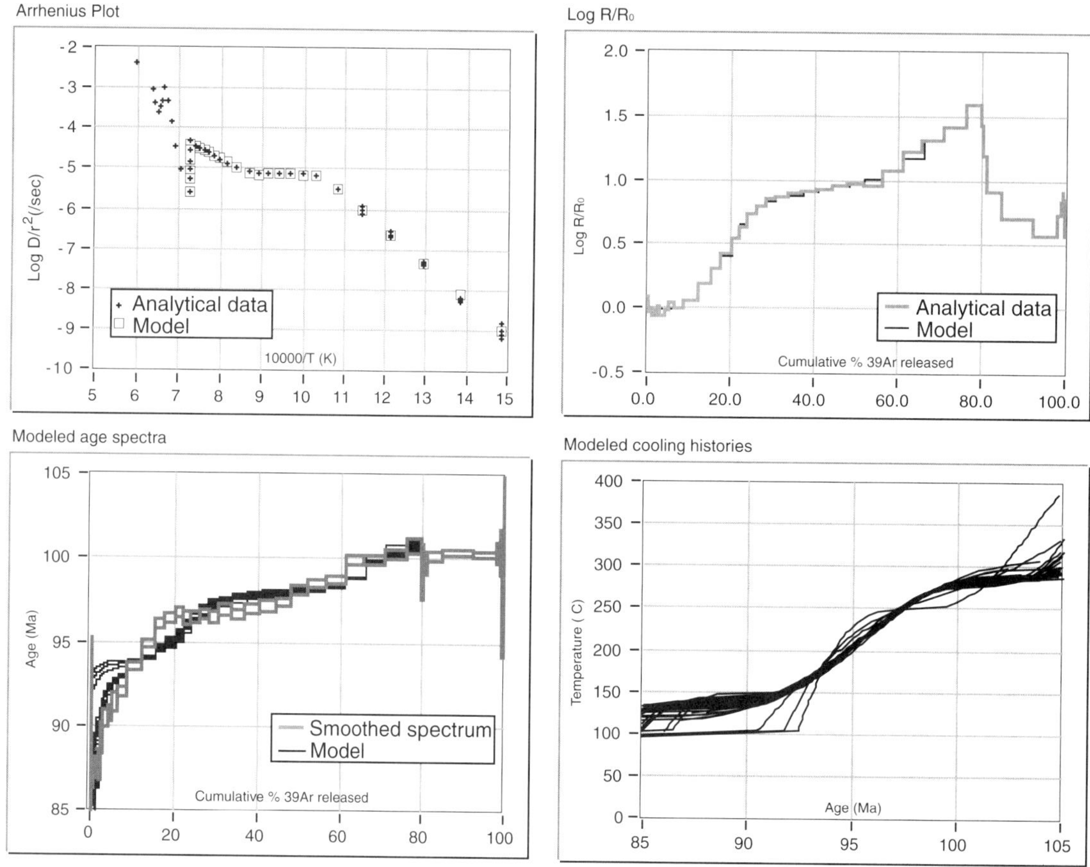

Figure 5. Examples of diffusion modeling of sample M22A. Arrhenius, log r/r$_o$ (Richter et al., 1991), age spectra with 2σ errors and modeled spectra, and modeled cooling history plots are shown.

Samples from across the map area record distinct temperature-time profiles reflecting significant deformation and thermal gradients in the Late Cretaceous. Metamorphic rocks collected from Lavrentia Bay cooled at a nearly linear rate between 15 and 35 °C/m.y., while rocks from deeper structural levels near Koolen Lake cooled later and much more rapidly (100–200 °C/m.y.) from high T to 300 °C and ~40 °C/m.y. below 300 °C. Brecciated granitoids above the low-angle fault in the western part of the map area are consistently 200 °C cooler than Lavrentia Bay samples, despite their present-day proximity (Fig. 2).

DISCUSSION

This study clearly defines segments of the geologic history in pressure-temperature and temperature-time space in the Cretaceous Koolen dome. All the samples analyzed have simple cooling histories that complement the metamorphic mineral assemblages documenting cooling during decompression. Linking these segments with field and petrographic analysis defines a significant portion of the peak and exhumation conditions and allows assessment of the mechanisms for exhumation. From our synthesis of mapping, petrography, and these analytical data we attempt to link pressure-temperature-time conditions to field relationships (Fig. 7).

1. Peak metamorphic conditions were >700 °C at 600–800 MPa in the core and ~540 °C at ~300 MPa on the periphery of the Koolen dome. Although it is unlikely that all levels achieved peak metamorphic conditions simultaneously, it is instructive to plot all data together (Fig. 8). Thermochronology indicates that these conditions occurred before or ca. 104 Ma, based on a U/Pb monazite age from deep structural levels (Bering Strait Geological Field Party, 1997), and before 103.1 Ma, based on an ^{40}Ar/^{39}Ar hornblende age from Lavrentia Bay. Peak pressures are inferred to result from crustal thickening related to the Brooks Range orogen (Dusel-Bacon et al., 1989) of Late Jurassic to Early Cretaceous age (Wirth et al., 1993; Calvert and Gans, 1996). We argue that the crust remained thick until Late Cretaceous arc magmatism and attendant metamorphism (Amato and Wright, 1997, 1998; Calvert and Gans, 1996). The mapped area includes deformed Late Cretaceous plutons (Natal'in et al., 1999) and we attribute peak temperatures in the field area to increased heat flow associated with this magmatism. The

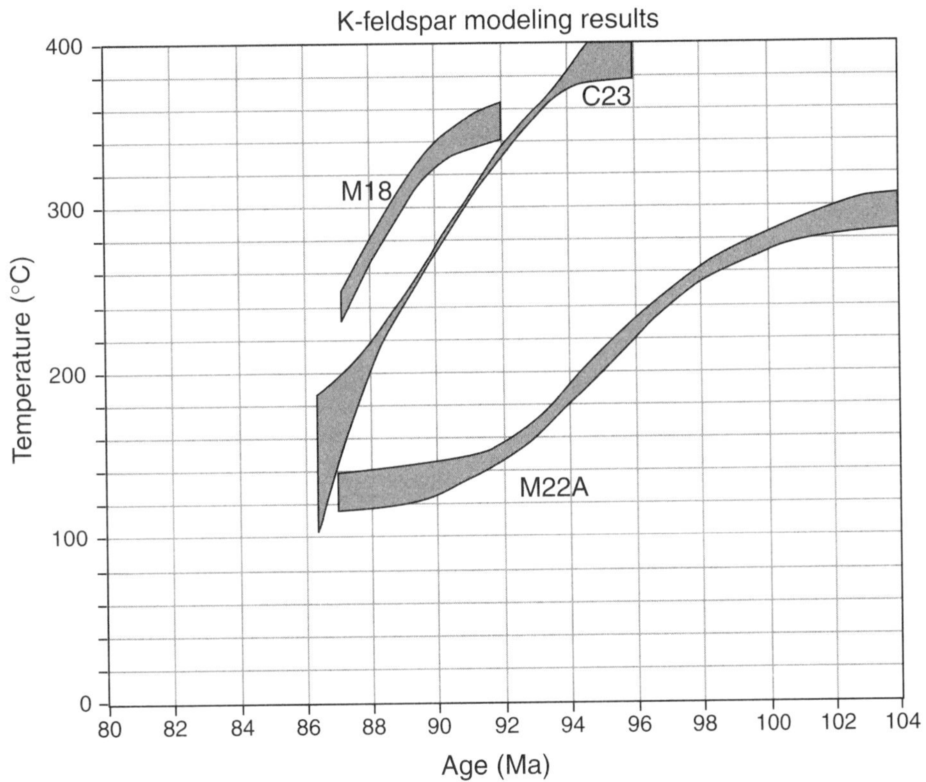

Figure 6. Modeled cooling histories derived from K-feldspar diffusion experiments. Each field shows range of modeled cooling histories derived for 50 optimized age spectra. Cooling histories are truncated at ages where there is no information from age spectra.

high metamorphic field gradient is an apparent gradient caused by subsequent ductile thinning during decompression.

2. Decompression of the complex involved no reheating, on the basis of preserved metamorphic mineral assemblages and textural relations. Although is unlikely that all levels achieved their decompressed metamorphic conditions simultaneously, it is likely bracketed in age between the 104 Ma peak and 94 Ma intrusion of undeformed epizonal granite and granodiorite (Bering Strait Geological Field Party, 1997), and is best resolved by F84 hornblende at 103 Ma. Hornblende closure (~535 °C) is within error of upper Lavrentiev Series decompression temperatures (Fig. 3; Table 2). Pressure determinations between core and rim vary more at deep structural levels than in the upper Lavrentiev Series. If the duration of crystal growth at the different structural levels was similar, the results may document ~20% internal vertical thinning during decompression, which helps explain the 500 MPa difference determined over the 10–12 km structural section. D_1 foliations and lineations in the peak assemblage likely continued to form during decompression, but annealed fabrics at all levels suggest that there was little deformation below ~500 °C. The decompression could have been in response to internal buoyancy caused by coeval magmatism or the significant melt fraction in the rocks, tectonically driven, or the result of simple erosion. Because the regional tectonics during this time are poorly understood, it is difficult to assess the decompression rate and therefore the mechanism.

3. Intrusion of granite and granodiorite throughout the area is best dated by a U/Pb monazite age of 94 Ma (Bering Strait Geological Field Party, 1997). Although these plutons are undeformed and epizonal, they mark a minimum age of penetrative deformation in the complex and likely an end to decompression recorded by the metamorphic petrology. The lack of contact-metamorphic aureoles suggests that the plutons did not markedly reheat the complex. There are a number of variably deformed and undeformed plutons, so additional U/Pb geochronology would improve resolution substantially.

4. We attribute the rapid cooling that occurred in the 6 m.y. following granite intrusion to doming. Simple tilting is a possibility that we cannot exclude because the entire dome was not mapped in this study. However, other workers (Shuldiner and Nedolmolkin, 1976; Zhulanova, 1990; Natal'in et al., 1999) have described the area as a simple structural dome, and the well-mapped adjacent Kigluaik dome on the Seward Peninsula (Amato and Miller, 1997) provides a potential analog. Cooling paths at all structural levels are well recorded by K-feldspar thermochronology and fit well with observed deformational data. When the complex is divided into three cooling paths from moderate temperatures to near-surface temperatures (Fig. 7), the paths are each initially separated by >200 °C reflecting a normal to steep geotherm across 15 km of section. The rapid cooling of deep levels illustrated in Figure 7 by the closely spaced isotherms was probably caused by doming of the complex (Fig.

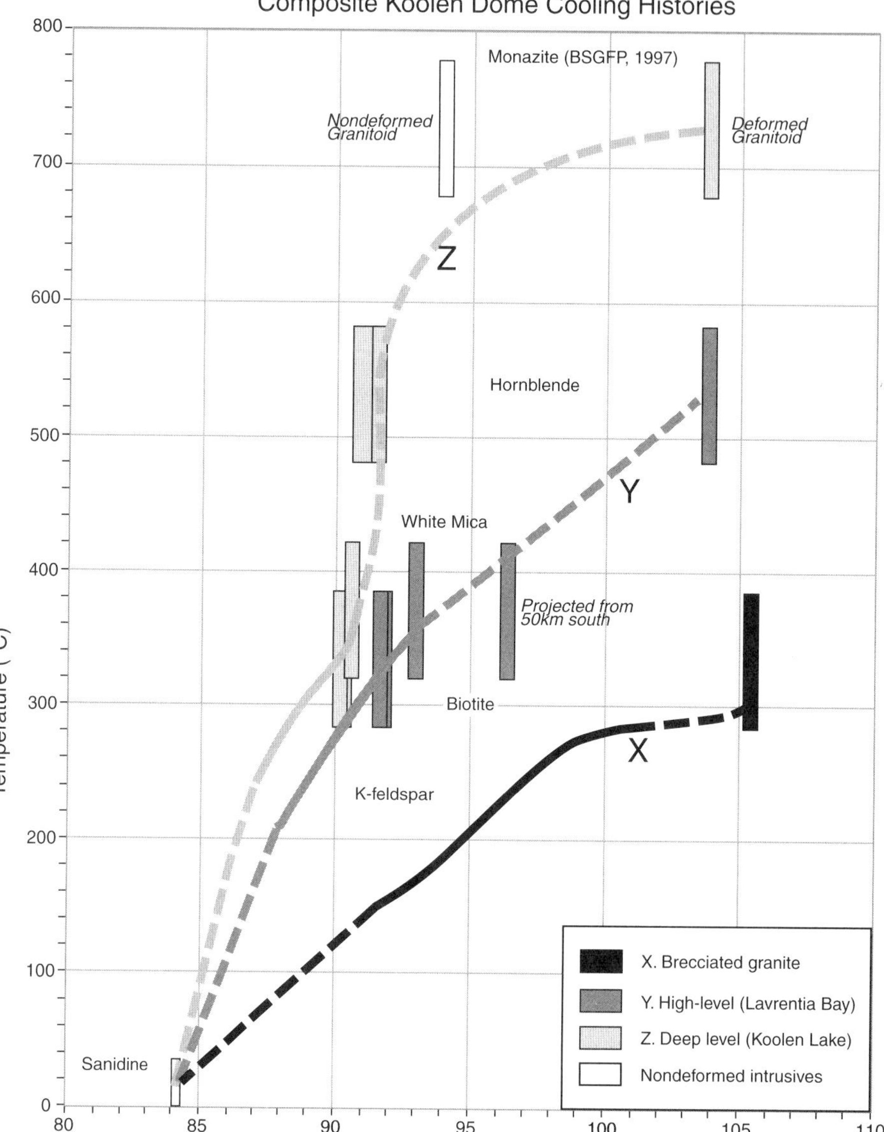

Figure 7. Cooling histories derived from all thermochronometers for three levels of complex. Line X is from brecciated granitic rocks above low-angle fault on west side of complex. Line Y is from shallow levels of Lavrentiev Series near Lavrentia Bay. Line Z is from deep levels near Koolen Lake. 84.2 Ma C5 sanidine was collected from subvolcanic rhyolite that crops out locally from deepest structural levels near Koolen Lake to high structural levels. BSGFP is Bering Strait Geological Field Party.

8). Doming is unlikely to have happened earlier due to the difficulty of maintaining dipping or folded isograds (Spear, 1993). There is no apparent penetrative strain associated with cooling at these low temperatures. The 60 °C difference in absolute temperatures after 91 Ma reflects reasonable temperature differences from the thermal lag of the system, as shown by our simple one-dimensional thermal modeling.

5. Final merging of all cooling histories pre–88 Ma is likely the result of motion on the subhorizontal mylonitic structure along the west side of the map area (Fig. 8). The fault appears to be extensional because it places significantly cooler granitoids on higher-temperature metamorphic rocks. If the fault is extensional, it cannot be older than the doming of the complex; marble mylonites in the footwall display consistent top-to-the-

south fabrics, and deeper structural levels are represented to the north of the faults. Biotite (105.6 ± 0.3 Ma) and K-feldspar (100–88 Ma) ages show cooling significantly before Lavrentiev Series samples and, given the convergence of low-temperature cooling at 88–84 Ma, this age may date motion on the fault that juxtaposes the units. The 200 °C difference between hanging wall and footwall across the fault between 90 and 100 Ma suggests that 5–10 km of section is missing across the fault.

6. Emplacement of volcanic and subvolcanic rhyolites at 84 Ma from Koolen Lake to near Lavrentia Bay indicates that the complex was at near-surface conditions.

The peak metamorphism and exhumation history of the flanks of the Koolen dome discussed here bear a remarkable similarity to the Arrigetch Peaks area of the south-central Brooks

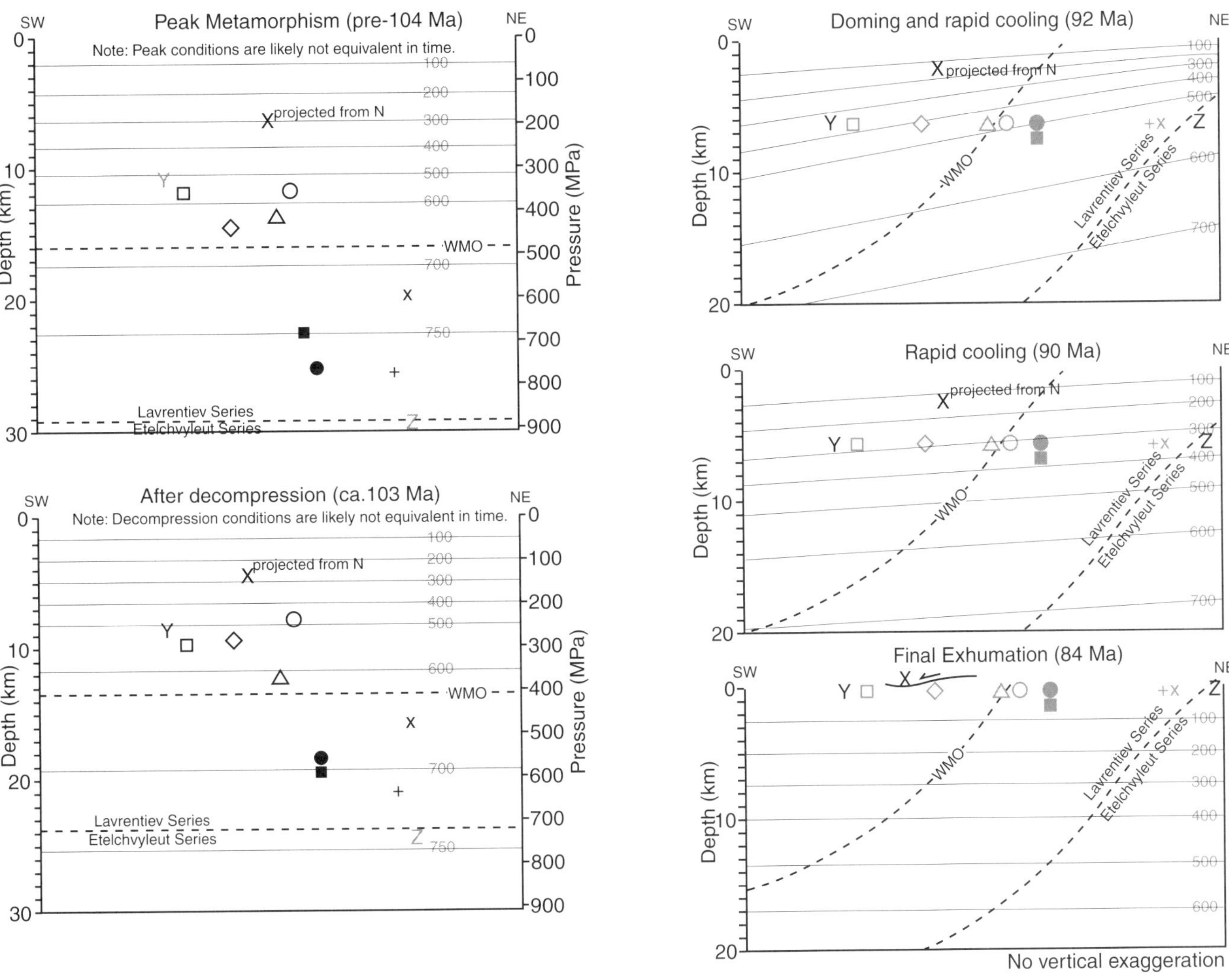

Figure 8. Hypothesized evolution of cross sections across southwest edge of Koolen dome at several times during Cretaceous. Symbols represent thermobarometry samples and letters represent three lines from Figure 7. Symbols and letters are gray when information is not relevant to individual time slices. WMO = white mica–out isograd.

Range (~154°W) of Alaska (Vogl et al., 1999). Peak metamorphic conditions in the core of the Koolen dome are higher temperature than the central Brooks Range, but the timing of exhumation is nearly identical. We attribute the temperature difference solely to the presence of arc magmatism in the Koolen dome. The timing and conditions of metamorphism in the Koolen dome bear great similarity to the Kigluaik, Bendeleben, Darby, and Neshkan domes (Table 1). Peak metamorphism of the Senyavin uplift (Calvert and Gans, 1999) preceded metamorphism in the Koolen dome by 30 m.y., but the volcanism and contact metamorphism of the Senyavin uplift between 90 and 110 Ma indicates similar events at shallower levels. This implies that peak metamorphism and exhumation following mid-Cretaceous crustal thickening occurred simultaneously along the orogen and suggests that peak conditions have regional significance.

CONCLUSIONS

Peak metamorphism in the Koolen dome was coeval with high-strain deformation and likely resulted from the flattening of depressed isotherms following an earlier crustal thickening event. Cooling and exhumation from the pre–104 Ma peak metamorphism in the Koolen dome occurred over the subsequent 20 m.y. and appear to have been the result of gentle, synplutonic exhumation from 103 to 92 Ma, rapid doming and exhumation between 92 and 88 Ma, and extensional (?) faulting between 88 and 84 Ma. The exhumation mechanism is difficult to determine due to a dearth of shear-sense indicators and a poor knowledge of regional events. The analytical data are compatible with anorogenic or magmatically driven doming followed by local extensional faulting.

ACKNOWLEDGMENTS

This project constituted a portion of Calvert's Ph.D. dissertation at University of California Santa Barbara and was funded by the Continental Dynamics Division of the National Science Foundation (NSF). NSF grant EAR-93-17087 to E. Miller and S. Klemperer funded the field and microprobe work and NSF grant EAR-93-17142 to P. Gans funded the $^{40}Ar/^{39}Ar$ analytical work. A portion of Akinin's work was funded by RFBR grant 01-05-65453 to Akinin. M. Gelman, B. Natal'in, E. Miller, J. Amato, J. Wright, R. Fantini, and J. Toro collaborated with field work. Akinin thanks E. Miller for arrangement of a travel grant, B. Jones for microprobe help at Stanford, and V. Fedkin for comments on an early draft of the manuscript. Detailed reviews by Clifford Todd, Barbara John, and Simon Klemperer improved the focus and content of the manuscript.

REFERENCES CITED

Akinin, V.V., 1995, Metaultramaphites of the crystalline basement of the Chukchi Peninsula, *in* Simakov, K.V., and Thurston, D.K., eds., Proceedings of the International Conference on Arctic Margins: Northeast Interdisciplinary Science Research Institute, Magadan, Russia, p. 214–219.

Akinin, V.V., and Kotlyar, I.N., 1997, GEOCHRON: The computer database of isotopic dating of minerals, rocks, and ores in Northeastern Siberia, *in* Byalobzeski, S.G., ed., Magmatism and mineralization in northeastern Siberia: Northeast Interdisciplinary Science Research Institute, Magadan, Russia, p. 314–318 (in Russian).

Amato, J.M., and Miller, E.L., 1997, Bedrock geologic map of the Kigluaik Mountains, Seward Peninsula, Alaska: State of Alaska Division of Geological and Geophysical Surveys, Public Data File 97–31, scale 1:63 360, 8 sheets.

Amato, J.M., and Wright, J.E., 1997, Potassic mafic magmatism in the Kigluaik gneiss dome, Northern Alaska: A geochemical study of arc magmatism in an extensional tectonic setting: Journal of Geophysical Research, B, Solid Earth and Planets, v. 102, no. 4, p. 8065–8084.

Amato, J.M., and Wright, J.E., 1998, Geochronologic investigations of magmatism and metamorphism within the Kigluaik Mountains gneiss dome, Seward Peninsula, Alaska, *in* Clough, J.G., and Larson, F., eds., Short notes on Alaska geology 1997: Alaska Division of Geological and Geophysical Surveys Professional Report 118, p. 1–21.

Amato J.M., Wright J.E., Gans P.B., and Miller E.L., 1994, Magmatically induced metamorphism and deformation in the Kigluaik gneiss dome, Seward Peninsula: Tectonics, v.13, no. 2, p. 515–527.

Aranovich, L.J., 1983, Biotite-garnet equilibrium in metapelites. 1. Thermodynamics of the solid solutions and mineral reactions, *in* Zarikov, V.A., and Fed'kin, V.V., eds., Sketches of physical-chemical petrology: Moscow, Nauka, v. 11, p.121–136 (in Russian).

Aranovich, L.J., Lavrent'eva, I.V., and Kosjakova, N.A., 1988, Biotite-garnet and biotite-orthopyroxene thermometers: The calibration with registration of variable content of Al in biotite: Geochemistry, v. 5, p. 668–676 (in Russian).

Avchenko, O.V., 1990, Mineral equilibrium in metamorphic rock and problems of the thermobarometry: Moscow, Nauka, 182 p. (in Russian).

Beikman, H.M., 1980, Geologic map of Alaska: U.S. Geological Survey Special Map, scale 1:2 500 000, 2 sheets.

Belyi, V.F., 1977, Structural-formational map of the Okhotsk-Chukotsk Volcanogenic Belt: Northeast Interdisciplinary Science Research Institute–Academy of Sciences, GKP "Sevvostgeologia," Magadan, scale 1:1 500 000, 1 sheet.

Bering Strait Geologic Field Party (BSGFP): Akinin, V.V., Gelman, M.L., Sedov, B.M., Amato, J.M., Miller, E.L., Toro, J., Calvert, A.T., Fantini, R.M., Wright, J.E., and Natal'in, B.A., 1997, Koolen metamorphic complex, NE Russia: Implications for the tectonic evolution of the Bering Strait region: Tectonics, v. 16, p. 713–729.

Bhattacharya, A., Mohanty, L., Maji, A., Sen, S.K., and Raith, M., 1992, Nonideal mixing in the phlogopite-annite binary: Constraints from experimental data on Mg-Fe partitioning and a reformulation of the biotite-garnet geothermometer: Contributions to Mineralogy and Petrology, v. 111, no. 1, p. 87–93.

Burg, J.P., Guiraud, M., Chen, G.M., and Li, G.C., 1984, Himalayan metamorphism and deformations in the North Himalayan Belt (southern Tibet, China): Earth and Planetary Science Letters, v. 69, p. 391–400.

Calvert, A.T., and Gans, P.B., 1996, Jurassic(?) kyanite-grade metamorphism and earliest Cretaceous exhumation of the Senyavin complex, Chukchi Peninsula, NE Russia [abs.]: Eos (Transactions, American Geophysical Union), v. 77, no. 46, p. 766.

Calvert, A.T., and Gans, P.B., 1999, Thermochronologic assessment of metamorphism and exhumation of mid-crustal gneiss domes in the Arctic Alaska terrane: Geological Society of America Abstracts with Programs, v. 31, no. 6, p. 42.

Calvert, A.T., Gans, P.B., and Amato, J.M., 1999, Diapiric ascent and cooling of a sillimanite gneiss dome revealed by $^{40}Ar/^{39}Ar$ thermochronology: The Kigluaik Mountains, Seward Peninsula, Alaska, *in* Ring, U., Brandon, M.T., Lister, G.S., and Willet, S.D., eds., Exhumation processes: Normal faulting, ductile flow, erosion: Geological Society [London] Special Publication 154, p. 205–232.

Crittenden, M.D., Coney, P.J., and Davis, G.H., editors, 1980, Cordilleran metamorphic core complexes: Boulder, Colorado, Geological Society of America Memoir 153, p. 485–490.

Dusel-Bacon, C., Brosgé, W.P., Till, A.B., Fitch, M.R., Mayfield, C.F., Reiser, H.M., and Miller, T.P., 1989, Distribution, facies, ages and proposed tectonic associations of regionally metamorphosed rocks in northern Alaska: U.S. Geological Survey Professional Paper 1497A, 44 p.

Ellis, D.J., and Green D.H., 1979, An experimental study of the effect of Ca upon garnet-clinopyroxene Fe-Mg exchange equilibria: Contributions to Mineralogy and Petrology, v. 71, no. 1, p. 13–22.

England, P.C., and Bickle, M.J., 1984, Continental thermal and tectonic regimes during the Archaean: Journal of Geology, v. 92, no. 4, p. 353–367.

Faure, G., 1986, Principles of isotope geology: New York, John Wiley & Sons, 589 p.

Ferry, J.M., and Spear, F.S., 1978, Experimental calibration of the partitioning of Fe and Mg between biotite and garnet: Contributions to Mineralogy and Petrology, v. 66, p. 113–117.

Gelman, M.L., 1973, Osnovnye osobennosti posleproterozoyskogo metamorfisma na severo-vostoke SSSR, *in* Smirnov, A.M., and Schuldiner, V.I., eds., Metamorficheskie kompleksy vostoka SSSR: Vladivostok, USSR, Academy of Science, p. 161–180 (in Russian).

Gelman, M.L., 1995, Phanerozoic granite-metamorphic domes of the Siberian Northeast. 1. Geological history of Paleozoic and Mesozoic domes: Tichoocenskaya Geologiya, v. 14, p. 102–115 (in Russian).

Gnibidenko, G.S., 1969, Metamorficheskie kompleksy v structurach severo-zapadnogo sektora Tokhookeanskogo poyasa: Moscow, Nauka, 135 p. (in Russian).

Grove, M., and Harrison, T.M., 1996, $^{40}Ar^*$ diffusion in Fe-rich biotite: American Mineralogist, v. 81, p. 940–951.

Hannula, K.A., and McWilliams, M.O., 1995, Reconsideration of the age of blueschist-facies metamorphism on the Seward Peninsula, Alaska, based on phengite $^{40}Ar/^{39}Ar$ results: Journal of Metamorphic Geology, v. 13, p. 125–139.

Hannula, K.A., Miller, E.L., Dumitru, T.A., Lee, J., and Rubin, C.M., 1995, Structural and metamorphic relations in the southwest Seward Peninsula, Alaska: Crustal extension and the unroofing of blueschists: Geological Society of America Bulletin, v. 107, no. 5, p. 536–553.

Hanson, G.N., and Gast, P.W., 1967, Kinetic studies in contact metamorphic zones: Geochimica et Cosmochimica Acta, v. 31, p. 1119–1153.

Harrison, T.M., 1981, Diffusion of ^{40}Ar in hornblende: Contributions to Mineralogy and Petrology, v. 78, no. 3, p. 324–331.

Harrison, T.M., Duncan, I., and McDougall, I., 1985, Diffusion of ^{40}Ar in biotite: Temperature, pressure and compositional effects: Geochimica et Cosmochimica Acta, v. 49, p. 2461–2468.

Harrison, T.M., Heizler, M.T., Lovera, O.M., Chen, W., and Grove, M., 1994, A chlorine disinfectant for excess argon released from K-feldspar during step heating: Earth and Planetary Science Letters, v. 123, nos. 1–4, p. 95–104.

Hill, E.J., Baldwin, S.L., and Lister, G.S., 1995, Magmatism as an essential driving force for formation of active metamorphic core complexes in eastern Papua New Guinea: Journal of Geophysical Research, v. 100, no. B7, p.10441–10451.

Holdaway, M.J., 1971, Stability of andalusite and the aluminum silicate phase diagram: American Journal of Science, v. 271, no. 2, p. 97–131.

Holland, T.J.B., and Powell, R., 1998, An internally consistent thermodynamic data set for phases of petrological interest: Journal of Metamorphic Geology, v. 16, p. 309–343.

Ivanov, O.N., and Kryukov, J.V., 1973, Precambrian of the eastern sector of Sov'et Arctic: Soviet Geology and Geophysics, no. 9, p. 50–59 (in Russian).

Kretz, R., 1983, Symbols for rock-forming minerals: American Mineralogist, v. 68, p. 277–279.

Lister, G.S., and Baldwin, S.L., 1993, Plutonism and the origin of metamorphic core complexes: Geology, v. 21, no. 7, p. 607–610.

Lister, G.S., and Baldwin, S.L., 1996, Modelling the effect of arbitrary P-T-t histories on argon diffusion in minerals using the MacArgon program for the Apple Macintosh: Tectonophysics, v. 253, nos. 1–2, p. 83–109.

Lister, G.S., and Davis, G.A., 1989, The origin of metamorphic core complexes and detachment faults formed during Tertiary continental extension in the northern Colorado River region, U.S.A.: Journal of Structural Geology, v. 11, nos. 1/2, p. 65–94.

Lovera, O.M., 1992, Computer programs to model ^{40}Ar/^{39}Ar diffusion data from multidomain samples: Computers in Geosciences, v. 18, no. 7, p. 789–813.

Lovera, O.M., Richter, F.M., and Harrison, T.M., 1989, The ^{40}Ar/^{39}Ar thermochronology for slowly cooled samples having a distribution of diffusion domain sizes: Journal of Geophysical Research, v. 94, no. B12, p. 17917–17935.

MacCready, T., Snoke, A.W., Wright, J.E., and Howard, K.A., 1997, Mid-crustal flow during Tertiary extension in the Ruby Mountains core complex, Nevada: Geological Society of America Bulletin, v. 109, no. 12, p. 1576–1594.

Miller, E.L., Calvert, A.T., and Little, T.A., 1992, Strain-collapsed metamorphic isograds in a sillimanite gneiss dome, Seward Peninsula, Alaska: Geology, v. 20, no. 6, p. 487–490.

Miller, T.P., and Bunker, C.M., 1976, A reconnaissance study of the uranium and thorium contents of plutonic rocks of the southeastern Seward Peninsula, Alaska: Journal of Research of the U.S. Geological Survey, v. 4, no. 3, p. 367–377.

Natal'in, B.A., 1979, Tectonic nature of a metamorphic complex of the Chukchi Peninsula: Soviet Geology and Geophysics, no. 6, p. 31–38 (in Russian).

Natal'in, B.A., Amato, J.M., Toro, J., and Wright, J.E., 1999, Paleozoic rocks of the Chegitun River valley, northern Chukotka Peninsula: Insights into the tectonic evolution of the eastern Arctic: Tectonics, v. 18, p. 977–1003.

Parrish, R.R., 1990, U-Pb dating of monazite and its application to geologic problems: Canadian Journal of Earth Science, v. 27, p. 1431–1450.

Patrick, B.E., and Lieberman, J.E., 1988, Thermal overprint on blueschists of the Seward Peninsula: The Lepontine in Alaska: Geology, v. 16, p. 1100–1103.

Perchuk, L.L., 1986, Evolution of metamorphism, *in* Zaricov, V.A., and Fed'kin, V.V., eds., Experiment in resolution of actual problems of geology: Moscow, Nauka, p. 151–173 (in Russian).

Perchuk, L.L., Lavrent'eva, I.V., Aranovich, L.J., and Podleskii, K.K., 1983, Biotite-garnet equilibrium and evolution of metamorphism: Moscow, Nauka, 197 p. (in Russian).

Ramberg, J., 1981, The role of gravity in orogenic belts, *in* McClay, K.R., and Price, N.J., eds., Thrust and nappe tectonics: Geological Society [London] Special Publication 9, p. 125–140.

Richter, F.M., Lovera, O, Harrison, T.M., and Copeland, P., 1991, Tibetan tectonics from ^{40}Ar/^{39}Ar analysis of a single K-feldspar sample: Earth and Planetary Science Letters, v. 105, nos. 1–3, p. 266–278.

Roddick, J.C., 1978, The application of isochron diagrams in ^{40}Ar-^{39}Ar dating: A discussion: Earth and Planetary Science Letters, v. 41, no. 2, p. 233–244.

Shilo, N.A., and Zagruzina, I.A., 1965, Magmaticheskie kompleksy i metallogeniya Vostochnoy Chukotki, *in* Shilo, N.A., ed., Pozdnemezozoiskie granitoidy Chukotki: Magadan, Russia, SVKNII, v. 12, p. 188–206 (in Russian).

Shuldiner, V.I., and Nedomolkin, V.F., 1976, Crystalline basement of Eskimoskii massif: Sov'etic Geology, no. 10, p. 33–47 (in Russian).

Snee, L.W., Sutter, J.F., and Kelly, W.C., 1988, Thermochronology of economic mineral deposits, dating the stages of mineralization at Panesqueira, Portugal by high-precision ^{40}Ar/^{39}Ar age spectrum techniques on muscovite: Economic Geology, v. 83, p. 335–354.

Spear, F.S., 1993, Metamorphic phase equilibria and pressure temperature time paths: Washington, D.C., Mineralogical Society of America Monograph 1, p. 799.

Till, A.B., and Dumoulin, J.A., 1994, Geology of Seward Peninsula and Saint Lawrence Island, *in* Plafker, G., and Berg, H.C., eds., The geology of Alaska: Boulder, Colorado, Geological Society of America, Geology of North America, v. G-1, p. 141–152.

Till, A.B., Dumoulin, J.A., Gamble, B., Kaufman, D., and Carroll, P.I., 1986, Preliminary geologic map and fossil data, Solomon, Bendeleben and southern Kotzebue quadrangles, Seward Peninsula, Alaska: U.S. Geological Survey Open-File Report 86–276, 60 p., 3 sheets.

Todd, C.S., 1998, Limits on the precision of geobarometry at low grossular and anorthite content: American Mineralogist, v. 83, p. 1161–1167.

Vogl, J.J., Gans, P.B., and Calvert, A.T., 1999, A thermochronologic transect across the south-central Brooks Range, Alaska: Exhumation of a metamorphic culmination: Geological Society of America Abstracts with Programs, v. 31, no. 6, p. 105.

Westcott, E., and Turner, D.L., 1981, Geothermal reconnaissance survey of the central Seward Peninsula, Alaska: University of Alaska Geophysical Institute Report 284, 123 p.

Wirth, K.R., Bird, J.M., Blythe, A.E., Harding, D.J., and Heizler, M.T., 1993, Age and evolution of western Brooks Range ophiolites, Alaska: Results from ^{40}Ar/^{39}Ar thermochronometry: Tectonics, v. 12, no. 2, p. 410–432.

Zhulanova, I.L., 1990, Earth crust of the northeast of Asia in Precambrian and Phanerozoic: Moscow, Nauka, 304 p. (in Russian).

MANUSCRIPT ACCEPTED BY THE SOCIETY MAY 15, 2001.

Evolution of crust and mantle beneath the Bering Sea region: Evidence from xenoliths and late Cenozoic basalts

Karl R. Wirth
Jeffrey Grandy*
Katherine Kelley*
Seth Sadofsky*
Macalester College, St. Paul, Minnesota 55105, USA

ABSTRACT

Late Cenozoic volcanic centers throughout the Bering Sea region of Alaska consist of small-volume eruptions of flows and cones of tholeiitic to alkaline basalt. Peridotite (spinel lherzolite, websterite, wehrlite, and pyroxenite) and gabbro xenoliths are abundant in some of the most alkaline flows and cones. Spinel lherzolite xenoliths are characterized by granuloblastic-equant and coarse-equant textures typical of mantle peridotites and yield equilibration temperatures of 1000–1200 °C (1.5 GPa). Relative to other spinel lherzolites worldwide, these xenoliths have low MgO and high $Fe_2O_3^T$ and are similar to some modeled compositions for fertile mantle. However, these xenoliths are relatively enriched in the light rare earth elements (REE), Th, U, K, Nb, and Ta. Rare amphibole peridotite xenoliths have textures and compositions that are consistent with an origin as pyroxenite veins in the mantle that were variably deformed and metasomatized with potassium-rich fluids. These xenoliths have concave-up light REE patterns similar to those of the host basalts, suggesting that the alkaline melts may have been produced by melting of amphibole-bearing mantle peridotites. Gabbroic xenoliths are variably deformed and recrystallized at relatively high pressure (0.4–0.6 GPa). Trace element ratios (e.g., Ba/Nb, La/Nb) indicate that these xenoliths crystallized from melts that were produced in a suprasubduction-zone setting and suggest that they represent cumulates that formed in the lower crust during Late Cretaceous subduction. We suggest that these cumulates are the source of the strong reflections that are observed in seismic reflection profile images of the Bering Sea region.

Volcanic flows consist of tholeiite, olivine tholeiite, alkali-olivine basalt, and basanite, whereas cones consist mostly of alkali-olivine basalt and basanite. Geochemical variations within each volcanic field largely reflect the effects of fractionation and melting (depth and fraction of melting), whereas between-field geochemical differences are largely due to source composition differences. The enriched incompatible trace element compositions (light REE, Ba, Th, Nb, Ta) and ratios (Th/Ta, Rb/Zr, Nb/La, La/Yb) are similar to intraplate basalts from ocean island and continental extensional settings. In particular, the compositions of Bering Sea basalts are similar to those of the diffuse volcanic province of Southeast Asia, a region where late Cenozoic volcanism has been related to tectonic extrusion along reactivated strike-slip faults.

*Current addresses: Grandy, Washington State University, Pullman, WA 99164; Kelley, Boston University, Boston, MA 02215; Sadofsky, Lehigh University, Bethlehem, PA 18015.
Data Repository item 2002077 contains additional material related to this article.

Wirth, K.R., Grandy, J., Kelley, K., and Sadofsky, S., 2002, Evolution of crust and mantle beneath the Bering Sea region: Evidence from xenoliths and late Cenozoic basalts, *in* Miller, E.L., Grantz, A., and Klemperer, S.L., eds., Tectonic Evolution of the Bering Shelf–Chukchi Sea–Arctic Margin and Adjacent Landmasses: Boulder, Colorado, Geological Society of America Special Paper 360, p. 167–193.

INTRODUCTION

Isolated volcanic centers are widespread throughout much of the northern Cordillera in Alaska and eastern Russia (Fig. 1), and are characterized by small-volume eruptions of tholeiitic to alkaline magmas with intraplate affinity (e.g., Hoare and Con-don, 1966, 1968, 1971a, 1971b; Patton and Csejtey, 1980; Swanson et al., 1981; Turner et al., 1981; Wirth, 1991; Moll-Stalcup, 1994, 1995; Feeley and Winer, 1999). The composition and character of this volcanism contrast sharply with the long history of subduction-related magmatism (calc-alkaline) and terrane accretion that characterized much of Alaska after the

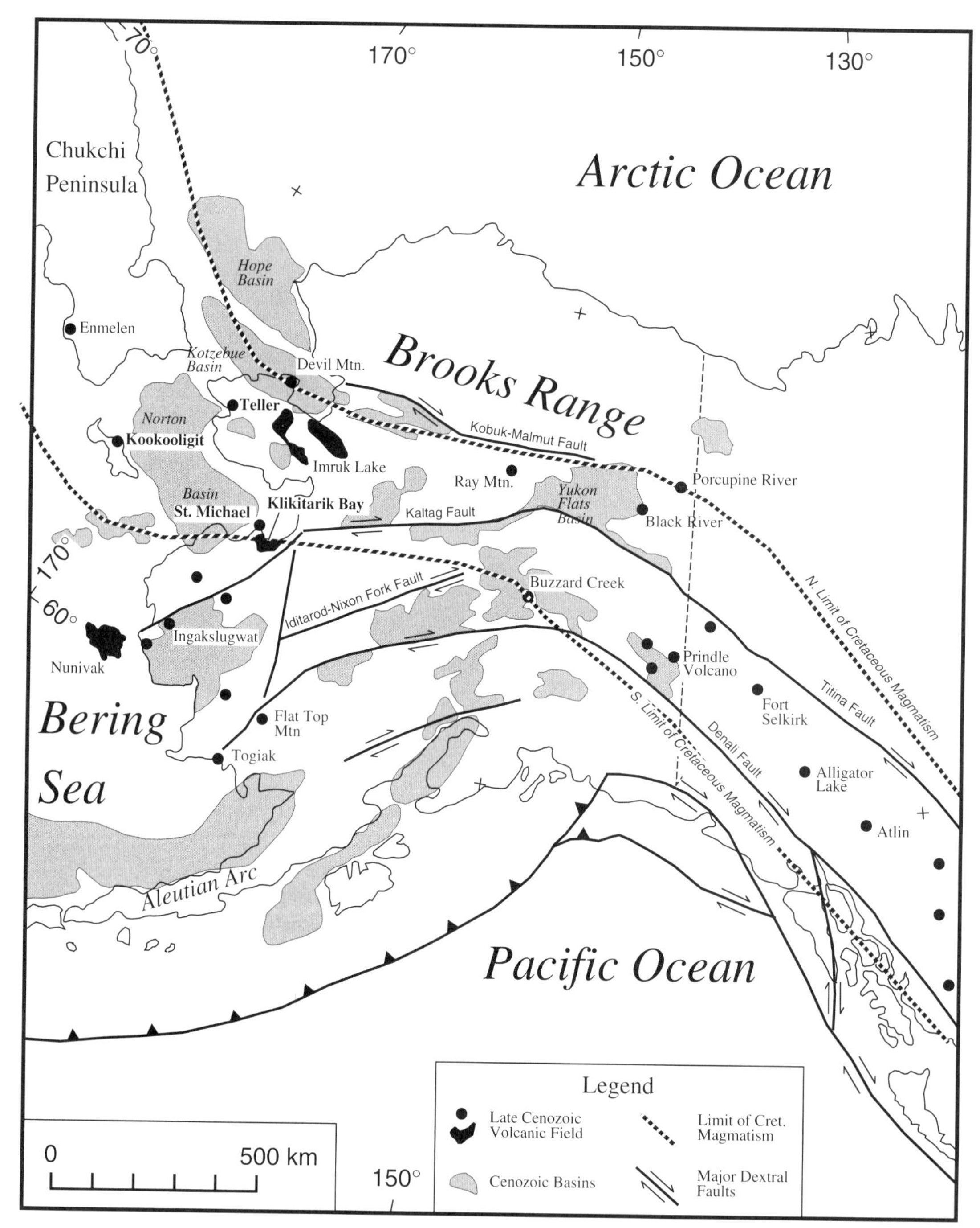

Figure 1. Map of Alaska showing locations of late Cenozoic volcanic fields, major dextral strike-slip faults, Cenozoic basins, and Cretaceous magmatic belt. Modified from Moll-Stalcup (1994), Kirschner (1994), Plafker and Berg (1994), and Amato and Wright (1997).

Mesozoic. The southern margin of Alaska has been the site of subduction since at least the Jurassic, resulting in the formation of the Brooks Range orogenic belt (Late Jurassic–Cretaceous) and the subsequent accretion of numerous terranes south of the Brooks Range. As the southern margin of Alaska stepped southward during Cretaceous terrane accretion, the locus of subduction-related magmatism also tracked to the south, resulting in a series of subparallel magmatic belts that extend westward from Alaska across the Bering Strait to eastern Russia (Hudson, 1979; Wallace and Engebretson, 1984). The Aleutian arc marks the present-day location of calc-alkaline magmatism related to northward subduction of the Pacific plate beneath the southern margin of Alaska. Widespread alkaline magmatism began suddenly ca. 6 Ma across a large region of Alaska 600–1100 km north of the present-day plate margin and the Aleutian arc (Fig. 1). This magmatism occurred significantly after the postorogenic collapse and extension of the southern Brooks Range orogen during the Late Cretaceous (Miller and Hudson, 1991). It also postdates the main phase of basin formation that occurred during the early Tertiary on the Bering platform (Fig. 1). The tectonic setting of Alaska has more recently been dominated by north-south convergence in eastern Alaska, with contemporaneous or tectonic extrusion and extension in the west (e.g., Plafker and Berg, 1994).

Alkaline volcanic provinces similar to those of the Bering Sea region occur in a wide range of tectonic settings. This type of volcanism has been related to mantle plumes (e.g., Chen and Frey, 1985), to regions of postsubduction magmatism (Gill and Whelan, 1989), rifting (Perry et al., 1987), and areas undergoing transtensional deformation (Hoang and Flower, 1998). The similar characteristics of alkaline magmas in all of these settings from both continental and oceanic plates (e.g., Fitton, 1987) have led some workers to suggest that these magmas must be derived from beneath the lithosphere-asthenosphere boundary. Others have argued that the source of alkaline melts is within the metasomatized lithosphere (e.g., Menzies and Murthy, 1980a; Bailey, 1987).

The goals of this study are (1) to study the character and origin of alkaline magmatism in the Bering Sea region, (2) to study the composition and evolution of the lower crust beneath the Bering Shelf using crustal xenoliths that were entrained in the alkaline volcanics, and (3) to examine the relationship between alkaline magmatism and the late Cenozoic tectonic evolution of western Alaska. We address this question by first describing the geochemistry and petrogenesis of several tholeiitic to alkaline volcanic suites in the Bering Sea region. The basalt geochemistry is dominated by melting processes and suggests that at least two different mantle sources are necessary to explain the diversity of magma types. Geochemical and pressure-temperature (*P-T*) data from mantle xenoliths indicate that at least some portions of the upper mantle beneath the Bering Shelf consist of relatively fertile peridotite that has been variably recrystallized, metasomatized, and melted. Studies of the crustal xenoliths, combined with the results of seismic and geochronologic studies, indicate that the lower crust beneath the Bering Strait includes significant amounts of mafic cumulates that were likely related to Late Cretaceous calc-alkaline magmatism and were subsequently deformed during Late Cretaceous extension. We argue that the alkaline volcanic rocks of Alaska fit the model of a diffuse volcanic province such as that proposed for southeastern Asia.

SAMPLING AND ANALYTICAL METHODS

More than 250 samples were collected in the field during a four-week field season during 1994. Of these, more than 150 samples of flows and xenoliths were selected for further analysis on the basis of geologic and petrographic evidence. Major element compositions of mineral grains were determined using a Zeiss DSM 960A scanning electron microscope (SEM) with an Oxford energy dispersive spectrometer (EDS) and Oxford Link Isis software. Operating conditions consisted of an accelerating voltage of 20 kV and an emission current of 70 µA. Elemental abundances were calculated using the ZAF technique and the EDS was standardized using mineral standards from the Smithsonian Institute.

All samples for whole-rock analysis were pulverized to fine powder with a shatterbox in an alumina container after crushing in a mild steel jaw crusher. Major elements and V, Cr, Ni, and Ba were determined by X-ray fluorescence (XRF) at McGill University (Ahmedali, 1983) with a Philips PW2400 using fused beads. Analytical precision for major elements is better than 0.1 wt% and better than 2% for the minor element abundances determined from fused beads. Rb, Sr, Nb, Zr, and Y were determined by XRF at McGill University on pressed powder pellets using an Rh-Kb Compton scatter matrix correction (analytical precision is estimated to be <2%). Major and trace element concentrations of Saint Lawrence samples were determined by XRF at Macalester College using a Philips PW 2400. Sample preparation (fused beads and pressed powder pellets) and analytical methods used are similar to those at McGill University. Based on replicate analyses, analytical precision is estimated to be better than 0.1 wt% for the major elements and better than 3 wt% relative for the minor and trace elements. Th, U, Ta, Ba, Hf, Cr, Ni, Co, Sc, and the rare earth element (REE) concentrations of Teller and Saint Michael basalts were determined by instrumental neutron activation analysis at Cornell University following the method of Kay and Kay (1988). Samples were irradiated in a flux of 10^8 neutrons $cm^{-2}s^{-1}$ in a TRIGA reactor at Cornell University and counted 7 and 40 days after irradiation. Analytical uncertainties are generally <5%, except for U (~ 8%). The REE, Ba, Th, Nb, Y, Hf, Ta, U, Pb, Rb, Cs, Sr, and Sc concentrations of all xenolith samples, as well the flows from Saint Lawrence Island, were analyzed by inductively coupled plasma–mass spectroscopy (ICP-MS) in the GeoAnalytical Laboratories at Washington State University. Samples were prepared by acid digestion of fused beads in HF, HNO_3, and $HClO_4$ followed by evaporation to dryness (Knaack et al., 1994). A final evaporation with $HClO_4$ converted any insoluble fluorides to soluble perchlorates. Samples were run on a Sciex Elan model 250 ICP-MS equipped with a Babington nebulizer, water-cooled spray chamber, and Brooks mass flow controllers. Analytical precision is better than 5%. A subset of samples was analyzed by two or more methods to ensure that the results are comparable between different laboratories and methods. Many of the trace elements were analyzed using two or more instrumental methods (XRF, INAA [instrumental neutron activation analysis], ICP-MS). When available, the results of ICP-MS and INAA methods are used because of their greater sensitivity and precision compared with those from XRF.

VOLCANIC FIELDS OF THE BERING SEA REGION

Relatively few of the volcanic fields of western Alaska and the Bering Shelf (Fig. 1) have been studied in detail. The most comprehensive descriptions of the character of the volcanism are found in review papers by Moll-Stalcup (1994, 1995). In this study we examined the field, petrographic, and geochemical characteristics of basalts from three volcanic fields in the Bering Sea region. These three volcanic centers are informally referred to as the Teller, Saint Michael, and Kookooligit volcanic fields (Moll-Stalcup, 1994).

Teller volcanic field

The Teller volcanic field, located north of Teller on the west end of the Seward Peninsula (Fig. 1), has received only brief mention in the literature (Swanson et al., 1981; Turner et al., 1981; Moll-Stalcup, 1994). It is a small volcanic field (~100 km[2]) that consists of several flows and numerous volcanic plugs. The volcanic plugs occur in the eastern part of the field and are steep-sided outcrops to several hundred meters in diameter that consist primarily of columnar-jointed basalt and dikes of various textural types. These features suggest that the plugs may be erosional remnants of former cinder cones. Flows are present throughout the field, but are best exposed in the western portions, where they are generally <15 m thick and typically have blocky surfaces. Massive, columnar-jointed flow interiors and vesicular margins are commonly present; however, some flows have been deeply weathered to rubble. Four samples from the cones have been dated as 2.5–2.8 Ma (data of Swanson and Turner, 1984, reported in Moll-Stalcup, 1994). The basement rocks of the region are weakly metamorphosed Precambrian (?) to Silurian clastic rocks and carbonates of the York terrane (Sainsbury, 1972; Silberling et al., 1994). The volcanic field is situated near the boundary of the York terrane and the Seward terrane (Till and Dumoulin, 1994).

The western flows are subophitic olivine tholeiites and are composed mainly of olivine, plagioclase, and clinopyroxene; minor opaque oxide minerals and glass are also present (Sad-

ofsky, 1995). Olivine phenocrysts (Fo_{70-90}) are frequently skeletal or resorbed. Plagioclase phenocrysts are intermediate in composition (An_{45-55}) and display normal and continuous compositional zoning. Many of the plagioclase grains occur as chadocrysts enclosed within pyroxene oikocrysts. Plagioclase microlites are also common constituents of the groundmass. Clinopyroxene (augite) is present both as large ophitic grains and as fine grains. Fresh volcanic glass was not observed in these flows, but several samples contain fine-grained interstitial alteration minerals that are likely devitrified glass. The petrography and mineral compositions of the eastern flows are similar to those of the western part of this volcanic field, except that pyroxene occurs only as small grains in the groundmass.

Samples from the eastern volcanic plugs vary in texture from holohyaline to intergranular and hypocrystalline. Olivine phenocrysts are more abundant and are frequently the only phenocryst phase in these samples. Plagioclase and clinopyroxene are present mainly as small microcrysts in the groundmass. Fresh and altered volcanic glass is a major constituent of the eastern plugs, and some secondary minerals (calcite and zeolites) are present lining vesicle walls.

Saint Michael volcanic field

The Saint Michael volcanic field is a fairly extensive (~3000 km[2]) region of flows, cones, and maar craters south of the Seward Peninsula and Norton Sound (Fig. 1). The extent and character of volcanism within the Saint Michael volcanic field was established by Hoare and Coonrad (1959, 1978a, 1978b), Hoare and Condon (1966, 1968, 1971a, 1971b), and Patton and Moll (1985). The timing of volcanism within the field is poorly defined and was summarized by Moll-Stalcup (1994). Magnetic polarity data suggest that volcanism occurred during both the Brunhes normal (0.7 Ma to the present) and Matuyama reversed (2.4–0.7 Ma) polarity chrons. Whole-rock K-Ar dates (Turner, 1986, reported in Moll-Stalcup, 1994) from several of the basal flows within the field range from 3.25 and 2.80 Ma. Relatively few geochemical analyses of Saint Michael volcanic field rocks have been published. In a study of Nunivak Island volcanic rocks, Mark (1971) reported Sr isotopic values ($^{87}Sr/^{86}Sr = 0.7026–0.7033$) for rocks from Nunivak Island and several other Bering Sea volcanic fields, including the Saint Michael volcanic field. Major element analyses from early mapping in the region were described by Moll-Stalcup (1994, 1995). The Saint Michael volcanic field is in the Yukon-Koyukuk terrane, which consists largely of subduction-related Jurassic and Cretaceous sedimentary and igneous rocks (Patton and Box, 1989).

The Saint Michael volcanic field extends from the Kogok and Golsovia Rivers in the east to the nearshore Saint Michael and Stuart Islands in the west. We studied three localities in the field, Saint Michael Island, Stuart Island, and Klikitarik Bay (east of Saint Michael Island). Volcanic rocks in the field were classified as tholeiite, olivine tholeiite, alkali basalt, alkali-olivine basalt, and basanite. In general, the flows and large, low-lying shields are characterized by less alkalic compositions, whereas the alkalic rocks form maar craters or steep-sided cones. Volcanic flows are typically <5 m thick and display pahoehoe or aa flow tops and columnar-jointed interiors. Steep-sided cones and flows with alkalic compositions contain abundant mantle and crustal xenoliths.

Rocks from the Saint Michael volcanic field vary from tholeiitic (tholeiite and alkali-olivine basalt; Stuart and Saint Michael Islands) to moderately alkaline (basanite and nephelinite of Klikitarik Bay). Igneous textures range from intersertal to intergranular and subophitic (Grandy, 1997). Most Saint Michael volcanic field samples are phenocryst poor; typically the groundmass and glass content is between 85 and 95 vol%. Olivine (Fo_{79} to Fo_{87}) is the most common mineral in rocks from the Saint Michael volcanic field. Olivine xenocrysts (Fo_{86} to Fo_{89}) are also present and are distinguished by abundant sub-grain boundaries and irregular shapes. Plagioclase (An_{72} to An_{60}) is present in the groundmass of rocks from all areas of the Saint Michael volcanic field and is the most common phenocryst phase along with olivine in samples from Stuart Island. Clinopyroxene is relatively rare (3–7 vol%) and fills interstices between plagioclase laths of various size in an intergranular mesostasis. Volcanic glass (1–4 vol%) is preserved in various stages of devitrification.

Kookooligit volcanic field (Saint Lawrence Island)

The Kookooligit volcanic field is located on north-central Saint Lawrence Island (Fig. 1) and consists of a single large shield volcano (>500 m elevation) that is dotted with more than 100 small (20–60-m-high) cones (Patton and Csejtey, 1980; Moll-Stalcup, 1994; Till and Dumoulin, 1994) that occur along an east-west ridge. Regional mapping and geochemical surveys indicate that the oldest flows are olivine tholeiite to alkali-olivine basalt, whereas younger flows and cones are primarily alkali-olivine basalt and basanite with minor hawaiite and nephelinite (Moll-Stalcup, 1994, 1995). Crustal (granite, granulite, mafic cumulates) and mantle (spinel lherzolite, dunite, wehrlite) xenoliths are abundant in the more alkaline cones and flows. Megacrysts of anorthoclase and amphibole (kaersutite) commonly occur with the xenoliths. Four flows from the Kookooligit volcanic field yielded K-Ar ages of 1.46–0.0238 Ma (Patton and Csejtey, 1980). The volcanic field overlies volcanic, plutonic, sedimentary, and metamorphic rocks of early Paleozoic to Tertiary age that are broadly correlative with those of the Seward and Arctic Alaska terranes (Till and Dumoulin, 1994). Petrographic features of the Kookooligit flows are similar to those of the Saint Michael volcanic field. Olivine and clinopyroxene phenocrysts are present in most flows and are joined by plagioclase in the tholeiites.

GEOCHEMISTRY OF VOLCANIC ROCKS

Analyses of representative samples from the three volcanic fields studied are presented in Table 1. Geochemical data for the complete sample suite are given in the Appendix.[1] The volcanic rocks are relatively unaltered, as evidenced by the general absence of alteration minerals in thin section and by the low loss-on-ignition values (LOI < 1.00 wt%) of most lavas. The rocks are classified mostly as basalt and rare basaltic andesite according to the total alkali versus silica system (Le Bas et al., 1986). Basaltic rocks are further subdivided according to normative mineral compositions. An Fe_2O_3/FeO ratio of 0.15 was assumed for calculation of magnesium numbers ($Mg\# = Mg/[Mg + Fe^{+2}]$) and normative mineral compositions (CIPW norms). Basalt having normative hypersthene is classified as tholeiite or as olivine tholeiite if it contains >10% normative olivine. Alkali-olivine basalt contains normative nepheline and >10% normative olivine. Basanite contains 10%–20% normative nepheline. Nephelinite (>20% normative nepheline) and hawaiite (<5 wt% MgO and >5 wt% $Na_2O + K_2O$) were not observed in this study.

The late Cenozoic volcanic rocks of the Teller, Saint Michael, and Kookooligit volcanic fields range from moderately hypersthene–normative tholeiite and basaltic andesite to nepheline–normative basanite. Flows from Saint Michael Island and Stuart Island (Saint Michael volcanic field) are the least undersaturated and are classified as tholeiite, olivine tholeiite, and alkali-olivine basalt (Fig. 2), whereas some basalts from the Kookooligit and Teller volcanic fields are the most undersaturated (olivine tholeiite, alkali-olivine basalt, and basanite). Flows in the Teller and Kookooligit volcanic fields are consistently less undersaturated (olivine tholeiite to alkali-olivine basalt) compared with volcanic cones (alkali-olivine basalt and basanite) in the same fields.

The compositions of the Bering Sea volcanic rocks form coherent trends on most major element plots. Titanium, magnesium, and iron decrease with increasing Si, whereas Al and Ca exhibit little variation or increase slightly with increasing Si (Fig. 2). Sodium, potassium, and phosphorous abundances exhibit two trends, one that is negatively correlated with increasing Si (Fig. 2), and one that is positively correlated with increasing silica. Silica-undersaturated basalts from the Saint Michael volcanic field (Klikitarik Bay region), Kookooligit, and Teller (cones) regions typically plot along the positively correlated trends, whereas olivine and nepheline-hypersthene normative basalts (Stuart Island, Saint Michael Island, and Teller volcanic field flows) plot along negatively correlated trends (Fig. 2). Plots of incompatible elements (Rb, Ba, Th, Nb, REE, Zr, Y) versus Si exhibit trends similar to those exhibited by Na, K, and P.

[1]GSA Data Repository item 2002077, Appendixes 1 and 2 are available on request from Documents Secretary, GSA, P.O. Box 9140, Boulder, CO 80301-9140, USA, editing@geosociety.org, or at www.geosociety.org/pubs/ft2002.htm, or on the CD-ROM accompanying this volume.

TABLE 1. MAJOR AND TRACE ELEMENT COMPOSITIONS OF REPRESENTATIVE BERING SEA VOLCANICS

	Teller volcanic field						Saint Michael volcanic field — Klikitarik Bay					Stuart Island			
	BM-06a	BM-06b	BM-06e	BM-09a2	BM-10	BM-11b	KB-03	KB-04c	KB-06	KB-07	KB-11a	SI-02a	SI-06a	SI-06f	SI-08a
	ol thol	aob	ol thol	aob	bas	bas	aob	bas	aob	bas	bas	thol	ol thol	thol	thol
Major Elements (wt%)															
SiO_2	49.21	46.69	48.74	45.17	44.79	46.01	48.04	46.44	45.07	44.78	43.17	50.68	49.88	53.54	52.56
TiO_2	1.97	2.44	2.05	2.52	2.77	2.75	2.68	2.83	2.95	3.01	3.33	1.88	1.93	1.38	1.62
Al_2O_3	14.71	14.28	14.55	14.46	14.52	15.36	16.19	16.30	14.70	15.16	15.08	15.43	15.07	15.29	15.19
$Fe_2O_3^{T\dagger}$	12.49	12.69	12.67	12.31	12.88	12.70	12.28	12.95	12.49	12.45	13.84	11.16	10.86	9.66	10.17
MnO	0.20	0.18	0.17	0.16	0.18	0.19	0.17	0.19	0.17	0.18	0.19	0.16	0.16	0.13	0.14
MgO	8.77	10.50	9.07	10.23	9.07	7.78	6.24	5.81	9.48	8.95	7.98	6.75	8.49	7.80	8.12
CaO	8.74	9.07	8.83	9.57	9.01	8.50	9.56	7.41	9.47	9.47	9.25	9.75	9.50	9.11	8.95
Na_2O	3.27	3.31	3.20	3.46	4.26	4.56	3.49	5.37	3.56	4.15	4.32	3.32	3.29	3.05	3.10
K_2O	0.75	1.19	0.88	1.32	2.03	2.12	1.29	2.44	1.50	1.84	1.93	0.93	1.16	0.36	0.46
P_2O_5	0.25	0.42	0.30	0.44	0.56	0.67	0.47	0.82	0.53	0.68	0.83	0.29	0.34	0.12	0.19
LOI	0.64	0.00	0.34	1.13	0.62	0.00	0.00	0.00	0.00	0.00	0.00	0.00	0.00	0.00	0.00
Total	101.06	100.82	100.86	100.83	100.72	100.67	100.43	100.57	99.96	100.70	99.95	100.39	100.72	100.48	100.54
Mg#††	0.61	0.65	0.61	0.65	0.61	0.57	0.53	0.50	0.63	0.61	0.56	0.57	0.63	0.64	0.64
CIPW Norms (wt%)[§]															
Ne		4.62		8.58	14.81	13.40	2.0	1	15.57	8.95	13.70	15.89			
Hy	6.12		3.13										8.72	1.09	23.30	24.56
Ol	16.08	22.22	18.63	20.64	18.12	16.63	14.38	14.44	19.09	17.45	16.98	8.17	16.54	0.00	0.00
Trace Elements (ppm)															
La	12.0	17.4	12.9	20.9	27.9	33.3	20.4	46.6	25.5	35.1	41.1	15.1	21.6	5.09	7.41
Ce	26.9	39.6	29.2	45.6	61.4	70.8	40.0	84.2	47.9	64.6	77.3	28.8	39.8	10.6	15.1
Pr							4.88	9.39	5.68	7.36	8.75	3.50	4.55	1.56	2.09
Nd	16.5	22.2	18.0	25.8	33.4	36.0	22.2	37.0	24.2	30.4	36.3	15.8	19.5	8.32	10.6
Sm	4.75	5.62	4.92	6.22	7.26	7.59	6.01	8.42	6.18	7.14	8.60	4.58	5.27	3.22	3.84
Eu	1.49	1.83	1.53	1.83	2.18	2.29	2.16	2.85	2.18	2.38	2.87	1.63	1.79	1.29	1.55
Gd							6.02	7.59	6.07	6.70	7.91	4.94	5.29	3.84	4.51
Tb	0.79	0.91	0.81	0.95	0.98	1.05	0.98	1.15	0.95	1.02	1.19	0.83	0.85	0.66	0.76
Dy							5.38	6.24	5.35	5.69	6.45	4.86	4.90	3.78	4.34
Ho							0.95	1.13	0.97	1.05	1.15	0.90	0.95	0.72	0.82
Er							2.33	2.78	2.37	2.57	2.77	2.29	2.31	1.76	2.05
Tm							0.30	0.38	0.32	0.34	0.37	0.31	0.32	0.24	0.27
Yb	1.98	1.96	1.93	2.26	1.65	2.26	1.69	2.27	1.82	1.97	2.03	1.76	1.93	1.36	1.56
Lu	0.26	0.26	0.26	0.31	0.21	0.30	0.26	0.34	0.27	0.29	0.31	0.27	0.30	0.20	0.23
Cs	0.09	0.19	0.22	0.28	0.39	0.25	0.12	0.34	0.20	0.25	0.27	0.39	0.43	0.30	0.25
Rb	9	11	11	14	22	21	11	33	16	20	25	16	20	7	7
Ba	151	109	132	182	219	252	188	401	200	269	340	166	265	93	103
Th	1.42	1.69	1.43	2.18	3.07	3.53	1.71	4.66	2.44	3.33	3.77	2.25	2.70	0.75	1.00
U	0.36	0.62	0.50	0.69	1.1	1.4	0.46	1.7	0.90	1.2	1.3	0.84	0.93	0.28	0.35
Nb	17	31	21	33	45	55	37	82	47	62	71	22	30	5	9
Ta	0.96	1.9	1.2	2.2	3.1	3.9	2.3	5.2	3.0	4.1	4.5	1.4	2.0	0.30	0.58
Sr	314	455	363	498	640	719	620	1080	619	781	883	337	445	200	226
Pb							1.2	2.7	1.5	2.0	2.0	2.4	1.6	1.0	1.3
Hf	3.1	4.0	3.3	4.2	5.5	5.9	4.2	6.7	4.4	5.2	5.7	3.2	3.8	2.3	2.8
Zr	127	164	134	172	220	244	174	334	192	238	258	138	156	83	111
Y	26	27	26	28	24	29	26	31	26	28	30	23	24	19	21
Cr	424	409	435	443	258	252	106	166	166	301	290	326	327	135	
Ni	252	244	279	227	205	140	70	73	211	183	117	111	195	193	230
Co	59	56	56	55	56	46	43								
Sc	23	24	23	27	18	19	25	17	27	27	23	28	26	24	24
V	177	205	182	249	241	216	245	203	262	268	278	209	205	134	141

TABLE 1. MAJOR AND TRACE ELEMENT COMPOSITIONS OF REPRESENTATIVE BERING SEA VOLCANICS (continued)

	Saint Michael volcanic field									Kookooligit volcanic field						
	Stuart Island			Saint Michael Island												
	SI-11a	SI-11b	SI-14	SM-02b	SM-03b	SM-06a	SM-06b	SM-08	SM-09	SV-05a	SV-08g	SV-08z	SV-15	SV-16	SV-20	SV-21a
	aob	aob	aob	aob	thol	thol	aob	aob	thol	bas	bas	aob	aob	aob	ol thol	bas
Major Elements (wt%)																
SiO_2	47.17	47.73	48.36	48.24	51.56	51.76	47.50	46.74	51.66	45.97	43.03	44.54	47.89	46.05	48.63	49.13
TiO_2	1.89	2.18	2.05	1.98	1.65	1.63	2.38	2.06	1.89	2.48	3.44	2.72	2.34	2.62	1.89	2.10
Al_2O_3	14.54	15.59	15.15	15.52	15.50	15.60	14.96	15.10	16.01	14.70	14.06	13.19	13.89	13.35	13.96	16.18
$Fe_2O_3^{T\dagger}$	11.37	11.10	11.78	11.68	10.22	9.75	11.75	11.64	10.44	11.96	13.73	12.93	12.82	12.84	13.09	11.23
MnO	0.17	0.17	0.17	0.17	0.15	0.14	0.17	0.18	0.15	0.18	0.16	0.17	0.16	0.17	0.17	0.16
MgO	11.21	8.21	9.00	8.27	7.78	7.39	8.92	9.63	5.98	8.15	9.02	10.65	9.35	10.38	9.62	5.36
CaO	9.25	9.75	9.25	9.19	9.33	9.48	9.38	9.56	9.73	8.24	8.65	9.46	8.82	8.91	8.69	6.55
Na_2O	2.83	3.15	3.24	3.51	3.22	3.22	3.45	3.31	3.44	4.36	4.10	3.45	3.31	3.27	3.20	5.41
K_2O	1.31	1.54	1.09	1.13	0.63	0.62	1.57	1.37	1.19	2.34	2.36	1.56	0.95	1.29	0.73	2.54
P_2O_5	0.40	0.46	0.36	0.36	0.20	0.19	0.48	0.40	0.31	0.64	0.62	0.57	0.38	0.47	0.26	0.86
LOI	0.00	1.00	0.00	0.00	0.00	0.19	0.00	0.00	0.00	0.00	0.00	0.00	0.00	0.00	0.00	0.00
Total	100.20	100.92	100.50	100.08	100.29	100.02	100.61	100.04	100.83	99.04	99.19	99.28	99.95	99.40	100.29	99.54
$Mg\#^{\dagger\dagger}$	0.68	0.62	0.63	0.61	0.63	0.63	0.63	0.65	0.56	0.60	0.59	0.64	0.62	0.64	0.62	0.51
CIPW Norms (wt%)[§]																
Ne	2.42	3.04	1.08	2.48			5.12	5.64		12.83	16.38	9.88	0.74	4.73		10.66
Hy					17.36	18.80			9.18						3.62	
Ol	23.42	17.01	19.63	18.37	4.09	1.80	18.24	19.99	5.78	17.05	18.37	21.06	20.74	21.79	19.73	14.00
Trace Elements (ppm)																
La	24.1	27.5		22.1	10.8	10.3	27.4	26.4	20.1	66.4	39.8	42.1	21.0	29.6	14.0	53.4
Ce	44.0	50.4		40.3	21.2	19.8	51.4	47.3	38.3	117.3	74.8	78.9	42.1	57.5	28.9	98.7
Pr	5.09	5.78		4.68	2.74	2.60	5.94	5.29	4.48	12.63	8.67	8.91	5.12	6.70	3.62	10.6
Nd	20.9	24.4		20.1	13.1	12.3	25.1	22.3	19.0	47.7	36.1	36.9	22.6	28.8	16.4	41.8
Sm	5.10	5.95		5.30	4.25	4.19	6.34	5.47	5.05	9.55	7.85	8.01	5.74	6.85	4.56	8.51
Eu	1.75	1.93		1.90	1.61	1.56	2.19	1.83	1.74	3.06	2.61	2.73	2.00	2.41	1.60	2.84
Gd	4.96	5.58		5.69	4.93	4.67	6.19	5.43	5.25	7.69	6.61	7.12	5.61	6.38	4.85	6.95
Tb	0.78	0.87		0.90	0.79	0.78	0.98	0.87	0.85	1.15	0.95	1.07	0.87	0.99	0.78	1.01
Dy	4.41	4.96		5.12	4.51	4.51	5.26	5.02	4.87	6.04	4.90	5.75	4.87	5.21	4.53	5.22
Ho	0.82	0.93		0.98	0.87	0.84	0.98	0.95	0.91	1.05	0.82	1.00	0.90	0.96	0.87	0.91
Er	2.02	2.29		2.46	2.15	2.00	2.37	2.36	2.33	2.32	1.72	2.26	2.10	2.16	2.16	2.07
Tm	0.28	0.32		0.33	0.29	0.27	0.31	0.32	0.31	0.29	0.20	0.28	0.27	0.27	0.28	0.27
Yb	1.67	1.80		1.98	1.68	1.65	1.84	1.94	1.87	1.57	1.07	1.52	1.60	1.54	1.67	1.47
Lu	0.25	0.28		0.30	0.25	0.24	0.27	0.29	0.28	0.22	0.15	0.22	0.22	0.23	0.24	0.22
Cs	0.34	0.58		0.22	0.38	0.39	0.22	0.40	0.81	0.64	0.36	0.32	0.14	0.23	0.09	0.49
Rb	18	21	16	17	12	11	20	21	23	56	36	29	17	23	13	54
Ba	291	333		264	152	151	307	299	284	747	589	475	271	370	193	782
Th	2.67	2.95		2.59	1.56	1.62	3.06	3.29	2.81	6.23	3.12	3.28	1.68	2.54	1.42	4.54
U	0.89	1.1		0.86	0.52	0.47	0.82	1.1	1.0	1.9	1.0	0.99	0.52	0.79	0.43	1.4
Nb	35	41	32	32	12	12	41	39	25	84	53	53	27	40	18	66
Ta	2.3	2.7		2.0	0.78	0.77	2.8	2.6	1.7	5.5	3.5	3.4	1.7	2.5	1.1	4.1
Sr	538	560	460	437	285	273	587	519	435	972	876	779	537	650	377	1050
Pb	2.7	3.6		2.5	2.3	2.0	2.9	3.7	3.5	5.5	3.1	3.2	1.9	2.5	1.7	4.2
Hf	3.7	4.3		3.6	2.9	3.0	4.3	3.9	3.7	7.8	5.5	5.3	3.9	4.6	3.1	6.5
Zr	161	188	163	152	112	114	178	166	148	362	252	231	167	195	134	314
Y	21	24	24	25	23	21	24	24	23	26	20	25	23	24	23	24
Cr										166	166	301	290	326	327	135
Ni	290	162	220	198	187	195	206	242	85	165	180	248	235	263	262	87
Co										47	62	56	57	59	59	33
Sc	26	29		28	26	22	24	27	25	17	15	22	23	24	24	14
V	212	230	189	195	165	163	224	225	212	211	277	253	227	240	214	154

Note: LOI is loss on ignition.

[†]$Fe_2O_3^T$ is all Fe as Fe_2O_3.

[††]Mg# is the cation wt % Mg/(Mg + Fe^{+2}) assuming $Fe^{+3}/Fe^{+2} = 0.15$.

[§]CIPW Norms calculated assuming assuming $Fe^{+3}/Fe^{+2} = 0.15$.

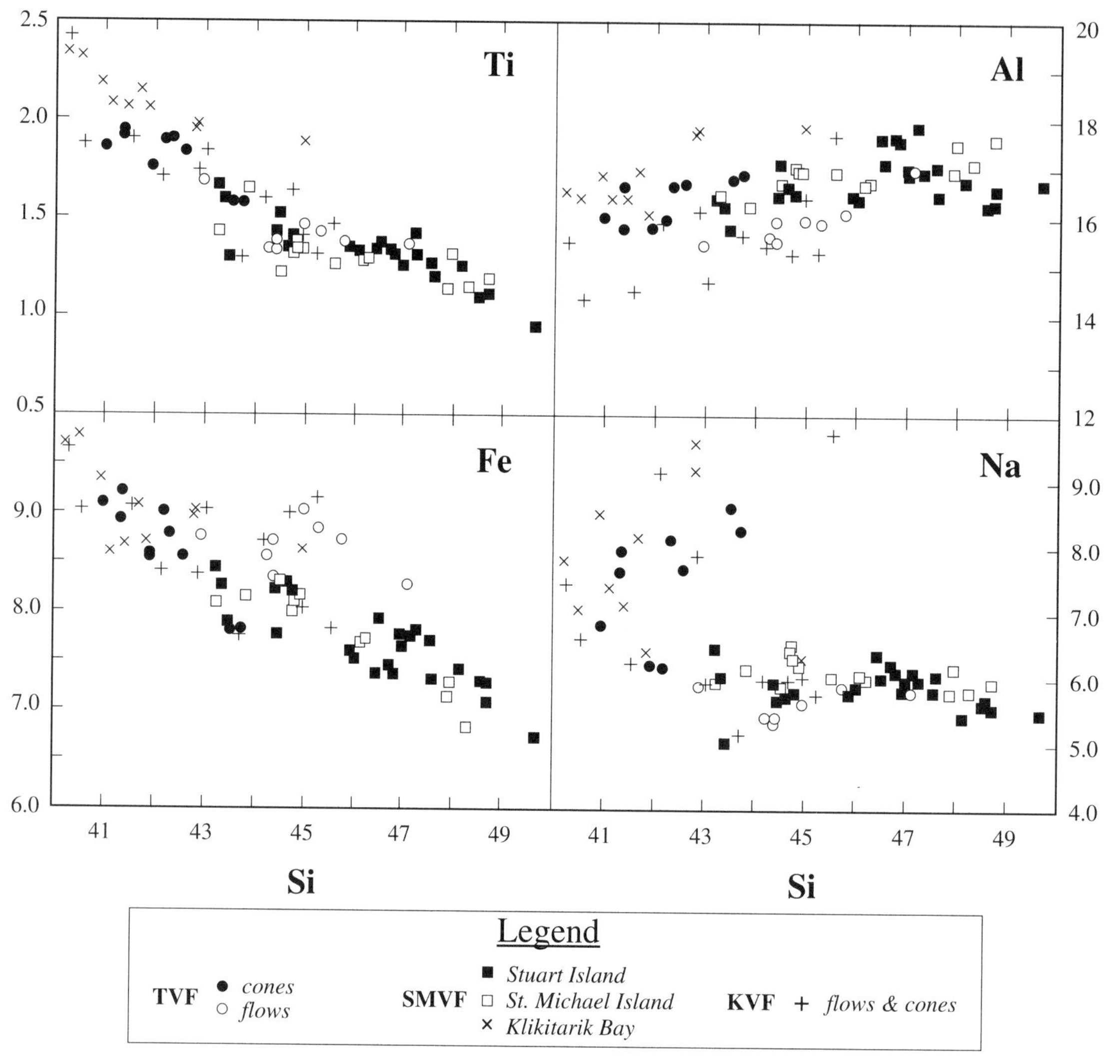

Figure 2. Harker-type diagrams of Si (cation wt%) vs. Ti, Al, Fe, and Na (cation wt%) from Teller volcanic field (TVF), Saint Michael volcanic field (SMVF), and Kookooligit volcanic field (KVF).

Compatible element concentrations (e.g., Cr, Ni) of Bering Sea volcanics are fairly well correlated with Mg#; however, many elements exhibit less coherent trends or plot along two sub-parallel trends (Fig. 3). Ti, Fe, K, and other incompatible elements (Rb, Ba, Th, Nb, REE, Zr, Y) form high (Teller, Klikitarik, and Kookooligit) and low (Stuart Island, Saint Michael Island) groups that exhibit increasing concentrations with decreasing Mg#. Furthermore, plots of ratios of highly incompatible trace elements with less incompatible elements (e.g., Zr/Y; Fig. 3) versus Mg# also exhibit similar trends.

Incompatible trace element abundances are well correlated with silica undersaturation in the Bering Sea volcanic rocks. Tholeiites and olivine tholeiites from Saint Michael Island, Stuart Island, and the Teller volcanic field have the lowest trace element abundances and the least light REE-enriched patterns (Fig. 4). In contrast, nepheline-normative basanites from Saint Lawrence Island and Klikitarik Bay exhibit the greatest light REE enrichment. La/Yb ratios also increase with increasing silica undersaturation. Many samples from Saint Michael Island, Stuart Island, Klikitarik Bay, and Saint Lawrence Island exhibit REE patterns with unusual sigmoidal shapes (Fig. 4); light REE patterns are concave up, whereas middle REE patterns are concave down. The shapes of these patterns were confirmed by analyzing a number of samples using both INAA and ICP-MS techniques. Flows and cones of the Teller region apparently lack the sigmoidal REE patterns that are characteristic of the other volcanic fields of the Bering Sea region (Fig. 4).

Trace element patterns of the Bering Sea province basalts indicate that significant differences exist within and between fields (Fig. 5). Most of the flows have broadly concave-downward patterns, and Nb and Ta have the highest mantle-normalized values. However, the degree of Ta and Nb enrichment, as well as the

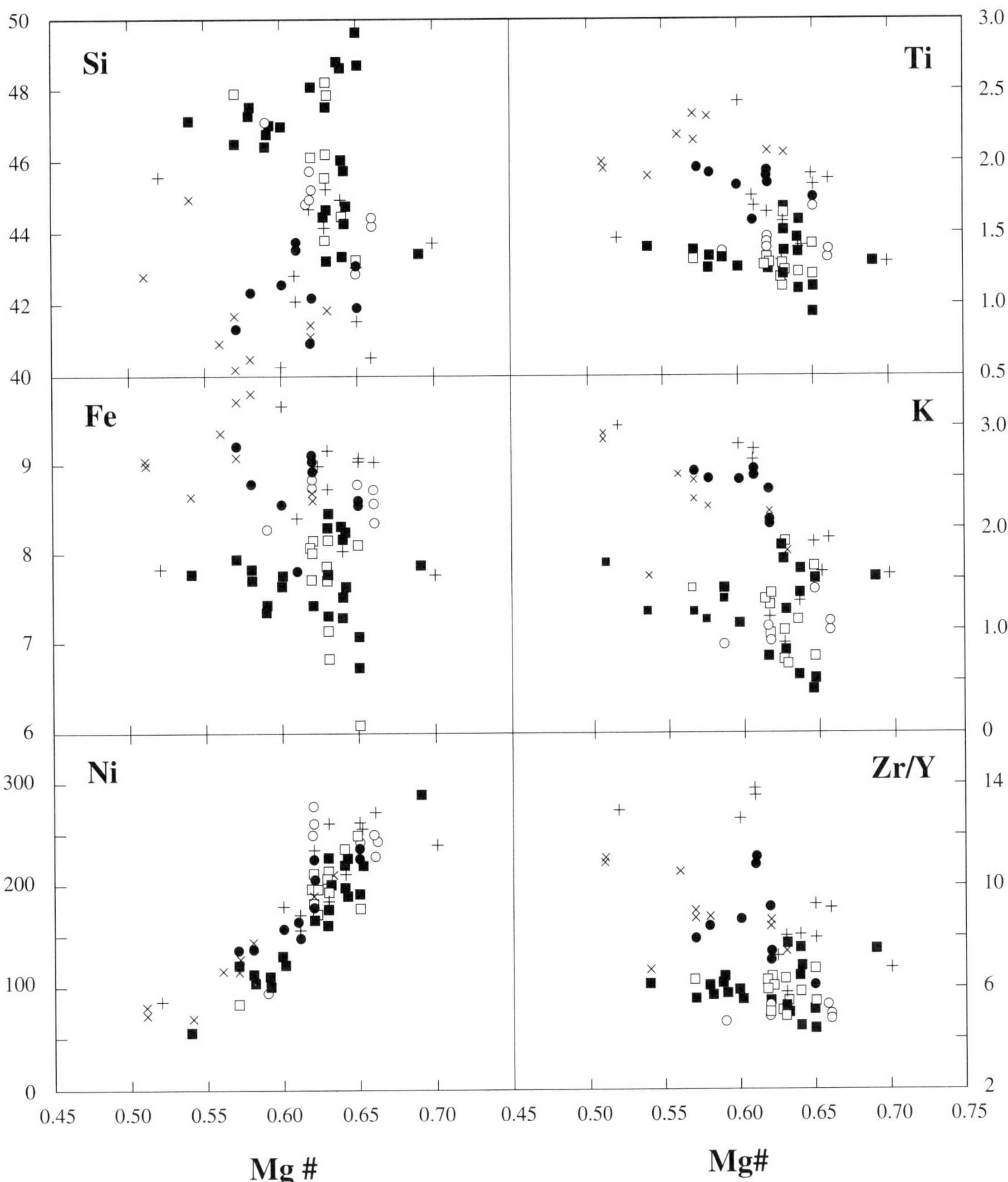

Figure 3. Plot of Mg# vs. Si, Ti, Fe, K, Ni, and Zr/Y in Bering Sea province volcanic rocks. Symbols as in Figure 2.

ratios of some other trace elements (e.g., Rb/U, Ba/Zr, Th/La, and Nb/La), varies considerably within and between volcanic fields. The least undersaturated flows (tholeiites and olivine tholeiites) from each volcanic field generally have the lowest Ba/Zr and Nb/La ratios, as do the flows relative to cones within the same field. Furthermore, lavas from the Kookooligit volcanic field have the highest Rb/U and Ba/Zr, and the lowest Th/La and Nb/La. In contrast, flows from Klikitarik Bay have low Rb/U, Ba/Zr, and Th/La ratios, and high Nb/La.

XENOLITHS

Abundant mantle and crustal xenoliths occur in the more silica-undersaturated flows and scoria cones of several of the Bering Sea volcanic fields. Mantle xenoliths from Nunivak Island were studied extensively by Francis (1976a, 1976b, 1978) and Roden et al. (1984, 1995). Swanson et al. (1987) reported equilibration temperatures for peridotites in their review of mantle xenoliths in the Bering Sea region. Xenoliths of the Enmelen

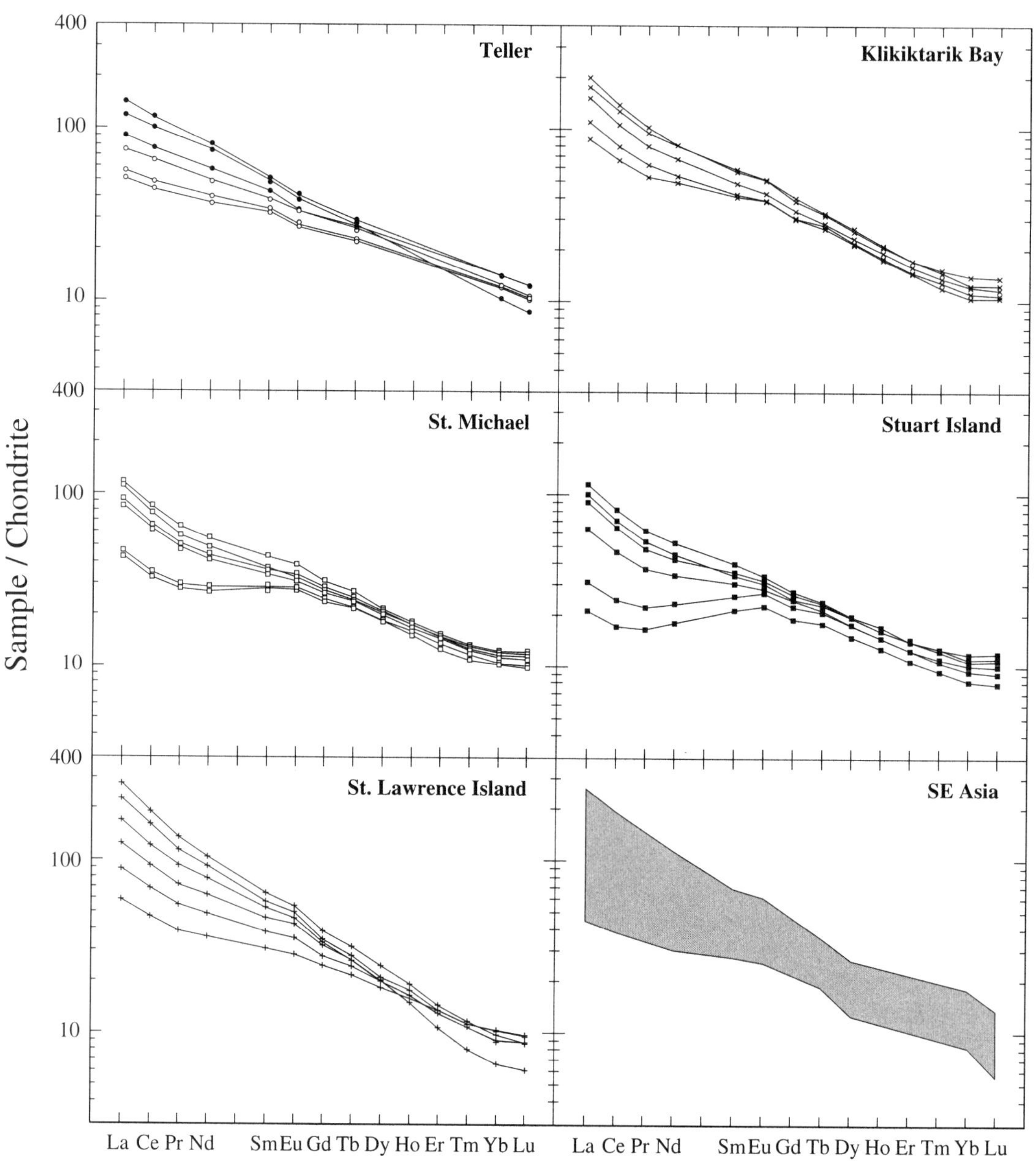

Figure 4. Rare earth element patterns of volcanic rocks from Teller, Saint Michael (Klikitarik Bay, Saint Michael, and Stuart Island regions), and Kookooligit (Saint Lawrence Island) volcanic fields. Chondrite normalization values are from McDonough and Sun (1995). Southeast Asia volcanic data are from Hoang et al. (1996).

volcanic field on the Chukchi Peninsula were described by Akinin et al. (1997). In this study we examined mantle and crustal xenoliths from the Kookooligit volcanic field on Saint Lawrence Island and the Klikitarik Bay region of the eastern Saint Michael volcanic field (Fig. 1).

Mantle xenoliths

Mantle xenoliths are abundant in silica-undersaturated flows and cones of the Kookooligit Mountains and near Klikitarik Bay (Saint Michael volcanic field). The xenoliths at both localities are generally small (<5 cm diameter) and consist mostly of spinel lherzolite of probable mantle origin. Other ultramafic xenoliths (harzburgite, dunite, olivine websterite, pyroxenite, websterite, and titaniferous pargasite and kaersutite-bearing pyroxenites) and megacrysts (anorthoclase, pyroxene, amphibole) likely originated within the upper mantle or lower crust (Wirth et al., 1995) and are not discussed further in this chapter.

The spinel lherzolites of Saint Lawrence Island and Klikitarik Bay (Saint Michael volcanic field) are fine to medium

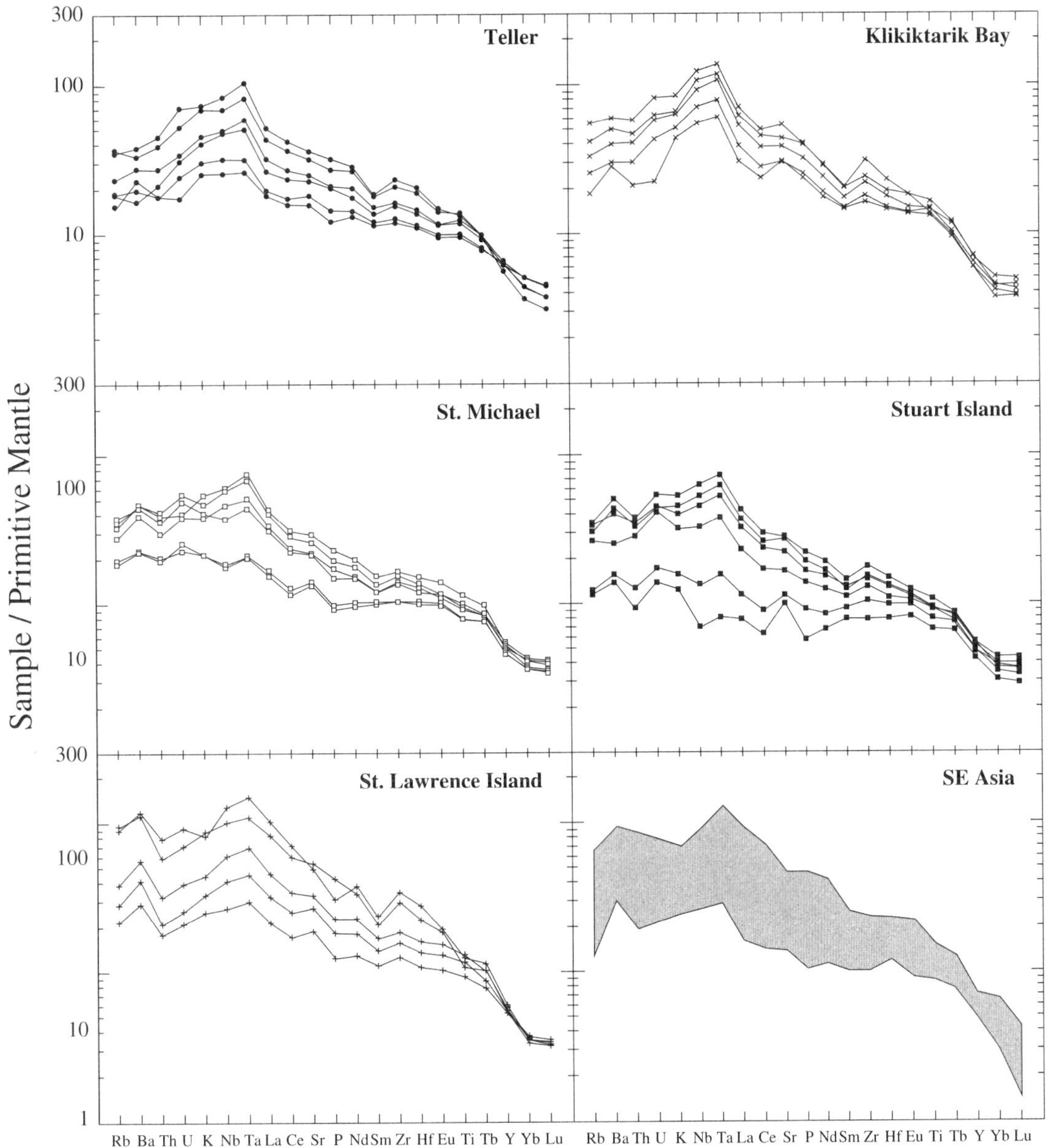

Figure 5. Plot of trace element patterns of volcanic rocks from Teller, Saint Michael (Klikitarik Bay, Saint Michael, and Stuart Island regions), and Kookooligit (Saint Lawrence Island) volcanic fields. Mantle normalization values are from McDonough and Sun (1995). Southeast Asia volcanic data are from Hoang et al. (1996).

grained and are characterized by two different textures: granuloblastic equant and coarse equant (Kelley, 1997). Granuloblastic-equant xenoliths collected from Klikitarik Bay are fine grained (0.5–1.0 mm) and exhibit evidence of deformation (foliation and alignment of elongate spinel grains) and recrystallization (equant olivine and pyroxene grains with polygonal boundaries). These xenoliths are also characterized by more abundant spinel and by the presence of large (<2 mm), deformed, and kinked porphyroclasts of orthopyroxene. This textural category corresponds to the equigranular mosaic texture of Mercier and Nicolas (1975). Lherzolite xenoliths with coarse-equant textures (protogranular texture of Mercier and Nicolas, 1975) occur primarily on Saint Lawrence Island and are characterized by coarse (2–5 mm), anhedral, olivine and pyroxene grains. The xenoliths generally lack a well-defined foliation, but olivine and orthopyroxene grains exhibit undulatory extinction and well-developed deformation bands. Spinel grains are generally less abundant and are rimmed by symplectic intergrowths of Cr-rich spinel and plagioclase, suggesting that recrystallization has occurred in some xenoliths.

The compositions of forsteritic olivine, chromian diopside, enstatite, and spinel in the xenoliths (Table 2) vary with texture (Kelley, 1997). Olivine compositions in granuloblastic-equant xenoliths range from Fo_{85} to Fo_{92}, whereas those with coarse-equant textures exhibit more limited compositions (Fo_{89} to Fo_{92}). Clinopyroxene grains of both texture types have compositions near end-member diopside, but clinopyroxenes in coarse-equant xenoliths have considerably higher Cr contents (0.97–1.66 wt% Cr_2O_3) than do those from the granuloblastic-equant xenoliths (0.19–0.86 wt% Cr_2O_3). Orthopyroxene in both texture groups have compositions near the enstatite-bronzite boundary ($En_{86-92}Fs_{7-13}Wo_1$) and those in coarse-equant xenoliths have greater alumina at a given magnesium content than in granuloblastic-equant xenoliths. Spinels are moderately aluminous in both texture groups (64.75–38.67 wt% Al_2O_3) and aluminum content is inversely correlated with chromium content (8.38–28.71 wt% Cr_2O_3, respectively); granuloblastic-equant xenoliths are characterized by high aluminum and low chromium, whereas those with coarse-equant textures have lower aluminum and higher chromium. All of the spinel analyses are similar to those of abyssal peridotite, and those from the granuloblastic xenoliths (Klikitarik Bay) are the least depleted. All of these relationships suggest that lherzolites with coarse-equant textures have undergone greater degrees of melting than granuloblastic-equant lherzolites. Although no hydrous minerals (e.g., mica or amphibole) were observed in the any of the lherzolite xenoliths, the symplectic intergrowths (plagioclase + Cr spinel) surrounding spinel in the coarse-equant lherzolites suggest that some metasomatism (Dautria and Girod, 1987; Ionov et al., 1994) may have occurred. Equilibration temperatures are similar for both texture groups, but there are systematic differences in calculated temperature depending on the method used. The spinel-olivine geothermometer of Sack and Ghiorso (1991) yields temperatures from 800 to 1030 °C, which are consistently lower than those calculated using the two-pyroxene method (870–1125 °C) of Brey and Köhler (1990) or the Al-in-clinopyroxene geothermometer (1000–1200 °C) of Gasparik (1984). All temperatures were calculated assuming a pressure of 1.5 GPa, representative of the spinel-lherzolite stability field. The temperatures calculated using pyroxene thermometers are similar to those calculated for xenoliths of other Bering Sea volcanic fields (e.g., Francis, 1978; Swanson et al., 1987; Akinin et al., 1997).

Most of the spinel-lherzolite xenoliths that we analyzed for bulk chemistry are from the Klikitarik Bay locality (Saint Michael volcanic field), and the number is limited because of their small size. The bulk compositions of the three granuloblastic-equant lherzolites (Table 2) are typical of spinel-lherzolite xenoliths from other northern Cordilleran alkaline volcanic centers (e.g., Francis, 1987) and similar to spinel lherzolites from continental and oceanic regions (Maaloee and Aoki, 1977), though they have noticeably lower MgO (36.84–37.23 wt% MgO compared with 42.21 wt% MgO in average spinel lherzolite) and higher $Fe_2O_3{}^T$ (9.39–12.07 wt%

$Fe_2O_3{}^T$ compared with 9.21 wt% $Fe_2O_3{}^T$ in average spinel). The compositions of these xenoliths closely resemble those of fertile upper mantle (pyrolite model; Ringwood, 1979) and models of primitive mantle (Hofmann, 1988; McDonough and Sun, 1995) (Table 2). The Bering Sea xenoliths differ from estimated compositions of fertile mantle by having slightly lower MgO, TiO_2, Al_2O_3, and Na_2O, and higher $Fe_2O_3{}^T$. REE patterns of the granuloblastic-equant lherzolites are characterized by nearly flat heavy REE patterns (Fig. 6), at about two times chondritic values, with enriched light REE (to six times chondritic values). Many of these lherzolites have normalized Pr/Tb values that are below chondritic values, suggesting that the lherzolites originally had light REE-depleted patterns and were subsequently enriched in the light REEs. Trace element patterns (normalized to primitive mantle) are relatively flat, but indicate that La, Ce, Nb, Ta, U, and Th are slightly enriched relative to the large-ion lithophile elements (LILE) and the less incompatible elements (Fig. 7). These REE and trace element patterns are similar to some of those described by Roden et al. (1984) for coarse-equant xenoliths from Nunivak Island. Geochemical studies (Roden et al., 1995) of fine- and coarse-grained lherzolites from Nunivak Island suggest that the xenoliths were derived from asthenospheric diapirs and metasomatized suboceanic lithosphere, respectively.

Amphibolite-pyroxenite xenoliths

Pyroxene-rich xenoliths, including olivine websterite, websterite, olivine clinopyroxenite, pyroxenite, and amphibole pyroxenite are also fairly common in the Bering Sea region, particularly at Klikitarik Bay (Saint Michael volcanic field; Fig. 1). The plagioclase- and pyroxene-rich xenoliths are typically medium grained and exhibit a wide range of textures. Most pyroxenite xenoliths have granular, interlocking or seriate textures, and mineral grains are anhedral and exhibit evidence of deformation and recrystallization (undulatory extinction, kinking, and deformation banding). A few xenoliths consist of small grains of orthopyroxene, olivine, and spinel enclosed within larger clinopyroxene in a poikilitic relationship. A small proportion of the xenoliths that we observed in the field (<2%) contain amphibole (to 50 vol%) that ranges in composition from pargasite to kaersutite. The amphibole occurs as small anhedral and interstitial grains to large (to 5-cm-diameter) poikilitic grains. Pyroxene grains in these xenoliths range from subhedral to anhedral, and are sometimes deeply embayed. Other pyroxenes are strongly zoned and have fine-grained reaction rims where they are in contact with amphibole. Amphibole also occurs along cleavage planes in clinopyroxene in some xenoliths that do not contain interstitial or poikilitic amphibole, suggesting that it formed as a result of a metasomatizing fluid. The amphibole pyroxenites have relatively elevated REE abundances (Table 2; La = 8–30 times chondrite values) and have flat to concave-up light REE patterns and sloping heavy REE patterns (Fig. 6). Trace element patterns of the amphibole pyroxenites

TABLE 2. MAJOR AND TRACE ELEMENT COMPOSITIONS OF MANTLE XENOLITHS FROM SAINT MICHAEL VOLCANIC FIELD (KLIKITARIK BAY)

	KB-10a lherz	KB-10b lherz	KB-10k lherz	KB-10n lherz	KB-10o lherz	KB-11b amph-pyx	KB-11j lherz	KB-11m amph-pyx	AVG	Primitive Mantle R	Primitive Mantle MS	Primitive Mantle H
SiO_2	44.64	44.14	44.55	44.79	44.93	42.89	45.99		44.56	45.1	44.92	45.56
TiO_2	0.10	0.10	0.11	0.12	0.16	2.78	0.11		0.50	0.2	0.20	0.18
Al_2O_3	3.15	2.36	3.41	3.56	3.46	8.59	3.64		4.02	3.3	4.44	4.06
$Fe_2O_3^T$	10.19	12.07	9.39	9.14	8.86	13.71	9.66		10.43	8.9	8.94	8.38
MnO	0.21	0.26	0.18	0.13	0.13	0.17	0.16		0.18	0.2	0.13	
MgO	37.67	36.87	37.23	38.15	37.53	19.11	36.84		34.77	38.1	37.80	37.78
CaO	2.85	3.14	3.31	2.95	3.33	11.12	3.16		4.27	3.1	3.54	3.21
Na_2O	0.30	0.28	0.30	0.28	0.36	1.64	0.29		0.49	0.4	0.36	0.33
K_2O	0.06	0.13	0.06	0.09	0.11	0.64	0.04		0.16	0.0	0.03	0.03
P_2O_5	0.02	0.02	0.02	0.02	0.03	0.07	0.01		0.03		0.02	
Cr_2O_3	0.25	0.30	0.27	0.26	0.22	0.07	0.00					
NiO	0.18	0.18	0.19	0.17	0.17	0.04	0.20					
LOI	0.00	0.00	0.00	0.00	0.00	0.00						
Total	99.62	99.84	99.02	99.66	99.29	100.82	100.10		99.41	99.3	100.38	99.53
Mg#	0.89	0.87	0.90	0.90	0.91	0.76	0.89					
La	1.21	1.43	1.07	0.85	1.44	6.60	0.58	1.94				
Ce	2.63	3.45	2.58	1.41	2.43	16.4	1.18	4.31				
Pr	0.33	0.47	0.32	0.17	0.27	2.46	0.16	0.60				
Nd	1.34	2.05	1.37	0.78	1.23	12.2	0.77	3.01				
Sm	0.36	0.54	0.40	0.27	0.38	3.89	0.27	1.07				
Eu	0.15	0.20	0.16	0.10	0.16	1.36	0.11	0.42				
Gd	0.44	0.50	0.47	0.41	0.49	4.18	0.37	1.36				
Tb	0.08	0.09	0.09	0.08	0.09	0.67	0.07	0.23				
Dy	0.55	0.57	0.62	0.55	0.64	3.65	0.53	1.44				
Ho	0.12	0.12	0.13	0.12	0.15	0.66	0.12	0.27				
Er	0.34	0.31	0.38	0.36	0.41	1.58	0.34	0.70				
Tm	0.05	0.05	0.06	0.05	0.06	0.20	0.05	0.09				
Yb	0.34	0.30	0.35	0.34	0.37	1.08	0.34	0.54				
Lu	0.05	0.05	0.06	0.06	0.06	0.16	0.06	0.09				
Cs	0.01	0.01	0.01	0.02	0.02	0.03	0.01	0.01				
Rb	0	1	0	1	1	2	0	1				
Ba	6	4	4	7	12	50	3	13				
Th	0.10	0.07	0.09	0.15	0.19	0.34	0.06	0.15				
U	0.05	0.02	0.04	0.08	0.09	0.12	0.03	0.06				
Nb	1	1	1	2	10	0.4	3					
Ta				0.08	0.12	0.67	0.03	0.22				
Sr	24	26	21	9	24	207	14	50				
Pb	0.2	0.2	0.2	0.	0.3	0.3	0.4	0.4				
Hf	0.14	0.16	0.17	0.19	0.25	2.2	0.22	0.65				
Zr		9	7				9					
Y	3	3	4	3	4	17	3	7				
Cr	2540	2990	2720	2560	2180	712						
Ni	1780	1780	1880	1720	1690	394	1960					
Co	104	106	105	87	93	74						
Sc	15	14	17	15	16	35	18	38				
V	70	67	76	62	71	272	84					

Note: See "Note" from Table 1. AVG = average Bering Sea spinel lherzolite; R = Ringwood (1979); pyrolite model; H = Hofmann (1988); primitive mantle; MS = McDonough and Sun (1995); silicate earth pyrolite model.

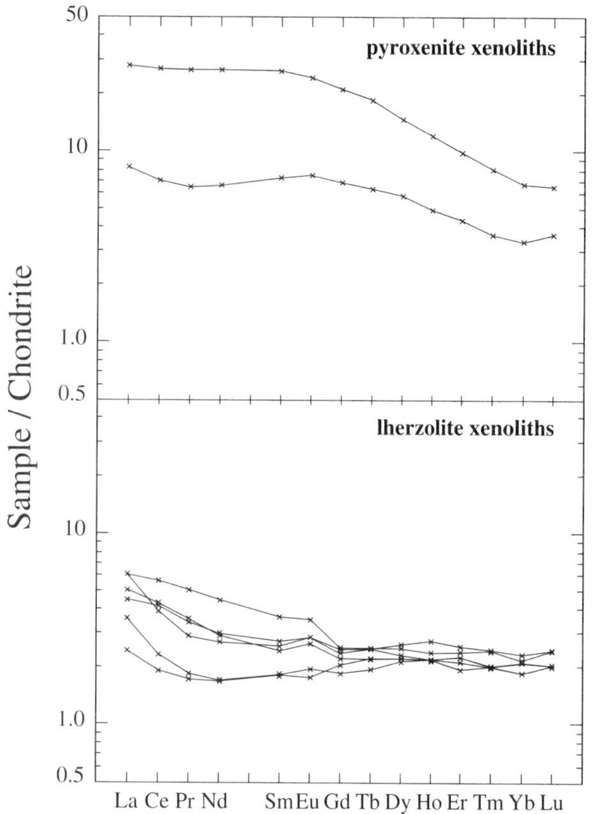

Figure 6. Rare earth element patterns of spinel lherzolite and amphibole-bearing pyroxenite xenoliths from Saint Michael volcanic field (Klikitarik Bay). Chondrite normalization values are from McDonough and Sun (1995).

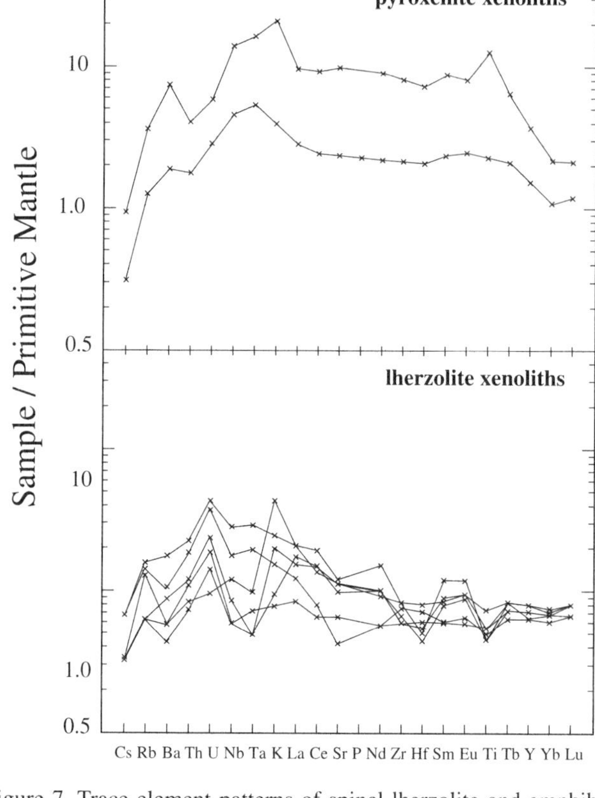

Figure 7. Trace element patterns of spinel lherzolite and amphibole-bearing pyroxenite xenoliths from Saint Michael volcanic field (Klikitarik Bay). Mantle normalization values are from McDonough and Sun (1995).

(Fig. 7) indicate that the xenoliths are enriched in the light REEs, K, Nb, Ta, U, Th, Ba, and Rb relative to primitive mantle.

The euhedral grain morphologies that are characteristic of the Klikitarik Bay pyroxenites and amphibole pyroxenites suggest they might be igneous in origin and that they are accidental xenoliths of ultramafic cumulates from the lower crust. This interpretation is supported by the presence of compositional layering that is present in one pyroxenite xenolith. A cumulate origin for these rocks would not be unreasonable given the considerable magmatism that has occurred in the region since the Cretaceous. However, several features of the xenoliths suggest that many of the pyroxene-rich xenoliths do not have a crustal origin. Pyroxene and olivine grains in some of the xenoliths exhibit evidence of deformation (undulatory extinction, kinking, deformation banding) and recrystallization (granoblastic textures), whereas interstitial and poikilitic amphibole is always undeformed. Pyroxene and olivine grains are commonly zoned and often have reaction rims with amphibole. Both of these features suggest that the amphibole postdates deformation and recrystallization of the olivines and pyroxenes. The common occurrence of crystallographically controlled zones of amphibole within pyroxene suggests that the amphibole originated by alteration of pyroxene with a hydrous fluid. If the xenoliths are of crustal origin (cumulates), then the deformed character of many pyroxene and olivine grains implies that they must have formed prior to the Late Cretaceous, the last major deformation event in the region. However, this seems unlikely because the trace element compositions of these xenoliths (Table 2) lack the signatures of suprasubduction-zone magmatism (e.g., high La/Nb) that are characteristic of the Cretaceous magmatism in the Bering Sea region (Moll-Stalcup and Arth, 1989; Miller, 1989; Moll-Stalcup, 1994; Amato and Wright, 1997). The trace element compositions of the pyroxenite xenoliths, especially those of the amphibole pyroxenites (Figs. 6 and 7), more closely resemble those of the late Cenozoic alkaline basalts of the Bering Sea province (Figs. 4 and 5). The textures and REE patterns of the Klikitarik Bay amphibole pyroxenite xenoliths are similar to those of coarse amphibole pyroxenite xenoliths from Nunivak Island (Roden et al., 1984). Textural and geochemical data from the Nunivak amphibole-bearing lherzolites and pyroxenites are interpreted to indicate that these xenoliths originated as precipitates from alkaline melts that were intruded into the lithospheric mantle (Francis, 1976a, 1976b; Roden et al., 1984, 1995).

Crustal xenoliths

Xenoliths of probable crustal origin are relatively rare in the Bering Sea province and were only observed at the Kookooligit volcanic field on Saint Lawrence Island (Fig. 1) during this study. These xenoliths consist mostly of medium- to coarse-grained plagioclase and orthopyroxene. Rare coarse-grained granite and fine-grained felsic volcanic rock are also present.

The plagioclase-orthopyroxene xenoliths range from medium to coarse grained and consist mostly of plagioclase (andesine) and lesser amounts of highly pleochroic orthopyroxene (hypersthene). Minor clinopyroxene (<15% diopside) is present with orthopyroxene in some xenoliths. Albite-rich perthite (mesoperthite) occurs with andesine in the most highly deformed xenoliths. Minor quartz, apatite, Fe-Ti oxides, and zircon are also present. The least-deformed xenoliths consist of coarse-grained plagioclase and pyroxene; some xenoliths also exhibit coarse compositional layering. Plagioclase grains in these xenoliths are subhedral and commonly enclose euhedral pyroxene in a poikilitic relationship. Other xenoliths are fine to coarse grained, heterogranular, and compositionally layered. Plagioclase grains are xenomorphic with undulatory extinction, and have kinked or bent twin lamellae. Pyroxenes are less commonly idiomorphic and are typically interstitial to feldspar grains. Small quartz and feldspar grains commonly occur along the boundaries of xenomorphic plagioclase grains. The most highly recrystallized xenoliths are medium grained and exhibit isogranular to polygonal granoblastic textures. The clinopyroxenes are characterized by relatively high Al^{VI} (0.03–0.3), a feature that is more commonly observed in clinopyroxenes from granulites (Aoki and Kushiro, 1968). Using the two-pyroxene geobarometer of Gasparik (1984), the Al contents of orthopyroxenes in the Saint Lawrence xenoliths indicate pressures of 0.4–0.6 GPa. On the basis of these textures and mineral composition data, we interpret the spectrum of igneous and metamorphic textures of the plagioclase-pyroxene xenoliths to record variable deformation of mafic igneous cumulates. U-Pb studies (SHRIMP) indicate crystallization ages ranging from 100 to 90 Ma, deformation and recrystallization occurring ca. 64 Ma (Miller et al., this volume, Chapter 10).

Pyroxene granulite xenoliths also occur on Nunivak Island (Francis, 1976c; Roden et al., 1994). These have corona textures and are olivine-plagioclase cumulates that recrystallized during isobaric cooling or increasing pressure in the lower crust (950 °C at >0.9 GPa; Francis, 1976c). The age of these xenoliths is unknown, but they are probably not related to middle to Late Cretaceous magmatism because they occur considerably south of the main axis of the surface expression of this belt. Major, trace element, and Sr and Nd isotopic data suggest that the Nunivak corona-bearing pyroxene granulites were derived from the underlying extended oceanic lithosphere (Roden et al., 1995).

The REE patterns of the mafic cumulate xenoliths are generally similar (e.g., subparallel and light REE enriched), but there are some significant differences (Table 3; Fig. 8). The xenoliths can be divided into high and low light REE groups. The two xenoliths of the high light REE group have steeper light REE patterns and less significant Eu anomalies when compared with the low light REE xenolith group. The differences between these two groups are even more pronounced on chondrite-normalized trace element plots (Fig. 9). The low light REE group is characterized by significant positive Sr and Ba anomalies, and by lower normalized Th, U, Nb, and Ta values. These differences do not appear to correlate with grain size, texture, modal proportions, deformation and recrystallization, zircon morphology, or age (Miller et al., this volume, Chapter 10). Compared with late Cenozoic volcanic rocks of the Bering Sea province, all of the mafic xenoliths have high La/Ta ratios and argue against a genetic relationship between the host lavas and these xenoliths. High La/Ta ratios typically indicate the involvement of continental crust, either as a component added by assimilation during crystallization in magma chambers or magma transport, or as a contaminant added directly in the mantle source (e.g., sediment subduction into the mantle). High La/Ta ratios are typical of the middle to Late Cretaceous magmatic rocks in the Seward Peninsula region that have been interpreted to be subduction related (Moll-Stalcup and Arth, 1989; Miller, 1989; Moll-Stalcup, 1994; Amato and Wright, 1997). The combined geochemical and geochronologic evidence argues favorably for the interpretation that these xenoliths are mafic cumulates that formed during middle to Late Cretaceous subduction-related magmatism, and were variably deformed and recrystallized during Late Cretaceous or early Tertiary extension. If these rocks are subduction related, it is not clear why their normalized Th and U values are so low relative to La. The low thorium and uranium abundances might be due to relatively low bulk distribution coefficients of these elements in the mafic cumulates, or they might have been lost from the mafic cumulates during Late Cretaceous–early Tertiary metamorphism and deformation.

The major element composition (Table 3) of a coarse-grained granitic xenolith indicates that it is a weakly peraluminous ($Al_2O_3/[K_2O + Na_2O + CaO] = 1.02$) granite. The composition of this granite plots along the compositional trends of mafic to felsic plutonic rocks from the Kigluaik gneiss dome on the Seward Peninsula (Amato and Wright, 1997). The REE and trace element patterns (Figs. 8 and 9) of this xenolith indicate that it is light REE enriched and that Nb and Ta are considerably depleted relative to the other more highly incompatible trace elements (e.g., Ba, Th, K, La, Ce). All of these features are consistent with crystallization of this granite from melts that were generated in a suprasubduction-zone setting similar to other Late Cretaceous granitic plutons of the Seward Peninsula (Amato and Wright, 1997).

PETROGENESIS OF BERING SEA VOLCANIC ROCKS

The late Cenozoic volcanic fields of the Bering Sea region consist almost entirely of basalt; only minor amounts of basaltic andesite are present on Stuart and Saint Michael Islands. Most

TABLE 3. MAJOR AND TRACE ELEMENT COMPOSITIONS OF GABBROIC AND GRANITIC XENOLITHS FROM THE KOOKOOLIGIT VOLCANIC FIELD (SAINT LAWRENCE ISLAND)

	SV-19b gabbro	SV-19d gabbro	SV-19i gabbro	SV-19z granite	SV-21a gabbro	SV-21b gabbro	SV-21c gabbro	SV-21e gabbro
SiO_2		50.03	51.46	73.07	49.13			63.41
TiO_2		0.63	0.42	0.05	2.10			0.15
Al_2O_3		20.03	21.05	14.50	16.18			18.22
$Fe_2O_3^T$		6.97	5.87	1.07	11.23			2.94
MnO		0.12	0.12	0.00	0.16			0.06
MgO		6.84	6.94	0.02	5.36			1.73
CaO		11.87	11.20	1.64	6.55			4.74
Na_2O		3.36	3.19	3.08	5.41			5.78
K_2O		0.37	0.23	5.72	2.54			2.61
P_2O_5		0.04	0.02	0.04	0.86			0.13
LOI		0.00	0.00	0.00	0.00			0.02
Total		100.26	100.51	99.19	99.52			99.80
Mg#		0.69	0.73	0.04	0.52			0.57
La	24.3	4.44	1.37	69.9	1.68	8.51	33.7	9.04
Ce	39.1	9.29	2.92	117.0	5.37	11.7	60.6	15.2
Pr	3.90	1.30	0.47	11.0	1.07	1.32	6.74	1.88
Nd	14.3	6.41	2.58	35.3	6.81	5.71	27.6	8.59
Sm	2.59	1.93	0.98	5.14	2.78	1.08	5.64	2.05
Eu	1.19	0.84	0.81	0.66	1.23	1.22	1.97	1.34
Gd	1.82	1.96	1.38	2.81	3.80	1.00	4.73	1.94
Tb	0.23	0.31	0.28	0.34	0.71	0.14	0.67	0.28
Dy	1.15	1.80	1.94	1.64	4.74	0.79	3.86	1.61
Ho	0.21	0.36	0.47	0.28	1.00	0.15	0.72	0.32
Er	0.50	0.89	1.40	0.69	2.74	0.42	1.84	0.79
Tm	0.07	0.12	0.22	0.10	0.38	0.05	0.25	0.11
Yb	0.46	0.72	1.43	0.61	2.35	0.33	1.53	0.68
Lu	0.08	0.11	0.23	0.10	0.36	0.06	0.25	0.11
Cs			0.01	0.44	0.02			
Rb	41	2	1	116	1	6	9	6
Ba	719	191	93	641	38	1200	728	1180
Th	0.52	0.05	0.41	9.83	0.09	0.05	0.75	0.03
U	0.21	0.02	0.01	0.52	0.02	0.01	0.24	0.01
Nb	7	1	0.1	2	0.5	0.5	15	0.2
Ta	0.3	0.07	0.01	0.24	0.05	0.02	0.86	0.01
Sr	397	1210	862	270	253	973	694	1080
Pb	16.5	1.2	0.5	20.9	0.7	7.3	6.1	6.5
Hf	2.6	0.93	0.47	5.3	2.0		5.5	
Zr				203				
Y	5	9	12	8	26	4	19	9
Cr				4				
Ni				1				
Co				2				
Sc	4	19	28		38	4	20	7
V				6				

Note: See "Note" from Table 1.

of the basalts have relatively low phenocryst contents (<10 vol%) and have high Mg#s (>0.60, Cr [>300 ppm], and Ni [>150 ppm]), suggesting that many of the basalts have undergone relatively minor crystal fractionation. However, compatible trace elements (e.g., Cr, Ni) are well correlated with Mg#, an effect predicted for crystallization of olivine ± clinopyroxene. Wall-rock assimilation can have a significant effect on magma composition whenever basaltic magmas are erupted through continental crust. However, these effects are often difficult to quantify without detailed knowledge of both parent magma and wall-rock compositions. Although crustal contamination has undoubtedly occurred in some Bering Sea magmas, several fea-

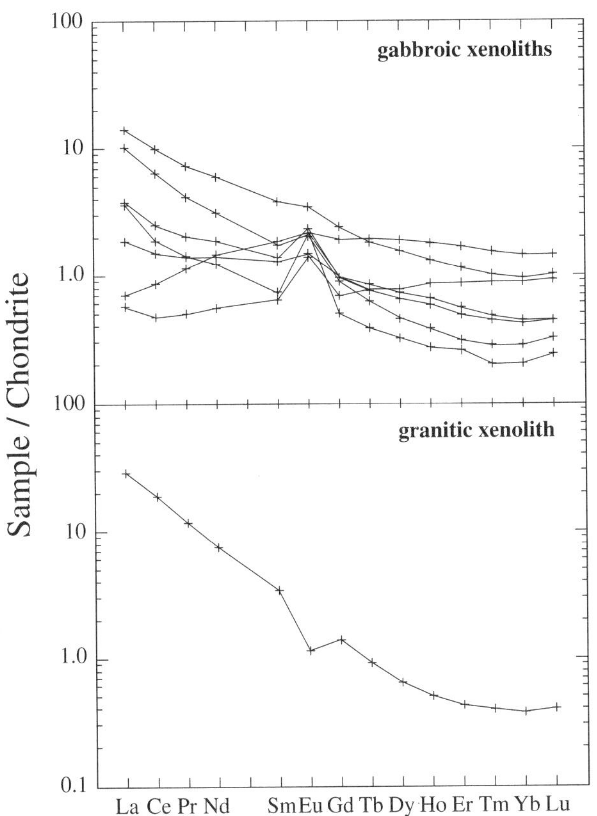

Figure 8. Rare earth element patterns of gabbroic and granitic xenoliths from Kookooligit volcanic field. Chondrite normalization values are from McDonough and Sun (1995).

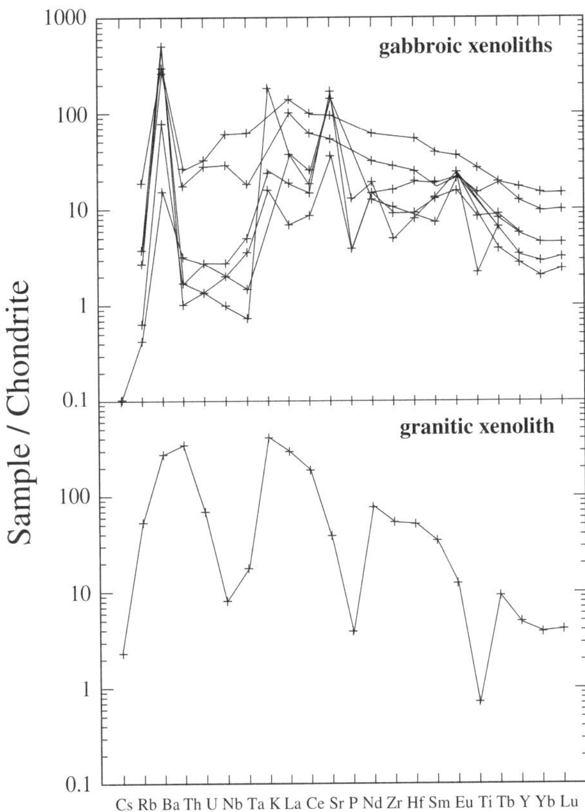

Figure 9. Trace element patterns of granitic xenolith from Kookooligit volcanic field. Chondrite normalization values are from McDonough and Sun (1995).

tures of the volcanic rocks suggest that this effect has been minimal. First, many of the lavas have relatively primitive compositions (e.g., Mg#s > 0.60; Cr > 300 ppm; Ni > 150 ppm), suggesting that they have undergone relatively little crystal fractionation and therefore probably did not have a long residence time in the crust. Second, the relatively common occurrence of mantle xenoliths in lavas from the Kookooligit and Saint Michael volcanic fields further supports the interpretation that the melts were transported directly from the mantle to the surface. Third, the high Nb/La and Nb/Ba ratios of all of the lavas preclude assimilation of significant amounts of crust (characterized by low Nb/La and Nb/Ba). Other trace element ratios that are sensitive to the addition of crustal materials to parent melts (e.g., Ce/Pb = 10–40; Nb/U = 20–90; Ba/La = 6–20; Rb/Zr = 0.06–0.16) are typical of oceanic values (e.g., Halliday et al., 1995). A few flows from the Saint Michael and Teller volcanic fields have high Pb (Ce/Pb = 8–10) and Ba (Ba/Nb = 12–21) that correlate with high K_2O/P_2O_5, Rb/Sr, and Th/Ta, and might have undergone significant contamination. With the exception of a few flows from the Saint Michael and Teller volcanic fields, we believe that crustal contamination played a relatively minor role in the petrogenesis of the Bering Sea volcanics.

Role of melting processes

The depth and extent of melting represented by the Bering Sea basalts can be qualitatively estimated on the basis of major and trace element compositions. The observation that many incompatible element abundances and ratios (e.g., K, Ti, Zr/Y) in the Bering Sea basalts increase with decreasing Si (Fig. 2) is not consistent with fractionation processes, which should lead to a correlation between silica and the incompatible trace elements. Furthermore, most primitive basalts (Mg# > 0.60) exhibit a wide range of incompatible element concentrations (e.g., K = 0.4–1.8 wt%; Fig. 3). These relationships can be understood in terms of the extent of melting; incompatible trace element abundances are approximately inversely proportional to extent of melting. Smaller melt fractions also tend to be more silica undersaturated, consistent with the inverse correlations between Si and Ti, K, Na, P, and other incompatible trace elements in the Bering Sea basalts (Fig. 2). In terms of the extent of melting, basalts from Saint Lawrence Island and Klikitarik Bay and the cones from Teller were produced by smaller degrees of melting relative to flows from Stuart Island, Saint Michael Island, and the Teller regions. Basalts from Saint Lawrence Island (Kookooligit volcanic field) have the lowest Yb (1.0–1.7 ppm; Fig. 5) and highest La/Yb (La/Yb = 8–45) relative to those from the Teller (Yb = 1.8–2.3; La/Yb = 5–15) and Saint Michael (Yb = 1.3–2.3; La/Yb = 3–22) volcanic fields, suggesting that residual garnet may have been present in the mantle source regions of some Kookooligit volcanic field basalts.

The most primitive Bering Sea basalts (Mg# > 0.6) are also characterized by a wide range of Fe compositions (Fe = 6.1–9.0 wt%; Fig. 3). Iron composition is relatively insensitive to the extent of melting at a given pressure, but increases with greater depths of melting (e.g., Jaques and Green, 1980; Takahashi, 1986). This implies that the high-Fe basalts from Saint Lawrence Island, Klikitarik Bay, and Teller region were produced by partial melting at greater depths than those from Stuart Island and Saint Michael Island. Subsequent fractionation of the high- and low-Fe parental magma groups then produced the two subparallel trends of increasing Fe with decreasing Mg# (Fig. 3). These relationships, coupled with the strong negative correlation between Fe and Si (Fig. 2), suggest that the spectrum of Bering Sea basalt compositions can be explained by crystal-fractionation processes superimposed on melts that were generated by varying degrees of melting over a range of depths in the mantle.

Role of mantle source compositions

Experimental studies have shown that bulk mantle composition exerts a strong control on the composition of partial melts (Hirose and Kushiro, 1993; Baker and Stolper, 1994). Relative to partial melts of fertile (asthenospheric) peridotite, melts of depleted peridotite have lower CaO, $Fe_2O_3^T$, TiO_2, Na_2O, and K_2O; the SiO_2 compositions of partial melts are relatively unaffected by starting peridotite composition (e.g., Turner and Hawkesworth, 1995, and references therein). Therefore, the wide range of major and trace element compositions exhibited by the less evolved Bering Sea basalts might also be explained by partial melting of variably depleted mantle peridotites. If variable source composition played a significant role in determining the composition of the Bering Sea volcanic fields, the basalts from Stuart Island and Saint Michael Island (Saint Michael volcanic field) were produced by melting of more depleted peridotite relative to those from the Teller volcanic field, Saint Lawrence Island (Kookooligit volcanic field), and Klikitarik Bay (Saint Michael volcanic field). Total iron content commonly increases with stratigraphic height in many continental rift sequences and is generally attributed to increased depths of melting during the history of magmatism (e.g., Turner and Hawkesworth, 1995; Lassiter and DePaolo, 1997). Low $Fe_2O_3^T$ and TiO_2 basalts commonly occur in the lowest portions of rift sequences and are interpreted to be derived from relatively depleted lithosphere mantle. In contrast, younger lavas in rift sequences commonly have higher $Fe_2O_3^T$ and TiO_2 and are interpreted to be melts of a plume (fertile asthenosphere). Although the alkaline volcanism of the Bering Sea region is not likely re-

lated to a plume, the variable major and trace element abundances suggest that the melts were generated from a spectrum of mantle compositions (depleted and fertile) and melt segregation depths (shallow to deep).

Plots of incompatible trace elements and trace element ratios also provide clues about mantle source composition and the extent of melting. In a plot of Zr/Nb versus Rb (Fig. 10A), the Bering Sea basalts plot along a curve from high Rb–low Zr/Nb (Saint Lawrence Island and Klikitarik Bay) to low Rb–high Zr/Nb (Stuart Island, Saint Michael Island, and Teller). Rubidium should behave as an incompatible trace element during partial melting and its abundance in melts should be approximately inversely proportional to the extent of melting. The Zr/Nb ratio can be used to infer source composition; depleted sources will have higher Zr/Nb ratios (e.g., normal mid-ocean ridge basalt [N-MORB], Zr/Nb ~30) than enriched (e.g., oceanic-island basalt,

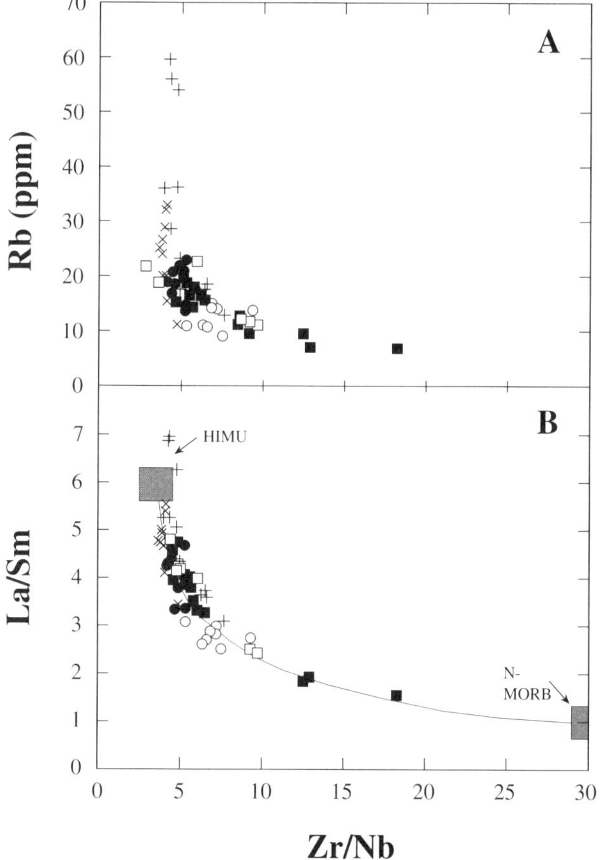

Figure 10. Plots of Bering Sea basalts on Zr/Nb vs. Rb (A) and La/Sm (B) diagrams. Basalts with high Zr/Nb were likely produced by larger degrees of melting of depleted source (normal mid-ocean ridge basalt [N-MORB]), whereas basalts with low Zr/Nb and high Rb were likely produced by small degrees of melting of enriched source. Mixing curve (tick marks are in 10% increments) is shown in B between estimated compositions of N-MORB and HIMU (high μ) (Sun and McDonough, 1989). Symbols as in Figure 2.

Zr/Nb = 6) sources. The relationships on this diagram imply that the Stuart Island and Saint Michael Island basalts (Saint Michael volcanic field) were produced by large degrees of melting of a relatively depleted (lithosphere?) source, whereas the Kookooligit volcanic field and Klikitarik Bay (Saint Michael volcanic field) basalts represent smaller melt fractions of a relatively enriched (asthenosphere?) source. These results are similar to those inferred from the major element compositions (Figs. 2 and 3). On a diagram of Zr/Nb versus La/Sm (Fig. 10b) the Bering Sea basalts follow a hyperbolic mixing trend between depleted (N-MORB) and enriched (high μ, HIMU) mantle sources. These relationships support the interpretation that Kookooligit volcanic field and Klikitarik Bay (Saint Michael volcanic field) basalts reflect melting of an enriched mantle source.

Studies of other alkaline volcanic centers have also indicated the involvement of end-member mantle compositions. Bering Sea region volcanic fields have Nd, Sr, and Pb isotopic ratios that range from N-MORB to near the composition of the bulk silicate earth (Bering Sea region, Mark, 1971; Pribilof Islands, Kay et al., 1978; Nunivak Island, Menzies and Murthy, 1980a, 1980b; Bering Sea region, von Drach et al., 1986; Fort Selkirk, Francis and Ludden, 1990; eastern Alaska, Wirth, 1991; Navarin basin, Davis et al., 1993; Bering Sea region, Moll-Stalcup, 1994, 1995). In general, basanites and nephelinites within each volcanic field have less radiogenic Nd and Sr isotopic compositions relative to tholeiites from the same field. Sr and Nd isotopic data also indicate a more radiogenic source for Saint Lawrence Island basalts relative to Saint Michael volcanic field basalts. Carignan et al. (1994) identified three mantle sources for alkaline centers in the northern Cordillera: (1) a source with relatively unradiogenic Pb and Sr that is amphibole bearing (source of nephelinites), (2) a source (continental lithospheric mantle?) with high $^{207}Pb/^{206}Pb$ (alkali-olivine basalts), and (3) a HIMU-type mantle (source of Mount Edziza lavas). Isotopic studies of spinel-lherzolite xenoliths from the Alligator Lake volcanic center (southern Yukon) suggest recent (<30 Ma) enrichment of the subcontinental lithosphere by subduction metasomatism (Carignan et al., 1996).

The nature of the mantle beneath the Bering Shelf can also be investigated using Ce and Pb abundances. Both elements have similar solid-silicate melt distribution coefficients, so they are not fractionated during melting processes. However, Pb is preferentially transferred to the continental crust along convergent margins, resulting in high Ce/Pb ratios in oceanic basalts and low Ce/Pb ratios in continental crust (Miller et al., 1994). Ce/Pb ratios of Bering Sea basalts range from 9 to 39 and are broadly correlated with Fe$_2$O$_3$ content (Fig. 11). Basalts from the Kookooligit volcanic field and Klikitarik Bay (Saint Michael volcanic field) plot entirely within the range of oceanic rocks. Other flows from the Saint Michael volcanic field (Stuart Island and Saint Michael Island) and Teller volcanic field have Ce/Pb ratios that range from 9 to 26; the Ce/Pb ratios of Teller volcanic field cones (Ce/Pb = 17–20) are consistently higher than Teller volcanic field flows (Ce/Pb = 10–12). These features suggest that the Kookooligit volcanic field and Klikitarik Bay (Saint Michael

volcanic field) rocks were produced by melting typical suboceanic mantle at relatively greater depths, whereas Teller volcanic field flows and some Saint Michael volcanic field flows (Stuart Island, Saint Michael Island) were produced from a shallower source that had been variably enriched in Pb. The Teller basalts have relatively high Fe and low Ce/Pb ratios and do not fit the trend of the other Bering Sea basalts. This relationship is not well understood, but might be related to the composition or evolution of the subcontinental lithosphere that underlies the Seward and York terranes.

In summary, the geochemical variations exhibited by individual volcanic suites largely reflect fractionation and melting effects, whereas variations between suites primarily reflect source composition differences. The Saint Michael Island, Stuart Island, and Teller flows were derived from relatively large degrees of melting of refractory mantle (N-MORB type) at shallow depths (subcontinental lithosphere?). The Teller cones were produced from a similar mantle source, but by smaller degrees of melting at greater depth. The mantle source regions of the Saint Michael and Stuart Island flows were modified (metasomatized?) and likely contained amphibole. Flows of the Kookooligit volcanic field and Klikitarik Bay (Saint Michael volcanic field) were apparently produced at greater depth by smaller degrees of melting of more fertile mantle (asthenosphere-like) that was relatively unmetasomatized.

Role of amphibole

Hydrous minerals (amphibole, phlogopite) are commonly present in some samples of the upper mantle (e.g., Nixon, 1987) and have been widely invoked to explain the chemical variations exhibited by suites of alkaline lavas (e.g., Sun and Hanson,

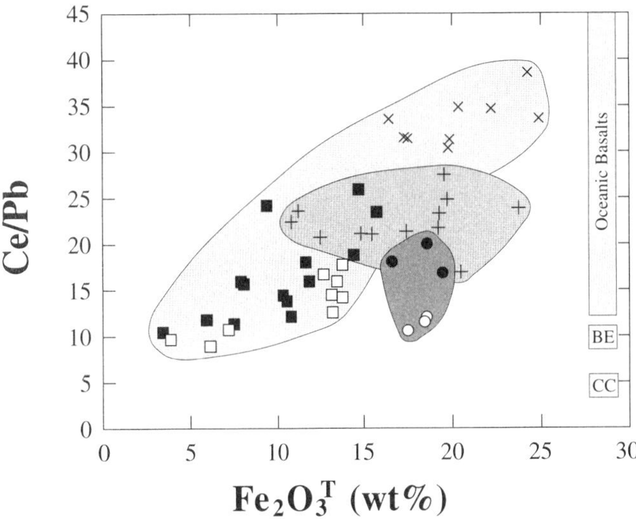

Figure 11. Plot of total iron (Fe$_2$O$_3^T$ wt%) vs. Ce/Pb of Bering Sea basalts. Estimated Ce/Pb ranges of bulk earth (BE) and oceanic and continental (CC) crust are from Miller et al. (1994). Symbols as in Figure 2.

1975; Clague and Frey, 1982). However, direct evidence of amphibole involvement in the generation of alkaline magmas has generally been lacking. Francis and Ludden (1995) suggested that the LILE, HFSE (high field strength elements), and light REE systematics of many intraplate alkaline suites exhibit evidence of residual amphibole in their genesis. Specifically, they argue that some alkaline lavas of the northern Canadian Cordillera formed by a two-stage melting process. In their model, the olivine nephelinites were produced during the first phase of melting when residual amphibole was present, and the hypersthene–normative olivine basalts through basanites were produced during a second stage of melting involving only the lherzolite host after the exhaustion of amphibole.

Several aspects of the geochemistry of the late Cenozoic volcanics of the Bering Sea region are similar to those of the northern Cordillera, suggesting that amphibole might also have been involved in the genesis of some of these lavas. In general, K/Na, K/La, and Nb/Zr are well correlated with silica undersaturation in the volcanics of the Bering Sea province. However, the most undersaturated basalts (basanites) exhibit relatively constant Nb/Zr values and lower K/Na and K/La ratios (i.e., constant K with increasing Na and La). These relationships are inconsistent with generation of these lavas by progressive melting of an anhydrous lherzolite mantle source (Francis and Ludden, 1995). Furthermore, the relatively high Rb/K of the basanites, compared with the tholeiites through alkali-olivine basalts, suggests the involvement of residual amphibole rather than phlogopite.

Further evidence of amphibole in the mantle source regions beneath the Bering Sea volcanic province comes from the REE patterns of the lavas. The REE patterns of basalts from the Saint Michael and Kookooligit volcanic fields exhibit a variety of shapes. In general, the slopes of the heavy REE patterns remain relatively constant within a given volcanic field. In contrast, the light REE patterns are concave up and rotate clockwise about Eu. The shapes of some of these patterns are similar to the amphibole-pyroxenite xenoliths from Klikitarik Bay (Fig. 6). These patterns are unusual and are not likely a result of fractionation of olivine, clinopyroxene, orthopyroxene, spinel, or plagioclase during melting or crystallization. The upturned La-Ce portions of the light REE patterns (Fig. 4) could have been produced by partial melting of a mantle source that had previously been enriched in La and Ce (Navon and Stolper, 1987; Johnson et al., 1990). If these enrichment effects (high La and Ce) are removed from the light REE patterns of the basalts, the patterns of the remaining REE would have smooth, concave-down shapes similar to those of amphibole. These relationships suggest that the REE patterns of basalts from the Saint Michael and Kookooligit volcanic fields might be affected by two processes. One process involved variable light REE enrichment (metasomatism?) of the mantle source; partial melting of these sources would result in melts with variably La- and Ce-enriched patterns. A second process involving residual amphibole might have resulted in melts that were variably enriched in the mid-REEs. The basalts with the strongest amphibole signatures (enriched mid-REEs) have the lowest total incompatible trace element abundances, and may have been produced by relatively large degree melts in the presence of amphibole. In contrast, basalts with the weakest amphibole signature are the most enriched in incompatible trace elements and represent relatively small degree melts of an amphibole-poor (unmetasomatized?) source. This model makes sense if the fraction of melt that can be produced is limited by the degree of metasomatism that is present; i.e., assuming relatively homogeneous temperature conditions in the mantle, the fraction of melt that might be directly related to the proportion of low-melting-temperature components (e.g., amphibole) that are present. In this model, tholeiite lavas are large-fraction melts of relatively depleted (shallow) mantle that was moderately metasomatized. In contrast, alkali-olivine basalts and basanites are smaller-fraction melts that were produced from relatively unmetasomatized mantle at greater depths. If portions of the mantle beneath the Bering Sea region are variably metasomatized, this might also explain the different Rb, Ba, Th, and K abundances that are exhibited by each of the volcanic fields (Fig. 5).

GEODYNAMIC IMPLICATIONS

A primary objective of this study was to examine the relationship between late Cenozoic volcanism and tectonic processes in Alaska. This is accomplished by comparing the style and composition of the volcanism with other volcanic provinces, and by examining the timing and distribution of the volcanic fields relative to major structural features and tectonic processes in Alaska.

Comparison with volcanic sequences in other regions

The petrographic, major and trace element, and isotopic compositions of the late Cenozoic tholeiitic to alkaline volcanics of the Bering Sea region are most similar to those from intraplate settings, including some ocean islands, continental rifts, and regions of continental extension. However, the small and isolated volcanic centers of the Bering Sea region are atypical of regions that have large igneous provinces (e.g., Coffin and Eldholm, 1994). Furthermore, with the exception of possible anomalous mantle beneath some regions of the northern Canadian Cordillera (Frederiksen et al., 1998; Shi et al., 1998), there is no evidence of a mantle plume beneath the northern Cordillera or Bering Sea region. Alkaline intraplate magmatism has also been described from some convergent plate margins that have undergone synsubduction or postsubduction extension (e.g., Antarctic Peninsula, Hole, 1988; Fiji, Gill and Whelan, 1989; Mexican Volcanic Belt, Luhr et al., 1989; New Zealand, Briggs and McDonough, 1990; western United States, Fitton et al., 1991), and this tectonic setting is not very dissimilar from the present situation in Alaska. However, with the exception of the Basin and Range Province, the alkaline volcanic rocks in these settings are restricted to relatively narrow belts and are intimately associated with subduction-related volcanics. Furthermore, volcanic sequences erupted in extensional settings (e.g., continental rifts, Basin and Range) typically include a significant proportion of felsic intrusive and extrusive rocks, whereas there is no evidence of felsic magmatism accompanying late Cenozoic basaltic volcanism in the Bering Sea region.

The widely dispersed tholeiitic volcanic centers of the Bering Sea region are chemically similar to those of Southeast Asia, which have been collectively termed a diffuse igneous province (Hoang et al., 1996). Volcanic flows in both Alaska and Southeast Asia consist mostly of tholeiitic to alkaline basalt and have very similar trace element patterns and ratios, despite being erupted through two continents with different compositions and histories. Specifically, the mantle-normalized trace element patterns of the Bering Sea basalts are most similar to basalts from the Dalat, Phyuoc Long, Song Cau, and Pleiku volcanic centers of Vietnam, Cambodia, and Laos (Fig. 5). Furthermore, both regions have volcanic sequences that exhibit a spectrum of basalt compositions ranging from high-Si, low-Fe tholeiites to low Si, high-Fe alkali basalts and basanites. During the Neogene to Quaternary, intraplate-type volcanism occurred in isolated centers that are widely distributed throughout eastern and southeastern Asia (e.g., Barr and McDonald, 1981; Whitford-Stark, 1987; Zhi et al., 1990, Hoang and Flower, 1998). The volcanism postdates the early Tertiary collision of India with Asia and occurs in a broad region of transtension that is east of a series of splaying strike-slip faults. The extensional basins (e.g., South China Sea, North China Sea, Andaman Sea) of this region are modeled as a consequence of tectonic extrusion of large crustal blocks along major strike-slip faults (Tapponier et al., 1982; Peltzer and Tapponier, 1988) resulting from the collision of India with Asia. In Southeast Asia the basins occur at the eastern ends of eastward-terminating left-lateral strike-slip faults.

Cretaceous to Holocene extension in the Bering Shelf region

There is considerable evidence for Cenozoic to Holocene crustal extension beneath western Alaska and the Bering Shelf region. The relatively low topography and thin crust of the Bering and Chukchi shelf regions is atypical of the Jurassic-Cretaceous Brooks Range fold and thrust belt in northern Alaska, the crustal root of which diminishes westward across the Bering Shelf (Barnes et al., 1994). The present-day crust beneath the Bering Sea is 30–35 km thick and suggests ~30% crustal extension (Klemperer et al., this volume, Chapter 1), assuming an initial crustal thickness of 50 km during the Early Cretaceous (Fuis et al., 1995). Miller et al. (1992), Amato et al. (1994), and Dumitru et al. (1995) recognized three periods of extension in the Bering Strait region. The earliest and most significant extension in the Seward Peninsula region, as recorded by the deformation, uplift, and cooling of deep crustal rocks, occurred during the middle to Late Cretaceous (100–70 Ma; Dumitru et al., 1995; Hannula and McWilliams, 1995) following regional magmatism and high-

grade metamorphism that peaked ca. 91 Ma (Amato and Wright, 1998; Miller et al., this volume, Chapter 10). Pervasive north-south mineral-stretching fabrics in the high-grade rocks (Miller and Hudson, 1991) and calcite-twin data (Grischkowsky et al., 1995) suggest that the region underwent north-south extension during Late Cretaceous collapse of the Jurassic–Early Cretaceous Brooks Range orogen. A second period of extension occurred during the early Tertiary (Eocene–early Oligocene) and resulted in regional tilting and erosion on the Seward Peninsula (Dumitru et al., 1995) and the formation of transtensional basins on the Bering Shelf (Fig. 1) (e.g., Norton and Hope basins; Worrall, 1991; Tolson, 1987). These Late Cretaceous–early Tertiary extension events may be represented by the deformation fabrics and mineral assemblages of the mafic crustal xenoliths (Miller et al., this volume, Chapter 10).

There is also considerable evidence for Pliocene to Holocene extension in the Bering Sea region; however, this extension might be unrelated to the Late Cretaceous–early Tertiary extension that is recorded in the crustal xenoliths. High heat flow, hot springs, late Neogene–Holocene normal faults, late Cenozoic sediment-filled grabens, and widely distributed intraplate magmatism have all been cited as evidence of recent extension in western Alaska (e.g., Nakamura et al., 1980; Turner et al., 1981; Biswas et al., 1986; Plafker et al., 1994). Focal mechanism solutions indicate that the region is currently undergoing northeast-southwest subhorizontal extension (Estabrook et al., 1988; Mackey et al., 1997). Fission-track data suggest only minor amounts of exhumation (<3 km) since the Pliocene (Dumitru et al., 1995). Extension during these later phases is probably related to westward extrusion of crustal fragments along major dextral strike-slip faults during rotation of terranes in northern Alaska and north-south shortening in eastern Alaska (e.g., Miller et al., 1993; Plafker and Berg, 1994). Mackey et al. (1997) further suggested that the Bering crustal block, which exhibits relatively little internal seismicity and deformation, and is undergoing clockwise motion relative to North America. Both of these scenarios are similar to the tectonic setting of Southeast Asia, except that the geometry is reversed and the extensional basins in Alaska occur at the west end of major right-lateral strike-slip faults (Whitney and Wallace, 1995; Mackey et al., 1997). The apparent disappearance of the major strike-slip faults in western Alaska beneath the Bering Shelf is an outstanding problem of Alaskan geology. Perhaps the motions along these faults are taken up in extensional-transtensional basins and provide a mechanism for the widely distributed mantle-derived lavas.

Many of the volcanic centers in western Alaska occur near the ends of the surface expressions of major strike-slip faults or along their projected traces (Fig. 1). For example, the Quaternary Togiak basalt field occurs at the terminus of the Denali fault and the Saint Michael volcanic field occurs near the end of the Kaltag fault. Other volcanic fields, such as at Imruk Lake and Teller, occur along major late Pleistocene to Holocene faults (e.g., Bering Strait fault zone and Bendeleben fault) or in association with swarms of lesser pre-Neogene faults in regions that are bounded by regional strike-slip faults. Several of the volcanic fields (Imruk Lake, Saint Michael, and Ingakslugwat) are closely associated with large Tertiary basins (Kotzebue, Norton, and Bethel basins, respectively). These large basins formed in three distinct phases, beginning in the middle to late Eocene, culminating in late Eocene to Miocene subsidence, and waning in the Pliocene to Holocene (Worrall, 1991; Kirschner, 1994).

The wide distribution of Cenozoic volcanic centers and sedimentary basins in eastern Alaska and northern British Columbia suggests that significant extension has also occurred in the northern Cordillera. However, the mechanism here cannot be one of tectonic extrusion. Extensive regions of interior Alaska are characterized by broad alluviated lowlands with subdued topography. Of significance to this study are those that are underlain by Cenozoic nonmarine sediments (e.g., Yukon Flats Basin, Tanana Basin, Northway lowlands). These basins typically contain fluvial and coal-bearing sedimentary sequences deposited in three Tertiary cycles (Paleocene to early Eocene, late Eocene to late Miocene, and late Miocene to Pliocene), and apparently developed as extensional basins along the major strike-slip faults in Alaska (Kirschner, 1994). Isolated volcanic centers of Cenozoic age occur throughout eastern Alaska between the Alaska Range and the Brooks Range. The oldest volcanics (Ray River region; 33.2 Ma) are predominantly tholeiitic and have radiogenic Nd and Sr isotopic compositions (ε_{Nd} = 4.0–4.7; $^{87}Sr/^{86}Sr$ = 0.70457–0.70468; Wirth, 1991). In contrast, progressively younger volcanic fields (e.g., Porcupine River, 15.9 Ma; Black River, 0.36 Ma; and Prindle Volcano, ca. 0.20 Ma) consist of lavas that are progressively more silica undersaturated, more highly enriched in trace elements, and have less radiogenic Nd and Sr isotopic compositions (e.g., Prindle Volcano: ε_{Nd} = 8.6–9.1; $^{87}Sr/^{86}Sr$ = 0.70291–0.70294). These trends are similar to those observed in the Bering Sea volcanic province and strongly suggest a progressive change in mantle sources and magmatic processes since the early Oligocene.

A number of late Tertiary to Holocene volcanic centers with intraplate characteristics also occur throughout the Canadian Cordillera (Fig. 1). These are generally considered to be related to local zones of transtension that developed along dextral transform faults between the North American and Pacific plates (Feisinger and Nicholls, 1977; Souther, 1977; Nicholls et al., 1982; Bevier, 1983; Souther et al., 1984; Eiché et al., 1987; Francis and Ludden, 1990; Carignan et al., 1994). Typically, these volcanic centers consist of hypersthene-normative basalt to nephelinite; intermediate and felsic rocks are rarely present. The volcanic stratigraphy typically consists of early silica-undersaturated lavas that become less silica undersaturated with decreasing age; several of the centers also display late-stage highly undersaturated cones and lavas. Mantle and crustal xenoliths occur in the nepheline-normative flows in many of the centers. Southeast of the subduction-related Wrangell volcanic belt, Cenozoic lavas that overlie transform faults are intermediate between calc-alkaline (suprasubduction) and intraplate lavas, indicating the change in mantle sources and magmatic processes

that occur in the transition from compressional to transpressional stress regimes in southeastern Alaska (Skulski et al., 1991). Edwards and Russell (1999, 2000) suggested that Neogene-Quaternary magmatism in the northern Cordilleran volcanic province is related to changes in relative plate motions between the Pacific and North American plates ca. 10 Ma. They argue that magmatism in the northern Cordilleran volcanic province continued through the Holocene due to local transtension along the Pacific–North American plate boundary.

Relationship between extension and volcanism in Alaska

The relationship between the timing of intraplate magmatism and extension in western Alaska is not well understood. As noted here, there have been several episodes of crustal extension in western Alaska since the Late Cretaceous. Magmatism associated with the main phase of extension in the Late Cretaceous is truly bimodal (Amato et al., 1994; Amato and Wright, 1997). Mafic and felsic plutonic (e.g., Seward Peninsula and Yukon-Koyukuk basin) and volcanic rocks (Saint Lawrence Island and Chukotka Peninsula) are exposed throughout the Cretaceous magmatic belt (southeast Alaska to northeast China). A volumetrically important mafic component of these widely distributed mafic and felsic magmatic systems was also apparently intruded as mafic sills in the lower crust, as evidenced by abundant variably deformed gabbroic xenoliths present in late Cenozoic basalts of the Bering Sea province. Both the gabbroic and granitic xenoliths have geochemical features consistent with formation in a suprasubduction setting. Magmatic rocks of the Yukon-Kanuti belt (Moll-Stalcup, 1994) indicate that subduction continued in this region until ca. 56 Ma, followed by a transition to postsubduction or possibly intraplate magmatism ca. 56–40 Ma. The earliest-known extension-related magmatism in the Bering Sea region is represented by middle to late Eocene basalt (42 ± 10 Ma; Tolson, 1986), recovered in the basal portions of drill core from the Kotzebue and Norton basins; however, the composition of these basalts is not known. The next younger intraplate volcanics in the Bering Sea region were erupted in the Kugruk volcanic field (28–26 Ma; Swanson et al., 1981); however, most of the mafic volcanic centers of the Bering Sea province are much younger (<6 Ma; Moll-Stalcup, 1994) and were erupted during the latest period of extension (Pliocene to Holocene). However, cooling and uplift studies suggest that most of the extension in the Bering Sea region occurred before the Late Cretaceous–early Tertiary and that only minor uplift (<2–3 km) has occurred since that time. Apparently, only minor amounts of magmatism accompanied the main period of basin formation during the late Eocene to Oligocene. It also seems that very little intraplate-type magmatism occurred during the main period (Late Cretaceous to early Tertiary; 85–50 Ma) of dextral motions along the major strike-slip faults in western Alaska (e.g., Tintina, Denali, Kaltag faults). Why was there a 40 m.y. hiatus between the main period of strike-slip faulting and basin formation and the onset of decompression melting in the mantle? Without detailed knowledge of the thermal structure and composition of the crust and mantle, and the timing and magnitude of motions along the strike-slip faults, it is difficult to answer this question definitively. The delay between extension and decompression melting might be a function of the thermal maturity of the crust or might indicate that strike-slip motions were more significant during the late Miocene to Holocene than is currently recognized. Alternatively, some models of decompression melting suggest that melt generation is favored during the later stages of multiple stretching events (e.g., Keen et al., 1994; Williamson et al., 1995).

The distribution of alkaline volcanic centers may provide some clues to the tectonic stresses in Alaska. With the exception of the small centers at Ray Mountain and Buzzard Creek (Fig. 1), there is a notable lack of intraplate volcanism in the region of central Alaska, where the trends of major strike-slip faults curve from northwest to west-southwest. If the present distribution is representative, the lack of volcanic fields in central Alaska might reflect the changing stress regimes between regions of transtension along strike-slip faults (east) and block rotations due to tectonic extrusion (west). It is interesting that Cenozoic basins are common in this region of central Alaska (Fig. 1). Dover (1994) attributed the formation of the Yukon Flats and Nenana basins at the apices of the Tintina and Denali faults to extension by rotation between active faults, or to extension along the convex sides of curving strike-slip faults.

SUMMARY

Late Cenozoic tholeiitic to alkaline volcanic fields occur as small isolated centers throughout the Bering Sea region. The erupted magmas consist mostly of basalt and range from voluminous tholeiite to volumetrically minor basanite. Tholeiites are characterized by high-Si, low-Fe, and low incompatible trace element abundances and were produced by large degrees of melting of depleted mantle at relatively shallow depth. In contrast, basanites have low-Si, high-Fe, and high trace element abundances and are small melt fractions of fertile mantle produced at greater depth. Major and trace element variations within each volcanic field reflect variable degrees of melting, crystal fractionation, and minor crustal contamination. At least two different mantle sources are required to explain the trace element compositions that characterize each of the volcanic fields. The REE element patterns of the basalts suggest that the mantle source regions of the tholeiites were enriched in the light REEs and that amphibole may have been present during melting.

Xenoliths are common in some of the silica-undersaturated flows and include mantle peridotites and gabbroic rocks. Mantle peridotites consist mostly of spinel lherzolites that have major element compositions similar to fertile mantle; incompatible trace element compositions of these xenoliths are enriched relative to primitive mantle. Amphibole-pyroxenite xenoliths contain pargasite or kaersutite and have textures suggesting that they originated within the upper mantle by metasomatic processes. However, it is unclear whether these xenoliths are the result, or were involved in the genesis, of the late Cenozoic alkaline magmas. Gabbroic xenoliths consist mostly of medium-grained plagioclase and orthopyroxene that have been variably deformed and recrystallized; thermobarometry studies suggest that the rocks equilibrated at 0.4–0.6 GPa. Compositional layering and the presence of igneous textures suggest that the rocks might be igneous cumulates from the lower crust. Combined geochronologic and trace element data from these xenoliths indicate an origin in a suprasubduction setting during the Late Cretaceous. The large number of mafic xenoliths in some lavas of the Bering Sea province suggest a significant component of mafic cumulates in the lower crust beneath the region. The inferred large volume of cumulates is the likely source of the highly reflective crust that is observed in seismic reflection profiles.

The composition and style of Bering Sea basaltic volcanism is similar to that of Southeast Asia. Both regions are characterized by small-volume and widely dispersed tholeiitic to alkaline eruptives. The magmatism in both regions is apparently a result of extension that is related to the tectonic extrusion along regional strike-slip faults. We propose that the late Cenozoic alkaline volcanic fields of the Bering Sea region fit the model of diffuse igneous provinces such as Southeast Asia.

ACKNOWLEDGMENTS

This study was supported by a National Science Foundation (NSF) Research Opportunity Award to E. Miller. Analytical studies at Macalester College were conducted on instruments that were purchased or upgraded with support from the NSF (grants NSF-9651385 and EAR-9601475). Neutron activation analyses were made possible by R. Kay with the support of the TRIGA Reactor Facility at Cornell University. Support for student summer research was provided by the Pew Midstates Science and Mathematics Consortium, the NASA Minnesota Space Grant Consortium, and Macalester College. The manuscript was significantly improved by reviews by J. Amato, E. Miller, and M. Roden. J. Craddock and J. Thole also reviewed early drafts of the manuscript. We are very grateful for the assistance and logistical support that we received from the many individuals of villages. Permission to work on native lands was granted by the Native Corporations of Brevig Mission, Saint Michael, Savoonga, and Stebbins. We thank D. Geist and E. Moll-Stalcup for thought-provoking discussions on alkaline magmatism. Editorial assistance was provided by T. Sandland.

REFERENCES CITED

Ahmedali, T., 1983, XRF procedures: Montreal, Canada, McGill University, Department of Geological Sciences, Circular no.1.

Akinin, V.V., Roden, M.F., Francis, D., Apt, J., and Moll-Stalcup, E., 1997, Compositional and thermal state of the upper mantle beneath the Bering Sea basalt province: Evidence from the Chukchi Peninsula or Russia: Canadian Journal of Earth Sciences, v. 34, p. 789–800.

Allegre, C.J., Lambert, B., and Richard, P., 1981, The subcontinental versus the sub-oceanic debate: Lead-neodymium-strontium isotopes in primary alkali basalts from a shield area: The Ahaggar volcanic field: Earth and Planetary Science Letters, v. 52, p. 85–92.

Amato, J.M., and Wright, J.E., 1997, Potassic mafic magmatism in the Kigluaik gneiss dome, northern Alaska: A geochemical study of arc magmatism in an extensional setting: Journal of Geophysical Research, v. 102, p. 8065–8084.

Amato, J.M., and Wright, J.E., 1998, Geochronologic investigations of magmatism and metamorphism with the Kigluaik Mountains gneiss dome, Seward Peninsula, Alaska, *in* Clough, J.G., and Larson, F., eds., Short notes on Alaska geology, 1977: State of Alaska, Department of National Resources, Division of Geological and Geophysical Surveys Professional Report 118, p. 1–21.

Amato, J.M., Wright, J.E., Gans, P.B., and Miller, E.L., 1994, Magmatically induced metamorphism and deformation in the Kigluaik gneiss dome, Seward Peninsula, Alaska: Tectonics, v. 13, p. 515–527.

Aoki, K., and Kushiro, I., 1968, Some clinopyroxenes from ultramafic inclusions in Dreiser Weiher, Eifel: Contributions to Mineralogy and Petrology, v. 18, p. 326–337.

Arndt, N.T., and Christensen, U., 1992, The role of lithospheric mantle in continental flood volcanism: Thermal and geochemical constraints: Journal of Geophysical Research, v. 97, p. 10967–10981.

Bailey, D.K., 1987, Mantle metasomatism, *in* Fitton, J.G., and Upton, B.G.J., eds., Alkaline igneous rocks: Geological Society [London] Special Publication 30, p. 1–13.

Baker, M.B., and Stolper, E.M., 1994, Determining the composition of high-pressure mantle melts using diamond aggregates: Geochimica et Cosmochimica Acta, v. 58, p. 2811–2827.

Barnes, D.F., Mariano, J., Morin, R.L., Roberts, C.W., Jachens, R.C., and Berg, H.C., 1994, Incomplete isostatic gravity map of Alaska, *in* Plafker, G., and Berg, H.C., eds., The geology of Alaska: Boulder, Colorado, Geological Society of America, Geology of North America, v. G-1, plate 9, 1 sheet, scale 1:2 500 000.

Barr, S.M., and MacDonald, A.S., 1981, Geochemistry and geochronology of late Cenozoic basalts of southeast Asia: Geological Society of America Bulletin, v. 92, p. 1069–1142.

Bering Strait Geologic Field Party, 1997, Koolen Metamorphic Complex, NE Russia: Implications for the tectonic evolution of the Bering Strait region: Tectonics, v. 16, p. 713–729.

Bevier, M.L., 1983, Implications of chemical and isotopic composition for petrogenesis of Chilcotin Group basalts, British Columbia: Journal of Petrology, v. 24, p. 207–226.

Biswas, N.N., Akim, K., Pulpan, H., and Tytgat, G., 1986, Characteristics of regional stresses in Alaska and neighboring areas: Geophysical Research Letters, v. 13, p. 177–180.

Box, S.E., and Patton, W.W., 1989, Igneous history of the Koyukuk Terrane, western Alaska: Constraints on the origin, evolution, and ultimate collision of an accreted island arc terrane: Journal of Geophysical Research, v. 94, p. 15843–15867.

Brey, G.P., and Köhler, T., 1990, Geothermometry in four-phase lherzolites 2: New thermometers, and practical assessment of existing thermometers: Journal of Petrology, v. 31, part 6, p. 1353–1378.

Briggs, R.M., and McDonough, W.F., 1990, Contemporaneous convergent margin and intraplate magmatism, North Island, New Zealand: Journal of Petrology, v. 31, p. 813–851.

Carignan, J., Ludden, J., and Francis, D., 1994, Isotopic characteristics of mantle sources for Quaternary continental alkaline magmas in the northern Canadian Cordillera: Earth and Planetary Science Letters, v. 128, p. 271–286.

Carignan, J., Ludden, J., and Francis, D., 1996, On the recent enrichment of subcontinental lithosphere: A detailed U-Pb study of spinel lherzolite xenoliths, Yukon, Canada: Geochimica et Cosmochimica Acta, v. 60, p. 4241–4252.

Chen, C.-Y., and Frey, F.A., 1985, Trace element and isotopic geochemistry of lavas from Haleakala volcano, East Maui, Hawaii: Implications for the origin of Hawaiian basalts: Journal of Geophysical Research, v. 90, p. 8743–8768.

Clague, D.A., and Frey, F.A., 1982, Petrology and trace element geochemistry of the Honolulu volcanoes, Oahu: Implications for the oceanic mantle below Hawaii: Journal of Petrology, v. 23, p. 447–504

Coffin, M.E., and Eldholm, O., 1994, Large igneous provinces: Crustal structure, dimensions, and external consequences: Reviews of Geophysics, v. 32, p. 1–36.

Dautria, J.M., and Girod, M., 1987, Cenozoic volcanism associated with swells and rifts, *in* Nixon, P.H., ed., Mantle xenoliths: Chichester, UK, John Wiley & Sons, p. 195–214.

Davis, A.S., Gunn, S.H., Gray, L.-B., Marlow, M.S., and Wong, F.L., 1993, Petrology and isotopic composition of Quaternary basanites dredged from the Bering Sea continental margin near Navarin Basin: Canadian Journal of Earth Sciences, v. 30, p. 975–984.

Dover, J.H., 1994, Geology of part of east-central Alaska, *in* Plafker, G., and Berg, H.C., eds., The geology of Alaska: Boulder, Colorado, Geological Society of America, Geology of North America, v. G-1, p. 153–204.

Dumitru, T.A., Miller, E.L., O'Sullivan, P.B., Amato, J.M., Hannula, K.A., Calvert, A.T., and Gans, P.B., 1995, Cretaceous to recent extension in the Bering Strait region, Alaska: Tectonics, v. 14, p. 549–563.

Edwards, B.R., and Russell, J.K., 1999, Northern Cordilleran volcanic province: A northern Basin and Range?: Geology, v. 27, p. 243–246.

Edwards, B.R., and Russell, J.K., 2000, Distribution, nature and origin of Neogene-Quaternary magmatism in the northern Cordilleran volcanic province, Canada: Geological Society of America Bulletin, v. 112, no. 8, p. 1280–1295.

Eiche, G.E., Francis, D.M., and Ludden, J.N., 1987, Primary alkaline magmas associated with the Quaternary Alligator Lake volcanic complex, Yukon Territory, Canada: Contributions to Mineralogy and Petrology, v. 95, p. 191–201.

Estabrook, C.H., Stone, D.B., and Davies, J.N., 1988, Seismotectonics of northern Alaska: Journal of Geophysical Research, v. 93, p. 12026–12040.

Feeley, T.C., and Winer, G.S., 1999, Evidence for fractionation of Quaternary basalts on St. Paul Island, Alaska, with implications for the development of shallow magma chambers beneath Bering Sea volcanoes: Lithos, v. 46, p. 661–676.

Feisinger, D.W., and Nichols, J., 1977, Petrography and petrology of Quaternary volcanic rocks, Quesnel Lake region, east-central British Columbia, *in* Barager, W.R.A., Coleman, L.C., and Hall, J.M., eds., Volcanic regimes in Canada: Geological Association of Canada Special Paper 16, p. 25–38.

Fitton, J.G., 1987, The Cameroon line, West Africa: A comparison between oceanic and continental alkaline volcanism, *in* Fitton, J.G., and Upton, B.G.J., eds., Alkaline igneous rocks: Geological Society [London] Special Publication 30, p. 273–291.

Fitton, J.G., James, D., and Leeman, W.P., 1991, Basic magmatism associated with Late Cenozoic extension in the western United States: Compositional variations in space and time: Journal of Geophysical Research, v. 96, p. 13693–13711.

Francis, D.M., 1976a, The origin of amphibole in lherzolite xenoliths from Nunivak Island, Alaska: Journal of Petrology, v. 17, p. 357–378.

Francis, D.M., 1976b, Corona-bearing pyroxene granulite xenoliths and the lower crust beneath Nunivak Island, Alaska: Canadian Mineralogist, v. 14, p. 291–298.

Francis, D.M., 1976c, Corona-bearing pyroxene granulite xenoliths and the lower crust beneath Nunivak Island, Alaska: Canadian Mineralogist, v. 14, p. 291–298.

Francis, D.M., 1978, The implications of the compositional dependence of texture in spinel lherzolite xenoliths: Journal of Geology, v. 186, p. 473–486.

Francis, D.M., 1987, Mantle-melt interaction recorded in spinel lherzolite xenoliths from Alligator Lake volcanic complex, Yukon, Canada: Journal of Petrology, v. 28, p. 569–597.

Francis, D., and Ludden, J., 1990, The mantle source for olivine nephelinite, basanite, and alkaline olivine basalt at Fort Selkirk, Yukon, Canada: Journal of Petrology, v. 31, p. 371–400.

Francis, D., and Ludden, J., 1995, The signature of amphibole in mafic alkaline lavas, a study in the northern Canadian Cordillera: Journal of Petrology, v. 36, p. 1171–1191.

Frederiksen, A.W., Bostock, M.G., VanDecar, J.C., and Cassidy, J.F., 1998, Seismic structure of the upper mantle beneath the northern Canadian Cordillera from teleseismic travel-time inversion: Tectonophysics, v. 294, p. 43–55.

Fuis, G.S., Levander, A.R., Lutter, W.J., Wissinger, E.S., Moore, T.E., and Christensen, N.I., 1995, Seismic images of the Brooks Range, Arctic Alaska, reveal crustal-scale duplexing: Geology, v. 23, p. 65–68.

Fujita, K., Cook, D.B., Hasegawa, H., Forsyth, D., and Wetmiller, R., 1990, Seismicity and focal mechanisms of the Arctic region and the North American plate boundary in Asia, *in* Grantz, A., Johnson, L., and Sweeney, J.F., eds., The Arctic Ocean region: Boulder, Colorado, Geological Society of America, Geology of North America, v. L, p. 79–99.

Gaspirik, T., 1984, Two-pyroxene thermobarometry with new experimental data in the system CaO-MgO-Al_2O_3-SiO_2: Contributions to Mineralogy and Petrology, v. 87, p. 87–97.

Gill, J., and Whelan, P., 1989, Postsubduction Ocean Island alkali basalts in Fiji: Journal of Geophysical Research, v. 94, p. 4579–4588.

Grandy, J., 1997, Petrogenesis of the Bering Sea Basalt Province: Evidence from St. Michael Volcanic Field, western Alaska [B.Sc. thesis]: St. Paul, Minnesota, Macalester College, 182 p.

Grischkowsky, B.A., Craddock, J.P., and Wirth, K.R., 1995, Progressive strain history from the Seward Peninsula: Evidence from twinned calcite preserved in Ordovician-Cretaceous sediments: 91st Annual Meeting of the Cordilleran Section, Fairbanks, Alaska, Geological Society of America Abstracts with Programs, v. 27, no. 5, p. 22.

Halliday, A.N., Lee, D.-E., Tommasini, S., Davies, G.R., Paslick, C.R., Fitton, J.G., James, D.E., 1995, Incompatible trace elements in OIB and MORB and source enrichment in the sub-oceanic mantle: Earth and Planetary Science Letters, v. 133, p. 379–395.

Hannula, K.A., and McWilliams, M.O., 1995, Reconsideration of the age of blueschist-facies metamorphism on the Seward Peninsula, Alaska, based on phengite $^{40}Ar/^{39}Ar$ results: Journal of Metamorphic Geology, v. 13, p. 125–139.

Hirose, K., and Kushiro, I., 1993, Partial melting of dry peridotites at high pressures: Determination of composition of melts segregated from peridotite using aggregate of diamonds: Earth and Planetary Science Letters, v. 114, p. 477–489.

Hoang, N., and Flower, M., 1998, Petrogenesis of Cenozoic basalts from Vietnam: Implication for origins of a "diffuse igneous province": Journal of Petrology, v. 39, p. 369–395.

Hoang, N., Flower, M.F.J., and Carlson, R.W., 1996, Major, trace element, and isotopic compositions of Vietnamese basalts: Interaction of enriched mobile asthenosphere with the continental lithosphere: Geochimica et Cosmochimica Acta, v. 60, p. 4329–4351.

Hoare, J.M., and Condon, W.H., 1966, Geologic map of the Kwiguk and Black Quadrangles, western Alaska: U.S. Geological Survey Miscellaneous Geological Investigations Series Map I-469, scale 1:250 000, 1 sheet.

Hoare, J.M., and Condon, W.H., 1968, Geologic map of the Hooper Bay Quadrangle, Alaska: U.S. Geological Survey Miscellaneous Geological Investigations Series Map I-523, scale 1:250 000, 1 sheet.

Hoare, J.M., and Condon, W.H., 1971a, Geologic map of the Marshall Quadrangle, western Alaska: U.S. Geological Survey Miscellaneous Geological Investigations Series Map I-668, scale 1:250 000, 1 sheet.

Hoare, J.M., and Condon, W.H., 1971b, Geologic map of the Saint Michael Quadrangle, Alaska: U.S. Geological Survey Miscellaneous Geological Investigations Series Map I-682, scale 1:250 000, 1 sheet.

Hoare, J.M., and Coonrad, W.L., 1959, Geology of the Russian Mission Quadrangle, Alaska: U.S. Geological Survey Miscellaneous Geological Investigations Series Map I-292, scale 1:250 000, 1 sheet.

Hoare, J.M., and Coonrad, W.L., 1978a, A tuya in Togiak Valley, southwest Alaska: U.S. Geological Survey Journal of Research, v. 6, p. 193–201.

Hoare, J.M., and Coonrad, W.L., 1978b, Geologic map of the Goodnews and Hagemeister Island Quadrangles region, southwestern Alaska: U.S. Geological Survey Open-File Report 78–9B, scale 1:250 000, 2 plates.

Hofmann, A.W., 1988, Chemical differentiation of the earth: The relationship between mantle, continental crust, and oceanic crust: Earth and Planetary Science Letters, v. 90, p. 297–313.

Hole, M.J., 1988, Post-subduction alkaline volcanism along the Antarctic Peninsula: Journal of the Geological Society of London, v. 145, p. 985–998.

Hudson, T., 1979, Mesozoic plutonic belts of southern Alaska: Geology, v. 7, p. 230–234.

Ionov, D.A., Hoffman, A.W., and Shimzu, N., 1994, Metasomatism-induced melting in mantle xenoliths from Mongolia: Journal of Petrology, v. 35, part 3, p. 753–785.

Jaques, A.L., and Green, D.H., 1980, Anhydrous melting of peridotite at 0–15 kb pressure and the genesis of tholeiitic basalts: Contributions to Mineralogy and Petrology, v. 73, p. 287–310.

Johnson, K.T.M., Dick, H.J.B., and Shimizu, N., 1990, Melting in the oceanic upper mantle: An ion microprobe study of diopsides in abyssal peridotites: Journal of Geophysical Research, v. 95, p. 2661–2678.

Kay, R.W., and Kay, S.M., 1988, Crustal recycling and the Aleutian arc: Geochimica et Cosmochimica Acta, v. 52, p. 1351–1359.

Kay, R.W., Sun, S.-s., and Lee-Hu, C.-N., 1978, Pb and Sr isotopes in volcanic rocks from the Aleutian Islands and Pribilof Islands, Alaska: Geochimica et Cosmochimica Acta, v. 42, p. 263–273.

Keen, C.F., Courtney, R.C., Dehler, S.A., and Williamson, M-C., 1994, Decompression melting at rifted margins: Comparison of model predictions with the distribution of igneous rocks on the eastern Canadian margin: Earth and Planetary Science Letters, v. 121, p. 403–416.

Kelley, K.A., 1997, Evolution of the mantle beneath the Bering Sea region: Evidence from ultramafic xenoliths of Klikitarik Bay and St. Lawrence Island, Alaska [B.Sc. thesis]: St. Paul, Minnesota, Macalester College, 92 p.

Kirschner, C.E., 1994, Interior basins of Alaska, *in* Plafker, G., and Berg, H.C., eds., The geology of Alaska: Boulder, Colorado, Geological Society of America, Geology of North America, v. G-1, p. 469–493.

Knaack, C., Cornelius, S., and Hooper, P., 1994, Trace element analyses of rocks and minerals by ICP-MS: Pullman, Washington, Washington State University, Department of Geology, GeoAnalytical Laboratory, Open-File Report.

Lassiter, J.C., and DePaolo, D.J., 1997, Plume/lithosphere interaction in the generation of continental and oceanic flood basalts: Chemical and isotopic constraints, *in* Mahoney, J.J., and Coffin, M.F., eds., Large igneous provinces: Continental, oceanic, and planetary flood volcanism: American Geophysical Union Geophysical Monograph 100, p. 335–355.

Le Bas, M.J., Le Maitre, R.W., Streckeisen, A., and Zanettin, B., 1986, A chemical classification of volcanic rocks based on the total alkali-silica diagram: Journal of Petrology, v. 27, p. 745–750.

Luhr, J.F., Aranda-Gomez, J.J., and Pier, J.G., 1989, Spinel-lherzolite-bearing Quaternary volcanic centers in San Luis Potos', Mexico. 1. Geology, mineralogy, and petrology: Journal of Geophysical Research, v. 94, p. 7916–7940.

Maaloee, S., and Aoki, K., 1977, The major element composition of the upper mantle estimated from the composition of lherzolites: Contributions to Mineralogy and Petrology, v. 63, p. 161–173.

Mackey, K.G., Fujita, K., Gunbina, L.V., Kovalev, V.M., Koz'min, B.M., and Imaeva, L.P., 1997, Seismicity of the Bering Strait region: Evidence for a Bering block: Geology, v. 25, p. 979–982.

Mark, R.K., 1971, Strontium isotopic study of basalts from Nunivak Island, Alaska [Ph.D. thesis]: Palo Alto, California, Stanford University, 50 p.

McDonough, W.F., and Sun, S.-s., 1995, The composition of the earth: Chemical Geology, v. 120, p. 223–253.

Menzies, M., and Murthy, V.R., 1980a, Mantle metasomatism as a precursor to the genesis of alkaline magmas: Isotopic evidence: American Journal of Science, v. 280–A, p. 622–638.

                                     K.R. Wirth et al.

Menzies, M., and Murthy, V.R., 1980b, Nd and Sr isotope geochemistry of hydrous mantle nodules and their host alkali basalts: Implications for local heterogeneities in metasomatically veined mantle: Earth and Planetary Science Letters, v. 46, p. 323–334.

Mercier, J-C.C., and Nicolas, A., 1975, Textures and fabrics of upper-mantle peridotites as illustrated by xenoliths from basalts: Journal of Petrology, v. 16, p. 454–487.

Miller, D.M., Goldstein, S.L., and Langmuir, C.H., 1994, Cerium/lead and lead isotope ratios in arc magmas and the enrichment of lead in the continents: Nature, v. 368, p. 514–520.

Miller, E.L., and Hudson, T.L., 1991, Mid-Cretaceous extensional fragmentation of a Jurassic–Early Cretaceous compressional orogen, Alaska: Tectonics, v. 10, p. 781–796.

Miller, E.L., Calvert, A.T., and Little, T.A., 1992, Strain-collapsed metamorphic isograds in a sillimanite gneiss dome, Seward Peninsula, Alaska: Geology, v. 20, p. 487–490.

Miller, E.L., Hannula, K.A., Dumitru, T., Amato, J.M., Calvert, A.T., Gans, P.B., and O'Sullivan, P., 1993, Cretaceous to recent N-S extension of the Bering Strait region: Constraints from geologic and thermochronologic studies on the Seward Peninsula, Alaska: Eos (Transactions, American Geophysical Union), v. 74, p. 614–615.

Miller, T.P., 1989, Contrasting plutonic rock suites of the Yukon-Koyukuk Basin and Ruby geanticline, Alaska: Journal of Geophysical Research, v. 94, p. 15969–15987.

Moll-Stalcup, E.J., 1994, Latest Cretaceous and Cenozoic magmatism in mainland Alaska, *in* Plafker, G., and Berg, H.C., eds., The geology of Alaska: Boulder, Colorado, Geological Society of America, Geology of North America, v. G-1, p. 589–619.

Moll-Stalcup, E., 1995, The origin of the Bering Sea basalt province, western Alaska, *in* Simakov, K.V., and Thurston, D.K., eds., Proceedings of the 1994 International Conference on Arctic Margins: Magadan, Russia, North East Science Center, Far East Branch, Russian Academy of Sciences, p. 113–123.

Moll-Stalcup, E.J., and Arth, J.G., 1989, The nature of the crust in the Yukon-Koyukuk Province as inferred from the chemical and isotopic composition of five Late Cretaceous to Early Tertiary volcanic fields in western Alaska: Journal of Geophysical Research, v. 94, p. 15989–16020.

Nakamura, K., Plafken, G., Jacob, K.H., and Davies, J.N., 1980, A tectonic stress trajectory map of Alaska using information from volcanoes and faults: Bulletin of the Earthquake Research Institute, v. 55, p. 89–100.

Navon, O., and Stopler, E., 1987, Geochemical consequences of melt percolation: The upper mantle as a chromatographic column: Journal of Geology, v. 95, p. 285–308.

Nicholls, J., Stout, M.Z., and Fiesinger, D.W., 1982, Petrologic variations in Quaternary volcanic rocks, British Columbia, and the nature of the underlying upper mantle: Contributions to Mineralogy and Petrology, v. 79, p. 201–218.

Nixon, P.A., editor, 1987, Mantle xenoliths: Chichester, UK, John Wiley & Sons, 844 p.

Patton, W.W., and Box, S.E., 1989, Tectonic setting of the Yukon-Koyukuk Basin and its borderlands, western Alaska: Journal of Geophysical Research, v. 94, p. 15807–15820.

Patton, W.W., Jr., and Csejtey, B., Jr., 1980, Geologic map of St. Lawrence Island, Alaska: U.S. Geological Survey Miscellaneous Geologic Investigation Series Map MF-1203, scale 1:250 000, 1 sheet.

Patton, W.W., Jr., and Moll, E.J., 1985, Geologic map of northern and central parts of Unalakleet Quadrangle, Alaska: U.S. Geological Survey Miscellaneous Field Studies Map MF-1749, scale 1:250 000, 1 sheet.

Peltzer, G., and Tapponnier, P., 1988, Formation and evolution of strike-slip faults, rifts, and basins during the India-Asia collision: An experimental approach: Journal of Geophysical Research, v. 93, p. 15085–15117.

Perry, F.V., 1987, Role of asthenosphere and lithosphere in the genesis of late Cenozoic basaltic rocks from the Rio Grande Rift and adjacent regions of the southwestern United States: Journal of Geophysical Research, v. 92, no. B9, p. 9193–9213.

Perry, F.V., Baldridge, W.S., and DePaolo, D.J., 1987, Role of asthenoscope and lithosphere in the genesis of late Cenozoic basaltic rocks from the Rio Grande Rift and adjacent regions of southwestern United States: Journal of Gephysical Research, v. 92, p. 9193–9213.

Plafker, G., and Berg, H.C., 1994, Overview of the geology and tectonic evolution of Alaska, *in* Plafker, G., and Berg, H.C., eds., The geology of Alaska: Boulder, Colorado, Geology of North America, v. G-1, p. 989–1021.

Plafker, G., Gilpin, L.M., and Lahr, J.C., 1994, Neotectonic map of Alaska, *in* Plafker, G., and Berg, H.C., eds., The geology of Alaska: Boulder, Colorado, Geological Society of America, Geology of North America, v. G-1, scale 1:2 500 000, 1 sheet.

Ringwood, A.E., 1979, Origin of the earth and moon: New York, Springer-Verlag, 295 p.

Roden, M.F., Frey, F.A., and Francis, D.M., 1984, An example of consequent mantle metasomatism in peridotite inclusions from Nunivak Island, Alaska: Journal of Petrology, v. 25, p. 546–577.

Roden, M.F., Francis, D.M., and Frey, F.A., 1995, Upper mantle composition beneath the eastern Bering Sea, *in* Simakov, K.V., and Thurston, D.K., eds., Proceedings of the 1994 International Conference on Arctic Margins: Magadan, Russia, North East Science Center, Far East Branch, Russian Academy of Sciences, p. 147–152.

Sack, R.O., and Ghiorso, M.S., 1991, Chromian spinels as petrogenetic indicators: Thermodynamics and petrological applications: American Mineralogist, v. 76, p. 827–847.

Sadofsky, S.J., 1995, Petrogenesis of the Bering Sea volcanic province: Evidence from lavas near Brevig Mission, Alaska [B.Sc. thesis]: St. Paul, Minnesota, Macalester College, 80 p.

Sainsbury, C.L., 1972, Geologic map of the Teller Quadrangle, western Seward Peninsula, Alaska: Miscellaneous Geologic Investigations Map I-0685, 4 p.

Shi, S., Francis, D., Ludden, J., Frederiksen, A., and Bostock, M., 1998, Xenolith evidence for lithospheric melting above anomalously hot mantle beneath the northern Canadian Cordillera: Contributions to Mineralogy and Petrology, v. 131, p. 39–53.

Silberling, N.J., Jones, D.L., Monger, J.W.H., Coney, P.J., Berg, H.C., and Plafker, G., 1994, Lithotectonic terrane map of Alaska and adjacent parts of Canada, *in* Plafker, G., and Berg, H.C., eds., The geology of Alaska: Boulder, Colorado, Geological Society of America, Geology of North America, v. G-1, scale 1:2 500 000, 1 sheet.

Skulski, T., Francis, D., and Ludden, J., 1991, Arc-transform magmatism in the Wrangell volcanic belt: Geology, v. 19, p. 11–14.

Souther, J.G., 1977, Volcanism and tectonic environments in the Canadian Cordillera: A second look, *in* Barager, W.R.A., Coleman, L.C., and Hall, J.M., eds., Volcanic regimes in Canada: Geological Association of Canada Special Paper 16, p. 3–24.

Souther, J.G., Armstrong, R.L., and Harakal, J., 1984, Chronology of the peralkaline, late Cenozoic Mount Edziza Volcanic Complex, northern British Columbia, Canada: Geological Society of America Bulletin, v. 95, p. 337–349.

Sun, S.-s., and Hanson, G.N., 1975, Evolution of the Mantle: Evidence from alkali basalt: Geology, v. 3, p. 297–302.

Sun, S.-s., and McDonough, W.F., 1989, Chemical and isotopic systematics of oceanic basalts: Implications for mantle composition and processes, *in* Saunders, A.D., and Norry, M.J., eds., Magmatism in the ocean basins: Geological Society [London] Special Publication 42, p. 313–345.

Swanson, S.E., Turner, D.L., Forbes, R.B., and Hopkins, D.M., 1981, Petrology and geochemistry of Tertiary and Quaternary basalts from the Seward Peninsula, western Alaska: Geological Society of America Abstracts with Programs, v. 13, p. 563.

Swanson, S.E., Kay, S.M., Brearley, M., and Scarfe, C.M., 1987, Arc and back-arc xenoliths in Kurile-Kamchatka and western Alaska, *in* Nixon, P.H., ed., Mantle xenoliths: Chichester, UK, John Wiley & Sons, p. 303–318.

Takahashi, E., 1986, Melting of a dry peridotite KLB-1 up to 14 GPa: Implication on the origin of peridotitic upper mantle: Journal of Geophysical Research, v. 91, p. 9367–9382.

Tapponnier, P., Peltzer, G., Le Dain, A.Y., Armijo, R., and Cobbold, P., 1982, Propagating extrusion tectonics in Asia: New insights from simple experiments with plasticine: Geology, v. 10, p. 611–616.

Till, A.B., and Dumoulin, J.A., 1994, Geology of Seward Peninsula and Saint Lawrence Island, *in* Plafker, G., and Berg, H.C., eds., The geology of Alaska: Boulder, Colorado, Geological Society of America, Geology of North America, v. G-1, p. 141–152.

Tolson, R.B., 1986, Structure and stratigraphy of the Hope Basin, southern Chukchi Sea, Alaska, *in* Scholl, D.W., Grantz, A., and Vedder, J.G., eds., Geology and resource potential of the continental margin of western North American and adjacent ocean basins: Beaufort Sea to Baja California: Circum-Pacific Council for Energy and Mineral Resources, Earth Science Series, v. 6, p. 59–72.

Turner, D.L., Swanson, S.E., and Wescott, E., 1981, Continental rifting: A new tectonic model for geothermal exploration of the central Seward Peninsula, Alaska: Geothermal Resources Council Transactions, v. 5, p. 213–216.

Turner, S., and Hawkesworth, C., 1995, The nature of the subcontinental mantle: Constraints from the major-element composition of flood basalts: Chemical Geology, v. 120, p. 295–314.

von Drach, V., Marsh, B.D., and Wasserburg, G.J., 1986, Nd and Sr isotopes in the Aleutians: Multicomponent parenthood of island-arc magmas: Contributions to Mineralogy and Petrology, v. 92, p. 13–34.

Wallace, W.K., and Engebretson, D.C., 1984, Relationship between plate motions and Late Cretaceous to Paleogene magmatism in southwestern Alaska: Tectonics, v. 3, p. 295–315.

Weaver, B.L., 1991, Trace element evidence for the origin of ocean-island basalts: Geology, v. 19, p. 123–126.

Weaver, B.L., Wood, D.A., Tarney, J., and Joron, J.L., 1987, Geochemistry of ocean island basalts from the South Atlantic: Ascension, Bouvet, St. Helena, Gough, and Tristan da Cunha, *in* Fitton, J.G., and Upton, B.G.J., eds., Alkaline igneous rocks: Geological Society [London] Special Publication 30, p. 253–267.

Whitford-Stark, J.L., 1987, A survey of Cenozoic volcanism on mainland Asia: Boulder, Colorado, Geological Society of America Special Paper 213, 71 p.

Whitney, J.W., and Wallace, W.K., 1995, Plate tectonic model for the neotectonics of northern Alaska: Geological Society of America Abstracts with Programs, v. 27, no. 5, p. 84.

Williamson, M-C., Courtney, R.C., Keen, C.E., and Dehler, S.A., 1995, The volume and rare earth concentrations of magmas generated during finite stretching of the lithosphere: Journal of Petrology, v. 36, p. 1433–1453.

Wirth, K.R., 1991, Processes of lithosphere evolution: Geochemistry and tectonics of mafic rocks in the Brooks Range and Yukon-Tanana region, Alaska [Ph.D. thesis]: New York, Cornell University, 367 p.

Wirth, K.R., Ronnback, C.M., Grandy, J.S., and Sadofsky, S.J., 1995, Mafic volcanism and xenoliths of the Northern Bering Sea region [abs.]: Eos (Transactions, American Geophysical Union) 1993 Fall Meeting, San Francisco, California, v. 76, no. 46, p. 589–590.

Wood, D.A., 1980, The application of a Th-Hf-Ta diagram to problems of tectonomagmatic classification and to establishing the nature of crustal contamination of basaltic lavas of the British Tertiary volcanic province: Earth and Planetary Science Letters, v. 50, p. 11–30.

Worrall, D.M., 1991, Tectonic history of the Bering Sea and the evolution of Tertiary strike-slip basins of the Bering Shelf: Boulder, Colorado, Geological Society of America Special Paper 257, 120 p.

Zhi, X., Song, Y., Frey, F.A., Feng, J., and Zhai, M., 1990, Geochemistry of Hannuoba basalts, eastern China: Constraints on the origin of continental alkalic and tholeiitic basalt: Chemical Geology, v. 88, p. 1–33.

MANUSCRIPT ACCEPTED BY THE SOCIETY MAY 15, 2001.

Geological Society of America
Special Paper 360
2002

Constraints on the age of formation of seismically reflective middle and lower crust beneath the Bering Shelf: SHRIMP zircon dating of xenoliths from Saint Lawrence Island

Elizabeth L. Miller and Trevor R. Ireland*
*Department of Geological and Environmental Sciences, Stanford University,
Stanford, California 94305-2115, USA*
Simon L. Klemperer
Department of Geophysics, Stanford University, Stanford, California 94305-2115, USA
Karl R. Wirth
Geology Department, Macalester College, Saint Paul, Minnesota 55105, USA
Vyacheslav V. Akinin
*Russian Academy of Science, Northeast Interdisciplinary Science Research Institute,
Portovaya Street, 16, Magadan, 685000, Russia*
Thomas M. Brocher
U.S. Geological Survey, 345 Middlefield Road, Menlo Park, California 94025, USA

ABSTRACT

Seismic reflection and/or refraction studies reveal reflective middle and lower crust and a sharp Moho (~32 km depth) beneath a broad region of the Bering Shelf between Alaska and northeast Russia. Basalt flows on Saint Lawrence Island of the late Cenozoic Bering Sea basalt province contain upper mantle and crustal xenoliths that include mafic cumulate rocks and lesser pyroxene-bearing gneisses that equilibrated at ~4–6 kbar. The gneissic xenoliths are interpreted as intrusive rocks that acquired deformation and/or recrystallization fabrics during granulite facies metamorphism.

Three gneissic xenoliths from two sites yielded zircons that were dated by the U-Pb method with the SHRIMP II (sensitive high resolution ion microprobe). Zoned prismatic zircons of magmatic origin yield ages mostly ca. 85–90 Ma. Rounded, non-zoned zircons from other samples are likely metamorphic in origin and yield mostly ~64 Ma ages. No older ages were obtained. More abundant gabbroic xenoliths are interpreted to represent mafic magmas emplaced into the middle to lower crust during this same approximate time span.

The oldest surface rocks on Saint Lawrence Island include Paleozoic-Mesozoic shelfal units of the Brooks Range (once deposited on Precambrian basement) but xenolith age data suggest that such older rocks, if ever volumetrically important in the deeper crust, could have been reconstituted and remobilized during younger thermal and/or magmatic events. Conversely, Late Cretaceous to Paleocene magmatic rocks are likely increasingly important with depth in the crust. A similarly young age is inferred for the development of seismically imaged reflective crust and (by inference) the Moho beneath the Bering Shelf.

*Current address: Research School of Earth Sciences, The Australian National University, 1 Mills Road, Canberra, ACT 0200, Australia.

Miller, E.L., Ireland, T.R., Klemperer, S.L., Wirth, K.R., Akinin, V.V., and Brocher, T.M., 2002, Constraints on the age of formation of seismically reflective middle and lower crust beneath the Bering Shelf: SHRIMP zircon dating of xenoliths from Saint Lawrence Island, *in* Miller, E.L., Grantz, A., and Klemperer, S.L., eds., Tectonic Evolution of the Bering Shelf–Chukchi Sea–Arctic Margin and Adjacent Landmasses: Boulder, Colorado, Geological Society of America Special Paper 360, p. 195–208.

INTRODUCTION

Deep seismic reflection profiling has brought new insight into the nature of tectonic, magmatic, and metamorphic processes occurring in the deep crust and mantle that are not easily inferred from surface geology alone. The results of such studies have had important implications for our understanding of the evolution and growth of continental crust and have also underscored the importance of magmatism and extension in modifying and stabilizing continental crust (e.g., Costa et al., 1994; Costa and Rey, 1995; Mooney and Meissner, 1992). Seismic reflection and refraction data were collected across the Bering Shelf–Chukchi Sea region as part of an international collabora-

tive effort; the effort included two parallel seismic reflection profiles that are the first to image the deep crust and mantle beneath the region (Klemperer et al., this volume, Chapter 1; Wolf et al., this volume, Chapter 2) (Fig. 1). Mantle- and crust-derived xenoliths are common in Neogene basalt fields of the Bering Sea basalt province, and their study provides a unique opportunity to place constraints on the nature and age of the seismically imaged crust. This chapter presents U-Pb age data on zircons from pyroxene-bearing gneissic xenoliths inferred to have originated in the middle crust beneath Saint Lawrence Island on the Bering Shelf (Fig. 1). These new data provide us with information on the age and history of the crust beneath this region, underscore the importance of magmatic processes in determining this his-

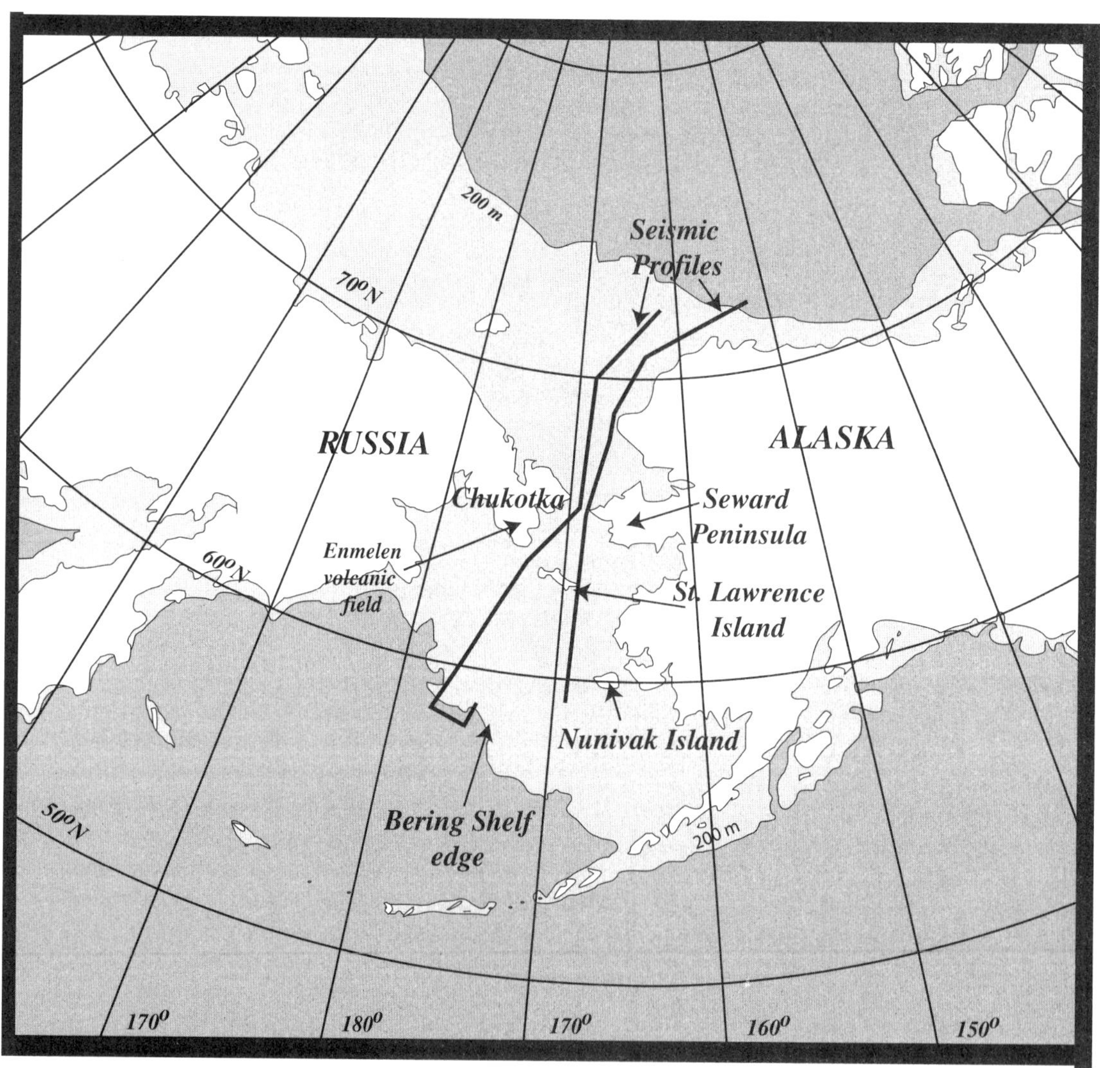

Figure 1. Location map of seismic reflection profiles of the Bering-Chukchi deep crustal transect and some of most important Neogene basalt fields of the Bering Sea basalt province: Saint Lawrence Island, Nunivak Island, the Candle and Immuruk basin volcanic fields in central Seward Peninsula of Alaska, and Emmelen volcanic field on Chukotka Peninsula, Russia. Base map is from U.S. Geological Survey and National Oceanographic and Atmospheric Administration.

tory, and emphasize the relative youthfulness of this history compared to the older accretionary evolution of the crust as recorded by its supracrustal geology.

GEOLOGIC SETTING

The Bering Shelf–Chukchi Sea region comprises over 50% of the total U.S. continental shelf and forms a broad isthmus of continental crust connecting the North American and Asian continents (Fig. 1). Overall, the geologic evolution of the Bering-Chukchi Shelf is only poorly understood, except for studies of the Tertiary basins of the Bering Shelf (e.g., Worral, 1991) and the Paleozoic and Mesozoic basins of the Chukchi Shelf (e.g., Sherwood et al., this volume, Chapter 3), which have been studied in detail.

The North American Cordillera is considered to be a prime example of crustal growth by the lateral accretion of terranes and island-arc systems, and the Alaskan portion of this belt is no exception. Alaska (and the Bering Shelf) is considered to have been shaped almost entirely by the process of terrane accretion since the Middle Jurassic (e.g., Howell et al., 1987; Nokleberg et al., 1998). This accretionary history was the fundamental process involved in creation of the Brooks Range foreland thrust belt in northern Alaska (Figs. 1 and 2) and in the assemblage of accreted oceanic and arc terranes along the southern margin of the range (e.g., Moore et al., 1994) (Fig. 2). Paleozoic and early Mesozoic autochthonous and allochthonous rock units of the Brooks Range orogen are found as far south as Saint Lawrence Island (Patton and Csejtey, 1980), but here the orogen is close to or beneath sea level, indicating that the processes of forma-

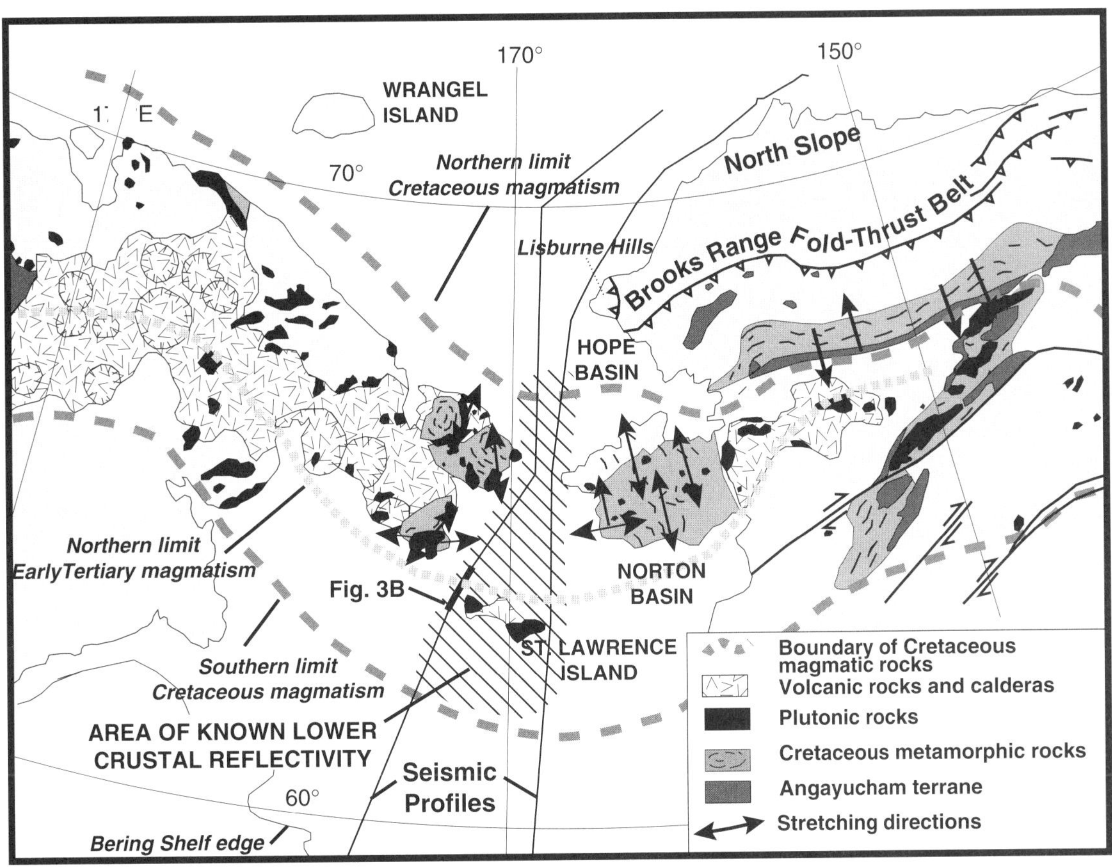

Figure 2. Location of seismic reflection lines of Bering-Chukchi deep seismic transect with respect to geologic features discussed in text. Region known to be underlain by reflective crust as imaged in seismic lines is shown by diagonal-line pattern, but probably underlies much broader region of inner Bering Shelf. Approximate northern and southern limits of mid–Late Cretaceous magmatism are shown by dark dashed lines. Approximate northern limit of very latest Cretaceous and early Tertiary magmatism is shown by light-gray dashed line. Its southern boundary is the Bering Shelf edge, based on distribution of Kuskokwim Mountains and Yukon-Kanuti magmatic belts of Alaska described by Moll-Stalcup (1994), and their inferred offshore continuation (Cooper et al., 1987; Worral, 1991). Stretching direction in metamorphic rocks within the belt of Cretaceous magmatism from Little et al. (1994), Law et al. (1994), Toro et al. (this volume), and Christiansen and Snee (1993) for Brooks Range; Till and Dumoulin (1999) and Dumitru et al. (1995) for Seward Peninsula; and Miller et al. (this volume, Chapter 17) for Russia.

tion and modification of continental crust in this region must differ from those characterizing the rest of the Brooks Range. Was terrane accretion the only mechanism of crustal growth across this portion of the orogen, and if so, why didn't this region undergo crustal thickening? Or, has subsequent crustal extension played a role here, and if so, at what time?

Cutting across the accreted terranes of Alaska and Russia are broad magmatic belts of Cretaceous age, the plutons of which span the time interval ca. 118–80 Ma (e.g., Nokleberg et al., 1997; Miller, 1994; Amato and Wright, 1997) (Fig. 2). Cretaceous plutons are commonly viewed as being of secondary importance to the mostly accretionary evolution of the Alaskan orogen (e.g., Nokleberg et al., 1998). Continuing work, however, has shown that Cretaceous magmatism was accompanied by significant regional metamorphism and deformation that may have played a fundamental role in the evolution of the deep crust beneath this region. This deformation is extreme near the Bering Strait, where it is responsible for the genesis of a series of metamorphic culminations or gneiss domes that expose sillimanite to granulite facies mid-crustal rocks (e.g., Bering Strait Geologic Field Party, 1997; Akinin and Calvert, this volume, Chapter 8; Amato et al., this volume, Chapter 7). Deformation associated with the formation of these gneiss domes is interpreted as related to the vertical rise of mid-crustal-level rocks to shallow crustal depths during regional extension. Data on the direction of crustal extension are compiled in Figure 2 and indicate north-south stretching within the Cretaceous magmatic belt at this latitude.

Miller (1994) and Amato and Wright (1997, 1998) summarized geochronologic, geochemical, and isotopic data from Cretaceous granites in Alaska, and Rowe (1998) summarized similar data from Russia. Isotopic values indicate a mantle component or origin for the magmas, modified by a variable but often significant crustal component. The variation of these values in given areas suggests a decreasing crustal component through time, at least for the Alaskan plutons (Amato and Wright, 1997). Amato and Wright (1997) suggested that plutonism is associated with northward subduction beneath the Bering Strait region and that the source region for mafic magmatism in the belt is inferred to have been enriched lithospheric mantle. The interpretation of these data from the Cretaceous plutonic belt, combined with regional geologic data, argues for significant heating (and melting and remobilization) of the crust in the Bering Strait region during the documented time span of mantle-derived magmatism (e.g., Bering Strait Geologic Field Party, 1997).

Less is known about the tectonic setting of the early Tertiary belt of volcanic and plutonic rocks that are inferred to underlie the outer part of the Bering Shelf (Fig. 2). On land, continental volcanic rocks and associated plutons form a broad belt defined by the Kuskokwim Mountains and the Yukon-Kanuti magmatic belts of Alaska, and are dated as very latest Cretaceous to early Tertiary (Paleocene) (Moll-Stalcup, 1994). The offshore continuation of this belt is inferred from magnetic anomalies along the outer edge of the Bering Shelf and by the dredging and coring of rocks of this age along the Bering Shelf

edge (Marlow et al., 1976; Worral, 1991). Equivalent rocks occur in the Anadyr basin, Russia (Agapitov et al., 1973), on Saint Matthew Island (Patton et al., 1976), and in the subsurface in the Saint George and Bristol basins (Burk, 1965; McLean, 1977, 1979). These rocks were referred to as the Anadyr-Bristol volcanogenic belt by Ivanov (1985).

SEISMIC DATA

Figure 3 summarizes seismic refraction (Fig. 3A) and reflection (Fig. 3B) data (Klemperer et al., this volume, Chapter 1) representative of the crust beneath Saint Lawrence Island and the Bering Strait. The approximate range of depths from which the studied xenoliths are inferred to have been derived is shown in Figure 3B (for pressure-temperature determinations, see Wirth et al., this volume, Chapter 9).

During collection of the seismic reflection data, seismic recorders were deployed on islands and on both sides of the Bering and Chukchi Seas (Fig. 3C). The crust and upper mantle in the Bering Sea between Saint Lawrence Island, Tin City, Alaska, and northeastern Russia was sampled using several recorders in Alaska and Russia. These instruments digitally recorded signals from the 137.7 L air-gun array towed behind the ship with large source-receiver offsets (Brocher et al., 1995). The ray coverage obtained by recording seismic arrivals onshore at large ranges (wide angles) from the seismic lines is shown schematically in Figure 3B. Seismic arrivals refracted from the upper mantle (Pn) begin to arrive before those from the upper crust (Pg arrivals) at ranges of ~140 km along our transect (Brocher et al., 1995) and are consistent with a crustal thickness of 30–35 km.

Traveltimes of first and second arrivals were forward modeled using one-dimensional velocity models (Luetgert, 1992) due to the lack of true ray path reversal along much of the transect (Fig. 3C) and the oblique geometry of the reflection lines to the receivers. Separate one-dimensional models were obtained for sources north and south of each receiver. These one-dimensional models were then organized by distance resulting in a two-dimensional velocity model (Fig. 3A). Subsequent two-dimensional modeling of the traveltime data agrees well with the model shown in Figure 3A (Wolf et al., this volume, Chapter 2). The oblique recording geometry means that there are no arrivals recorded at ranges <20–30 km, greatly limiting our knowledge of velocities in the uppermost few kilometers of the crust.

The wide-angle data indicate lower crust velocities on the order of 6.1–6.4 km/s at 16–17 km depth (Conrad discontinuity) increasing to 6.7–6.9 km/s at ~32 km depth (Moho) (Fig. 3). There is a fairly clear demarcation of velocities, with velocities ≤6 km/s above the Conrad and ≥7.9 km/s below the Moho.

Seismic reflection data that imaged the deep crust and mantle (Klemperer et al., this volume, Chapter 1) indicate several unique aspects of the crust beneath Saint Lawrence Island that must be considered when interpreting the significance of the U-Pb ages we have obtained from crustal xenoliths. The seismic

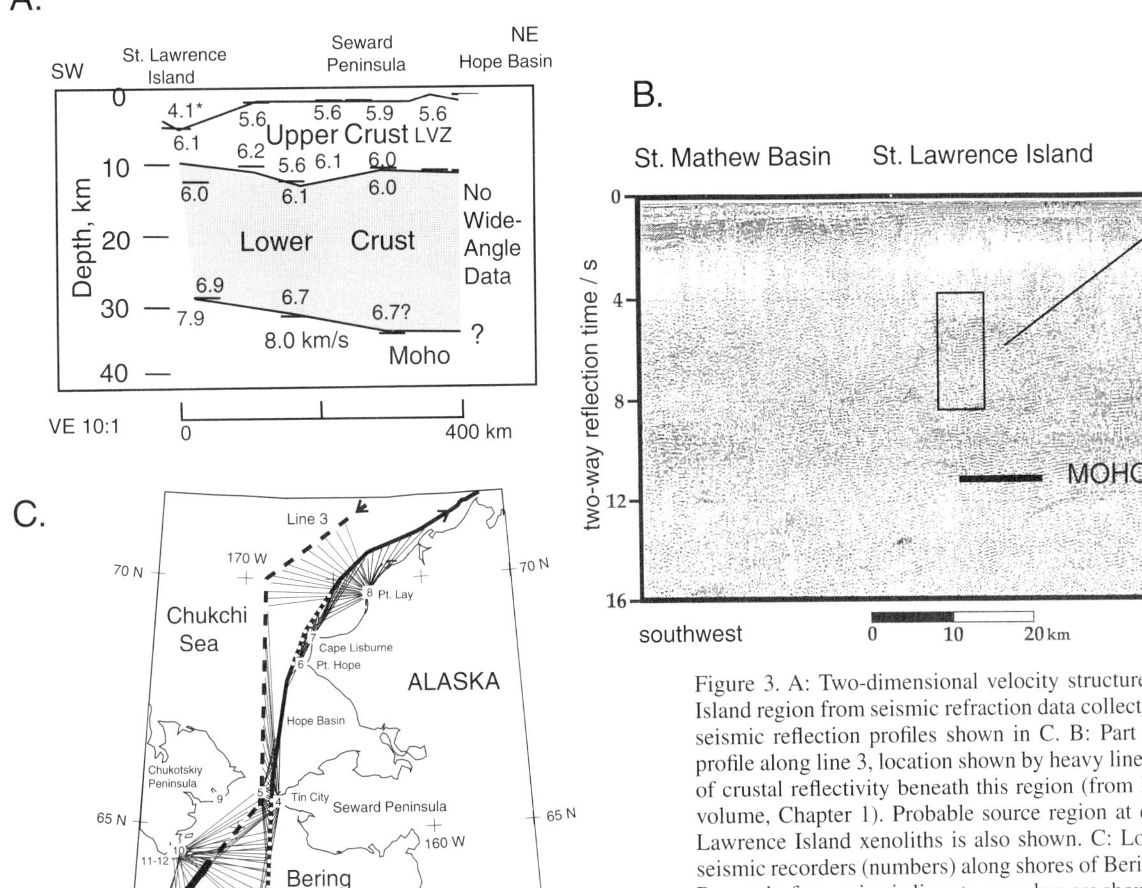

Figure 3. A: Two-dimensional velocity structure for Saint Lawrence Island region from seismic refraction data collected during shooting of seismic reflection profiles shown in C. B: Part of seismic reflection profile along line 3, location shown by heavy line in C, showing nature of crustal reflectivity beneath this region (from Klemperer et al., this volume, Chapter 1). Probable source region at depth for dated Saint Lawrence Island xenoliths is also shown. C: Location of wide-angle seismic recorders (numbers) along shores of Bering and Chukchi Seas. Ray paths from seismic lines to recorders are shown schematically (thin straight lines) to illustrate variations in sampling of crust resulting from this investigation.

reflection data (Klemperer et al., this volume, Chapter 1) indicate a gradual but significant change in the character of lower-crustal reflections and the reflection Moho just south of the Hope Basin (Fig. 3A). South from there, the reflection Moho is normally picked as the base of a sequence of many reflections that occupy the lower one- to two-thirds of the crust, marking the transition to a usually nonreflective upper mantle (Fig. 3B). The layered character of the crust is most prominent from the Bering Strait to Saint Lawrence Island and southward (Fig. 3B), a distance of ~500 km, and disappears near Saint Matthew Basin. Across this same distance, the crust is also a uniform thickness, estimated as 32 km (Klemperer et al., this volume, Chapter 1). It is noteworthy that the shallowest reflections of this nature are observed and occur within the uppermost 5–10 km of crust just north of Saint Lawrence Island (Fig. 3B) (and near the Seward Peninsula) (Klemperer et al., this volume, Chapter 1). In general, however, as with many if not most crustal-scale reflection profiles, the uppermost crystalline crust contains few reflections, in large part due to technical reasons involved in the data collection (Klemperer et al., this volume, Chapter 1). It is not surprising, there-

fore, that the data do not image specific plutons, metamorphic complexes, or gneiss domes along our line of transect.

Examples of similarly reflective crust were first observed to be widespread in the thin crust of young extensional provinces such as the North Sea and the Basin and Range Province of the western United States (e.g., Matthews and Cheadle, 1986; Allmendinger et al., 1987). In general, this type of reflectivity is thought to represent tectonically imposed layering formed by subhorizontal flow and vertical shortening during crustal thinning or extension, and/or primary igneous layering produced by mafic intrusions into the lower crust (e.g. Green et al., 1990; Warner, 1990). Crustal collapse or crustal thinning following an episode of tectonic thickening is the most likely sequence of events leading to the formation of such highly reflective crust (Warner, 1990; Nelson, 1992; Rey, 1993). In the Bering Strait–Saint Lawrence region, we attribute the reflectivity to the postulated collapse and extension of crust previously thickened during Brookian shortening in the Late Jurassic to Early Cretaceous. Collapse and extension of the crust took place during plutonism associated with the middle to Late Cretaceous magmatic belt, when mafic underplating of the crust is also likely to have occurred (Miller and Hudson, 1991; Rubin et al., 1995; Bering Strait Field Geology Party, 1997; Amato and Wright, 1997, 1998). The region of uniform crustal thickness and high reflectivity coincides with the locus of Cretaceous magmatism at the Earth's surface (Fig. 2). The lower part of the crust beneath Saint Lawrence Island has velocities that range from 6.1 to 6.7 km/s (Fig. 3B). These velocities are consistent with some interplating of mafic sills into lower crust, leading to a bimodal velocity structure at a small scale generated by ~6.5 km/s felsic to intermediate igneous and metamorphic lithologies and ~7 km/s gabbroic intrusions (Klemperer et al., this volume, Chapter 1), but are not consistent with large volumes of gabbroic intrusives, which would have velocities of ~7 km/s. Mafic sills are a likely source for the gabbroic xenoliths with cumulate textures collected from the Neogene basalt flows on Saint Lawrence Island (Wirth et al., this volume, Chapter 9). The middle part of the crust (16–24 km depth) is characterized by fairly low velocities of 6.1–6.4 km/s (Wolf et al., this volume, Chapter 2), precluding a large component of new mafic intrusions at these depths. Thus, the reflectivity observed higher in the crust must represent ductile extensional fabrics and transposed compositional layering in metamorphic (metasedimentary and metaigneous) rocks. Examples of these rocks are present in the large gneiss domes or metamorphic culminations of the Chukotka and Seward Peninsulas (Akinin and Calvert, this volume, Chapter 8; Amato et al., this volume, Chapter 7; Bering Strait Field Geology Party, 1997).

PETROLOGY OF LAVAS AND XENOLITHS FROM SAINT LAWRENCE ISLAND

The late Cenozoic Bering Sea basalt province (Moll-Stalcup, 1994; Akinin and Apt, 1994, 1997; Akinin et al., 1997) consists of more than 17 volcanic fields on islands in the Bering Sea, on the west coast of Alaska, and on the northeastern coast of Russia. Several of these are shown in Figure 1. These volcanic fields represent the most significant volume of magma erupted in the Arctic region within the past 10 m.y. and consist primarily of tholeiitic and alkali-olivine basalt flows with subordinate basanite and nephelinite cones, flows, and maars. Alkali basalts and nephelinite from the province contain upper mantle and lower and/or upper crustal xenoliths, generally a common occurrence in alkali basalt provinces worldwide. Most of the volcanic fields are dominated by intraplate basalts that compositionally resemble those erupted in oceanic islands and in some continental settings (e.g., Wood, 1980; Clague and Frey, 1982). Pb, Sr, and Nd isotopic differences between individual volcanic fields in the Bering Sea basalt province are greater than the difference within volcanic fields, which suggests different mantle sources for the lavas (Davis et al., 1995; Akinin and Apt, 1997).

The volcanic field on Saint Lawrence Island consists of a large Paleocene to Quaternary shield volcano of alkali-olivine and tholeiitic basalt flows and more than 70 small younger cones and short flows of mostly basanite and nephelinite (Moll-

Stalcup, 1994). The xenolith population in these lavas consists of spinel lherzolite, harzburgite, wehrlite, pyroxenite, amphibole-bearing pyroxenite, gabbro-norite, anorthosite, and crustal rocks including granites, gneisses, felsic volcanic, and sedimentary rock (Wirth et al., 1995). Ultramafic xenoliths are typically foliated and contain mineral grains that exhibit deformation interpreted to be due to flow in the mantle. Many of the plagioclase-bearing xenoliths have cumulate textures and are variably deformed, suggesting they originated in magma chambers in the lower crust (Wirth et al., 1995).

Many of the plagioclase- and pyroxene-bearing xenoliths are coarse-grained gabbro and gabbronorite with cumulate and poikolitic textures suggesting a magmatic origin. Deformation in these xenoliths is largely limited to undulatory extinction and slight bending of twin lamellae in plagioclase. In contrast, many of the medium-grained and feldspar-rich xenoliths exhibit a more pronounced metamorphic layering and foliation (Fig. 4). In thin section, these xenoliths exhibit a range of textures from polygonal granoblastic equigranular to interlobate and seriate; some relict igneous textures are also present. The primary pyroxene present in all of these xenoliths is hypersthene; minor diopside is also present in some xenoliths. Feldspars typically exhibit variable undulatory extinction and twin lamellae that are sometimes bent or broken. Plagioclase feldspars are typically andesine in composition, whereas alkali feldspars are sodium rich (anorthoclase composition) and exhibit coarse exsolution lamellae (mesoperthite). Minor quartz (1%–10%), Fe-Ti oxides, and apatite are also typically present. Hydrous minerals are absent, except as rare alteration products around the margins of pyroxene.

The compositions and textures of the plagioclase-pyroxene–bearing xenoliths are interpreted to indicate that they are largely of igneous origin. This interpretation is consistent with the lack of convincing evidence that any of the observed xenoliths are metasedimentary. The aluminum compositions of the clinopyroxenes plot in the field of granulites (Al [IV] versus Al [VI]; Aoki and Kushiro, 1968). Furthermore, the Al contents of clinopyroxene and orthopyroxene pairs (Gasparik, 1984) suggest equilibration pressures of 4–6 kbar. Therefore, although we cannot unambiguously demonstrate that these xenoliths are granulite facies metamorphic rocks, we believe that the observed textures and compositions are consistent with an origin in which mafic to intermediate magmas were intruded into the mid-crust and were subsequently deformed and recrystallized at relatively high temperatures and pressures.

Xenoliths provide us with the only means of directly sampling the lower crust and mantle. Although detailed studies of xenoliths are numerous, relatively few studies have determined the ages of xenoliths that sample lower crust and mantle (e.g., Rudnick and Williams, 1987; Rudnick and Fountain, 1995; Costa and Rey, 1995; O'Reilly et al., 1995; Davis, 1996). Our initial work suggests that there are plentiful zircon populations in the probable granulite facies gneissic xenoliths brought to the surface by young basalts of the Bering Sea basalt province. The descriptions and data discussed in the following are preliminary

Figure 4. Pyroxene-bearing gneissic xenoliths collected from basalt flows on Saint Lawrence Island by Karl Wirth in 1994, illustrating nature of conspicuous gneissic foliation observed at hand-specimen scale.

and few in number compared to the information potentially available from other such rocks across this vast region.

SHRIMP DATING OF ZIRCONS

Four pyroxene-bearing gneissic xenoliths from Saint Lawrence Island were processed for zircon. Three of these yielded zircons that were dated. Sample SV19B has a visible foliation in hand specimen. In thin section it is fine grained (~0.5 mm) and granoblastic, with foliation defined by layers or concentrations of pyroxene versus plagioclase. It is composed of hypersthene (20%–30%), andesine (70%–80%) with some granophyric overgrowths on feldspars, minor quartz, and ilmenite-magnetite. Brown glass is associated with, and rims, some pyroxenes. Sample 21C is medium to coarse grained (2–4 mm) and granoblastic. Foliation or layering is present in hand specimen, but not visible in thin section due to the coarse-grained nature of the rock (Fig. 5). It is composed of diopside (trace), hypersthene (20%–30%), andesine (70%–80%), minor perthite, and oxides (2%–5%). Plagioclase grains are often strained and bent and sometimes broken. Sample SV21E is similar to SV21C and is a coarse-grained (2–4 mm), granoblastic, pyroxene-bearing gneiss. Although foliation or layering is present in hand specimen, it is not obvious in thin section due to the coarse-grained nature of the sample. It is composed of mesoperthite (70%–80%), andesine (5%–10%), diopside (10%–15%), hypersthene (2%–5%), and quartz (5%–10%). Feldspars are often strained and bent.

Zircons were mounted in epoxy and polished to reveal midsections. Prior to ion microprobe analysis, the zircons were photographed in transmitted light (e.g., Fig. 6) and imaged by cathodoluminescence for internal structures largely related to trace element zoning (e.g., Fig. 7). U-Pb analyses were carried out on the SHRIMP II at the Australian National University (Table 1). Owing to the youthfulness of the zircons, most effort was placed on obtaining a precise U-Pb age; $^{207}Pb/^{206}Pb$ ages are imprecise owing to the small amount of radiogenic ^{207}Pb present and the uncertainties propagated from the $^{204}Pb/^{206}Pb$ correction. As such, the fraction of common Pb in an analysis was estimated by assessing the measured $^{207}Pb/^{206}Pb$ ratio in relation to the projected position, from the common Pb isotopic composition, on to the Tera-Wasserburg concordia, which also thus yields the $^{206}Pb/^{238}U$ age (e.g., see Muir et al., 1996a,b). The common Pb composition used was a model Pb composition based on the inferred radiogenic $^{206}Pb/^{238}U$ age for each analysis. A general feature of analyses of the granulite zircons is instability in the ^{206}Pb count rate, not correlated with U concentration. The uncertainty in the $^{206}Pb/^{238}U$ ratio is often dominated by variation in the ratio, rather than statistical variations related to counting errors. Although a unique interpretation cannot be given for any given analysis, it appears that the variability is related to fine-scale Pb loss, particularly in SV19B. Data are presented in Figure 8 in the form of probability density functions by assigning unit area Gaussian curves to each $^{206}Pb/^{238}U$ age datum and then summing all data through the range in 1 Ma bins. The operation of summing the individual Gaussian curves should result in a Gaussian curve if the data are distributed about the mean in a manner consistent with the individual uncertainties. All of the data from the xenoliths show large scatter. This preliminary data set is somewhat imprecise because it does not constitute a statistically cohesive data set.

Sample SV19B contains a fairly homogeneous zircon pop-

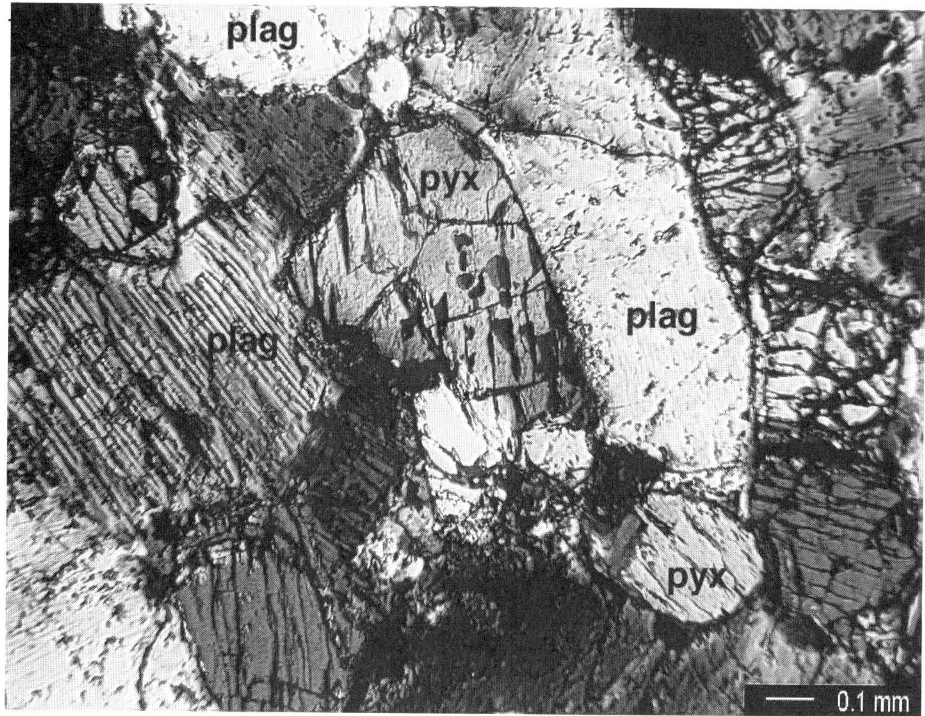

Figure 5. Thin-section photograph of sample SV21C (plag is plagioclase, pyx is pyroxene). Note coarse equigranular textures, despite strong foliation in hand specimen, indicating slow cooling or elevated temperatures long after deformation had ceased, compatible with a mid-crustal origin for xenoliths inferred from metamorphic assemblages.

ulation consisting of elongate subhedral to euhedral zircons that generally have inclusion-crowded, uranium-rich cores and clear, inclusion-free, uranium-poor rims (Figs. 6 and 7). The zircon rims typically show oscillatory zoning (Fig. 7). We infer from these characteristics that the zircons are most likely igneous in origin and, given the metamorphic fabric of the rock, that the plutonic protolith was subsequently metamorphosed and deformed at probable granulite facies conditions. We found that

zircon cores from sample SV19B are systematically, but only marginally, older than the rims (Table 1), suggesting a slightly earlier magmatic precursor for this rock. The main age represented in the peak of the distribution suggests a magmatic age of ca. 85 Ma, although two of the analyses are distinctly younger (ca. 74 Ma) and suggest that Pb loss may have occurred. The suggestion that Pb loss may explain the younger ages is based on the reasoning that superimposed granulite facies metamor-

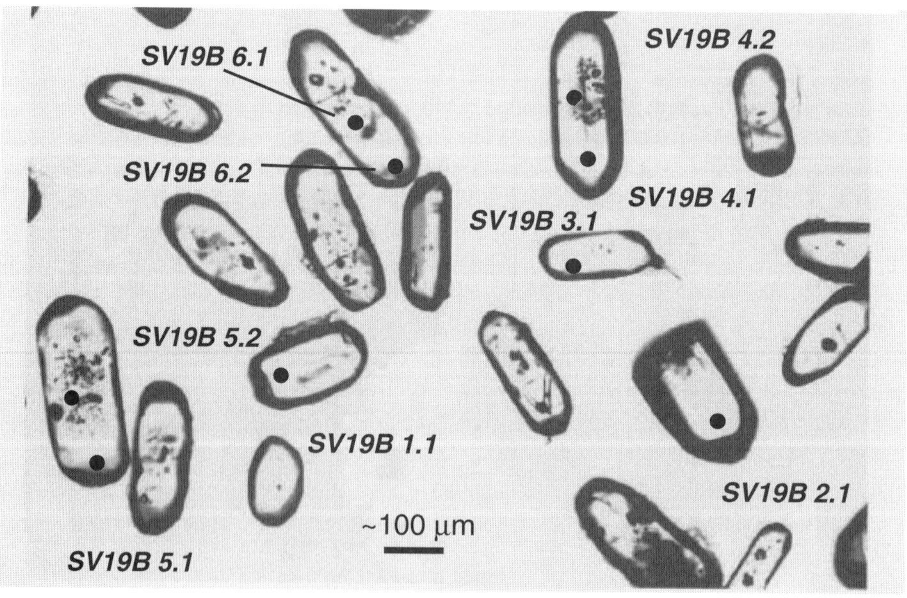

Figure 6. Transmitted light photo of zircons from sample SV19B showing spot localities dated with SHRIMP II.

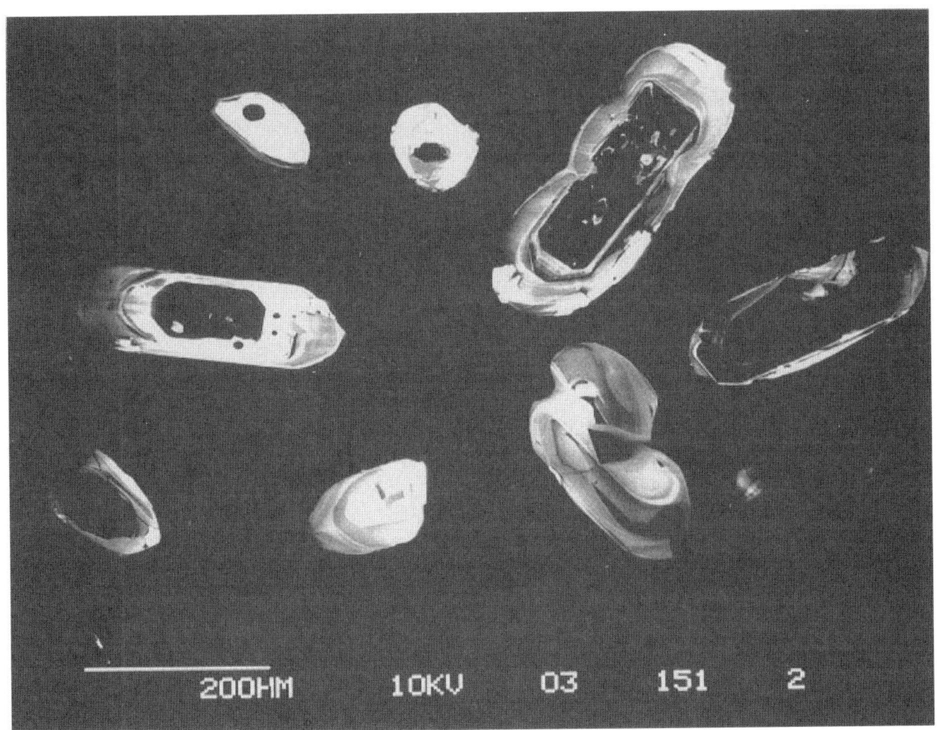

Figure 7. Cathodoluminescence scan of typical zircons from sample SV19B showing oscillatory zoning. Note that publication-quality cathodoluminescent scans shown here and in Figure 10 imaged a different part of the zircon population than those dated.

TABLE 1. U-PB ISOTOPIC DATA, SAINT LAWRENCE ISLAND XENOLITHS

Labels	U	Th	Th/U	^{207}Pb/^{206}Pb	^{238}U/^{206}Pb	Age (Ma)
SV19B						
1.1	419	121	0.29	0.0456 ± 0.0035	75.95 ± 5.66	84.5 ± 6.3
2.1	460	253	0.55	0.0502 ± 0.0029	71.19 ± 2.36	89.7 ± 3.0
3.1	390	102	0.26	0.0496 ± 0.0035	86.13 ± 5.05	74.2 ± 4.3
4.1c	422	112	0.26	0.0491 ± 0.0025	70.70 ± 6.40	90.4 ± 8.1
4.2r	2335	139	0.06	0.0499 ± 0.0015	73.69 ± 1.32	86.7 ± 1.5
5.1r	393	170	0.43	0.0501 ± 0.0024	87.65 ± 3.04	72.9 ± 2.5
5.2c	4050	760	0.19	0.0471 ± 0.0012	60.57 ± 1.71	105.7 ± 3.0
6.1r	2685	571	0.21	0.0479 ± 0.0009	76.68 ± 0.82	83.5 ± 0.9
6.2c	373	143	0.38	0.0513 ± 0.0022	70.48 ± 3.67	90.5 ± 4.7
SV21C						
1.1	31	20	0.64	0.0985 ± 0.0071	91.31 ± 3.62	66.3 ± 2.7
2.1	44	32	0.72	0.0489 ± 0.0048	113.77 ± 6.56	56.3 ± 3.2
3.1	548	202	0.37	0.0488 ± 0.0011	90.82 ± 2.07	70.5 ± 1.6
4.1	206	68	0.33	0.0487 ± 0.0025	93.88 ± 3.52	68.2 ± 2.5
6.1	357	229	0.64	0.0468 ± 0.0026	103.34 ± 2.51	62.1 ± 1.5
7.1	297	205	0.69	0.0536 ± 0.0031	99.39 ± 1.59	64.1 ± 1.0
8.1	28	21	0.75	0.0362 ± 0.0057	99.49 ± 4.21	65.3 ± 2.8
SV21E						
1.1	479	277	0.58	0.0481 ± 0.0013	70.12 ± 2.84	91.3 ± 3.7
1.2	28	6	0.21	0.0315 ± 0.0042	94.20 ± 3.07	69.2 ± 2.3
2.1	607	453	0.75	0.0465 ± 0.0011	64.06 ± 0.73	100.0 ± 1.1
3.1	9	3	0.25	0.0756 ± 0.0104	94.96 ± 2.44	63.7 ± 1.8

Note: All errors are 1σ, unless otherwise stated.
Standard error of 11 AS3 analyses = 0.33% (with 1.0% error analysis, mean square of weighted deviates = 1.15).
c, r designates analyses of cores and rims, respectively.

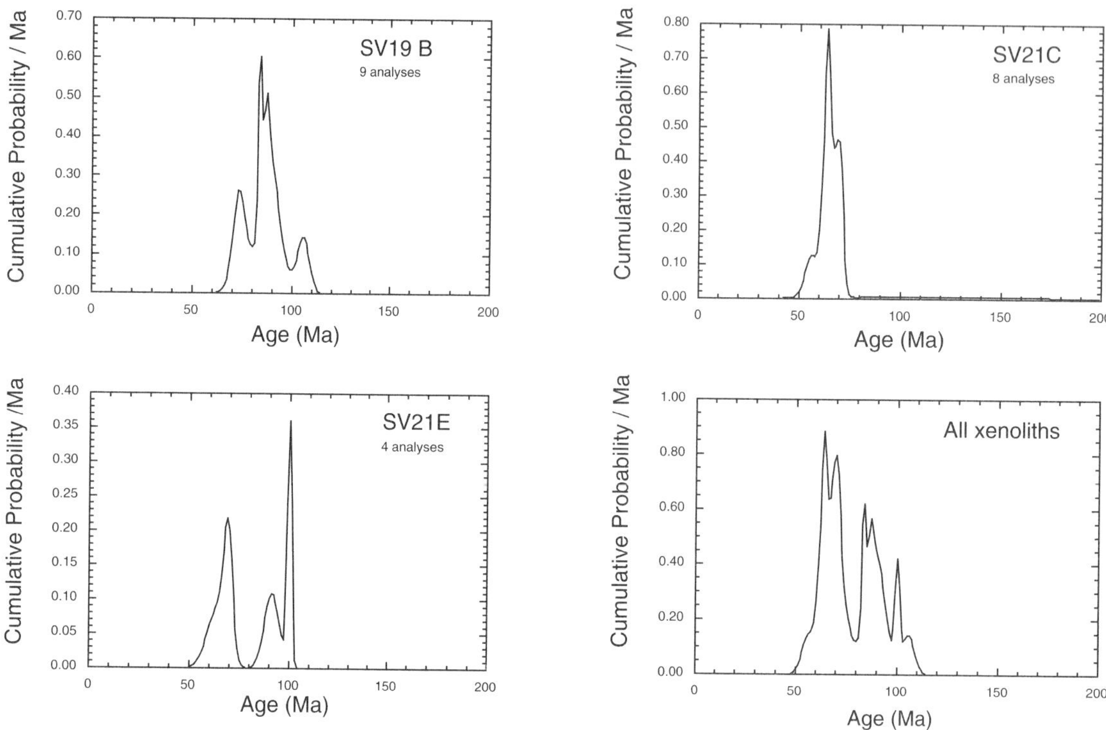

Figure 8. Cumulative probability plots of U-Pb zircon ages from samples dated.

phism did not appear to have precipitated any new zircon in this rock; the delicate oscillatory zoning observed in the zircon rims is more consistent with magmatic crystallization. Thus the ca. 74 Ma ages are likely the result of Pb loss, at this or at a younger time of heating and metamorphism. The proposed protolith age of 85 Ma may underestimate the age of magmatic crystallization of this rock, although it is unlikely, based on the data, that this magmatic age would be much older than ca. 90 Ma.

Samples SV21C and SV21E have similar zircon populations, distinct from those of sample SV19B (Figs. 9 and 10). The majority of zircons from these two samples are rounded and equidimensional to stubby in shape and are smaller than those from sample SV19B. They have poorly defined crystal faces, are adamantine in luster, and mostly very clear (Fig. 9). Few of the zircons are elongate or prismatic, but these grains, in contrast to those of sample SV19B, have poorly developed crystal faces and some grains have small rounded overgrowths. Cathodoluminescence images of zircons from sample SV21C (Fig. 10) indicate that the smaller rounded zircons are not zoned in any systematic fashion, and vary in luminescence from grain to grain and commonly across the grain (Fig. 10). The larger grains commonly show relict oscillatory zoning, and some have oscillatory-zoned cores surrounded by clear, uranium-poor rims. We interpret these characteristics as indicating that most of the zircon grew in a metamorphic environment, but that the protolith of the rock was likely plutonic. The seven analyses of zircons from sample SV21C are fairly consistent and indicate an age of ca. 64 Ma,

which is significantly younger than the age of zircons in sample SV19B, 85–90 Ma. One of the analyses from a small rounded grain (SV21C 3.1), with no apparent core, gives an age of ca. 70 Ma, detectably older than the mean in sample SV21C. Grain 7, which is a long prismatic grain of inferred magmatic origin (Fig. 9) yields an age (64 Ma) no different from the rounded, inferred metamorphic zircons. Although the age of inferred magmatic grain 7 may have been reset during granulite facies metamorphism at 64 Ma, the slightly older age of grain 3 may suggest a slightly earlier metamorphic event ca. 70–75 Ma. In this respect, the ages of the younger grains of SV19B are consistent with such an age, but more data are required to assess the presence of a distinct population of 70–75 Ma metamorphic zircons.

Ages of the four zircon analyses from sample SV21E are mixed; two younger ages are similar to the metamorphic ages obtained from SV21C and two older ages of 100 and 91 Ma are similar to the magmatic populations of SV19B. The two younger samples are likely metamorphic in origin, whereas the older grains may be relict magmatic zircons of middle to Late Cretaceous age. The metamorphic zircons in SV21E are noteworthy for their low U concentrations, which are characteristic of metamorphic zircons that have grown under granulite facies conditions (Gibson and Ireland, 1995). However, these zircons do not show the extreme fractionation in Th/U (often <0.01) found in some granulite zircons.

In summary, zircons from the pyroxene-bearing crustal xenoliths indicate that magmatic protolith rocks of 85–100 Ma

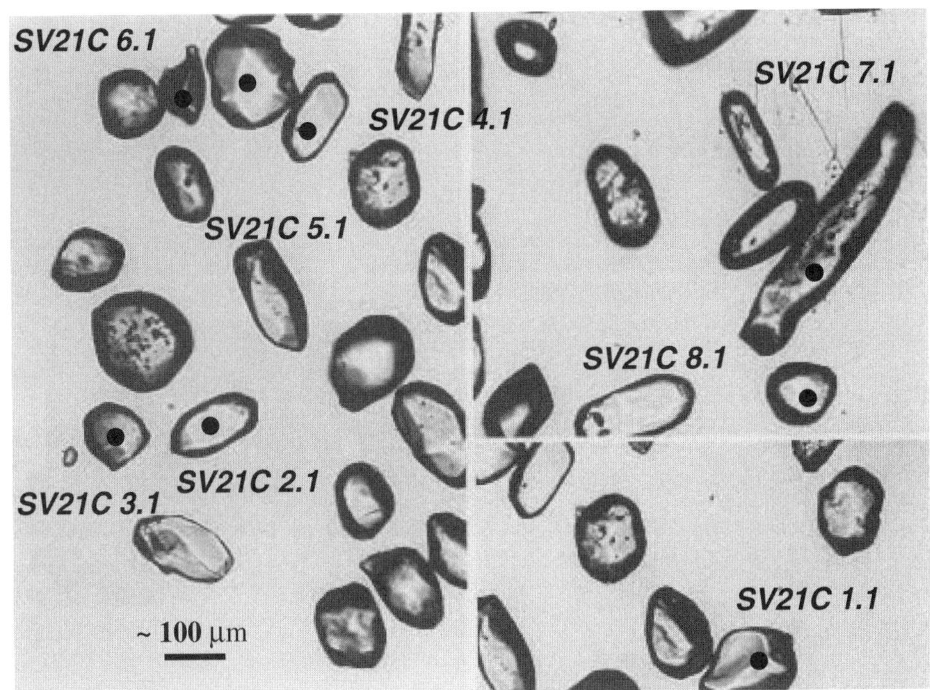

Figure 9. Transmitted light photo of zircons from samples SV21C and SV21E showing spot localities dated with SHRIMP II.

were involved in a younger thermal event between 60 and 75 Ma; a peak event ca. 64 Ma affected the deeper crust beneath the inner Bering Shelf. This younger thermal event coincided with the emplacement of the Yukon-Kanuti and Kuskokwim Mountains magmatic belts between 76 and 55 Ma (Moll-Stalcup, 1994). The northernmost exposures of volcanic rocks associated with the Yukon-Kanuti and Kuskokwim magmatic belts occur on Saint Lawrence Island, where volcanic rocks have been dated by the K-Ar method on sanidine (64.4 Ma), hornblende (62.2 Ma), and whole rock (64 Ma) (Wilson et al., 1994); these ages

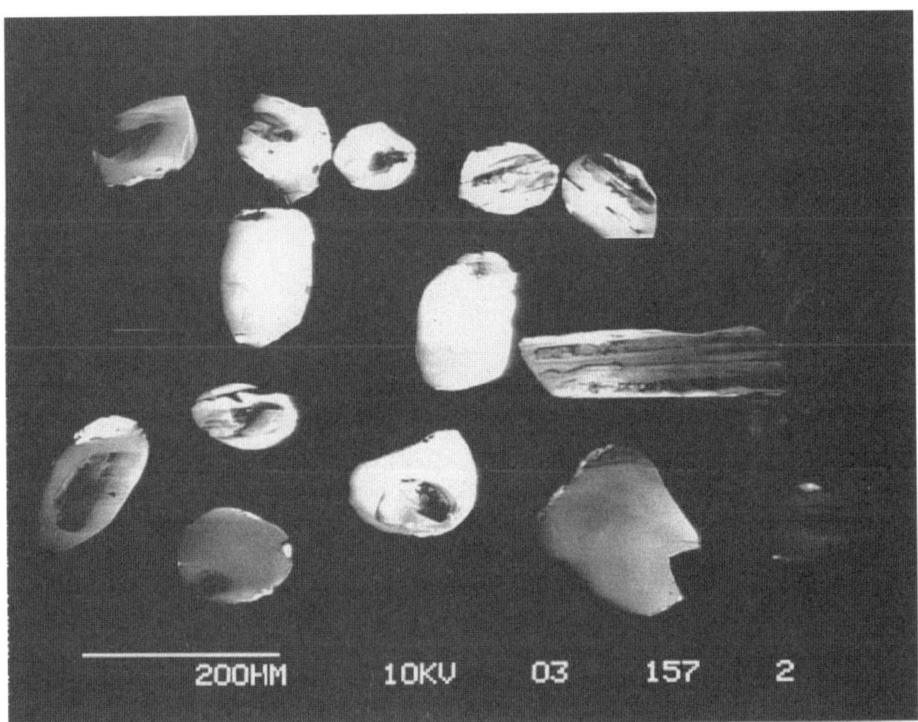

Figure 10. Cathodoluminescence scan of typical zircons from sample SV21C.

are indistinguishable from the age of the mid-crustal metamorphic zircons we dated. The ages of the older, magmatic zircons coincide with ages obtained from the broad belt of Cretaceous granitoid plutons that crosses the Bering Strait from Alaska to Russia and extends from the Seward Peninsula to Saint Lawrence Island (Fig. 2).

CONCLUSIONS

The crustal and mantle xenoliths in Neogene basalt flows on Saint Lawrence Island provide an unusual opportunity to study the nature and history of the crust and underlying mantle beneath the inner Bering Shelf. These xenoliths also provide insight for the interpretation of the deep crustal seismic reflection data collected across this region (Klemperer et al., this volume, Chapter 1).

On the basis of refraction studies, the crust beneath Saint Lawrence Island exhibits normal velocities and a sharply defined Moho between 30 and 35 km, which is fairly typical of continental crust worldwide. Seismic reflection data indicate that the middle and lower crust is characterized by well-developed subhorizontal reflectivity down to the Moho, where the underlying mantle is transparent. Worldwide, this type of reflectivity has been interpreted as a product of modification of continental crust by extensional processes (e.g., Mooney and Meissner, 1992).

This fairly persistent crustal reflectivity has been mapped seismically over a broad part of the inner Bering Shelf and southern Hope Basin and coincides in a general way with the locus of middle to Late Cretaceous magmatism in Alaska and Russia (Fig. 2). The northernmost outcrops of Paleocene volcanic rocks are on Saint Lawrence Island and continue on land in Alaska as the Kuskokwim Mountains and Yukon-Kanuti magmatic belts of latest Cretaceous to Paleocene age (Moll-Stalcup, 1994). At the scale of the Bering Shelf, it appears that magmatism youngs southward and that the locus of latest Cretaceous to Paleocene magmatism in part overprints the locus of mid–Late Cretaceous magmatism (Fig. 2). Geochronologic and isotopic data from middle to Late Cretaceous magmatic rocks (Miller, 1994; Amato and Wright, 1997, 1998; Bering Strait Geologic Field Party, 1997; Rowe, 1998) suggest that magmas that intruded the crust during this event were mantle derived, and mixed, to variable extent, with crustal rocks. The mantle-derived magmatism is inferred to be at least in part related to subduction beneath the ancient Bering margin. Magmatism likely occurred in an extensional tectonic environment as the active subduction zone migrated south with time (Amato and Wright, 1997, 1998). An extensional tectonic setting for magmatism is also inferred from geologic studies on either side of the Bering Strait that outline the relationship of Cretaceous magmatism to metamorphism, melting, and wholesale flow of the middle and lower crust within the broad metamorphic culminations or gneiss domes within the magmatic belt (Fig. 2). Less is known about the tectonic setting of the younger latest Cretaceous to Paleocene magmatic belt, except that it covers an extensive region from Saint Lawrence Island south to the edge of the present Bering Shelf, and that volcanic rocks in the belt were erupted in a subaerial setting (Cooper et al., 1987; Worral, 1991).

Pyroxene-bearing gneissic xenoliths derived from the midcrust beneath Saint Lawrence Island most likely represent highly deformed intrusive rocks that have undergone granulite facies metamorphism. These xenoliths have zoned prismatic zircons that appear to be magmatic and yield preliminary Late Cretaceous U-Pb ages, which we interpret as intrusive ages between ca. 85 and 90 Ma with one age as old as 100 Ma. Some of these xenoliths have rounded, nonzoned zircons that appear to be metamorphic, and these yield Late Cretaceous to early Tertiary metamorphic ages between 60 and 75 Ma with a peak of ages at 64 Ma. Nonfoliated gabbroic xenoliths are the more common crustal xenolith represented in basalt flows on Saint Lawrence Island. Although these have not yet been dated, we interpret them to be derived from Late Cretaceous or early Tertiary mafic sills and plutons in the middle and lower crust. Together, these preliminary data suggest that the igneous and metamorphic events responsible for development of the observed reflective crust and sharp Moho beneath Saint Lawrence and the inner Bering Shelf occurred during the Late Cretaceous and early Tertiary.

The Alaskan mainland south of the Brooks Range has long been portrayed as an orogenic collage of allochthonous terranes presumed to continue westward beneath the Bering Shelf (e.g., Nokleberg et al., 1998). Some of these terranes have been identified on Saint Lawrence Island, where Paleozoic rocks similar to shelfal sequences in the Brooks Range occur together with a Triassic sequence of chert, shale, and greenstone similar to that in the Angayucham terrane of the southern Brooks Range (Patton and Dutro, 1969; Patton and Csejtey, 1980). Although one might have expected the deep crustal seismic reflection data collected as part of this project to image the orogenic collage of allochthonous terranes presumed to underlie the Bering Shelf, the data actually provide little to no insight into this Albian and older Mesozoic accretionary history. Seismic reflection data in conjunction with the preliminary U-Pb zircon ages of deep crustal xenoliths indicate that the most important events that formed the present-day middle and lower crust postdate the accretionary history of the region and are linked to the Cretaceous to early Tertiary magmatic history of the region.

ACKNOWLEDGMENTS

The Continental Dynamics Program, National Science Foundation (NSF), award EAR-93-17087 funded most of the research presented here. In addition, an NSF Research Opportunity Award tied to EAR-93-17087 funded filed work by K. Wirth and students who collected the xenoliths on Saint Lawrence Island (see Chapter 9 of this volume). The Office of Polar Programs, NSF, grant OPP-99-05790 supported data compilation and interpretation. V. Akinin acknowledges support by RFBR grant 01-05-65453. We thank our reviewers Walter Mooney, Betsy

Moll-Stalcup, James Mortensen, and an anonymous reviewer for improvements on our original manuscript.

REFERENCES CITED

Agapitov, D.I., Ivanov, V.V., and Krainov, V.G., 1973, New data on geology and perspective of oil and gas resources in Anadyr basin, *in* The problems of oil and gas potential in the North-East of USSR (Novye dannye po geologii i perspektivam neftegasonosnosti Anadyrskogo basseina. V kn: Problemy neftegazonosnosti Cevero-Vostoka SSSR): Magadan, Russia, Northeast Interdisciplinary Science Research Institute, v. 49, p. 23–39 (in Russian).

Akinin, V.V., 1995, Petrology of alkali lavas and deep-seated inclusions of Enmelen volcanoes, Chukchi Peninsula, *in* Simakov, K.V., and Thurston, D.K., Proceedings of the 1994 International Conference on Arctic Margins: Magadan, Russia, North East Science Center, Far Eastern Branch of the Russian Academy of Science, p. 138–146.

Akinin, V.V., and Apt, J.E., 1994, Enmelen volcanoes, Chukchi Peninsula: Petrology of alkaline lavas and deep seated inclusions: Magadan, Russia, Northeast Interdisciplinary Science Research Institute, Far Eastern Branch of the Russian Academy of Science, 97 p. (in Russian).

Akinin, V.V., and Apt, Y.E., 1997, Late Cenozoic alkaline basaltic volcanism of the North East of Russia, *in* Byalobzhesky, S.G., ed., Magmatism and ores of north east of Russia: Magadan, Russia, Northeast Interdisciplinary Science Research Institute, Far Eastern Branch of the Russian Academy of Science, p. 153–172.

Akinin, V.V., Roden, M.F., Francis, D.M., Apt, J.E., and Moll-Stalcup, E., 1997, Compositional and thermal state of the upper mantle beneath the Bering Sea basalt province: Evidence from the Chukchi Peninsula of Russia: Canadian Journal of Earth Sciences, v. 34, p. 789–800.

Allmendinger, R.W., Hauge, T.A., Hauser, E.C., Potter, C.J., Klemperer, S.L., Nelson, K.D., Knuepfer, P.L.K., and Oliver, J. 1987, Overview of the CO-CORP 40° transect, western United States, the fabric of an orogenic belt: Geological Society of America Bulletin, v. 98, p. 308–319.

Amato, J.M., and Wright, J.E., 1997, Potassic mafic magmatism in the Kigluaik gneiss dome, Northern Alaska: A geochemical study of arc magmatism in an extensional tectonic setting: Journal of Geophysical Research, v. 102, p. 8065–8084.

Amato, J.M., and Wright, J.E., 1998, Geochronologic investigations of magmatism and metamorphism within the Kigluaik Mountains gneiss dome, Seward Peninsula, Alaska, *in* Clough, J.G., and Larson, F., eds., Short notes on Alaska geology: State of Alaska, Department of National Resources, Division of Geological and Geophysical Surveys Professional Report 118, p. 1–21.

Amato, J.M., Wright, J.E, Gans, P.B., and Miller, E.L., 1994, Magmatically induced metamorphism and deformation in the Kigluaik gneiss dome, Seward Peninsula, Alaska: Tectonics, v. 13, p. 515–527.

Aoki, K., and Kushiro, I., 1968, Some clinopyroxenes from ultramafic inclusions in Dreiser Weiher, Eifel: Contributions to Mineralogy and Petrology (former title: Beitraege zur Mineralogie und Petrologie), v. 18, no. 4, p. 326–337.

Bering Strait Geologic Field Party, 1997, Koolen metamorphic complex, NE Russia: Implications for the tectonic evolution of the Bering Strait region: Tectonics, v. 16, p. 713–729.

Brocher, T.M., Allen, R.M., Stone, D.B., Wolf, L.W., and Galloway, B.K., 1995, Data report for onshore-offshore wide-angle seismic recordings in the Bering-Chukchi Sea, western Alaska and eastern Siberia: U.S. Geological Survey Open-File Report 95–650, 57 p.

Burk, C.A., 1965, The geology of the Alaska peninsula–island arc and continental margin: Boulder, Colorado, Geological Society of America Memoir 99, 250 p.

Choukroune, P., and the ECORS Pyrenees Team, 1989, The ECORS Pyrenean deep seismic profile: Reflection data and the overall structure of an orogenic belt: Tectonics, v. 8, p. 23–39.

Christiansen, P.P., and Snee, L.W., 1993, Structure, metamorphism and geochronology of the Cosmos Hills and Ruby Ridge, Brooks Range schist belt, Alaska: Tectonics, v. 14, p. 193–213.

Clague, D.A., and Frey, F.A., 1982, Petrology and trace element geochemistry of the Honolulu Volcanics, Oahu: Implications for the oceanic mantle below Hawaii: Journal of Petrology, v. 23, p. 447–504.

Cooper, A.K., Marlow, M.S., and Scholl, D.W., 1987, Geologic framework of the Bering Sea Crust, *in* Scholl, D.W., Grantz, A., and Vedder, J.C., eds., Geology and resource potential of the continental margin of western North America and adjacent ocean basins: Beaufort Sea to Baja California: Houston, Texas, Circum-Pacific Council for Energy and Mineral Resources, Earth Science Series, v. 6, p. 73–102.

Costa, S., and Rey, P., 1995, Lower crustal rejuvenation and growth during post-thickening collapse: Insights from a crustal cross section through a Variscan metamorphic core complex: Geology, v. 23, p. 905–908.

Costa, S., Rey, P., Todt, W., and Goldstein, S.L, 1994, Evolution of the lower continental crust during post-thickening collapse: Mineralogical Magazine, v. 58A, no. A-K, p. 197–198.

Davis, A.S., Marlow, M.S., and Wong, F.L., 1995, Petrology of Quaternary basalt from the Bering Sea continental margin, *in* Simakov, K.V., and Thurston, D.K., eds., Proceedings of the 1994 International Conference on Arctic Margins: Magadan, Russia, North East Science Center, Far Eastern Branch of the Russian Academy of Science, p. 124–137.

Davis, W.J., 1996, U-Pb geochronology of lower crustal xenoliths from the Archean Slave Province: Evidence for protracted Archean metamorphism and Proterozoic mafic magmatism in the lower crust: Eos (Transactions, American Geophysical Union), v. 77, no. 46, supplement, p. 821.

Dumitru, T.A., Miller, E.L., O'Sullivan, P.B., Amato, J.M., Hannula, K.A., Calvert, A.T., and Gans, P.B., 1995, Cretaceous to recent extension in the Bering Strait region, Alaska: Tectonics, v. 14, p. 549–563.

Gasparik, T., 1984, Experimentally determined stability of clinopyroxene + garnet + corundum in the system CaO-MgO-Al_2O_3-SiO_2: American Mineralogist, v. 69, no. 11–12, p. 1025–1035.

Gibson, G.M., and Ireland, T.R., 1995, Granulite formation during continental extension in Fiordland, New Zealand: Nature, v. 375, p. 479–482.

Green, A.G., Milkereit, B., Percival, J.A., Davidson, A., Parrish, R.R., Cook, F.A., Geis, W.T., Cannon, W.F., Hutchinson, D.R., West, G.F., and Clowes, R.M., 1990, Origin of deep crustal reflections: Seismic profiling across high-grade metamorphic terranes in Canada: Tectonophysics, v. 173, p. 627–638.

Holliger, K., and Levander, A., 1994, Lower crustal reflectivity modelled by rheological controls on mafic intrusions: Geology, v. 22, p. 367–370.

Howell, D.G., Moore, G.W., and Wiley, T.J., 1987, Tectonics and basin evolution of western North America: An overview, *in* Scholl, D.W., Grantz, A., and Vedder, J.G., eds., Geology and resource potential of the continental margin of western North America and adjacent ocean basins: Beaufort Sea to Baja California: Houston, Texas, Circum-Pacific Council for Energy and Mineral Resources, Earth Science Series, v. 6, p. 1–15.

Ivanov, V.V., 1985, Sedimentary basins of North-Eastern Asia (comparative oil and geological analysis): Moscow, Nauka, 208 p. (in Russian).

Law, R.D., Little, T.A., Miller, E.L., and Lee, J., 1994, Extensional origin of ductile fabrics in the Schist Belt, central Brooks Range, Alaska. 2. Quartz fabric studies: Journal of Structural Geology, v. 16, no. 7, p. 919–940.

Little, T.A., Miller, E.L., Lee, J., and Law, R.D., 1994, Extensional origin of ductile fabrics in the Schist Belt, Central Brooks Range, Alaska. 1. Geologic and structural studies: Journal of Structural Geology, v. 16, no. 7, p. 900–918.

Luetgert, J.H., 1992, MacRay: Interactive two-dimensional seismic ray tracing for the Macintosh: U.S. Geological Survey Open-File Report 92–356, 43 p., 1 diskette.

Marlow, M.S, Scholl, D.W., Cooper, A.K., and Buffington, E.C., 1976, Structure and evolution of Bering Sea shelf south of St. Lawrence Island: American Association of Petroleum Geologists Bulletin, v. 60, no. 2, p. 161–183.

Matthews, D.H., and Cheadle, M.J., 1986, Deep reflections from the Caledonides and Variscides west of Britain and comparison with the Himalayas, *in* Barazangi, M., and Brown, L.D., eds., Reflection seismology: A global perspective: American Geophysical Union Geodynamics Series, v. 13, p. 5–19.

Matthews, D.H., and the BIRPS Group, 1990, Progress in BIRPS deep seismic reflection profiling around the British Isles: Tectonophysics, v. 173, p. 387–396.

McLean, H., 1977, Organic geochemistry, lithology and paleontology of Tertiary and Mesozoic rocks from wells on the Alaska Peninsula: U.S. Geological Survey Open-File Report N 77–813, 63 p.

McLean, H., 1979, Tertiary stratigraphy and petroleum potential of Cold Bay–False Pass Area, Alaska Peninsula: American Association of Petroleum Geologists Bulletin, v. 63, no. 9, p. 1522–1526.

Miller, E.L., and Hudson, T.L., 1991, Mid-Cretaceous extensional fragmentation of a Jurassic–Early Cretaceous compressional orogen, Alaska: Tectonics, v. 10, p. 781–796.

Miller, E.L., Calvert, A.T., and Little, T.A., 1992, Strain-collapsed metamorphic isograds in a sillimanite gneiss dome, Seward Peninsula, Alaska: Geology, v. 20, p. 487–490.

Miller, T.P., 1994, Pre-Cenozoic plutonic rocks in mainland Alaska, *in* Plafker, G., and Berg, H.C., eds., The geology of Alaska: Boulder, Colorado, Geological Society of America, Geology of North America, v. G-1, p. 589–620.

Moll-Stalcup, E.J., 1994, Latest Cretaceous and Cenozoic magmatism in mainland Alaska, *in* Plafker, G., and Berg, H.C., eds., The geology of Alaska: Boulder, Colorado, Geological Society of America, Geology of North America, v. G-1, p. 589–619.

Mooney, W.D., and Meissner, R., 1992, Multi-genetic origin of crustal reflectivity: A review of seismic reflection profiling of the continental lower crust and Moho, *in* Fountain, D.M., Arculus, R., and Kay, R.W., eds., Continental lower crust: New York, Elsevier, p. 45–71.

Moore, T.E, Wallace, W.K., Bird, K.J., Karl, S.M., Mull, C.G., and Dillon, J.T., 1994, Geology of northern Alaska, *in* Plafker, G., and Berg, H.C., eds., The geology of Alaska: Boulder, Colorado, Geological Society of America, Geology of North America, v. G-1, p. 49–140.

Muir, R.J., Ireland, T.R., Weaver, S.D., and Bradshaw, J.D., 1996a, Ion microprobe dating of Paleozoic granitoids: Devonian magmatism in New Zealand and correlations with Australia and Antarctica: Chemical Geology, v. 127, no. 1–3, p. 191–210.

Muir, R.J., Weaver, S.D., Bradshaw, J.D., Eby, G.N., Evans, J.A., and Ireland, T.R., 1996b, Geochemistry of the Karamea Batholith, New Zealand and comparisons with the Lachlan fold belt granites of SE Australia: Lithos, v. 39, no. 1–2, p. 1–20.

Nelson, K.D., 1992, Are crustal thickness variations in old mountain belts like the Appalachians a consequence of lithospheric delamination?: Geology, v. 20, p. 498–502.

Nokleberg, W.J., Parfenov, L.M., Monger, J.W.H., Baranov, B.V., Byalobzhesky, S.G., Bundtzen, T.K., Fenney, T.D., Fujita, K., Gordey, S.P., Grantz, A., Khanchuk, A.I., Natal'in, B.A., Natapov, L.M., Norton, I.O., Patton, W.W., Jr., Plafker, G., Scholl, D.W., Stone, D.B., Tabor, R.W., Tsukanov, N.V., and Vallier, T.L., 1997, Summary: Circum-Pacific tectonostratigraphic terrane map: U.S. Geological Survey Open-File Report OF 96–0727, scale 1: 10 000 000, 1 sheet.

Nokleberg, W.J., Parfenov, L.M., Monger, J.W.H., Norton, I.O., Khanchuk, A.I., Stone, D.B., Scholl, D.W., and Fujita, K., 1998, Phanerozoic tectonic evolution of the circum-North Pacific: U.S. Geological Survey Open-File Report 98–0754, 125 p.

O'Reilly, S.Y., et al., 1995, Special issue: The xenolith window to the lower crust: Lithos, v. 36, p. 155–306.

Parsons, T., Christiansen, N.I., and Wilshire, H., 1995, Velocities of southern Basin and Range xenoliths: Insights on the nature of lower crustal reflectivity and composition: Geology, v. 23, p. 129–132.

Patton, W.W., and Csejtey, B., 1980, Geologic map of Saint Lawrence Island: U.S. Geological Survey Miscellaneous Investigation Series Map I-203, scale 1:250 000, 1 sheet.

Patton, W.W., Jr., and Dutro, J.T., Jr., 1969, Preliminary report on Paleozoic and Mesozoic sedimentary sequence on Saint Lawrence Island, Alaska, *in* Geological Survey Research 1969: U.S. Geological Survey Professional Paper 650–D, p. D138–D143.

Patton, W.W., Lanphere, M.A., Miller, T.P., and Scott, R.A., 1976, Age and tectonic significance of volcanic rocks in St. Matthew Island, Bering Sea Alaska: U.S. Geological Survey Journal of Research, v. 11, no. 1, p. 67–73.

Rey, P., 1993, Seismic and tectonometamorphic characters of the lower continental crust in Phanerozoic areas, a consequence of post-thickening extension: Tectonics, v. 12, p. 580–590.

Rowe, H., 1998, Petrogenesis of plutons and hypabyssal rocks of the Bering Strait Region, Chukotka, Russia [M.S. thesis]: Houston, Texas, Rice University, 90 p.

Rubin, C.M., Miller, E.L., and Toro, J., 1995, Deformation of the northern circum-Pacific margin: Variations in tectonic style and plate-tectonic implications: Geology, v. 23, p. 897–900.

Rudnick, R.L., 1992, Xenoliths-samples of the lower continental crust, *in* Fountain, D.M., Arculus, R., and Kay, R.W., eds., Continental lower crust: New York, Elsevier, p. 269–308.

Rudnick, R.L., and Fountain, D.M., 1995, Nature and composition of the continental crust: A lower crustal perspective: Reviews of Geophysics, v. 33, p. 267–309.

Rudnick, R.L., and Williams, I.S., 1987, Dating the lower crust by ion microprobe: Earth and Planetary Science Letters, v. 85, p. 145–161.

Till, A.B., and Dumoulin, J.A., 1994, Geology of Seward Peninsula and St. Lawrence Island, *in* Plafker, G., and Berg, H.C., eds., The geology of Alaska: Boulder, Colorado, Geological Society of America, Geology of North America, v. G-1, p. 141–152.

Warner, M.R., 1990, Basalts, water or shear zones in the lower continental crust?: Tectonophysics, v. 173, p. 163–174.

Wilson, F.H., Shew, N., and DuBois, G.D., 1994, Map and table showing isotopic age data in Alaska, *in* Plafker, G., and Berg, H.C., eds., The geology of Alaska: Boulder, Colorado, Geological Society of America, Geology of North America, v. G-1, scale 1:2 500 000, with tables, 1 sheet.

Wirth, K.R., Ronnback, C.M., Grandy, J.S., and Sadofsky, S.J., 1995, Mafic volcanism and xenoliths of the northern Bering Sea Region [abs.]: Eos (Transactions, American Geophysical Union), v. 76, no. 46, supplement, p. 589–590.

Wood, D.A., 1980, The application of a Th-Hf-Ta diagram to problems of tectonomagmatic classifications and to establishing the nature of crustal contamination of basaltic lavas of the British Tertiary volcanic province: Earth and Planetary Science Letters, v. 50, p. 11–30.

Worral, D.M., 1991, Tectonic history of the Bering Sea and the evolution of Tertiary strike-slip basins of the Bering Shelf: Boulder, Colorado, Geological Society of America Special Paper 257, 120 p.

MANUSCRIPT ACCEPTED BY THE SOCIETY MAY 15, 2001.

Geological Society of America
Special Paper 360
2002

South Anyui suture, northeast Arctic Russia: Facts and problems

Sergei D. Sokolov
Grigoriy Ye. Bondarenko
Odin L. Morozov
Geological Institute, Russian Academy of Sciences, Pyzhevsky pereulok 7, 109017 Moscow, Russia
Vladimir A. Shekhovtsov
Sergei P. Glotov
Anyui Governmental Mining and Geological Enterprise, Ministry of Geology of Russia,
Geologov Street 13, 686720 Bilibino, Chukotka, Russia
Alexander V. Ganelin
Igor R. Kravchenko-Berezhnoy
Geological Institute, Russian Academy of Sciences, Pyzhevsky pereulok 7, 109017 Moscow, Russia

ABSTRACT

The South Anyui suture zone extends from the eastern part of the Laptev Sea into western Chukotka (northeastern Arctic Russia) and is everywhere marked by Mesozoic terrigenous turbidite and fragments of ophiolitic sequences. Ophiolitic sequences of both Paleozoic and Mesozoic age are believed to be present. The units flanking the suture are known to host suprasubduction-zone arc complexes of late Paleozoic–Late Jurassic age on the south and of Late Jurassic–Early Cretaceous age on the north.

The Paleozoic Anyui oceanic basin is proposed to have been connected with the Taimyr and Polar Urals ocean basins. The relationship of the Anyui ocean basin with the paleo-Pacific remains debatable. In the Late Jurassic–Early Cretaceous (or, possibly, beginning in the late Paleozoic), the South Anyui basin became separated from the Pacific by a convergent boundary. The accretionary mélange and approximately coeval suprasubduction-zone complexes are regarded as indicators of convergent plate motion and boundaries between the oceanic plates of the South Anyui basin and the Siberian continent, and the Chukotka microcontinent.

Four principal stages of deformation are distinguished in the South Anyui zone (Natal'in, 1984). The first two stages gave rise to fold and thrust–type structures. The two later stages involved strike-slip faulting. The most important interaction between the Chukotka microcontinent and Siberia during latest Jurassic and Early Cretaceous time was likely related to oblique collision between the two continental masses that involved lengthwise dextral strike-slip motion along the South Anyui suture.

INTRODUCTION

In northeastern Russia, Mesozoic orogenic structures (Mesozoides) occupy a vast area extending from the Siberian craton on the west to the Okhotsk-Chukotka continental-margin volcanic belt to the east (Fig. 1). Farther east, the Koryak-Kamchatka fold belt occurs along the Pacific margin of northeastern Russia. Collisional processes played an important role in the formation of the Mesozoic orogenic belts of northeastern Russia (Parfenov, 1984; Natal'in, 1984; Parfenov et al., 1993; Zonenshain et al., 1990; Pushcharovsky et al., 1992; and others). One of these collisional belts is the South Anyui suture zone (Fig. 1). Since the

Sokolov, S.D., Bondarenko, G.Ye., Morozov, O.L., Shekhovtsov, V.A., Glotov, S.P., Ganelin, A.V., and Kravchenko-Berezhnoy, I.R., 2002, South Anyui suture, northeast Arctic Russia: Facts and problems, *in* Miller, E.L., Grantz, A., and Klemperer, S.L., eds., Tectonic Evolution of the Bering Shelf–Chukchi Sea–Arctic Margin and Adjacent Landmasses: Boulder, Colorado, Geological Society of America Special Paper 360, p. 209–224.

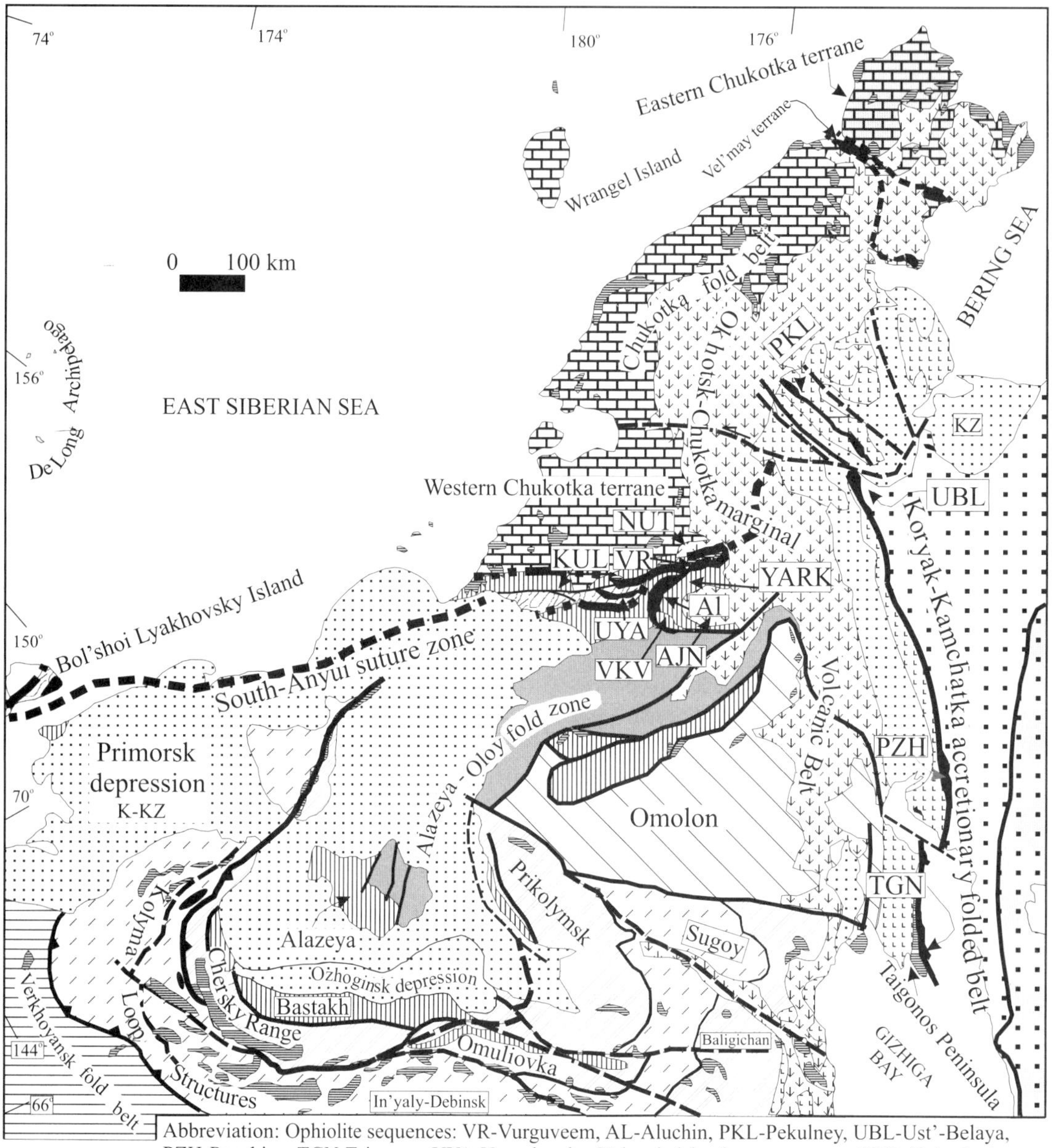

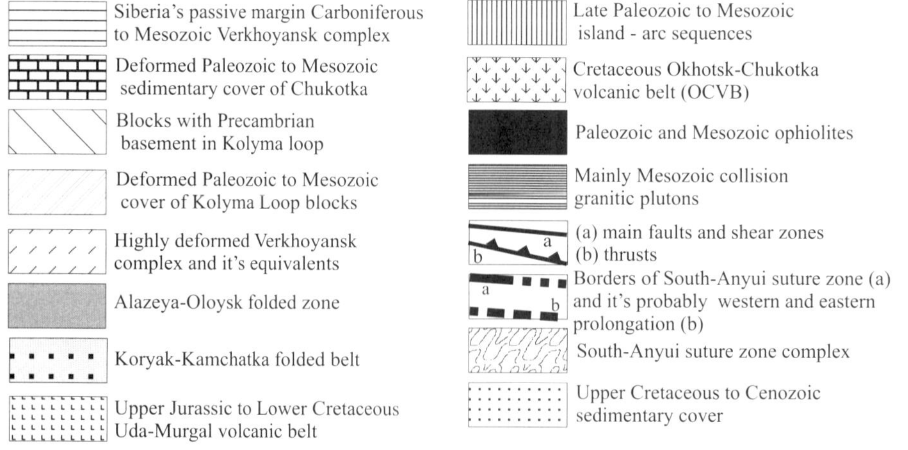

Abbreviation: Ophiolite sequences: VR-Vurguveem, AL-Aluchin, PKL-Pekulney, UBL-Ust'-Belaya, PZH-Penzhina, TGN-Taigonos, UYA-Uyamkanda. Volcanic island arc sequences: KUL-Kulpolnei, NUT-Nutesyn, AJN-Ajnakhkurgen, VKV-Vukvaam. YARK-Yarakvaam terrane

Siberia's passive margin Carboniferous to Mesozoic Verkhoyansk complex

Late Paleozoic to Mesozoic island - arc sequences

Deformed Paleozoic to Mesozoic sedimentary cover of Chukotka

Cretaceous Okhotsk-Chukotka volcanic belt (OCVB)

Blocks with Precambrian basement in Kolyma loop

Paleozoic and Mesozoic ophiolites

Deformed Paleozoic to Mesozoic cover of Kolyma Loop blocks

Mainly Mesozoic collision granitic plutons

Highly deformed Verkhoyansk complex and it's equivalents

(a) main faults and shear zones (b) thrusts

Alazeya-Oloysk folded zone

Borders of South-Anyui suture zone (a) and it's probably western and eastern prolongation (b)

Koryak-Kamchatka folded belt

South-Anyui suture zone complex

Upper Jurassic to Lower Cretaceous Uda-Murgal volcanic belt

Upper Cretaceous to Cenozoic sedimentary cover

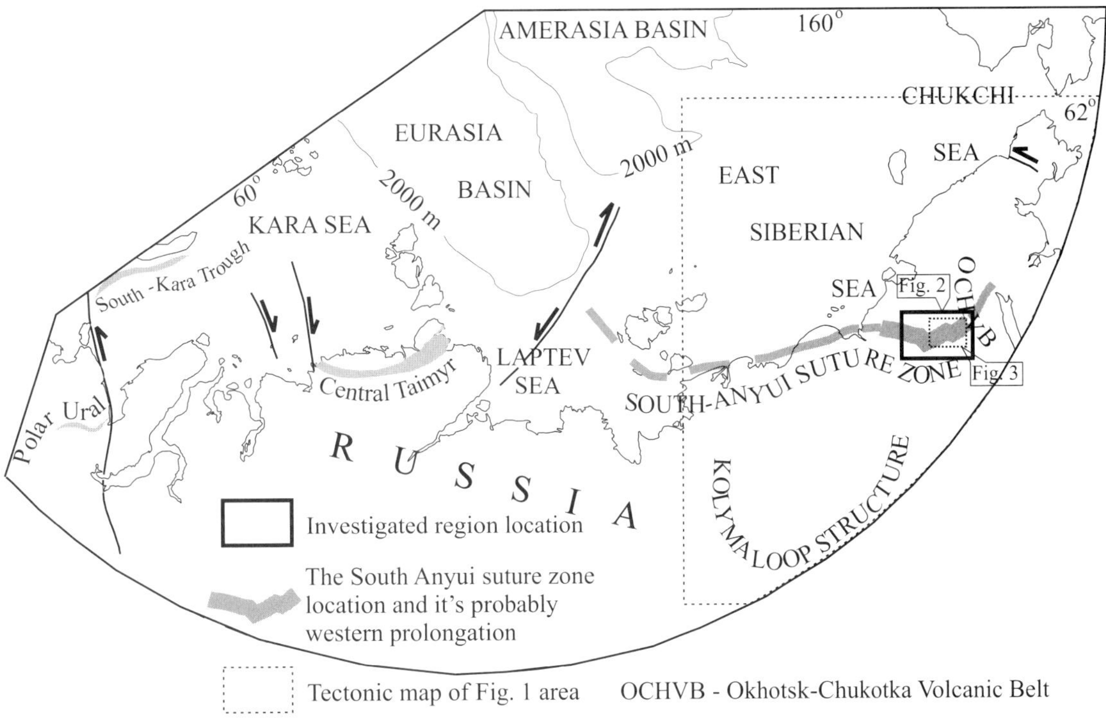

Figure 1 (*continued*)

works by Seslavinskiy (1970) and Til'man (1973), the South Anyui suture zone has customarily been viewed as a highly deformed zone that formed during and after the closure of a rift-related late Mesozoic oceanic basin. Later closing of the South Anyui ocean basin was linked to the collision between the Siberian continent and the Chukotka microcontinent, which belonged either to the hypothetical Arctida (or Hyperborea) continent (Seslavinskiy, 1979; Zonenshain and Natapov, 1988, 1990) or to the North American plate (Parfenov, 1984; Natal'in, 1984). At the same time, Parfenov (1984) indicated that accretionary processes may have also taken place along this boundary. Thus, it is more accurate to consider the South Anyui suture zone as one of the collisional-accretionary belts that are widespread around the edges of the Siberian craton (Berzin et al., 1994).

The collisional nature of the South Anyui suture zone is not questioned, yet the age of formation of the ocean basin, its life span, and size prior to collision are open to discussion. These issues are critical to paleotectonic reconstructions of the Arctic and to understanding the interaction between the Eurasian and North American plates. Unfortunately the South Anyui suture zone remains poorly known, and the available data do not in all cases agree, giving rise to diverse, often contradictory views on the structure and history of the South Anyui suture zone. The most controversial issues regarding the geology of the South Anyui suture zone are (1) the timing of the inception and existence of the South Anyui oceanic basin and its original size; (2) the age of the sedimentary complexes involved in the formation of the South Anyui suture zone; (3) the age and genesis of the ophiolites; (4) the inner structure of the South Anyui suture zone and the age and sequence of deformations; (5) the spatial distribution and age of island-arc complexes in and adjacent to the South Anyui suture zone, and what they tell us about the convergent boundaries that flank the basin; and (6) the dynamics of the collision of the continental masses of Chukotka (North America) and Siberia (Asia) that developed during the evolution of this zone.

This chapter is intended to summarize and analyze existing and new geological data in order to reach a more reasoned interpretation of the evolution of the South Anyui suture zone, and discusses the challenges and prospects that face future studies in the South Anyui suture zone.

STRUCTURAL SETTING AND PRINCIPAL TECTONIC UNITS

The South Anyui suture zone is an important tectonic unit of Northeast Asia (Fig. 1). It terminates structures of the Kolyma loop, including the arcuate belt of allochthonous terranes (Zonenshain et al., 1990) as well as island-arc terranes of the Alazeya-Oloy fold zone (Til'man, 1973), and separates the Anyui-Chukotka fold belt from the Siberian craton (Fujita and

Figure 1. Main tectonic features of northeastern Eurasia and the Arctic regions of Eurasia showing location of Polar Urals, Taimyr, and South Anyui suture zone, modified from Parfenov et al. (1983) and Sokolov et al. (1997).

Newberry, 1983; Parfenov, 1984; Pushcharovsky et al., 1992). Structures of the Kolyma loop and the Anyui-Chukotka fold belt are briefly described in the following.

The Kolyma loop structures are thrust over the Verkhoyansk fold belt (Parfenov, 1984; Parfenov et al., 1993; Oxman et al., 1995), which is composed of thick shelfal and continental-slope sediments deposited along the margin of the Siberian craton and is referred to as the Verkhoyansk complex. The Verkhoyansk complex is Mississippian to Early Cretaceous in age. Important constituents of the Kolyma loop include microcontinental fragments with an early Paleozoic to Cretaceous sedimentary cover, such as the Okhotsk, Omolon, Prikolymsk, and Omuliovka blocks. These are separated by zones of imbricate thrust sheets composed of late Paleozoic and Mesozoic deposits similar to the Verkhoyansk complex (Fig. 1) (Gagiyev, 1992; Natapov and Shul'gina, 1991; Savostin et al., 1994).

Over a significant portion of its length (~700 km), the South Anyui suture zone is bounded by the structures of the Alazeya-Oloy fold zone (Til'man, 1973; Parfenov, 1984) (Fig. 1). This zone consists of a number of terranes composed of Paleozoic metamorphic rocks, Paleozoic to Mesozoic clastic and sedimentary-volcanic complexes, and ophiolites (Parfenov et al., 1993; Nokleberg et al., 1994). Late Paleozoic to Mesozoic island-arc complexes are believed to be widespread within the Alazeya-Oloy zone (Seslavinskiy, 1979; Parfenov, 1984; Natal'in, 1984; Natapov and Surmilova, 1986; Natapov and Shul'gina, 1991).

The Anyui-Chukotka fold belt of Til'man (1973), Parfenov et al. (1993), and Nokleberg et al. (1994) includes (1) the Western Chukotka terrane, (2) the Vel'may terrane, and (3) the Eastern Chukotka terrane (Fig. 1). The Western Chukotka terrane is composed of Paleozoic and Mesozoic, largely clastic strata deposited in a passive continental-margin setting. Triassic turbidites show a southerly direction of sediment transport (modern reference frame) (Til'man, 1973; Morozov, 1996). The Vel'may terrane is composed of volcanic, clastic, and chert-bearing sequences that have yielded Upper Triassic fauna, as well as minor ultramafic, gabbro, and plagiogranite bodies (Tynankergav and Bychkov, 1987). The Eastern Chukotka terrane is composed of Proterozoic metamorphic rocks and their Paleozoic sedimentary cover (Zhulanova, 1990); these are markedly similar to the Seward Peninsula complexes (Alaska) and are regarded as a single terrane (Nokleberg et al., 1994).

SOUTH ANYUI SUTURE

The South Anyui suture zone can be traced for more than 1600 km from the eastern Laptev Sea in a southeasterly direction to the north part of the Primorsk depression (Rusakov and Vinogradov, 1969; Spektor et al., 1981; Sekretov, 1993) into the upper reaches of the Bol'shoi Anyui River (Fig. 2). The South Anyui suture zone is not yet clearly defined. Some workers define the South Anyui suture zone to comprise only late Mesozoic clastic deposits and the associated ophiolitic sequences of the Polyarninsk block and the Bol'shoi Lyakhovsky Island

(Radziwill, 1964; Til'man, 1973; Dovgal' et al., 1975; Seslavinskiy, 1970; Parfenov, 1984; Til'man and Bogdanov, 1992; Drachev and Savostin, 1993). Others believe the South Anyui suture zone (sensu lato) to be much broader and to include the Paleozoic to lower Mesozoic sedimentary and volcanic sequences and Paleozoic ophiolites found in the adjacent Yarakvaam terrane (Ged'ko et al., 1991; Lychagin et al., 1991). Natal'in (1984) suggested that the South Anyui suture zone also includes the Late Jurassic–Early Cretaceous complexes of the Nutesyn and Oloy zones.

To the southeast, rocks of the South Anyui suture zone are unconformably overlain by volcanic rocks of the Albian–Upper Cretaceous Okhotsk-Chukotka volcanic belt, and its southward continuation from there is debatable (Natal'in, 1984; Til'man and Bogdanov, 1992; Zonenshain et al., 1990). In general, its continuation is believed to be defined by sporadic outcrops of ultramafic rocks, gabbro, and clastic-chert-volcanic sequences in the vicinity of the Vel'may terrane in eastern Chukotka (Natal'in, 1984; Parfenov, 1984).

The South Anyui suture zone consists chiefly of highly deformed clastic deposits (flysch) that were previously referred to as Upper Jurassic to Lower Cretaceous (Radziwill, 1964; Dovgal' et al., 1975; Seslavinskiy, 1970, 1979; Til'man, 1973; Natal'in, 1984; Parfenov, 1984). In some localities, however, lithologically similar rocks have yielded Upper Triassic faunal assemblages (Bychkov and Solov'yev, 1992; Glotov, 1995). These fossil finds bring into question the assignment of all of these rocks to the Late Jurassic to Early Cretaceous.

In the South Anyui suture zone, rocks mapped as terrigenous clastic deposits (flysch) locally contain exposures of ultramafic rocks, gabbro, and chert that have been previously interpreted as fragments of dismembered ophiolites (Pinus and Sterligova, 1973; Dovgal' et al., 1975; Natal'in, 1984; Parfenov, 1984; Lychagin, 1985; Lychagin et al., 1991). Most workers (Pinus and Sterligova, 1973; Til'man, 1973; Seslavinskiy, 1979; Natal'in, 1984; Lychagin et al., 1991; Parfenof, 1984) assign the mafic rocks and cherts to the Upper Jurassic to Lower Cretaceous sequence because of the presumed age of the enclosing clastic deposits. However, no direct evidence exists for the age of the ophiolitic rocks. The first direct ages for the ophiolitic sequences were determined from rocks that make up south-vergent thrust slices in the Polyarninsk block (Fig. 2). Here, ophiolitic sequences comprise serpentinized oceanic lherzolite and harzburgite, orthoamphibolite, cumulate gabbro, anorthosite, plagiogranite, sheeted diabase dikes, and oceanic pillow basalts that are associated with Middle to Late Jurassic radiolarian

Figure 2. Tectonic map of eastern part of South Anyui suture zone and Alazeya-Oloy fold zone (modified from Lychagin et al., 1991) with schematic stratigraphic columns for Kulpolney island arc (A) and Yarakvaam terrane (B). Schematic cross sections include transect from Aluchin ophiolite to Kulpolney island arc (1-1′–2-2′) and across southern part of Polyarninsk block (cross-section 3–4) (modified from Ged'ko et al., 1991).

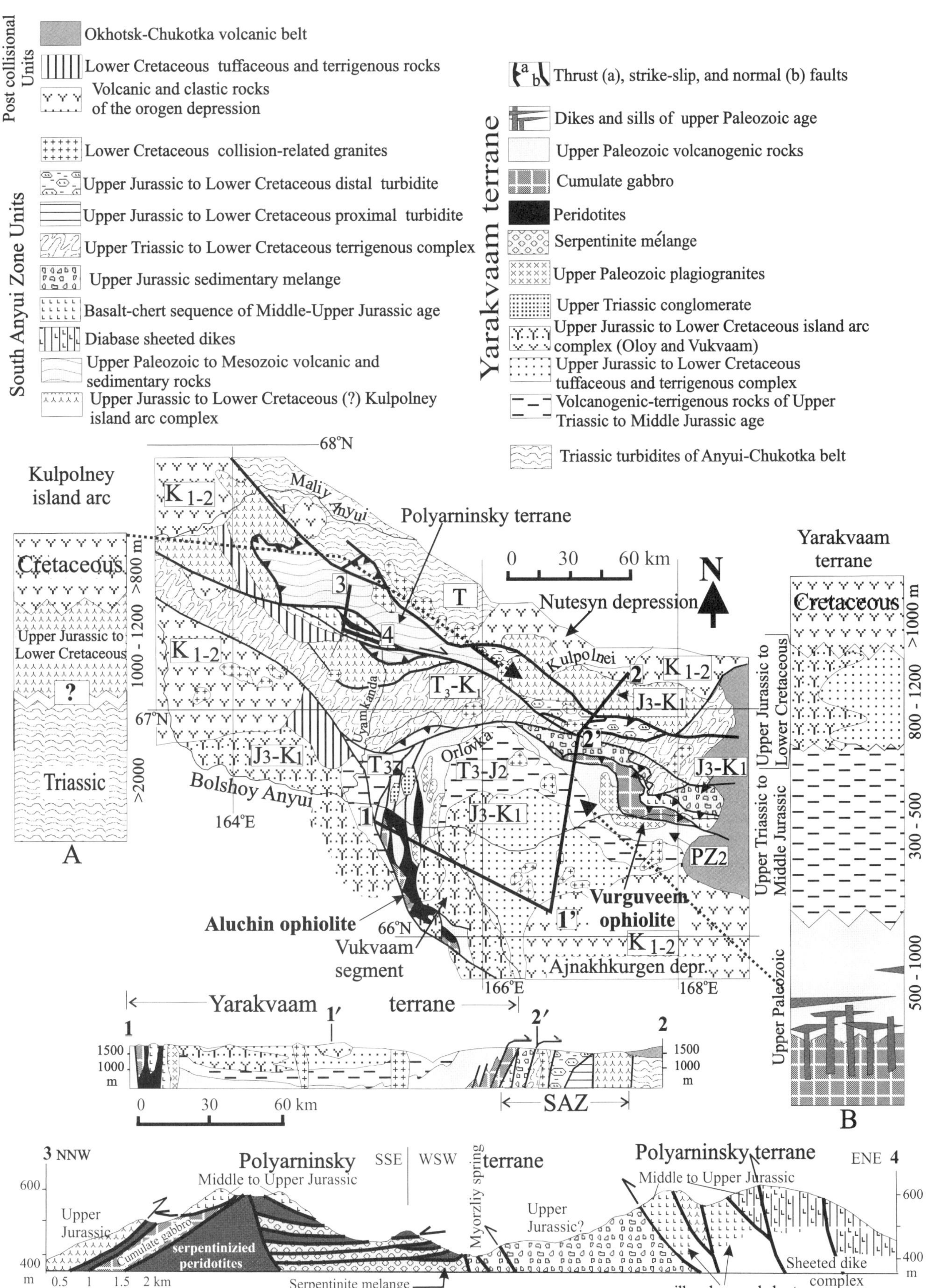

Schematic without scale cross section along the 3-4 line (modified from Ged'ko,1991)

cherts (Ged'ko et al., 1991) (Fig. 2, cross section 3–4). Another ophiolite is associated spatially with upper Paleozoic sedimentary-volcanic rocks of probable island-arc affinity (Lychagin et al., 1991).

Subsequent studies have provided evidence for a Paleozoic age for some of the ophiolitic rocks of the South Anyui suture zone (Lychagin, 1985; Lychagin et al., 1991; Ged'ko et al., 1991). In Bol'shoi Lyakhovsky Island, dismembered ophiolites are thrust to the northwest over Permian-Triassic shale and sandstone similar to those of the Chukotka terrane (Drachev and Savostin, 1993). The ophiolites comprise orthoamphibolite (473 Ma; K-Ar whole-rock dating, Drachev and Savostin, 1993), lherzolite, harzburgite, and pillow basalt (291 ± 62 Ma; Sm-Nd mineral and whole-rock dating, Drachev and Savostin, 1993). The ophiolites are overthrust by Early Cretaceous island-arc andesites and crosscut by granitoids (100–110 Ma; K-Ar whole-rock dating, Drachev and Savostin, 1993).

The Aluchin and Vurguveem ophiolites were included by Ged'ko et al. (1991) and Lychagin et al. (1991) as parts of the South Anyui suture zone (sensu lato). These workers suggested that these ophiolites (Fig. 2) are late Paleozoic, based on the oldest K-Ar data from plagiogranite intrusions of the Vurguveem ophiolite (250 Ma, whole rock, Palymskaya and Palymsky, 1975) and Aluchin ophiolite (231 Ma, K-Ar data, whole rock, Dovgal' et al., 1975; and 374 Ma, K-Ar data, whole rock, Glotov, 1995). Upper Triassic conglomerates and sandstones were deposited unconformably on the Aluchin ophiolite and older complexes of the Yarakvaam terrane (Orlovka River), and contain spinel, chromite, and clasts of ultramafic rocks, gabbro, and plagiogranite (Dovgal' et al., 1975; Cecile et al., 1991). There are pebbles of Vurguveem gabbro and plagioganite in Upper Triassic and Middle Jurassic rocks of the Yarakvaam terrane (Palymsky and Palymskaya, 1975; Shekhovtsov, 1991). However, assignment of the Aluchin and Vurguveem ophiolites to the South Anyui suture zone remains debatable.

GEOLOGY AND STRATIGRAPHY OF THE EASTERN PART OF THE SOUTH ANYUI SUTURE

In 1997 and 1998, we carried out geological studies in the eastern part of the South Anyui suture zone. We focused our efforts in a transect (Fig. 2) that crosses the South Anyui suture zone from the Alazeya-Oloy folded zone to the south to the Anyui-Chukotka fold belt to the north. A preliminary summary of these studies, in which we describe the structural elements of the zone from south to north, is given in the following. These elements include the Yarakvaam terrane, the South Anyui suture-zone complex, and the Kulpolney island-arc complex.

The Yarakvaam terrane, which is the most structurally intact terrane of the South Anyui suture zone, is bounded by shear zones on the south, north, and west and is overlapped by Okhotsk-Chukotka volcanic belt volcanic rocks on the east (Figs. 1 and 2). The Vurguveem (Gromadnensky) ophiolite massif occurs along the boundary between the Yarakvaam terrane and the South Anyui suture zone (Figs. 2 [segment 1'–2'] and 3). The ophiolitic rocks make up a thrust package that verges northward and consists of layered gabbro, ferrogabbro, and gabbronorite with a minor amount of ultramafic rocks and orthoamphibolite (Lychagin, 1985; Lychagin et al., 1991). The mafic rocks are cut by plagiogranites and granodiorites that yield two groups of K-Ar whole-rock dates (Palymskaya and Palymsky, 1975; Shekhovtsov, 1991): 252–207 Ma (latest Permian to Late Triassic) and 179–147 Ma (Middle to Late Jurassic).

During field work in the south part of the Vurguveem massif in 1998, we mapped a suite of compositionally variable (diabase-andesite-dacite) sheeted dikes. The dikes are subvertical and trend chiefly 100°–140°SE. The lower part of the dike complex contains gabbro and gabbro-diabase screens, and the upper part contains screens of extrusive and pyroclastic rocks. Compositionally similar dikes and sills and their extrusive counterparts are widespread in a thick Carboniferous-Permian sedimentary-volcanic sequence of island-arc affinity belonging to the Yarakvaam terrane (Parfenov et al., 1993). This implies that the late Paleozoic sedimentary-volcanic sequence unconformably overlies the plutonic rocks of the Vurguveem massif. Basal conglomerates of Early Permian (?) age were described by Palymskaya and Palymsky (1975). Therefore, the plutonic rocks of the Vurguveem massif together with the overlying island-arc volcanic rocks might compose a single late Paleozoic arc and suprasubduction-zone ophiolitic complex. We interpret the gabbro to be a fragment of lower crust associated with an island arc, perhaps similar to the Tonsina assemblage in Alaska (De Bari and Coleman, 1989). The island-arc volcanic rocks are among the oldest stratified rocks of the Yarakvaam terrane. These stratified rocks consist of shallow-marine and continental sedimentary and volcanic rocks (mafic through felsic volcanic rocks, and pyroclastic and clastic rocks with intercalated limestone) and contain Carboniferous and Permian faunas and floras (Radziwill, 1964; Palymskaya and Palymsky, 1975; Yegorov, 1990; Shekhovtsov, 1991; Glotov, 1995) (Fig. 2, column B). The Carboniferous-Permian sequence is riddled with mafic, intermediate, and felsic dikes and sills and is cut by minor plagiogranite, diorite, and gabbro-diabase intrusions. This late Paleozoic island-arc volcanic sequence is overlain unconformably by transgressive Upper Triassic clastic rocks. From the center to the northern boundary of the South Anyui suture zone, shallow-marine facies give way to deeper water turbidites and submarine-slump deposits (Shekhovtsov, 1991; Yegorov, 1990; Bychkov and Solov'yev, 1992). Faunal assemblages in these deposits include Tethyan species (Afizkiy, 1970). In the west, Upper Triassic conglomerate unconformably overlaps the Aluchin ophiolite (Dovgal' et al., 1975; Cecile et al., 1991). These relationships imply a pre–Late Triassic accretion or amalgamation of the Aluchin ophiolite to the Yarakvaam terrane (possibly between the end of the Permian and the beginning of the Late Triassic).

Higher in the section, Lower to Middle Jurassic clastic rocks and Oxfordian to Valanginian pyroclastic-terrigenous rocks occur (Afizkiy, 1970; Shekhovtsov, 1991). The southwest part of the Yarakvaam terrane (Vukvaam zone, after Gulevich,

1975) displays a widespread sequence of compositionally variable island-arc volcanic rocks of Oxfordian to Volgian age (Natal'in, 1984). Along the boundary with the South Anyui suture zone west of the Yarakvaam terrane, the Volgian to Valanginian part of the succession comprises a volcanic sequence with layers of intercalated tuff, conglomerate, and sandstone. Basalts and andesites within the volcanic sequence are alkalic in composition. The island-arc kinship of these volcanic rocks is thus not readily apparent (Gulevich, 1975; Parfenov, 1984), and further studies are required to elucidate their geodynamic setting.

It has long been proposed that the eastern part of the Alazeya-Oloy zone is underlain by older continental basement (Yablonsky massif) that is in turn part of the Omolon massif (Til'man, 1973; Grinberg and Rudich, 1981; Natal'in, 1984). However, the relationships just described between the Vurguveem gabbro and the intruding dikes and sills with the unconformably overlying Paleozoic island-arc suites are at variance with this hypothesis; hence, the interpretation that the Yarakvaam terrane is of island-arc affinity (Parfenov et al., 1993; Nokleberg et al., 1994) is more reasonable.

LITHOLOGIC ZONATION

A system of high-angle fault-bounded tectonic slices that verge chiefly north make up the central part of the South Anyui suture zone (Figs. 2 and 3). Their constituents, from south to north, are as follows.

Basalt-chert complex

The basalt-chert complex makes up the southern, tectonically bounded slices in the South Anyui suture zone and is overridden by the gabbro of the Vurguveem massif to the south (Figs. 2 and 3). The sequences within individual tectonic slices are fragmentary and differ in the number and thickness of basalt flows, presence or absence of chert (including red radiolarian chert), and the presence or absence of interpillow carbonate lenses and layers. The basalt-chert sequences in the tectonic slices are separated by mélanges of tuff and terrigenous material. The matrix of olistostrome includes some clasts of felsic tuff. Geochemically, the basalts are similar to oceanic or backarc basin basalts (Natal'in, 1984; Lychagin et al., 1991). The cherts yield the radiolaria *Haliodictya* cf. *Hojnosi* Riedel and Sanfilippo, *Stichocapsa convexa* Yao, *Williriedellum* sp., and *Zhamoidellum* sp., which is a typical Bajocian-Kimmeridgian fauna (according to N.Yu. Bragin, Geological Institute, Russian Academy of Sciences, Moscow).

Sedimentary mélange

The sedimentary mélange consists of tuffaceous clastic rock exhibiting various degrees of tectonization. The mélange contains olistoliths and blocks of various sizes composed of basalt, andesite, chert, gabbroic rocks, and plagiogranite that resemble parts of the Vurguveem ophiolite. The extrusive rocks comprise both mid-ocean ridge basaltlike rocks and medium-Ti, high-Al tholeiitic basalts, andesites, and andesitic dacites of island-arc affinity. The blocks of the mélange are arranged in lenses and boudinaged layers within which the rocks are lithologically uniform. The matrix of the mélange displays sedimentary structures, such as rocks within the layers showing submarine-slump structures, evidence of redeposition by bottom currents, and horizons of grain-flow deposits and pebbly mudstone or diamictite and turbidites. Some tectonic slices within the clastic terrigenous mélange that are composed of turbidites yield Oxfordian to Volgian and Berriasian to Valanginian faunas (Radziwill, 1964; Radziwill and Radziwill, 1975; Glotov, 1995). Similar lithologies are found in the matrix of the mélange, suggesting that the fragments in the matrix are of the same age as the rocks that compose the tectonic slices. Direct evidence for the age of the matrix is absent, however, and it may contain fragments of various ages. The sedimentary mélange shows various degrees of tectonization, and in places it is metamorphosed to greenschist facies. Textures common to accretionary mélanges, such as C-S tectonites, broken formations, and block-in-matrix structures, are widespread in the sedimentary mélange unit and are comparable to those found in accretionary mélanges elsewhere around the world (Cowan, 1985).

Terrigenous complex

The terrigenous complex comprises alternating terrigenous turbidite and shale with sulfide-carbonate concretions and calcareous sandstone intercalations. Sandstone grains consist of quartz (80%–90%), feldspar, and less abundant sedimentary rocks and pyroxene. Measurements of the persistent system of bottom current marks suggest a north-to-south clastic supply (present reference frame). These deposits were previously assigned to the Upper Jurassic to Lower Cretaceous (Radziwill, 1964; Seslavinskiy, 1979; Natal'in, 1984). Geologists from the Bilibino Geological Survey, however, established that the rocks that yield the Late Jurassic to Early Cretaceous faunas are distinguished by a more polymictic sandstone composition. These turbidites are lithologically similar to the Triassic deposits of the Anyui-Chukotka fold belt, which are tentatively dated as Triassic (Glotov, 1995). One turbidite sample, collected in 1997, yielded sporadic, poorly preserved conodont remains (as determined by V.A. Aristov, Geological Institute, Russian Academy of Sciences). Farther west along the strike of the transect, similar terrigenous rocks within the South Anyui suture zone have yielded Upper Triassic faunas (Glotov, 1995). The terrigenous clastic complex is thus more likely to be Triassic than Upper Jurassic to Lower Cretaceous in age.

Turbidite complex

This complex of subvertical to steeply northward-dipping tectonic slices consists of distal and proximal turbidites with interbeds of high-Ti subalkalic pillow basalt and picritic basalt. The clasts in the sandstones show evidence of explosive volcanism of felsic composition. Faunal remains in the terrigenous rocks yield Volgian ages (Glotov, 1995). Preliminary geochem-

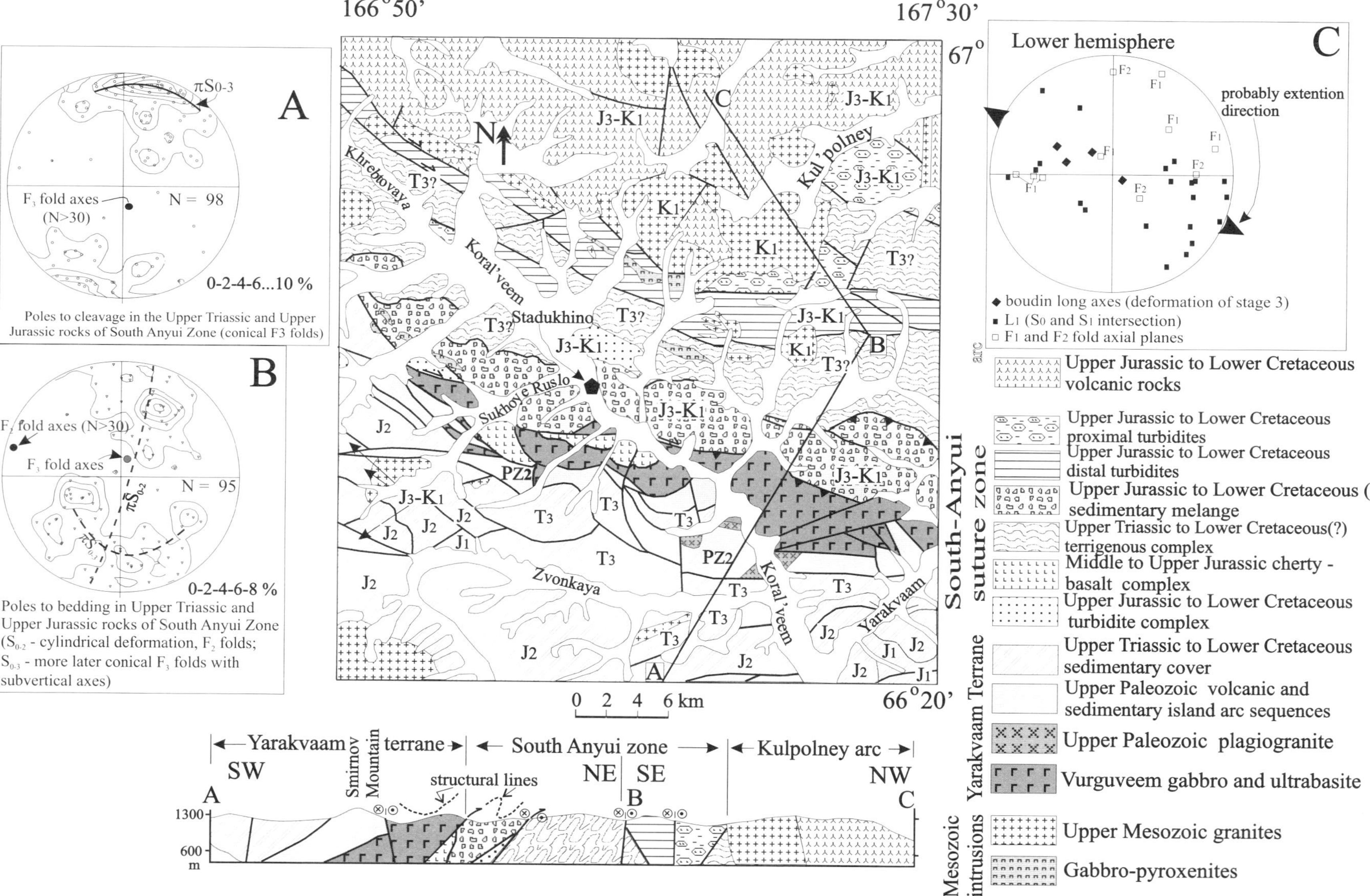

Figure 3. Geological map of Koralveem River basin in eastern South Anyui zone showing location of cross section A-B-C. Stereograms (Wolf, lower-hemisphere projection) show (A) plot of poles to cleavage deformed during stage 3 of deformation dextral strike-slip faulting; (B) two stages of bedding deformation: S_0, 2, belt of poles for stage 2 cylindric thrust folds; $S_0$3, belt of poles for stage 3 conical strike-slip deformation (see text for details); (C) orientation of fold axes and long boudin axes, and L1 lineation (the result of cleavage and bedding intersection during stage 2). Most F2 folds have northerly vergence.

ical analysis suggests that the picritic basalts formed in a supra-subduction-zone setting that was undergoing extension (Sharaskin et al., 1983). This rock assemblage may have formed in the frontal (distal) part of an island arc. Tectonic slices composed of volcaniclastic turbidites with an admixture of tuffaceous material occur farther north. The amount of volcanic material increases toward the north, and grain-flow and clastic gravity-flow deposits such as olistostromes with olistoliths and pebbles of tuffs and volcanic rocks of mafic and intermediate to felsic composition appear. The turbidites yield Volgian to Valanginian faunas (Glotov, 1995). The coarsening of the turbidite clastics and the abundant volcanic clasts suggest that these deposits were derived from an arc and were deposited in a proximal forearc setting. The relationships between the turbidite and terrigenous complexes described here are tectonic.

Fault-bounded wedges of shallow-water pebble conglomerate are present within the tectonic slices of Upper Jurassic to Lower Cretaceous polymict turbidites. Clast composition is diverse, but is dominated by lithologies derived from the erosion of the South Anyui suture zone units. These deposits can be regarded as deformed neoautochthonous strata. The age of the shallow-water conglomerate is Valanginian to Hauterivian according to Yegorov (1990), and Hauterivian to Barremian according to Dovgal' et al. (1975). The new ages (Yegorov, 1990) pose further limitations on the final time of closing of the South Anyui basin.

Kulpolney island-arc complex

This complex in the northern part of the South Anyui suture zone (Fig. 2) is dominated by various types of volcanic rocks that include pillow lavas and massive amygdular basalts, andesites, dacites, and their subvolcanic and pyroclastic counterparts. Volcanic flows are interbedded with thin lenses of tuff, ash beds, siliceous shale, and interpillow jasperoid chert. Field evidence indicates that, in places, the volcanic complex is in stratigraphic continuity with the underlying turbidite sequence. The upper part of the turbidite sequence includes tuff and lava horizons, and the lower part of the volcanic sequence contains tuffaceous layers. Lateral transitions and facies changes between the upper part of the turbidite section and the lower part of the volcanic section might also be present. These relationships imply a Late Jurassic to Early Cretaceous (?) age for the volcanic complex. The volcanic rocks display calc-alkaline and subalkaline trends and may have formed in an island-arc setting.

Structurally upsection in the Kulpolney island arc, intermediate-felsic extrusive rocks and associated tuffaceous and terrigenous deposits have stratigraphic positions with respect to underlying rocks that remain unclear. They either cap the Upper Jurassic to Lower Cretaceous island-arc sequence (Glotov, 1995) or make up the volcanic fill of the younger, and mostly overlying, Nutesyn depression of Early Cretaceous age (Shekhovtsov, 1991).

The issue of what constitutes the basement of the Kulpolney island arc remains open to discussion. Most workers (Natal'in, 1984; Parfenov, 1984; Til'man and Bogdanov, 1992; Morozov,

1996) believe that this island-arc complex formed along the margin of the Chukotka microcontinent. Our investigations, however, have not found any direct relationships between the Kulpolney complex and the sedimentary cover of the Chukotka terrane. Furthermore, the thrust-bounded wedges of Lower Jurassic sandstone and siltstone occur along the boundary between the Chukotka terrane and the Kulpolney island-arc sequence. These clastic rocks have polymict compositions and differ from that of the cover of the Chukotka terrane (Glotov, 1995).

STRUCTURAL STUDIES

The first detailed structural studies in the South Anyui suture zone were carried out by Natal'in (1984), who identified thrust- and strike-slip-related deformation of various ages. Through our field work, we collected additional structural data and recognized four main stages of tectonic deformation of Mesozoic age, which we describe in the following. From oldest to youngest, these include thrusting of probable pre–Late Triassic age, Late Jurassic to Early Cretaceous thrusting and dextral strike-slip faulting, and post-Albian sinistral strike-slip faulting.

Stage 1

Deformation associated with stage 1 is best developed in the gabbroic rocks of the Vurguveem ophiolite (Fig. 4, A and B). It is represented by isoclinal F_1 folds confined to dynamothermally metamorphosed zones and amphibolite schists (metamorphosed gabbro (Fig. 4B). Fold axes trend roughly east-west. Tectonic transport was directed from south to north. Figure 4A shows plagiogranite veins affected by late-stage deformation supporting this transport direction. The age of plagiogranite is late Paleozoic (Palymskaya and Palymsky, 1975). This deformation took place later. We suppose that the deformation accompanied the late Paleozoic ophiolite accretion. This accretionary event is marked by Upper Triassic stratigraphic unconformity at the edge of the Yarakvaam terrane. In terms of geometry, later deformational structures in the Vurguveem ophiolite are similar to fabrics of stages 2, 3, and 4 (described in the following). Among these structures are the thrust faults of north and south vergence and the strike-slip faults that were documented by Natal'in (1984). According to Natal'in (1984), the tectonic history of the Vurguveem ophiolite plutonic rocks was increasingly complicated. We have observed very complicated fold structures of magmatic layering and the presence of high-temperature shear zones, but we believe that these are related to the magmatic and metamorphic history of the Paleozoic plutonic rocks of the Vurguveem ophiolite. We do not address these deformations, but focus instead on the identification and interpretation of the main tectonic fabrics of Mesozoic age.

Stage 2

This stage of deformation is characterized by two substages of deformation (a and b) of similar age (Figs. 3C and 4, C and

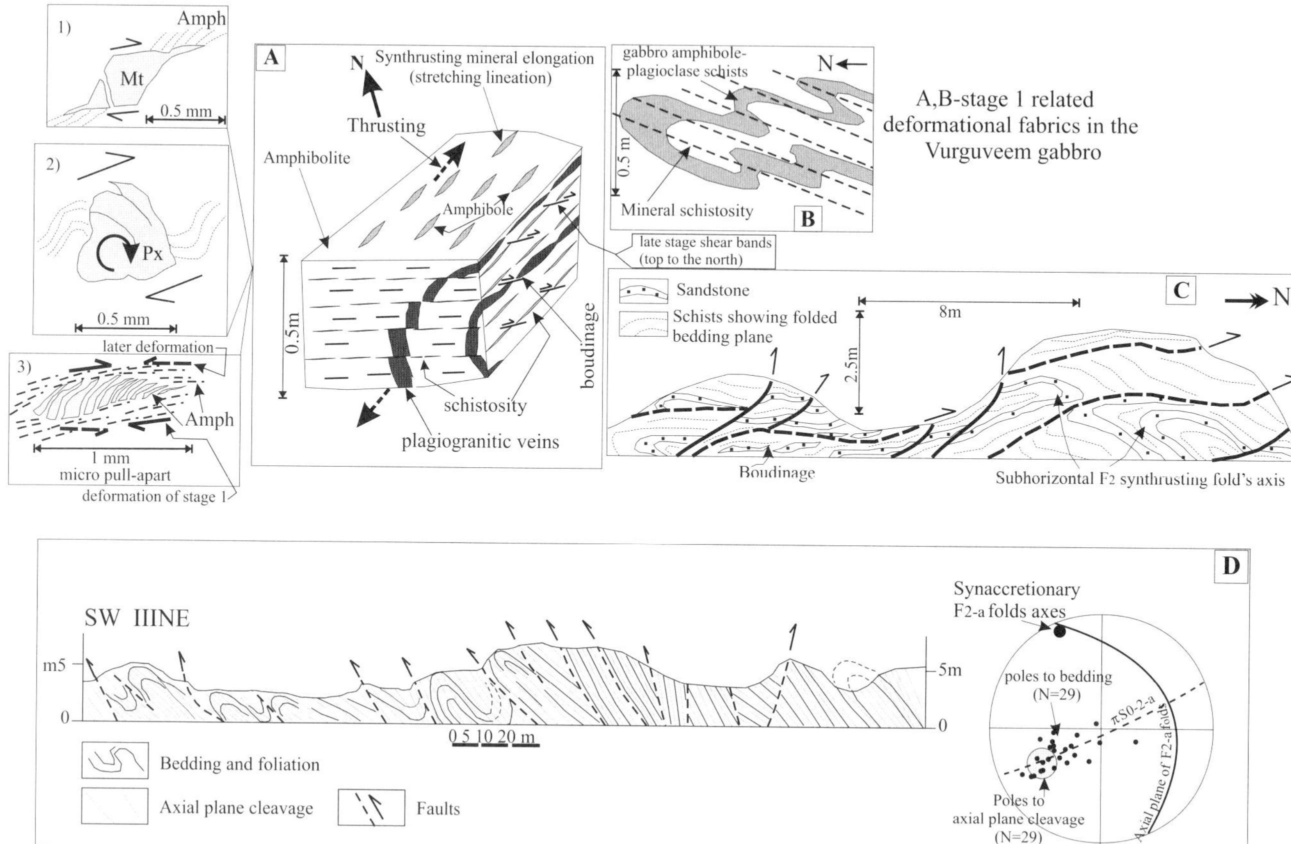

Figure 4. Examples of fold and thrust–related stage 1 deformation in Vurguveem ophiolite (A and B), and examples of stage 2 deformation in Mesozoic terrigenous turbidite of South Anyui suture zone (C). (A) Thrust-related boudinage of plagiogranitic veins and mineral stretching lineation in Vurguveem metamorphosed gabbro show north-vergent tectonic transport during stage 1 deformation. Photomicrograph drawings show structural fabrics related to stage 1 deformation (north-vergent thrusting): 1, magnetite (Mt) grain clockwise rotation around subhorizontal axis; 2, pyroxene (Px) grain clockwise rotation; 3, microscopic pull-apart structure along schistosity. Thin section cuts foliation and parallels stretching lineation. (B) F1 isoclinal folds in amphibole-plagioclase schists (metagabbro); metamorphic schistosity is parallel to fold axial planes. (C) F2–a isoclinal folds and thrust faults of northerly vergence in Upper Jurassic to Lower Cretaceous terrigenous turbidite sequence, southern part of South Anyui suture zone. (D) F2–b isoclinal folds and thrust faults of southerly vergence in Late Triassic (?) terrigenous clastic complex, northern part of South Anyui suture zone.

D). Both substages are manifest in the accretionary mélange, the Upper Triassic terrigenous complex, and the Upper Jurassic to Lower Cretaceous turbidites, and are characterized by isoclinally folded bedding (Fig. 4, C and D) and an axial-plane S_1 penetrative cleavage. The fold axes trend north-northwest–south-southeast (Fig. 3, stereogram B), and the cleavage is strongly deformed by subsequent events.

Substage 2a. The characteristic feature of substage 2a is the southward vergence of folds and an axial-plane cleavage (Fig. 4D). We have identified substage 2a deformation only in some outcrops of the Late Triassic terrigenous complex and the Upper Jurassic–Lower Cretaceous clastic turbidites. The south-vergent folds of this substage are located in the northern part of the South Anyui suture zone. We propose that substage 2a deformation was synchronous with subduction beneath the Kulpolney island arc, in the forearc region or accretionary prism. The deformation of substage 2a is the result of the formation of an accretionary wedge in Late Jurassic–Early Cretaceous time.

Substage 2b. The deformation of substage 2b is characterized by northward vergence of the folds and development of an axial-plane cleavage (Figs. 3 [stereograms B and C] and 4C). This deformation is similar to the F_3, F_2, and F_4 folds described by Natal'in (1984). These features are more widespread than substage 2a deformational features, especially in the southern part of the South Anyui suture zone, and the zones of dynamothermal metamorphism are characterized by more intensive development and a greater strain. These zones locally also develop the ductile C-S tectonite fabrics. Substage 2b deformation may be due to compression directed roughly north-south. The lower age limit of this deformation is delimited by the fact that the folds affect the Upper Jurassic to Valanginian rocks of the terrigenous mélange; hence deformation must be Valanginian or younger in age. We propose that deformation occurring in substage 2b is associated with the earliest phase of collision of the Siberian and Chukotka microcontinents. As a result, the deformation of substage 2b is superposed on the synaccretion deformation of substage 2a. This deformation began in the southern part of the South Anyui oceanic paleobasin and then migrated to the north.

Stage 3

Structures associated with stage 3 deformation are widespread in the South Anyui suture zone and its immediate vicinity. We believe that these structures are related to motion along the east-west– and west-northwest–east-southeast–trending subvertical dextral strike-slip faults. Conical folds of bedding with a new axial-plane cleavage are developed chiefly in this event and characterized by subvertical axes (Fig. 3, stereograms A and B). Lense-shaped zones of tectonic mélange are also present and are thought to be related to this event. Boudin axes in these tectonic mélange zones are subvertical (Fig. 3, stereogram C). C-S structures indicating dextral strike-slip fabrics are ubiquitous at all scales of observation (Fig. 5A). Z-shaped folds of bedding, foliation including the boundaries of tectonic slices, and boudins with subvertical long axes in the tectonic mélange zones are common at all scales of observation. This is illustrated by the Z-shaped form of the Vurguveem ophiolite massif (Fig. 5B). The upper age limit for the dextral strike-slip faults is delimited by the fact that this deformation affects rocks as young as those of the Upper Jurassic–Lower Cretaceous turbidite complex.

Stage 4 (postcollisional)

Structures related to stage 4 deformation are developed in conjunction with roughly east-west–trending brittle sinistral strike-slip faults, the locations of which may be inherited from preexisting fault planes. Sinistral strike-slip faulting is demonstrated by the local development of pressed conical folds with subvertical axes within zones of high deformation only as thick as a few tens of meters. The domains between these zones show only open folds of bedding and cleavage and widely spaced tension fractures. The sinistral strike-slip faults offset the Albian-Cenomanian rocks of the Okhotsk-Chukotka volcanic belt, and are conjugate with north-northeast–south-southwest–trending strike-slip faults, to which belts of Upper Cretaceous dikes are confined.

Paleotectonic interpretation of the South Anyui suture zone

Previous studies together with new data establish the presence in the southern part of the South Anyui suture zone of chert-basalt complexes of Bajocian to Kimmeridgian age, which implies the former presence in the region of a South Anyui oceanic basin (scnsu stricto). The sedimentary mélange complex can be interpreted as the relic of an accretionary prism. Furthermore, the ophiolite clastics in the mélange, derived from the erosion of the Yarakvaam terrane and Vurguveem ophiolite, together with the northerly vergence of the tectonic slices, suggest a spatial association of the accretionary prism with the Vukvaam segment of the island arc. This segment developed along the northern margin of the amalgamated terranes of the Alazeya-Oloy fold belt (Fig. 6B).

Our studies also document the existence of clastic and volcanic deposits of the Kulpolney island arc developed in the

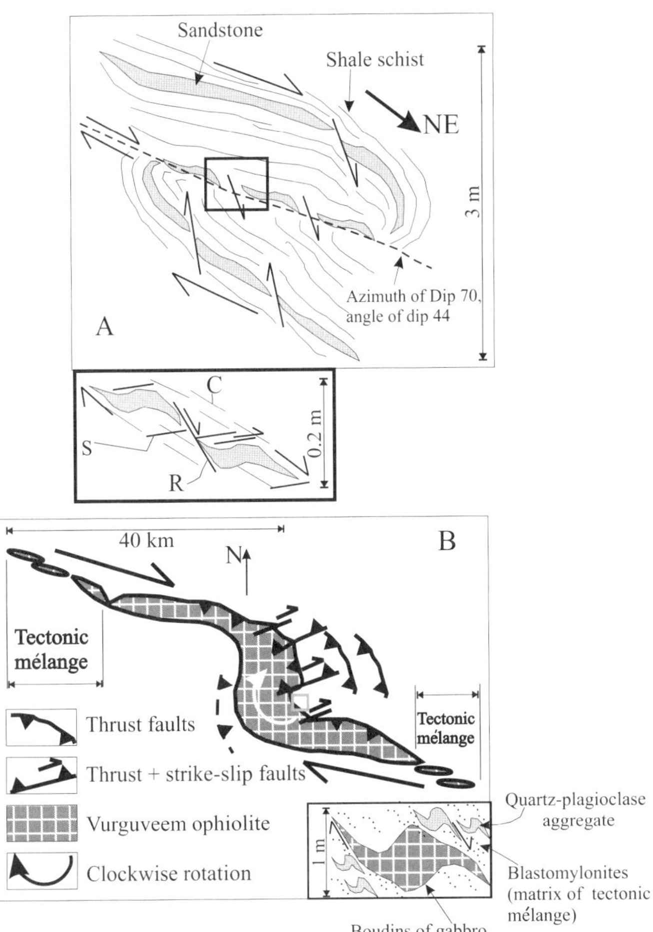

Figure 5. Examples of dextral strike-slip faults related to stage 3 deformation in Mesozoic sedimentary rocks of South Anyui suture zone (A) and Vurguveem ophiolite (B). Letters in A denote C-S tectonite structures. R and S—Riedel shears: R is extensional normal fault (in picture plane); S is compressional reverse fault (in picture plane) congruated with schistosity; C is shear surface.

northern part of the South Anyui suture zone (Fig. 6C). The Late Jurassic Epoch is characterized by the north-dipping subduction of the South Anyui oceanic crust beneath the Chukotka microcontinent (Natal'in, 1984) (Fig. 6C). The spatial relationships of the Upper Jurassic island-arc complexes with the turbidite deposits of the South Anyui suture zone further support this view. The main axis of volcanism related to the Kulpolney island arc was to the north. To the south, these volcanic rocks are inferred to interfinger with proximal and distal turbidites deposited in a forearc (Fig. 6C). Picritic basalts probably were erupted in local zones of extension in the frontal part of the forearc, similar to picrite-basalt volcanism in the Tonga arc (Sharaskin et al., 1983). The location of this arc along the margin of the Chukotka microcontinent (Fig. 6C) is only an inference. We cannot exclude the possibility that the Kulpolney island arc developed in an intraoceanic setting similar to the Koyukuk island arc of Alaska (Moore et al., 1994). Note that our data support the notion (Parfenov, 1984; Natal'in, 1984; Nokleberg et al., 1998) of

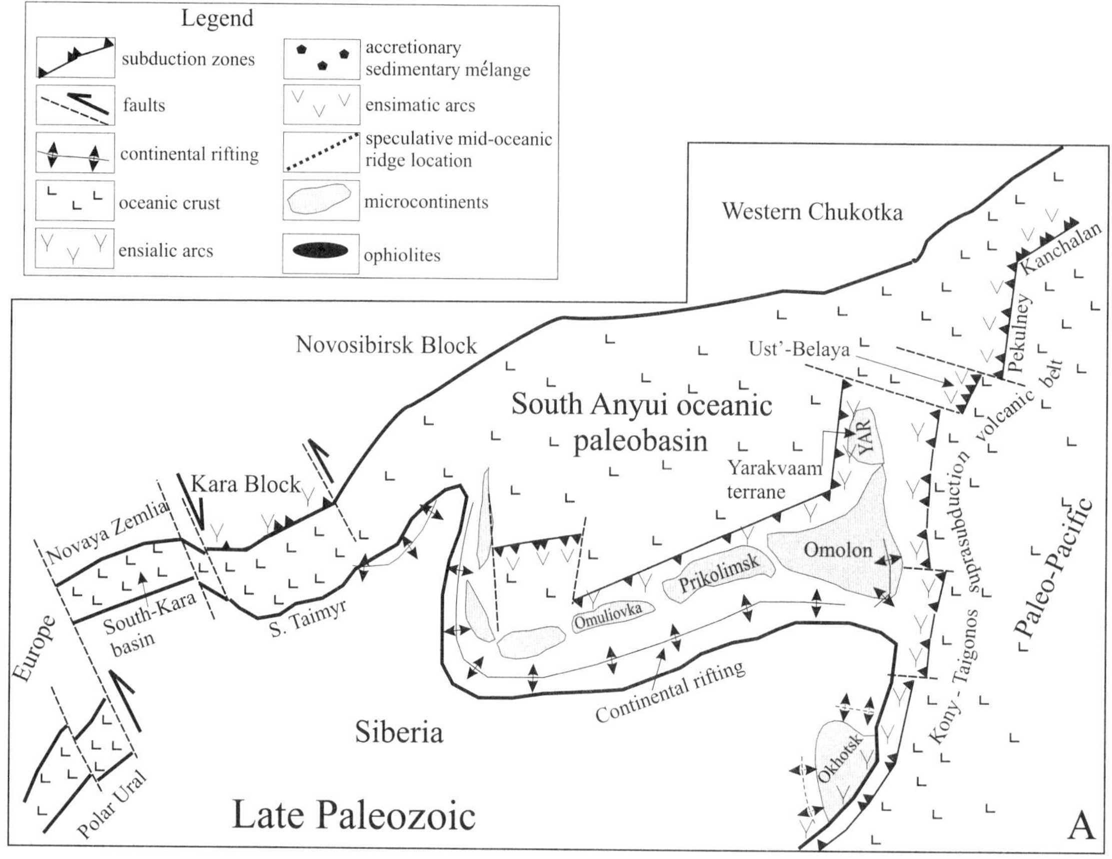

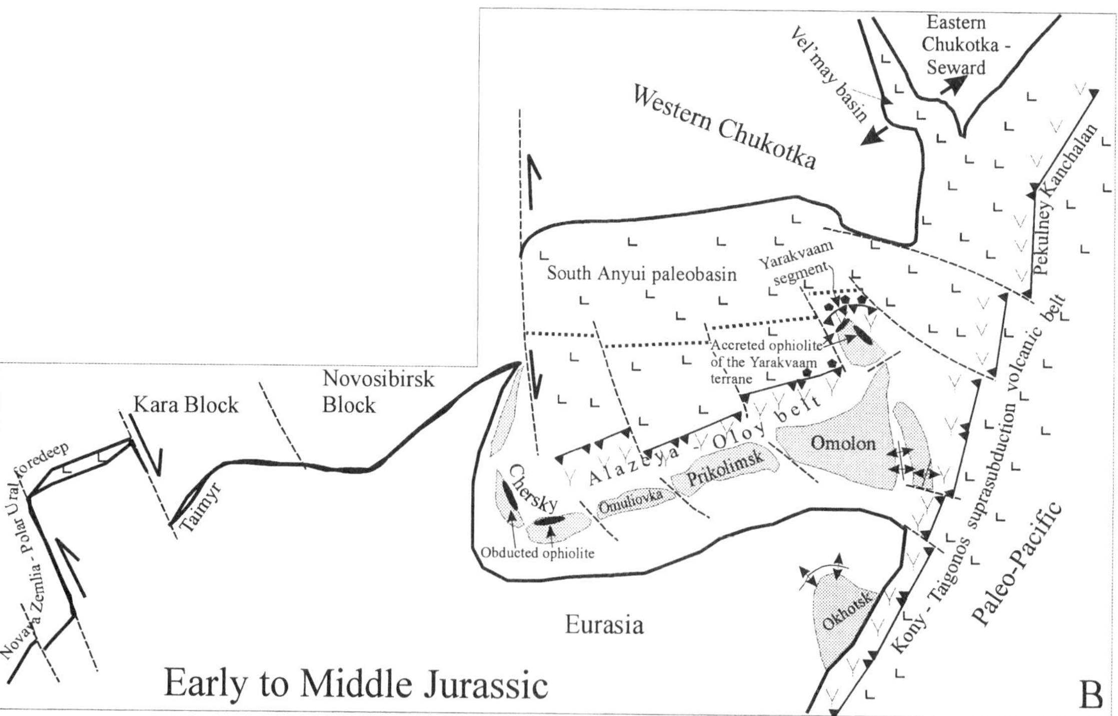

Figure 6. Interpretative paleotectonic schemes for late Paleozoic (A), Early to Middle Jurassic (B), and Late Jurassic to Early Cretaceous (C) stages of evolution of Anyui oceanic basin.

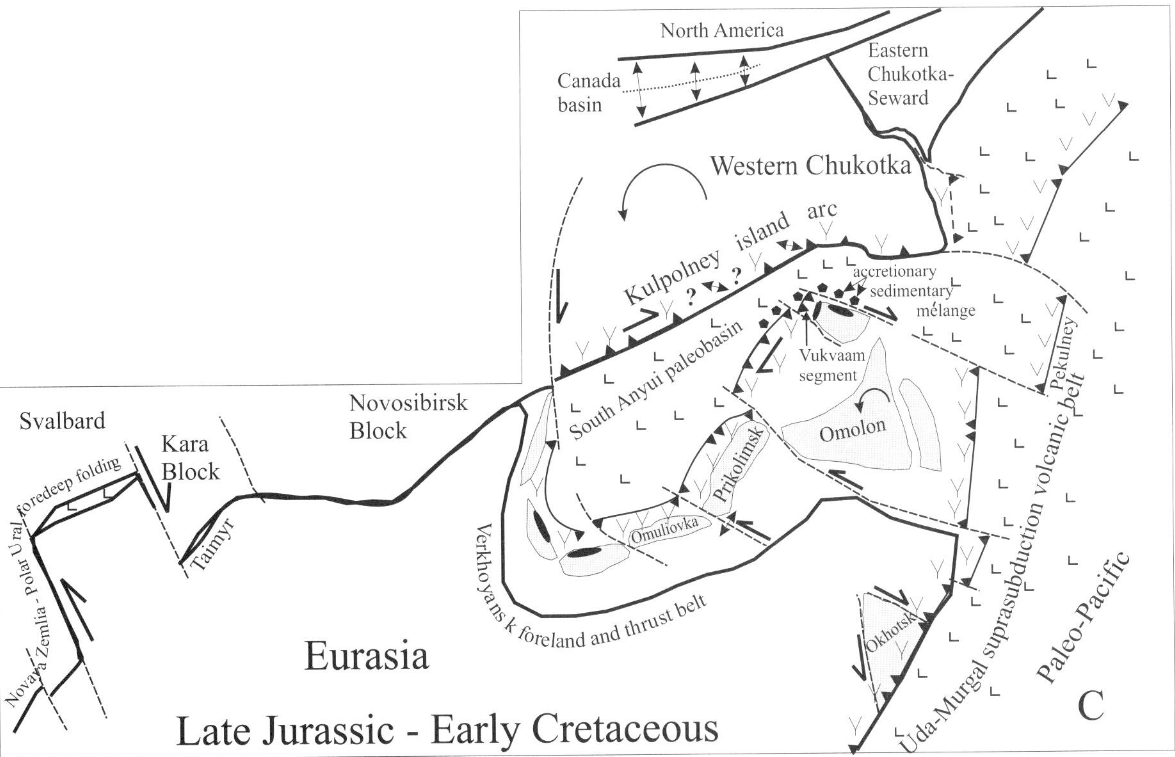

Figure 6 (*continued*)

DISCUSSION AND PRINCIPAL ISSUES TO BE RESOLVED

a degree of similarity between the South Anyui suture zone complexes and the Angayucham terrane (Moore et al., 1994).

The new data on the age of the ophiolites, pelagic cherts, and basalts in the South Anyui suture zone and Alazeya-Oloy zones testify to the existence of an Anyui (South Anyui) oceanic basin between the Siberian and North American plates in Paleozoic and early Mesozoic time (e.g., Fujita and Newberry, 1983; Zonenshain and Natapov, 1988; Zonenshain et al., 1990; Golonka et al., 1994; Nokleberg et al., 1998). The age of formation and early history of this basin remain relatively poorly known. Its youngest deposits are pelagic chert and shale of Bajocian to Kimmeridgian age.

In Paleozoic time the Anyui oceanic basin may have been connected, via Taimyr, with the paleo-Urals Ocean. The presence of late Paleozoic ophiolites in the Polar Urals region (Ruzhentsev and Aristov, 1998) (Fig. 1) suggests that the Polar Urals Ocean closed in late Paleozoic time. Simultaneously, through the collision of Siberia and the Kara microcontinent, the Taimyr basin ceased to exist (Vernikovsky, 1996). Various viewpoints exist concerning the eastward continuation of the Anyui oceanic basin. Most workers (Zonenshain et al., 1990; Parfenov, 1984; Til'man and Bogdanov, 1992) suggest that the Anyui

basin was a reentrant of the paleo-Pacific. We believe that, starting at least in the late Paleozoic, a convergent boundary separated the Anyui basin from the northwest Pacific (Morozov, 1996; Sokolov et al., 1997) (Fig. 6A). At the end of the Paleozoic and in the early Mesozoic, after the folding in the western (present reference frame) branch of the Anyui basin within the Mesozoides of northeastern Russia, the Anyui oceanic basin was still in part preserved, but had a rather complex configuration (Fig 6B). The widespread record of late Paleozoic to early Mesozoic island-arc magmatism in the Alazeya-Oloy zone together with evidence for tectonic deformation of the ophiolite sequences of the Chersky Range (Oxman et al., 1995) suggest a shrinking of the Anyui basin. In pre–Late Triassic time, an amalgamation of island-arc terranes (e.g., Yarakvaam, Alazeya) and oceanic terranes (e.g., Aluchin) took place. The newly formed heterogeneous basement provided the site for the development of the younger Alazeya-Oloy island arc (Late Triassic to Lower Jurassic), which rimmed the southern margin of the South Anyui oceanic basin (present reference frame) (Seslavinskiy, 1979; Parfenov, 1984; Natal'in, 1984).

Paleomagnetic data analyzed by A.N. Didenko (Sokolov et al., 1997) suggest that the Kolyma loop terranes once formed a single megablock together with the Siberian plate, beginning at least in Middle Devonian time. Before the Middle Jurassic these terranes had not been rigidly connected with Siberia, yet their motions were characterized by similar paths. In post–Middle

Jurassic time these blocks finally became permanently incorporated into the Eurasian plate. Geologic data (Parfenov et al., 1993) suggest that during this time, in the Middle Jurassic, the Kolyma loop terranes became a single Kolyma-Omolon superterrane, and ophiolites were obducted in the Chersky Range (Fig. 6B) (Oxman et al., 1995).

In Late Jurassic time new convergent boundaries developed along the modified margin of the northeastern Asian continent. At the Pacific boundary, the Uda-Murgal convergent margin developed upon a heterogeneous basement (Sokolov et al., 1997, 1998) (Fig. 6C). Geologic data suggest that this margin continued into the southern part of the Chukotka Peninsula (Morozov, 1996) and into the southern part of Alaska, probably into the Togiak terrane (Nokleberg et al., 1994), or the Chitina arc (Parfenov and Spector, 1997). This convergent boundary separated the South Anyui basin from the Pacific (Fig. 6C).

In Late Jurassic to Early Cretaceous time the Svyatoi Nos–Oloy island arc developed along the boundary between the Asian continent and the South Anyui basin (Til'man et al., 1977; Natapov and Surmilova, 1986). The Vukvaam island arc is a representative part of this larger arc (Natal'in, 1984). The Kulpolney arc developed along the north side of the South Anyui basin. It can therefore be inferred that oceanic crust of the South Anyui basin was consumed by subduction along both its northern and southern boundaries, bringing about its rapid closure and demise. The Chukotka-Eurasia collision that resulted from this closure culminated at the end of the Early Cretaceous, approximately synchronous with the onset of seafloor spreading in the Canada basin (Embry and Dixon, 1990; Grantz and May, 1983; Grantz et al., 1990). After the collision the Nutesyn orogenic basin developed north of the South Anyui suture zone, and the Aynakhkurgen orogenic basin formed south of it. Both of these basins filled with Hauterivian to Albian sedimentary deposits and volcanic rocks.

Our paleotectonic interpretation, which is presented in Figure 6, is a very general one and is offered as a working hypothesis that requires further refinement and verification. Future studies must focus on the ophiolites in order to (1) elucidate the time of inception of the South Anyui basin and (2) compare these ophiolites with coeval ophiolites in the Chersky Range, Polar Urals, Taimyr, and the northwest Pacific margin. The Paleozoic ophiolites (e.g., the Aluchin ophiolite) lack demonstrable volcanic members, whereas the South Anyui suture zone Mesozoic (Bajocian to Kimmeridgian) oceanic assemblages are not known to contain ultramafic-gabbroic complexes. This might be due to a poor knowledge of the geologic details of this region or to the poor degree of preservation of these ophiolites. Otherwise, a satisfactory explanation must be found for this fact, which appears to support the idea of the rift-related origin of the South Anyui suture zone (Seslavinskiy, 1970, 1979; Dovgal' et al., 1975; Til'man et al., 1977). It remains unclear whether the Anyui basin was a vast ocean or a minor basin. Tackling this issue requires a paleomagnetic study of the terranes north and south of the South Anyui suture zone.

The Late Jurassic to Early Cretaceous syncollisional deformational events that affected rocks of the South Anyui suture zone were closely related to thrust structures, which likely reflect the same event. The thrusting must have marked the initial stage of a continent-to-continent collision: during this stage collision developed orthogonally. The final stage of continental collision developed by oblique slip and was accompanied by transpression and dextral strike-slip faults along the South Anyui suture zone. Some thrusts may have formed through a symmetric or asymmetric flower structure-type arrangement at right angles to the suture. Such phenomena are widespread in zones of transpression (Sylvester, 1988; Woodcock and Fischer, 1986; Bondarenko, 1996).

The map distribution of Upper Jurassic to Early Cretaceous island-arc complexes along the South Anyui suture zone (Fig. 1) shows gaps. These gaps might be due to segments where the continental and oceanic plates interacted along strike-slip faults. Some of the volcanic complexes currently categorized as arc related might have formed in pull-apart structures adjacent to strike-slip faults. That dextral displacements played an important role suggests that the collision involved a counterclockwise rotation of the Chukotka microcontinent relative to Siberia. This rotation may have been due to differential rifting in the Canada basin (Embry and Dixon, 1990). Specifying the dynamics of collision between the continental masses will require the determination of reliable timing of the deformation stages identified herein.

CONCLUSIONS

The tectonic interpretation proposed for the South Anyui suture zone remains a working hypothesis because many structural details and fundamental issues about the evolutionary history of this basin are not yet understood. At the same time, this geological summary highlights the key role of the South Anyui suture zone to our understanding of the tectonic history and palinspastic reconstructions of northeast Asia and the Arctic. Understanding this history would further refine the scenario for the interaction between the Eurasian, North America, and Pacific plates.

In discussing the existing viewpoints and in criticizing the available geological data, our purpose has been to suggest approaches that will be lucrative for future South Anyui suture zone studies. We believe that the first priorities for study are (1) a more detailed study of the ophiolites, (2) a geochemical study of the Late Jurassic to Early Cretaceous volcanic rocks, (3) study of the stratigraphy and sedimentology of the terrigenous sequences, and (4) a paleomagnetic study of the terranes that flank the South Anyui zone.

ACKNOWLEDGMENTS

We thank V.T. Burchenkov, director of the Anyui Governmental Mining and Geological Enterprise (Bilibino) for his help in our field work. A. Ishiwatari (Kanazawa University, Japan) assisted in the field study of the Aluchin and Vurguveem ophiolites. This work was supported by the Russian Foundation for Basic Research (projects 97-05-65711, 98-07-90015, 99-05-65649, and 00-07-90000), the International Association for the Promotion of Cooperation with Scientists from the New Independent States of the Former Soviet Union (INTAS) (grant 96-1880), the North Atlantic Treaty Organization (NATO) (grant 97-5739), and Young Scientists Russian Academy of Science grant 1999 to O. Morozov. We are grateful to the reviewers R.G. Coleman, M. Cecile, and B. Natal'in for well-disposed criticism. We also thank Elizabeth Miller and Arthur Grantz for their fruitful suggestions, which helped to improve this text.

REFERENCES CITED

Afizkiy, A.I., 1970, Biostratigraphy of Triassic and Jurassic sediments of Bolshoi Anyui River Basin (Western Chukotka): Moscow, Nauka, 146 p. (in Russian).

Berzin, N.A., Coleman, R.G., Dobretsov, N.L., Zonenshain L.P., Xiao Xuchan, and Chang, E.Z., 1994, Geodynamic map of the western Paleoasian Ocean: Geologia i Geofizika, v. 35, nos. 7/8, p. 8–28 (in Russian).

Bondarenko, G.Ye., 1995, The structural evolution of the Eastern part of the Kolyma Loop: 5th Zonenshain Conference on Plate Tectonics, Moscow, Abstracts, p. 142.

Bondarenko, G.Ye., 1996, Dextral strike-slip faults in the southwestern part of the Omolon Massif: Byulleten' Moskovskogo Obshchestva Ispytatelei Prirody, Seriya Geologicheskaya, v. 71, no. 5, p. 18–24 (in Russian).

Bondarenko, G.Ye., and Didenko, A.N., 1997, New geological and paleomagnetic data on the Jurassic-Cretaceous history of the Omolon Massif: Geotektonika, no. 2, p. 14–26 (in Russian).

Bychkov, Yu.M., and Solov'yov, G.I., 1992, New data on the stratigraphy and lithology of Triassic sediments in the upper reaches of the Bol'shoi Anyui River, Lower Mesozoic deposits of the right side of the Kolyma River and northwestern Kamchatka: Magadan, Russia, Severo-Vostochnyi Kompleksnyi Nauchno-Issledovatel'skyi Institut Rossiyiskoi Akademii Nauk, no. 1, p. 3–24 (in Russian).

Cecile, M.P., Lane, L.S., Kos'ko, M.K., Bychkov, Yu.M., Vinogradova, O.N., and Gorodinsky, M.E., 1991, Joint visit of Canadian and Soviet scientists to the northeastern Soviet Union: Episodes, v. 14, no. 2, p. 125–130.

Cowan, D.S., 1985, Structural style in Mesozoic and Cenozoic mélanges in the Western Cordillera of North America: Geological Society of America Bulletin, v. 96, p. 451–462.

DeBari, S.M., Coleman, R.G., 1989, Examination of the deep levels of an island arc: Evidence from the Tonsina ultramafic-mafic assemblage, Tonsina, Alaska: Journal of Geophysical Research, v., 94, no. B4, p. 4373–4391.

Dovgal', Yu.M., Gorodinsky, M.Ye., and Sterligova, M.Ye., 1975, The Aluchinskyi ultrabasite complex, *in* Magmatism of Northeast Asia: Magadan, Russia, Knizhnoe Izdatel'stvo, v. 2, p. 59–70 (in Russian).

Drachev, S.S., and Savostin, L.A., 1993, Ophiolite of Bolshoi Lyakhovsky Island (Novosibirsk Archipelago): Geotektonika, no. 3, p. 98–107 (in Russian).

Embry, A.F., and Dixon, J., 1990, The breakup unconformity of the Amerasia Basin, Arctic Ocean: Evidence from Arctic Canada: Geological Society of America Bulletin, v. 102, no. 11, p. 1526–1534.

Fujita, K., Newberry, J.T., 1983, Accretionary terranes and tectonic evolution of Northeast Siberia, *in* Hashimoto, M., and Yeda, S., eds., Accretion tectonics in the circum-Pacific regions: Tokyo, Terrapub, p. 43–58.

Gagiyev, M.Kh., 1992, Stratigraphy and conodonts of Early–Middle Devonian sediments of northeastern USSR: Novosibirsk, Russia, Nauka, 40 p. (in Russian).

Ged'ko, M.I., Postnikov, S.N., and Svirina, M.A., 1991, A geological report on cosmoaerogeological mapping Quadrangles R-58, 59, 60, Q-57, 58, between 1988–1990: Moscow, "Aerogeologia," Ministry of Geology of Russia, 310 p. (in Russian).

Glotov, S.P., 1995, A report on the 1:50 000 geological mapping and general exploration, Quadrangle Q-58, between the Orlovka and Nutesyn Rivers between 1990–1995: Bilibino, Anuyuiskoe Gosudarstvennoe Gorno-Geologicheskoe Predprivatie Press, Ministry of Geology of Russia, 250 p. (in Russian).

Golonka, J., Ross, M., and Scotese, C.R., 1994, Phanerozoic paleogeographic and paleoclimatic modelling maps, *in* Embry, A.F., Beauchamp, B., and Glass, D., eds., Pangaea: Global environments and resources: Canadian Society of Petroleum Geologists Memoir 17, p. 1–47.

Grantz, A., and May, S.B., 1983, Rifting history and structural developments of the continental margin north of Alaska, *in* Watkins, J.S., and Drake, C.L., eds., Studies in continental margin geology: American Association of Petroleum Geologists Memoir 34, p. 7–100.

Grantz, A., May, S.D., and Lawyer, L.A., 1990, Canada Basin, *in* Grantz, A., Johnson, G.L., and Sweeney, J.F., eds., The Arctic Ocean region: Boulder, Colorado, Geological Society of America, Geology of North America, v. L, p. 379–402.

Grinberg, G.A., and Rudich, K.N., editors, 1981, Tectonics, magmatic, and metamorphic complexes of the Kolyma-Omolon Massif: Moscow, Nauka, 359 p. (in Russian).

Gulevich, V.V., 1975, Late Jurassic volcanism of the upper reaches of the Bol'shoi Anyui River, *in* Magmatism of northeast Asia: Magadan, Russia, Knizhnoe Izdatel'stvo, v. 2, p. 83–89 (in Russian).

Lychagin, P.P., 1985, The Aluchinsky Massif and the problem of ophiolitic ultrabasites and gabbro in the Mesozoides of northeast USSR: Tikhookeanskaya Geologiya, p. 33–41 (in Russian).

Lychagin, P.P., 1997, The volcanic complexes of the South-Anyui fold zone, *in* Byalobzhesky, S.G., ed., Magmatism and mineral deposits of north-east Russia: Magadan, Russia, Northeast Interdisciplinary Science Research Institute, Far East Branch, Russian Academy of Science (SVKNII RAN), p. 17–33.

Lychagin, P.P., Byalobzhessky, S.G., Kolyasnikov, Yu.A., Korago, Ye.A., and Likman, V.B., 1991, The magmatic history of the South Anyui folded zone, *in* Byalobzhessky, S.G., ed., Geology of continent-ocean transition zone in northeast Asia: Magadan, Russia, Severo-Vostochnyi Kompleksnyi Nauchno-Issledovatel'skyi Institut Rossiyiskoi Akademii Nauk (SVKNII RAN), p. 140–157 (in Russian).

Merzlyakov, V.M., Terekhov, M.I., Lichagin, P.P., and Dylevsky, Ye.F., 1982, Tectonics of the Omolon Massif: Geotektonika, no.1, p. 74–81 (in Russian).

Moore T.E., Wallace, W.K., Bird, K.J., Karl, S.M., Mull, C.G., and Dillon, J.T., 1994, Geology of northern Alaska, *in* Plafker, G., and Berg, H.C., eds., The geology of Alaska: Boulder, Colorado, Geological Society of America, Geology of North America, v. G-1, p. 49–140.

Morozov, O.L., 1996, Geological composition and tectonic evolution of Central Chukotka: Moscow, GIN RAN, Abstract of Kandidatskaya Dissertaciya, 29 p. (in Russian).

Natal'in, B.A., 1984, Early Mesozoic eugeosynclinal systems in the northern part of the circum-Pacific Belt: Moscow, Nauka, 136 p. (in Russian).

Natapov, L.M., and Shul'gina, V.S., editors, 1991, Geological map of the USSR, Quadrangles Q-56, 57, Srednekolymsk, New Series: Leningrad, Nedra, scale 1:1 000 000, explanatory note, 111 p. (in Russian).

Natapov, L.M., and Surmilova, Ye.P., editors, 1986, Geological map of the USSR, Quadrangle Q-54, 55, Khonuu, New Series): Leningrad, Nedra, explanatory notes, 120 p. (in Russian).

Nokleberg, W.J., Parfenov, L.M., Monger, J.W.H., Baranov, B.V., Byalobzhesky, S.G., Bundtzen, T.K., Feeney, N.D., Fujita, K., Gordey, S.P., Grantz, A., Khanchuk, A.I., Natalin, B.A., Natapov, L.M., Norton, I.O., Patton, W.W., Plafker, G., Jr., Scholl, D.W., Sokolov S.D., Sosunov, G.M., Stone, D.B., Tabor, R.W., Tzukanov, N.V., Vallier, T.L., and Wakita, K., 1994,

Circum-north Pacific tectonostratigraphic terrane map: U.S. Geological Survey Open-File Report 94–714, scale 1:5 000 000, 1 sheet.

Nokleberg, W.J., Parfenov, L.M., Monger, J.W.H., Norton, I.O., Khanchuk, A.I., Stone, D.B., Scholl, D.W., and Fujita, K, 1998, Phanerozoic tectonic evolution of the circum-north Pacific: U.S. Geological Survey Open-File Report 98–754, 125 p.

Oxman, V.S., Parfenov, L.M., Prokopiev, A.V., Timofeev, V.F., Tretyakov, F.F., Nedosekin, Y.D., Lawyer, P.W., and Fujita, K., 1995, The Chersky Range ophiolite belt, northeast Russia: Journal of Geology, v. 103, no. 5, p. 539–556.

Palymskaya, Z.A., and Palymsky, B.F., 1975, Late Paleozoic intrusive magmatism of the eastern part of the Anyui-Oloy Block (Western Chukotka), *in* Egiazarov, B.X., ed., Magmatism of northeast Asia: Magadan, Russia, Knizhnoe Izdatel'stvo, v. 2, p. 51–58 (in Russian).

Parfenov, L.M., 1984, Continental margins and island arcs in the Mesozoides of northeastern Asia: Novosibirsk, Russia, Nauka, 192 p. (in Russian).

Parfenov, L.M., and Spector, V.B., editors, 1997, Geological monuments of the Sakha Republic (Yakutia): Novosibirsk, Nauka, 80 p

Parfenov, L.M., Natapov, L.M., Sokolov, S.D., and Tsukanov, N.V., 1993, Terrane analysis and accretion in northeast Asia: The Island Arcs, v. 2, p. 35–54.

Pinus, G.V., and Sterligova, V.Ye., 1973, A new alpine-type hyperbasite belt in the northeast USSR: Geologiya i Geofizika, v. 14, p. 109–111 (in Russian).

Pushcharovsky, Yu.M., Sokolov, S.D., Tielman, S.M., and Krylov, K.A., 1992, The northwest circum-Pacific: Tectonics and geodynamics, *in* Borukaev, Ch.B., and Vrublevskiy, A.A., eds., Problems of tectonics, mining, and energy resources of the northwestern Pacific: Khabarovsk, Russia, Insitut Tektoniki I, Geofiziki, Dal'nevostochnoe otdelenie, Akademiya Nauk, USSR, v. 1, p. 128–137 (in Russian).

Radziwill, A.Ya., 1964, New data on the geology of eastern Anyui Range, *in* Materials on geology and mining of the northeastern USSR: Magadan, Russia, Knizhnoe Izdatel'stvo, v. 17, p. 57–62 (in Russian).

Radziwill, A.Ya., and Radziwill, V.Ya., 1975, Late Jurassic magmatic associations of the south Anyui depression, *in* Magmatism of Northeastern Asia: Magadan, Russia, Knizhnoe Izdatel'stvo, v. 2, p. 71–80 (in Russian).

Rusakov, I.M., and Vinogradov, V.A., 1969, Eugeosynclinal and myogeosynclinal areas in the northeastern USSR, *in* Egizarov, B.X., Rusakov, I.M., eds., Trudy Nauchno-Issledovatel'skogo Instituta Geologii Arktiki, Regional Geology, v. 15, p. 5–27 (in Russian).

Ruzhentsev, S.V., and Aristov, V.A., 1998, New data on the Polar Ural geology, *in* Knipper, A.L., and Kurenkov, S.A., eds., The Ural: The basic problems of geodynamics and stratigraphy: Moscow, Nauka, Geologicheskiy Institut Reports, v. 500, p. 25–41 (in Russian).

Savostin, L.A., Bondarenko, G.Ye., Safonov, V.G., and Pavlov, V.E., 1994, Structural evolution of Omolon Massif's northwestern frame in the Jurassic: Geotektonika, no. 5, p. 46–62 (in Russian).

Sekretov, S.B., 1993, Geological structure of the Laptev Sea Shelf on seismic data: St. Petersburg, Shirshov Institute of Oceanology, RAS, Abstract of Kandidatskaya Dissertaciya, 24 p. (in Russian).

Seslavinskiy, K.B., 1970, The structure and development of the South Anyui sutural trough (Western Chukotka): Geotektonika, no. 5, p. 56–68 (in Russian).

Seslavinskiy, K.B., 1979, The South Anyui Suture (Western Chukotka): Akademiya Nauk SSSR, Doklady, v. 249, no. 5, p. 1181–1185 (in Russian).

Sharaskin, A.Ya., Puschin, I.K., Zlobin, S.K., and Kolesov, G.M., 1983, Two ophiolite sequences from the basement of the Northern Tonga Arc: Ofioliti, v. 8, no. 3, p. 411–430.

Shekhovtsov, V.A., 1991, A report on the geological mapping and general exploration, Scale 1:50 000, Quadrangle Q-58, Gremuchaya-Ainakhkurgen Rivers, between 1986–1991: Bilibino, Russia, Anuyuiskoe Gosudarstvennoe Gorno-Geologicheskoe Predprivatie Press, 312 p. (in Russian).

Sokolov, S.D., Didenko, A.N., Grigoriyev, V.N., Alexyutin, M.V., Bondarenko, G.Ye., and Krylov, K.A., 1997, Paleotectonic reconstructions of Northeastern Russia: Problems and uncertainties: Geotektonika, no. 6, p. 72–90 (in Russian).

Sokolov, S.D., Bondarenko, G.Ye., Morozov, O.L., Aleksyutin, M.V., Chamov, N.P., Khudoley, A.K., Layer, P., Lutchitskaja, M.V., and Silantyev, S.A., 1998, The evolution of the east Siberia Mesozoic convergent margin (the Taigonos Segment, NE Russia): 6th Zonenshain International Conference on Plate Tectonics, Moscow, Abstracts, p. 95–96.

Spektor, V.B., Andrusenko, A.M., Dudko, E.A., and Kareva, N.F., 1981, Continuation of the South-Anyui Suture in the Priimorsk Island: Akademiya Nauk SSSR, Doklady, v. 260, p. 1447–1450 (in Russian).

Surnin, A.A., and Okrugin, V.A., 1989, Basite-ultrabasite magmatism of the South Anyui Suture: Tikhookeanskaya Geologiya, no. 5, p. 10–18 (in Russian).

Sweeney, J.F., 1985, Comments on the age of the Canada Basin: Tectonophysics, v. 114, p. 1–10.

Sylvester, A.G., 1988, Strike-slip faults: Geological Society of America Bulletin, v. 100, no. 11, p. 1666–1703.

Til'man, S.M., 1973, Comparative tectonics of the North Pacific Ring Mesozoides: Novosibirsk, Nauka, 325 p. (in Russian).

Til'man, S.M., and Bogdanov, N.A., 1992, Tectonic map of northeast Asia, *in* Pushcharovsky, Yu.M., ed., Tectonic map of northeat Asia: Moscow, Institute of the Lithosphere, scale 1:500 000, 1 sheet (in Russian).

Til'man, S.M., Afizky, A.I., and Chekhov, A.D., 1977, Comparative tectonics of the Alazeya and Oloi zones (north-east of the USSR) and the Kolyma massif problem: Geotektonica, no. 4, p. 6–17.

Tynankergav, G.A., and Bychkov, J.M., 1987, Upper Triassic chert-volcanic-terrigeneous assemblages of Western Chukotka: Akademiya Nauk SSSR, Doklady, v. 296, p. 698–700 (in Russian).

Vernikovsky, V.A., 1996, Geodynamic evolution of Taimyr folded area: Novosibirsk, Nauka, 202 p. (in Russian).

Woodcock, N.H., and Fischer, M., 1986, Strike-slip duplexes: Journal of Structural Geology, v. 8, no. 7, p. 725–735.

Yegorov, V.V., 1990, A report on geological mapping and general exploration, Quadrangle Q-58, 59, Left side of the Yarakvaam River, between 1985–1990: Bilibino, Russia, Anuyuiskoe Gosudarstvennoe Gorno-Geologicheskoe Predprivatie Press, scale 1:50 000, 1 sheet, 367 p. (in Russian).

Zhulanova, I.L., 1990, The Earth crust of NE Asia in the Precambrian and Paleozoic: Moscow, Nauka, 302 p. (in Russian).

Zonenshain, L.P., and Natapov, L.M., 1988, Tectonic history of the Arctic, *in* Pushcharovsky, Yu.M., ed., Topical problems of the oceans and continents: Moscow, Nauka, p. 31–57 (in Russian).

Zonenshain, L.P., Kuzmin, M.I., and Natapov, L.M., 1990, Plate tectonics of the USSR Territory, Book 2: Moscow, Nedra, 334 p. (in Russian).

MANUSCRIPT ACCEPTED BY THE SOCIETY MAY 15, 2001.

Geological Society of America
Special Paper 360
2002

Jurassic-Cretaceous history of the Omolon massif, northeastern Russia: Geologic and paleomagnetic evidence

Alexei N. Didenko*
Institute of the Physics of the Earth, Russian Academy of Sciences, Moscow
Grigoriy Ye. Bondarenko
Sergei D. Sokolov
Igor R. Kravchenko-Berezhnoy
Geological Institute, Russian Academy of Sciences, Moscow

ABSTRACT

Geologic and paleomagnetic research in the western part of the Omolon massif and in the Sugoi folded zone, northeastern Russia, suggests significant lateral displacements of these units relative to one another and with respect to the Siberian continent in late Mesozoic time. In the northern part of the Sugoi folded zone folds and thrust faults exhibit vergence in various directions, which may be related to a history of transpression. Lower Jurassic fine-grained terrigenous and calcareous deposits were studied, yielding two ancient paleomagnetic directions: (1) a postfolding one with declination (dec) = 30°, inclination (inc) = 83°, precision parameter (K) = 117, and confidence limit (α_{95}) = 4° and (2) a probable prefolding one with dec = 229°, inc = 58°, K = 43, and α_{95} = 9.4°.

Fault offsets (to a few hundreds of kilometers) occur mainly on sinistral strike-slip faults in the Sugoi zone, and were formed as the Omolon terrane and separate tectonic blocks in the Sugoi zone rotated 30°–70° counterclockwise relative to Siberia. In Early Jurassic time, the Omolon massif was located at lat 60°N, and in Early Cretaceous time, at lat 76°N.

INTRODUCTION

Two major regional belts of deformation are recognized in northeastern Russia (Fig. 1): the Verkhoyansk-Kolyma-Chukotka Mesozoic fold and thrust belt and the Koryak-Kamchatka system. They are separated by the extensive Okhotsk-Chukotka volcanic belt. The Verkhoyansk-Kolyma-Chukotka belt is a composite of many structural units that are mostly northwest trending and have formed through several stages of deformation of the Asian continental margin, the final stage taking place in the Late Jurassic–Early Cretaceous. During this deformation, the continental margin developed and grew mainly through collisional processes. Tectonic blocks or fragments underlain by continental crust (such as the Omolon, Okhotsk, Prikolymsk, and Omuliovka terranes) (Fig. 1) played an important role in orogenesis. There are two viewpoints about the origin of these continental blocks. Some researchers view them as fragments of the Siberian continent (Merzlyakov et al., 1982; Shapiro and Ganelin, 1988), and others view them as exotic terranes that are considerably farther traveled (Ustritsky and Khramov, 1984; Zonenshain et al., 1990; Talent, 1990). The existence and size of the oceans that separated these blocks from the continental margin remain topics of discussion (Ustritski and Khramov, 1984; Shapiro and Ganelin, 1988; Zonenshain et al., 1990; Parfenov et al., 1993).

The first viewpoint, that the continental blocks represent

*E-mail: didenko@uipe-ras.scgis.ru

Didenko, A.N., Bondarenko, G.Ye., Sokolov, S.D., and Kravchenko-Berezhnoy, I.R., 2002, Jurassic-Cretaceous history of the Omolon massif, northeastern Russia: Geologic and paleomagnetic evidence, *in* Miller, E.L., Grantz, A., and Klemperer, S.L., eds., Tectonic Evolution of the Bering Shelf–Chukchi Sea–Arctic Margin and Adjacent Landmasses: Boulder, Colorado, Geological Society of America Special Paper 360, p. 225–241.

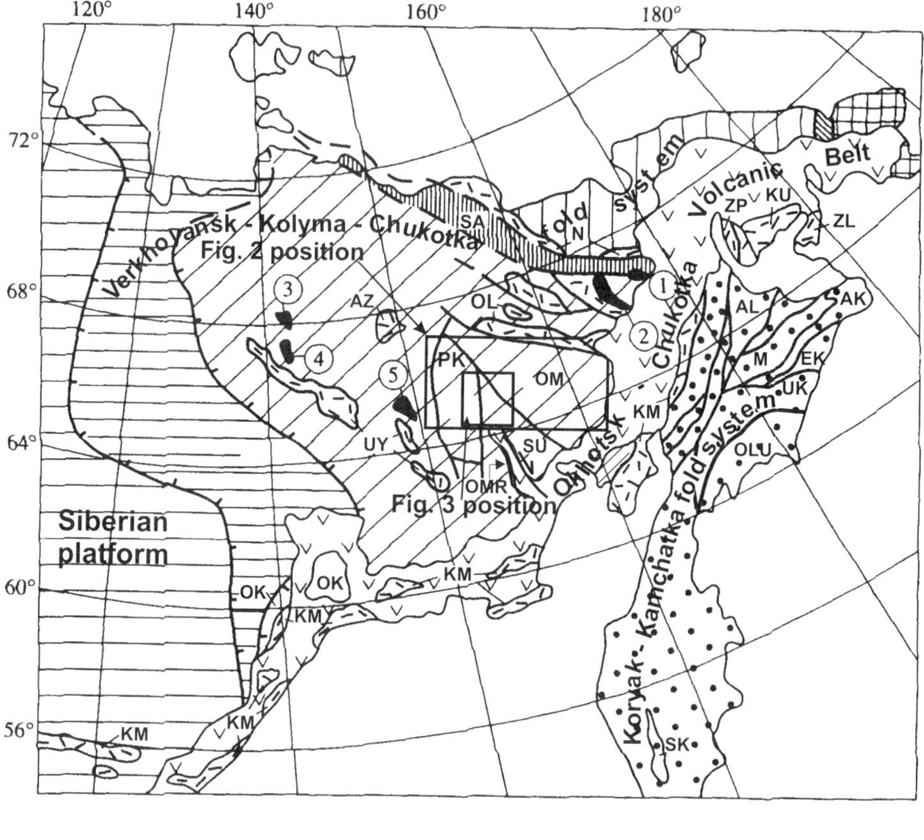

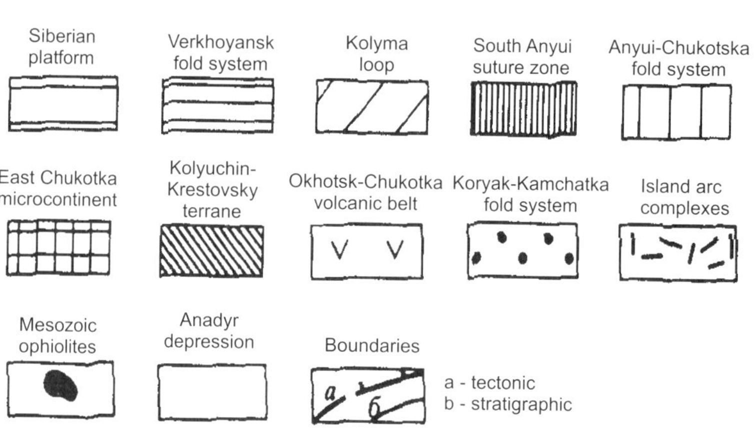

Figure 1. Regional map of northeastern Russia showing principal tectonic zones. Terranes and other tectonic elements: AZ, Alazeya; AL, Algansk; AK, Alkatvaam; ZL, Zolotogorsk; ZP, west Pekulnei; KU, Kanchalan and Ushkanegorsk; M, Mainits; N, Nutesyn; OL, Oloi; OM, Omolon; PK, Prikolymsk; OMR, Omsukchan Range; OK, Okhotsk; UK, Ukelayat; OLU, Olyutorsk; SK, Sredinno-Kamchatsky; EK, Ekonai; SU, Sugoi; UY, Uyandina-Usachnaya volcanic belt; SA, South Anyui suture zone; KM, Koni-Taigonos and Uda-Murgal island arcs. Circled numbers denote ophiolite terranes: 1, Vurguveem; 2, Aluchin; 3, Uyandina; 4, Tas-Khayakhtakh; 5, Arga-Tas. Larger boxed area is Figure 2, and smaller is Figure 3.

earlier rifted fragments of the continental margin, is based on detailed stratigraphic and faunal data (e.g., Gagiev, 1991; Zhuravlev, 1988; Terekhov, 1979; Shapiro and Ganelin, 1988). The second viewpoint, that these are far-traveled, is based primarily on paleomagnetic data (Kolesov, 1981; Lozhkina, 1982; Khramov, 1988; Khramov and Ustritsky, 1990). The two alternate hypotheses describing the Mesozoic geodynamic evolution of northeastern Russia advocated by these two groups of researchers compel one to either question the reliability of their arguments or to search for a better compromise between the two

data sets. The goal of this study is to develop a working model capable of integrating the broadest range of paleomagnetic and geologic data available.

To this end, we have carried out a geologic and structural transect across the west slope of the Prikolymsk uplift and the Sugoi folded zone along the Kolyma, Korkodon, Bulun, and Tokur-Yuriakh Rivers in the western part of the Omolon massif (Figs. 2 and 3). Structural and geometric analysis of folds and kinematic analysis of faults were carried out following the methods of Gonchar and Drachev (1993) and Kazakov (1980).

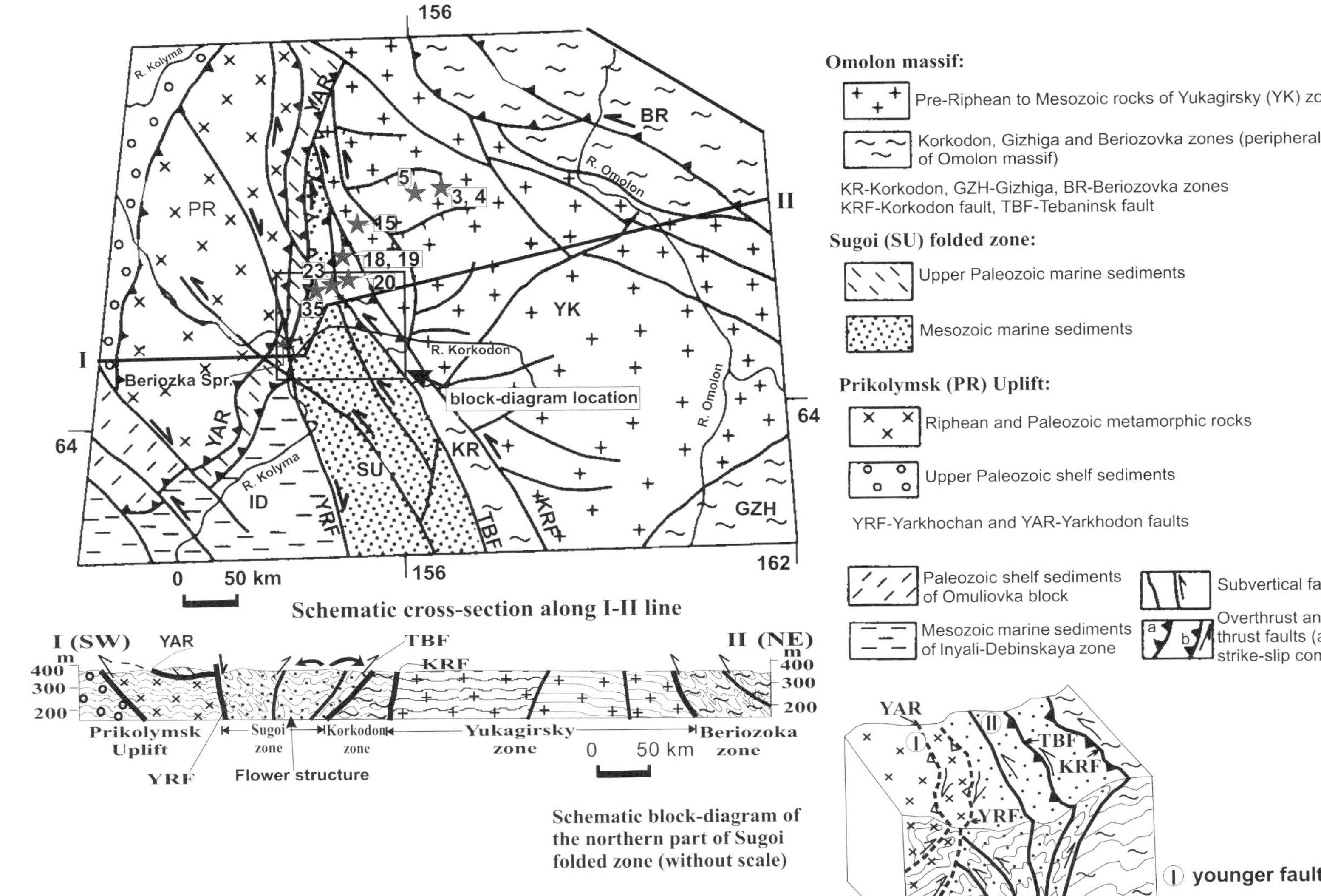

Figure 2. Geological map showing Omolon massif and surrounding areas (Bondarenko and Didenko, 1997). OL, Oloi fold zone; BR, Berezovka fold zone; GZH, Gizhiga fold zone; YK, Yukagirsky block; KR, Korkodon fold zone; SU, Sugoi fold zone; ID, Inyali-Debin fold zone. Asterisks mark our paleomagnetic sampling localities. Principal shear zones: KRF, Korkodon fault, TBF, Tebaninsky fault, YAR, Yarkhodon fault, YRF, Yarkhochan fault. Bottom left, cross section across Prikolymsk uplift, Sugoi fold zone, and Omolon massif (line I–II). Bottom right, block diagram showing flower structure in northern part of Sugoi fold zone.

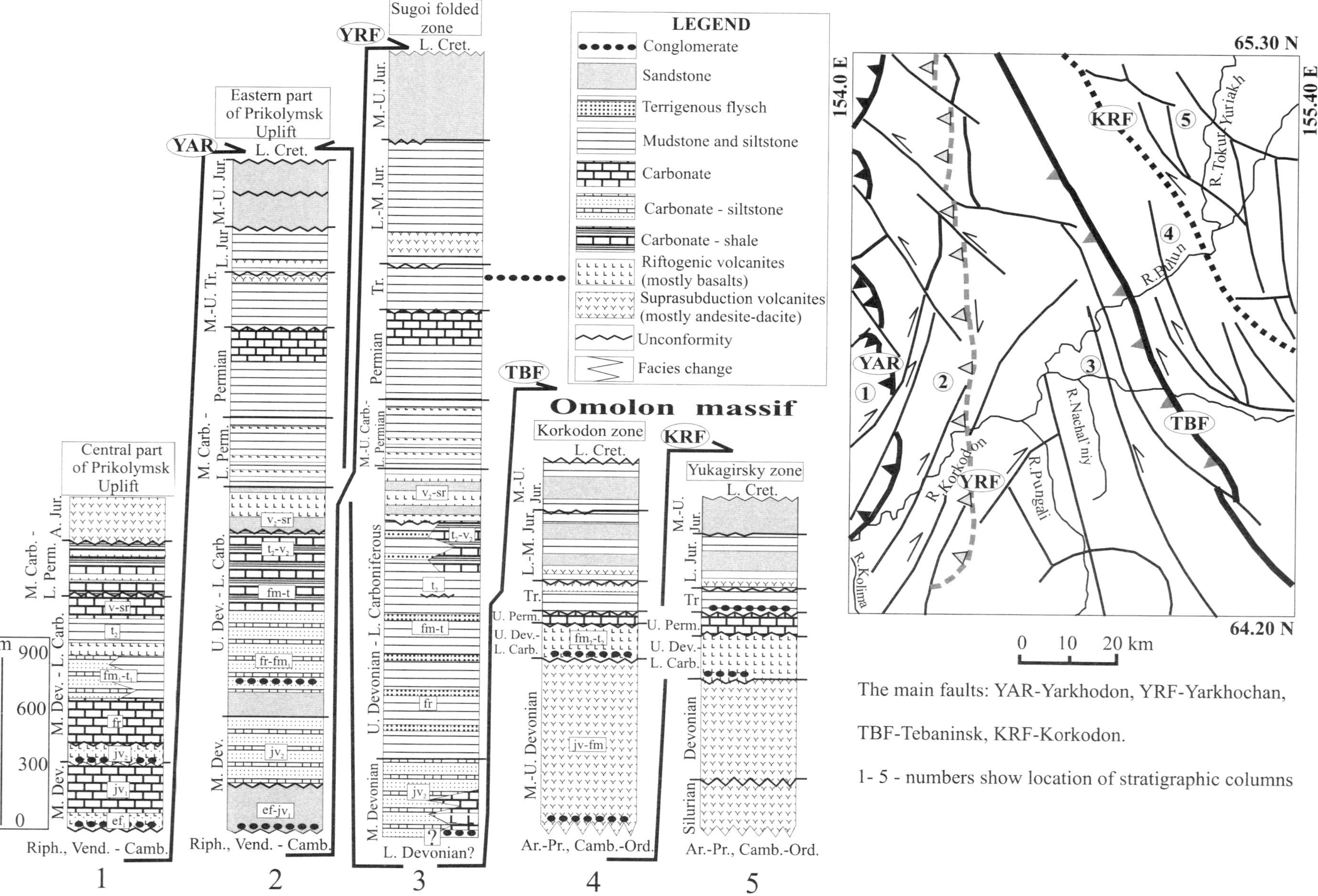

Figure 3. Map showing principal faults in northern part of Sugoi fold zone and schematic stratigraphic sections for main tectonic units of western part of Omolon massif and northern part of Sugoi fold zone. Arrows indicate relationships between stratigraphic sections and modern structure (e.g., overthrusts, strike-slips faults). Abbreviations: ef_1, lower Eifelian; jv_1, lower Givetian; jv_2, upper Givetian; fr, Frasnian; fm_1, lower Fammenian; t_1, lower Tourneisian; t_2, upper Tourneisian; v_2, upper Visean; sr, Serpukhovian.

Samples for paleomagnetic studies were collected from Lower and Middle Jurassic terrigenous strata; they are suitable for paleomagnetic study and well dated by their faunal assemblages.

GEOLOGIC FRAMEWORK

The study area includes the western part of the Omolon massif, the northern part of the Sugoi folded zone, and the eastern part of the Prikolymsk uplift (Figs. 2 and 3).

The Omolon massif consists of Archean-Proterozoic crystalline basement overlain by a Riphean-Phanerozoic sedimentary and volcanic cover sequence. The Lower Jurassic bathyal to shallow-marine terrigenous and calcareous deposits (Terekhov, 1979; Savostin et al., 1993) sampled for this study conformably overlie Upper Triassic strata on the Omolon massif, and were deposited in a shelf to continental slope setting. Beginning in the Middle Jurassic (Bajocian), clastic material derived from the east, and possibly from the west, was deposited in the basin and increased markedly in grain size with time, supposedly due to the first orogenic movements in this region (Savostin et al., 1993). Toward the western limit of the Omolon, Jurassic strata become progressively thicker, more continuous, and coarser grained. Following Shulgina and Natapov (1991), we define this marginal, relatively depressed part of the Omolon as the Korkodon zone (KR, Fig. 2). In the Korkodon zone the Mesozoic cover of the Omolon is notably thicker (300–700 m) and more highly deformed than to the east (Savostin et al., 1993, 1994). The rest of the Omolon (Yukagirsky zone 2) is separated from the Korkodon zone by a system of subvertical sinistral northwest-trending strike-slip faults, recognized as the Korkodon fault zone (KRF, Fig. 2). The Sugoi zone is separated from the Korkodon zone by the Tebaninsky fault (TBF, Fig. 2) (Shulgina and Natapov, 1991; Kuznetsov, 1975; Savostin et al., 1994). West of the Tebaninsky fault in the Sugoi zone, Lower Jurassic and Mesozoic strata as a whole increase abruptly in thickness to more than 1000 m. In the Sugoi zone the grain size of these sediments decreases, and turbidites appear in the section (Savostin et al., 1993).

The western part of the Prikolymsk uplift (Fig. 2) exposes widespread Mississipian to Permian volcanic-carbonate-terrigenous deposits that are lithologically similar to coeval sequences in the Sugoi zone, and laterally transitional to Paleozoic strata in the central part of the Prikolymsk uplift (Gagiev, 1991; Bondarenko, 1995). This transition is evidenced by lensoidal debris-flow bodies, interpreted as submarine canyon facies within Mississippian fine-grained, rhythmic terrigenous turbidites found on the middle reaches of the Kolyma River on its north side at the mouth of Berezka Creek. The clasts included in these deposits (dolomite, marble, conglomerate, quartzite) are surrounded by a fine-grained terrigenous matrix. The clasts are typical of the rocks of the central part of the Prikolymsk uplift (Gagiev, 1991), which lacks Mesozoic deposits. In the study area, Mesozoic deposits of the western part of the Sugoi zone are separated from Paleozoic strata by the Yarkhochan fault (YRF, Fig. 2) (Shulgina and Natapov, 1991; Savostin et al., 1994).

The Lower Jurassic terrigenous deposits that are the focus of our research are represented by lithologic associations of three types.

Type 1 deposits are widespread in the inner parts of the Omolon massif (YK, Fig. 2) and consist of terrigenous rocks (sandstone, siltstone, mudstone) deposited in shallow-marine shelf environments.

Type 2 deposits are widespread in the Korkodon zone of the Omolon massif (KR, Fig. 2). These terrigenous deposits differ from type 1 deposits in that they are thicker and have a greater proportion of mud, and they are interpreted to have been deposited in a deeper-shelf and continental slope setting (proximal turbidites).

Type 3 deposits are widespread in the Sugoi zone (between the Tebaninsky and Yarkhochan faults) and consist mostly of silt and mud with coarse-grained beds and lenses of sand deposited in even greater water depths (distal turbidites) (Savostin et al., 1994).

Here we present new data on the Mesozoic paleogeodynamic history of this region, which build on detailed structural and paleomagnetic studies (Savostin et al., 1994; Bondarenko, 1995; Bondarenko and Didenko, 1997). We briefly summarize the structural setting and history of the study region in the following.

STRUCTURAL STUDIES

In the central parts of the Omolon massif (the Yukagirsky zone; YK, Fig. 1) the Jurassic sequences are mostly flat lying or gently dipping and involved in minor flexure-like monoclines. These monoclines are mostly restricted to the location of mapped subvertical northwest-trending faults that often show a sinistral strike-slip component of motion.

The Yukagirsky zone (Fig. 3) is bounded on the southwest by a series of en echelon subvertical sinistral transpressive strike-slip faults that trend northwest and are associated with conjugate strike-slip faults of lesser displacement. We define this assemblage of faults as the Korkodon fault zone (KRF, Fig. 2). Southwest of this zone (in the Korkodon folded zone), the Jurassic deposits become more highly folded and faulted (Savostin et al., 1993, 1994). Bedding is folded into map-scale, mainly northeast-vergent folds that become increasingly more compressed to the southwest. Faults dip mostly southwest at high angles and are dominated by sinistral strike-slip and transpressional senses of motion. Folding is strongest near the Tebaninsky fault (TBF, Fig. 2), where folds are tight, occasionally isoclinal, and where observed, verge to the northeast.

The Tebaninsky fault zone has a long and complicated kinematic history (Savostin et al., 1994) and is characterized by intense northeast-vergent folds modified by southwest-dipping overthrust and transpressional strike-slip faults as well as more highly disrupted zones of tectonic mélange.

West of the Tebaninsky fault in the Sugoi folded zone, the degree of deformation in the Jurassic deposits varies significantly from place to place. The most intense deformation (isoclinal folding, ductile flow, boudinage, and tectonic mélange) is found on the western and eastern flanks of the Sugoi zone, near the Yarkhochan and Tebaninsky faults, respectively (Fig. 2). Within the Sugoi zone strongly deformed areas are confined to fault zones and alternate with less deformed domains. The structure of the Sugoi zone was detailed by Bondarenko (1995) and Savostin et al. (1994). The west part of the Sugoi zone exhibits west-vergent compressed to isoclinal folds in Jurassic deposits.

Structural studies in the main tectonic units of the study region and in the intervening fault zones suggest four stages of compressional deformation (Savostin et al., 1994). Two of these stages played principal roles in the evolution of the region.

The first main stage of compressional deformation gave rise first to overthrusts that are usually concordant with bedding or bedding parallel, and then to folding accompanied by the development of axial-plane cleavage. Geologic relationships (Shulgina and Natapov, 1991) tentatively date this first stage of deformation as Middle Jurassic (Savostin et al., 1994). Structures likely associated with this stage of deformation are widespread and involve upper Paleozoic strata between the Yarkhodon and Yarkhochan faults in the eastern part of the Prikolymsk uplift, and the western and central parts of the Sugoi folded zone (Savostin et al., 1994).

The second main stage of compressional deformation formed a system of sinistral strike-slip faults that trend northwest-southeast and roughly east-west (the Korkodon and Tebaninsky faults). Displacements of this stage are also documented along the Yarkhochan fault, which separates the Sugoi zone from the Prikolymsk uplift. The Yarkhochan fault was later reactivated as a dextral strike-slip fault, and the preexisting sinistral strike-slip–related deformation was mostly obscured by the superimposed deformation. Sinistral strike-slip motions of the second main stage are also recorded in the Sugoi and Korkodon zones, when subvertical sinistral strike-slip and transpressional faults formed and moved. These motions were west vergent in the west of the Sugoi zone and east vergent in the eastern part adjacent to the Korkodon zone. Associated, chiefly sinistral, strike-slip faults are also documented, as reported by Kuznetsov (1991).

The compressional sinistral strike-slip faults near the western and eastern boundaries of the Sugoi zone are associated with tectonic mélange. The boudins in the mélange have undergone counterclockwise rotation around the subvertical axes. The kinematic history of this type of fault has been well determined based on measurements of grooves, striations, and slickensides; fractures with calcite and quartz fillings; and apparent displacements of various preexisting structural features (Savostin et al., 1994).

Sinistral strike-slip faults are associated with folds of various scale (centimeters to many tens of meters), the hinges of which generally plunge more steeply than 35°–40°. Near the western limit of the Sugoi zone these folds are west vergent, and near the eastern limit they are east vergent. This doubly vergent structure resembles flower structures, in which vergence is outward, toward both of the adjacent centers of the more rigid basement blocks—the Prikolymsk uplift and the Omolon massif (Fig. 2, block diagram). Folded structures of this type, with doubly vergent limiting faults, are typical of zones of transpressional deformation (Sylvester, 1988; Woodcock and Fischer, 1986).

The axes of the folds that are associated with the sinistral strike-slip faults trend north-northeast, at an acute angle to the boundaries of the Sugoi zone. The axes of preexisting folds have been rotated by strain in the same manner so they are now approximately parallel to the young folds axes (Bondarenko, 1995); this further supports the sinistral kinematic interpretation of the faults to which they are related (Sylvester, 1988; Woodcock and Fischer, 1986).

There is no direct evidence for the age of deformation of the sinistral strike-slip faults in the northern part of the Sugoi zone or along its margins. These faults, however, are overprinted by younger dextral strike-slip faults, which are similar to those associated with Aptian-Albian magmatism of the Omsukchan Range (Fig. 1). Therefore, the older sinistral strike-slip faults described here are post–Middle Jurassic to pre-Aptian in age. We believe that they most likely formed in the first half of the Late Jurassic.

The sinistral strike-slip faults are mapped southwest of the Yukagirsky zone, in the Korkodon folded zone, and at the boundary between the Korkodon and Sugoi zones. Their origin can be attributed to the rotation of tectonic blocks bounded by sinistral strike-slip faults. We believe that the distribution and sense of offset of these faults require that the Omolon massif rotated counterclockwise.

Therefore, in the Late Jurassic, the area between the Omolon massif and the Siberian craton (the Sugoi zone) became the site of a system of sinistral strike-slip faults, which evolved until the end of the Neocomian. Strike-slip motion was accompanied by overall shortening, which resulted in the observed doubly vergent structure of the Sugoi zone.

PALEOMAGNETISM OF THE LOWER JURASSIC DEPOSITS

Paleomagnetic samples were collected from the Lower Jurassic fine-grained terrigenous and calcareous deposits of the Omolon and Sugoi structural zones. In the western part of the Omolon massif, we sampled the rocks of the Yukagirsky zone at localities 3–5; those of the Korkodon zone at localities 15, 18–20; and rocks of the Sugoi zone at localities 23 and 35 (Fig. 2).

At each locality, 15–30 hand samples were collected. The sampled sections are at least 50–100 m thick; hence the secular variation in geomagnetic field was averaged in the overall mean results for each section. The samples are tied to particular stratigraphic levels to within 1–2 stages by paleontological studies (Karago and Shpetnyi, 1982; Kudley, 1978). The orientations of samples and bedding attitudes were measured by magnetic compass and are precise within 5°. Magnetic declination in the study area is ~11°W, and was accounted for during measurement of the mean directions of components.

Bedding attitudes (overturned versus normal) were determined using sedimentological criteria and structural methods. As a rule, the strike and dip of the rocks varied negligibly at localities with nearly monoclinal bedding, but across the study area strike and dip varied significantly from locality to locality. This enabled us to analyze the magnetic components we isolated by a modified fold test (Bazhenov and Shipunov, 1988).

We collected 217 hand samples from nine localities (Fig. 2); 2–3 cubes, 2 cm on a side, were cut with a diamond saw from each hand sample. The cubes were subjected to thermal cleaning to 560–400 °C, as a rule, at 8–12 thermal steps. The thermal demagnetizer was shielded from the external magnetic field, and was <20 nT in the cooling chamber. After each stage of heating, sample remanence was measured on a JR-4 two-component spin magnetometer (sensitivity ~1 × 10^{-4} A/m). The measuring region was placed in shielding Helmholtz coils ensuring a 300–400 factor reduction of the laboratory magnetic field.

After each heating step, the magnetic susceptibility of the samples was measured on a KLY-2 kappa-meter in a constant magnetic field of 3 nT. The uniformity of the laboratory magnetic field permitted measurements to within 1 × 10^{-7} SI units, which enabled us to determine any changes in magnetic mineralogy in the samples during heating.

For nearly all the samples original natural remanent magnetization (NRM) intensities ranged between 0.2 and 10 mA/m, and the range in magnetic susceptibilities was 70–700 × 10^{-7} SI units, resulting in low Koenigsberger ratios (<1) for the entire collection. After thermal cleaning to 400–450 °C, the NRM values of a significant number of samples increased by a factor of 1.5–2 relative to the original values. This implies that the heating of these samples gave rise to a new magnetic carrier. The newly formed phase might have been magnetite formed through sulfide decomposition on heating. The presence of sulfides in the rocks under study was noted in hand samples and thin sections. Samples in which the NRM increased abruptly during thermal cleaning and those in which the NRM intensity was near the sensitivity limit of the magnetometer (usually NRM dropped 50%–70% through thermal cleaning to 450 °C) were rejected (127 of 217). These 90 samples were used in further thermal cleanings and for complete principal component analysis following Kirshvink's (1980) method.

Analyzed samples with recognizable NRM during heating suggest the existence of at least three magnetic components in the samples (Fig. 4). These components are (1) a low-temperature (*T*) component of NRM removed by heating to 200–300 °C and virtually coinciding with the present local geomagnetic field; (2) a medium-*T* component of NRM, isolated between 300 and 500 °C and showing a very steep inclination in both geographic and stratigraphic reference frames; and (3) a high-*T* component of NRM, having both polarities and isolated by heating above 500 °C. Typical demagnetization behavior is illustrated by six selected samples.

Sample 3–7 (Fig. 4A) from the Yukagirsky zone lost the modern field overprint vector (declination [dec] = 3°, inclination [inc] = 74°) on heating to 300 °C. All Zijderveld plots (Zijderveld, 1967) in Figure 4 are given in a geographic reference frame. This vector accounts for 50% of the NRM intensity. Further heating to 450 °C isolates a vector with a roughly similar dec of 345° and a steeper inc of 86°, accounting for ~40% of the overall intensity. Above 475 °C, a vector with dec = 43° and inc = –56° is isolated, accounting for <10% of the overall NRM intensity. The stereogram in Figure 4B shows clearly how the NRM vector migrates from the lower to the upper hemisphere. Unfortunately only 4 of 90 samples from our collection from the Yukagirsky and Korkodon zones behaved so ideally through the thermal cleaning.

Sample 19–12 (Fig. 4D) from the Korkodon zone yields no low-*T* component of NRM close to the modern magnetic field, but the medium-*T* and high-*T* components of NRM were observed. The medium-*T* component, with blocking temperatures in the range 200–470 °C, has a dec of 205° and an inc of 75°. The high-*T,* shallower component is isolated in the range 500–560 °C and has a vector direction of dec = 236°, inc = 62°. Sample 19–12 (Fig. 4D) differs from 3–7 (Fig. 4A) mainly in that the high-*T* polarity is normal (Fig. 4, B and E). A positive inclination for the high-*T* component is recorded for 36 of the 90 samples. Such samples were found in all three tectonic zones, i.e., Yukagirsky, Korkodon, and Sugoi, the latter showing samples of this type only (Fig. 4F, sample 23–11). For sample 20–8 (Fig. 4G), also from the Korkodon zone, the sample magnetization behaved very much like sample 3–7 during thermal cleaning. The same three magnetic components of NRM were isolated: (1) low-*T* components with blocking temperatures to 300 °C directed at dec = 24° and inc = 63°; (2) a medium-*T* component in the range 350–500 °C directed at dec = 334° and inc = 70°; and (3) a high-*T* component observed above 500 °C, with dec = 34° and inc = –49°.

The difference lies in the fact that the high-*T* component for samples 3–7 and 15–3 includes the origin (Fig. 4, A and C), whereas for samples 20–8 and 18–1 this component is incomplete, and does not include the origin of coordinates (Fig. 4, G and I). Note also the significantly lower inclination of NRM for these samples: dec = 28°, inc = –47° (Fig. 4H).

There are more than 20 samples having NRMs that behave in this manner in the high-*T* range, but they are found only in the Yukagirsky and Korkodon zones. The latter zone shows a significant proportion of such samples (Table 1).

The rocks dip nearly monoclinally at each locality, hence a fold test can only be valid for the entire data set. Table 1 shows directions for the medium-*T* and high-*T* components for both individual exposures and for the entire data set. The change in the Fisher estimate of kappa (K$_g$/K$_s$ = 2.65; Table 1) clearly shows the postfolding nature of the medium-*T* component of NRM.

A stepwise unfolding test (Shipunov, 1993) was applied for medium-*T* component resulting in a maximum K of 30%, dec = 14°, and inc = 82° (Table 1). These data suggest a postfolding

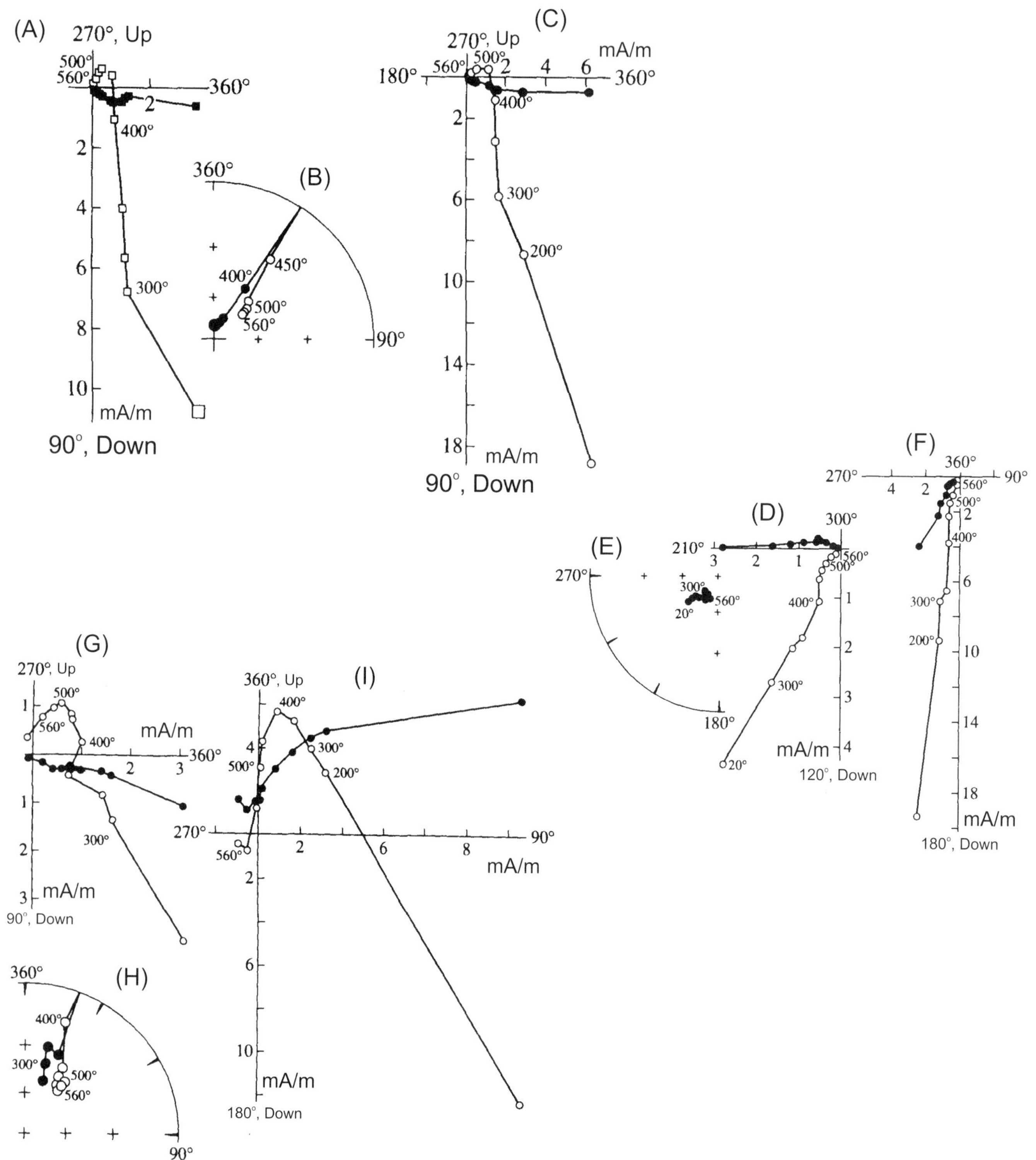

Figure 4. Orthogonal vector diagrams (Zijderveld, 1967) of representative samples that show three natural remanent magnetization (NRM) components (A, C), and distribution on equal-area projection (B) of those components that show NRM projections migrating from lower to upper hemisphere during progressive thermal demagnetization; two NRM components (D, F) and distribution of these components on equal-area projection (E); and selected samples showing that characteristic directions do not pass through zero point (G, I) and distribution of these components on equal-area projection (H). A, B: Siltstone sample (3–7) from Yukagirsky zone. C: Siltstone sample (15–3) from Korkodon zone. D, E: Fine-grained sandstone sample (19–12) from Korkodon zone. F: Calcareous sandstone sample (15–3) from Sugoi zone. G, H: Sandstone sample (20–8) from Korkodon zone. I: Fine-grained sandstone sample (18–1) from Korkodon zone. Symbols on Zijderveld plots show end points of magnetization vector during progressive thermal demagnetization. Magnetization intensities are indicated along axes (mA/m). Empty symbols represent projection onto vertical plane on Zijderveld plots and upper hemisphere on equal-area projections; filled symbols represent projection onto horizontal plane on Zijderveld plots and lower hemisphere on equal-area projections.

TABLE 1. MEAN PALEOMAGNETIC DIRECTIONS FOR LOWER JURASSIC ROCKS FROM THE OMOLON MASSIF AND SUGOI ZONE

No.		n	Medium-T component				n	High-T component				n	Great circles	
			dec_g	inc_g	dec_s	inc_s		dec_g	inc_g	dec_s	inc_s		dec_s	inc_s
Local.	Lithology and attitude		K/α_{95}		K/α_{95}			K/α_{95}		K/α_{95}			K/α_{95}	
	Yukagirsky zone (65°31′N, 155°38′E)													
3	Siltstones	8	346	85	35	82	8	222	60	218	65	8	200	75
	Az. 30-90 ∠ 3-13		74/6		78/6			60/7		82/5			15/25	
4	Silty sandstones	16	160	85	25	81	11	30	−27	36	−37	13	21	−69
	Az. 340-30 ∠ 3-30		46/5		47/5			10/13		11/13			6/12	
5	Siltstones	8	22	82	8	76	8	259	61	269	59	7	272	72
	Az. 320-350 ∠ 5-10		62/6		73/6			17/13		16/13			15/25	
	Mean 3, 4, 5	3	26	87	20	80	3	226	51	231	56	3	223	75
			136/7		405/4			11/24		13/22			43/12	
	Korkodon zone (65°15′N, 155°26′E)													
15	Siltstones	15	18	81	319	80	5	30	−52	38	−45	13	91	−78
	Az. 180-270 ∠ 4-15		123/3		157/3			13/18		13/17			5/10	
18	Silty sandstones	17	98	80	290	82	17	9	−63	44	−49	17	91	−71
	Az. 240-320 ∠ 7-50		36/5		68/4			5/15		8/12			9/13	
19	Silty sandstones	8	347	86	344	82	3	209	63	216	64	7	253	71
	Az. 240-360 ∠ 4-10		59/7		61/6			17/2		25/16			9/15	
20	Sandstones	5	33	78	14	74	4	42	−51	50	−54	6	87	−73
	Az. 320-10 ∠ 4-9		58/8		73/7			18/17		17/17			15/31	
	Mean 15, 18, 19, 20	4	42	83	340	81	4	209	58	222	53	4	265	73
			134/6		116/7			65/9		84/7			332/4	
	Sugoi zone (64°59′N, 155°4′E)													
23	Silty sandstones	6	14	72	294	60	6	127	84	230	60	6	248	62
	Az. 220-270 ∠ 20-50		21/12		48/8			42/9		43/9			7/8	
35	Silty sandstones	7	30	80	345	62	—	—	—	—	—	—	—	—
	Az. 290-320 ∠ 15–35		6/22		5/24					—			—	
	Mean 23, 35	2	20	76	319	63	—	—	—	—	—	—	—	—
			169/19		22/55					—			—	
General mean 3, 4, 5, 15, 18,		9	30	83	340	78	8	217	64	229	58	8	248	73
19, 20, 23, 35			117/4		44/7			14/14		39/8			64/6	
	Kmax-30%				14	82				Kmax-80%			244	75
					121/4								74/6	
Mean direction for samples with positive inclination:							36	224	69	234	73			
								10/8		17/6				
Mean direction for samples with negative inclination:							26	27	−49	41	−46			
								6/10		10/8				

Note: Local. is locality; Az and ∠ are azimuth and dip of bedding; T = temperature; n is number of sites or localities; dec_g, dec_s are declination in geographic and stratigraphic coordinates, respectively; inc_g, inc_s are inclination in geographic and stratigraphic coordinates, respectively; K, α_{95} are Fisher (1953) statistical parameters (dec, inc, and α_{95} are given in grades).

(or very late synfolding) origin for this component. The secondary nature of this magnetic component is shown by the high concentration of medium-T directions both within individual localities and for all exposures (Table 1). The main orogeny in the region is dated as Late Jurassic–Neocomian (Bondarenko, 1995; Shulgina and Natapov, 1991; Parfenov and Natalin, 1977). We suggest a similar age for the acquisition of the medium-T component, tentatively estimating its age as mid-Neocomian (140 ± 10 Ma). Because the maximum precision for this component was obtained at a 30% tilt correction (Table 1), we consider this magnetization to be synfolding.

Similar magnetization directions are recorded for Upper Jurassic–Lower Cretaceous rocks (Savostin et al., 1993). Note that the medium-T paleomagnetic pole ($\Phi = 75°$, $\Lambda = 183°$, $A_{95} = 8°$, N = 9; Table 2) is a very close fit to the 134 Ma paleomagnetic pole for Eurasia (Khramov, 1991), suggesting that the Omolon massif has been an integral part of the Eurasian plate since at least the Late Jurassic–Early Cretaceous.

Zijderveld plots (Fig. 4) imply clearly that the studied NRMs include high-T components, which we successfully isolated in 62 of 217 samples (Table 2). Analyzing mean directions by locality, the overall mean for this component (Fig. 5, A and

TABLE 2. COMPARISON OF THE OBSERVED PALEOMAGNETIC DIRECTIONS AND THE EXPECTED ONES FROM NORTH EURASIA

Component	Age	N (n)	D (°)	I (°)	K	α_{95} (°)	φ_a (°)	$\Delta\varphi_a$ (°)	R (°)	Φ (°)	Λ (°)	A_{95} (°)
1. Postfolding	J_3-K_1	9 (90)	30	83	117	4	76 ± 8	2 ± 12	8 ± 60	75	183	8
			38	84			78 ± 9			73	182	9
2a. Prefolding	J_1	8 (62)	229	58	43	9	39 ± 12	37 ± 13	43 ± 30	21	116	11
			272	83			76 ± 6			62	124	6
2b. Prefolding, with N-polarity samples only	J_1	(36)	234	73	17	6	59 ± 10	17 ± 12	38 ± 33	40	121	11
			272	83			76 ± 6			62	124	6
2c. Synfolding—80%	J_1 (?)	8 (80)	244	75	74	6	62 ± 11	14 ± 12	28 ± 35	45	118	11
			272	83			76 ± 6			62	124	6

Note: N is number of sites; n is number of samples; D is declination; I is inclination; K is precision parameter; α_{95} is confidence limit around paleomagnetic direction; φ_a is paleolatitude; $\Delta\varphi_a$ is difference between observed and expected paleolatitudes; R is rotation of the Omolon terrane with respect to Siberia; Φ, Λ is latitude and longitude of paleomagnetic pole; α_{95} is confidence limit around paleomagnetic pole. First row is observed data and the second row, expected from North Eurasia (Khramov, 1991) on point 65°N and 155°E. Confidence limits for flattening and rotation calculated according to Demarest (1983) and Butler (1992).

B; Table 1) shows internal consistency values of as much as 10–20, typical of detrital and postdetrital magnetization, and a significant increase in precision for the overall mean in the stratigraphic reference frame ($K_s/K_g = 2.78$). In addition, this magnetic component records both normal and reverse polarities, implying that it represents prefolding magnetization.

The Zijderveld plot for sample 20–8 (Fig. 4G) shows an incomplete demagnetization during thermal cleaning. The high-T component behaves in this manner in many cases, and it is thus doubtful that direct calculations of reverse high-T directions (with negative inclinations) will yield true values. This behavior during demagnetization may be caused by overlapping ranges of blocking temperatures for the high-T component and/or an overprinted medium-T component with a positive inclination. The same is suggested by the systematic understatement for the localities dominated by negative inclinations (Table 1). The mean poles for high-T components (Fig. 5B) follow a great-circle distribution.

In this way we have estimated the mean paleomagnetic directions for each locality using the method of converging remagnetization circles (Halls, 1976), which calculates great circles for each sample using three to four values obtained in thermal cleaning.

For locality 4 in Figure 2, the mean obtained from remagnetization circles is much steeper (inc = −69°) than for the mean inclinations ($inc_s = -37°$; Table 1). Note the shape of distribution of the means from remagnetization circles in Figure 5C: it is rather similar to the Fisher distribution (Fig. 5D), unlike the one in Figure 5B, which shows the locality means obtained by a direct calculation of isolated components on Zijderveld plots. The reversal test is negative because the observed angular difference between the positive and negative directions is 15.4°, and the critical angular difference for 95% probability is only 12.2° (McFadden and McElhinny, 1990).

We have also assessed the true mean by separate calculations for samples with normal and reverse polarities (Table 1). The difference in mean inclinations is 27°. Taken together, these data show that the true direction is best approximated by the means from remagnetization circles and from samples with normal polarity only. Nevertheless the values summarized in Table 2 and Figure 5D include both normal and reverse polarities.

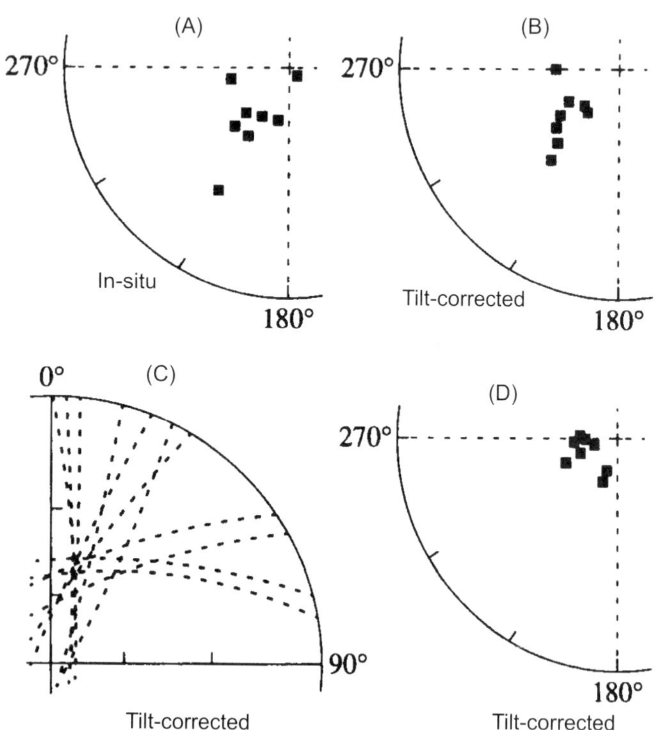

Figure 5. Distribution of mean directions by locality from three zones before (A) and after (B and D) tilt correction, and converging remagnetization circles (C). D: Mean directions for localities calculated from converging remagnetization circles. A, B, and D: Lower-hemisphere projections. C: Upper-hemisphere projections.

All the rocks studied are Jurassic in age; i.e., their deposition spanned hardly more than 60 m.y. However, because they are reliably dated as Hettangian and Sinemurian (Karago and Shpetnyi, 1982; Kudley, 1978), the interval of deposition was only about 10–15 m.y.

DISCUSSION

Thus, paleomagnetic study of the Lower Jurassic sedimentary rocks in the west of the Omolon massif and in the north of the Sugoi folded zone yielded two ancient NRM components. One of them is prefolding, with dec = 234°, inc = 73°, K = 17, and α_{95} = 6°, and the other is overprint, with dec = 30°, inc = 83°, K = 117, and α_{95} = 4°, in the Early Cretaceous. According to the stratigraphic and structural geologic data, the main tectonic deformation took place from Upper Jurassic to Early Cretaceous. We suggest that the age of the prefolding component is not younger than upper part of Lias, but probably Hettangian to Pliensbachian, because we were studying the lower part of the Lias sedimentary sequence. According to our interpretation the Omolon massif underwent translation from 60° ± 10°N in the Western Hemisphere via the polar region to 76° ± 8°N in the Eastern Hemisphere between the Middle Jurassic and Early Cretaceous, while rotating 30°–40° counterclockwise relative to Siberia. The data we obtained in this study confirm that the Omolon massif and the Sugoi folded zone were part of the Eurasian plate from the Early Cretaceous, as suggested by Bogdanov and Tielman (1992), Parfenov et al. (1993), and Sokolov et al. (1997).

In early–Middle Jurassic time the Omolon massif was in the Northern Hemisphere, because our entire set of geological, paleontological, and paleomagnetic data suggests that it was then in northeast Asia. Hence, the pole with a northeasterly direction and negative inclination can be treated as a reverse polarity, while the pole with a southwesterly direction and positive inclination can be treated as a normal polarity. Based on the main paleomagnetic tenet, that the magnetic paleomeridian coincides with the geographic paleomeridian while magnetization is being acquired, we can analyze the absolute motion of the Omolon massif in two ways.

1. After the acquisition of the Early Jurassic NRM component the Omolon terrane moved northward from latitudes of 55°–65° to ~80°N (Fig. 6B); it rotated counterclockwise through an angle of 150° ± 70° in the mid-Neocomian (Fig. 6A), and descended to 60°–70°N in post-Neocomian time. Note that for the steep inclinations we are dealing with, the errors in paleodeclinations and, hence, in determining the rotation angle, increase greatly.

2. From the Early Jurassic to the Neocomian, the Omolon terrane passed through the polar region; i.e., it must have migrated from high latitudes in the Western Hemisphere to high latitudes in the Eastern Hemisphere, rather than traveling first from low latitudes in the Western Hemisphere into the polar region and then back again. This motion would have resulted in a much smaller (to 30°–40°) clockwise rotation than would be required by alternative 1.

We prefer the second model, especially because the northeast margin (in the geographic reference frame) of the Eurasian plate was proposed to have migrated in a similar manner (Zonenshain et al., 1990; Khramov, 1991).

Analysis of the paleomagnetic data obtained 10–20 yr ago for middle Paleozoic and Mesozoic rocks defining the large crustal blocks of northeastern Russia (Gondwana Consultants, 1995) reveals a wide scatter of their paleomagnetic poles and, accordingly, of directions of paleomeridians relative to the Siberian poles for the Paleozoic, and Eurasian poles for the Mesozoic. The best evidence for the exotic nature of these blocks with respect to Siberia has been provided by the directions of their paleomeridians. In the middle to late Paleozoic the Siberian paleomeridians were oriented ~180° (normal polarity), changing orientation through 270° to 360° in the first half of the Mesozoic, before the coeval Omolon and Aluchinsk paleomeridians acquired an azimuth of 75°–345° (Fig. 6A).

The latitude gap between the Omolon block and Siberia is as much as 12° in Late Devonian and Carboniferous time; to 38° in Late Permian time; and to 36° in Triassic–Early Jurassic time (Gondwana Consultants, 1995; measurements 5049, 4733, 4434, respectively). A similar gap of 40° in the latitudes between the Aluchinsk block and Siberia in the late Paleozoic to Triassic was also reported (Gondwana Consultants, 1995; measurements 6073, 6065). The gap between Siberia and the basement of the Okhotsk-Chukotka volcanic belt in Late Devonian–Early Permian time may have been ~16° (Gondwana Consultants, 1995; measurement 4734). Note that in this case the Siberian plate must have been south of the Okhotsk-Chukotka basement, whereas in the preceding instances Siberia was north of the Omolon block. In Middle to Late Triassic time, these blocks might already have been 50° south of Siberia (Gondwana Consultants, 1995; measurement 4642). The averaged paleolatitudes for individual directions from Omolon rocks are not widely separated from the Siberian paleolatitudes in the middle to late Paleozoic and Mesozoic (Fig. 6B). In the Early Permian the gap was ~30°, and was insignificant in the early Mesozoic.

These data provide convincing evidence that the blocks of the Kolyma loop are exotic with respect to Siberia, yet Bondarenko and Didenko (1997) showed that virtually all of the magnetic components were isolated through low-T and medium-T magnetic cleaning, and some were obtained only through alternating field or single-axis field magnetic cleaning.

New paleomagnetic data we analyze in this chapter refer to the Arga-Tas Range (Iosifidi, 1988), Tas-Khayakhtakh Range (L'vov and Neustroev, 1991), and Omolon terrane, and are listed in Table 3 (abbreviations also used in Fig. 6).

Note that nearly all the paleomagnetic directions obtained for this region have northeast and, in part, southeast inclinations (Fig. 6A), and that directions that depart from the majority of data (Table 3) appear to be due to local rotations about vertical

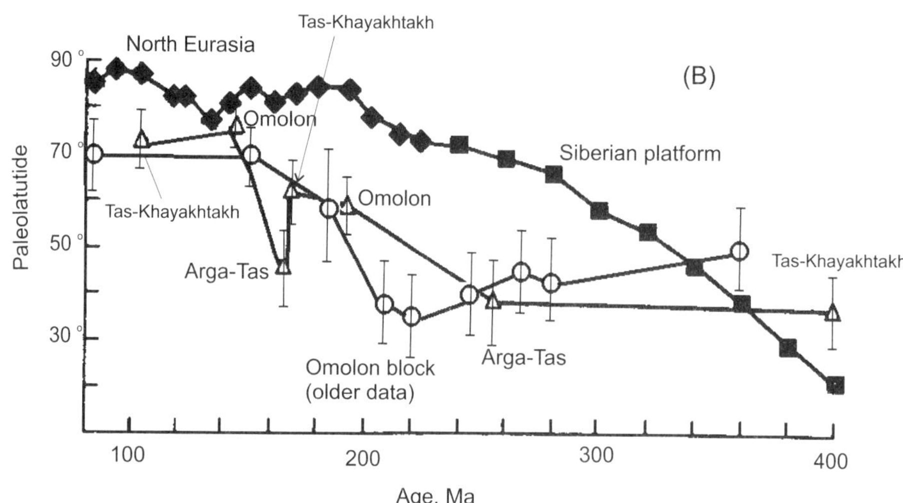

Figure 6. Comparison of new paleomagnetic results from Omolon block and Kolyma loop (triangles) and previous studies (circles) with expected declinations (A) and paleolatitudes (B) from Paleozoic Siberian platform (square) (Pechersky and Didenko, 1995) and stable Mesozoic North Eurasia continent (diamond) (Khramov, 1991). Abbreviations as in Table 3.

TABLE 3. PALEOMAGNETIC DATA FOR SOME OBJECTS OF THE VERKHOYANSK-KOLYMA COLLISIONAL ZONE

Terrane	Site coord. N	Site coord. E	Age rock	Age ChRM	Test	D (°)	I (°)	α_{95} (°)	Φ (°)	Λ (°)	A_{95} (°)	φ_a (°)	Reference
Arga-Tas	65.6	145.5	J_3	J_3?	?	187	60	9	17	140	12	41	Iosifidi, 1988
Arga-Tas	65.0	149.0	P_2	P_2?	?	226	57	12	19	112	15	38	Iosifidi, 1988
Omolon	65.0	155.4	J_1	145 ± 15	F-	30	83	4	75	184	8	76	this work
Omolon	65.0	155.4	J_1	J_1	F+, R	234	73	6	40	122	11	59	this work
Tas-Khayakhtakh	67.5	139.5	J_{2-3}	K	F- (25%)	8	79	7	87	208	13	68	L'vov and Neustroev, 1991
Tas-Khayakhtakh	67.4	139.3	J_2	J_2	F+	235	75	7	46	106	13	62	L'vov and Neustroev, 1991
Tas-Khayakhtakh	67.3	139.5	D_1	D_1?	F+?	239	56	9	23	91	11	37	L'vov and Neustroev, 1991

Note: ChRM is characteristic paleomagnetic direction; in column "Test": F is fold test, R is reversal test; D is declination of ChRM; I is inclination of ChRM; α_{95} is confidence limit around paleomagnetic direction; Φ is latitude of paleomagnetic pole; Λ is longitude of paleomagnetic pole; A_{95} is confidence limit around paleomagnetic pole; φ_a is paleolatitude.

axes, as is the case with the Upper Permian rocks of the Arga-Tas Range (Iosifidi, 1988).

Our study has established a high-T (>520 °C) prefolding component for the Lower Jurassic rocks of the Omolon massif and Sugoi zone (Tables 1–3) and has shown the secondary na-

ture of the medium-T paleomagnetic component, formerly interpreted as prefolding. Between the Early Jurassic and the Late Jurassic to Early Cretaceous the high-T component suggests a counterclockwise rotation of 30°–40°, rather than the clockwise rotation that was concluded from previous paleomagnetic data

(Fig. 7). During the same time interval the Omolon terrane, like the northeast margin of Siberia (which was not yet amalgamated into a single entity), passed through the polar region and continued its drift as a constituent of the newly consolidated Eurasian continent.

In Late Jurassic–Early Cretaceous time, the Omolon terrane became the site of vigorous magmatism (Bogdanov and Tielman, 1992; Parfenov et al., 1993). We believe that the directions of a significant part of the early paleomagnetic measurements from this region are the result of an Early Cretaceous overprint, rather than reflecting the true directions of the pre-Cretaceous geomagnetic field. The passage of the Omolon and Tas-Khayakhtakh terranes through the polar region in post-Triassic time was proposed by Rodionov (1991), but his data led him to conclude that the Omolon terrane rotated clockwise, and the Tas-Khayakhtakh (Fig. 1) terrane rotated counterclockwise (Lvov and Neustroev, 1991). On the basis of our data, the Omolon massif underwent a counterclockwise rotation between the Early Jurassic and Early Cretaceous.

The mismatch between (1) the previous paleomagnetic determinations for the Omolon terrane and (2) paleobiogeographic and structural geologic observations as well as the new paleomagnetic directions for the blocks of the Kolyma loop appearing in more recent studies (Table 3) (Bondarenko and Didenko, 1997; Iosifidi, 1988; Lvov and Neustruev, 1991) call for reinterpretation of the behavior of these terranes with respect to Siberia based on these more recent measurements alone, rather than on the older database.

Note the similar behavior of the paleomagnetic declinations (paleomeridians) of the new NRM component from the Kolyma terranes and Siberia between the middle Paleozoic and Mesozoic (Fig. 6A). Whereas the Paleozoic paleomeridians differ by 30°–50°, beginning ca. 240–230 Ma the measured paleomeridians virtually coincide with those calculated from the Siberian plate.

The interpretation that the Kolyma loop blocks were 15°–25° farther south than between Pennsylvanian and Middle Jurassic time (Fig. 6B) is somewhat equivocal, however. Mean directions for the Jurassic (Bondarenko and Didenko, 1997; Lvov and Neustroev, 1991) and Devonian (Lvov and Neustroev, 1991) magnetic components have been obtained from converging remagnetization circles, a method known to occasionally yield a systematically biased mean (Shipunov, 1993).

The more recent direction obtained by more rigorous methods suggests that the Kolyma loop blocks formed a continuous superterrane that was within the confines of the Siberian plate since at least the Middle Devonian. From the middle Paleozoic to the Middle Jurassic, the Kolyma loop terrane was no longer connected rigidly with Siberia, yet their motions had some features in common: they included a clockwise rotation and a northward drift, with Omolon, Tas-Khayakhtakh (Bondarenko and Didenko, 1997; Rodionov, 1991), and the northeast margin of the Siberian plate passing almost simultaneously across the polar region. The Omolon terrane, however, underwent a coun-

terclockwise rotation relative to Siberia at this time because of a time lag in its rotation. In post-Jurassic time, these terranes were finally deformed and incorporated into the Eurasian plate.

On the basis of these data we offer three time-slice reconstructions for the geodynamic evolution of northeast Eurasia: Early Jurassic, Late Jurassic–Neocomian, and Barremian (Fig. 7). The North American and Eurasian blocks are positioned using the paleomagnetic poles of Besse and Courtillot (1990) and Khramov (1991). The positions of the smaller continental blocks and island-arc systems are based on our data (this work and Sokolov et al., 1997), and the position of the Omuliovka terrane is after Lvov and Neustroev (1991).

In Early Jurassic time (200 ± 10 Ma), the Omolon terrane was at 60°–70° and was separated from Eurasia by a marine basin (Inyali-Debin and Sugoi) (Fig. 7A). The Omolon terrane was part of an extensive system of continental terranes (e.g., Omuliovka, Prikolymsk, Okhotsk) that was between Eurasia and the Pacific Ocean and the Anyui oceanic basin to the south. Behind this system of continental terranes, a convergent boundary existed between Eurasia and the oceanic plates of the Anyui basin and the Pacific Ocean. This boundary was marked by the Oloi and Koni-Taigonos island arcs (Fig. 7A), which are represented by the Upper Triassic to Lower Jurassic volcanic complexes of the Oloi and Koni-Taigonos zones in the present framework of northeast Eurasia (Fig. 1; Parfenov et al., 1993; Sokolov et al., 1997). South of this convergent margin, within the Pacific Ocean, a system of ensimatic island arcs existed (Kanchalan [KN] and Pekulney [PK], Fig. 7A) (Morozov, 1992), the paleolatitudinal positions of which are after Sokolov et al. (1997).

The next reconstruction spans Late Jurassic to Neocomian time 155 ± 5 Ma (Fig. 7B). It indicates that, beginning in the Early Cretaceous, the Eurasian plate rotated clockwise while migrating 15°–20° southward. The system of continental blocks that incorporated the Omolon terrane (the Kolyma-Omolon superterrane of Parfenov et al., 1993) migrated in the same direction as Eurasia, but all the while rotated counterclockwise with respect to Eurasia. During that time interval, the Omolon terrane underwent sinistral strike-slip displacements and rotated 30°–40° counterclockwise (Table 2) with respect to Eurasia, and reached subpolar latitudes (Fig. 7B). The shelf basin that separated the Omolon terrane from Eurasia shrank by partial subduction and was rotated by strike-slip faulting (Bondarenko, 1995). The crust of the Pacific and the Anyui basins (Uyandina-Yasachny [UYA, Fig. 7B] and Uda-Murgal island arcs) continued to subduct beneath Eurasia. The ensimatic island arcs (Kanchalan and Pekulney) migrated from 30°N to 60°–70°N, toward Eurasia. The Anyui (Arctic) oceanic basin shrank, possibly in response to the onset of opening of the Canada basin (Grantz et al., 1990) and the beginning of the motion of the Chukotka blocks (WCH and ECH, Fig. 7B) toward Eurasia.

By the Hauterivian-Barremian (130 ± 5 Ma), owing to completion of the collision between the Omolon terrane and Eurasia, the Kolyma loop had formed (Fig. 7C) and Eurasia and the

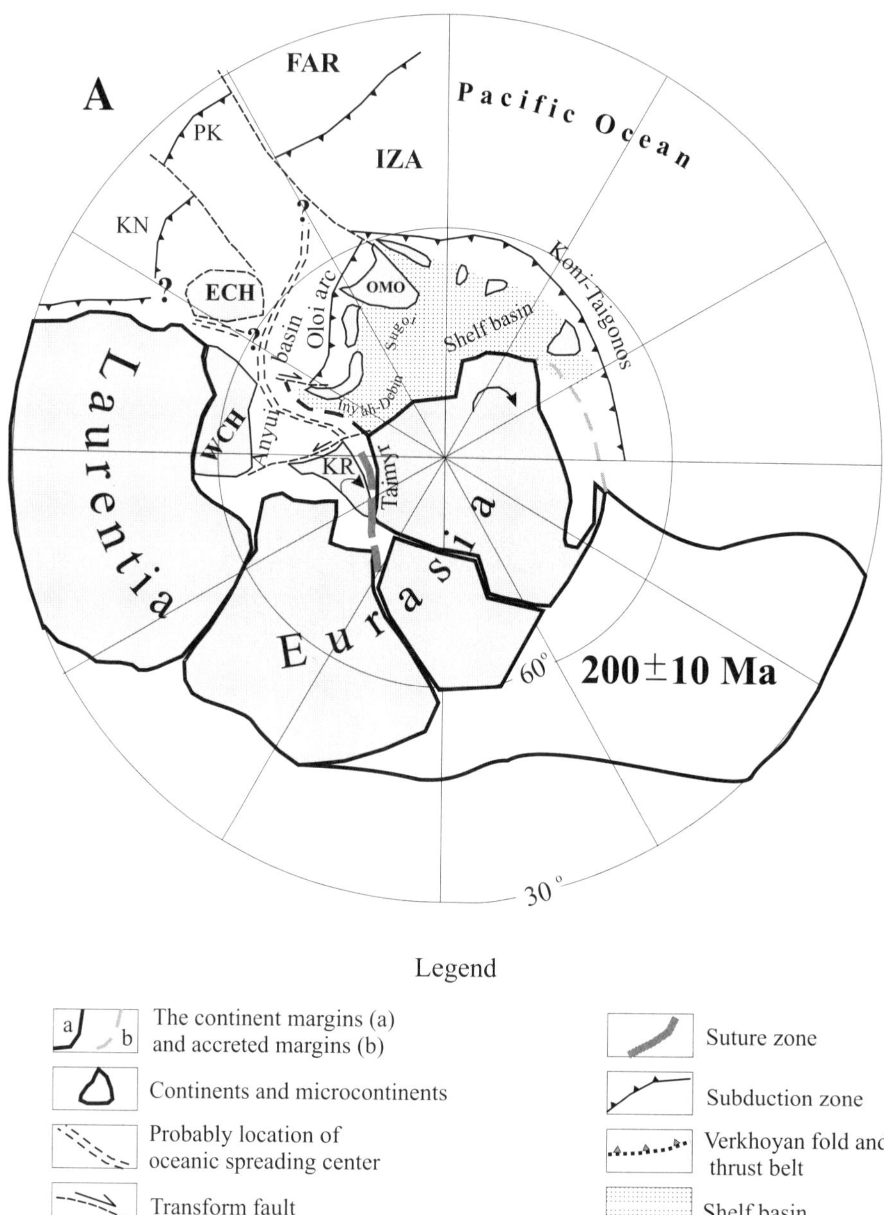

Figure 7 (on this and facing page). Plate tectonic reconstructions for northeastern Eurasia for 200 Ma (A), 155 Ma (B), and 130 Ma (C). Positions of Laurentia and Eurasia are after Besse and Courtillot (1990) and Khramov (1991). Position of Omuliovka block is after Lvov and Neustroev (1991). Positions of island-arc units are after Sokolov et al. (1997). Position of Omolon block is after this study. Main microcontinents: KR, Kara; WCH, western Chukotka; ECH, eastern Chukotka; OMO, Omolon; OM, Omuliovka; PR, Prikolymsk; OK, Okhotsk. Main subduction zones: PK, Pekulney; KN, Kanchalan; UYA, Uyandina-Yasachny. Oceanic plates of Pacific: FAR, Farallon; IZA, Izanagi. SAZ is South Anyui suture zone.

Legend

continental terranes of the Kolyma loop migrated another 15°–20° south. Structural data show that the Omolon terrane continued to undergo minor displacements on sinistral strike-slip faults while rotating clockwise with respect to Eurasia. The available paleomagnetic data (according to which the Omolon block rotated 8° ± 60° with respect to Siberia; Table 2) do not permit a precise assessment of this rotation because both the Omolon block and adjacent parts of Eurasia were at high latitudes. On the Pacific side of Eurasia, the Uda-Murgal arc continued to develop. The Anyui basin closed, the Canada basin reached its maximum size, and the Chukotka terrane migrated toward Eurasia during the Hauterivian to Barremian. Dextral

strike-slip faults formed along the boundary between the Chukotka terrane and adjoining Eurasia, and the development of the South Anyui suture was complete. Along the Pacific-Chukotka transition, ensimatic island arcs continued to exist, but became extinct between the end of Barremian and the beginning of Aptian (Morozov, 1992).

We propose the following mechanism to explain the rotation of the region's tectonic blocks between the Early Jurassic and Lower Cretaceous and to constrain the timing of this event. In the second half of the Late Jurassic (starting from the Kimmeridgian), the southeast margin (present reference frame) of the Omolon terrane became the site of a subduction zone and

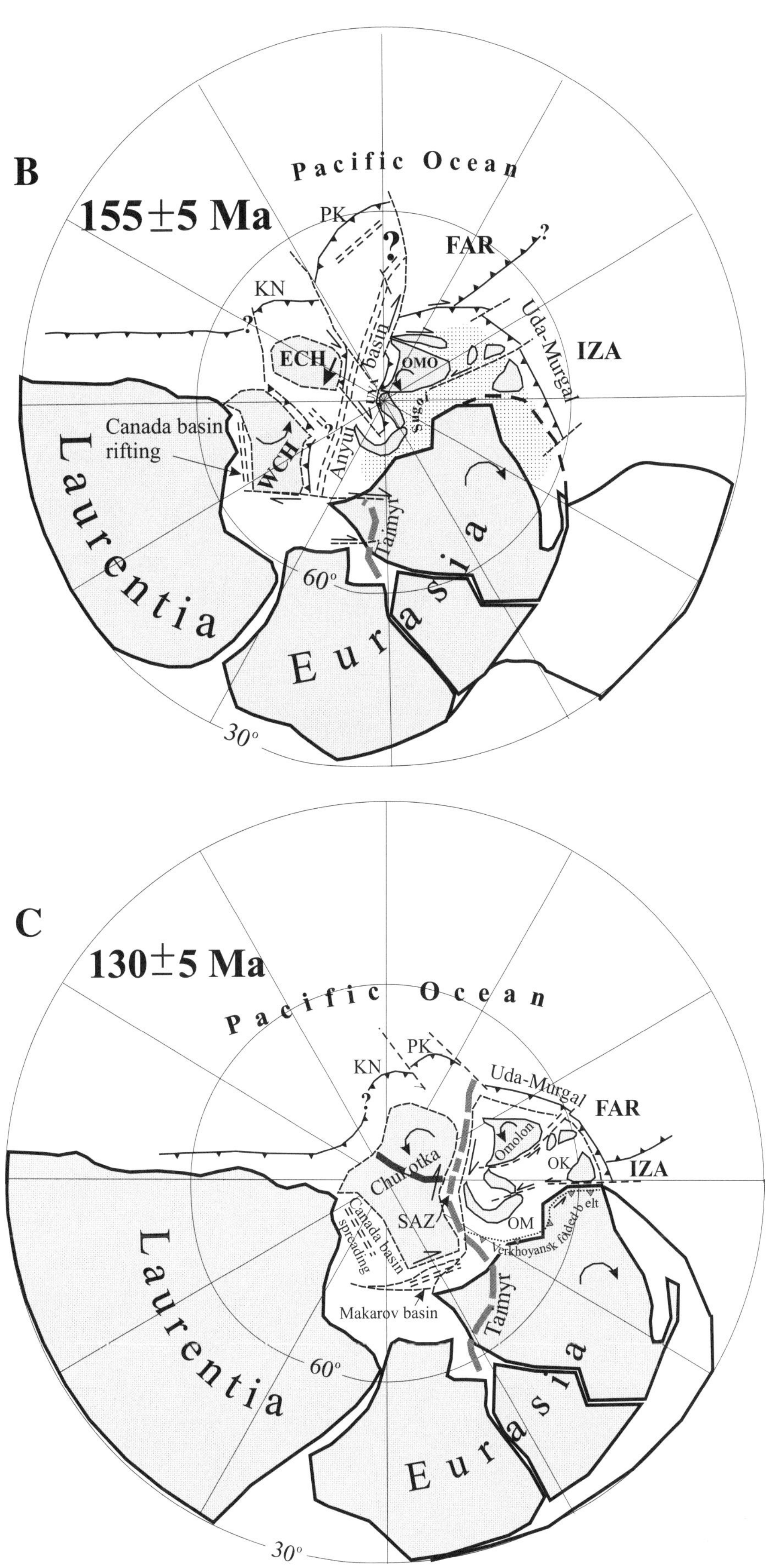

B
155±5 Ma
Pacific Ocean
PK
?
FAR
?
KN
?
Uda-Murgal
IZA
ECH
OMO
Canada basin
rifting
WCH
Anyui basin
Sigo
Taimyr
60°
Laurentia
Eurasia
30°

C
130±5 Ma
Pacific Ocean
KN
PK
?
Uda-Murgal
FAR
Chukotka
Omolon
OK
IZA
Canada basin
spreading
SAZ
OM
Verkhoyansk folded belt
Makarov basin
Taimyr
60°
Laurentia
Eurasia
30°

volcanic belt, the Uda-Murgal island arc of Late Jurassic to Neocomian age (Nekrasov, 1976; Zaborovskaya, 1978). The subduction responsible for this belt might have impeded the southerly drift of the Omolon terrane. Simultaneously (Kimmeridgian to Berriasian), the Oloi island arc became established on the northeast margin of the Omolon terrane (Shulgina and Natapov, 1991; Natal'in, 1984). The subduction related to this arc might also have played a decelerating role on the southerly drift of the Omolon terrane. The velocity of the absolute southward drift of the Omolon terrane in that time interval appears to have been slower than that of Siberia.

The disparity in southward drift rates between the Omolon terrane and Siberia may have given rise to a sinistral strike-slip boundary between continental blocks (Bondarenko, 1995). Displacement on this boundary must have induced the counterclockwise rotation, with respect to Siberia, of the Omolon and intervening tectonic blocks in the Sugoi zone (Bondarenko, 1995). This would have resulted in dextral strike-slip faulting with coeval shortening on the southeast margin of the Omolon massif (Bondarenko, 1995; Khudoley and Sokolov, 1994) at the Koni-Taigonos and Uda-Murgal island arcs (KM, Fig. 1). Geologic field observations supporting dextral strike-slip motion on the northern Omolon margin were given in Natal'in (1984) and Sokolov et al. (this volume).

The results obtained in our study suggest major lateral displacements between tectonic blocks in the study region during the Late Jurassic and Neocomian; in particular, the displacement of Omolon relative to Siberia. At the same time there is no need, at least for the Mesozoic, to regard Omolon as an exotic microcontinent with respect to Siberia or to invoke a special model for its orthogonal collision with the Siberia margin. Most likely, as proposed by Savostin et al. (1994), the Omolon terrane, jointly with the Prikolymsk and Omuliovka uplifts, held a marginal position with respect to the Siberian paleocontinent during the Mesozoic, perhaps migrating a short distance from it as a consequence of several phases of rifting during the late Paleozoic and Mesozoic.

PRINCIPAL CONCLUSIONS

In the Late Jurassic and Early Cretaceous the Sugoi paleobasin, which separated the Omolon terrane from the Siberian paleocontinent, became the site of sinistral strike-slip faults with attendant shortening and transpression, which led to the northward translation of the Omolon terrane with respect to Siberia. In the present framework, these strike-slip faults are located mainly within the Sugoi folded zone. Simultaneously, motions on systems of dextral strike-slip faults took place along the boundary between Omolon and the Uda-Murgal and, possibly, Oloi island arcs. Lower Jurassic sedimentary rocks in the west of the Omolon terrane and in the north of the Sugoi folded zone yield two ancient NRM components: (1) prefolding (dec = 234°, inc = 73°, K = 17), with a respective paleomagnetic pole (Φ = 40°, Λ = 121°, A_{95} = 11°), and (2) overprinted Early Cretaceous

(dec = 30°, inc = 83°, K = 117), with a respective paleomagnetic pole (Φ = 75°, Λ = 183°, A_{95} = 8°). Between the end of the Jurassic and earliest Cretaceous, the Omolon terrane underwent translation from 60° ± 10°N in the Western Hemisphere via the polar region to 76° ± 8°N in the Eastern Hemisphere, while rotating 30°–40° counterclockwise relative to Siberia. In the region's structural grain, this motion is manifest by a system of sinistral strike-slip faults with shortening and auxiliary strike-slip faults. Starting from the Early Cretaceous, the Omolon terrane and the Sugoi folded zone appear to have been integral to the Eurasian plate.

ACKNOWLEDGMENTS

This work was supported by the Russian Foundation for Basic Research (grants 99-05-64857, 97-05-65711, and 98-07-90015). We thank M. Bazhenov and D. Pechersky for their advice and critical remarks on our paleomagnetic results. We are grateful to A. Varvarov, V. Pavlov, Z. Sharonova, and G. Yanova for their assistance in field work and laboratory research. Special thanks are due to the 3rd International Conference on Arctic Margins organizers and sponsors, who made it possible for us to present these data at the meeting in Celle, and E. Miller (Stanford University), A. Grantz, and W. Harbert (University of Pittsburgh) for their invaluable remarks and help in improving this paper.

REFERENCES CITED

Bazhenov, M.L., and Shipunov, S.V., 1988, Fold test in paleomagnetism: Izvestiya Akademii Nauk SSSR, Fizika Zemli, no. 7, p. 89–101.

Besse, J., and Courtillot, V., 1990, Revised and synthetic apparent polar wander paths of the African, Eurasian, North American and Indian Plates and true polar wander paths since 200 Ma: Journal of Geophysical Research, v. 96, p. 4029–4050.

Bogdanov, N.A., and Tielman, S.M., 1992, Tectonics and geodynamics of northeast Asia: Moscow, ILSAN, 80 p., explanatory note (in Russian).

Bondarenko, G.E., 1995, Sinistral strike slips in the structure of the southwest rim of the Omolon Massif: Byulleten' Moskovskogo Obshchestva Ispytatelei Prirody, Seriya Geologicheskaya, v. 71, no. 5, p. 18–24 (in Russian).

Bondarenko, G.E., and Didenko, A.N., 1997, New geologic and paleomagnetic data on the Jurassic-Cretaceous history of the Omolon Massif: Geotektonika, no. 2, p. 14–27.

Butler, R.F., 1992, Paleomagnetism: Magnetic domains to geologic terranes: Cambridge, Blackwell, 319 p.

Demarest, H.H., 1983, Error analysis for the determination of tectonic rotation from paleomagnetic data: Journal of Geophysical Research, v. 88, p. 4321–4328.

Fisher, R.A., 1953, Dispersion on sphere: Proceedings, Royal Society of London Philosophical Transactions, ser. A, v. 217, p. 295–305.

Gagiev, M.K., 1991, Stratigraphy and conodonts of the Middle Paleozoic of the northeastern USSR [Ph.D. thesis abstract]: (Geol.-Miner.) Magadan, Severo-Vostochnyi Kompleksnyi Nauchno-Issledovatel'skyi Institut Rossiyiskoi Akademii Nauk (in Russian).

Gonchar, V.V., and Drachev, S.S., 1993, The reconstruction of the fields of tectonic strains and stresses of various ages using kinematic analysis of breakdown structures: Fizika Zemli, no. 12, p. 22–28.

Gondwana Consultants, 1995, Global paleomagnetic database, Ver. 3.1, in P.O. Box S, Hat Head NSW 2440, Australia, (e-mail: midemce@ozemail .com.au).

Grantz, A., May, S.D., and Lawyer, L.A., 1990, Canada Basin, *in* Grantz, A., Johnson, G.L., and Sweeney, J.F., eds., The Arctic Ocean region: Boulder, Colorado, Geological Society of America, Geology of North America, v. L, p. 379–402.

Halls, H.C., 1976, A least-squares method to find a remanence direction from converging remagnetization circles: Geophysical Journal of the Royal Astronomical Society, v. 45, p. 297–304.

Iosifidi, A.G., 1988, New paleomagnetic data on the Kolyma Massif (Arga-Tas Range), *in* Khramov, A.N., ed., Paleomagnetism and accretionary tectonics: Leningrad, VNIGRI, p. 104–123 (in Russian).

Karago, E.A., and Shpetnyi, A.P., 1982, State geologic map of the USSR, Quadrangle Q-56–24: Leningrad, VSEGEI, scale 1:200 000, explanatory note (in Russian).

Kazakov, A.N., 1980, Geometrical analysis of fold structures using stereograms, *in* Shul'ts, S.S., ed., Geologic survey of strongly deformed complexes: Leningrad, Nedra, p. 2–14 (in Russian).

Khramov, A.N., 1988, Paleomagnetism and issues of accretionary tectonics of the northwest segment of the Pacific Mobile Belt, *in* Khramov, A.N., ed., Paleomagnetism and accretionary tectonics: Leningrad, VNIGRI, p. 141–153 (in Russian).

Khramov, A.N., 1991, Standard series of paleomagnetic poles for the plates of North Eurasia: Connection with the issues of geodynamics of the USSR territory: Leningrad, VNIGRI, p. 135–149 (in Russian).

Khramov, A.N., and Ustritsky, V.I., 1990, Paleopositions of some northern Eurasian tectonic blocks: Paleomagnetic and paleobiologic constraints: Tectonophysics, v. 184, p. 101–109.

Khudoley, A.K., and Sokolov, S.D., 1994, On the role of strike slips in the tectonics of the southwestern Koryak Highlands, Doklady Rossiiskoi Akademii Nauk (Reports of the Russian Academy of Science), v. 335, p. 74–76.

Kirshvink, J.L., 1980, The least-square line and plane and the analysis of paleomagnetic data: Geophysical Journal of the Royal Astronomical Society, v. 2, p. 699–718.

Kolesov, E.V., 1981, Paleomagnetic characteristics of the Middle Paleozoic deposits of the Omolon Massif, *in* Lin'kova, T.I., ed., Magnetism of the rocks and paleomagnetic stratigraphy of eastern and northeastern Asia: Magadan, Russia, Severo-Vostochnyi Kompleksnyi Nauchno-Issledovatel'skyi Institut Rossiyiskoi Akademii Nauk (SVKNII), p. 68–74 (in Russian).

Kudley, E.M., 1978, State geologic map of the USSR, Middle Kolyma Series, Quadrangle Q-56–29, 30: Leningrad, VSEGEI, scale 1:200 000, explanatory note (in Russian).

Kuznetsov, V.M., 1975, The southwest boundary of the Omolon Massif, *in* Shilo, E., ed., The materials on geology and minerals of the northeastern USSR: Magadan, Russia, Knizhnoe Izdatel'stvo, no. 22, p. 42–48 (in Russian).

Kuznetsov, V.M., 1991, Arcuate faults of the northeastern USSR, *in* Shilo, E., ed., The materials on geology and minerals of the northeastern USSR: Magadan, Russia, Knizhnoe Izdatel'stvo, no. 27, p. 39–46 (in Russian).

Lozhkina, N.V., 1982, The NRM in the Upper Triassic deposits of the Omolon Massif, *in* Lin'kova, T.I., ed., Paleomagnetic methods used to solve geologic problems (a case study from the Far East): Vladivostok, Russia, DVO DVIN AN SSSR, p. 97–106 (in Russian).

Lvov, M.Yu, and Neustroev, A.P., 1991, The data from paleomagnetic studies in some Mesozoic and Paleozoic rocks of the Tas-Khayakhtakh Range (Eastern Yakutia), *in* Khramov, A.N., ed., Paleomagnetism and paleogeodynamics of the USSR Territory: Leningrad, VNIGRI, p. 95–113 (in Russian).

McFadden, P.L., and McElhinny, M.W., 1990, Classification of the reversal test in paleomagnetism: Geophysical Journal International, v. 103, p. 725–729.

Merzlyakov, V.M., Terekhov, M.I., Lychagin, L.L., and Dylevski, E.F., 1982, Tectonics of the Omolon Massif: Geotektonika, no. 1, p. 74–86.

Morozov, O.L., 1992, Paleoisland arc system of Pekulney Range (central Chukotka), *in* Mezhelovsky, N., ed., Regional geodynamics and stratigraphy of Asian part of USSR: Leningrad, VSEGEI, p. 120–172.

Natal'in, B.M., 1984, Early Mesozoic geosynclinal systems of the northern part of the Pacific Rim: Moscow, Nauka, 136 p. (in Russian).

Nekrasov, G.E., 1976, Tectonics and magmatism of the Taigonos Peninsula and Northwestern Kamchatka: Moscow, Nauka, 160 p. (in Russian).

Parfenov, L.M., and Natal'in, B.M., 1977, Tectonic evolution of northeast Asia in the Mesozoic and Cenozoic: Dokl. Akad. Nauk SSSR, v. 235, p. 1132–1135.

Parfenov, L.M., Natapov, L.M., Sokolov, S.D., and Tsukanov, N.V., 1993, Terranes and accretionary tectonics of northeast Asia: Geotektonika, no. 1, p. 68–78.

Pechersky, D.M., and Didenko, A.N., 1995, The paleo–Asian Ocean: Petromagnetic and paleomagnetic data on its lithosphere: Moscow, OIFZ RAN, 298 p. (in Russian).

Rodionov, V.P., 1991, The kinematic models of relationships between the Siberian craton and the blocks of the Verkhoyansk-Kolyma region, *in* Khramov, A.N., ed., Paleomagnetism and paleogeodynamics of the USSR Territory: Leningrad, VNIGRI, p. 113–119 (in Russian).

Savostin, L.A., Pavlov, V.E., Safonov, V.G., and Bondarenko, G.E., 1993, The Lower and Middle Jurassic deposits in the west of the Omolon Massif: Dokl. Ross. Akad. Nauk, v. 333, p. 481–486.

Savostin, L.A., Bondarenko, G.E., Safonov, V.G., and Pavlov, V.E., 1994, The structural evolution of the southwest surroundings of the Omolon Massif in the Jurassic: Geotektonika, no. 5, p. 46–62.

Shapiro, M.N., and Ganelin, V.G., 1988, The paleotectonic relationships of major blocks in the Mesozoides of the northeastern USSR: Geotektonika, no. 5, p. 94–104.

Shipunov, S.V., 1993, The fundamentals of paleomagnetic analysis: Theory and practice: Moscow, Nauka, 159 p.

Shulgina, V.S., and Natapov, L.M., 1991, editors, Geologic map of the USSR, Quadrangle Q-56, 57–Srednekolymsk, New Series: Leningrad, VSEGEI, scale 1 000 000, explanatory note (in Russian).

Sokolov, S.D., Didenko, A.N., Grigoriyev, V.N., Alexyutin, M.V., Bondarenko, G.Ye., and Krylov, K.A., 1997, Paleotectonic reconstructions of northeastern Russia: Problems and uncertainties: Geotektonika, no. 6, p. 72–90 (in Russian).

Sylvester, A.G., 1988, Strike-slip faults: Geological Society of America Bulletin, v. 100, p. 1666–1703.

Talent, J.A., 1990, Relationships between lithospheric blocks in the northeast USSR: Autochtons or wanderers from afar?: Geotektonika, no. 2, p. 123–125.

Terekhov, M.I., 1979, Stratigraphy and tectonics of the south part of the Omolon Massif: Moscow, Nauka, 114 p. (in Russian).

Tielman, S.M., 1973, Comparative tectonics of the Mesozoides of the north part of the Pacific Rim: Novisibirsk, Russia, Nauka, 325 p. (in Russian).

Ustritsky, V.I., and Khramov, A.N., 1984, Geologic history of the Arctic regions from a plate-tectonic standpoint, *in* Geologic structure of the USSR: Regularities in the distribution of mineral resources: Leningrad, Nedra, Soviet Arctic Seas, v. 5, p. 253–265 (in Russian).

Woodcock, N.H., and Fischer, M., 1986, Strike-slip duplexes: Journal of Structural Geology, v. 8, p. 725–735.

Zaborovskaya, N.B., 1978, The inner zone of the Okhotsk-Chukotka Belt in the Taigonos Peninsula: Moscow, Nauka, 198 p.

Zhuravlev, A.Yu., 1988, Archaeocyatha from the Lower Cambrian of the extreme northeast of the USSR, *in* Gagiev, M., ed., The Cambrian of Siberia and Central Asia: Moscow, Nauka, p. 97–110 (in Russian).

Zijderveld, J.D.A., 1967, A.C. demagnetization of rocks: Analysis of results, *in* Collinson, D.W., and Creer, K.M., eds., Methods in paleomagnetism: Amsterdam, Elsevier, p. 254–286.

Zonenshain, L.P., Kuzmin, M.M., and Natapov, L.M., 1990, Plate tectonics of the USSR Territory, Book 2: Moscow, Nedra, 334 p. (in Russian).

MANUSCRIPT ACCEPTED BY THE SOCIETY MAY 15, 2001.

Geological Society of America
Special Paper 360
2002

Paleomagnetic paleolatitudes for Upper Devonian rocks of the Omolon massif, northeastern Russia

Evgenii Kolesov
Northeast Interdisciplinary Scientific Research Institute, 16 Portovaya Street, Magadan, 685010, Russia
David B. Stone
Geophysical Institute, University of Alaska, Fairbanks, Alaska 99775, USA, and
Departments of Physics and Geology, Rhodes University, Grahamstown 6140, South Africa

ABSTRACT

The Omolon massif in northeastern Russia forms the core of the Kolyma-Omolon superterrane. The massif has an early Precambrian crystalline basement and is surrounded by the Paleozoic shelf deposits of the Omulevka terrane and various other terrane fragments. Geologic models range from the superterrane being a fragment of the Siberian craton that rifted away in late Precambrian time and traveled several tens of degrees to the south relative to the craton, to models that never allow it to move very far (of the order of 500 km) from Siberia. To test the various hypotheses, paleomagnetic data were obtained from a suite of volcanic rocks of Late Devonian age (Famennian) from the Omolon massif. These results pass fold ($K_s/K_g = 3.5$), reversal, and conglomerate tests. The combined data (N = 44) give a paleolatitude of $30.3° \pm 4.5°$, and a virtual geomagnetic pole located at latitude 20.6°N and longitude 224.0°E. When compared with poles of the same age from the Siberian platform, this result indicates that the massif could have been in roughly its present position, although rotated about a vertical axis, with respect to the Siberian platform in Devonian time. This assumes that both the Siberian platform and the massif were in the same (northern) hemisphere. Additional paleomagnetic data for the Kolyma-Omolon superterrane selected from the literature indicate that in post-Devonian time, the Kolyma-Omolon superterrane was farther south with respect to the Siberian platform, and subsequently drifted back north to its present location in Jurassic time.

INTRODUCTION

There are two alternative viewpoints on the history of tectonic development of the Northeast Asia. Some researchers consider the area east of the stable Siberian platform to consist of amalgamated allochthonous terranes similar to those proposed for Alaska (e.g., see Jones et al., 1972, 1977; Coney et al., 1980). In this way, the Siberian platform and the terranes to the east of it are a mirror image of the Canadian shield and the amalgamated terranes of the Cordillera and Alaska to the west (Fig. 1). Together these terranes fill the vast area separating the Arctic and Pacific Oceans and the Siberian and Canadian continental blocks. The major amalgamation and accretion of the terranes of northeastern Russia took place in Mesozoic and Tertiary time (Burke, 1984; Karasik et al., 1984; Khramov et al., 1985; Khramov and Ustritsky, 1990; Nokleberg et al., 1998; Parfenov, 1984, 1991; Scotese et al., 2001; Talent, 1990; Ustritsky and Khramov, 1987; Zonenshain, 1984; Zonenshain et al., 1990a, 1990b).

The other viewpoint claims that the large tectonic units forming much of northeastern Russia, particularly those now known as the Kolyma-Omolon superterrane, are fragments of the Siberian continent that have remained essentially in place

Kolesov, E., and Stone, D.S., 2002, Paleomagnetic paleolatitudes for Upper Devonian rocks of the Omolon massif, northeastern Russia, *in* Miller, E.L., Grantz, A., and Klemperer, S.L., eds., Tectonic Evolution of the Bering Shelf–Chukchi Sea–Arctic Margin and Adjacent Landmasses: Boulder, Colorado, Geological Society of America Special Paper 360, p. 243–257.

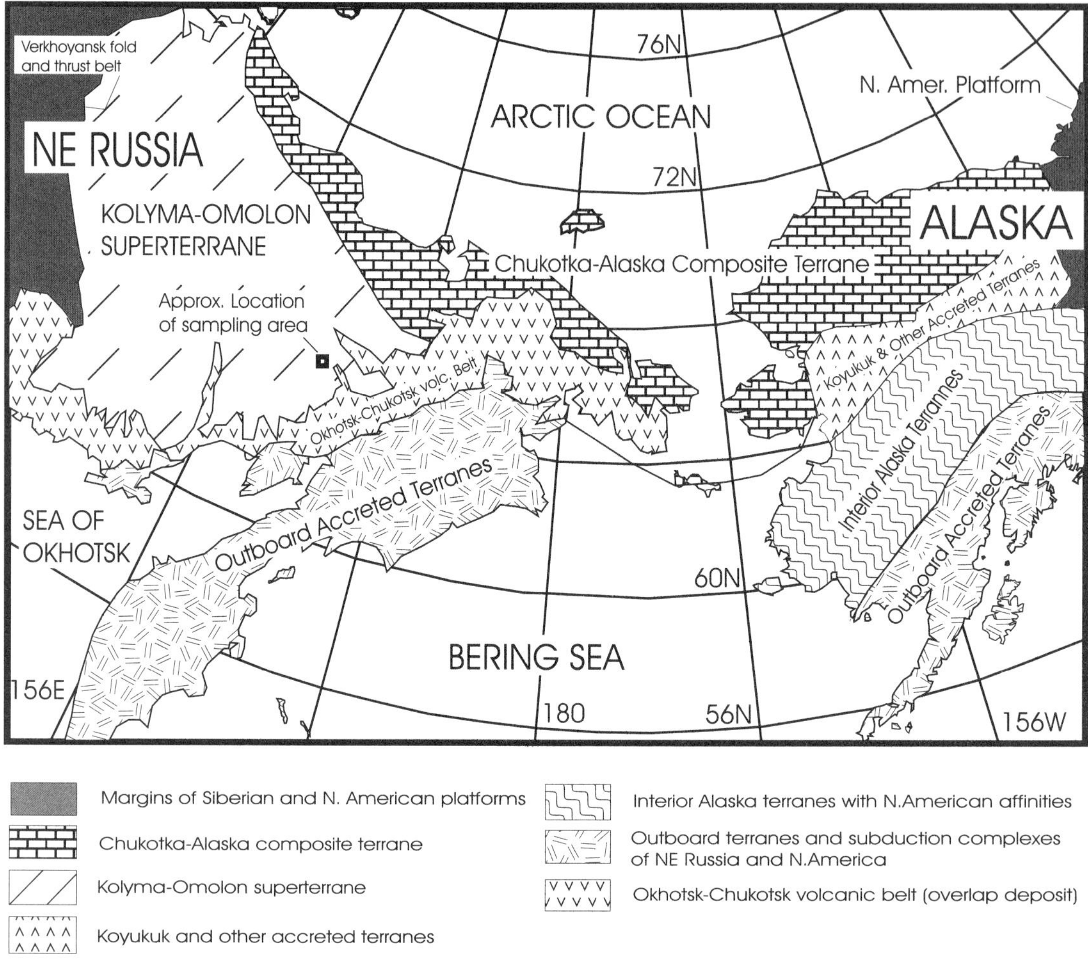

Figure 1. Schematic representation of tectonostratigraphic terranes between Siberian and North American platforms, and approximate location of study area. Chukotka-Alaska composite terrane is thought to have rotated away from Canadian Arctic margin, and outboard accreted terranes were rafted northward on plates of Pacific and Interior Alaska terranes translated northward along North American margin. Kolyma-Omolon superterrane is amalgamation of terranes that accreted to Siberian platform.

with respect to the Siberian craton throughout their Phanerozoic history (Chekhov, 1990; Merzlyakov et al., 1982; Shapiro and Ganelin, 1988; Sharkovsky, 1975; Til'man, 1973).

The data presented here support the view that the Omolon massif plus the Prikolyma and Omulevka terranes, which together form a large part of the Kolyma-Omolon superterrane (Figs. 1 and 2), behaved independently from the Siberian craton, though they may never have been very far distant from it.

GEOLOGIC SETTING

The terranes and continental fragments that make up northeastern Russia include many different paleoenvironments and rock types, but can be divided into three general groups on the basis of the general timing of amalgamation of the terranes and their accretion to the Siberian platform.

In Figure 1, the northwestern Bering Sea and Okhotsk Sea margins are labeled as outboard accreted terranes and include subduction-related complexes. These terranes bear a striking resemblance to the terranes of southern Alaska in terms of the rock types and ages represented, as well as in their paleomagnetic paleolatitudes through time. These latter data (e.g., see Stavsky et al., 1990; Kovalenko and Remizova, 1997) indicate that the terranes forming this margin have been transported northward to their present locations, presumably by the generally northward motions of the plates of the Pacific basin (e.g., see Engebretson et al., 1985). In northeastern Russia the boundary between these and the more inboard terranes is largely covered by the dominantly Late Cretaceous Okhotsk-Chukotsk volcanic belt (Nokleberg et al., 1994) (Figs. 1 and 2), which is presumed to have been generated by northward subduction of the North Pacific prior to and during accretion of the outboard terranes.

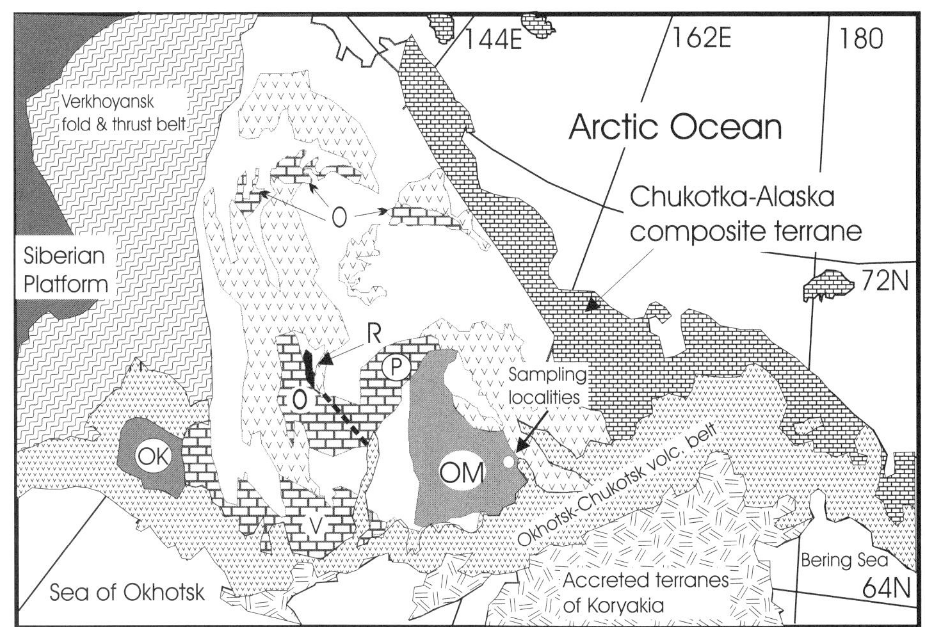

Figure 2. Selected terranes of Kolyma-Omolon superterrane. General horseshoe shape of terranes within superterrane represents Kolyma structural loop of Zonenshain (1990a, 1990b). Dominantly Late Cretaceous Okhotsk-Chukotsk volcanic belt forms major overlap sequence. Details of terranes can be found in Nokleberg et al. (1994).

The northern part of northeastern Russia, bounded to the south by the South Anuyi suture (Seslavinskiy, 1979; Churkin and Trexler, 1981; Sokolov et al., this volume) is generally considered to be closely related to, and commonly presumed to be an integral part of, the Arctic Alaska composite terrane often referred to as the Chukotka-Alaska terrane (Fig. 1). There are many models for the origin and history of the Chukotka-Alaska terrane (Lawver and Scotese, 1990; Lawver et al., 1996), but the most acceptable model today involves rotating the entire block away from the Canadian Arctic margin in Early Cretaceous time about a pivot located to the south of the Mackenzie Delta near the Alaska-Yukon border. The South Anuyi suture is poorly defined and poorly understood, but its geology is consistent with a collisional margin, and it may thus represent the zone of collision between the Chukotka-Alaska terrane and the terranes to the south.

The area enclosed by the South Anuyi suture, the Okhotsk-Chukotsk volcanic belt, and the eastern margin of the Siberian platform consists of many terranes representing many different geologic environments collectively known as the Kolyma-Omolon superterrane. The individual terranes have been given various labels and subdivided and collected together in many ways. In general they consist of an ancient terrane with crystalline basement, the Omolon terrane; the Omulevka, Prikolyma, and Viliga terranes, three terranes dominated by continental margin rocks; and many terranes representing the remains of island arcs, collapsed oceanic basins, and fragments of continental shelf (Fig. 2). These terranes have been intruded by the major granitic belts that outline the arms of the Kolyma structural loop of Zonenshain et al. (1990b), and partially covered by overlap deposits (Parfenov, 1991; Fujita et al., 1997). The Omulevka, Prikolyma, and Omolon terranes have many features in common, and are frequently referred to as the Omulevka-Omolon composite terrane, whereas the Kolyma-Omolon superterrane includes the Omulevka-Omolon composite terrane together with surrounding terranes that amalgamated prior to their accretion to the Siberian platform in Jurassic time. The boundaries of the Kolyma-Omolon superterrane have been located differently by different authors (Nokleberg et al., 1994, 1998). They are outlined here as encompassing the entire area enclosed by the Verkhoyansk fold and thrust belt, the South Anuyi suture, and the Okhotsk-Chukotsk volcanic belt (Figs. 1

and 2). The new paleomagnetic data discussed in this report come from the Omolon terrane.

GEOLOGY OF THE STUDY AREA

The study area is located on the eastern end of the Perevalny anticline in the northeastern part of the Omolon massif (representative lat 65.00°, long 162.12°, Figs. 2 and 3). The core of the anticline consists of mainly volcanogenic rocks of the Pylkatveem suite, which includes red lava, tuff, and ignimbrite of subalkaline rhyolites (Simakov, 1974; Simakov and Shevchenko, 1974). The samples we studied are from the Upper Devonian–Lower Carboniferous section of volcaniclastic and other sedimentary rocks exposed in the northwest limb of this structure. The volcanic sequences are interpreted to be rift related (Gagiev, 1995, 1996).

From bottom to top this section consists of the Upper Devonian Khelon, Perevalny, and Elergetkhyn suites, and the Lower Carboniferous Mol and Sikambrian suites (Fig. 4). The sampling for this study was confined to the volcanic rocks of the Khelon suite, where they crop out on Perevalny Creek, a tribu-

tary of the Molandzha River; however, published paleomagnetic data are available from all but the Mol suite (Kolosev, 1981) and are discussed in a later section.

Exposures of the sampled section are not continuous, but all parts sampled are within a distance of ~200 m. The uppermost units of the volcanic section are not exposed in the sampled section.

Five units were recognized in the Khelon suite in the field (Fig. 5). From unit 1 at the base, they consist of the following.

Unit 1 is ~22 m thick, and composed of gray-greenish to black andesite and basaltic andesite lavas, which are massive in the lower parts and amygdaloidal near the top. The middle portion of the flow has large (3–5 cm) spherical nodules with sharply defined crusts and clear amygdule zoning. The amygdules are calcite filled, and become elongate near the top of the unit.

Unit 2 is ~5 m and composed of poorly bedded friable psammitic tuff of basic composition. It is generally brown with a hint of lilac color, and contains sparse fragments of andesite and basaltic andesite that are 2–2.5 cm in diameter.

Unit 3 is ~55 m thick and consists of two andesite to basaltic andesite layers that are very similar in appearance to those of

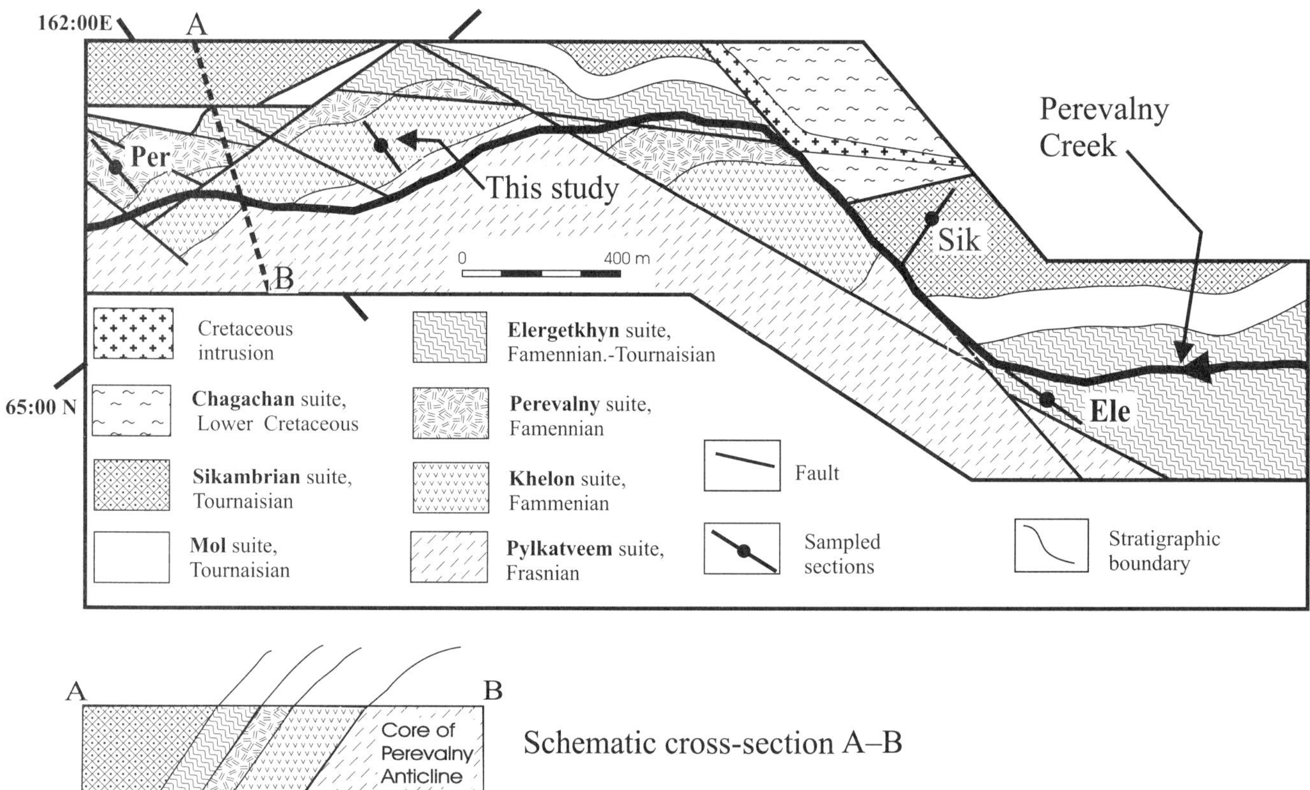

Figure 3. Sketch map of geology around sampling area near Perevalny Creek modified from Gagiev (1982). Axis of Perevalny anticline is south of map area. Khelon suite sampling sites described herein are labeled "this study." Data from Kolosev (1981), listed in Table 4 and shown in Figure 8, are from labeled sections. The Per (Perevalny), Ele (Elergetkhyn), Sik (Sikambian) suites, and Khe (Khelon) suite were sampled ~10 km west of map area.

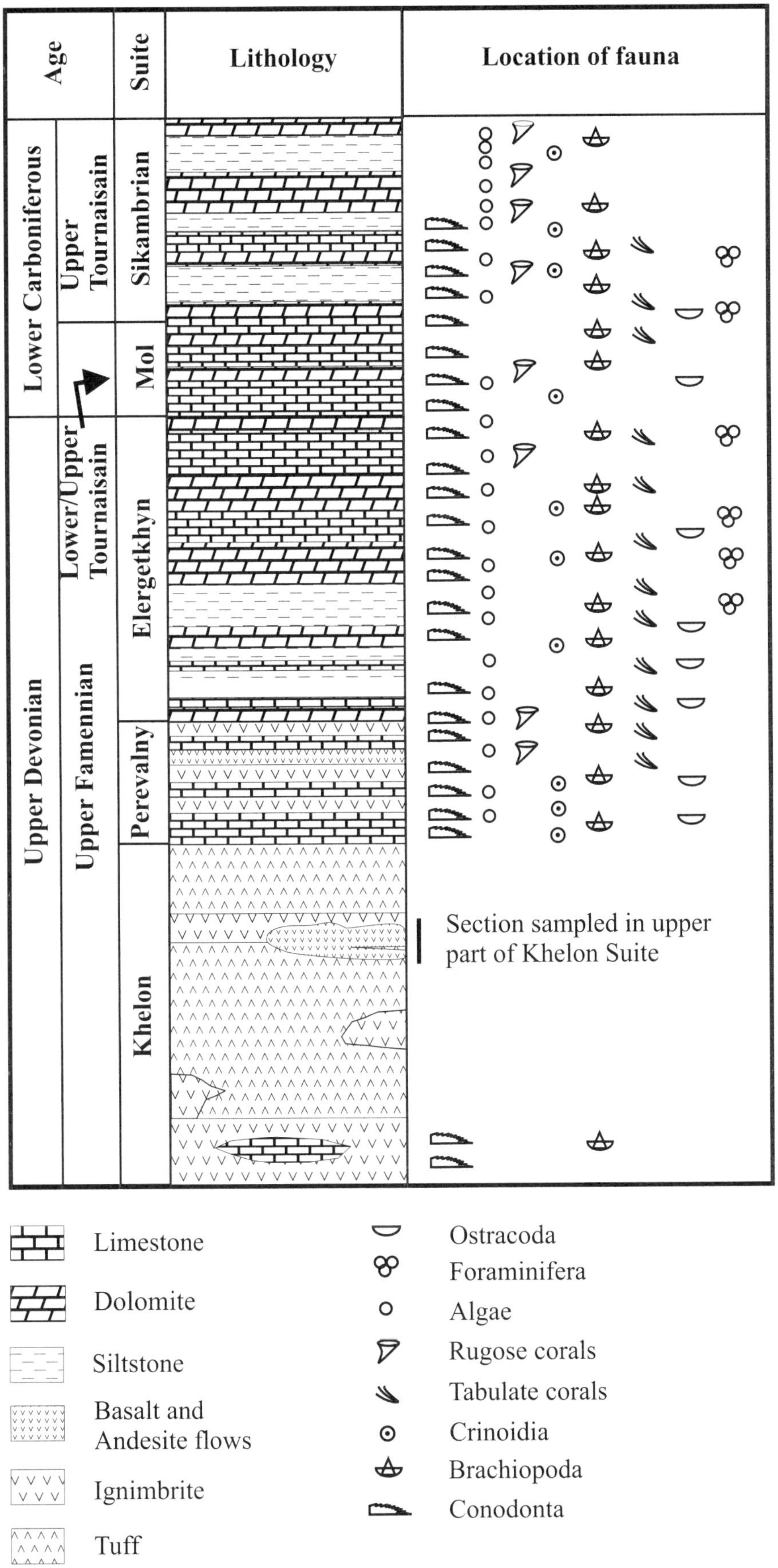

Figure 4. Composite lithostratigraphic column for Mol River basin (which includes Perevalny Creek area) modified from Gagiev (1996). Section sampled was from volcanic rocks of Khelon suite; age control was from fossil assemblages collected from underlying, overlying, and laterally equivalent sections (see text). Paleomagnetic data from Khelon suite (this paper and Kolosev, 1981) and from Perevalny, Elgergetkhyn, and Sikambrian suites (Kolosev, 1981) are listed in Table 4 and shown in Figure 8.

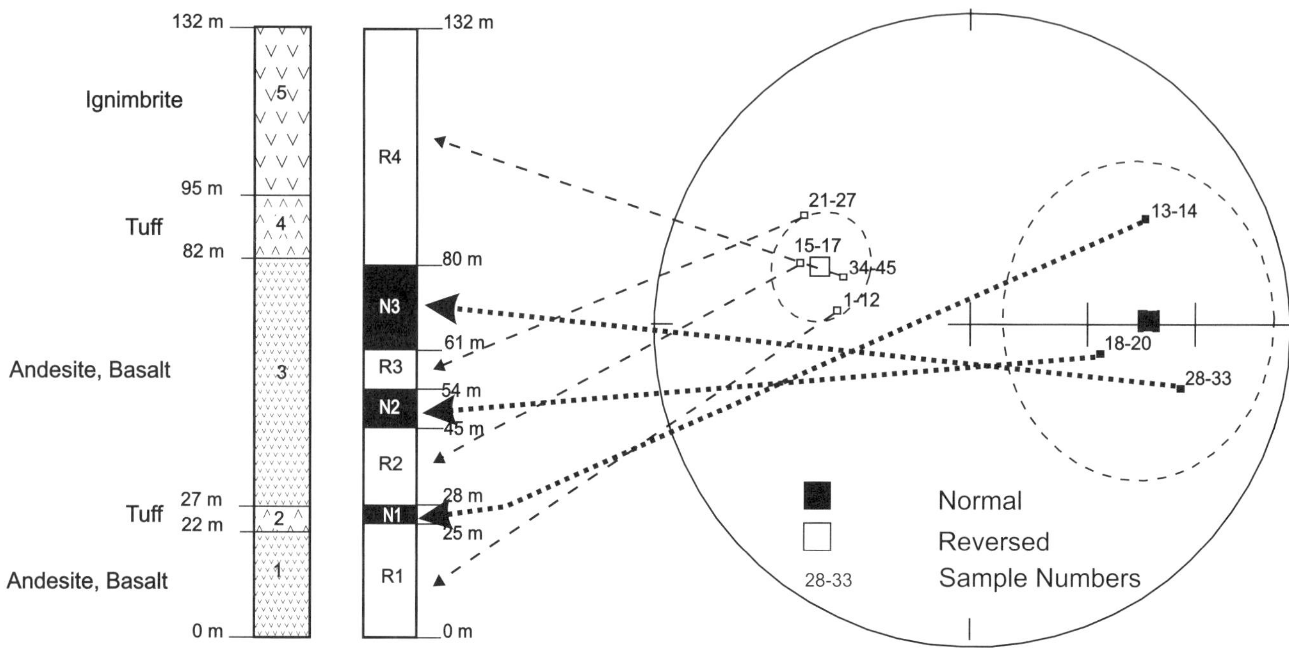

Figure 5. Mean paleomagnetic vector directions in stratigraphic (tilt-corrected) coordinates for reverse (R) and normal (N) sections from Khelon suite of Famennian age (at Perevalny Creek); with R and N means and their circles of confidence (α_{95}), and correlation between magnetostratigraphy and rock type are shown.

unit 1. The lower layer is 4–5 m thick and consists of dark brown to dark green basaltic andesite. The lowermost part of this flow is scoriaceous, but it becomes more massive and amygdaloidal upward. The amygdules increase in size and become more abundant toward the top. The upper layer is ~50 m thick. Its lowermost portion (0.5 m) consists of black scoriaceous loose rock that grades upward into dark green to reddish rock of basic (?) composition. The middle portion consists of dark brown to greenish massively amygdaloidal basaltic andesite with spheroidal joint sets to 5 m in diameter. Amygdules make up 30%–35% by volume of the top of the unit.

Unit 4 is ~13 m thick and composed of psammitic tuff that grades upward into psephitic tuff of mostly basic composition. The weathered surfaces of these rocks are light green, but dark brown where fractured. Angular rock fragments in the tuff are brown andesite, more rarely pink-red rhyolite with inclusions of hydrous quartz and pink feldspar.

Unit 5 is ~37 m thick and composed of andesite-dacite ignimbrites that are light brown on weathered surfaces and dark brown where fractured. These rocks are mostly vitreous and contain pieces of welded tuff as long as a few centimeters. They also contain 2–3-mm-sized clasts of andesite and less commonly rhyolite and granitoids that make up 5%–15% of the rock volume. The rock fragments are commonly concentrated in interbeds.

The uppermost parts of the volcanogenic section are not exposed in the sampling area, but can be seen in neighboring sections. They consist of intercalated tuffaceous sandstone and con-

glomerate. The conglomerates contain pebbles with the same red coloring as the rocks of the sampled section of the Khelon suite, together with material similar to the underlying Pylkatveem suite, and unaltered fragments of other rock types. Petrographic analyses of the rock samples from the Khelon suite indicate that the red color was due to significant alteration that occurred during or shortly after emplacement, thus the red pebbles were considered suitable for a conglomerate test.

The age of the Khelon suite was determined from conodonts obtained from the base of the Perevalny suite and from laterally equivalent sections in which sediments are exposed above and below the stratigraphic level of the volcanic suite within the Khelon suite. These conodonts place the sampled section in the standard conodont *trachytera* zone (Gagiev, 1996), of early–late Famennian age (362.5–365 Ma; Harland et al., 1990).

The depositional environment for the sediments of the Perevalny suite, which overlies the volcanic rocks of the Khelon suite, is nearshore; both shallow-water and terrestrial sediments are represented. These clastic rocks are in turn overlain by Upper Devonian through Mississippian carbonates of the Elergetkhyn, Mol, and Sikambrian suites. Age control for the sampled section is based on poorly preserved brachiopods of the lower–upper Famennian *trachytera* zone recovered from calcareous sandstones within the Khelon suite (Gagiev, 1996), consisting of *Productella subaculeata* (Murchison), *Productellana binaries* (Stanbrook), *Gastrodetoechia utahensis muolensis* (Rzon), *Mucrospirifer posterus* (Hall), *Euritatospirifer tionesta*, and *E. spicatus, Valisodpirifer validus* (Simakov, 1979), accompanied by the tabulate corals *Syringopora intermixta, S. fragilis, and S. hecheri* (Smirnova, 1979). Conodonts were identified in the sediments of the Perevalny suite, which concordantly overlies, or overlaps, the predominantly volcanogenic Khelon suite. These include *Scaphignathus vilefer, Polygnathus semicostatus, P. obliquicoatatus, Neoicriodus terminalis,* and *Icriodus* sp., which characterize the upper Famennian *postera* zone (Gagiev, 1982, 1996).

Samples from both extrusive and the pyroclastic rocks of the Khelon suite were dated by the ^{40}Ar/^{39}Ar method in the geochronology laboratory of the University of Alaska. Good plateau ages of ca. 120 Ma were obtained; occasional samples yielded ages ranging to 230 Ma (P. Layer, 1999, personal commun.). The samples dated show evidence that they contained considerable amounts of glass when they were emplaced, and it is thought that a low-level metamorphic event at 120 Ma was sufficient to release the trapped argon from the glass, thus resetting the argon clock. A remagnetization signature removed by thermal demagnetization between ~200 and 300 °C suggests that the event was low temperature. The 120 Ma age may be an indicator of the timing of the pervasive remagnetization event seen throughout much of northeastern Russia (Stone et al., 1994).

PALEOMAGNETIC RESULTS

The results described in the following sections were all obtained by one of us (Kolesov) and colleagues using the paleomagnetic laboratory of the Northeast Interdisciplinary Scientific Research Institute in Magadan. The equipment used includes a JR-4 spinner magnetometer and KLY-2 kappa bridge made by Geofyzika (now AGICO), together with several standard thermal and alternating field (AF) demagnetizing devices set up in magnetic field free spaces maintained by Helmholtz coil systems. Oriented block samples collected in the field were cut into 2 cm cubes in the laboratory. Because several samples were available from each block, it was possible to carry out both thermal and AF demagnetizations on most of the samples. Isothermal remanent magnetizations (IRM) were obtained in fields to ~0.8 T (8000 G), examples of which are shown in Figure 6.

The intensity of magnetization (J_n) of the samples is in the range $10–140 \times 10^{-3}$ A/m; the average value is 26.6×10^{-3} A/m. The magnetic susceptibility is in the range $10–150 \times 10^{-5}$ SI with an average value of 34×10^{-5} SI. The mean magnetic susceptibility of the Khelon suite volcanic rocks is much smaller than that found in other regions of the Omolon massif (Pechersky and Yakupov, 1967). The presence of hematite in the rocks is indicated by the thermal demagnetization and IRM acquisition curves shown in Figure 6. On the basis of petrographic analyses and the observation that similar red-colored rock fragments are found in nearby conglomerates (Lychagin, 1976), the hematite formed synchronously with emplacement of the volcanic rocks.

To measure the direction of the original magnetization, the samples were demagnetized using both thermal and AF demagnetization techniques. Viscous magnetization accounts for

~20% of the initial intensity of magnetization (J_n). Optimal destruction of the magnetization took place in the range of 10–30 mT (100–300 G) and 300–400 °C. Possible changes in the magnetic minerals during heating were checked by measuring the magnetic susceptibility after each stage of heating. Line-fitting techniques were used to determine the vectors representing the high-temperature components.

Thermal demagnetization was more effective than alternating magnetic fields. The thermomagnetic analyses show the presence of hematite with a blocking temperature, T_b, of 675 °C, and an applied magnetic field in excess of 100 Am^{-1} failed to reach saturation magnetization. As can be seen from Figures 6 and 7, the rocks also contain hydrous iron (50–150 °C) and titanomagnetite (T_b 350–550 °C). Titanomagnetite was also confirmed by conventional mineralogic analysis as an accessory mineral with octahedral and rhombic dodecahedral crystals (Lychagin, 1978).

Analysis of the magnetic components seen during thermal demagnetization indicates that there are at least three components of magnetization in most of the samples (Fig. 7).

1. A low-temperature component is destroyed by thermal demagnetization at temperatures <~200–300 °C. This component appears to represent an overprint common to the entire sampling area and is characterized by a generally steeply dipping magnetization in the geographic reference frame with widely dispersed azimuths.

2. An intermediate-temperature component is sometimes seen between 300 and 450 °C, and has generally low, and commonly negative magnetic inclination and a general northeast declination in stratigraphic coordinates. It represents an overprinting

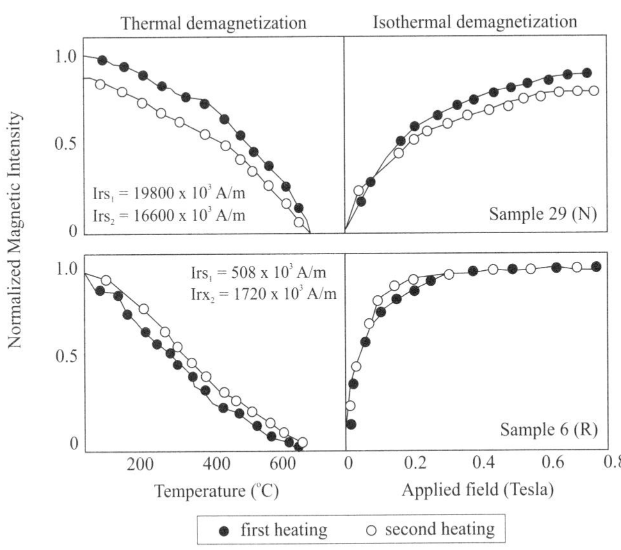

Figure 6. Data from two samples of volcanic flows (samples 6 and 29) showing acquisition of isothermal remanent magnetization followed by thermal demagnetization (dots); entire process is repeated second time (circles). Upper sample (29) shows clear evidence of hematite, and lower sample (6) shows range of titanomagnetites with some hematite.

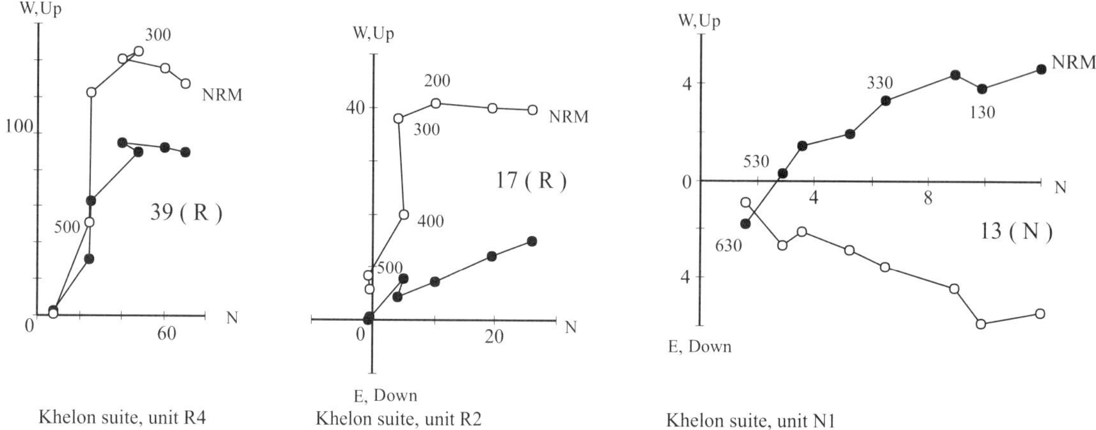

Figure 7. Zijderveld orthogonal plots of thermal demagnetization of samples 39 and 17 from reversely (R) magnetized units R_4 and R_2, and normally (N) magnetized sample 13 from unit N_1 (Table 1). All plots are in stratigraphic coordinates. Dots represent projection of declination onto horizontal plane, and circles projection of inclination onto north-south vertical plane. X and Y axes both show intensities of magnetization in mA/m, and labels on individual points indicate demagnetizing temperature (in °C). Both overprint and characteristic remanent magnetizations (ChRM) can be seen in these plots. All samples were subjected to similar demagnetization regimes.

field with shallow to intermediate dips and azimuths that are generally in the northwest quadrant in geographic coordinates.

3. A high-temperature component is commonly seen between 500 and 630 °C. This component is considered to represent the characteristic magnetization (ChRM) (see Table 1).

The overprinting magnetic vectors are not defined well enough to calculate a useful mean, but the low-temperature component averages to a steep inclination in geographic coordinates.

Examples of the magnetic components are shown in Zijderveld plots (Fig. 7). Thermal demagnetization of sample 39 from the R_4 reversely magnetized interval gives a component below 300 °C with a direction (declination, $D_s = 8°$, inclination, $I_s = 24°$; $D_g = 74°$, $I_g = 64°$, where the subscripts s and g represent data in stratigraphic [s] and geographic [g] coordinate systems) that is characteristic of the R_4 magnetic zone overprint. Demagnetization between 300 and 400 °C gives a component with $D_s = 303°$, $I_s = -31°$; $D_g = 305°$, $I_g = 24°$, and the intensity drops to ~80% of the initial NRM. For temperatures between 400 and 600 °C, a component with $D_s = 283°$ and $I_s = -63°$ is seen. Sample 17, from the R_2 magnetic interval, displays similar behavior. Within the temperature interval 20–300 °C, a magnetic direction $D_s = 332°$, $I_s = -2°$; $D_g = 323°$, $I_g = 62°$ is found. Between 300 and 500 °C, the trend of the successive demagnetizations displayed on the Zijderveld plot turns sharply toward the origin ($D_s = 319°$, $I_s = -79°$), and is thus considered the characteristic magnetization.

Samples from magnetic zones with normal polarity display a generally similar behavior under thermal demagnetization. Thus, in sample 13 (N_1 magnetic interval) increasing demagnetization temperatures to 330 °C give a component (shown in stratigraphic coordinates in Fig. 7) that is similar to that seen in

the reversely magnetized intervals, i.e., it is close to the overprinting field ($D_s = 348°$, $I_s = 21°$; $D_g = 49°$, $I_g = 82°$). An intermediate-temperature component ($D_s = 301°$, $I_s = -23°$; $D_g = 298°$, $I_g = 30°$) is isolated between 430–530 °C and a high-temperature component is seen between 530–630 °C ($D_s = 302°$ and $I_s = 36°$). The final intensity after demagnetization at 630 °C is still almost 20% of the initial NRM.

Essentially all samples investigated are characterized by a so-called incomplete demagnetization because they are not completely demagnetized at the maximum temperatures applied (600–630 °C). This indicates the presence of a high-temperature ferromagnetic mineral in the rocks. This is presumed to be the syngenetic hematite that has been detected petrographically and is in the host rock.

The paleomagnetic measurements show that both normal and reversed polarities are present in the Khelon suite, and that they occur in a pattern of three normal and four reversed zones in the section sampled (Table 1; Fig. 5). Each polarity interval except N_1 is represented by several samples. N_1 is represented by a normally magnetized flow unit (sample 14) and an underlying normally magnetized psammitic tuff (sample 13). The characteristic magnetization vector (ChRM) for the tuff is displaced from the mean, and is of opposite polarity from the tuffs immediately below it (samples 9–12). We therefore interpret these data as representing a remagnetization of the uppermost tuff by the overlying flow, and have not included it in the final data set. If this interpretation is correct, it provides a (poorly defined) contact test.

The mean directions of the reversed and normal sites are antipodal within the 95% confidence limits, but the circle of confidence for the normal samples is rather large, in part because

TABLE 1. DEMAGNETIZED PALEOMAGNETIC DATA FOR PEREVALNY CREEK SAMPLES SORTED BY STRATIGRAPHY AND POLARITY

Polarity	Sample	N	D_g	I_g	K_g	α_g	D_s	I_s	K_s	α_s
R_4	45–34	12	313.7	1.9	9.1	15.2	290.6	−55.1	9.7	14.7
N_3	33–28	6	111.0	−16.2	5.4	31.7	107.1	31.7	8.0	25.2
R_3	27–21	7	307.8	20.9	6.0	27.0	303.6	−37.7	5.8	27.4
N_2	20–18	3	92.0	−32.8	1.3	90+	103.0	55.7	5.9	56.2
R_2	17–15	3	302.8	11.5	6.4	53.5	290.1	−42.9	6.3	54.1
N_1	14–13	1	113.2	12.5	—	—	74.0	50.0	—	—
R_1	12–1	12	300.2	−22.6	5.3	20.9	276.0	−55.6	18.8	10.3
Mean of samples (corrected to normal (N) polarity)										
	1–45	44	303.2	−1.9	5.6	10.1	106.2	49.5	8.5	7.8
Mean of Polarity Zones										
		7	297.8	7.2	12.4	17.8	104.9	47.9	33	10.7

Virtual Geomagnetic Poles	N	Lat	Long	d_p	d_m
Mean of samples	44	20.6	224.0	6.9	10.4
Mean of Polarity Zones	7	20.1	226.2	9.1	14.0

Note: Figure 4 gives the locations of the sampling points. R_1 to R_4 and N_1 to N_3 are the polarity zones based on the samples listed. These were block samples from which multiple specimens were cut. N = the number of Characteristic Remanent Magnetization (ChRM) values used. D_g, I_g, K_g, α_g are the mean declination, inclination, kappa value and alpha 95 (Fisher, 1953) in geographic coordinates; D_s, I_s, K_s, α_s are the equivalent values in stratigraphic (tilt-corrected) coordinates. d_p and d_m represent the oval of uncertainty in the determination of the virtual geomagnetic poles.

it only contains three localities. When a reversal test based on McFadden and McElhinny (1990) is applied to all the samples, they show a positive reversal test (see Table 2).

The combined samples also pass a whole sampling area tilt test with a ratio of geographic (with respect to present-day horizontal and north) to stratigraphic (tilt-corrected) kappa confidence parameters of $K_s/K_g = 1.52$. Two fold tests (McFadden and Jones, 1981; McElhinny, 1964) were also applied to the characteristic remanent magnetizations from the lower part of the section ($K_s/K_g = 3.5$), the rest of the section being essentially monoclinal. Both tests were positive (Table 3).

The rocks pass a conglomerate test based on the study of pebbles from a conglomerate unit in the Perevalny suite, which overlies the Khelon suite. This conglomerate consists of debris from the Khelon suite rocks and the characteristic magnetizations have a distribution not significantly different from random ($K = 1.4$). In contrast, the mean of the secondary magnetizations has a distribution represented by $K = 9$.

The data derived from this study are listed in Tables 1–3, and are compared with data from the same region (Fig. 4) (Kolosev, 1981) in Table 4. These data sets show clear similarities. When compared with apparent polar wander paths for the Siberian platform (Didenko and Pechersky, 1993), a counterclockwise rotation of ~90° with little or no relative latitude change is indicated (Fig. 8). Using the 360 Ma pole for Siberia (Didenko and Pecherski, 1993) (Table 5), the displacement between the observed paleolatitude of the sampling area and the paleolatitude it would have had it had remained in its present position with respect to Siberia is −6.7° ± 7.6°.

Although the similarity between the paleolatitudes obtained for the Omolon terrane and the Siberian platform suggests that they may have been adjacent to one another in Famennian time (364 Ma), there is no longitude control on their relative positions, and thus several different paleogeographies are still possible. However, the simplest reconstructions place the Omolon massif adjacent to Siberia in Devonian time, which leaves the question of whether a Devonian rifting event (Gagiev, 1995, 1996) was followed by significant southward motion, as implied by the data from the Rassokha terrane (Iosofidi, 1989), or by an apparently continuous southward trend that ended in latest Triassic–earliest Jurassic time (Fig. 9).

TABLE 2. REVERSAL TEST FOR PEREVALNY CREEK SAMPLES

	R	N	D	I	K
Normal	8.76	10	97.3	36.07	7.2
Reversed	30.47	34	288.8	51.3	9.3

Note: R—resultant of N unit vectors; N—the number of ChRM vectors used; D, I, K represent the declination, inclination, and kappa value for each mean. Angle between reversed and normal directions: $Y_o = 17.4$. Critical angle between reversed and normal directions: $Y_c = 18.5$. $Y_c > Y_o$ thus reversal test is positive.

TABLE 3. FOLD TEST FOR PEREVALNY CREEK SAMPLES

MCELHINNY (1964)

Bedding Azimuth = 340E, Dip = 65					Bedding Azimuth = 20E, Dip = 20				
Sample*	D_g	I_g	D_s	I_s	Smpl.#	D_g	I_g	D_s	I_s
8	293.8	−18.3	247.1	−52.8	12	306.7	−69.1	254.1	−68.2
7	271.1	10.8	266.1	−20.1	11	325.1	−39.8	306.9	−51.2
6	329.2	−0.5	321.8	−65.8	10	305.1	−54.1	275.8	−56.1
5	302.9	−16.1	254.2	−60.5	9	301.5	−55.8	269.9	−57.1
4	305.1	7.9	289.8	−38.1					
3	235.7	−14.1	256.1	−53.8					
2	324.2	−2.1	311.8	−65.1					
1	300.8	−3.1	273.9	−51.8					

MCFADDEN AND JONES (1981)

R_a[†]	N_a[§]	R_b[†]	N_b[§]	R_{ab}[#]	Observed	Critical
7.514	8	3.916	4	11.414	0.029	0.349

Note: Fold Test: D_g = 300.2; I_g = −22.7; N = 12; K_g = 5.3; $\alpha95_g$ = 20.9; D_s = 275.9; I_s = −55.6; N = 12; K_s = 18.9; $\alpha95_s$ = 10.3; K_s/K_g = 3.5. For a positive fold test with N = 12 and a 95% confidence level, K_s/K_g must exceed 2.05. Thus this test is positive.

*Sample numbers used for the fold test.

[†]R_a and R_b are the lengths of the resultant vectors from each limb of the fold.

[§]N_a and N_b are number of sites from each limb

[#]R_{ab} is total length of the resultant vector from both limbs of the fold.

The observed value does not exceed critical value, thus the test is positive.

COMPILATION AND EVALUATION OF PREVIOUS PALEOMAGNETIC DATA FROM THE KOLYMA-OMOLON SUPERTERRANE

A number of paleomagnetic studies have been reported for the Kolyma-Omolon superterrane, and many data have been taken from the Omulevka and Omolon terranes. These data have been interpreted using the assumption that the Omolon crystalline basement originated as part of the Siberian craton that rifted away, perhaps as early as late Precambrian, or perhaps as late as Devonian time. The earlier interpretations of the paleo-

TABLE 4. VIRTUAL GEOMAGNETIC POLES FOR PALEOZOIC ROCKS FROM THE PEREVALNY CREEK AREA (LONG 162°E, LAT 65°N)

Age	Suite	N	VGP Lat	VGP Long	d_m	d_P
C_1	Sikambrian	20	25	224	10.5	7.3
D_3	Elergetkhyn	10	34	232	12.5	9.2
D_3	Perevalny	10	21	222	13.7	9.3
D_3	Khelon	10	26	212	25.3	18.7
D_3 (this study)	Khelon	44	20.6	224	10.4	6.9

Note: See Figure 3 for locations of sample sets, Figure 4 for generalized stratigraphy. C_1, Early Carboniferous; D_3, Late Devonian. All data shown (excluding the present study) are from Kolosev (1981). VGP is virtual geomagnetic pole.

magnetic data suggest that the Omolon terrane moved several thousand kilometers south with respect to the craton, then drifted back to roughly its present relative position in Early to Middle Jurassic time (e.g., Khramov, 1991; Neustroev et al., 1993). However, much work has been published during the past decade on the lithofacies, tectonic structure, biostratigraphy, and paleomagnetism of northeastern Russia; these works appear to tie the Omolon massif, and Prikolyma and Omulevka terranes to the Siberian platform from Ordovician or earlier time until at least Devonian time (Bondarenko and Didenko, 1997; Gagiev, 1990, 1995, 1996; Rodionov, 1991; Rodionov and Shemyakin, 1988; Savostin et al., 1993, 1994; Sokolov et al., 1997).

The earlier models (Khramov, 1991) were based on matching APWPs, but gave few details of the magnetic data from which the APWPs were drawn. A later model (Neustroev et al., 1993) based the relative latitudinal motions on paleomagnetic measurements emphasizing paleolatitudes, but the details of the data were sparse. A compilation (Stone et al., 1994) derived from data sets for which there is sufficient background information to allow a determination of the overall quality of the results was included in Nokleberg et al. (1998). (For the tabulated data set, see http://www.gi.alaska.edu/TSRG/Research/NPacificPaleomag/index.html.)

The individual results were divided into four groups, the A group being the most reliable.

A. Multiple demagnetization steps must have been applied to most or all of the samples, and positive fold or reversal tests quoted together with the relevant statistics. In some cases other

Figure 8. Apparent polar wander path for Siberian platform from 480 to 240 Ma from Didenko and Pechersky (1993). Devonian pole for Khelon suite reported in this chapter is labeled "this study." Data from same area in Omolon massif (Kolosev, 1981) are labeled C_1 Sik (Sikambrian suite), D_3 Ele (Elergetkhyn suite), D_3 Per (Perevalny suite), and D_3 Khe (Khelon suite). D_3 represents Upper Devonian and C_1 represents Mississippian. All circles of confidence are based on α_{95} values quoted by original authors (Tables 1, 4, and 5).

evidence of positive stability tests was accepted (e.g., a conglomerate test or a baked contact test) provided there was an adequate statistical analysis to validate the test.

B. Multiple demagnetization steps have been applied to a significant number of the samples, and the author claims to have established positive stability for the magnetization.

C. Either single-level blanket demagnetization techniques have been applied and the author claims positive stability tests but has supplied no supporting data, or normal and reversely magnetized samples were combined without any quantitative evidence that the magnetizations are antipodal. Group C was also assigned in cases where ancient horizontal was poorly delimited, e.g., as in data from intrusive rocks.

D. The author claims to have data that represent the paleomagnetic magnetic field, but gives no supporting evidence. Group D is also used for cases where the data are apparently clean, but normal stability tests are not possible.

All other data in the literature were rejected until such time as positive magnetic stability tests of some type are available.

Paleolatitudes meeting the foregoing criteria are plotted against geologic time for the individual terranes of the Kolyma-Omolon superterrane in Figure 9. The face value interpretation of the data shown in this figure is that the Omolon terrane was at a latitude of ~40°N in latest Devonian–earliest Carboniferous time (360 Ma). The choice of a location in the Northern Hemisphere is based on similarities of the fossil assemblages of Devonian age in the Omulevka terrane with those found in rocks of similar age that are clearly part of the Siberian platform (Gagiev, 1990, 1995, 1996). However, a tie between the Omulevka and Omolon terranes for Devonian time is difficult to verify. By one interpretation of the data, the Kolyma-Omolon superterrane moved rapidly southward to a much lower paleolatitude in Mississippian time (ca. 350 Ma), then moved slowly northward to its present latitude by Middle Jurassic time. However, the very rapid latitudinal change in this scenario is largely controlled by one data point representing Carboniferous time (Iosofidi 1988, 1989). This point comes from samples from the Rassokha terrane (Nokleberg et al., 1994) and passes the selection criteria in the B group, but we cannot at this time support or refute the data. One interpretation is that the sampled sections from the Ras-

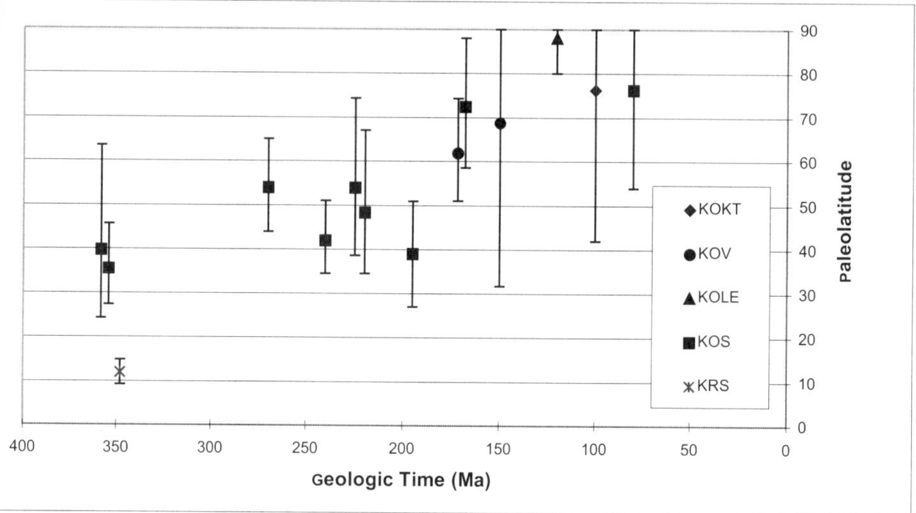

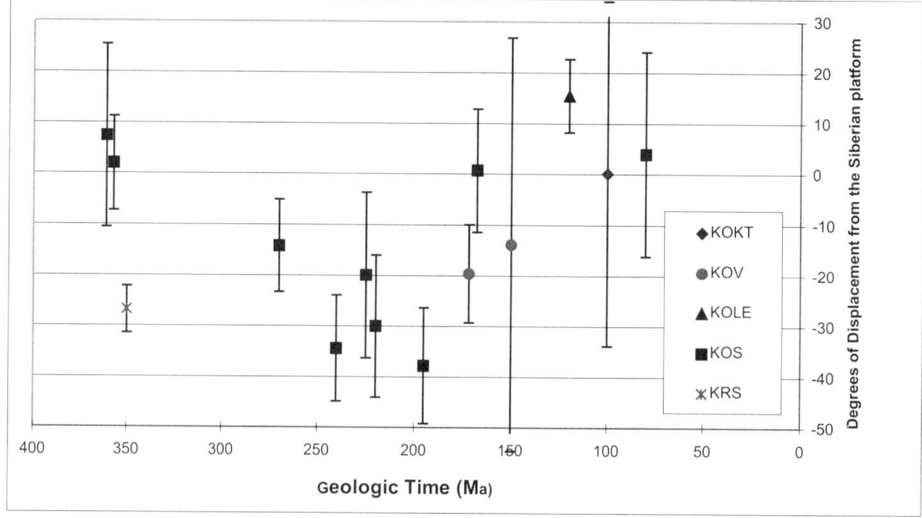

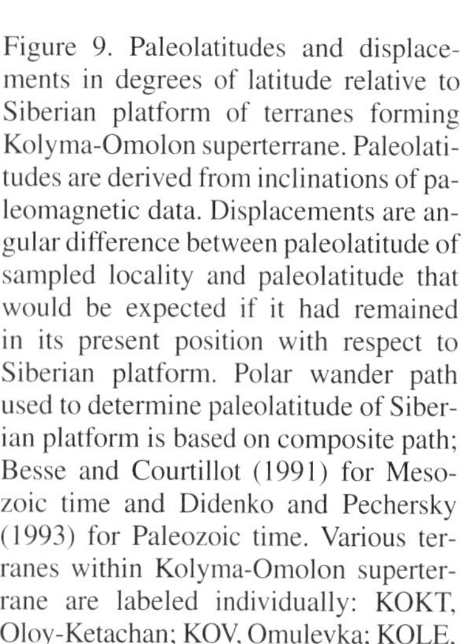

Figure 9. Paleolatitudes and displacements in degrees of latitude relative to Siberian platform of terranes forming Kolyma-Omolon superterrane. Paleolatitudes are derived from inclinations of paleomagnetic data. Displacements are angular difference between paleolatitude of sampled locality and paleolatitude that would be expected if it had remained in its present position with respect to Siberian platform. Polar wander path used to determine paleolatitude of Siberian platform is based on composite path; Besse and Courtillot (1991) for Mesozoic time and Didenko and Pechersky (1993) for Paleozoic time. Various terranes within Kolyma-Omolon superterrane are labeled individually: KOKT, Oloy-Ketachan; KOV, Omulevka; KOLE, Eropol; KOS, Omolon; KRS, Rassokha (Nokleberg et al., 1994).

sokha terrane have acquired a significant unrecognized secondary magnetization. Another is that the Rassokha terrane had a different and separate travel history from the rest of the Omulevka-Omolon composite terrane. This is suggested by some basic differences between the Ordovician stratigraphy of the Rassokha terrane and the rest of the Omulevka-Omolon composite terrane.

Merslyakov (1971) reported that a complete Ordovician section is present in the Rassokha block. This section is composed mainly of volcanic rocks (trachybasalts and their tuffs and trachyrhyolites) up to 3000 m thick that may have originated in a rifting event (Natapov et al., 1977). There are as yet no reliable paleomagnetic data from these rocks. Zonenshain et al. (1990a) noted that this Ordovician section is very different from the carbonate and carbonate-schist strata of the same age in the Omulevka block. In contrast, other geologic data have been interpreted to indicate that the Rassokha terrane was part of the Kolyma-Omolon superterrane as it developed in close associa-

tion with the Siberian craton (Bondarenko and Didenko, 1997; Gagiev, 1990, 1995, 1996; Rodionov, 1991; Savostin et al., 1993, 1994; Sokolov et al., 1997).

Unlike the data from the Rassokha terrane, those from the Devonian of the Omolon massif reported in this chapter clearly describe a primary magnetization. If both data sets represent primary magnetizations, and both were in their present relative positions, then the combined data sets require a very rapid post-Devonian southward migration of the composite terrane.

An alternative way of displaying the paleolatitude data is to compare them with the paleolatitudes that would be expected if the terranes in question had remained fixed with respect to the Siberian platform, and interpret the differences as displacements (Fig. 9). To calculate the displacement we use a composite APWP derived from the Eurasian path for Mesozoic and later time (Besse and Courtillot, 1991), and the compilation of Didenko and Pechersky (1993) for Paleozoic time. The displacement curve indicates that the Omolon block was not latitudinally displaced with respect to Siberia in Devonian time. If the data from the Rassokha terrane are excluded, then a steady southward displacement of the superterrane with respect to the Siberian platform can be inferred. This displacement reached ~40° in latest Triassic–earliest Jurassic time, and was followed by northward relative motion of the Kolyma-Omolon superterrane with respect to the Siberian platform until Middle Jurassic time, after which there was little or no relative displacement. The inclusion of the Rassokha terrane gives extraordinary rates of displacement in latest Devonian–earliest Carboniferous time, followed by little displacement until Permian time, after which it followed the northward migration just described.

There is considerable geological evidence indicating extension in Siberia in mid-Paleozoic (Devonian?) time (Khramov, 1988; Parfenov, 1984; Zonenshain et al., 1990a). It seems probable that the eastern margin of the Siberian craton was involved in the pervasive Devonian rifting recognized in many parts of the world. These observations suggest that the Kolyma-Omolon superterrane may have started to rift away from the Siberian craton in Devonian time and thus has paleolatitudes that match those determined for the craton.

CONCLUSIONS

A paleomagnetic study of a suite of volcanic and volcanogenic rocks from the Omolon-Omulevka composite terrane gives apparently good results. The data obtained pass the standard fold, reversal, and conglomerate tests. The mean directions obtained indicate a paleolatitude of ~30°N, which is very similar to that determined for nearby parts of the Siberian platform. This fits well with, but does not prove, the models that claim that the Omolon (and by implication the Okhotsk) continental fragment terranes were once part of the Siberian craton, were rifted away in Devonian time, and had an independent history until accretion to the Siberian platform in Middle Jurassic time. More data are needed for late Paleozoic time before questions relating to the extent of the travels of these terranes can be answered with confidence.

ACKNOWLEDGMENTS

We thank Pavel Minyuk in Magadan and Stephen Crumley in Fairbanks for their help in many aspects of this paper; Paul Layer for the ^{40}Ar/^{39}Ar age data; and W. Harbert, J. Hillhouse, and the editors of this GSA Special Paper for their helpful comments on the manuscript. This work was supported by Russian Fund of Basic Research (RFBR) grant 98-05-65472, North Atlantic Treaty Organization grant SA.12-2-02 (ENVIR.LG 930919), and U.S. National Science Foundation grant OPP-9500241.

REFERENCES CITED

Besse, J., and Courtillot, V., 1991, Revised and synthetic polar wander paths of the African, Eurasian, North American and Indian plates, and true polar wander since 200 Ma: Journal of Geophysical Research, v. 96, p. 4029–4050.

Bondarenko, G.Ye., and Didenko, A.N., 1997, New geological and paleomagnetic data on Jurassic-Cretaceous history of the Omolon massif: Geotectonics, no. 2, p. 14–27 (in Russian).

Burke K., 1984, Plate tectonic history of the Arctic: 27th International Geological Congress, Arctic Geology section 4, Papers, Volume 4: Moscow, Nauka, p. 159–167 (in Russian).

Chekhov, A.D., 1990, Structure and development of the Mesozoides of the northeast USSR, Part 2: Magadan, Russia, Northeast Interdisciplinary Scientific Research Institute (SVKNII), 79 p. (in Russian).

Churkin, M., and Trexler, J.H., 1981, Continental plates and accreted oceanic terranes in the Arctic, *in* Nairn, A.E.M., Churkin, M., and Stehli, F.G., eds., The Arctic Ocean: New York, Plenum Press, The Ocean Basins and Margins, v. 5, p. 1–20.

Coney, P.C., Jones, D.L., and Monger, J.W.H., 1980, Cordilleran suspect terranes: Nature, v. 288, p. 329–333.

Didenko, A., and Pechersky, D., 1993, Revised Paleozoic apparent polar wander paths for E. Europe, Siberia, N. China and Tarim Plates: Program with Abstracts for L.P. Zonenshain Memorial Conference on Plate Tectonics, Moscow, p. 47.

Engebretson, D., Cox, A., and Gordon, R.C., 1985, Relative motions between oceanic and continental plates in the Pacific Basin: Boulder, Colorado, Geological Society of America Special Paper 106, 59 p.

Fisher, R.A., 1953, Dispersion on a sphere: Proceedings of the Royal Society, London, v. A217, p. 295–305.

Fujita, K., Stone, D.B., Layer, P.W., Parfenov, L.M., and Koz'min, B.M., 1997, Cooperative program helps decipher tectonics of northeastern Russia: Eos (Transactions, American Geophysical Union), v. 78, p. 245, 252–253.

Gagiev, M.Kh., 1982, Conodonts of Upper Famennian and Tournaisian sediments of the northeastern part of the Omolon massif: Summary of Kandidat in Geological and Mineralogical Sciences dissertation: Moscow, Moscow State University, 25 p. (in Russian).

Gagiev, M.Kh., 1990, Stratigraphy of the Middle Paleozoic and the concept of heterogeneity of the modern structure of the northeast USSR: Tectonics and mineralogeny of the northeast USSR: Magadan, Russia, Northeast Interdisciplinary Scientific Research Institute (SVKNII), p. 45–47 (in Russian).

Gagiev, M.Kh., 1995, Stratigraphy of the Devonian and Lower Carboniferous of the Omulevka uplift (north-east Asia): Magadan, Russia, Northeast Interdisciplinary Scientific Research Institute (SVKNII), 196 p. (in Russian).

Gagiev, M.Kh., 1996, Middle Paleozoic of northeastern Asia: Magadan, Russia, Northeast Interdisciplinary Scientific Research Institute (SVKNII), 120 p. (in Russian).

Harland, W.B., Armstrong, R.L., Cox, A.V., Craig, L.E., Smith, A.G., and Smith, D.G., 1990, A geologic timescale 1989: Cambridge, Cambridge University Press, p. 263.

Iosifidi, A.G., 1988, New paleomagnetic data on the Kolyma massif (Arga-Tas Range), *in* Khramov, A.N., ed., Paleomagnetism and accretionary tectonics: Leningrad, All Union Oil Scientific Research Institute for Geology and Exploration (VNIGRI), p. 104–123 (in Russian).

Iosifidi, A.G., 1989, Paleomagnetic direction and pole positions: Data for the USSR: Moscow, Soviet Geophysical Committee: World Data Center-B, Catalog, Issue 7 (in Russian).

Jones, D.L., Irwin, W.P., and Ovenshine, A.T., 1972, Southeastern Alaska: A displaced continental fragment?: U.S. Geological Survey Professional Paper 800–B, p. 211–217.

Jones, D.L., Silberling, N.J., and Hillhouse, J., 1977, Wrangellia: A displaced terrane in northwestern North America: Canadian Journal of Earth Sciences, v. 14, p. 2565–2577.

Karasik, A.M., Ustritsky, V.I., and Khramov, A.N., 1984, The history of formation of the Arctic Ocean, *in* Proceedings of the 27th International Geologic Congress, Arctic Geology, Colloquium 4, Reports, Volume 4: Moscow, Nauka, p. 179–189.

Khramov, A.N., 1988, Paleomagnetism and problems of accretionary tectonics of the northwestern segment of the Pacific mobile belt, *in* Khramov, A.N., ed., Paleomagnetism and accretionary tectonics: Leningrad, All Union Oil Scientific Research Institute for Geology and Exploration (VNIGRI), p. 141–153 (in Russian).

Khramov, A.N., 1991, Standard series of paleomagnetic pole for plates of northern Eurasia: Connection with problems of paleogeodynamics for the territory of the USSR, *in* Paleomagnetism and paleogeodynamics of the territory of the USSR: Leningrad, All Union Oil Scientific Research Institute for Geology and Exploration (VNIGRI), p. 135–149 (in Russian).

Khramov, A.N., and Ustritsky V.I., 1990, Paleopositions of some northern Eurasian tectonic blocks: Paleomagnetic and paleobiologic constraints, *in* Van der Voo, R., and Schmidt, P.W., eds., Reliability of paleomagnetic data: Tectonophysics, v. 184, p. 101–109.

Khramov, A.N., Gurevich, E.L., Komissarova, R.A., Osipova, E.P., Pisarevsky, S.A., Radionov, V.P., and Slautsitais, I.P., 1985, Paleomagnetism, microplates and Siberian plate consolidation: Journal of Geodynamics, v. 2, p. 127–139.

Kolesov, E.V., 1981, Paleomagnetic characteristics of Middle Paleozoic sediments of the Omolon Massif, *in* Magmatism of rocks and paleomagnetic stratigraphy of East and Northeast Asia: Magadan, Russia: Northeast Interdisciplinary Scientific Research Institute (SVKNII), p. 68–74 (in Russian).

Kovalenko, D.V., and Remizova, L.L., 1997, Paleomagnetism of the northwestern Olyutorskii zone, southern Koryak upland: Izvestia, Physics of the Solid Earth, v. 33, p. 589–598.

Lawver, L.A., and Scotese, C.R., 1990, A review of tectonic models for the evolution of the Canada Basin, *in* Grantz, A., Johnson, G.L., and Sweeney, J.F., eds., The Arctic Ocean region: Boulder, Colorado, Geological Society of America, Geology of North America, v. L, p. 593.

Lawver, L.A., Srivastava, S.P., Fujita, K., Stone, D.B., and Embry, A., 1996, The tectonic development of the Canada Basin and surrounding basins: GSA Today, v. 6, p. 17–18.

Lychagin, P.P., 1976, Middle Paleozoic volcanism of Omolon massif: Summary of Kandidat in Geological and Mineralogical Sciences dissertation: Novosibirsk, Russia, Institute of Geology and Geophysics, 26 p. (in Russian).

Lychagin, P.P., 1978, Middle Paleozoic magmatism of Omolon massif: Moscow, All Union Institute for Scientific and Technical Information (VINITI), Deposition no. 496–78, 195 p. (in Russian).

McElhinny, M.W., 1964, Statistical significance of the fold test in palaeomagnetism: Geophysical Journal of the Royal Astronomical Society, v. 8, p. 338–340.

McFadden, P.L., and Jones, D.L., 1981, The fold test in paleomagnetism: Geophysical Journal of the Royal Astronomical Society, v. 67, p. 53–58.

McFadden, P.L., and McElhinny, M.W., 1990, Classification of the reversal test in paleomagnetism: Geophysical Journal International, v. 103, p. 725–729.

Merzlyakov, V.M., 1971, Stratigraphy and tectonics of the Omulevka uplift: Moscow, Nauka, 152 p. (in Russian).

Merzlyakov, V.M., Terekhov, M.I., Lychagin, P.P., and Dulevsky, Ye.F., 1982, Tectonics of the Omolon massif: Geotectonics, no. 1, p. 74–86 (in Russian).

Natapov, L.M., Zonenshain, L.P., and Shulgina, V.M., 1977, Geological development of the Kolyma-Indigirka region and the problem of the Kolyma massif: Geotectonics, no. 4, p. 18–31 (in Russian).

Neustroev, A.P., Parfenov, L.M., and Rodionov, V.P., 1993, Paleomagnetic data and nature of the Tas Kayakhtakh terrane, Verkhoyansk-Kolyma region: Geology and Geophysics, v. 34, no. 3, p. 25–37 (in Russian).

Nokleberg, W.J., Parfenov, L.M., Monger, J.W.H., Baranov, B.V., Byalobzhesky, S.G., Bundtzen, T.K., Feeney, T.D., Fujita, K., Gordey, S.P., Grantz, A., Khanchuk, A.I., Natal'in, B.A., Natapov, L.M., Norton, I.O., Patton, W.W., Jr., Plafker, G., Scholl, D.W., Sokolov, S.D., Sosunov, G.M., Stone, D.B., Tabor, R.W., Tsukanov, N.V., Vallier, T.L., and Wakita, K., 1994, Circum-North Pacific tectono-stratigraphic terrane map: U.S. Geological Survey Open-File Report 94–714, scale 1:5 000 000, 2 sheets, and scale 1:10 000 000, 2 sheets, 211 p.

Nokleberg, W.J., Parfenov, L.M., Monger, J.W.H., Norton, I.O., Khanchuk, A.I., Stone, D.B., Scholl, D.W., and Fujita, K., 1998, Phanerozoic tectonic evolution of the circum-north Pacific: U.S. Geological Survey Open-File Report 98–754, p. 125.

Parfenov, L.M., 1984, Continental margins and island arcs of the Mesozoides of northeast Asia: Novosibirsk, Russia, Nauka, 192 p. (in Russian).

Parfenov, L.M., 1991, Tectonics of the Verkhoyansk-Kolyma Mesozoides in the context of plate tectonics: Tectonophysics, v. 199, p. 319–342.

Pechersky, D.M., and Yakupov, V.S., 1967, Magnetization of magmatic rocks of northeast USSR: Scientific and Technical Information Division (ONTI), All-Union Scientific Research Institute for Economics of Mineral Raw Materials and Geologic Exploration Work (VIEMS), ser. 3, p. 22 (in Russian).

Rodionov, V.P., 1991, Kinematic models of the interrelation of Siberian platform and blocks of the Upper Kolyma region, *in* Khramov, A.N., ed., Paleomagnetism and paleogeodynamics of the territory of the USSR: Leningrad, All Union Oil Scientific Research Institute for Geology and Exploration (VNIGRI), p. 113–119 (in Russian).

Rodionov, V.P., and Shemyakin, Ye.V., 1988, Interrelations among some blocks of Siberian in the early Paleozoic and Silurian according to paleomagnetic data, *in* Khramov, A.N., ed., Paleomagnetism and accretionary tectonics: Leningrad, All Union Oil Scientific Research Institute for Geology and Exploration (VNIGRI), p. 123–127 (in Russian).

Savostin, L.A., Pavlov, V.E., Safonov, V.G., and Bondarenko, G.Ye., 1993, Lower and middle Jurassic deposits west of the Omolon massif: Conditions of sediment formation and paleomagnetism: Proceedings of the Russian Academy of Sciences, v. 333, p. 481–486 (in Russian).

Savostin, L.A., Bondarenko, G.Ye., Safonov, V.G., and Pavlov, V.E., 1994, Structural evolution of the south-western frame of the Omolon mass in Jurassic: Geotectonics, no. 5, p. 46–62 (in Russian).

Scotese, C.R., Nokleberg, W.J., Monger, J.W.H., Norton, I.O., Parfenov, L.M., Buntdzen, T.K., Dawson, K.M., Eremin, R.A., Frolov, Y.F., Fujita, K., Goryachev, N.A., Kanchuk, A.I., Podzeev, A.I., Ratkin, V.V., Rodinov, S.M., Rozenblum, I.S., Scholl, D.W., Shpikerman, V.I., Sidorov, A.A., and Stone, D.B., 2001, *in* Nokleberg, W.J., and Diggles, M.F., eds., U.S. Geological Survey Open File Report 01–261, CD-ROM.

Seslavinskiy, K.B., 1979, The South Anyuy geosture (western Chukotka): Proceedings, Earth Science Sections, Russian Academy of Sciences, v. 249, p. 78–81.

Shapiro, M.N., and Ganelin, V.G., 1988, Paleotectonic correlations of the large blocks in mesozoides of the USSR North-East: Geotectonics, no. 5, p. 94–104 (in Russian).

Sharkovsky, M.B., 1975, Tectonics of the Kolyma-Indigirka interfluve: Geotectonics, no. 6, p. 44–60 (in Russian).

Simakov, K.V., 1974, Stratigraphy of Middle Paleozoic sediments of the upper course of the Omolon River, *in* Simakov, K.V., ed., Basic problems of biostratigraphy and paleogeography of the northeast USSR: Magadan, Russia, Northeast Interdisciplinary Scientific Research Institute (SVKNII), p. 234–270 (in Russian).

Simakov, K.V., 1979, Separation and correlation of the Upper Famennian and Lower Tournaisian sediments of the Omolon massif after brachiopods *in* Simakov, K.V., ed., 14th Pacific Science Congress Guide of Scientific Excursion, Route 9: Appendix 8: Magadan, Russia, Northeast Interdisciplinary Scientific Research Institute (SVKNII), p. 183–214 (in Russian).

Simakov, K.V., and Shevchenko, V.M., 1974, Kedon series: Size, composition, age and conditions of formation, *in* Simakov, K.V., ed., Basic problems of biostratigraphy and paleogeography of the northeast USSR: Magadan, Russia, Northeast Interdisciplinary Scientific Research Institute (SVKNII), Issue 62, p. 189–233 (in Russian).

Smirnova, L.V., 1979, Separation and correlation of the Upper Famennian and Lower Tournaisian sediments of the Omolon massif after tabulates: 14th Pacific Science Congress, Guide of Scientific Excursion, Route 9: Appendix 8: Magadan, Russia, Northeast Interdisciplinary Scientific Research Institute (SVKNII), p. 215–235 (in Russian).

Sokolov, S.D., Didenko, A.N., Grogoryev, V.N., Aleksyutin, M.V., Bondarenko, G.Ye., and Krylov, K.A., 1997, Paleotectonic reconstructions of north-east Russia: Problems and unclear facts: Geotectonics, no. 6, p. 72–90 (in Russian).

Stavsky, A.P., Chekhovitch, V.D., Kononov, M.V., and Zonenshain, L.P., 1990, Plate tectonics and palinspastic reconstructions of the Anadyr-Koryak region, northeast USSR: Tectonics, v. 9, p. 81–101.

Stone, D.B., Crumley, S.G., Neustroev, A., and Parfenov, L.M., 1994, Paleomagnetic data relating to the paleolatitudes of the Omolon and Omulevka terranes [abs.]: International Conference on Arctic Margins, Magadan, Russia, p. 113.

Talent, J.A., 1990, Interrelations among lithological blocks on north-east of the USSR: Autochthonous or newcomer from far away?: Geotectonics, no. 2, p. 123–125 (in Russian).

Til'man, S.M., 1973, Comparative tectonics of the Mesozoides of the northern Pacific ring: Novosibirsk, Russia, Nauka, 326 p. (in Russian).

Ustritsky, V.I., and Khramov, A.N., 1987, On the history of formation of northern part of Pacific ocean and the Pacific mobile belt (comparison of paleomagnetic and paleobiogeographical data from the adjacent land), *in* Pushcharovsky, Yu.M., ed., Essays on the geology of the Northwestern Sector of Pacific Tectonic Belt: Moscow, Nauka, p. 239–276 (in Russian).

Zonenshain, L.P., 1984, Tectonics of the inner continental folded belts *in* Proceedings of the 27th International Geologic Congress, Tectonics, Colloquium 7, Papers, Volume 7: Moscow, Nauka, p. 48–59 (in Russian).

Zonenshain, L.P., Kuzmin, M.I., and Natapov, L.M., 1990a, Tectonics of the lithosphere plates of the USSR, Book 2: Moscow, Nedra, 334 p. (in Russian).

Zonenshain, L.P., Kuzmin, M.I., and Natapov, L.M., 1990b, Geology of the USSR: A plate-tectonic synthesis, *in* Page, B.M., ed., American Geophysical Union Geodynamic Series, v. 21, 242 p.

MANUSCRIPT ACCEPTED BY THE SOCIETY MAY 15, 2001.

Geological Society of America
Special Paper 360
2002

Seismicity of Chukotka, northeastern Russia

Kazuya Fujita
Kevin G. Mackey
Robert C. McCaleb
Department of Geological Sciences, Michigan State University, East Lansing, Michigan 48824-1115, USA
Larissa V. Gunbina
Valentin N. Kovalev
Magadan Experimental Methodological Seismological Division, Skuridina 6b,
Magadan 685000, Russian Federation
Valery S. Imaev
Yakutsk State University, 56 Ulitsa Belinskogo, Yakutsk 677000, Russian Federation
Vladimir N. Smirnov
Northeast Interdisciplinary Scientific Research Institute, Portovaya 16,
Magadan 685000, Russian Federation

ABSTRACT

The seismicity of Chukotka forms the central part of a Trans-Bering seismic belt that extends from the Koryak highlands through the Seward Peninsula to central Alaska. The seismicity of the Chukotka section has been recorded teleseismically (1928 to present), using a temporary network (1964–1966), a multiinstrument single station (1966–1982), and a regional network (1980–1993). Three seismic zones are identified within the Chukotka section of the Trans-Bering seismic belt: the Kolyuchin Gulf–Eastern Chukotka zone, a western extension of the rift system of Seward Peninsula, Alaska; the Koryak-Provideniya-Seward zone, representing the transpressional southern boundary of the Trans-Bering seismic belt; and the Anadyr–Amguema–Chukchi Sea zone, a weak, transtensional zone of seismicity forming the northern edge of the Trans-Bering seismic belt. The rest of Chukotka is relatively aseismic; the seismicity reported from the Polyarnyi area is anthropogenic. The Kolyuchin Gulf–Eastern Chukotka zone is a highly active transtensional zone (earthquakes to magnitude 7) with three northeast-striking segments of seismicity that are presumed to be transform faults offsetting rift segments. The Koryak-Provideniya-Seward zone is characterized by thrusting in its southern end, which changes to transform motion in the Gulf of Anadyr, and connects with the east-west–striking rift systems in Chukotka and the Seward Peninsula. The seismicity and focal mechanisms are consistent with the existence of an independent Bering Sea block rotating clockwise with respect to North America about a pole in western Chukotka. The Trans-Bering seismic belt forms the northern border of the Bering Sea block.

Fujita, K., Mackey, K.G., McCaleb, R.C., Gunbina, L.V., Kovalev, V.N., Imaev, V.S., and Smirnov, V.N., 2002, Seismicity of Chukotka, northeastern Russia , *in* Miller, E.L., Grantz, A., and Klemperer, S.L., eds., Tectonic Evolution of the Bering Shelf–Chukchi Sea–Arctic Margin and Adjacent Landmasses: Boulder, Colorado, Geological Society of America Special Paper 360, p. 259–272.

INTRODUCTION

Chukotka is located in northeasternmost Asia across the Bering Strait from Alaska; its seismicity is in the central part of a diffuse belt of earthquakes that extends from northwestern Kamchatka through the Seward Peninsula to central Alaska (Fig. 1; e.g., Mackey et al., 1997; Imaev et al., 1999), and includes the Bering Strait region. We refer to the belt as the Trans-Bering seismic belt, and it is among the most seismically active regions within what is usually considered to be part of the North American plate. The Chukotka part of the Trans-Bering seismic belt has undergone five magnitude 6, and eight magnitude 5, events during the past 75 yr. The region is sparsely populated and, as a result, the area was generally considered only weakly seismic. No macroseismic reports exist from Chukotka prior to 1971, thus seismicity prior to the early twentieth century is unknown.

Traditionally, the seismicity of western Alaska has been interpreted either as intraplate activity (Biswas et al., 1980) or as backarc effects of the Alaska subduction zone (e.g., Nakamura et al., 1980; Biswas et al., 1986a; Estabrook et al., 1988), and not related to the area being part of an extensive seismic belt. The combination of Russian and U.S. data, however, indicates that the diffuse seismicity of western Alaska is continuous with that in northeastern Russia (Fig. 1); thus the tectonics of the Trans-Bering seismic belt should be examined as a whole. Based on the combined data set, Lander et al. (1996) and Mackey et al. (1997) suggested that the Trans-Bering seismic belt forms the northern boundary of an independent Bering Sea plate or block that is rotating clockwise relative to North America about an Euler pole in western Chukotka (Fig. 1).

This chapter summarizes the seismicity of the previously poorly studied Chukotka segment of the Trans-Bering seismic belt. The seismicity of the other segments of the belt was described by Biswas et al. (1983, 1986b) for western Alaska and

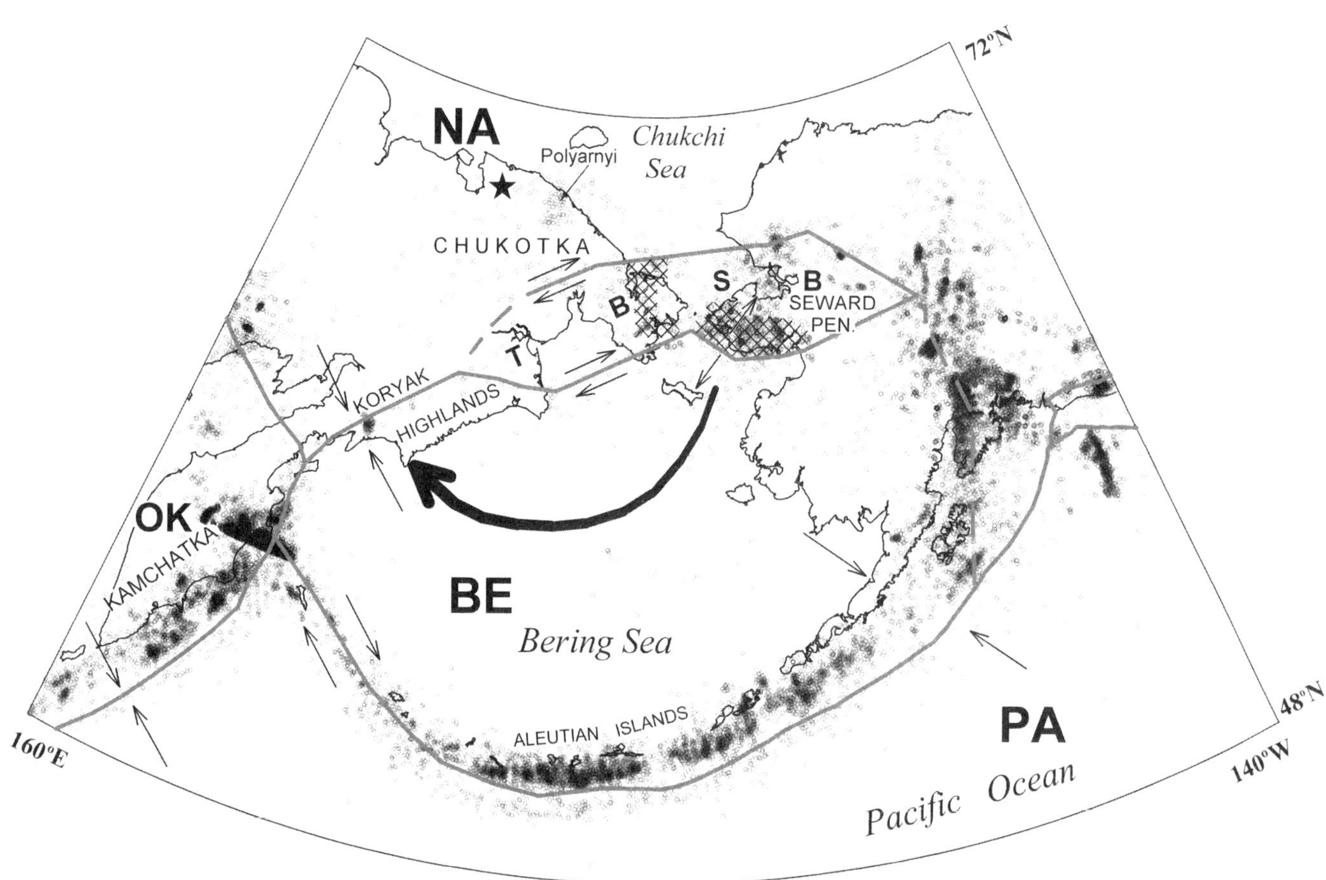

Figure 1. Seismicity map of Trans-Bering seismic belt (TBSB). Major plates and blocks are labeled (NA, North America; OK, Okhotsk block; BE, Bering Sea block; PA, Pacific). Star denotes pole of rotation of Bering block relative to North America (Mackey et al., 1997). Cross hachures indicate extensional areas on Seward Peninsula and in Chukotka. Seismicity around Polyarnyi is anthropogenic in origin (see Appendix). Plate and block boundaries are shown in gray, dashed where uncertain or diffuse. Arrows show relative motions between plates and blocks and large bold arrow shows motion of Bering block. Only teleseismic seismicity is shown south of 56°N in Kamchatka, in the Aleutians, and in Alaska outside of Seward Peninsula.

the Seward Peninsula; by Estabrook et al. (1988) and Page et al. (1991) for north-central Alaska; and by Lander et al. (1996) for the southern Koryak highlands.

SEISMICITY

Teleseismic records

Prior to 1964, all records of Chukotkan seismicity were teleseismic. Figure 2 shows teleseismically located seismicity from 1900 to the present based on data from the International Seismological Summary (ISS), the International Seismological Center (ISC), and the Preliminary Determination of Epicenters (PDE). Two prominent clusters of seismicity are noted, one near the Kolyuchin Gulf, Chukotka, and the other near Cape Prince of Wales, Alaska. Viewed on a broader scale (Fig. 1), however, most of the seismicity occurs in a band, the northern boundary of which passes through the Kolyuchin Gulf and the De Long Mountains, and the southern boundary of which extends from Cape Navarin to the southern Seward Peninsula. When combined with the microseismicity discussed in the following, this defines the Chukotka segment of the Trans-Bering seismic belt.

Teleseismic detection capability for this region has ranged from about magnitude (M) = 6 in the 1920s to M = ~4 today. Lo-

cation accuracy was ~±100 km in the 1920s to 1950s due to the lack of stations and poor timing, and has decreased to ±15 km today. Based on more recent seismicity, the three large (M = 6–7) events that occurred east of the Kolyuchin Gulf (Fig. 2) in 1928 are probably mislocated from the northeast-striking trend ~50 km to the north. In addition, a number of other early twentieth-century events attributed to this area are also mislocated or spurious; the most significant of these is a spurious event listed in the ISS for May 24, 1927, north of Wrangel Island. There are also two teleseismic events (labeled X in Fig. 2) listed in the ISC bulletin for 1971 that do not appear in any Russian bulletins; these may be a result of misassociated arrivals from events that occurred off the east coast of Kamchatka. All events are assumed to be crustal (10–33 km depth).

NEISRI test network

From late 1964 to early 1966, short-period sensors to record local and regional events were placed in towns throughout Chukotka as part of a larger temporary deployment by the Northeast Interdisciplinary Scientific Research Institute (NEISRI; Andreev et al., 1967). Although a large number of events were recorded, only 12 events were located in Chukotka due to the small number of simultaneously operating stations (Fig. 3).

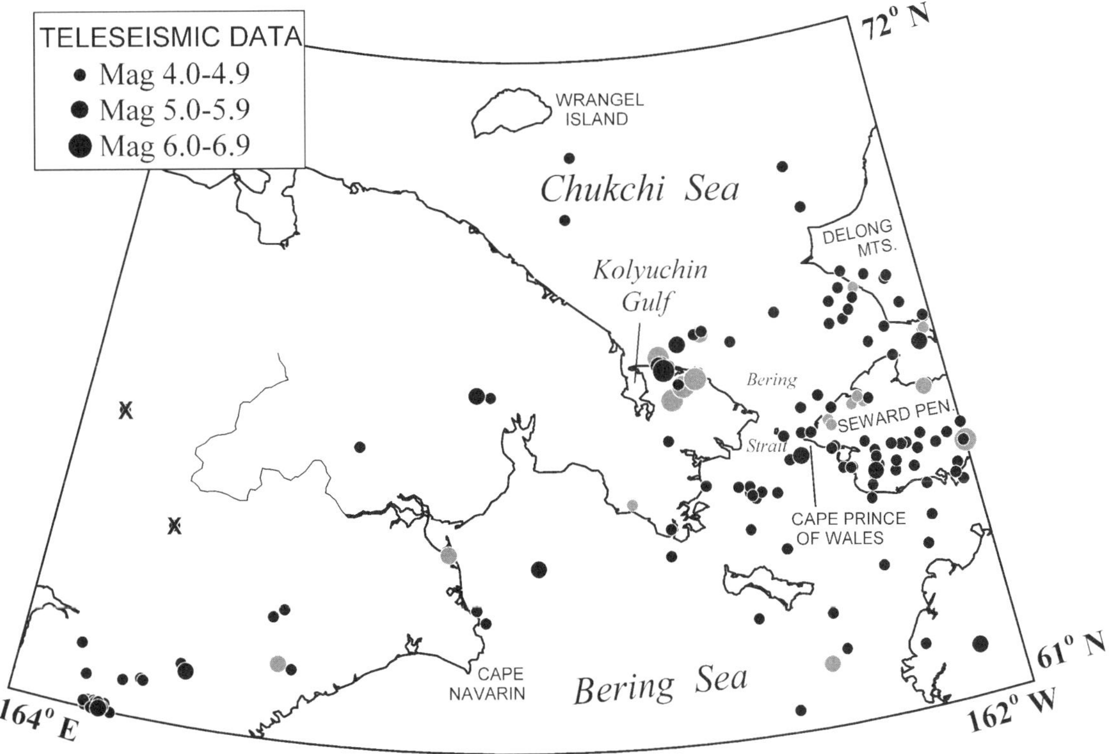

Figure 2. Teleseismically determined epicenters for Chukotka and westernmost Alaska. Events prior to 1964 are shown in gray and have greater location uncertainties. Two events marked X are discussed in text. Magnitude is indicated by size of circle.

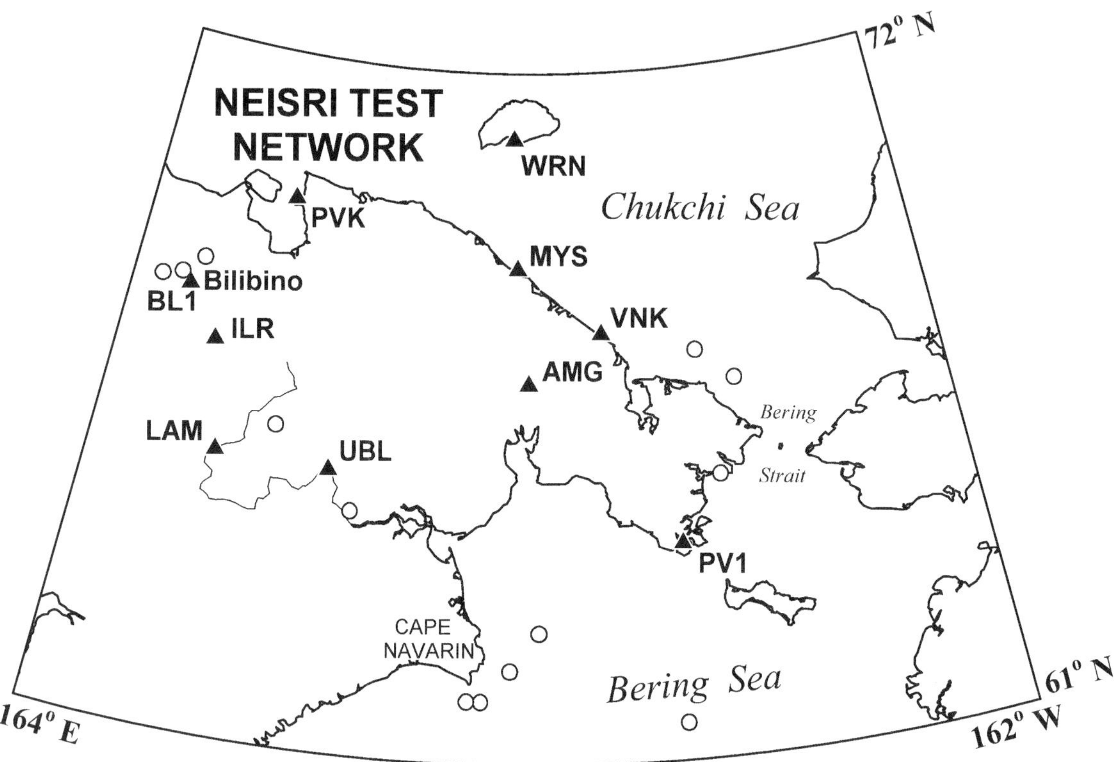

Figure 3. Northeast Interdisciplinary Scientific Research Institute (NEISRI) test network located epicenters (open circles) and station locations (triangles; see text).

Based on time of day and year, the events around Bilibino (BL1) are almost certainly explosions (Andreev et al., 1967; see Appendix). Seismicity was also identified off Cape Navarin, in the Bering Sea, and in the Bering Strait region. As a result of this, a permanent seismic station in Chukotka was deployed.

Iul'tin station

A permanent seismic station, equipped with three-component long- and short-period instruments, was deployed at Iul'tin (Table 1; Fig. 4) by the Institute of Physics of the Earth, Moscow, in March 1966 (Lazareva, 1970, 1973). Epicenters taken from Kondorskaya and Shebalin (1982) and the Arctic section of the annual *Earthquakes in the USSR* (annual, 1961–1989, Nauka, Moscow) are shown in Figure 4. A high level of seismicity was discovered south of Polyarnyi and southwest of the Kolyuchin Gulf. The seismicity south of Polyarnyi is anthropogenic in origin (see Appendix).

From 1966 to 1982, epicentral coordinates were determined for these regional events using the S-P time to obtain the origin time and distance while the azimuth was based on the polarization of the P wave (Lazareva, 1975). All earthquakes were assumed to be shallow (15 km). Only ~7% of the events recorded by Iul'tin were locatable (Avetisov, 1996). The epicentral accuracy was presumed to be ~±50 km, although comparison of lo-

cations using Iul'tin data alone and multistation relocations from the western Alaska network, operated by the University of Alaska, for five events in 1981 and 1982 in eastern Chukotka yield differences of as much as 50–100 km. These events are far from Iul'tin, thus events closer in may be better located. There is an abnormal north-south lineation of epicenters directly north and south of the station, suggesting that some locations are poor.

Godzikovskaya and Lander (1991), on the basis of energy envelopes from Iul'tin records that they interpret as lacking crustal phases (i.e., Pg, Sg), suggested that some events in Chukotka have a subcrustal focus. However, these events are at or close to the crossover distance for crustal and mantle phases, thus only one set of phases is likely to be observed. In addition, relocation of an event in 1984, cited as being subcrustal, yields a focal depth of 8 ± 11 km and is probably an explosion in the Polyarnyi mining district. Thus, there is no convincing evidence for subcrustal earthquakes in Chukotka.

NEISRI regional network

In October 1979, the Magadan Experimental-Methodological Seismic Division (EMSD) was organized under NEISRI to operate a regional seismic network. In the fall of 1980, the Magadan EMSD began the deployment of a regional network in Chukotka (Table 1; Fig. 5) with short-period sensors.

TABLE 1. MAGADAN EMSD CHUKOTKA STATION COORDINATES

Code	Station	Lat	Long	Elev	Open	Closed
ANSS*	Anadyr-1	64.77	177.57	40	11.80	1.89
ANYS*	Anadyr	64.734	177.496	55	4.89	7.93
					9.96	Open
BIL*	Bilibino	68.058	166.449	282.6	8.81	4.92
BILL	Bilibino GSN	68.065	166.452	299	8.95	Open
EGV*	Egvekinot	66.323	179.127W	18	1.90	—.94
ILT	Iul'tin	67.87	178.74W	235	3.66	7.93
MKI*	Maiskii	68.97	173.71	261	8.82	7.91
MKV*	Markovo	64.684	170.412	25.2	10.85	4.92
OMO*	Omolon	65.23	160.54	260	6.82	7.93
PVD*	Provideniya	64.427	173.224W	25.5	9.80	12.93
ULN*	Uelen	66.16	169.84W	5	—.81	—.82

Note: ILT was operated by the Institute of Physics of the Earth, Moscow. Asterisk denotes unofficial code. Elev is elevation in meters. Lat and Long are latitude (°N) and longitude (°E unless noted), respectively.
Sources: Yugova et al., 1997; Magadan EMSD.

Epicenters located by the NEISRI network shown in Figure 5 were obtained from the *Earthquakes in the USSR* and the Magadan EMSD network catalog. The seismicity is diffuse, but is concentrated in eastern Chukotka, northwest of Provideniya (PVD, Fig. 5), and in bands that trend from the Khatyrka River basin across the Gulf of Anadyr to the Bering Strait, and from north of Egvekinot (EGV, Fig. 5) to the eastern Chukchi Sea.

The 15 events from the Polyarnyi area that are believed to be anthropogenic in origin (see Appendix) are omitted from Figure 5.

From 1983 to 1993, earthquakes were located by graphical methods (Vorob'eva and Yugova, 1988) using the traveltime curve of Andreev (1984), which uses P- and S-wave velocities for the crust of 6.1 km/s and 3.51 km/s and mantle velocities of 8.1 km/s and 4.66 km/s, respectively. Starting in 1982, some

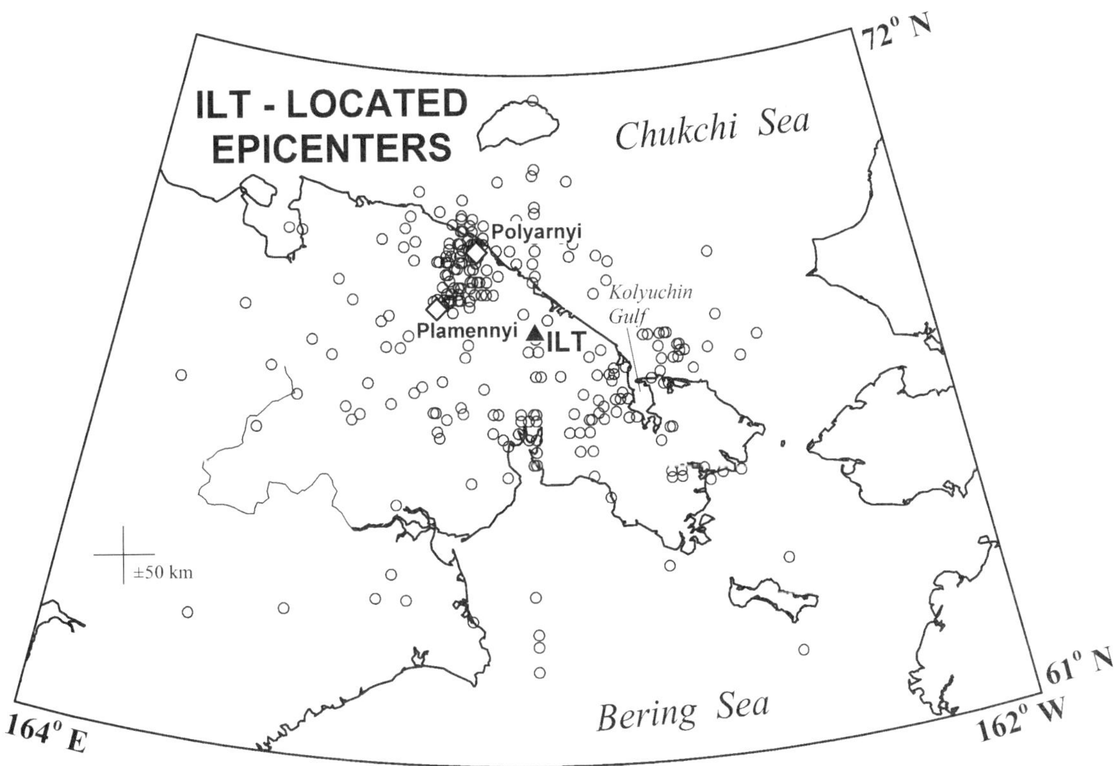

Figure 4. Epicenters (open circles) located using three-component station at Iul'tin (ILT, triangle). Location error shown near lower left.

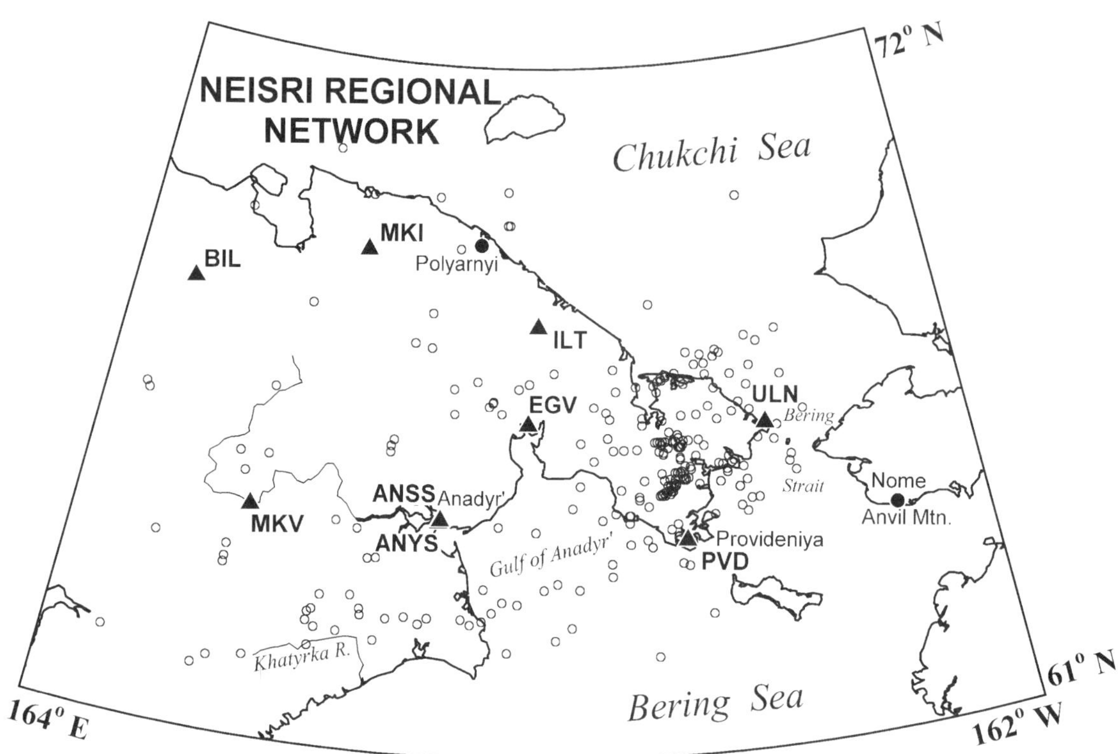

Figure 5. Epicenters (open circles) located by Northeast Interdisciplinary Scientific Research Institute (NEISRI) regional network and station locations (triangles; see text). See Table 1 for station codes. Note that 15 events believed to be of anthropogenic origin near Polyarnyi have been omitted.

larger earthquakes were located using personal computers, and computer-based locations became standard in the early 1990s. Formal errors are ±25–50 km in the early 1980s and ±10–15 km in the late 1980s and early 1990s. The detection level in Chukotka with three to four stations was about M = 3 (Artamonov and Mishina, 1984) and possibly slightly lower during the late 1980s. With the collapse of the Soviet Union, funding decreases resulted in the rapid collapse of the Chukotka regional network in 1992–1993 (Table 1) and an increase in detection levels to about M = 4 throughout the region. All stations were closed by the end of 1993.

About 50 events were relocated by Mackey (1999; mainly from easternmost Chukotka in the mid-1980s) by supplementing the phase data from the Magadan network bulletin with arrivals picked from the develocorder records for Anvil Mountain (Nome), Alaska. The mean relocation distance was ~25 km. Some clear trends become visible in the relocated epicenters (Fig. 6).

SEISMIC ZONES OF CHUKOTKA

The combination of the teleseismic and regional data make the Trans-Bering seismic belt in Chukotka much more prominent (Fig. 7) and suggests there are three zones of elevated seismicity: Kolyuchin Gulf–Eastern Chukotka (zone A, exten-

sional), Koryak-Provideniya-Seward (zone B, transpressional), and Anadyr–Amguema–Chukchi Sea (zone C, transtensional).

Because the seismicity in the Polyarnyi region appears to be strongly contaminated with, if not entirely the result of, explosions, and because of the poor quality and quantity of the NEISRI test network data, these events have been omitted. Epicenters from the western Alaska network within map areas are from Biswas et al. (1983). Iul'tin epicenters are shown in gray due to their greater uncertainties. It is possible that the diffuse seismicity in Norton Sound is part of the Trans-Bering seismic belt; however, detailed studies on this seismicity are lacking.

Kolyuchin Gulf–Eastern Chukotka seismic zone

The Kolyuchin Gulf–Eastern Chukotka zone (Fig. 7, zone A), which we suggest is extensional in nature, is the most active and enigmatic of the seismic zones of Chukotka. The zone extends from the Kolyuchin Gulf in the northwest to the Provideniya area in the southeast. We propose that this zone is a continuation of the rift system of the Seward Peninsula (Fig. 7, zone D) into Chukotka. Zone A is truncated by zone B in the southeast and by zone C in the northwest. The seismicity within zone A defines three short segments, all oriented southwest-northeast, along with additional scattered events (arrows, Fig. 7).

The northernmost segment extends northeastward from the

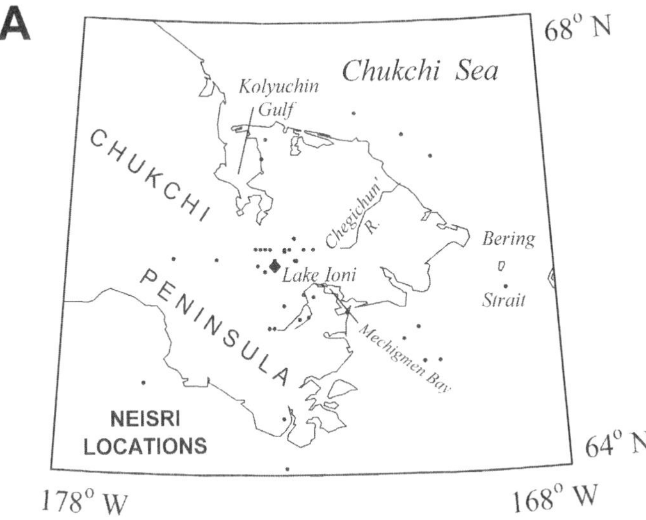

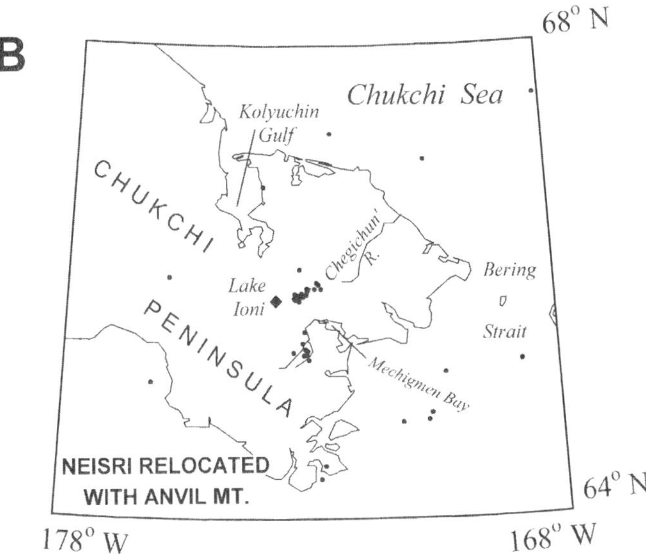

Figure 6. A: Northeast Interdisciplinary Scientific Research Institute (NEISRI) locations. B: Relocated epicenters using additional data from Anvil Mountain, Alaska. Lake Ioni shown by solid diamond.

Kolyuchin Gulf into the Chukchi Sea. The segment is also parallel to the edge of the topographic highlands of eastern Chukchi Peninsula formed by exposures of Precambrian crystalline rocks. Events of magnitude 5–6 occurred along this segment in 1928, 1962, 1971, 1996, and 1997. Three other 1928 events, with magnitudes of 6.2–6.9, are located to the southeast; however, given location accuracy in the 1920s and the lack of present-day activity, it is likely that they also were along this northern segment.

Two moment tensor solutions have been determined for this segment (Fig. 8; Dziewonski et al., 1997b, 1998). Both solutions indicate right-lateral transtension, assuming that the fault plane is the nodal plane (strike, $\varphi = 230°–249°$) striking approximately parallel to the strike of the segment ($\varphi \approx 240°$). The other nodal plane ($\varphi = 127°–134°$) is roughly perpendicular to the segment and parallel to the trend of older tectonic features found ~100 km to the east (e.g., north coast of Chukchi Peninsula and Kotzebue arch, Eittreim et al., 1979). A focal mechanism for the 1971 event based on teleseismic first motions and short-period body-wave modeling using the method of Kroeger (1978) yields a similar solution, although rotated about 20° (Table 2; $\varphi = 267°$). Previous solutions for the 1971 event (Biswas et al., 1986a; Fujita and Koz'min, 1994) are erroneous, probably due to the emergent first motions used from Alaskan stations. Teleseismic P-wave first-motion data and waveform characteristics are similar between the 1971 and 1962 events, thus we presume that they had similar mechanisms; no mechanisms can be constructed for the 1928 events (Fujita and Koz'min, 1994). Thus the northern segment is under north-south-directed extension with a right-lateral strike-slip component.

The central segment of zone A is within the Kolyuchin-Mechigmen graben in a topographically low region filled with Quaternary deposits (Gorodinsky, 1982). Our relocated epicenters define a northeast-southwest trend (Figs. 6 and 7) near Lake Ioni. This trend is colinear with the valley of the Chegichun' River along which Natal'in et al. (1999) mapped a fault with possible right-lateral motion of unknown age. No large events have been recorded from this zone.

The southern segment is partly along a lineament visible on Landsat images that extends southwest from Mechigmen Bay and cuts the strike of the regional topography (Figs. 6 and 7). Relocated events define only the northern end of this trend (cf. Figs. 6 and 7). The seismicity of the northern end of this segment is under Quaternary deposits filling the Kolyuchin-Mechigmen graben, while the southern end is under eroding Cretaceous extrusive volcanics.

On the basis of the focal mechanisms from the northern segment and the topographic depressions linking the northern, central, and southern segments, we propose that zone A represents an extension of the rift system of the Seward Peninsula (Turner and Swanson, 1981) into Chukotka (Mackey et al., 1997). This rift consists of three northwest-southeast–striking rift segments parallel to the overall strike of zone A, offset by strike-slip faults represented by the three northeast-southwest–striking seismically active segments (Fig. 8). As in the Seward Peninsula (Biswas et al., 1986b), the extension is oriented north-south to northeast-southwest. We suggest that the current locus of the rift system follows the Kolyuchin-Mechigmen graben, a fault-bounded lowland (Pol'kin, 1984) that extends from the southern end of the Kolyuchin Gulf to Mechigmen Bay. The graben is asymmetic with a steeper and higher northern side, and the bounding faults are expressed in geophysical fields (Pol'kin, 1984). Based on the southern segment of seismicity of zone A, the southeastern end may be offset farther to the south, but there is no clear topographic basin. However, little seismicity is observed in areas between the segments, i.e., the extensional areas.

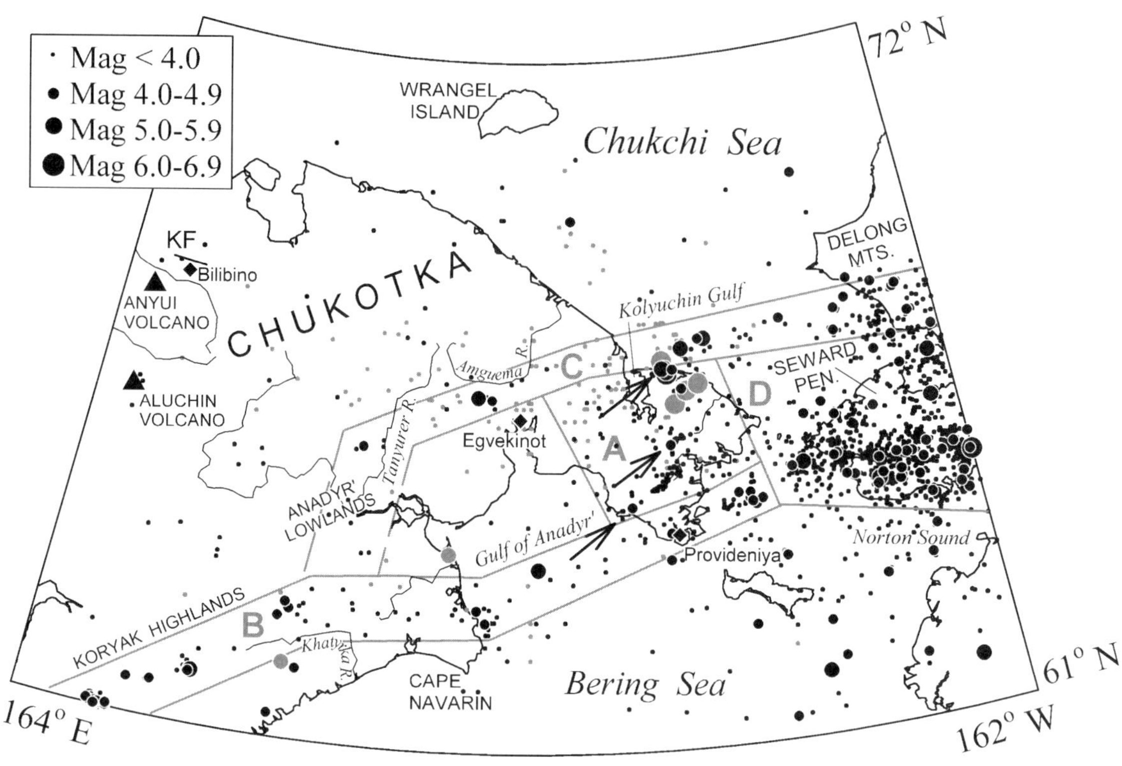

Figure 7. Seismicity map of Chukotka using best available data. Seismic zones defined in text are outlined in gray. A—Kolyuchin Gulf–Eastern Chukotka zone. B—Koryak-Provideniya-Seward zone. C—Anadyr–Amguema–Chukchi Sea zone. Iul'tin determined epicenters and pre-1964 teleseismic epicenters are shown in gray due to lower location accuracy. Arrows in zone A highlight segments discussed in text. Triangles denote Quaternary (?) volcanoes. KF is Keperveyem fault. D—seismic zone of the Seward Peninsula.

This may be due to the fact that the Kolyuchin-Mechigmen graben is an established structure and a zone of relative weakness. In the Seward Peninsula the rift is breaking through an intact Precambrian block (Till and Dumoulin, 1994).

Late Cenozoic alkaline basalts found in the southern Chukchi Peninsula are also indicative of extension (Enmelen volcanics, EV, Fig. 8; Akinin and Apt, 1994), as are hydrothermal areas near Provideniya and in the eastern part of the Kolyuchin-Mechigmen graben (Imaev et al., 1999).

Russian geophysical data suggest that there was a continuation of the rift system into the Chukchi Sea, north of the Kolyuchin Gulf. Shipilov et al. (1989) suggested that there are north-northwest–striking fault-bounded troughs, which may be part of a much larger, now inactive, rift system that extended north from the Kolyuchin Gulf toward Wrangel Island, as well as along the northeast coast of the Chukchi Peninsula. There are no published data to suggest that the Kolyuchin-Mechigmen graben extends to the west of the Kolyuchin Gulf.

Turner and Swanson (1981) summarized the geologic evidence for extension and rifting in the Seward Peninsula. Combined with focal mechanisms determined for the Seward Peninsula by Biswas et al. (1986b), it seems apparent that the Seward Peninsula has been a region of Quaternary extension, which continues to this day. This extensional regime appears to have dominated the Bering Strait region for the entire Cenozoic and perhaps the latest Mesozoic. Dumitru et al. (1995) suggested that extension began as early as 120 Ma. Offshore seismic studies in the Hope basin (Tolson, 1987) and basaltic volcanic deposits in the Seward Peninsula and west-central Alaska (Moll-Stalcup, 1994; Till and Dumoulin, 1994) suggest the continuation of this extension throughout the Cenozoic (Mackey et al., 1997). This extensional regime may also be evidenced by deep reflectors and thinned crust underneath the Bering Strait recorded by the 1994 cruise of the R/V *Ewing* (Brocher et al., 1995; Klemperer et al., this volume, Chapter 1; Wolf et al., this volume).

In Chukotka and the western Chukchi Sea, the timing of extension is less well defined. However, the formation of Kolyuchin-Mechigmen graben is believed to have initiated in Late Cretaceous to Paleogene time (Pol'kin, 1984) and the Enmelen and associated volcanics have been dated as 4–10 Ma based on the K-Ar method and palynology (Akinin and Apt, 1994; Belyi, 1995). Shipilov et al. (1989) also suggested a Late Cretaceous origin for extension in the Chukchi Sea. Thus, the

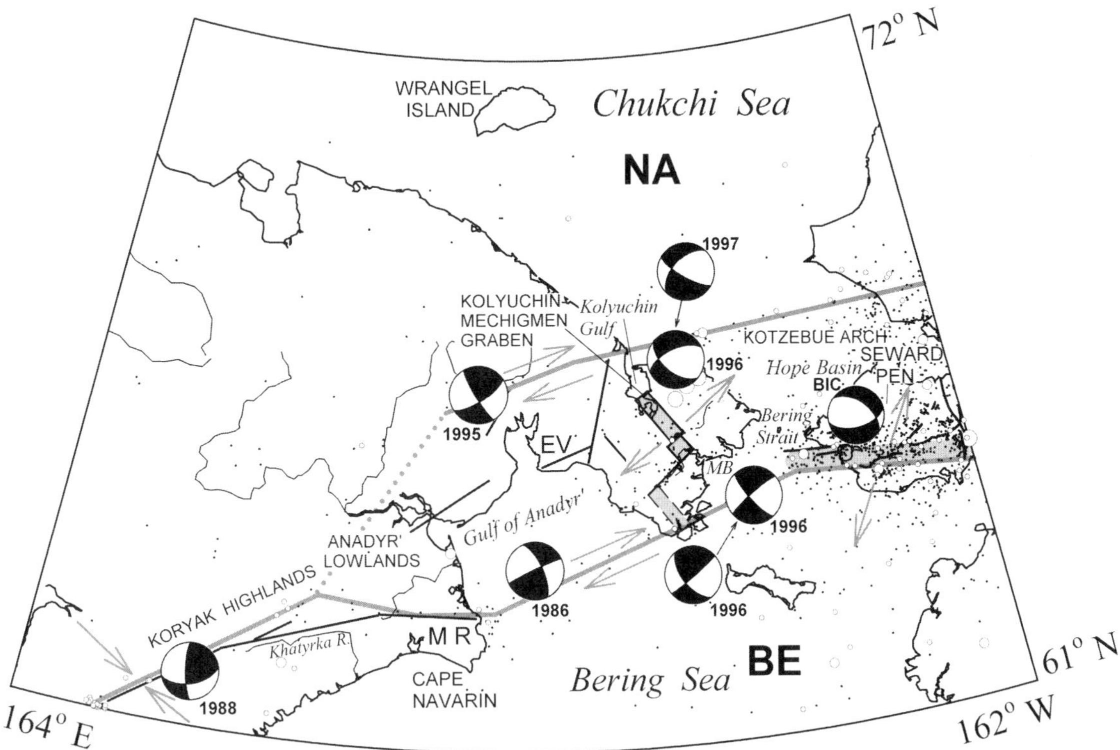

Figure 8. Tectonic map of Chukotka and western Alaska showing proposed tectonic boundaries (gray), earthquakes (open circles), focal mechanisms with year (compressional quadrant solid; see Table 2 for Chukotka mechanisms), and faults visible on space imagery or on topographic maps (black lines). Proposed rift areas are shaded in gray. Focal mechanism labeled BIC is composite mechanism from Biswas et al. (1983) and is representative of Seward Peninsula. MB—Mechigmen Bay. EV—Enmelen volcanics. MR—Meingypil'gyn Range. Large arrows show approximate relative motions. NA is North American plate, BE is Bering Sea block.

extension appears to have started contemporaneously in both Chukotka and the Seward Peninsula. Because the regions of rifting adjoin, we suggest that they form two segments of a unified extensional system, which Dumitru et al. (1995) and Mackey et al. (1997) suggested is related to interactions between the Pacific plate and Alaska, possibly far-field effects of the accretion of terranes in southeastern Alaska.

Koryak-Provideniya-Seward zone

Zone A is truncated on the south by zone B, which extends southwestward from the Bering Strait to the northern Koryak highlands (Fig. 7, zone B). The southern extension of zone B (Lander et al., 1996) traverses the Koryak highlands and connects with the North America–Okhotsk plate boundary (Cook et al., 1986) in northern Kamchatka. Our discussion is restricted to the part of zone B north of 62°N.

The northern end of zone B merges into the seismicity of the Seward Peninsula (zone D, Fig. 7) across the Bering Strait from Provideniya. The northernmost part of the segment is defined by a cluster of seismicity striking northeast. Moment tensors have been determined for three earthquakes that oc-

curred in late 1996 within this cluster; the two larger events (Fig. 8) have solutions consistent with right-lateral strike-slip faulting on a plane parallel to the strike of the cluster ($\varphi = 235°$–$238°$), and the smallest event is right-lateral transpression along a similar plane ($\varphi = 249°$; Dziewonski et al., 1997b, 1998). To the north of this cluster, zone B trends into the rifts and seismicity of the southern Seward Peninsula described by Biswas et al. (1986b) and Turner and Swanson (1981), and then further into central Alaska (Page et al., 1991). On the basis of the orientation of the seismicity and the focal mechanisms, we suggest that zone B near Provideniya acts as a strike-slip transform fault between the southern part of zone A and the Seward Peninsula rift system (Fig. 8).

To the southwest, in the Gulf of Anadyr, zone B is traced by a band of microearthquakes, with a few teleseisms, that trends toward Cape Navarin and the Koryak highlands. The largest event ($M = 5.3$) in this part of the zone occurred in 1986. The moment tensor solution (Fig. 8; Table 2; Dziewonski et al., 1987) is also a right-lateral strike slip on a plane striking parallel to the seismicity ($\varphi \approx 245°$), although the P-wave first-motion data (Gunbina et al., 1987) admit a more transpressional solution.

No seismic stations have been deployed in the Koryak high-

TABLE 2. FOCAL MECHANISMS OF THE CHUKOTKA-BERING STRAIT REGION

Date			Origin			Lat	Long	Mag	Plane 1			Plane 2			Meth	Ref
88	10	13	00	32	13.	61.85	169.65	5.7	070	53	164	170	77	39	CMT	1
86	10	19	18	30	57.	63.90	−178.69	5.3	336	68	4	245	86	158	CMT	1
									323	50	28	216	69	137	PB	2
95	10	02	01	35	48.3	66.69	179.14	5.1	055	71	170	148	81	19	CMT	1
97	03	24	06	56	13.3	67.07	−173.31	5.1	230	53	−160	127	74	−39	CMT	1
96	10	24	19	31	56.2	67.13	−172.84	6.0	251	60	−138	136	54	−38	CMT	1
71	10	05	01	40	41.6	67.38	−172.57	5.2	267	55	−145	142	50	−40	SYN,P	2
96	08	09	18	33	25.6	64.91	−170.34	4.7	001	50	32	249	66	135	CMT	1
96	08	09	18	45	43.4	64.93	−170.45	4.9	148	58	4	055	86	148	CMT	1
96	11	03	23	24	30.7	64.84	−170.41	5.1	146	76	−8	238	82	−166	CMT	1

Note: Epicenters after ISC, PDE, or Kondorskaya and Shebalin (1982). References: 1—Harvard CMT Catalog (Dziewonski et al., 1987, 1989, 1997a, 1997b, 1998); 2—this paper. Origin time in UTC; Lat and Long are latitude and longitude (north and east positive); Mag is body wave magnitude, Plane 1 and Plane 2 are given in strike, dip, and slip according to standard seismological convention; Meth is method (CMT—centroid moment tensor; PB—P waves from seismograms and bulletins; SYN—short-period synthetics; P—P-wave first motions from seismograms.

lands, thus the detection threshold is higher than in Chukotka, and fewer events have been located. In the Meingypil'gyn Range (MR, Fig. 8; also called the Ukvushvuynen Mountains), the seismicity is approximately coincident with a zone of east-west-striking faults; the faults offset river valleys that can be seen on Landsat images and 1:200 000 topographic maps (Fig. 9, areas A and B). Toward Cape Navarin (Fig. 9, area C), younger faults are superposed on, and offset, a series of faults and the large-scale structural trend of the region (dashed lines in Fig. 9). River valleys are offset as much as 2 km in a right-lateral sense across many of these faults. A total offset of 5–7 km across the entire fault system is estimated. Landslides and similar features, most likely generated by strong earthquakes, have also been mapped in this area by Russian field parties, suggesting that significant faulting continues in this area.

West of 177.3°E, the faults and seismicity curve to a west-southwest strike and the larger river valleys are bent or offset by them in a right-lateral sense (Fig. 9, area D). Farther west, the seismicity continues into a large basin near 175°E filled with Quaternary deposits, and then into the Khatyrka River valley. A magnitude 5.5–5.7 event (March 9, 1934) was located in the Khatyrka River valley, although the epicentral error may be large. A focal mechanism here indicates transpression (Fig. 8; Table 2), which becomes thrusting south of 61°N (Lander et al., 1996).

The neotectonics of this area have not been extensively studied. Examination of presumed peneplanation (accordant summit) surfaces and river terraces has been used by Russian authors to suggest Holocene uplift of as much as 500–1000 m in the seismically active region of the Koryak highlands (Smirnov, 1995), although such surfaces may simply represent equilibrium surfaces (Keller and Pinter, 1996). Based on re-leveling surveys, the Anadyr lowlands, to the north of zone B, are subsiding at rates of ~5–7 mm/yr, while the coastline to the south is subsiding at ~2 mm/yr (Zolotarskaya et al., 1987). The Meingypil'gyn Range is undergoing uplift at 2–5 mm/yr. Uplifted terraces along river valleys show uplift of ~1 mm/yr (Glushkova et al., 1987). These data suggest overthrusting of the Koryak highlands over the Anadyr lowlands. This is consistent with the large number of thrust focal mechanisms from earthquakes farther southwest in the southern Koryak highlands (Lander et al., 1996; Mackey et al., 1997) and with the southward movement of the Gulf of Anadyr region (Fig. 8). Therefore, based on the river offsets and the uplift data, we suggest that the area is under right-lateral transpression.

We also suggest that zone B represents the southern edge of the Trans-Bering seismic belt; right-lateral strike-slip displacement, possibly with a small amount of compression, is occurring in the Gulf of Anadyr and the Bering Strait, and increasing transpression is occurring in the Koryak highlands. This is consistent with the clockwise motion of the Bering Sea plate or block relative to North America about a pole in western Chukotka, as proposed by Lander et al. (1996) and Mackey et al. (1997).

Anadyr–Amguema–Chukchi Sea zone

The Anadyr–Amguema–Chukchi Sea zone (Fig. 7, zone C) represents a band of elevated seismicity that marks the northern edge of the Trans-Bering seismic belt. It extends to the east and west of the northern segment of zone A. Its southern end, near the Anadyr River, is diffuse and has only a little seismicity. It is possible that the zone extends southward to join the seismicity of the Koryak highlands. Teleseisms and microseismicity, largely defined by the Iul'tin data, suggest that zone C extends from the Tanyurer River valley to the north of Egvekinot. From there, zone C is more clearly defined as the northern edge of scattered seismicity across Chukotka to the Kolyuchin Gulf, where it joins the northern segment of zone A. The seismicity then crosses the Chukchi Sea, where there are several teleseismic events, to the De Long Mountains of Alaska.

The moment tensor solution for a magnitude 5.1 event

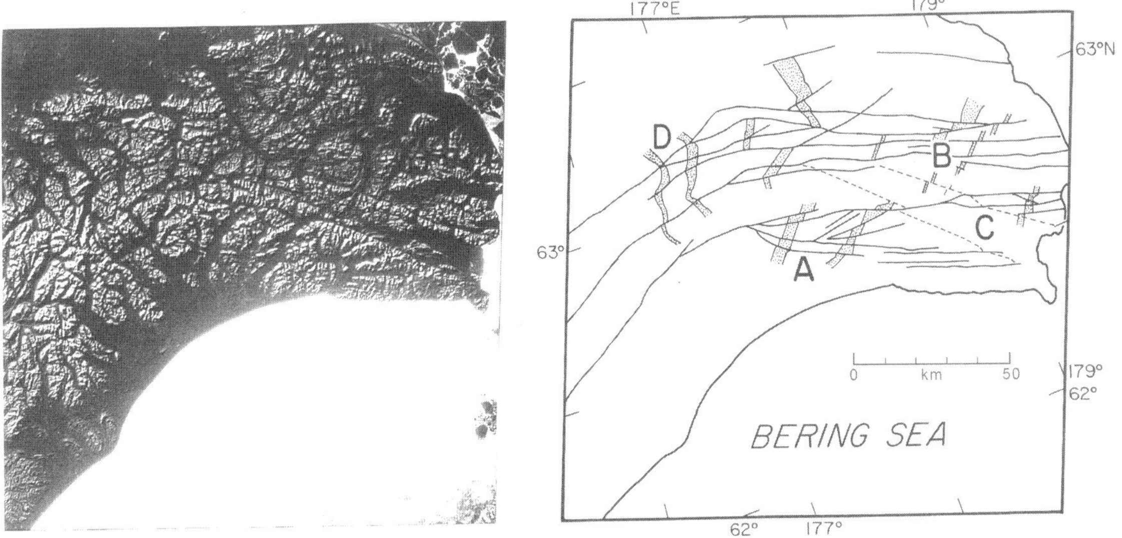

Figure 9. Left: Landsat image of western Koryak highlands and (right) interpreted faults (solid lines) and river offsets (valleys stippled). Dashed lines represent older tectonic grain. A–D: Area discussed in text. The line drawing covers the same area as the photograph.

northwest of Egvekinot in 1995 shows right-lateral strike slip (Fig. 8; Dziewonski et al., 1997a), assuming that the northeast-trending plane ($\varphi = 235°$), which roughly parallels the strike of the zone, is the fault plane. The zone passes through several wide river valleys that are filled with Quaternary sediments; thus few lineaments can be clearly suggested to be of tectonic origin.

Although the continuity of this zone is difficult to assess, this may represent the northwestern and northern edges of the currently active rift and transtensional zone and, therefore, the edge of the rigid North American plate. This zone bounds the Kolyuchin Gulf–Eastern Chukotka seismic zone (zone A) on the northwest. The larger transtensional events from the northern segment of that zone may be along this "cold," newly rifting margin, explaining their greater magnitude.

Other regions

The remainder of Chukotka is aseismic with only a weak background of microearthquakes (M < 3; Fig. 7). However, there are several volcanoes believed to be young (e.g., Anyui and Aluchin volcanoes, thought to be younger than 1–2 ka; Rudich, 1985; Fig. 7) 100–200 km south of Bilibino. While volcanoes are particularly well preserved in the Arctic (cf. Balagan-Tas, dated as 286 ka by Layer et al. [1993], but originally thought to be a few thousand years old), this does leave open the possibility that there may have been significant tectonic activity in the region thousands to hundreds of thousands of years ago. On the basis of inferred peneplanation surfaces, Kruglyakov et al. (1987) inferred neotectonic offsets of as much as 500 m on the Keperveyem (KF, Fig. 7) and other faults near Bilibino. No detailed evidence of Holocene geologic movements in the Bili-

bino area has been published. However, recent field work is reported to have identified landslide features that could have been seismically generated. In 1989–1991, a small local network was deployed around Bilibino to investigate the potential seismic risk to the atomic reactor by the Baikal EMSD (Pavlov et al., 1994). No events with M > 0.7 were recorded. We cannot, however, exclude the possibility of significant seismic activity in western Chukotka.

CONCLUSIONS

We suggest that the seismicity in Chukotka forms the central part of a Trans-Bering seismic belt and is characterized by three seismic zones: the Anadyr–Amguema–Chukchi Sea zone (zone C) with earthquakes generally of M < 5, representing the edge of rigid North America and the northern boundary of the seismic belt; the Koryak-Provideniya-Seward zone (zone B), representing the transtensional northern boundary of the Bering block and the southern edge of the seismic belt, with events, generally strike slip, of M < 6; and the Kolyuchin Gulf–Eastern Chukotka rift zone (zone A) within the seismic belt, with extensional to transtensional events of M < 7. The seismicity near Polyarnyi is anthropogenic in origin and should not be included in tectonic analyses (see Appendix).

The continuity of seismicity from the southern Koryak highlands, through Chukotka, to the Seward Peninsula and west-central Alaska strongly suggests that the Bering Sea, which is generally aseismic (Fig. 1), represents a rigid block in motion with respect to North America (Mackey et al., 1997). The eastern boundary of this Bering Sea block is considerably more diffuse, as shown by the widely scattered seismicity in southwest continental Alaska (Fig. 1). Focal mechanisms indicate extension or transtension in Chukotka and the Seward Peninsula and transtension and compression in the Koryak highlands. This is consistent with the clockwise rotation of the Bering Sea block, relative to North America, about a pole in northwest Chukotka. The driving force for the rifting in Chukotka and the Seward Peninsula could be either backarc extension (e.g., Nakamura et al., 1970) or terrane accretion in southeast Alaska (e.g., Page et al., 1991; Mackey et al., 1997), or some combination of both. However, the backarc spreading model cannot explain the thrusting in the southern Koryak highlands. Thus, we prefer the Bering block model. The magnitude of the relative motion across the various segments of the Trans-Bering seismic belt is unknown, but given the levels of seismicity and tectonic activity, it is presumed to be relatively small (millimeters per year), except possibly in the southern Koryak highlands.

The Chukotka rift zone (zone A) is completely within the Trans-Bering seismic belt and is bounded by strike-slip zones B and C. The Trans-Bering seismic belt between zones B and C to the southwest of zone A must be moving more rapidly to the west than the eastern part in order for the rifting in zone A to occur. This is likely due to the drag from the more rapidly moving Bering Sea block slivering the edge of North America proper.

To the east of zone A, the Trans-Bering seismic belt is dominated by north-south extension (Page et al., 1991). If this region is an analog of the extrusion tectonics of Southeast Asia, as proposed by Mackey et al. (1997), zones B and C are the equivalent of some of the large strike-slip faults of Southeast Asia with a zone of extension between them (e.g., Peltzer and Tapponier, 1988). Alternatively, zone A may represent a region of weakness where differential motions of small blocks and slivers within the Trans-Bering seismic belt result in extension.

Studies are currently in progress to remove the artificial data barrier that existed between far-eastern Russia and Alaska until the 1990s; this will greatly improve our knowledge of how the seismicity of Chukotka relates to that of the Seward Peninsula. Additional phase data will be required before meaningful relocations of western Alaskan and Trans-Bering events can be obtained; previously determined epicenters are known to have significant location errors (Page et al., 1991; Mackey, 1999). The reopening of Bilibino as an Incorporated Research Institutions for Seismology Global Seismic Network station in 1995, and of Anadyr (ANYS, Fig. 5) in 1997 by the Magadan EMSD, and future cooperation between Alaska and northeastern Russia in data exchange will, undoubtedly, improve our understanding of the seismicity of this complex region.

ACKNOWLEDGMENTS

This study was conducted as part of a cooperative research program between the University of Alaska, Michigan State University, the Northeast Interdisciplinary Scientific Research Institute (Magadan, Russia), and the Yakutian Institute of Geological Sciences (Yakutsk, Russia). We thank David Cook, Alexandra DeJong, Dimitry Gunbin, Boris Koz'min, Paul Layer, Erin Lynch, Melissa McLean, Elizabeth Miller, Ken Ridgway, David Stone, Jaime Toro, Guy Tytgat, Bruce Watson, Robert Young, the late Lev Zonenshain, and two anonymous reviewers for helpful discussions, comments, and criticisms on various parts of this work. This research was supported by National Science Foundation grants OPP- 94–24139 and OPP-98–06130. Data acquisition was supported by the Incorporated Research Institutions for Seismology Joint Seismic Program.

APPENDIX: CONTAMINATION OF THE SEISMICITY CATALOG BY EXPLOSIONS

With the opening of the Iul'tin station, considerable seismic activity was located near the Polyarnyi placer gold mining district and in the Ekvyvatap Range, ~200 km northwest of Iul'tin (Fig. 4). The events were located from 1969 to 1982. Based on these data, this region was considered the most seismically active in Chukotka. However, this activity is coincident with the expansion of mining at Polyarnyi starting in 1968 and is likely to be anthropogenic.

The anthropogenic nature of the events near Polyarnyi can be demonstrated by examining the temporal variation in seismicity (Godzikovskaya, 1995; Mackey and Fujita, 1999). Explosions in placer mines are normally conducted in daytime hours in the late winter and spring to break up frozen ground for summer extraction activities; a few night explosions also occur. Figure A1a shows the temporal distribution of Iul'tin-located events for the Polyarnyi region and clearly shows a bias similar to that for known explosions from the NEISRI network period (Fig. A1b). First motions from a sequence of events in 1978 yielded compressional first motions at Iul'tin (Lazareva, 1982), also supporting an explosion origin. Given the location error associated with the single-station Iul'tin epicenters, all of the events in the Ekvyvatap Range can be ascribed to the Polyarnyi gold or the Plammenyi mercury mines (Fig. 4). Explosions here generally yield magnitudes of M = ~2.0 or less (Godzikovskaya, 1995).

The NEISRI network data show very little seismicity in this region. A total of 15 earthquakes are cataloged between Polyarnyi and Plamennyi in the same areas as the Iul'tin-located events. These events are also the same size (M = 2.0–2.3) and have the same temporal distribution as known explosions. Other events with explosion characteristics within a few hundred kilometers of a station are identified as such by NEISRI network operators and excluded from seismicity catalogs.

Extension of the temporal analysis to other parts of Chukotka shows no significant biases in the Iul'tin or NEISRI data sets except for a few explosions associated with exploratory drilling for the construction of a hydroelectric station across the Amguema River (Godzikovskaya, 1995) in 1981–1984.

REFERENCES CITED

Akinin, V.V., and Apt, Y.E., 1994, Enmelen volcanoes (Chukchi Peninsula): Petrology of alkaline lavas and deep-seated inclusions: Magadan, Russia, Severo-Vostochnyi Kompleksnyi Nauchno-Issledovatel'skii Institut, 97 p. (in Russian).

Andreev, T.A., 1984, Calculation of parameters of weak earthquakes on an electronic calculating machine, *in* Izmailov, L.I., and Lin'kova, T.I., eds., Seismic processes in the northeast USSR: Magadan, Russia, Severo-Vos-

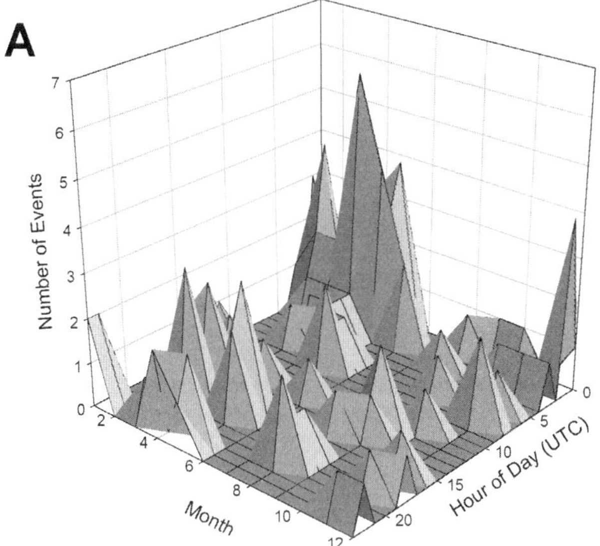

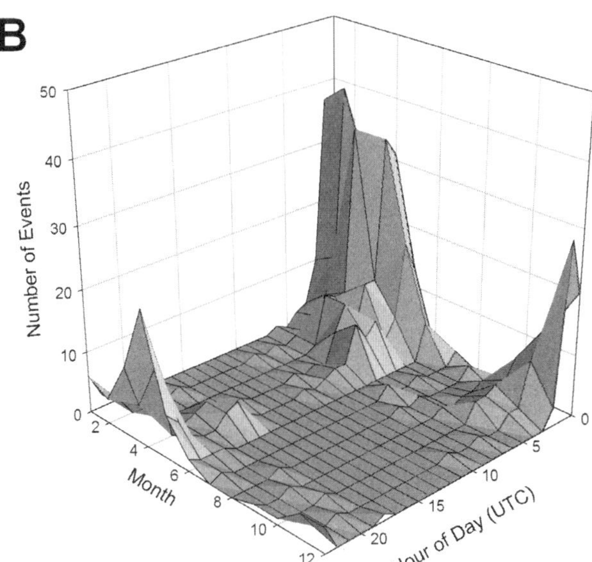

Figure A1. Temporal distribution of (A) events from Polyarnyi region located by Iul'tin station (1966–1982) and (B) explosions identified by Northeast Interdisciplinary Scientific Research Institute (NEISRI) station operators (1983–1996).

tochnyi Kompleksnyi Nauchno-Issledovatel'skii Institut, p. 116–127 (in Russian).

Andreev, T.A., Kravets, I.F., and Mishin, S.V., 1967, On the seismic activity of the Northeast: Magadan, Russia, Trudy Severo-Vostochnyi Kompleksnyi Nauchno-Issledovatel'skii Institut, v. 30, p. 159–166 (in Russian).

Artamonov, V.V., and Mishina, L.V., 1984, Recording possibilities of the network of seismic stations in the northeast USSR, *in* Izmailov, L.I., and Lin'kova, T.I., eds., Seismic processes in the northeast USSR: Magadan, Russia, Severo-Vostochnyi Kompleksnyi Nauchno-Issledovatel'skii Institut, p. 99–115 (in Russian).

Avetisov, G.P., 1996, Seismoactive zones of the Arctic: St. Petersburg, Vserossiiskii Nauchno-Isseldovatal'skii Institut Okeanolgiya, 185 p. (in Russian).

Belyi, V.F., 1995, Bering volcanic province: Tikhookeanskaya Geologiya, v. 14, no. 4, p. 50–59 (in Russian).

Biswas, N.N., Gedney, L., and Agney, J., 1980, Seismicity of western Alaska: Bulletin of the Seismological Society of America, v. 70, p. 873–883.

Biswas, N.N., Pujol, J., Tytgat, G., and Dean, K., 1983, Synthesis of seismicity studies for western Alaska, Final report for contract NA81–RAC00112: Fairbanks, University of Alaska, Geophysical Institute, 74 p.

Biswas, N.N., Aki, K., Pulpan, H., and Tytgat, G., 1986a, Characteristics of regional stresses in Alaska and neighboring areas: Geophysical Research Letters, v. 13, p. 177–180.

Biswas, N.N., Pujol, J., Tytgat, G., and Dean, K., 1986b, Synthesis of seismicity studies for western Alaska: Tectonophysics, v. 131, p. 369–392.

Brocher, T.M., Allen, R.M., Stone, D.B., Wolf, L.W., and Galloway, B.K., 1995, Data report for onshore-offshore wide-angle seismic recordings in the Bering-Chukchi Sea, western Alaska and eastern Siberia: U.S. Geological Survey Open-File Report 95–0650, 57 p.

Cook, D.B., Fujita, K., and McMullen, C.A., 1986, Present-day plate interactions in northeast Asia: North American, Eurasian, and Okhotsk plates: Journal of Geodynamics, v. 6, p. 33–51.

Dumitru, T.A., Miller, E.L., O'Sullivan, P.B., Amato, J.M., Hannula, K.A., Calvert, A.T., and Gans, P.B., 1995, Cretaceous to recent extension in the Bering Strait region, Alaska: Tectonics, v. 14, p. 549–563.

Dziewonski, A.M., Ekström, G., Woodhouse, J.H., and Zwart, G., 1987, Centroid-moment tensor solutions for October–December 1986: Physics of the Earth and Planetary Interiors, v. 48, p. 5–17.

Dziewonski, A.M., Ekström, G., Woodhouse, J.H., and Zwart, G., 1989, Centroid-moment tensor solutions for October–December 1988: Physics of the Earth and Planetary Interiors, v. 54, p. 10–21.

Dziewonski, A.M., Ekström, G., and Salganik, M.P., 1997a, Centroid-moment tensor solutions for October–December 1995: Physics of the Earth and Planetary Interiors, v. 101, p. 1–12.

Dziewonski, A.M., Ekström, G., Maternovskaya, N.N., and Salganik, M.P., 1997b, Centroid-moment tensor solutions for July–September 1996: Physics of the Earth and Planetary Interiors, v. 102, p. 133–143.

Dziewonski, A.M., Ekström, G., and Maternovskaya, N.N., 1998, Centroid-moment tensor solutions for October–December 1996: Physics of the Earth and Planetary Interiors, v. 105, p. 95–108.

Eittreim, S., Grantz, A., and Whitney, O.T., 1979, Cenozoic sedimentation and tectonics of Hope Basin, southern Chukchi Sea, *in* Sisson, A., ed., The relationship of plate tectonics to Alaskan geology and resources: Anchorage, Alaska Geological Society, p. B1–B11.

Estabrook, C.H., Stone, D.B., and Davies, J.N., 1988, Seismotectonics of northern Alaska: Journal of Geophysical Research, v. 93, p. 12026–12040.

Fujita, K., and Koz'min, B.M., 1994, Seismicity of the Amerasian Arctic shelf and its relationship to tectonic features, *in* Thurston, D.K., and Fujita, K., eds., Proceedings of the 1992 International Conference on Arctic Margins: Anchorage, Alaska, Minerals Management Service Outer Continental Shelf MMS Report 94-0040, p. 307–312.

Glushkova, O.Y., Degtyarenko, Y.P., and Prokhorova, T.P., 1987, Pleistocene freezing and features of sedimentation and reconstruction of the hydrologic network in the Koryak Highlands, *in* Pokhialainen, V.P., ed., The Quaternary period in northeast Asia: Magadan, Russia, Severo-Vostochnyi Kompleksnyi Nauchno-Issledovatel'skii Institut, p. 33–54 (in Russian).

Godzikovskaya, A.A., 1995, Local explosions and earthquakes: Moscow, Institut Gidroproekt, 98 p. (in Russian).

Godzikovskaya, A.A., and Lander, A.V., 1991, Are mantle earthquakes possible in Chukotka?: Academy of Sciences of the USSR, Izvestiya, Earth Physics, v. 27, p. 1084–1086.

Gorodinsky, M.E., editor, 1982, Geologic map of the northeast USSR (dated 1980): Leningrad, Vsesoyuznyi Nauchnyi-Isseldovatel'skii Geologichskii Institut, scale 1:1 500 000, 9 sheets (in Russian).

Gunbina, L.V., Vorob'eva, E.A., and Bobkov, A.O., 1987, Anadyr' earthquake of 19 October, 1986: Yuzhno Sakhalinsk, Russia, Institut Morski Geologii i Geofiziki, Dal'nevostochnoe Nauchnoe Otdelenie, Akademii Nauk SSSR, 22 p. (in Russian).

Imaev, V.S., Imaeva, L.P., Koz'min, B.M., Gunbina, L.V., Mackey, K., and Fujita, K., 1999, Modern geodynamics of lithospheric plates of northeastern Asia: Doklady Earth Sciences, v. 369, p. 1105–1110.

Keller, E.A., and Pinter, N., 1996. Active tectonics, earthquakes, uplift, and landscape: Upper Saddle River, New Jersey, Prentice-Hall, 338 p.

Kondorskaya, N.V., and Shebalin, N.V., editors, 1982, New catalog of strong earthquakes in the USSR from ancient times through 1977: Boulder, Colorado, World Data Center A for Solid Earth Geophysics, Report SE-31, 608 p.

Kroeger, G.C., 1978, Synthesis and analysis of teleseismic body wave seismograms [Ph.D. thesis]: Palo Alto, California, Stanford University, 135 p.

Kruglyakov, M.I., Vladimiriv, V.G., and Kharlov, E.M., 1987, Estimate of seismic conditions in area of the Bilibino Atomic Energy Station: Geologiya i Geofizika, v. 28, no. 8, p. 84–90 (in Russian).

Lander, A.V., Bukchin, B.G., Kiryushin, A.V., and Droznin, D.V., 1996, The tectonic environment and source parameters of the Khailino, Koryakiya earthquake of March 8, 1991: Does a Beringia plate exist?: Computational Seismology and Geodynamics, v. 3, p. 80–96.

Layer, P., Parfenov, L.M., Surnin, A.A., and Timofeyev, V.F., 1993, First ^{40}Ar-^{39}Ar ages of igneous and metamorphic rocks from the Upper Yana–Kolyma Mesozoides: Transactions (Doklady) of the Russian Academy of Sciences, Earth Science Sections, v. 330, p. 130–134.

Lazareva, A.P., 1970, Earthquakes of the Arctic, *in* Vvedenskaya, N.A., and Kondorskaya, N.V., eds., Earthquakes in the USSR in 1966: Moscow, Nauka, p. 259–261 (in Russian).

Lazareva, A.P., 1973, Earthquakes of the Arctic, *in* Vvedenskaya, N.A., Kondorskaya, N.V., and Shebalin, N.V., eds., Earthquakes in the USSR in 1969: Nauka, Moscow, p. 209–211 (in Russian).

Lazareva, A.P., 1975. Earthquakes of the Arctic in 1970 and 1971, *in* Gorbunova, I.V., Kondorskaya, N.V., and Shebalin, N.V., eds., Earthquakes in the USSR in 1971: Moscow, Nauka, p. 145–149 (in Russian).

Lazareva, A.P., 1982, Earthquakes of Chukotka and the Arctic Basin, *in* Kondorskaya, N.V., ed., Earthquakes in the USSR in 1978: Moscow, Nauka, p. 71–72 (in Russian).

Mackey, K.G., 1999, Seismological studies in northeast Russia [Ph.D. thesis]: East Lansing, Michigan State University, 346 p.

Mackey, K.G., and Fujita, K., 1999, The northeast Russia seismicity database and explosion contamination of the Russian earthquake catalog, *in* Proceedings of the 21st Seismic Research Symposium: Technologies for monitoring the Comprehensive Nuclear-Test-Ban Treaty: U.S. Department of Defense, LA-UR-99–4700, v. 1, p. 151–161.

Mackey, K.G., Fujita, K., Gunbina, L.V., Kovalev, V.N., Imaev, V.S., Koz'min, B.M., and Imaeva, L.P., 1997, Seismicity of the Bering Strait region: Evidence for a Bering block: Geology, v. 25, p. 979–982.

Moll-Stalcup, E.J., 1994, Latest Cretaceous and Cenozoic magmatism in mainland Alaska, *in* Plafker, G., and Berg, H.C., eds., The geology of Alaska: Boulder, Colorado, Geological Society of America, Geology of North America, v. G-1, p. 589–620.

Nakamura, K., Plafker, G., Jacob, K.H., and Davies, J.N., 1980, A tectonic stress trajectory map of Alaska using information from volcanoes and faults: Bulletin of the Earthquake Research Institute, v. 55, p. 89–100.

Natal'in, B.A., Amato, J.M., Toro, J., and Wright, J.E., 1999, Paleozoic rocks of northern Chukotka Peninsula, Russian Far East: Implications for the tectonics of the Arctic region: Tectonics, v. 18, p. 977–1003.

Page, R.A., Biswas, N.N., Lahr, J.C., and Puplan, H., 1991, Seismicity of continental Alaska, *in* Slemmons, D.B., Engdahl, E.R., Zoback, M.D., and Blackwell, D.D., eds., Neotectonics of North America: Boulder, Colorado, Geological Society of America, Decade Map Volume, p. 47–68.

Pavlov, O.V., Chernykh, E.N., and Tabulevich, V.N., 1994, Estimates of areas of weak seismicity where important objects are to be constructed (an example of northeast Russia): Geologiya i Geofizika, v. 35, no. 10, p. 116–123 (in Russian).

Peltzer, G., and Tapponier, P., 1988, Formation and evolution of strike-slip faults, rifts, and basins during the India-Asia collision: An experimental approach: Journal of Geophysical Research, v. 93, p. 15085–15117.

Pol'kin, Y.I., 1984, Chukchi Sea, *in* Gramberg, I.S., and Pogrebitsky, Y.E., eds., Geologic structure of the USSR and regularities in the distribution of mineral resources: Leningrad, Nedra, Seas of the Soviet Arctic, v. 9, p. 67–79 (in Russian).

Rudich, N.N., 1985, The hot and cold north: Moscow, Nauka, 80 p. (in Russian).

Shipilov, E.V., Senin, B.V., and Yunov, A.Y., 1989, Sedimentary cover and basement of the Chukchi Sea from seismic data: Geotectonics, v. 23, p. 456–463.

Smirnov, V.N., 1995, Morphotectonics of mountain-building regions of northeast Asia: Author's summary of doctoral dissertation: Moscow, Moscow State University, 40 p. (in Russian).

Till, A.B., and Dumoulin, J.A., 1994, Geology of Seward Peninsula and Saint Lawrence Island, *in* Plafker, G., and Berg, H.C., eds., The geology of Alaska: Boulder, Colorado, Geological Society of America, Geology of North America, v. G-1, p. 141–152.

Tolson, R.B., 1987, Structure and stratigraphy of the Hope Basin, southern Chukchi Sea, Alaska, *in* Scholl, D.W., Grantz, A., and Vedder, J.G., eds., Geology and resource potential of the continental margin of western North America and adjacent ocean basins: Beaufort Sea to Baja California: Houston, Texas, Circum-Pacific Council for Energy and Mineral Resources, Earth Science Series, v. 6, p. 59–71.

Turner, D.L., and Swanson, S.E., 1981, Continental rifting: A new tectonic model for the central Seward Peninsula, *in* Wescott, E., and Turner, D., eds., Geothermal reconnaissance survey of the central Seward Peninsula, Alaska: Fairbanks, University of Alaska, Geophysical Institute Report 284, p. 7–36.

Vorob'eva, L.A., and Yugova, R.S., 1988, Earthquakes of the Northeast, *in* Kondorskaya, N.V., ed., Earthquakes in the USSR in 1985: Moscow, Nauka, p. 174–177 (in Russian).

Yugova, R.S., Efremova, L.V., and Gunbina, L.V., 1997, Earthquakes of the Northeast, *in* Starovoit, O.E., ed., Earthquakes in Northern Eurasia in 1992: Moscow, Geoinfommark, p. 101–105.

Zolotarskaya, S.B., Nikitenko, Y.P., and Ufimtsev, G.F., 1987, Contemporary vertical movements of the Earth's crust, Eastern Siberia and the Far East, *in* Logachev, N.A., ed., Processes in the formation of the relief of Siberia: Novosibirsk, Russia, Nauka, p. 116–121 (in Russian).

Manuscript Accepted by the Society May 15, 2001.

Paleozoic links among some Alaskan accreted terranes and Siberia based on megafossils

Robert B. Blodgett
Department of Zoology, Oregon State University, Corvallis, Oregon 97331, USA
David M. Rohr
Department of Geology, Sul Ross State University, Alpine, Texas 79832, USA
A.J. Boucot
Department of Zoology, Oregon State University, Corvallis, Oregon 97331, USA

ABSTRACT

Biogeographically distinctive Paleozoic fossil biotas from the Farewell terrane (which includes three previously defined subterranes, Nixon Fork, Dillinger, and Mystic) of southwestern Alaska document close ties to Siberia and the Urals. Close ties are especially notable in Middle Cambrian trilobites, Upper Ordovician brachiopods, Upper Silurian brachiopods and aphrosalpingid sponges, Lower Devonian (Emsian) brachiopods, and Permian plants. These biogeographic links suggest that the Farewell terrane most likely represents a continental margin sequence that rifted away from the Siberian continent. Siberian origins are also indicated for other major Alaskan terranes (i.e., the Alexander terrane of southeastern Alaska, the Arctic Alaska terrane, and York terrane of northern Alaska) based on their close faunal ties with the Farewell terrane and Siberia. Faunally, the only truly North American part of Alaska is the triangular area within east-central Alaska, bounded roughly by the Porcupine River to the northwest, and the Yukon River on the southwest. Paleozoic faunas from this area are closely similar to those found in Laurentian (North American) miogeoclinal and cratonic strata of northwestern Canada to the east, and are quite distinct from Farewell terrane faunas at the species level throughout most of the Paleozoic.

INTRODUCTION

In recent years the concept that nearly all of Alaska, as well as much of the western Cordillera of North America, is composed of numerous discrete, accreted tectonostratigraphic terranes (Coney et al., 1980; Jones et al., 1981, 1982, 1986, 1987; Jones and Silberling, 1982; Nokleberg et al., 1994) has gained general acceptance. The only exception to this in Alaska is the triangular area of east-central Alaska, bounded on the northwest by the Porcupine River and on the southwest by the Yukon River.

This region, which includes the Nation Arch, represents a continuation of Precambrian and Paleozoic rocks exposed farther to the east in the Ogilvie Mountains of Yukon Territory.

In this chapter we review the newly emerging published literature on Paleozoic (primarily Ordovician-Devonian) megafaunas, and to a lesser extent megafloras, from Alaskan accreted terranes in order to better determine their origins. With the exception of the southeastern part of the state (Alexander terrane), Paleozoic faunas and floras were poorly documented in the paleontological literature and it has only been since the 1960s that

Blodgett, R.B., Rohr, D.M., and Boucot, A.J., 2002, Paleozoic links among some Alaskan accreted terranes and Siberia based on megafossils, *in* Miller, E.L., Grantz, A., and Klemperer, S.L., eds., Tectonic Evolution of the Bering Shelf–Chukchi Sea–Arctic Margin and Adjacent Landmasses: Boulder, Colorado, Geological Society of America Special Paper 360, p. 273–290.

there has been significant documentation of their past life forms in other parts of Alaska. This lack of knowledge has been even more acute in southwestern and west-central Alaska (Farewell terrane), for which publications on fossils have appeared only since the early 1980s.

The terranes discussed in this chapter include those in which early and middle Paleozoic strata form a significant part of the stratigraphic framework, most notably the Farewell, York, Livengood, and Alexander terranes, as well as the Arctic Alaska terrane (see Fig. 1 for locations of these terranes). In addition, there is limited discussion of the fossil biotas of the only nonaccreted portion of Alaska, which is the triangular area of east-central Alaska that is roughly bounded on the north by the Porcupine River and on the south by the Yukon River and Tintina fault.

FAREWELL TERRANE

The Farewell terrane of southwestern and west-central Alaska (Fig. 1) was established by Decker et al. (1994) as a tectonostratigraphic entity incorporating three previously named, genetically related terranes (Nixon Fork, Dillinger, and Mystic). The latter were relegated at that time to subterranes of the former. Gilbert and Bundtzen (1984) regarded the Dillinger and Mystic terranes (now subterranes; see Fig. 1 for their distribution) to represent a single stratigraphic succession of Paleozoic to Triassic age, and preferred to apply the term "sequence" to each terrane. They considered the Dillinger sequence to be a Cambrian to Lower Devonian deep-water succession and the Mystic to consist of laterally variable Devonian to Triassic (?) shallow-water to nonmarine sedimentary rocks with intrusive and extrusive mafic and ultramafic rocks. Strata as young as Early Jurassic are now also known from the Mystic sequence or subterrane from the Shellabarger Pass region, Talkeetna (C-6) Quadrangle (Sandy and Blodgett, 2000). The close stratigraphic relationship between the Dillinger and Mystic sequences (or subterranes) was supported by stratigraphic studies by Blodgett and Gilbert (1992a) to the southwest in the Lime Hills Quadrangle. The Nixon Fork subterrane (Fig. 1) was originally defined as a full-rank terrane by Patton (1978), and is characterized primarily by a dominantly carbonate succession of early and middle Paleozoic age, and some interbeds of deep-water, basinal rocks. This dominantly shallow-water succession was recognized to be essentially equivalent to deeper-water Cambrian to Lower Devonian rocks of the Dillinger subterrane (Blodgett, 1983; Blodgett and Gilbert, 1983; Bundtzen and Gilbert, 1983; Gilbert and Bundtzen, 1983). Rocks as old as probable Late Proterozoic are known in the lowermost part of the Nixon Fork subterrane (Babcock et al., 1993, 1994; Blodgett, in Nokleberg et al., 1994).

Until several years ago, most prior interpretations of the gross lithostratigraphy and faunal affinities of the best-studied subterrane, the Nixon Fork subterrane, had indicated either fragmentation from or lithic continuity with northwestern Canada (Blodgett, 1983; Churkin et al., 1984; Blodgett and Clough, 1985; Rohr and Blodgett, 1985; Abbott, 1995). Newly acquired biogeographic data of Farewell terrane biotas from time intervals of heightened global endemism, most notably the Ashgillian (late Late Ordovician) and Emsian (late Early Devonian), in conjunction with affinities of faunas and floras of other time intervals, now indicate that this continental margin sequence originated from the Siberian continent (see Fig. 2 for localities referred to from within the former Soviet Union), rather than northwestern North America (Blodgett, 1998; Blodgett and Brease, 1997; Blodgett and Boucot, 1999).

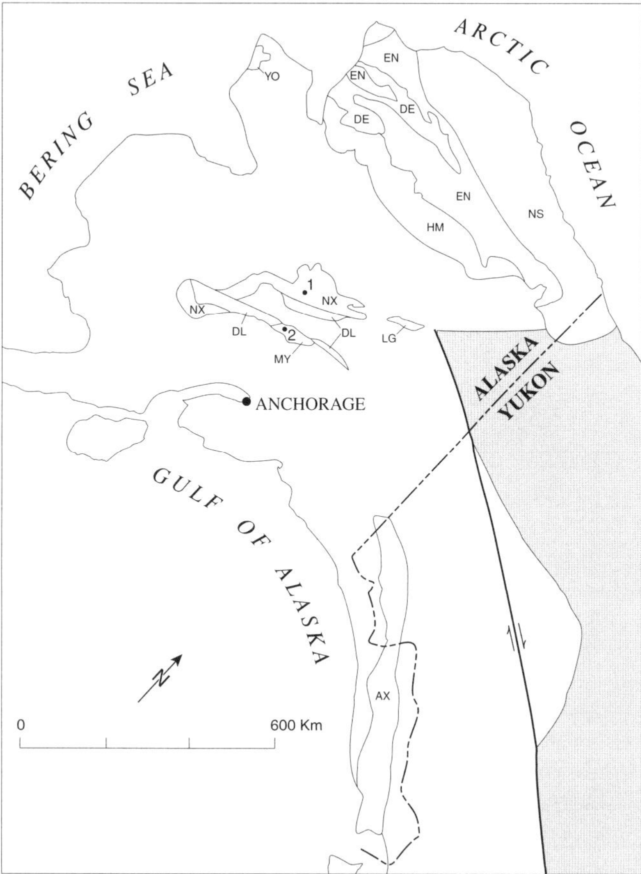

Figure 1. Location of cited tectonostratigraphic terranes and component subterranes in Alaska (modified from Coney et al., 1980). Farewell terrane: NX—Nixon Fork subterrane, MY—Mystic subterrane, DL—Dillinger subterrane, LG—Livengood terrane. Arctic Alaska terrane: HM—Hammond subterrane, DE—De Long Mountains subterrane, EN—Endicott Mountains subterrane, NS—North Slope subterrane, YO—York terrane, AX—Alexander terrane. Stipple pattern is North American autochthonous basement. Localities: 1, Reef Ridge area, Medfra (B-3) Quadrangle. 2, Shellabarger Pass, Talkeetna (C-6) Quadrangle.

Figure 2. Map of former Soviet Union showing localities referred to in text that show close faunal relationships with Alaskan accreted terranes during early and middle Paleozoic time. See Figure 1 for abbreviations.

NIXON FORK SUBTERRANE

Middle Cambrian faunas

Two stratigraphically distinct, rich, diverse Middle Cambrian trilobite faunas were discovered in 1984 in the Nixon Fork subterrane strata in the Sleetmute (A-2) Quadrangle. These were initially discussed by Palmer et al. (1985), who considered both faunas to be representative of an outer shelf environment that is biogeographically of Siberian aspect, with faunal elements previously unrecognized in North America. Alternatively, it was suggested by Babcock and Blodgett (1992) and Babcock et al. (1993) that these southwestern Alaskan trilobite faunas of Siberian aspect may have dispersed in cool waters below the thermocline because they show strong similarity to autochthonous outer shelf faunas of north Greenland. Additional discussion of these trilobite faunas can be found in St. John and Babcock (1997) and Kingsbury and Babcock (1998). However, A.R. Palmer (1999, oral commun. to Blodgett) reiterated his views that the Middle Cambrian trilobite faunas of southwest Alaska were of Siberian aspect and most closely similar to faunas from the eastern part of the Siberian platform.

Late Ordovician (Ashgillian) fauna

Upper Ordovician (Ashgillian) strata of the Lone Mountain area (McGrath [C-4] Quadrangle, west-central Alaska), also part of the Nixon Fork subterrane, yield a rich, diverse silicified fauna of brachiopods, gastropods, and corals (Rohr and Blodgett, 1985; Blodgett et al., 1992a). The most abundant faunal element is the pentameroid brachiopod *Tcherskidium* (Rohr and Blodgett, 1985; Potter et al., 1988a, 1988b; Potter and Blodgett, 1992; Blodgett et al., 1992a). This genus has been identified at only two sites of undoubted North America origin (Black River

Quadrangle of east-central Alaska and north Greenland), but is characteristic and abundant in Ashgillian faunas from Siberia (Kolyma, Taimyr, and Chukotka). The Lone Mountain form represents a new species of this genus, which is also present in the Shublik Mountains (Fig. 3, 1–5) of the northeastern Brooks Range in the Arctic Alaska terrane of Moore (1992). Gastropods from the Lone Mountain section are most similar to species from the York Mountains of the Seward Peninsula (Blodgett et al., 1992a), which belongs to the York terrane. The new genus *Alaskadiscus* Rohr, Frýda, and Blodgett, 2002 is known only from the York terrane and Lone Mountain. *Siskiyouspira* is an uncommon gastropod genus reported only from Lone Mountain, the Brooks Range (see Fig. 4, 4–6), and the Klamath Mountains, California. Older Ordovician shelly faunas are known from the Nixon Fork subterrane, but their biogeographic affiliations are more generalized, containing taxa of both North American and Siberian affinities (Rohr and Blodgett, 1988; Rohr et al., 1992; Rohr and Gubanov, 1997; Rohr and Yochelson, 1999; Measures et al., 1992).

Late Silurian–early Early Devonian (Lochkovian) fauna and flora

Farewell terrane Late Silurian and early Early Devonian (Lochkovian) rocks dramatically differ from contemporaneous North American rocks in the widespread and extensive buildup of algal reef mound complexes along the outer carbonate platform margin of the Nixon Fork subterrane of the Farewell terrane (Clough and Blodgett, 1985, 1988, 1992; Blodgett et al., 1984; Blodgett and Clough, 1985; Blodgett and Gilbert, 1992b). These southwestern Alaskan buildups, along with their associated brachiopods, aphrosalpingids (sphinctozoan sponges restricted to the Upper Silurian, see Rigby et al., 1994; also Fig. 5), and calcareous green algal flora are very similar to equivalent, coeval buildups in the Urals. The only other part of present-day North America with similar algal buildups of equivalent age is the Alexander terrane of southeastern Alaska (Soja, 1994; Soja and Antoshkina, 1997), which is also of accretionary origin.

Late Early Devonian (Emsian) fauna

A biogeographically distinctive brachiopod- and coral-rich fauna is present in the Soda Creek Limestone of late Early Devonian age (early Emsian [*Polygnathus dehiscens* zone]) (Blodgett et al., 1995, 2000; Blodgett, 1998) in the Medfra (B-3) Quadrangle, west-central Alaska. Blodgett et al. (1995) and Chalmers et al. (1995) published abstracts on the biostratigraphic and depositional environments of this distinctive 150 m unit, named the Soda Creek Limestone by Blodgett et al. (2000). This unit composes a distinctive marker unit within the Whirlwind Creek Formation of Dutro and Patton (1982). We believe the Whirlwind Creek Formation should be raised in level to group rank, rather than formation.

Figure 3. Biogeographically significant Upper Ordovician (Ashgillian) brachiopods from North Slope subterrane of Arctic Alaska terrane and Kolyma Region of northeastern Siberia. 1–5: *Tcherskidium* n. sp. from upper part of member 8 of Nanook Limestone, Shublik Mountains. 1–3 are ventral, lateral, and posterior views of ventral valve, ×1.5. 4 and 5 are dorsal and posterior views of dorsal valve, ×1.5. 6: *Tcherskidium unicum* (Nikolaev, 1968) from Ina River, Kolyma Region, northeastern Siberia, ventral view of ventral valve, ×1.5. This species is larger and more coarsely ribbed than *Tcherskidium* n. sp. It also occurs in Baird Mountains Quadrangle of northwestern Brooks Range (in rocks of Hammond subterrane).

The Whirlwind Creek Formation was described by Dutro and Patton (1982, p. H19) as "an Upper Silurian to Upper Devonian sequence of predominantly shallow-water carbonate rocks, 1,000–1,500 m thick." The type section of the formation was designated as a south-dipping sequence exposed on the ridge between Whirlwind and Soda Creeks in sec. 23, T. 24 S., R. 23 E., Medfra (B-3) Quadrangle. A supplemental section was designated for the higher beds of the formation on the north flank of the syncline in sec. 29 and 30, T. 23 S., R. 25 E., Medfra (B-3) Quadrangle. The Soda Creek Limestone corresponds to the Emsian-Siegenian age interval of limestones with coral biostromes shown in Figure 5 of Dutro and Patton (1982). Although not mentioned in their paper, this distinctive unit is repeated once by a thrust fault along their designated type section of the Whirlwind Creek Formation. The Whirlwind Creek Formation of Dutro and Patton (1982) includes several distinctive stratigraphic units, and is considered here to possibly represent a stratigraphic unit of group rank. Its uppermost strata, as recognized by us, are stratigraphically equivalent to the Cheeneetnuk Limestone of Blodgett and Gilbert (1983) of the McGrath (A-4) and (A-5) Quadrangles of west-central Alaska. The upper part (early Eifelian) of the Cheeneetnuk Limestone contains the

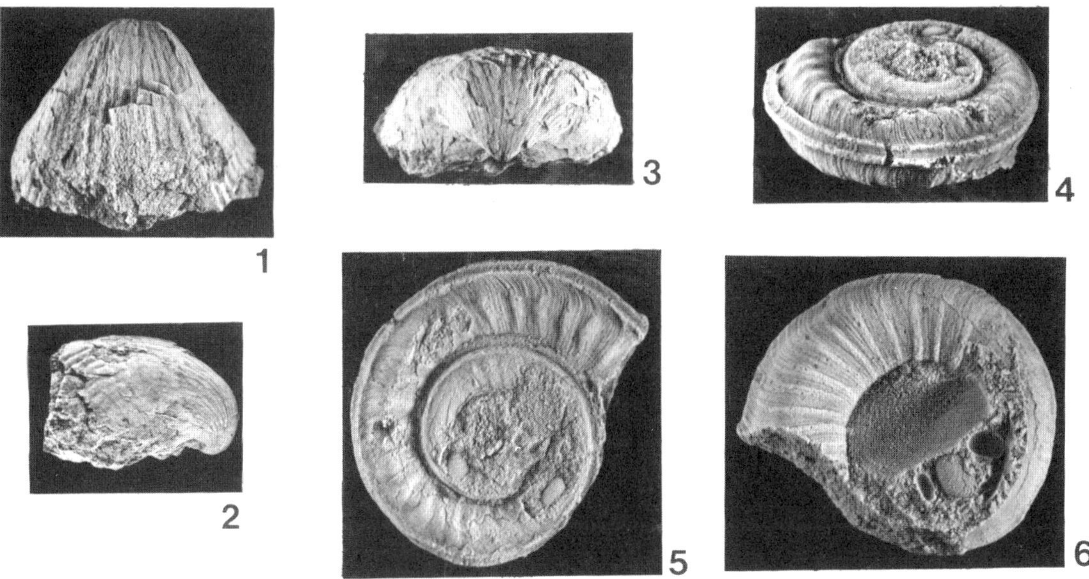

Figure 4. Biogeographically significant Upper Ordovican (Ashgillian) fauna from North Slope subterrane of Arctic Alaska terrane. Both specimens from upper part of member 8 of Nanook Limestone, Shublik Mountains. 1–3: *Eoconchidium* sp., ventral, lateral, and posterior views of ventral valve, ×1.5. 4–6: *Siskiyouspira* sp., oblique side view, apical, and basal views, ×4.

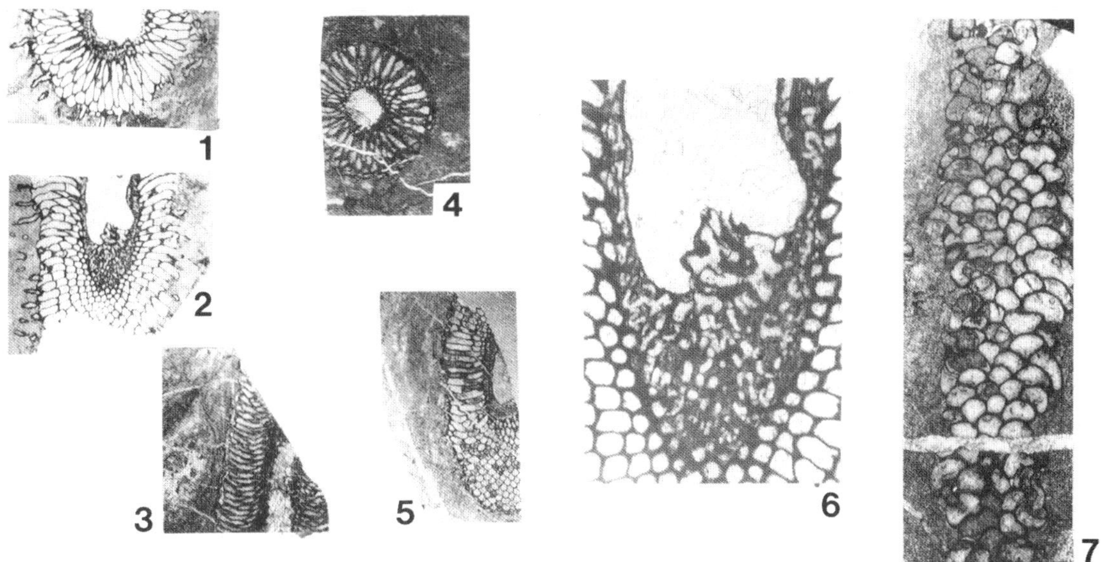

Figure 5. 1–7: *Aphrosalpinx textilis* Miagkova, 1955, a distinctive sphinctozoan sponge associated with Upper Silurian algal reef buildups in Urals, Novaya Zemlya, and Farewell and Alexander terranes of Alaska. Specimens shown in 1, 2, and 6 are from White Mountain area, McGrath (A-4) Quadrangle, west-central Alaska (Nixon Fork subterrane of Farewell terrane); specimens in 3 and 4 are from Shellabarger Pass area, Talkeetna (C-6) Quadrangle, south-central Alaska (from isolated algal reef buildup in Dillinger subterrane of Farewell terrane); those in 5 and 7 are from Tuxekan Island, Craig (D-4) Quadrangle, southeastern Alaska (Alexander terrane). 1, transverse section, ×3; 2, tangential section, ×3; 3, vertical section, ×3; 4, transverse section, ×3; 5, diagonal section in upper part grading to tangential in lower part, ×3; 6, tangential section, ×10; 7, tangential section, ×5.

same fauna and similar lithology as found in the uppermost strata of the type section of the Whirlwind Creek Formation. The Whirlwind Creek was indicated by Dutro and Patton (1982) to be of early Late Devonian (Frasnian) age in its uppermost part. Examination by one of us (Blodgett) of the site indicated by W.W. Patton Jr. as bearing Frasnian fauna shows that these strata belong to the Soda Creek Limestone of Emsian age. No rugose corals belonging to *Smithiphyllum,* the age indicator for the Frasnian noted in Dutro and Patton (1982, p. H21), or the reported accompanying sticklike stromatoporoid *Amphipora,* were recovered. Instead, a diverse fauna of Emsian-age corals was found at the site where *Smithiphyllum* reportedly occurred.

The largest examined collections from the Soda Creek Limestone were derived from a north-trending ridge containing the type section, which is located in sec. 23, T. 24 S., R. 23 E. of the Medfra (B-3) Quadrangle. The unit is repeated once by thrust faulting along this ridge, long known informally as Reef Ridge by mineral company geologists (Fig. 1, locality 1). Brachiopods from this unit are represented by more than 30 species, the most common of which are rhynchonellids (notably uncinulids). Genera present include *Plicogypa, Stenorhynchia, Taimyrrhynx, "Uncinulus," Nordotoechia?, Spinatrypa, Nucleospira, Protathyris, Howellella,* and *Aldanispirifer.* Biogeographically restricted species from this fauna include *Plicogypa* cf. *kayseri* (Peetz), *Taimyrrhynx taimyrica* (Nikiforova), *"Uncinulus" polaris* Nikiforova, and *Howellella yacutica* Alekseeva. These species are known from various parts of Siberia (Kolyma, Taimyr, and Kuznetsk Basin) and Arctic Russia (Novaya Zemlya), but are unknown in equivalent-age strata of east-central Alaska (Nation Arch area), or northwestern Canada. Rugose corals identified by A.E.H. Pedder, formerly of the Canadian Geological Survey of Canada, consist mostly of solitary forms, including *Pseudoamplexus altaicus, Lithophyllum* spp., rare *Zonophyllum, Rhizophyllum schischkaticum,* new species of *Zelophyllum,* and *Acanthophyllum.* Pedder noted that this fauna is related to early Emsian coral faunas from the Kolyma region of northeastern Russia.

Early Middle Devonian (Eifelian) fauna

The Cheeneetnuk Limestone, defined by Blodgett and Gilbert (1983), consists of 457 m of well-bedded limestone exposed in the McGrath (A-4) and (A-5) Quadrangles of west-central Alaska. The upper part of the formation contains several thick beds crowded with the remains of a richly diverse, open-marine biota of Eifelian (early Middle Devonian) age. Biotic elements include brachiopods, rugose and tabulate corals, gastropods, bivalves, rostroconchs, nautiloids, goniatites (House and Blodgett, 1982), tentaculitids, trilobites, ostracodes, crinoid ossicles, sphinctozoan and siliceous sponges (Rigby and Blodgett, 1983), stromatoporoids, and calcareous algae (Poncet and Blodgett, 1987). The same fauna and a similar lithology are found in the uppermost strata of the Whirlwind Creek Formation of Dutro and Patton (1982).

The Cheeneetnuk sponge fauna (Rigby and Blodgett, 1983) is dominated by sphinctozoan sponges, a peculiar calcareous sponge group that has previously not been noted in Devonian strata of cratonic North America. Nonetheless, these sponges are extremely common in the Farewell terrane, where two occurrences of Emsian age are also known, on the southern flank of Limestone Mountain, Medfra (B-4) Quadrangle (Nixon Fork subterrane), and in the Talkeetna (C-6) Quadrangle (Mystic subterrane). The goniatite fauna from the Cheeneetnuk is characterized by the presence of the index fossil *Pinacites jugleri* Roemer, which was previously unknown from North America but is typical of Eifelian faunas of Eurasia (House and Blodgett, 1992). The Cheeneetnuk brachiopod fauna consists of ~25 species, several of which have been described (Baxter and Blodgett, 1994; Blodgett and Johnson, 1994; Johnson and Blodgett, 1993). Gastropods include more than 50 species, of which only ~7 species have been described (Blodgett, 1992, 1993; Blodgett and Johnson, 1992; Blodgett and Rohr, 1989). These gastropods are especially notable for their extremely ornate shell forms. The numerous spines and nodes have been interpreted to indicate that these faunas lived in warm, tropical waters, and they are the most ornate warm-water biota known among Middle Devonian gastropod faunas of North America (e.g., see 9 in Fig. 6). The Cheeneetnuk gastropod fauna is distinct at the species level from correlative gastropod faunas known from the Northwest Territories and Nevada, which probably constituted a single faunal province during Eifelian time. Although the Cheeneetnuk gastropod fauna contains some of the same genera found in the Eifelian strata of the Northwest Territories and Nevada, not one species is as yet known to occur in both areas. It would appear that the gastropod fauna of the Cheeneetnuk Limestone (and its equivalent in the uppermost beds of the Whirlwind Creek Formation, as well as the Cascaden Ridge unit of the Livengood terrane [see following]) composed a distinct, separate faunal province from the Eifelian seaway of the Western Cordillera of cratonic North America. Comparison of the Eifelian gastropod fauna of the Farewell terrane with that of Siberia is precluded by the virtual lack of described species from Siberia. Butusova (1960) described only a few Middle Devonian gastropod species from the Kuznetsk Basin of Siberia.

The upper part of the Cheeneetnuk Limestone (Eifelian) also contains a rich and diverse algal flora, including the dasyclad genus *Coelotrochium,* the charophyte genus *Sycidium,* and the udoteacean alga *Lancicula sergaensis* Shuysky. *Sycidium* is widely reported in the Devonian of the Russian platform and Germany, but is not known elsewhere in North America. *L. sergaensis* was described from the Cheeneetnuk by Poncet and Blodgett (1987), but was originally described from the early Emsian of the Urals by Shuysky (1973). None of these algal taxa have been reported in nonaccreted, cratonic North American Devonian strata. These observations on the general lack of sphinctozoans and calcareous green algae in Early and Middle Devonian strata of the Western Cordillera of North America are supported by the examination (by Blodgett) of the large paleon-

Figure 6. Biogeographically significant Lower and Middle Devonian gastropods from Farewell and Alexander terranes. 1–3: *Chlupacispira spinosa* Blodgett and Rohr, 1989, Emsian (late Early Devonian), south flank of Limestone Mountain, Medfra (B-4) Quadrangle, west-central Alaska (Nixon Fork subterrane of Farewell terrane), apical, apertural, and basal views, ×12. 4: *Euomphalopterus* sp., Emsian (late Early Devonian), south flank of Limestone Mountain, Medfra (B-4) Quadrangle, west-central Alaska (Nixon Fork subterrane of Farewell terrane), lateral view, ×3. 5–6: *Tubina* sp., Emsian (late Early Devonian), Kasaan Island, southeastern Alaska (Alexander terrane), apical and oblique abapertural views, ×4. 7–8: *Oriostoma* sp., Emsian (late Early Devonian), Kasaan Island, southeastern Alaska (Alexander terrane), apical and lateral views, ×4. 9: *Spinulrichospira cheeneetnukensis* Blodgett and Rohr, 1989, Eifelian (early Middle Devonian), Cheeneetnuk Limestone, McGrath (A-5) Quadrangle, west-central Alaska (Nixon Fork subterrane of the Farewell terrane), apertural view, ×5.

tological collections of the Geological Survey of Canada (Calgary, Alberta), and the United States Geological Survey (Washington, D.C.).

MYSTIC SUBTERRANE

Late Early Devonian (Emsian) fauna

A biogeographically distinctive fauna is present in an unnamed limestone of late Emsian (*Polygnathus serotinus* zone) age in the Shellabarger Pass area (Fig. 1, locality 2), Talkeetna (C-6) Quadrangle, south-central Alaska. This lithologic unit is widespread in the Mystic subterrane, and late Emsian conodonts have been described from it, to the west in the Lime Hills (D-5) Quadrangle (Savage and Blodgett, 1995). This unit locally forms the base of the Mystic sequence (Blodgett and Gilbert,

1992b; Gilbert and Bundtzen, 1984). Typically the unit appears to be ~50 m thick throughout the central Alaska Range. In the Shellabarger Pass area, the limestone unit was examined at a single large hilltop exposure, just slightly above 2300 feet in elevation, in the center of the NE1/4 sec. 15, T. 28 N., R. 18 W., Talkeetna (C-6) Quadrangle. It consists of 38.7 m of lime mudstone- and wackestone-bearing moderately abundant megafossils (mostly brachiopods, rugose and tabulate corals, trilobites, and sphinctozoan sponges in order of decreasing abundance). These beds appear to stratigraphically or structurally overlie bedded cherts. Conodonts from this section were identified by Norman M. Savage (1996, personal commun.) as belonging to the *Polygnathus serotinus* zone, indicating a late Emsian age. Distinctive two-hole cirral crinoid ossicles commonly ascribed to the species *Gasterocoma? bicauli* Johnson and Lane 1969, are common throughout the section, as they are in most shallow-water, open-marine platform carbonate environments in Emsian strata of Alaska and northwestern Canada. Brachiopod taxa, in addition to the eospiriferinids reported in the following, also include *Opsiconidon, Teichertina, Sibirirhynchia alata* (Khodalevich), *Gypidula, Ivdelinia* sp. (Fig. 7), clorindinid, *Atrypa, Spinatrypa, Carinatina, Punctatrypa* (*Undatrypa*), and the eospiriferinid genera *Myriospirifer* (Fig. 8, 1–5) and *Janius* (Fig. 8, 6–8). Fossils were noted earlier at this locality (Reed and Nelson, 1980, locality 17; U.S. Geological Survey locality 9589–SD), which they denoted as a locally derived slump block of their Dl unit, limestones of Middle and Late Devonian age. On the basis of its fauna, this locality was indicated to be either late Early Devonian (Emsian) or early Middle Devonian (Eifelian) in age. Detailed field examination of the surrounding area (by Blodgett in late July 1996) indicates that the exposure is actually an in situ outcrop of an Emsian-age limestone that is separated by a thick section of siliciclastic beds from a much more prominent, thicker limestone unit of Frasnian age.

The eospiriferinid genera *Myriospirifer* and *Janius,* as well as the subfamily that includes them, the Eospiriferinae, are nearly unknown anywhere in the Early Devonian or Eifelian strata of the Cordilleran Region of the Old World Realm. The only exception is the citation of the genus *Janius* (Perry and Lenz, 1978, p. 142) or "*Janius*" (Savage et al., 1979, p. 196) from the Royal Creek area, Yukon Territory. No illustrations of this taxon are provided in either publication. Whereas eospiriferinids are only a minor element in the Royal Creek fauna, they form a major portion of the Shellabarger Pass Emsian brachiopod fauna, due to the abundance of specimens belonging to *Myriospirifer.* Thus, this taxon is the second-most abundant brachiopod recovered from an interval 14.9–15.8 m above the base of the Emsian limestone at Shellabarger Pass measured by Blodgett in 1996. However, in marked contrast to their general absence in the Cordilleran Region, eospiriferinid brachiopods are common elements in Emsian strata of Siberia (Kolyma and the Kuznetsk Basin), at numerous localities throughout the Urals and Novaya Zemlya, and in the Central Asia part of the former Soviet Union (Blodgett, 1998).

The Eurasian aspect of the Shellabarger late Emsian fauna is also indicated by other brachiopods. The gypidulinid genus *Ivdelinia* is a typical Old World Realm taxon that is widely reported from Early Devonian (Lochkovian-Emsian) and early Middle Devonian (Eifelian) rocks of the Rhenish-Bohemian and Uralian Regions, but that is almost unknown in the Cordilleran Region of the Old World Realm, which in the Emsian included Arctic and western Canada and Nevada. Only two brachiopod species belonging to this genus have been described from Emsian-Eifelian strata of the Cordilleran Region. These are *Ivdelinia grinnellensis* Brice, 1982, and *Ivdelinia* (*Ivdelinella*) *ellesmerensis* Brice, 1982, both of which occur in the Canadian Arctic Islands. The Shellabarger Pass *Ivdelinia* is distinct in its external morphology from the previously mentioned species and is more closely related to species described from the Ural Mountains. The rhynchonellid brachiopod *Sibirirhynchia alata* (Khodalevich, 1951), which occurs in the late Emsian beds of Shellabarger Pass, was previously unreported in North America, but occurs in Emsian-Eifelian strata in the Urals and Kolyma. As with the Nixon Fork subterrane (Blodgett, 1998; Blodgett and Brease, 1997), the Emsian brachiopods of the Mystic subterrane, as typified by the Shellabarger Pass fauna, also appear

Figure 7. 1–5: *Ivdelinia* sp., biogeographically distinctive brachiopod from Lower Devonian (Emsian) of Shellabarger Pass, Talkeetna (C-6) Quadrangle, south-central Alaska (Mystic subterrane of Farewell terrane), ventral, anterior, lateral, ventral, and dorsal views, ×3.

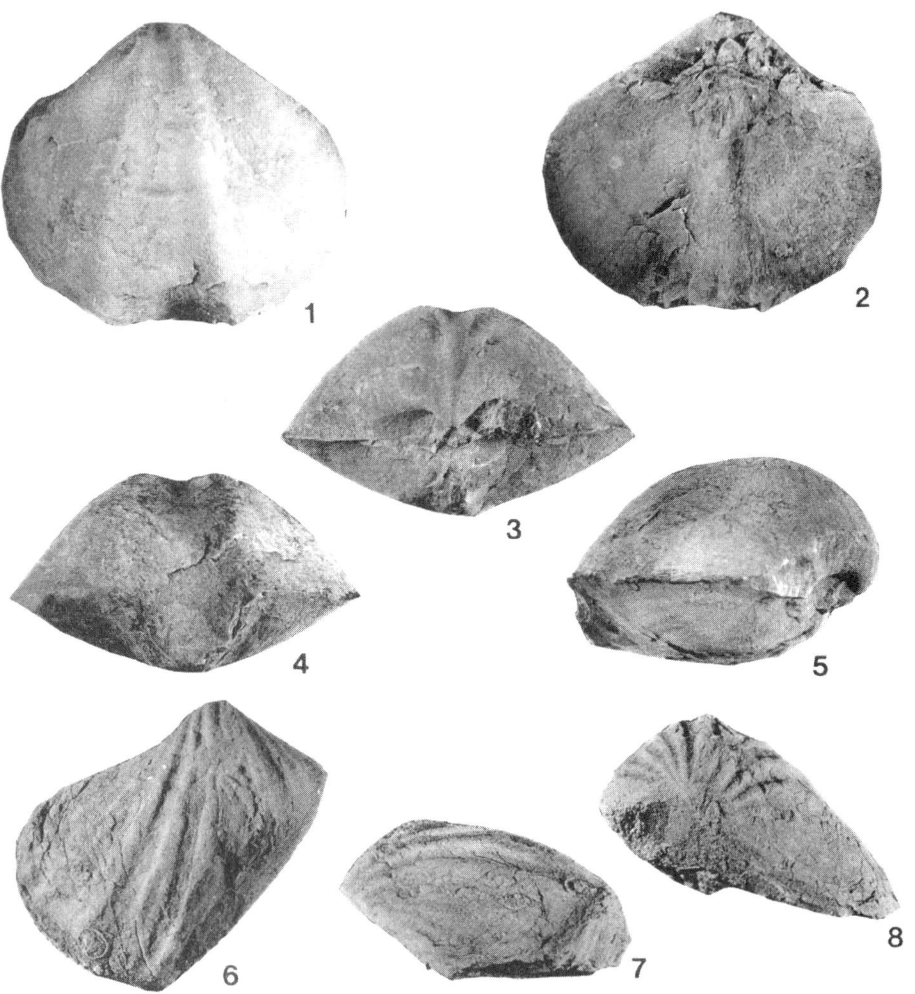

Figure 8. Biogeographically significant Lower Devonian (Emsian) brachiopods from Lower Devonian (Emsian) of Shellabarger Pass, Talkeetna (C-6) Quadrangle, south-central Alaska (Mystic subterrane of Farewell terrane). 1–5: *Myriospirifer breasei* Garcia-Alcalde and Blodgett, 2001; ventral, dorsal, posterior, anterior, and lateral views, ×1.5. 6–8: *Janius* cf. *vetulus* (Eichwald); ventral, posterior, and lateral views of ventral valve, ×1.5.

to biogeographically belong to the Uralian, and not to the Cordilleran Region of the Old World Realm.

Early Late Devonian (Frasnian) fauna

Mamet and Plafker (1982) described and discussed some elements of a Frasnian microbiota (mostly foraminifers) collected from limestones of the Mystic subterrane exposed along both sides of the Farewell fault in two areas: (1) two localities on the north side of the Farewell fault ~6.4 km south-southwest of Farewell in the McGrath (B-2) Quadrangle; and (2) from three closely associated localities on the south side of the fault in the McGrath (A-5) Quadrangle. Close faunal relationships were noted by Mamet and Plafker (1982) between the Farewell assemblage and coeval faunas from the Russian Platform and the Urals, and to some degree with the Southesk Formation of Jasper National Park, Alberta, Canada.

Pennsylvanian fauna. Pennsylvanian-age trilobites from the Farewell terrane in both the Nixon Fork and Mystic subter-

ranes (Hahn et al., 1985; Hahn and Hahn, 1985) showed affinities with Eurasian, rather than North American, taxa.

Permian flora. Mamay and Reed (1984) described and illustrated Permian plants from the conglomerate of Mount Dall in the Talkeetna (C-5) Quadrangle, central Alaska Range, that they considered to be derived from the Mystic terrane of Jones et al. (1981). Although the flora shares some elements with localities in the southwestern part of the United States, its most distinctive element is the genus *Zamiopteris,* which is characteristic of Angaraland (Siberia) and is not known elsewhere in North America. On this basis, they suggested that the Mystic terrane possibly was exotic and accreted to Alaska between post–Early Cretaceous and early Tertiary time.

MIDDLE DEVONIAN FAUNA
OF THE LIVENGOOD TERRANE

The Livengood terrane of east-central Alaska (Fig. 1) was defined by Churkin et al. (1982). Few shelly fauna-bearing lithologic units are present within the stratigraphic succession of this terrane, notable exceptions being the informally named Lost Creek unit of Late Silurian age and the Cascaden Ridge unit of Middle Devonian age (Blodgett et al., 1988c; Weber et al., 1992, 1994). The Cascaden Ridge unit contains a rich gastropod fauna consisting of 35 species of Eifelian (early Middle Devonian) age (Blodgett, 1992). The species of this mid-Eifelian-age fauna were noted to be most similar, and in many cases conspecific, with gastropods from Eifelian-age beds of the Cheeneetnuk Limestone of the McGrath (A-5) Quadrangle and the uppermost part of the Whirlwind Creek Formation in the Medfra (B-3) Quadrangle, both of the Nixon Fork subterrane of the Farewell terrane. Like these two occurrences, the Cascaden Ridge unit gastropod beds also contain abundant remains of the calcareous green algal genus *Coelotrochium* (a dasycladacean, originally described as a foraminifer by Schlüter, 1879). This genus is known from a number of localities of late Emsian–late Eifelian age in Germany (Siegfried Rietschel, 1981, written commun.). It also occurs in the late Eifelian (late Couvinian) of Belgium (Mamet and Preat, 1992) as well as many localities in Eifelian-age rocks of the Farewell terrane. The gastropod fauna is notable in that, although it shares many genera with Eifelian and Givetian localities in the western and mid-continent regions of North America, it contains only one species that is closely related to a species occurring in contemporaneous rocks of undoubted North American origin. Comparison with Siberian Devonian gastropod faunas is precluded due to the nearly total lack of described material from the latter area. However, one of the bellerophontid gastropods present in the Cascaden Ridge unit fauna, *Bucanopsis sullivani* Blodgett, 1992, shows the closest resemblance to one of the few gastropods described from the Middle Devonian of Siberia, i.e., *Bucanopsis salairica* Butusova, 1960, from the Mamontov beds (Eifelian) of the Kuznetsk Basin.

ARCTIC ALASKA TERRANE

The Arctic Alaska terrane, equivalent here to the Arctic Alaska superterrane of Moore (1992), extends the entire length of the Brooks Range in northern Alaska (Fig. 1). It incorporates a number of previously defined terranes (i.e., North Slope, Endicott Mountains, De Long Mountains, Hammond, and Coldfoot terranes). Although its early and middle Paleozoic-age megafossils are poorly characterized in comparison to other parts of Alaska, biogeographically distinctive shelly faunas of that age are now known from a number of localities within this terrane.

HAMMOND SUBTERRANE

Late Ordovician (Ashgillian) fauna

The pentameroid brachiopod *Tcherskidium unicum* (Nikolaev, 1968) is present in Late Ordovician (Ashgillian) carbonates exposed near Hub Mountain in the northwestern Brooks Range. These specimens are from locality 66ATr82.3, collected

by I.L. Tailleur in 1966 from ~60.8 m above the base of a carbonate unit in the northeastern part of the Baird Mountains Quadrangle (lat 67°44.15′N, long 159°27.5′W). The occurrence of the genus *Tcherskidium* Nikolaev and Sapelnikov, 1969, was also noted from this locality in Blodgett et al. (1992a). In the late 1960s, these specimens had been referred to the pentameroid brachiopod genus *Conchidium,* resulting in the miscorrelation of these beds with the Silurian. *Tcherskidium unicum* (Nikolaev, 1968), the type species of the *Tcherskidium,* originally was described from specimens collected in the Kolyma River basin of northeastern Siberia (6 in Fig. 3) and subsequently reported from the basin of the Lower Taimyr River in central Taimyr (Nikiforova, 1989; Cocks and Modzalevskaya, 1997).

DE LONG MOUNTAINS SUBTERRANE

Early Middle Devonian (Eifelian) fauna

Baxter and Blodgett (1994) illustrated and described a new species of *Droharhynchia, D. rzhonsnitskayae,* which is present in Eifelian-age beds of the Cheeneetnuk Limestone of the McGrath Quadrangle (Nixon Fork subterrane of the Farewell terrane) and in the Baird Group in the Howard Pass (B-5) Quadrangle in the northwestern Brooks Range. As noted in Baxter and Blodgett (1994, p. 1235), several other brachiopod and coral species from the Howard Pass locality also are present in the Cheeneetnuk fauna of the Farewell terrane.

Late Middle Devonian (Givetian) fauna

Blodgett and Dutro (1992) illustrated stringocephalid brachiopods (belonging to the subgenus *Stringocephalus* [*Stringocephalus*] and questionably to the genus *Geranocephalus*) from Givetian carbonate rocks of the Baird Group exposed at Punupkahkroak Mountain, De Long Mountains (A-3) Quadrangle in the northwestern Brooks Range. Unfortunately for the purposes of biogeographic analysis, these taxa primarily indicate only a generalized Old World Realm tropical nature of the fauna.

Mississippian flora

Fossil plants were recovered from a number of localities in the Mississippian Kurupa Sandstone of the central Killik River Quadrangle (Spicer and Thomas, 1987). Lycopods dominate the described flora, which includes taxa previously known only from Mississippian floras in Siberia. On this basis, Spicer and Thomas suggested that northern Alaska and Siberia were joined together during Mississippian time.

ENDICOTT MOUNTAINS SUBTERRANE

Late Late Devonian (Famennian) fauna

Dutro et al. (1994) illustrated and discussed late Late Devonian brachiopods from the Noatak Sandstone of the Endicott

Group in the Killik River (A-5) Quadrangle. Both the brachiopods and conodonts from this locality indicate an early Famennian age. Unfortunately, for the purposes of biogeographic analysis, Famennian faunas tend to be very cosmopolitan, and hence little can be said about the brachiopod fauna other than that the species illustrated appear to be closely related to taxa from western Canada and Nevada.

NORTH SLOPE SUBTERRANE

Late Ordovician (Ashgillian) fauna

Ashgillian fossils from the uppermost part of the Nanook Limestone (as revised by Blodgett et al., 1992b) in the North Slope subterrane in the Shublik Mountains, northeastern Brooks Range, were discussed and illustrated in Blodgett et al. (1988b). These fossils included a different species of *Tcherskidium, T.* n. sp. (Fig. 3, 1–5), identical with the species found in the Lone Mountain area (McGrath [C-4] Quadrangle) of the Farewell terrane. Another distinctive pentameroid brachiopod, *Eoconchidium* (Fig. 4, 1–3), is also present in these beds. This biogeographically restricted genus was previously reported only from the Ashgillian of northeastern Siberia (type species, *E. indigiricum,* from the Kolyma Region) and Kazakhstan (Sapelnikov, 1985), although Sapelnikov noted a questionable occurrence from the lower Ludlovian of southern Kazakhstan.

Early Middle Devonian (Eifelian) fauna

Popov et al. (1994) described a new species of the lingulid brachiopod genus *Bicarinitina, B. kongakutensis,* from the informally named Ulungarat unit of Anderson (1991) from the Demarcation Point (A-4) Quadrangle, northeastern Brooks Range. This occurrence is considered to be Eifelian (early Middle Devonian), on the basis of the presence of the dasycladacean alga *Coelotrochium* and a distinctive new genus of murchisoniid gastropod, known elsewhere in Alaska only in the Eifelian. The genus *Bicarinatina* previously was known in the Devonian only from Eifelian and Givetian beds of the East Baltic and Russian Platform. The dasycladacean *Coelotrochium* is a seemingly widespread algal element of shallow, nearshore or inner platform facies rocks in all Alaskan accreted terranes (Farewell, Alexander, Livengood, and Arctic Alaska) that contain well-developed Eifelian-age stratigraphic successions. Despite its extreme abundance at a number of sites in these disparate, widely separated Alaskan terranes, *Coelotrochium* has never been observed in Middle Devonian localities of the North American craton. This observation has been confirmed by the examination by Blodgett of the large collections of Middle Devonian fossils in the collections of the U.S. Geological Survey and the Geological Survey of Canada, as well as by inquiry and shipment of *Coelotrochium* specimens by Blodgett to nearly all paleontologists and stratigraphers working on Devonian problems in western North America. Outside of its occurrence in Alaska's ac-

creted terranes, the genus has previously been noted only in late Emsian–late Eifelian strata of Germany, and in the Givetian of Belgium.

ORDOVICIAN FAUNA OF THE YORK TERRANE

The York terrane consists primarily of unmetamorphosed carbonate rocks of Ordovician to Mississippian age, and is situated at the western end of the Seward Peninsula (Fig. 1). Siberian and/or Russian Platform affinities are well demonstrated by shelly faunas from the Ordovician of the York terrane. Early Ordovician trilobites from the Lost River area were first described by Ross (1965). Subsequently, Ormiston and Ross (1979, p. 57) noted ". . . trilobite fauna of Nova Zemlya includes all the Lost River genera except *Triarthrus* and suggests a close faunal tie between Novaya Zemlya, Alaska and eastern Balto-Scandia in the Early Ordovician." This evidence was suggested, as for the monorakid trilobites noted in the following, to indicate that the Seward Peninsula was not part of Alaska until after Ordovician time. Ormiston and Ross (1979) also pointed out that the occurrence of the trilobite genus *Monorakos* in Upper Ordovician strata of the western Seward Peninsula (York terrane) suggested close linkage of the latter area with what they termed the *Siberia-Kolyma continent* because this genus was reported previously only from the northern part of Russia (Siberian platform, Taimyr, New Siberian Islands, Kolyma, and Chukotka). The Seward Peninsula *Monorakos* was recognized to represent a new species, named *Monorakos kledos* by Ormiston (1978), based on material he collected along the Don River.

Siberian and/or Russian affinities also are indicated for Upper Ordovician corals from the York terrane (Robert J. Elias, 1998, written commun.). Elias wrote the following (our brackets):

With regard to Seward [Peninsula] biogeography, the only possibly useful information I can provide at this time is based on collections 68–ADu-16 (USGS 6746–CO) and 66–ASn-442 (USGS 6026–CO), both from the same horizon situated near the base of the Ordovician-Silurian section at Hill 760, about 1 mile west of Don River, Seward Peninsula, Alaska (Sainsbury et al., 1971, p. C55–C56, fig. 3). Oliver (in Oliver et al., 1975, p. 23, table 2) identified *Bighornia* sp.; in my opinion, this is an undescribed new species of *Bighornia*. It differs from all known species of the genus in having a circular cross-sectional shape throughout ontogeny, rather than subcalceoloid or modified subcalceoloid form. It resembles three species, which also have a cardinal septum that becomes thin and decreases in length during intermediate ontogenetic stages: *Bighornia patella* (Wilson, 1926), which occurs widely in Richmondian strata within the Red River-Stony Mountain Province of North America, *Bighornia orvikui* Kaljo, 1960, from the Ashgill of Estonia, and *Bighornia* sp. of Sytova, 1979, from the Boorsky Subhorizon of the Ketsky Horizon (Ashgill) in the Morkoka River basin of the Siberian platform. It may be closest to *Bighornia* sp. of Sytova, in having a very large columella.

Rohr (1988) described a diverse Late Ordovician gastropod fauna from the York terrane. Many of the taxa are cosmopolitan, but the species cited by Rohr (1988, p. 552) as "*Tropidodiscus* aff. *T. subactus* (Ulrich)" occurs only in the York terrane and near Lone Mountain (McGrath [C-4] Quadrangle, west-central Alaska) in the Nixon Fork subterrane of the Farewell terrane. Recently, this species has been redescribed and formally established under the name *Alaskadiscus donensis* as the type species of new bellerophontoidean genus by Rohr et al. (2002).

ALEXANDER TERRANE

Silurian faunas

Despite having been better studied paleontologically than the remainder of Alaska, the Alexander terrane of southeastern Alaska (Fig. 1) remains relatively poorly known in comparison with cratonic North America. Nevertheless, the Silurian-Devonian shelly faunas of the Alexander terrane clearly demonstrate non-North American affinities and close affinities with Alaskan accreted terranes that have Siberian and/or Uralian affinities, notably the Farewell terrane. Late Silurian brachiopods are reasonably well documented from the Heceta Limestone of the western part of Prince of Wales Island (Kirk, 1922, 1925, 1926; Kirk and Amsden, 1952; Savage, 1989). One of the collections listed by Kirk and Amsden (1952), USGS (U.S. Geological Survey) locality 2689–SD, was determined by Savage (1981) on the basis of conodonts to be early Pragian (middle Early Devonian) in age, and he reassigned it to the Karheen Formation. Savage also suggested that the citation of *Atrypoidea* from this locality is an error. The overall aspect of the Silurian brachiopod faunas is Late Silurian, post-mid-Wenlock, pre-Pridolian.

Several of the Silurian taxa listed by Kirk and Amsden (1952), such as *Delthyris?* and *Atrypa* cf. *reticularis,* are biogeographically relatively cosmopolitan. However, *Brooksina, Harpidium,* and *Cymbidium* are relatively abundant in the Uralian-Cordilleran Region, with *Harpidium* also occurring in the North American Province of the North Atlantic Region. *Atrypoidea* is widespread in the Uralian-Cordilleran Region, but also occurs in the North Silurian Region (Gotland, the Baltic States, mid-continent North America) and South China plus eastern Australia (Sino-Australian Province). Kirk and Amsden's (1952) *Conchidium alaskense* has been reassigned to the relatively cosmopolitan, extra-Malvinokaffric Realm genus *Kirkidium.* Savage (1989) recognized Jones's (1979) pentameroid genus *Nanukidium* in southeast Alaska. This genus was originally described from Arctic Canada (North American Province). Except for the presence of *Nanukidium,* however, the southeast Alaska brachiopod fauna has Uralian-Cordilleran affinities. The significance of *Nanukidium* is problematic, however, because it occurs at only two localities globally and requires good serial sections for its recognition (it is an external homeomorph of *Kirkidium*).

Notable among the Silurian brachiopods are such large, distinctive pentameroid genera as *Brooksina* Kirk, *Harpidium* Kirk, and *Cymbidium* Kirk, all based on specimens found in the Heceta Limestone. These three genera even occur together in one collection, USGS locality 1005–SD, from the south shore of Kosciusko Island (Kirk and Amsden, 1952, p. 53). In terms of overall brachiopod taxic composition, the faunal association from Kosciusko Island closely matches the *Brooksina* community of Sapelnikov et al. (1999) based on collections from the Lozva River, Ivdel Region of the eastern slope of the northern Urals. As noted by Kirk and Amsden (1952, p. 54), the Heceta faunas most closely resemble Late Silurian faunas described by Khodalevich (1939) from the eastern slope of the Urals. The genus *Brooksina,* according to Sapelnikov (1985), is known from late Wenlock to Ludlovian strata and occurs in both the eastern and western slope of the Urals, Gornyi Altai, southern Tian Shan, Kazakhstan, and northeastern Siberia, as well as in Ludlovian strata of southeastern Alaska and Nevada (Johnson et al., 1976). The genus *Cymbidium,* according to Sapelnikov (1985), occurs in Ludlovian strata of southeastern Alaska and central Nevada and questionably in the Omulevsk Mountains of northeastern Siberia. The genus *Harpidium,* according to Sapelnikov (1985), ranges in age from late Llandoverian to Ludlovian and is much more widespread in distribution than the first two genera, occurring in the Urals, the Central Asia portion of the former USSR, Kazakhstan, the Goryni Altai, northeastern Siberia, southeastern Alaska, the mid-continent region of the United States, and Greenland. The overall affinities of these three distinctive Alexander terrane genera still appear to be closest to the Urals. Although the genus *Cymbidium* has not been reported from the Urals, it is questionably reported from the Kolyma Region of Siberia.

Another interesting Silurian spiriferoid genus, *Alaskospira* Kirk and Amsden, 1952, was also established from, and appears to be restricted to, southeastern Alaska. The type species is *Alaskospira dunbari* Kirk and Amsden from the south shore of Kosciusko Island. According to Kirk and Amsden (1952, p. 55), another point of similarity between the southeastern Alaskan Silurian brachiopod faunas and those of the Urals is the presence of a species described by Khodalevich (1939) as *Spirifer (Martinia) pseudopentameriformis.* Kirk and Amsden were of the opinion that this species was probably congeneric with *Alaskospira.* It is also interesting to note that a species nearly identical externally to *A. dunbari* is present in strata transitional between deep-water, thin-bedded platy limestones and an overlying massive algal reef complex (similar to equivalent-age reefal buildups in the Alexander terrane) on the southeastern flank of White Mountain, in the McGrath (A-4) Quadrangle in the Farewell terrane of southwestern Alaska.

Like contemporaneous rocks in the Farewell terrane, the Heceta Limestone of the Alexander terrane contains extensive Late Silurian algal reef mound complexes with an algal flora and associated aphrosalpingid sphinctozoan sponges (Fig. 5) that are also known in the Urals (Soja, 1994; Soja and Antoshkina, 1997; Soja and Riding, 1993; Riding and Soja, 1993; Rigby et al., 1994). Similar complexes are unknown from rocks of equivalent age in the North America craton.

Devonian faunas

An unusually rich fossil fauna is known from Emsian (late Early Devonian) strata of Kasaan Island on the eastern side of Prince of Wales Island, southeastern Alaska. Kindle (1907) provided a fauna list from these beds and noted that the entire fauna was more strongly allied with Asiatic and European faunas than with contemporaneous North American faunas outside of Alaska. The foreign affinities were particularly emphasized by the presence of the bivalve *Hercynella* in a single horizon in the lower part of the Devonian section on Kasaan Island. Chapin and Kirk (*in* Buddington and Chapin, 1929) provided a revised faunal list for the Kasaan Island faunas. Forney et al. (1981, p. 131) noted the presence of typical Bohemian molluscan elements in the Emsian strata of Kasaan Island, on the eastern side of Prince of Wales Island, southeastern Alaska, including the distinctive bivalve *Hercynella* cf. *bohemica* Barrande, the oriostomatid gastropod *Oriostoma* sp. aff. *princeps* Oehlert, and the tremanotid gastropod *Boiotremus* cf. *fortis* Frech. Lower Devonian Bohemian (or Barrandian) faunas have close biogeographic affinities with Uralian faunas of Russia. We examined newly acquired gastropod collections from Kasaan of C.M. Soja and N.M. Savage, as well as the original collection of E.M. Kindle. Our perusal of these collections clearly demonstrates their non-North American character, which is unlike contemporaneous faunas from Nevada, western Canada, or Arctic Canada.

Blodgett et al. (1988a) illustrated two gastropod species from the Kasaan beds: *Oriostoma* sp. (7 and 8 in Fig. 6) and *Tubina* sp. (5 and 6 in Fig. 6). The oriostomatid genus *Tubina* is foreign to North America; there are no previous reports of the taxon from the Western Hemisphere. In a broad survey of all known Emsian gastropod collections known from western or Arctic North America, not one species of the richly diverse Kasaan Island fauna is shared with faunas of undoubted North American origin. The closest related fauna from western North America is that from the Farewell terrane on the south flank of Limestone Mountain, Medfra (B-4) Quadrangle, west-central Alaska. Like the Kasaan fauna, the Limestone Mountain Emsian gastropod fauna (see Fig. 6, 1–4 for representatives) is also distinctly of Bohemian cast (Frýda and Blodgett, 1998). The Emsian brachiopod fauna of Kasaan Island was described by Soja (1988). Several of the taxa described therein show Russian or Siberian ties, notably the genus *Sibirotoechia* and two species of *Desquamatia,* which she compared with species from the Urals. In addition, the subspecies *Gypidula pelagica alaskaforma* Soja, 1988, appears to belong to the genus *Plicogypa* Rzhonsnitskaya, 1975, previously reported from the Kuznetsk Basin of southwestern Siberia, the western slope of the Urals, Novaya Zemlya, the Harz Mountains of Germany, and the Farewell terrane of southwestern Alaska. It is noteworthy that none of the species described in that paper are conspecific with previously described species from North American rocks in the Great Basin or western Canada.

Eifelian (early Middle Devonian) strata of the Alexander terrane also have distinctly non-North American faunal affinities. Several collections from the Wadleigh Limestone on a small island off the western side of Prince of Wales Island, gathered by USGS geologists, contain a moderately diverse molluscan fauna (gastropods, bivalves, and ammonoids). The gastropods and ammonoids are in many cases conspecific with taxa in the Cheeneetnuk Limestone of the Farewell terrane. The most common gastropod in the Wadleigh Limestone collections is a distinctive new murchisoniid genus, apparently the same as a species found in the Cheeneetnuk Limestone. In 1958, Donald J. Miller of the USGS recovered another Eifelian collection dominated by gastropods from the Chilkat Mountains of southeastern Alaska (Juneau [D-6] Quadrangle). This small collection contained several brachiopod species, as well as a number of gastropod species and the dasyclad alga genus *Coelotrochium.* Many of these species were conspecific with species in the Cheeneetnuk Limestone of the Farewell terrane. As noted elsewhere, the alga genus *Coelotrochium* is almost a hallmark for the shallow-water, inner carbonate platform facies of Eifelian-age rocks of the Farewell terrane and is also known from northern Europe (Baltica), but it has never been reported from cratonic or miogeoclinal rocks of North America. Oliver et al. (1975, p. 19) also noted that the Middle Devonian coral fauna from Wadleigh Limestone of southeastern Alaska was distinct from that of western North America, stating that (our brackets) "none of these corals from southeastern Alaska seems conspecific or very closely related to Hume [Shale of Northwest Territories] or Great Basin species."

In summary, Silurian and Devonian megafossils from the Alexander terrane consistently show virtually no affinities with North American faunas from the western Cordillera of the United States or western or Arctic Canada, but are closely allied with faunas described from the Urals, Central Asia, and the Kuznetsk Basin of Siberia. Their closest consistent biotic ties, however, are with the Farewell terrane of southwestern Alaska. This suggests that although they are dominated by differing sedimentologic regimes (eugeoclinal for the Alexander terrane; miogeoclinal for the Farewell), these terranes were close to one another during Silurian-Devonian time, thus resulting in commonality among a great portion of their fossil biota.

EAST-CENTRAL ALASKA (CRATONIC NORTH AMERICA)

The triangular area in east-central Alaska (corresponding to the area shown in dashed pattern in Fig. 1), bounded roughly on the north by the Porcupine River, on the south by the Yukon River, and the east by the Alaska-Yukon border, is thought to be the only portion of Alaska that has not been tectonically accreted (Coney et al., 1980; Blodgett, 1998; Blodgett and Boucot, 1999). The Paleozoic rocks of this region appear to represent a western continuation of the North American miogeocline from

Yukon and the Northwest Territories, and the megafossils of the early and middle Paleozoic strata of east-central Alaska are very similar to those of northwestern Canada. Nonetheless, this triangular area was divided by Silberling et al. (1994) into two tectonic elements: (1) the Porcupine terrane, which includes rocks northwest of the Kandik Basin, and (2) North American rocks equivalent to the Nation Arch of the older literature. The Porcupine terrane includes a thick succession of Paleozoic rocks exposed in the Black River and southern part of the Coleen Quadrangles, and extends eastward into Yukon Territory. These rocks were referred to by Morrow and Geldsetzer (1989) as the Porcupine platform, an isolated structurally high region. Few papers have appeared on the Paleozoic fossil fauna of the Porcupine platform despite the fact that rich faunas are present in its Ordovician-Devonian strata as well as in later Paleozoic rocks.

The early Middle Ordovician (Whiterockian, or late Arenigian) gastropod genus *Palliseria,* which was reported by Rohr and Blodgett (1994) in the Porcupine River canyon (Black River [D-4] Quadrangle), is known only from North America. It has been documented in Whiterockian strata of southeastern Yukon Territory and adjacent District of Mackenzie, Alberta, British Columbia, Nevada, California (Death Valley), Oklahoma, and State of Chihuahua, Mexico. The Early Devonian–Eifelian Salmontrout Limestone, the type section of which is near the confluence of the Salmontrout and Porcupine Rivers in the Coleen (A-2) Quadrangle, contains a rich and diverse fauna of corals and brachiopods. Brachiopods from this formation (Lane and Ormiston, 1975; Blodgett, 1978, 1998) are closely allied with faunas found to the east in Yukon and Northwest Territories (Lenz, 1977a, 1977b; Perry, 1984), and to the south in the Ogilvie Formation in the Charley River (A-1) Quadrangle of the Nation Arch (Blodgett, 1978).

The Nation Arch area (southeastern Charley River and northeastern Eagle Quadrangles) includes a thick succession of Paleozoic rocks that represents the probable western terminus of North American miogeoclinal stratigraphy in Alaska. Close biogeographic ties can be noted between many of the Paleozoic faunas of this area, especially in the Early Devonian Ogilvie Formation faunas (Pragian-Emsian) of the Charley River (A-1) Quadrangle (Blodgett, 1978). These faunas share a majority of their contained species with faunas described to the east in rocks of Yukon and Northwest Territories (Lenz, 1977a, 1977b; Perry, 1984). The biogeographic distinctiveness of the late Early Devonian brachiopod fauna of east-central Alaska becomes even more obvious when it is compared to the now well-sampled Farewell terrane faunas of southwestern and west-central Alaska. Although some genera occur in both east-central Alaska and the Farewell terrane, not one species appears to be common to both areas.

CONCLUSIONS

Review of numerous biogeographically significant, primarily Ordovician-Devonian fossil collections from the Farewell, Arctic Alaska, York, Livengood, and Alexander accreted terranes of Alaska indicates that their closest biotic ties are with Eurasia, notably the Siberian platform and nearby areas (Kolyma, Chukotka, Taimyr) considered to represent peri-Siberian terranes, the Kuznetsk Basin of Siberia, and the Urals. In the Devonian they also have close ties with parts of Europe (Germany, Czech Republic, northwestern France, and the Carnic Alps of Austria and Italy). These affinities suggest that these Alaskan terranes originated as fragments rifted from Siberia during the Devonian. Another plausible, but apparently less likely, origin would be derivation from the modern-day northeastern margin of Baltica (Ural Mountains). The close faunal affinities between the Alaskan terranes and between the Siberian and Uralian biotas suggest they were all in close proximity from Cambrian to Devonian time. The only truly North American part of Alaska is east-central Alaska (including the Nation Arch), which contains Ordovician and Devonian faunas that are closely related to those found in Laurentian (North American) miogeoclinal and cratonic strata of northwestern and Arctic Canada. These east-central Alaskan faunas share virtually no species with the accreted terranes that make up most of Alaska.

ACKNOWLEDGMENTS

Blodgett and Rohr thank the Committee for Research and Exploration of the National Geographic Society for providing funds that permitted field work in a number of field areas of southwestern and west-central Alaska. Blodgett also thanks Phil F. Brease of the U.S. National Park Service, Denali National Park, for his able assistance during field work in the Shellabarger Pass, Talkeetna (C-6) Quadrangle. Norman M. Savage, University of Oregon, Eugene, identified most of the conodonts recovered from the acidization of limestones from this part of Alaska. Blodgett is indebted to the Alaska Division of Geological and Geophysical Surveys for involving him in field studies of southwestern Alaska during the past two decades, and thanks Sohio, ARCO Alaska, Union Oil Company of California, and Louisiana Land and Exploration Company for providing logistical field support in many remote areas of Alaska. We thank Thomas E. Moore, Arthur Grantz, David Brew, Warren Nokleberg, Ray Sullivan, James G. Clough, and David L. LePain for useful discussions during the writing of this paper. Special thanks to Robert J. Elias for providing us with his valuable comments on the biogeographic affinities of Ordovician corals from the York terrane. Boucot thanks the National Science Foundation for grants over the years. We appreciate the efforts of Arthur Grantz, Alfred C. Lenz, and George D. Stanley Jr. for their thoughtful reviews of this manuscript. This paper is dedicated to the paleontologists of the former Soviet Union, whose untiring efforts for the past 40 years have greatly illuminated our knowledge of the early and middle Paleozoic biostratigraphy and biogeography of the present-day circum-Arctic region.

REFERENCES CITED

Abbott, G., 1995, Does Middle Cambrian rifting explain the origin of the Nixon Fork terrane?: Geological Society of America Abstracts with Programs, v. 27, no. 5, p. 1.

Anderson, A.V., 1991, Ulungarat Formation type section of a new formation, headwaters of the Kongakut River, eastern Brooks Range, Alaska: Alaska Division of Geological and Geophysical Surveys, Public Data File 91-4, 27 p.

Babcock, L.E., and Blodgett, R.B., 1992, Biogeographic and paleogeographic significance of Middle Cambrian trilobites of Siberian aspect from southwestern Alaska: Geological Society of America Abstracts with Programs, v. 24, no. 5, p. 4.

Babcock, L.E., Blodgett, R.B., and St. John, J., 1993, Proterozoic and Cambrian stratigraphy and paleontology of the Nixon Fork terrane, southwestern Alaska: Proceedings of the First Circum-Pacific and Circum-Atlantic Terrane Conference, Guanajuato, Mexico, p. 5–7.

Babcock, L.E., Blodgett, R.B., and St. John, J., 1994, New Late (?) Proterozoic formations in the vicinity of Lone Mountain, McGrath Quadrangle, west-central Alaska, *in* Till, A.B., and Moore, T.E., eds, Geologic studies in Alaska by the U.S. Geological Survey, 1993: U.S. Geological Survey Bulletin 2107, p. 143–155.

Baxter, M.E., and Blodgett, R.B., 1994, A new species of *Droharhynchia* from the lower Middle Devonian (Eifelian) of west-central Alaska: Journal of Paleontology, v. 68, p.1235–1240.

Blodgett, R.B., 1978, Biostratigraphy of the Ogilvie Formation and limestone and shale member of the McCann Hill Chert (Devonian), east-central Alaska and adjacent Yukon Territory [M.S. thesis]: Fairbanks, University of Alaska, 142 p.

Blodgett, R.B., 1983, Paleobiogeographic implications of Devonian fossils from the Nixon Fork terrane, southwestern Alaska, *in* Stevens, C.H., ed., Pre-Jurassic rocks in western North America suspect terranes: Los Angeles, Pacific Section, Society of Economic Paleontologists and Mineralogists, p. 125–130.

Blodgett, R.B., 1992, Taxonomy and paleobiogeographic affinities of an early Middle Devonian (Eifelian) gastropod faunule from the Livengood Quadrangle, east-central Alaska: Palaeontographica Abteilung A, v. 221, p. 125–168.

Blodgett, R.B., 1993, *Dutrochus,* a new microdomatid (Gastropoda) genus from the Middle Devonian (Eifelian) of west-central Alaska: Journal of Paleontology, v. 67, p. 194–197.

Blodgett, R.B., 1998, Emsian (Late Early Devonian) fossils indicate a Siberian origin for the Farewell terrane, *in* Clough, J.G., and Larson, F., eds., Short notes on Alaska geology 1997: Alaska Division of Geological and Geophysical Surveys Professional Report 118, p. 53–61.

Blodgett, R.B., and Boucot, A.J., 1999, Late Early Devonian (late Emsian) eospiriferinid brachiopods from Shellabarger Pass, south-central Alaska, and their biogeographic importance: Further evidence for a Siberian origin of the Farewell and allied Alaskan accreted terranes: Senckenbergiana Lethaea, v. 79, no. 1, p. 209–221.

Blodgett, R.B., and Brease, P.F., 1997, Emsian (late Early Devonian) brachiopods from Shellabarger Pass, Talkeetna C-6 Quadrangle, Denali National Park, Alaska indicate Siberian origin for Farewell terrane: Geological Society of America Abstracts with Programs, v. 29, no. 5, p. 5.

Blodgett, R.B., and Clough, J.G., 1985, The Nixon Fork terrane: Part of an in-situ peninsular extension of the North American Paleozoic continent: Geological Society of America Abstracts with Programs, v. 17, no. 6, p. 342.

Blodgett, R.B., and Dutro, J.T., Jr., 1992, *Stringocephalus* (Brachiopoda) from Middle Devonian (Givetian) rocks of the Baird Group, western Brooks Range, Alaska, *in* Chaplin, J.R., and Barrick, J.E., eds., Special papers in paleontology and stratigraphy: A tribute to Thomas W. Amsden: Oklahoma Geological Survey Bulletin 145, p. 91–111.

Blodgett, R.B., and Gilbert, W.G., 1983, The Cheeneetnuk Limestone; A new Early(?) to Middle Devonian formation in the McGrath A-4 and A-5 quadrangles, west-central Alaska: Alaska Division of Geological and Geophysical Surveys Professional Report 85, scale 1:63 360, 1 sheet, 6 p.

Blodgett, R.B., and Gilbert, W.G., 1992a, Upper Devonian shallow-marine siliciclastic strata and associated fauna and flora, Lime Hills D-4 Quadrangle, southwest Alaska, *in* Bradley, D.C., and Dusel-Bacon, C., eds., Geologic studies in Alaska by the U.S. Geological Survey, 1991: U.S. Geological Survey Bulletin 2041, p. 106–115.

Blodgett, R.B., and Gilbert, W.G., 1992b, Paleogeographic relations of Lower and Middle Paleozoic strata of southwest and west-central Alaska: Geological Society of America Abstracts with Programs, v. 24, no. 5, p. 8.

Blodgett, R.B., and Johnson, J.G., 1992, Early Middle Devonian (Eifelian) gastropods of central Nevada: Palaeontographica, Abteilung A, v. 222, p. 85–139.

Blodgett, R.B., and Johnson, J.G., 1994, First recognition of the genus *Verneuilia* Hall and Clarke (Brachiopoda, Spiriferida) from North America (west-central Alaska): Journal of Paleontology, v. 68, p. 1240–1242.

Blodgett, R.B., and Rohr, D.M., 1989, Two new Devonian spine-bearing pleurotomariacean gastropod genera from Alaska: Journal of Paleontology, v. 63, p. 47–52.

Blodgett, R.B., Clough, J.G., and Smith, T.N., 1984, Ordovician–Devonian paleogeography of the Holitna Basin, southwestern Alaska: Geological Society of America Abstracts with Programs, v. 16, no. 5, p. 271.

Blodgett, R.B., Rohr, D.M., and Boucot, A.J., 1988a, Lower Devonian gastropod biogeography of the Western Hemisphere, *in* McMillan, N.J., Embry, A.F., and Glass, D.J., eds., Devonian of the world: Canadian Society of Petroleum Geologists Memoir 14, v. 3, p. 285–305.

Blodgett, R.B., Rohr, D.M., Harris, A.G., and Rong Jia-yu, 1988b, A major unconformity between Upper Ordovician and Lower Devonian strata in the Nanook Limestone, Shublik Mountains, northeastern Brooks Range, *in* Hamilton, T.D., and Galloway, J.P., eds., Geologic studies in Alaska by the U.S. Geological Survey during 1987: U.S. Geological Survey Circular 1016, p. 18–23.

Blodgett, R.B., Ning Zhang, Ormiston, A.R., and Weber, F.R., 1988c, A Late Silurian age determination for the limestone of the "Lost Creek Unit," Livengood C-4 Quadrangle, east-central Alaska, *in* Hamilton, T.D., and Galloway, J.P., eds., Geological studies in Alaska by the U.S. Geological Survey during 1987: U.S. Geological Survey Circular 1016, p. 54–56.

Blodgett, R.B., Rohr, D.M., and Clough, J.G., 1992a, Late Ordovician brachiopod and gastropod biogeography of Arctic Alaska and Chukotka: International Conference on Arctic Margins Abstracts with Program, Anchorage, Alaska, p. 11.

Blodgett, R.B., Clough, J.G., Harris, A.G., and Robinson, M.S., 1992b, The Mount Copleston Limestone, a new Lower Devonian formation in the Shublik Mountains, northern Brooks Range, Alaska, *in* Bradley, D.C., and Ford, A.B., eds., Geologic studies in Alaska by the U.S. Geological Survey, 1990: U.S. Geological Survey Bulletin 1999, p. 3–7.

Blodgett, R.B., Savage, N.M., Pedder, A.E.H., and Rohr, D.M., 1995, Biostratigraphy of an Upper Lower Devonian (Emsian) limestone unit at "Reef Ridge," Medfra B-3 Quadrangle, west-central Alaska: Geological Society of America Abstracts with Programs, v. 27, no. 5, p. 6.

Blodgett, R.B., Rohr, D.M., Measures, E.A., Savage, N.M., Pedder, A.E.H., and Chalmers, R.W., 2000, The Soda Creek Limestone, a new upper Lower Devonian formation in the Medfra Quadrangle, west-central Alaska, *in* Pinney, D.S., and Davis, P.K., eds., Short notes on Alaska geology 1999: Alaska Division of Geological and Geophysical Surveys Professional Report 119, p. 1–9.

Brice, D., 1982, Brachiopodes du Dévonien Inférieur et Moyen des Formations de Blue Fiord et Bird Fiord des Iles Arctiques Canadiennes: Geological Survey of Canada Bulletin 326, 175 p.

Buddington, E.F., and Chapin, T., 1929, Geology and mineral deposits of southeastern Alaska: U.S. Geological Survey Bulletin 800, 398 p.

Bundtzen, T.K., and Gilbert, W.G., 1983, Outline of geology and mineral resources of the upper Kuskokwim region, Alaska: Journal of the Alaska Geological Society, v. 3, p. 101–119.

Butusova, I.P., 1960, Nekotorye gastropody mamontovskikh sloev srednego devona kuznetskogo basseina: Leningrad, Vsesoyuznyi nauchno-issledovatelskii geologicheskii institut (VSEGEI), Informatsionnyi sbornik, v. 35, p. 81–89.

Chalmers, R.W., Measures, E.A., Rohr, D.M., and Blodgett, R.B., 1995, Depositional environments of an upper Lower Devonian (Emsian) limestone unit at "Reef Ridge," Medfra B-3 Quadrangle, west-central Alaska: Geological Society of America Abstracts with Programs, v. 27, no. 5, p. 6.

Churkin, M., Jr., Foster, H.L., Chapman, R.M., and Weber, F.R., 1982, Terranes and suture zones in east central Alaska: Journal of Geophysical Research, v. 87, p. 3718–3730.

Churkin, M., Jr., Wallace, W.K., Bundtzen, T.K., and Gilbert, W.G., 1984, Nixon Fork–Dillinger terranes: A dismembered Paleozoic craton margin in Alaska displaced from Yukon Territory: Geological Society of America Abstracts with Programs, v. 16, no. 5, p. 275.

Clough, J.G., and Blodgett, R.B., 1985, Comparative study of the sedimentology and paleoecology of middle Paleozoic algal and coral-stromatoporoid reefs in Alaska, *in* Gabrie, G., and Salvat, B., eds., Proceedings of the Fifth International Coral Reef Congress, Papeete, Tahiti, v. 2 (abstract), p. 78, v. 3 (text), p. 593–598.

Clough, J.G., and Blodgett, R.B., 1988, Silurian-Devonian algal reef mound complex of southwest Alaska, *in* Geldsetzer, H.H.J., James, N.P., and Tebbutt, G.E., eds., Reefs, Canada and adjacent areas: Canadian Society of Petroleum Geologists Memoir 13, p. 404–407.

Clough, J.G., and Blodgett, R.B., 1992, A southwest Alaska Late Silurian–Early Devonian algal reef-rimmed carbonate ramp: Depositional cycles and regional significance: Geological Society of America Abstracts with Programs, v. 24, no.5, p. 16.

Cocks, L.R.M., and Modzalevskaya, T.L., 1997, Late Ordovician brachiopods from Taimyr, Arctic Russia, and their palaeogeographical significance: Palaeontology, v. 40, p. 1061–1093.

Coney, P.J., Jones, D.L., and Monger, J.W.H., 1980, Cordilleran suspect terranes: Nature, v. 288, p. 329–333.

Decker, J., Bergman, S.C., Blodgett, R.B., Box, S.E., Bundtzen, T.K., Clough, J.G., Coonrad, W.L., Gilbert, W.G., Miller, M.L., Murphy, J.M., Robinson, M.S., and Wallace, W.K., 1994, Geology of southwestern Alaska, *in* Plafker, G., and Berg, H.C., eds., The geology of Alaska: Boulder, Colorado, Geological Society of America, Geology of North America, v. G-1, p. 285–310.

Dutro, J.T., Jr., and Patton, W.W., Jr., 1982, New Paleozoic formations in the northern Kuskokwim Mountains, west-central Alaska: U.S. Geological Survey Bulletin 1529–H, p. 13–22.

Dutro, J.T., Jr., Blodgett, R.B., and Mull, C.G., 1994, *Cyrtospirifer* from Upper Devonian rocks of the Endicott Group, west-central Alaska, *in* Till, A.B., and Moore, T.E., eds., Geologic studies in Alaska by the U.S. Geological Survey, 1993: U.S. Geological Survey Bulletin 2107, p. 133–142.

Forney, G.G., Boucot, A.J., and Rohr, D.M., 1981, Silurian and Lower Devonian zoogeography of selected molluscan genera, *in* Gray, J., Boucot, A.J., and Berry, W.B.N., eds., Communities of the past: Stroudsberg, Pennsylvania, Hutchinson Ross Publishing Company, p. 119–164.

Frýda, J., and Blodgett, R.B., 1998, Two new cirroidean genera (*Vetigastropoda, Archaeogastropoda*) from the Emsian (late Early Devonian) of Alaska with notes on the early phylogeny of Cirroidea: Journal of Paleontology, v. 72, p. 265–273.

Garcia-Alcalde, J., and Blodgett, R.B., 2001, New Lower Devonian (Upper Emsian) *Myriospirifer* (Brachiopoda, Eospiriferinae) species from Alaska and northern Spain and the paleographic distribution of the genus *Myriospirifer:* Journal of the Czech Geological Society, v. 46, no. 3/4, p. 59–68.

Gilbert, W.G., and Bundtzen, T.K., 1983, Paleozoic stratigraphy of Farewell area, southwest Alaska Range, Alaska, *in* Alaska Geological Society Symposium: New Developments in the Paleozoic Geology of Alaska and Yukon, Anchorage, Alaska, 1983: Alaska Geological Society Program and Abstracts, p. 10–11.

Gilbert, W.G., and Bundtzen, T.K., 1984, Stratigraphic relationship between Dillinger and Mystic terranes, western Alaska Range, Alaska: Geological Society of America Abstracts with Programs, v. 16, no. 5, p. 286.

Hahn, G., and Hahn, R., 1985, Trilobiten aus dem hohen Ober-Karbon oder Unter-Perm von Alaska: Senckenbergiana Lethaea, v. 66, p. 445–485.

Hahn, G., Blodgett, R.B., and Gilbert, W.G., 1985, First recognition of the Gshelian (Upper Pennsylvanian) trilobite *Brachymetopus pseudometopina* Gauri and Ramovs in North America and a description of accompanying trilobites from west-central Alaska: Journal of Paleontology, v. 59, p. 27–31.

House, M.R., and Blodgett, R.B., 1982, The Devonian goniatite genera *Pinacites* and *Foordites* from Alaska: Canadian Journal of Earth Sciences, v. 19, p. 1873–1876.

Johnson, J.G., and Blodgett, R.B., 1993, Russian Devonian brachiopod genera *Cyrtinoides* and *Komiella* in North America: Journal of Paleontology, v. 67, p. 952–958.

Johnson, J.G., and Lane, N.G., 1969, Two new Devonian crinoids from central Nevada: Journal of Paleontology, v. 43, p. 69–73.

Johnson, J.G., Boucot, A.J., and Murphy, M.A., 1976, Wenlockian and Ludlovian age brachiopods from the Roberts Mountains of central Nevada: University of California Publications in Geological Sciences, v. 115, 102 p., 55 pls.

Jones, B., 1979, *Nanukidium,* a new name for *Rossella* Jones, 1978: Journal of Paleontology, v. 53, p. 1261.

Jones, D.L., and Silberling, N.J., 1982, Mesozoic stratigraphy: The key to tectonic analysis of southern and central Alaska, *in* Leviton, A.E., Rodda, P.U., Yochelson, E., and Adrich, M.L., eds., Frontiers of geological exploration of western North America: American Association for the Advancement of Science, Pacific Division, 60th Annual Meeting, 1979, Proceedings (papers from two meeting symposiums), University of Idaho, Moscow, Idaho, p. 139–153.

Jones, D.L., Silberling, N.J., Berg, H.C., and Plafker, G., 1981, Map showing tectonostratigraphic terranes of Alaska, columnar sections, and summary description of terranes: U.S. Geological Survey Open-File Report 81–792, scale 1:2 500 000, 2 sheets, 20 p.

Jones, D.L., Silberling, N.J., Gilbert, W.G., and Coney, P.J., 1982, Character, distribution, and tectonic significance of accretionary terranes in the central Alaska Range: Journal of Geophysical Research, v. 87, p. 3709–3717.

Jones, D.L., Silberling, N.J., and Coney, P.J., 1986, Collision tectonics in the Cordillera of western North America: Examples from Alaska, *in* Coward, M.P., and Ries, A.C., eds., Collision tectonics: Geological Society [London] Special Publication 19, p. 367–387.

Jones, D.L., Silberling, N.J., Coney, P.J., and Plafker, G., 1987, Lithotectonic terrane map of Alaska (west of the 141st Meridan): U.S. Geological Survey Map MF-1874–A, scale 1:2 500 000, 1 sheet.

Kaljo, D., 1960, Nekotorye voprosy razvitiya ordovikskikh tetrakorallov: Eesti NSV Teaduste Akadeemia Geoloogia Instituudi Uurimused, v. 5, p. 245–258.

Khodalevich, A.N., 1939, Verkhnesiluriiskie brakhiopody vostochnogo sklona urala: Sverdlovsk, Trudy uralskogo geologicheckogo upravleniya, Izdanie Uralgeoupravleniya, 135 p.

Khodalevich, A.N., 1951, Nizhnedevonskie i eifelskie brakhiopody ivdelskogo i serovskogo raionov sverdlovskoioblasti: Trudy sverdlovskogo gornogo instituta imemi V.V. Vakhrusheva, Vypusk 18: Moscow, Gosgeolotekhizdat, 169 p.

Kindle, E.M., 1907, Notes on the Paleozoic faunas and stratigraphy of southeastern Alaska: Journal of Geology, v. 15, p. 314–337.

Kingsbury, S.A., and Babcock, L.E., 1998, Biogeography and paleogeographic implications of early Middle Cambrian trilobites and enigmatic fossils from the Farewell terrane, southwestern Alaska: Geological Society of America Abstracts with Programs, v. 30, no. 2, p. 27.

Kirk, E., 1922, *Brooksina,* a new pentameroid genus from the Upper Silurian of southeastern Alaska: Proceedings of the United States National Museum, v. 60, article 19, p. 1–8.

Kirk, E., 1925, *Harpidium,* a new pentameroid brachiopod genus from southeastern Alaska: Proceedings of the United States National Museum, v. 66, article 32, p. 1–7.

Kirk, E., 1926, *Cymbidium,* a new genus of Silurian pentameroid brachiopods from Alaska: Proceedings of the United States National Museum, v. 69, article 23, p. 1–5.

Kirk, E., and Amsden, T.W., 1952, Upper Silurian brachiopods from southeastern Alaska: U.S. Geological Survey Professional Paper 233–C, p. 53–66.

Lane, H.R., and Ormiston, A.R., 1979, Siluro-Devonian biostratigraphy of the Salmontrout River area, east-central Alaska: Geologica et Palaeontologica, v. 13, p. 39–96.

Lenz, A.C., 1977a, Upper Silurian and Lower Devonian brachiopods of Royal Creek, Yukon, Canada. 1. Orthoida, Strophomenida, Pentamerida, Rhynchonellida: Palaeontographica Abteilung A: v. 159, p. 37–109.

Lenz, A.C., 1977b, Upper Silurian and Lower Devonian brachiopods of Royal Creek, Yukon, Canada. 2. Spiriferida: Atrypacea, Dayiacea, Athyridacea, Spiriferacea: Palaeontographica Abteilung A: v. 159, p. 111–138.

Mamay, S.H., and Reed, B.L., 1984, Permian plant megafossils from the conglomerate of Mount Dall, central Alaska Range, *in* Coonrad, W.L., and Elliott, R.L., eds., The United States Geological Survey: Accomplishments during 1981: U.S. Geological Survey Circular 868, p. 98–102.

Mamet, B., and Preat, A., 1992, Algues du Dévonien Moyen de Wellin (Synclinorium de Dinant, Belgique): Revue de Micropaléontologie, v. 35, p. 53–75.

Mamet, B.L., and Plafker, G., 1982, A Late Devonian (Frasnian) Microbiota from the Farewell–Lyman Hills area, west-central Alaska: U.S. Geological Survey Professional Paper 1216–A, p. 1–12.

Measures, E.A., Rohr, D.M., and Blodgett, R.B., 1992, Depositional environments and some aspects of the fauna of Middle Ordovician rocks of the Telsitna Formation, northern Kuskokwim Mountains, Alaska, *in* Bradley, D.C., and Dusel-Bacon, C., eds., Geologic studies in Alaska by the U.S. Geological Survey, 1991: U.S. Geological Survey Bulletin 2041, p. 186–201.

Miagkova, E.I., 1955, Novye predstaviteli tipa Archaeocyatha; Doklady Akademiya Nauk SSSR, v. 104, p. 638–641.

Moore, T.E., 1992, The Arctic Alaska superterrane, *in* Bradley, D.C., and Dusel-Bacon, C., eds., Geologic studies in Alaska by the U.S. Geological Survey, 1991: U.S. Geological Survey Bulletin 2041, p. 238–244.

Morrow, D.W., and Geldsetzer, H.H.J., 1989, Devonian of eastern Canadian Cordillera, *in* McMillan, N.J., Embry, A.F., and Glass, D.J., eds., Devonian of the world: Canadian Society of Petroleum Geologists Memoir 14, v. 1, p. 85–121.

Nikiforova, O.I., 1989, Pozdneordovikskie pentameridy (brakhiopody) tsentralnogo taimyra, *in* Kolobova, I.M., and Khozatkii, L.J., eds., Ezhegodnik vsesoyuznogo paleontologicheskogo obshechestva, Tom 32: Leningrad, Nauka, p. 77–87.

Nikolaev, A.A., 1968, *Conchidium? unicum* sp. nov., p. 47–48, *in* Balashov, Z.G., ed., Polevoi atlas ordovikskoi fauny severo-vostoka SSSR: Magadan, Magadanskoe Knizhnoe Izdatelstvo, p. 47–48.

Nikolaev, A.A., and Sapelnikov, V.P., 1969, Dva novykh roda pozdneordovikskikh Virgianidae, *in* Trifonov, V.P., and Shabalina, N.S., eds., Voprosy geologii i gidrogeologii urala: Sverdlovsk, Trudy Sverdlovskogo ordena trudovogo krasnogo znameni gornogo instituta, Vypusk 63, p. 11–17.

Nokleberg, W.J., Moll-Stalcup, E.J., Miller, T.P., Brew, D.A., Grantz, A., Reed, J.C., Jr., Plafker, G., Moore, T.E., Silva, S.R., and Patton, W.W., Jr., with contributions on specific regions by Blodgett, R.B., Box, S.E., Bradley, D.C., Bundtzen, T.K., Dusel-Bacon, C., Gamble, B.M., Howell, D.G., Foster, H.L., Karl, S.M., Miller, M.L., and Nelson, S.W., 1994, Tectonostratigaphic terrane and overlap assemblage map of Alaska: U.S. Geological Survey Open-File Report 94–194, scale 1:2 500 000, 1 sheet, 53 p.

Oliver, W.A., Jr., Merriam, C.W., and Churkin, M., Jr., 1975, Ordovician, Silurian, and Devonian corals of Alaska: U.S. Geological Survey Professional Paper 823–B, p. 13–44.

Ormiston, A.R., 1978, *Monorakos* (Trilobita) from the Ordovician of the Seward Peninsula, Alaska: Journal of Paleontology, v. 52, p. 345–352.

Ormiston, A.R., and Ross, R.J., Jr., 1979, *Monorakos* in the Ordovician of Alaska and its zoogeographic significance, *in* Gray, J., and Boucot, A.J., eds., Historical biogeography, plate tectonics, and the changing environment: Corvallis, Oregon State University Press, p. 53–59, 5 figs.

Palmer, A.R., Egbert, R.M., Sullivan, R., and Knoth, J.S., 1985, Cambrian trilobites with Siberian affinities, southwestern Alaska [abs.]: American Association of Petroleum Geologists Bulletin, v. 69, p. 295.

Patton, W.W., Jr., 1978, Juxtaposed continental and ocean-island arc terranes in the Medfra Quadrangle, west-central Alaska, *in* Johnson, K.M., ed., The United States Geological Survey in Alaska: Accomplishments during 1977: U.S. Geological Survey Circular 772–B, p. 38–39.

Perry, D.G., 1984, Brachiopoda and biostratigraphy of the Silurian-Devonian Delorme Formation in the District of Mackenzie, the Yukon: Royal Ontario Museum Life Sciences Contributions, no. 138, 243 p., 46 pls.

Perry, D.G., and Lenz, A.C., 1978, Emsian paleogeography and shelly fauna biostratigraphy of Arctic Canada, *in* Stelck, C.R., and Chatterton, B.D.E., eds., Western and Arctic Canadian biostratigraphy: Geological Association of Canada Special Paper 18, p. 133–160.

Poncet, J., and Blodgett, R.B., 1987, First recognition of the Devonian alga *Lancicula sergaensis* Shuysky in North America (west-central Alaska): Journal of Paleontology, v. 61, p. 1269–1273.

Popov, L.Y., Blodgett, R.B., and Anderson, A.V., 1994, The first occurrence of the genus *Bicarinatina* (Brachiopoda, Inarticulata) from the Middle Devonian of North America: Journal of Paleontology, v. 68, p. 1214–1218.

Potter, A.W., and Blodgett, R.B., 1992, Paleobiogeographic relations of Ordovician brachiopods from the Nixon Fork terrane, west-central Alaska: Geological Society of America Abstracts with Progams, v. 24, no. 5, p. 76.

Potter, A.W., Blodgett, R.B., and Rohr, D.M., 1988a, Paleobiogeographic relations and paleogeographic signficance of Late Ordovician brachiopods of Alaska: Geological Society of America Abstracts with Programs, v. 20, no. 7, p. 339.

Potter, A.W., Blodgett, R.B., and Rohr, D.M., 1988b, Paleobiogeographic relations and paleogeographic significance of Late Ordovician brachiopods from Alaska: Fifth International Symposium on the Ordovician System, Memorial University of Newfoundland, St. John's, Newfoundland, Canada, Abstracts Volume, p. 74.

Reed, B.L., and Nelson, S.W., 1980, Geological map of the Talkeetna Quadrangle, Alaska: U.S. Geological Survey Miscellaneous Investigation Series Map I-1174, scale 1:250 000, 1 sheet, 15 p.

Riding, R., and Soja, C.M., 1993, Silurian calcareous algae, cyanobacteria, and microproblematica from the Alexander terrane: Journal of Paleontology, v. 67, p. 710–728.

Rigby, J.K., and Blodgett, R.B., 1983, Early Middle Devonian sponges from the McGrath Quadrangle of west-central Alaska: Journal of Paleontology, v. 57, p. 773–786.

Rigby, J.K., Nitecki, M.H., Soja, C.M., and Blodgett, R.B., 1994, Silurian aphrosalpingid sphinctozoans from Alaska and Russia: Acta Palaeontologica Polonica, v. 39, p. 341–391.

Rohr, D.M., 1988, Upper Ordovician gastropods from the Seward Peninsula, Alaska: Journal of Paleontology, v. 62, p. 551–565.

Rohr, D.M., 1993, Middle Ordovician carrier shell *Lytospira* (Mollusca, Gastropoda) from Alaska: Journal of Paleontology, v. 67, p. 959–962.

Rohr, D.M., and Blodgett, R.B., 1985, Upper Ordovician Gastropoda from west-central Alaska: Journal of Paleontology, v. 59, p. 667–673.

Rohr, D.M., and Blodgett, R.B., 1988, First occurrence of *Helicotoma* Salter (Gastropoda) from the Ordovician of Alaska: Journal of Paleontology, v. 62, p. 304–306.

Rohr, D.M., and Blodgett, R.B., 1994, *Palliseria* (Middle Ordovician Gastropoda) from east-central Alaska and its stratigraphic and biogeographic significance: Journal of Paleontology, v. 68, p. 674–675.

Rohr, D.M., and Gubanov, A.P., 1997, *Macluritid opercula* (Gastropoda) from the Middle Ordovician of Siberia and Alaska: Journal of Paleontology, v. 71, p. 394–400.

Rohr, D.M., and Yochelson, E.L., 1999, Life association of shell and operculum of Middle Ordovician gastropod *Maclurites:* Journal of Paleontology, v. 73, p. 1078–1080.

Rohr, D.M., Dutro, J.T., Jr., and Blodgett, R.B., 1992, Gastropods and brachiopods from the Ordovician Telsitna Formation, northern Kuskokwim Mountains, west-central Alaska, *in* Webby, B.D., and Laurie, J.R., eds., Global perspectives on Ordovician Geology, Proceedings of the Sixth International Symposium on the Ordovician System: Sydney, Australia, Balkema Press, p. 499–512.

Rohr, D.M., Frýda, J., and Blodgett, R.B., 2002, *Alaskadiscus,* a new bellephontoidean gastropod from the Upper Ordovician of the York and Farewell terranes of Alaska, *in* Clautice, K.H., ed., Short notes on Alaska geology 2001: Alaska Division of Geological and Geophysical Surveys Professional Report (in press).

Ross, R.J., Jr., 1965, Early Ordovician trilobites from the Seward Peninsula, Alaska: Journal of Paleontology, v. 39, p. 17–20.

Rzhonsnitskaya, M.A., 1975, Biostratigrafiya devona okrain kuznetskogo basseina. Opisanie brakhiopod: Leningrad, Nauka, Tom 2, 232 p.

Sainsbury, C.L., Dutro, J.T., Jr., and Churkin, M., Jr., 1971, The Ordovician-Silurian boundary in the York Mountains, western Seward Peninsula, Alaska: U.S. Geological Survey Professional Paper 750–C, p. 52–57.

Sandy, M.R., and Blodgett, R.B., 2000, Early Jurassic spiriferid brachiopods from Alaska and their paleogeographic significance: Geobios, v. 33, p. 319–328.

Sapelnikov, V.P., 1985, Sistema i stratigraficheskoe znachenie brakhiopod podotryada pentameridin: Moscow, Nauka, 208 p.

Sapelnikov, V.P., Bogoyavlenskaya, O.V., Mizens, L.I., and Shuysky, V.P., 1999, Silurian and Early Devonian benthic communities of the Ural–Tien Shan region, *in* Boucot, A.J., and Lawson, J.D., eds., Paleocommunities: A case study from the Silurian and Lower Devonian: New York, Cambridge University Press, p. 510–544.

Savage, N.M., 1981, A reassignment of the age of some Paleozoic brachiopods from southeastern Alaska: Journal of Paleontology, v. 55, p. 353–369.

Savage, N.M., 1989, The occurrence of the brachiopods *Nanukidium* and *Atrypoidea* in the Late Silurian of southeastern Alaska, Alexander terrane: Journal of Paleontology, v. 63, p. 530–533.

Savage, N.M., and Blodgett, R.B., 1995, A new species of the conodont *Amydrotaxis* from the Lower Devonian of southwestern Alaska, *in* Combellick, R.A., and Tannian, F., eds., Short notes on Alaska geology 1995: Alaska Division of Geological and Geophysical Surveys Professional Report 117, p. 69–73.

Savage, N.M., Perry, D.G., and Boucot, A.J., 1979, A quantitative analysis of Lower Devonian brachiopod distribution, *in* Gray, J., and Boucot, A.J., eds., Historical biogeography, plate tectonics and the changing environment: Corvallis, Oregon State University Press, p. 169–200.

Schlüter, C., 1879, *Coelotrochium decheni,* eine Foraminifere aus dem Mitteldevon: Zeitschrift der Deutschen Geologischen Gesellschaft, v. 1879, p. 668–675.

Shuysky, V.P., 1973, Vodorosli roda *Lancicula* iz nizhnego devona Urala, *in* Sapelnikov, V.P., and Chuvashov, B.I., eds., Materialy po paleontologii srednego paleozoya Uralo-Tyanshanskoi oblasti: Sverdlovsk, Akademiya Nauk SSSR, Uralskii Nauchnii Tsentr, Trudy Instituta Geologii i Geokhimii, Vypusk 99, p. 3–12.

Silberling, N.J., Jones, D.L., Monger, J.W.H., Coney, P.J., Berg, H.C., and Plafker, G., 1994, Lithotectonic terrane map of Alaska and adjacent parts of Canada, *in* Plafker, G., and Berg, H.C., eds., The geology of Alaska: Boulder, Colorado, Geological Society of America, Geology of North America, v. G-1, scale 1:2 500 000, 1 sheet, plate 3.

Soja, C.M., 1988, Lower Devonian (Emsian) brachiopods from southeastern Alaska: Palaeontographica. Abteilung A, v. 201, p. 129–193.

Soja, C.M., 1994, Significance of Silurian stromatolite-sphinctozoan reefs: Geology, v. 22, p. 355–358.

Soja, C.M., and Antoshkina, A.I., 1997, Coeval development of Silurian stromatolite reefs in Alaska and the Urals Mountains: Implications for paleogeography of the Alexander terrane: Geology, v. 25, p. 539–542.

Soja, C.M., and Riding, R., 1993, Silurian microbial associations from the Alexander terrane: Journal of Paleontology, v. 67, p. 728–738.

Spicer, R.A., and Thomas, B.A., 1987, A Mississippian Alaska-Siberia connection: Evidence from plant megafossils, *in* Tailleur, I., and Weimer, P., eds., Alaska North Slope geology, Volume 1: Bakersfield, California, Pacific Section, Society of Economic Paleontologists and Mineralogists and the Alaska Geological Society, Book 50, p. 355–358.

St. John, J.M., and Babcock, L.E., 1997, Late Middle Cambrian trilobites of Siberian aspect from the Farewell terrane, southwestern Alaska, *in* Dumoulin, J.A., and Gray, J.E., eds., Geologic studies in Alaska by the U.S. Geological Survey, 1995: U.S. Geological Survey Professional Paper 1574, p. 269–281.

Sytova, V.A., 1979, Rugozy mangazeiskogo, dolborskogo i ketskogo gorizontov, *in* Rozman, Kh.S., Stukalina, G.A., Krasilova, I.N., and Sytova, V.A., eds., Fauna ordovika srednei sibiri: Akademiya Nauk SSSR, Ordena trudovogo krasnogo znameni geologicheskii institut, Trudy, Vypusk 330: Moskow, Izdatelstvo Nauka, p. 159–176.

Weber, F.R., Wheeler, K.L., Rinehart, C.D., Chapman, R.M., and Blodgett, R.B., 1992, Geologic map of the Livengood Quadrangle, Alaska: U.S. Geological Survey Open-File Report 92–562, scale 1:250 000, 1 sheet, 20 p.

Weber, F.R., Blodgett, R.B., Harris, A.G., and Dutro, J.T., Jr., 1994, Paleontology of the Livengood Quadrangle, Alaska: U.S. Geological Survey Open-File Report 94–215, scale 1:250 000, 1 sheet, 24 p.

Wilson, A.E., 1926, An Upper Ordovician fauna from the Rocky Mountains, British Columbia: Geological Survey of Canada, Museum Bulletin 44, p. 1–34.

MANUSCRIPT ACCEPTED BY THE SOCIETY MAY 15, 2001.

Geological Society of America
Special Paper 360
2002

Lithostratigraphic, conodont, and other faunal links between lower Paleozoic strata in northern and central Alaska and northeastern Russia

Julie A. Dumoulin*
U.S. Geological Survey, 4200 University Drive, Anchorage, Alaska 99508-4667, USA
Anita G. Harris
U.S. Geological Survey, 926A National Center, Reston, Virginia 20192, USA
Mussa Gagiev†
Russian Academy of Sciences, Portovaya Street 16, Magadan, 685010, Russia
Dwight C. Bradley
U.S. Geological Survey, 4200 University Drive, Anchorage, Alaska 99508-4667, USA
John E. Repetski
U.S. Geological Survey, 926A National Center, Reston, Virginia 20192, USA

ABSTRACT

Lower Paleozoic platform carbonate strata in northern Alaska (parts of the Arctic Alaska, York, and Seward terranes; herein called the North Alaska carbonate platform) and central Alaska (Farewell terrane) share distinctive lithologic and faunal features, and may have formed on a single continental fragment situated between Siberia and Laurentia. Sedimentary successions in northern and central Alaska overlie Late Proterozoic metamorphosed basement; contain Late Proterozoic ooid-rich dolostones, Middle Cambrian outer shelf deposits, and Ordovician, Silurian, and Devonian shallow-water platform facies, and include fossils of both Siberian and Laurentian biotic provinces. The presence in the Alaskan terranes of Siberian forms not seen in well-studied cratonal margin sequences of western Laurentia implies that the Alaskan rocks were not attached to Laurentia during the early Paleozoic.

The Siberian cratonal succession includes Archean basement, Ordovician shallow-water siliciclastic rocks, and Upper Silurian–Devonian evaporites, none of which have counterparts in the Alaskan successions, and contains only a few of the Laurentian conodonts that occur in Alaska. Thus we conclude that the lower Paleozoic platform successions of northern and central Alaska were not part of the Siberian craton during their deposition, but may have formed on a crustal fragment rifted away from Siberia during the Late Proterozoic. The Alaskan strata have more similarities to coeval rocks in some peri-Siberian terranes of northeastern Russia (Kotelny, Chukotka, and Omulevka). Lithologic ties between northern Alaska, the Farewell terrane, and the peri-Siberian terranes diminish after the Middle Devonian, but Siberian affinities in northern and central Alaskan biotas persist into the late Paleozoic.

*E-mail: dumoulin@usgs.gov.
†See Acknowledgments.
Data Repository item 2002078 contains additional material related to this article.

Dumoulin, J.A., Harris, A.G., Gagiev, M., Bradley, D.C., and Repetski, J.E., 2002, Lithostratigraphic, conodont, and other faunal links between lower Paleozoic strata in northern and central Alaska and northeastern Russia, *in* Miller, E.L., Grantz, A., and Klemperer, S.L., eds., Tectonic Evolution of the Bering Shelf–Chukchi Sea–Arctic Margin and Adjacent Landmasses: Boulder, Colorado, Geological Society of America Special Paper 360, p. 291–312.

INTRODUCTION

Lower Paleozoic platform carbonate strata crop out discontinuously across northern Alaska, from westernmost Seward Peninsula to the northeastern Brooks Range (Figs. 1 and 2). Lithologic and faunal similarities between these carbonate successions strongly suggest that they formed along a single continental margin (Dumoulin and Harris, 1994). Microfossil and megafossil assemblages from the Alaskan successions have Siberian elements not found in coeval strata of the Canadian Cordillera or the Canadian Arctic Islands (Dumoulin et al., 1998c, 2000a), but they also contain Laurentian elements. This distinctive co-occurrence of Siberian and Laurentian faunal elements is also documented in coeval carbonate rocks of the Farewell terrane in central Alaska (Dumoulin et al., 1998b; Blodgett, 1998), in successions that have many lithologic similarities with those of northern Alaska. In this chapter we compare lower Paleozoic stratigraphies and faunas from northern and central Alaska to coeval stratigraphies and faunas from northeastern Russia. These comparisons provide critical data that constrain models of Arctic paleogeography and tectonic evolution.

NORTHERN ALASKA

Lower Paleozoic carbonate strata in northern Alaska crop out on the Seward Peninsula and discontinuously throughout the Brooks Range (Fig. 2). Their present configuration is primarily the result of Mesozoic and Tertiary tectonism, but may also reflect Paleozoic tectonic events (Moore et al., 1994, 1997). Original facies relations, such as those between shallow- and deep-water strata, have been disrupted and obscured by widespread thrust and perhaps extensional faulting. In the Brooks Range, lower Paleozoic rocks have been assigned to the Arctic Alaska terrane, whereas coeval strata on Seward Peninsula have been assigned to the York and Seward terranes (Nokleberg et al., 1994, 1998; Silberling et al., 1994). The Arctic Alaska terrane includes numerous subterranes; lower Paleozoic carbonate rocks in the western and central Brooks Range belong to the Hammond subterrane, whereas those in the northeastern Brooks Range are part of the North Slope subterrane (Moore et al., 1994). Stratigraphic and faunal evidence suggests that lower Paleozoic carbonate successions in all three terranes were once part of a contiguous carbonate platform (herein called the North Alaska carbonate platform) that was disrupted by later tectonic events (Dumoulin and Harris, 1994).

Paleogeographic reconstruction of northern Alaska is further complicated by extensive exposures, particularly in the southern Brooks Range, of undated, deformed, chiefly metasiliciclastic and lesser metavolcanic rocks that could include both shallow- and deep-water facies of early Paleozoic age (parts of the Hammond and Coldfoot subterranes of the Arctic Alaska terrane of Moore et al., 1994). The stratigraphic and depositional positions of these strata are uncertain. They have been interpreted as largely Devonian deposits that accumulated in block-

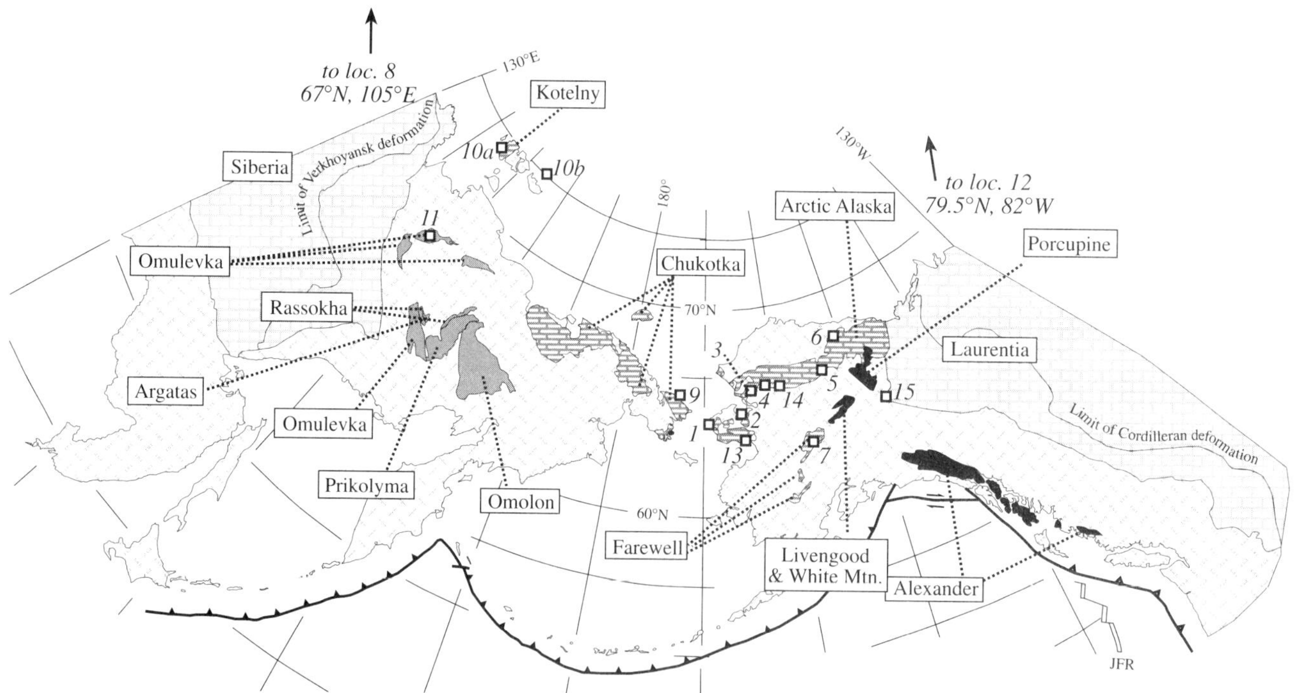

Figure 1. Terranes of northeastern Russia and Alaska, modified from Nokleberg et al. (1998), as follows: Chukotka includes Russian part of Seward terrane; Arctic Alaska consists of parts of Arctic Alaska, Seward (Alaskan part), and York terranes that contain lower Paleozoic carbonate rocks. Numbers 1–12 are areas of stratigraphic sections shown in Figure 3; 13–15 are localities mentioned in text: 13, southeast Seward Peninsula; 14, northeast Ambler River Quadrangle; 15, Jones Ridge.

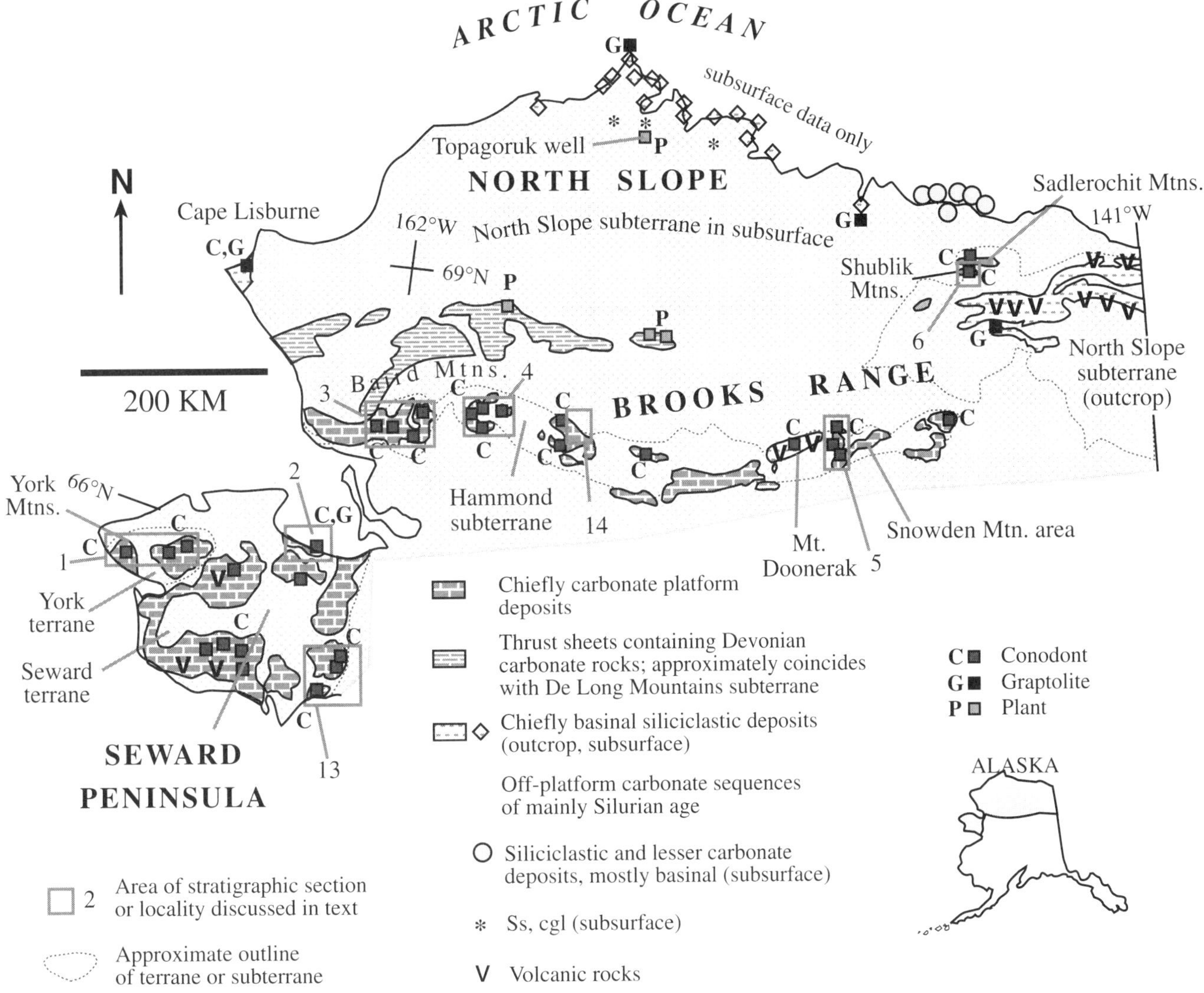

Figure 2. Distribution of lower Paleozoic lithofacies and selected fossil localities across northern Alaska. Patterns and symbols refer to outcrop data unless specified as subsurface.

faulted basins within the North Alaska carbonate platform during inferred middle Paleozoic extension (Moore et al., 1997), or alternatively, as Devonian and older basinal successions that formed along the southern margin of the carbonate platform (Oldow et al., 1998).

In northernmost Alaska (north of lat 68°N), lower Paleozoic strata are largely deep-water, siliciclastic and lesser volcanic deposits included in the North Slope subterrane of the Arctic Alaska terrane (Moore et al., 1994). These rocks crop out to the west near Cape Lisburne, in the central Brooks Range at Mount Doonerak, and in the far northeastern Brooks Range, and have been encountered by numerous exploratory wells drilled for petroleum beneath the North Slope (Fig. 2). The lower Paleozoic deep-water strata of northernmost Alaska could have formed along the northern margin of the North Alaska carbonate plat-

form, but they cannot be depositionally tied to that platform until Devonian time or later (Moore et al., 1994); this point is further discussed in the following.

Lithofacies

Carbonate successions of early Paleozoic age have been studied in detail in the York Mountains (western Seward Peninsula), the central and eastern Seward Peninsula, the western and eastern Baird Mountains (western Brooks Range), the Snowden Mountain area (central Brooks Range), and the Shublik and Sadlerochit Mountains (eastern Brooks Range) (Figs. 2 and 3, columns 1–6; Dumoulin and Harris, 1994; Harris et al., 1995, and references therein). Strata in the York, Shublik, and Sadlerochit Mountains are unmetamorphosed. The other successions

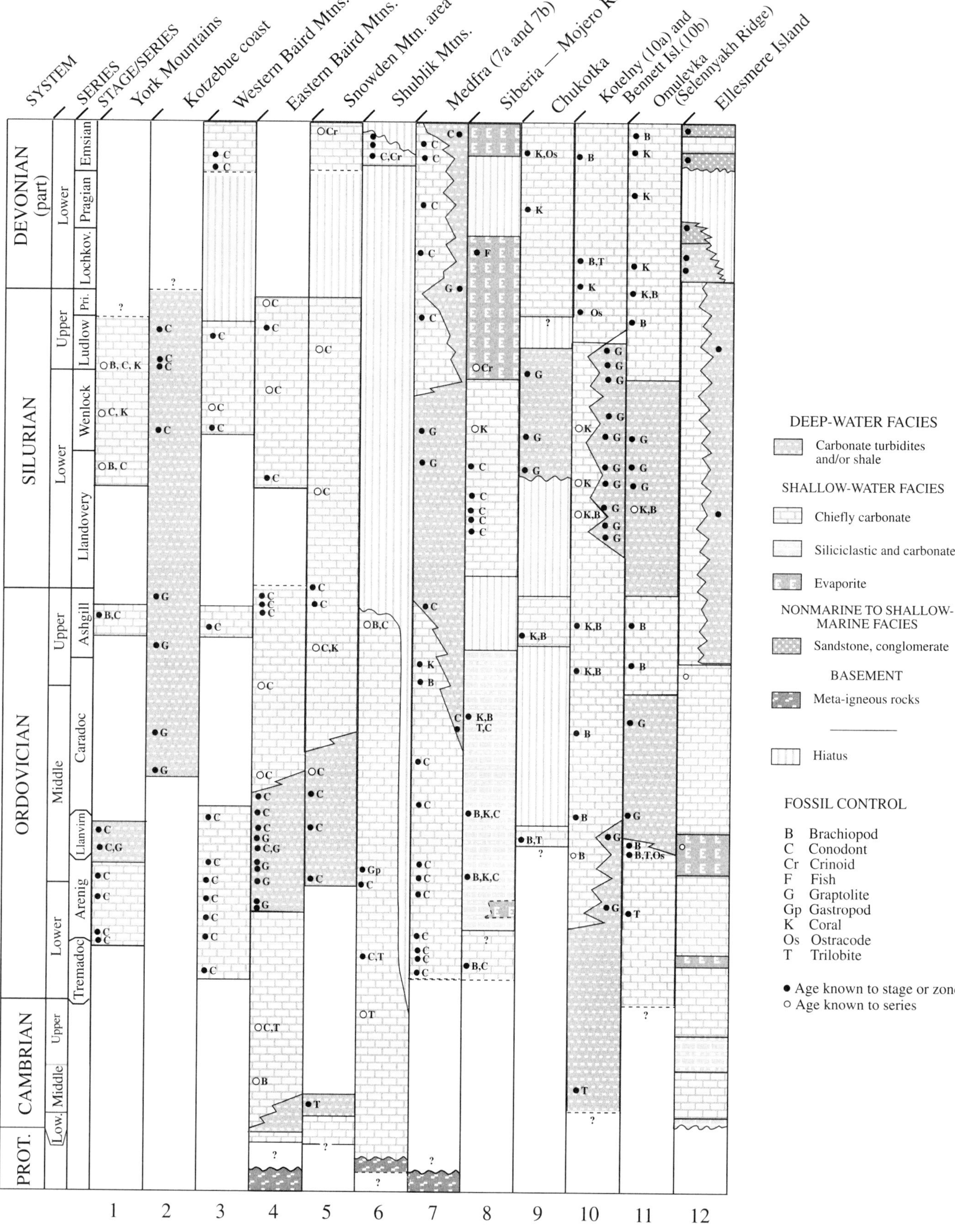

SYSTEM
SERIES
STAGE/SERIES
York Mountains
Kotzebue coast
Western Baird Mtns.
Eastern Baird Mtns.
Snowden Mtn. area
Shublik Mtns.
Medfra (7a and 7b)
Siberia — Mojero River
Chukotka
Kotelny (10a) and Bennett Isl.(10b)
Omulevka (Selennyakh Ridge)
Ellesmere Island

DEVONIAN (part)
Lower
Emsian
Pragian
Lochkov.
SILURIAN
Upper
Pri.
Ludlow
Wenlock
Lower
Llandovery
ORDOVICIAN
Upper
Ashgill
Middle
Caradoc
Llanvirn
Lower
Arenig
Tremadoc
CAMBRIAN
Upper
Middle
Low.
PROT.

DEEP-WATER FACIES
Carbonate turbidites and/or shale
SHALLOW-WATER FACIES
Chiefly carbonate
Siliciclastic and carbonate
Evaporite
NONMARINE TO SHALLOW-MARINE FACIES
Sandstone, conglomerate
BASEMENT
Meta-igneous rocks
Hiatus

FOSSIL CONTROL
B   Brachiopod
C   Conodont
Cr Crinoid
F   Fish
G   Graptolite
Gp Gastropod
K   Coral
Os Ostracode
T   Trilobite

● Age known to stage or zone
○ Age known to series

1   2   3   4   5   6   7   8   9   10   11   12

are metamorphosed to greenschist and blueschist facies, but sedimentary features and faunal assemblages are locally well preserved.

Rocks interpreted as basement to the carbonate platform succession are not widely exposed and have been little studied. In the eastern Baird Mountains, this basement consists of amphibolite facies metasedimentary rocks that have calculated Nd crustal residence ages of 2.0 Ga and are intruded by metagranites with Late Proterozoic (750 ± 6 Ma) U-Pb zircon crystallization ages (Nelson et al., 1993). On the central Seward Peninsula, schists intruded by orthogneiss bodies with U-Pb crystallization ages of 676 ± 15 Ma and 681 ± 3 Ma structurally underlie lower Paleozoic carbonate rocks (Patrick and McClelland, 1995). In the Shublik and Sadlerochit Mountains, carbonate strata overlie metasedimentary rocks intruded by a mafic dike with an Rb-Sr age of 801 ± 20 Ma (Clough and Goldhammer, 1995).

Carbonate strata of Late Proterozoic and Cambrian age in northern Alaska compose a chiefly shallow-water succession of at least 100 m in the western and central Brooks Range and as much as 3300 m in the eastern Brooks Range. Dolostones of known or inferred Late Proterozoic age crop out in the eastern Baird Mountains, Snowden Mountain area, and Shublik-Sadlerochit Mountains and contain abundant stromatolites and coated grains (both oncoids and ooids). These rocks are overlain by peritidal cycles of limestone and dolostone of Middle and Late Cambrian age in the eastern Baird and Shublik-Sadlerochit Mountains; fossils from these strata include protoconodonts, acrotretid brachiopods, and trilobites. Rare Lower (and lower Middle?) Cambrian carbonate rocks on the central and eastern Seward Peninsula contain protoconodonts and the problematic microfossil *Lapworthella* sp., which indicates a shallow-water setting (Till et al., 1986). Somewhat deeper-water (outer shelf and slope) deposits of Middle Cambrian and older (?) age occur locally in the eastern Baird Mountains and in the Snowden Mountain area. Pre-Ordovician strata are not recognized in the York or western Baird Mountains, but redeposited Late Cambrian conodonts are found in basal parts of the Ordovician sections in the western Baird and Snowden areas.

Shallow-water carbonate rocks of Early and Middle Ordovician age in the York, western Baird, Shublik, and Sadlerochit Mountains are widely distributed, at least 250–450 m thick, and formed in a range of supratidal to subtidal, locally restricted, inner and middle shelf environments. The York and western Baird sections are differentiated into two lithofacies: yellow- to orange-weathering carbonate that contains abundant argillaceous material and quartz silt and local ooids and intraclasts, and relatively pure, gray peloidal micrite. Decameter-scale shallowing-upward cycles characterize both lithofacies in both areas. The Shublik-Sadlerochit section consists chiefly of peloidal, locally oolitic, pack-grainstone.

An intraplatform basin or basins developed within the North Alaska carbonate platform during late Early Ordovician time and persisted throughout most of the Middle Ordovician. Basinal strata are longest lived in the eastern Baird and Snowden Mountain sections, where they overlie shallower-water strata of Cambrian and possibly Early Ordovician age. However, deep-water rocks also crop out to the west, in the western Baird Mountains and on the Seward Peninsula, where they are underlain by and intercalated with shallow-water carbonate rocks of Early and Middle Ordovician age. Basinal lithofacies range from at least 50 to more than 200 m thick, are of Arenig to early Caradoc (?) age, and consist of graptolitic shale or phyllite grading upward into carbonate turbidites. On the central and eastern Seward Peninsula, a continental platform succession was punctuated by at least one episode of Ordovician rifting during which iron- and titanium-rich metagabbros that are compositionally similar to modern ridge tholeiites intruded a thick sequence of calcareous and volcanogenic sediments (Till and Dumoulin, 1994).

Deposition of neritic carbonate resumed in the Late Ordovician across most of Seward Peninsula and the Brooks Range and persisted through the Silurian and locally into the Middle Devonian. Upper Ordovician through Devonian carbonate rocks are 150–300 m thick and generally similar across northern Alaska; variations mainly reflect differential erosion during Early Silurian and/or post–Early Devonian time. Meter- to decameter-scale shallowing-upward cycles characterize Upper Ordovician strata in the eastern Baird and Snowden Mountain areas, Silurian rocks in all northern Alaskan sections that contain neritic carbonate, and Lower–Middle Devonian strata in the western Brooks Range.

Shallow-water carbonate platform deposits of late Early and Middle Devonian age are widely distributed across southern and central Seward Peninsula and the western Baird Mountains, but are of more limited extent elsewhere in northern Alaska. The youngest strata known in the York Mountains are a single outcrop of coralline carbonate rocks of probable Late Silurian (late Ludlow) to Early Devonian age (Dumoulin and Harris, 1994). The youngest parts of the carbonate succession elsewhere consist of rare outcrops of upper Lower and/or lower Middle Devonian metalimestone in the Snowden Mountain area

---

Figure 3. Correlation of Upper Proterozoic to Lower Devonian rocks in selected areas of northern Alaska, interior Alaska, northeastern Russia, and Ellesmere Island, Arctic Canada. See Figures 1, 2, and 4 for locations of columns. Only fossil groups that most narrowly restrict age of collection or unit are listed; fossil groups that provide age control for column 12 not given by sources. Data sources: columns 1–6, Dumoulin and Harris (1994), Harris et al. (1995); column 7, Dutro and Patton (1982), Dumoulin et al. (1999, 2000b); columns 8–11, stratigraphic columns compiled in 1992 by Valentyn Bondarev and colleagues, VNII-Okeangeologiya, Saint Petersburg, for Geological Survey of Canada Paleozoic Circumpolar Correlation Project, with additional information from Kanygin et al. (1988) (column 8), Oradovskaya (1988) (columns 9–11) and Kos'ko et al. (1990) (column 10); column 12, de Freitas et al. (1997), stratigraphic column compiled in 1992 for Geological Survey of Canada Paleozoic Circumpolar Correlation Project. Former Llandeilo Series (Middle Ordovician) is now considered a stage of Llanvirn Series (Fortey et al., 1995).

and Lower Devonian (Emsian) limestone that disconformably overlies Upper Ordovician limestone in the Shublik Mountains.

Devonian shallow-water carbonate rocks also occur in a series of thrust sheets north of the lower Paleozoic carbonate platform in the western Brooks Range (Kelly River and Ipnavik River allochthons of Mayfield et al., 1988; approximately equivalent to the De Long Mountains subterrane of the Arctic Alaska terrane of Moore et al., 1994), but the relationship of these strata to the North Alaska carbonate platform is uncertain. The thrust sheets contain 50–700 m of chiefly unmetamorphosed limestone and lesser dolostone that is Middle to Late Devonian (Eifelian, Givetian, Frasnian, and Famennian) (Mayfield et al., 1990; Dumoulin and Harris, 1992; Baxter and Blodgett, 1994). The lower boundary of this unit is invariably a thrust fault and the upper boundary is a thrust fault or a conformable contact with Mississippian carbonate rocks.

Devonian carbonate strata in the thrust sheets are lithologically similar to, but largely younger than, the Devonian part of the North Alaska carbonate platform; no carbonate rocks definitively older than Eifelian are known in the thrust sheets, and none definitively younger than Givetian are found in the North Alaska platform. Carbonate rocks in the thrust sheets may be the structurally detached uppermost part of the North Alaska platform, or they may be part of a discrete Devonian-Mississippian carbonate platform that was not originally contiguous with the North Alaska carbonate platform, as suggested by some stratigraphic evidence (discussed in the following). Faunal assemblages, however, discussed in the section "Paleobiogeography," imply ties between the two successions.

Siliciclastic rocks of Devonian and Mississippian age that accumulated chiefly in shallow-marine and nonmarine settings overlie older carbonate and noncarbonate strata across much of northern Alaska. In the western Baird and Snowden sections, Devonian platform carbonates grade up into interbedded limestones and quartz-rich sandstones of Middle and early Late Devonian age. These rocks in turn grade up into the siliciclastic Endicott Group (Upper Devonian and Lower Mississippian), which contains abundant detrital chert and is locally as much as 2600 m thick (Moore and Nilsen, 1984). The Endicott Group is largely absent from thrust sheets of the Kelly River and Ipnavik River allochthons, however (Mayfield et al., 1988). This difference in stratigraphic succession suggests the possibility that Devonian carbonate rocks in these thrust sheets were not originally part of the North Alaska carbonate platform.

The deep-water siliciclastic successions of early Paleozoic age in northern Alaska are generally overlain by Devonian-Mississippian siliciclastic strata (Endicott Group and related rocks). These younger rocks are Mississippian near Cape Lisburne and Mount Doonerak, but Middle Devonian to Early Mississippian in the northeastern Brooks Range (Tailleur, 1965; Armstrong et al., 1976; Anderson et al., 1994). Chert-rich sandstone and conglomerate were also penetrated in several wells beneath the coastal plain in the north-central North Slope; in the Topagoruk well (Fig. 2), this sequence contains plant fragments of proba-

ble late Early–early Middle Devonian age (Witmer et al., 1981). These strata are compositionally similar to, but somewhat older than, exposures of the Endicott Group in the thrust sequences of the Brooks Range to the south.

Thus, both pre-Carboniferous platform carbonate strata and coeval deep-water siliciclastics now exposed to the north are overlain by chert-rich, siliciclastic shallow-marine and nonmarine rocks of Devonian and Mississippian age. These chert-rich younger siliciclastic strata may have been produced by a single depositional system that overlapped older Paleozoic rocks throughout northern Alaska, but no simple age progression for this overlap has been established.

Fault-bounded, off-platform carbonate sequences of primarily Silurian age and limited geographic extent are recognized at several localities on Seward Peninsula and in the Brooks Range, primarily along the present-day northern margin of the North Alaska carbonate platform. Basinal strata on the northern Seward Peninsula (Kotzebue coast; Fig. 2, location 2; Fig. 3, column 2) are the best documented of these occurrences, and consist of Middle and Upper Ordovician graptolitic shales and radiolarian cherts that grade upward into Silurian (middle Wenlock and lower to middle Ludlow) limestone turbidites and debris flows (Till et al., 1986; Ryherd and Paris, 1987; Ryherd et al., 1995). Off-platform deposits that are lithologically similar to and generally coeval with those on the Kotzebue coast are also found on the southeastern Seward Peninsula (Fig. 2, location 13; Till et al., 1986) and in the central Brooks Range (Ambler River Quadrangle; Fig. 2, location 14; Dumoulin and Harris, 1988); these rocks are largely carbonate turbidites and lesser graptolite- and radiolarian-bearing phyllites of Silurian age. The paleogeographic significance of these deep-water carbonate strata is uncertain. Broad lithologic and faunal correlations suggest that they were derived from the North Alaska carbonate platform; they may be intraplatform deposits like those of Middle Ordovician age described here, or (less likely) remnants of continental slope and rise deposits that accumulated along the carbonate platform margin.

Paleobiogeography

Both Siberian and North American affinities have been reported for various fossils from the lower Paleozoic carbonate platform successions in northern Alaska (Dumoulin and Harris, 1994, and references therein). In this paper we use "Laurentian" in place of "North American" to more faithfully represent Paleozoic paleogeography. Specifically, Laurentia includes the originally contiguous cratonic cores of North America and Greenland and excludes terranes accreted to these cratons in Phanerozoic time. The paleobiogeographic data we present here are derived primarily from conodonts, supplemented where possible by information from other fossil groups. Conodonts have numerous advantages for such studies, including abundance and preservation in a wide variety of lithofacies and diagenetic and metamorphic regimes. In addition, conodonts are the principal micropaleontologic index fossils in marine rocks of Late Cambrian through Triassic age. In this chapter we rely heavily on Ordovician conodonts for paleobiogeographic comparisons between Alaska and northeast Russia because conodont provinciality worldwide is strongest during this period (Bergström, 1990b). Cambrian conodonts are comparatively rare and Silurian and Devonian conodonts are relatively cosmopolitan. Our conodont database for Ordovician rocks in northern Alaska and the Farewell terrane of central Alaska consists of 284 samples (Table 1) that have yielded ~20 000 specimens. Megafossils of various types also provide useful paleobiogeographic data (e.g., Blodgett et al., this volume), but the use of these data is complicated for several reasons. Paleobiogeographic realms defined by one fossil group may differ from those outlined by another group during the same time period, and realms defined by the same fossil group may shift through time.

In the following analysis we distinguish conodonts representative of two realms and two provinces; in addition, some other conodonts have mixed provincial affinities (Table 1). The cosmopolitan realm includes conodont genera in which most species have worldwide distribution (includes the relatively high latitude conodont faunal realm of Sweet and Bergström, 1986). Taxa of this realm are generally abundant in most depositional settings in relatively high paleolatitudes, but in low paleolatitudes they are found chiefly in basin and continental margin deposits. The tropical cosmopolitan realm includes genera in which most species have worldwide distribution in the tropics or are known from at least three large, distinct tropical cratons and their margins (see Scotese, 1997, for Ordovician paleogeographic reconstructions). Some tropical cosmopolitan species, however, may occur as rare components in cosmopolitan faunas (e.g., rare specimens of *Rossodus manitouensis,* an Early Ordovician tropical cosmopolite, were reported by Löfgren et al., 1998, from a relatively high paleolatitude cosmopolitan assemblage in Scandinavia).

Conodont provinces are more limited in geographic extent than realms. The Laurentian province includes conodont genera in which some or most species are restricted to or are most common in the tropical shallow-water, cratonic, platform, and shelf deposits of Laurentia and peri-Laurentia (the North American midcontinent conodont province or faunal succession of Ethington and Clark, 1971; Sweet et al., 1971; Sweet, 1984, among others). Laurentia encompassed much of present-day North America and Greenland, but Ordovician conodont faunas with Laurentian affinities may occur in other tropical areas that were near, or were part of, the Ordovician Laurentian plate (e.g., Bergström, 1990a; Lehnert, 1995). From time to time, plate movements and/or changes in oceanic circulation allowed some Laurentian taxa to invade higher-paleolatitude shelf or platform areas. The Siberian province includes conodont genera in which some or most species are restricted to, or most common in, the tropical, shallow-water cratonic, platform, and shelf deposits of Siberia and peri-Siberia (e.g., Abaimova, 1975; Moskalenko, 1970, 1973, 1983). Ordovician conodont faunas with Siberian affinities may also occur in other tropical areas that were near the Ordovician Siberian plate (Dumoulin and Harris, 1994).

We recognize in addition some conodonts of mixed provincial affinity. Species that are limited to Siberia and northern and/or central Alaska are designated Siberian-Alaskan forms; these forms have not been recognized in areas established as parts of Laurentia. Similarly, species found in Ordovician Laurentia, western and central Alaska, Siberia, and peri-Siberia but unknown elsewhere are designated Laurentian-Siberian forms. In all our collections, we have recognized only two conodont species representative of the low-latitude, relatively shallow water North Chinese province (Wang et al., 1996). *Tangshanodus tangshanensis* An and *Tasmanognathus sishuiensis* Zhang both occur in our northeast Russian collections, and *T. tangshanensis* was also found in a collection from the western Brooks Range (Tables 1 and 2).

In northern Alaska, fossils with Siberian affinities occur in the same successions and commonly at the same stratigraphic levels as forms with Laurentian provinciality. Conodonts illustrate this pattern of mixed affinities particularly well. Tropical cosmopolitan and cosmopolitan forms dominate most Ordovician conodont collections from northern Alaska (Table 1), but four Ordovician time intervals are characterized by an important component of Siberian-Alaskan conodonts. These intervals—early Arenig, late Arenig–early Llanvirn, latest Llanvirn-early Caradoc, and middle (?) Ashgill—coincide approximately with global sea-level highstands (Ross and Ross, 1988). Siberian-Alaskan conodonts (Table 2) include *Fryxellodontus?* n. sp. (Dumoulin and Harris, 1994, Fig. 24, no. 99; = *Acodina? bifida* of Abaimova, 1975) in early Arenig time, acanthodinids and acanthocordylodids in late Arenig–early Caradoc time, *Stereoconus corrugatus* Moskalenko and *Plectodina?* cf. *P.? dolboricus* (Moskalenko) in early Caradoc time, and *Belodina? repens* Moskalenko in middle (?) Ashgill time. Our lower Paleozoic conodont database for northern Alaska encompasses several hundred collections from more than 100 localities. Siberian elements occur in several tens of collections, and are strikingly abundant in a few. However, typical Laurentian conodonts—such as species of *Clavohamulus* in the early Arenig—occur along with the Siberian forms and are locally abundant. In our Ordovician collections, Laurentian province species outnumber Siberian-Alaskan species throughout northern Alaska (Table 1).

Northern Alaskan megafossils also have mixed affinities. Fossils with Siberian affinities include Middle Cambrian trilobites from the central Brooks Range (both from the carbonate platform rocks and from the siliciclastic-volcanic sequence at Mount Doonerak), Early and Late Ordovician trilobites from the Seward Peninsula, and Late Ordovician brachiopods and gastropods from the Seward Peninsula and the western and eastern Brooks Range (Dumoulin and Harris, 1994, and references therein; Ormiston and Ross, 1976; Blodgett et al., 1992, and this volume). Laurentian forms include Early and Late Cambrian trilobites from the eastern Brooks Range, and Late Ordovician corals, stromatoporoids, and brachiopods from the Seward

TABLE 1. FAUNAL AFFINITIES AND DISTRIBUTION OF LATEST CAMBRIAN AND ORDOVICIAN CONODONT TAXA IN NORTHEASTERN RUSSIA AND THE FAREWELL TERRANE, SEWARD PENINSULA, AND BROOKS RANGE, ALASKA

Taxon	Northeastern Russia	Farewell terrane, central Alaska	Seward Peninsula	Western Brooks Range	Central and eastern Brooks Range
Acanthodina sp. indet.	SA	SA			
Acanthocordylodus sp. indet.	SA	SA	SA	SA	
"Acanthodus" lineatus (Furnish)	TC	TC	TC	TC	
"Acanthodus" uncinatus Furnish		TC			
Acodus deltatus Lindström			C	C	
Acodus cf. *A. deltatus* Lindström				C	
Aloxoconus iowensis (Furnish)		L		L	
Amorphognathus sp. indet.	C	C	C	C	
Ansella nevadensis (Ethington & Clark)		L?			
Ansella sp.		C		C	
Aphelognathus divergens Sweet				L	L
Baltoniodus aff. *B. variabilis* Bergström				C	
Belodinids				TC	TC
Belodina compressa (Branson & Mehl)		TC		TC	
Belodina monitorensis Ethington & Schumacher?		L		L	
Belodina? repens Moskalenko	SA	SA	SA		
Bergstroemognathus extensus Serpagli	TC				
Cahabagnathus sweeti (Bergström)		C			
Chosonodina rigbyi Ethington & Clark				L	
Clavohamulus densus Furnish					L
Clavohamulus n. sp. of Dumoulin & Harris, 1987			L	L	
Clavohamulus n. sp. cf. *C.* n. sp. A of Repetski, 1982	LS				
Colaptoconus quadraplicatus (Branson & Mehl)	TC		TC	TC	
Coleodus sp.	TC				
Complexodus sp. indet.	C				
Cordylodus angulatus Pander	C			C	
Cordylodus intermedius Furnish	C	C			C
Cordylodus proavus Müller	C	C			C
Cordylodus sp.	C	C			
Cornuodus longibasis (Lindström)		C			
Cornuodus? sp.	C	C			
Culumbodina occidentalis Sweet		L	L		L
Diaphorodus sp.			LS		
Drepanodus arcuatus Pander			C	C	C
Drepanodus concavus (Branson & Mehl)		TC		TC	
Drepanodus sp.	C	C			
Drepanoistodus basiovalis (Sergeeva)		C		C	
Drepanoistodus forceps Lindström			C	C	
Drepanoistodus pervetus Nowlan		L?	L?	L?	
Drepanoistodus suberectus (Branson & Mehl)		C	C	C	
Drepanoistodus spp.	C	C	C		
Eoplacognathus elongatus Bergström transitional to *Polyplacognathus* sp.					C
Erraticodon balticus Dzik		C	C		C
Erraticodon patu Cooper		C?			
Erraticodon sp. indet.	C				
Eucharodus parallelus (Branson & Mehl)		TC	TC	TC	
Eucharodus toomeyi Ethington & Clark				L	
Evencodus sibericus Moskalenko		SA			
Fahraeusodus marathonensis (Bradshaw)	TC		TC	TC	
Fryxellodontus? n. sp. (= *Acodina? bifida* Abaimova)		SA	SA	SA	
Gapparodus bisulcata Müller				C	
Hirsutodontus hirsutus Miller	TC	TC			
Histiodella donnae Repetski		TC		TC	
Histiodella holodentata Ethington & Clark	C				
Histiodella n. sp. 2 of Harris et al., 1979	LS			LS	
Histiodella n. sp. (Early Ordovician)			L	L	
Iapetognathus sprakersi Landing	C				
Juanognathus jaanussoni Serpagli	C				
Juanognathus variabilis Serpagli	TC			TC	TC

Taxon	Northeastern Russia	Farewell terrane, central Alaska	Seward Peninsula	Western Brooks Range	Central and eastern Brooks Range
Juanognathus? sp. indet.	TC				
Jumudontus gananda Cooper	C			C	
Loxodus? spp.		LS			
Macerodus n. sp.				TC	
Oepikodus communis (Ethington & Clark)	TC		TC		TC
Oepikodus evae (Lindström)	C				
Oistodus bransoni Ethington & Clark		L			
Oistodus lanceolatus Pander			C		
Oistodus? lecheguillensis Repetski				L	
"Oistodus" mehli Furnish s.f.				L	
Oistodus multicorrugatus Harris	LS	LS			
Oistodus cf. *O multicorrugatus* Harris	TC	TC			
Oneotodus costatus Ethington & Brand				L	
"Oneotodus" gracilis (Furnish)		TC			
Oneotodus simplex (Furnish) as used by Ethington & Brand, 1981		L			
"Oneotodus" variabilis Lindström		C	C	C	
Ozarkodina hassi (Pollock, Rexroad, & Nicoll)		C			
Ozarkodina sesquipedalis Nowlan & McCracken		L			
"Paltodus" acuminatus (Pander)				C	
Paltodus cf. *P. deltifer* (Lindström)				C	
Paltodus inaequalis (Pander)				C	
Paltodus spurius Ethington & Clark				L	
Paltodus subequalis Pander			C	C	
Panderodus gracilis (Branson & Mehl)	C	C	C	C	
Panderodus sp.	C	C	C	C	C
Paracordylodus gracilis Lindström			C	C	
Parapanderodus? consimilis (Moskalenko)	LS			LS	
Parapanderodus paracornuformis (Ethington & Clark)		TC			
Parapanderodus striatus (Graves & Ellison)			C	C	C
Paraprioniodus costatus (Mound)?				L	
Paraserratognathus abruptus (Repetski)		TC	TC		
Paroistodus? horridus (Barnes & Poplawski)				C	
Paroistodus? mutatus (Rhodes)	C	C	C	C	C
Paroistodus cf. *P. neumarcuatus* Lindström		C		C	
Paroistodus originalis (Lindström)		C			
Paroistodus parallelus (Pander)			C	C	
Paroistodus proteus (Lindström)		C	C	C	
Paroistodus sp. indet.	C		C	C	
Periodon aculeatus Hadding	C	C	C	C	C
Periodon flabellum (Lindström)	C		C		C
Periodon sp. indet.	C	C	C		
Phakelodus tenuis (Müller)				C	C
Phragmodus flexuosus Moskalenko	TC			TC	
Phragmodus n. sp. of Barnes, 1974					L*
Plectodina? cf. *P.? dolboricus* (Moskalenko)				SA	SA†
Plectodina? lunguskaensis (Moskalenko)		TC	TC	TC	TC
Polonodus tablepointensis Stouge	C				
Polonodus sp. indet.	C			C	
Polycostatus aff. *Po. minutus* Ji & Barnes		L			
Prattognathus rutriformis (Sweet & Bergström)					L
Prioniodus elegans Pander				C	
Prioniodus sp. indet.	C				
Proconodontus muelleri Miller				C	
Protopanderodus graeai (Hamar)					C
Protopanderodus insculptus (Branson and Mehl)	TC	TC		TC	
Protopanderodus leei Repetski		TC	TC	TC	
Protopanderodus liripipus Kennedy, Barnes, & Uyeno	C			C	C
Protopanderodus elongatus Serpagli		C	C	C	

TABLE 1. (continued)

Taxon	Northeastern Russia	Farewell terrane, central Alaska	Seward Peninsula	Western Brooks Range	Central and eastern Brooks Range
Protopanderodus gradatus Serpagli				C	
Protopanderodus rectus (Lindström)	C		C	C	
Protopanderodus robustus (Hadding)	C			C	
Protopanderodus varicostatus (Sweet and Bergström)					C
Protopanderodus cf. *P. varicostatus* (Sweet and Bergström)	C	C		C	
Protopanderodus sp. indet.	C	C	C	C	C
Protoprioniodus aranda Cooper	TC	TC	TC		
Pseudobelodina adentata Sweet			L		
Pseudobelodina dispansa (Glenister)			TC	TC	
Pseudooneotodus mitratus (Moskalenko)	C	C			
Pygodus anserinus Lamont & Lindström		C		C	C
Pygodus serra (Hadding)					C
Rossodus manitouensis Repetski and Ethington	TC	TC			TC
Rossodus n. sp.			L	L	
Scalpellodus sp.		C	C	C	
Scandodus brevibasis (Sergeeva) s.f.				C	
Scandodus sinuosus Mound				L	
Scolopodus bolites Repetski			TC	TC	
Scolopodus *cornutiformis* Branson & Mehl s.f.				L	
"Scolopodus" filosus Ethington & Clark		TC	TC	TC	
Scolopodus floweri Repetski	TC	TC	TC	TC	
Scolopodus kelpi Repetski	LS		LS		
Scolopodus rex Lindström		C	C	C	
Scolopodus sulcatus Furnish		TC	TC	TC	
Semiacontiodus nogamii (Miller)	TC				
Spinodus spinatus (Hadding)	C	C		C	C
Strachanognathus parvus Rhodes		C			
Staufferella aff. *S. brevispinata* Nowlan & Barnes			L		
Stereoconus corrugatus Moskalenko		SA	SA	SA	
Striatodontus prolificus Ji & Barnes			TC	TC	
Tangshanodus tangshanensis An	NC-SA			NC-SA	
Tasmanognathus sishuiensis Zhang	NC-S				
Teridontus nakamurai (Nogami)	C	C			C
Triangulodus cf. *T. brevibasis* (Sergeeva)				C	
Tripodus laevis Bradshaw	TC			TC	TC
Tripodus sp.	C				
Tropodus comptus (Branson & Mehl)			TC		
Tropodus sp.			TC	TC	
Ulrichodina deflexa Furnish				L	
Ulrichodina n. sp. 1 of Repetski, 1982		L		L	
Utahconus longipinnatus Ji & Barnes		TC			
Variabiloconus bassleri (Furnish)	TC	TC	TC	TC	TC
Walliserodus australis			TC	TC	
Walliserodus ethingtoni (Fåhraeus)	C		C	C	
Walliserodus sp.	C			C	

Note: Conodont taxa in this list are from collections distributed as follows: northeastern Russia, 88; Farewell terrane, 80; Seward Peninsula, 50; western Brooks Range, 132; and central and eastern Brooks Range, 22. Abbreviations of conodont faunal affinities: C, cosmopolitan realm; TC, tropical cosmopolitan realm; L, Laurentian province (species occur in Laurentia and peri-Laurentia); LS, Laurentian-Siberian (species occur in Laurentia, peri-Laurentia, Siberia, and peri-Siberia); SA, Siberian-Alaskan (species restricted to Siberia, peri-Siberia, and Alaska); NC-S, North Chinese–Siberian (species restricted to North China, Siberia, and peri-Siberia); NC-SA, North Chinese–Siberian–Alaskan (species known from North China, Siberia, peri-Siberia, and Alaska). L?, taxon known from province but may be more widespread paleogeographically.

*Dumoulin and Harris (1994) considered phragmodid elements assigned by Moskalenko (1983, fig. 4W, X) to *P. undatus* Branson and Mehl, likely equivalent to *Phragmodus* n. sp. of Barnes, 1974. Consequently, they characterized the faunal affinities of this species as Siberian–northern North American (Laurentian-Siberian as used herein). Because Moskalenko (1983) did not figure the complete appartus of the species, we are uncertain if her specimens are the same as *P.* n. sp. of Barnes, 1974. Thus, we herein consider the faunal affinities of the Barnes (1974) and Dumoulin and Harris (1994, pl. 2, figs. 20–24) *Phragmodus* as Laurentian.

†Species not found east of the central Brooks Range.

TABLE 2. DISTRIBUTION OF LATEST CAMBRIAN AND ORDOVICIAN CONODONT TAXA OF MIXED FAUNAL AFFINITIES IN NORTHEASTERN RUSSIAN AND THE FAREWELL TERRANE, SEWARD PENINSULA, AND BROOKS RANGE, ALASKA

Taxon	Northeastern Russia	Farewell terrane, central Alaska	Seward Peninsula	Western Brooks Range	Central and eastern Brooks Range
Acanthodina sp. indet.	SA	SA			
Acanthocordylodus sp. indet.	SA	SA	SA	SA	
Belodina? repens Moskalenko	SA	SA	SA		
Clavohamulus n. sp. cf. *C.* n. sp. A of Repetski, 1982	LS				
Diaphorodus sp.			LS		
Evencodus sibericus Moskalenko		SA			
Fryxellodontus? n. sp. (= *Acodina? bifida* Abaimova)		SA	SA	SA	
Histiodella n. sp. 2 of Harris et al., 1979	LS			LS	
Loxodus? spp.		LS			
Oistodus multicorrugatus Harris	LS	LS			
Parapanderodus? consimilis (Moskalenko)	LS			LS	
Plectodina? cf. *P.? dolboricus* (Moskalenko)				SA	SA (central Brooks Range only)
Scolopodus kelpi Repetski	LS		LS		
Stereoconus corrugatus Moskalenko		SA	SA	SA	
Tangshanodus tangshanensis An	NC-SA			NC-SA	
Tasmanognathus sishuiensis Zhang	NC-S				

Note: Thus far, latest Cambrian and Ordovician conodonts with Siberian affinities have not been found in the eastern Brooks Range. Abbreviations of conodont faunal affinities: LS, Laurentian-Siberian (species occur in Laurentia, peri-Laurentia, Siberia, and peri-Siberia); SA, Siberian-Alaskan (species restricted to Siberia, peri-Siberia, and Alaska); NC-S, North Chinese–Siberian (species restricted to North China, Siberia, and peri-Siberia); and NC-SA, North Chinese–Siberian–Alaskan (species known from North China, Siberia, peri-Siberia, and Alaska).

Peninsula and the central Brooks Range (Dumoulin and Harris, 1994, and references therein; Potter, 1984). Carbonate strata in thrust sheets of the De Long Mountains subterrane contain Middle Devonian brachiopods with Siberian affinities (Baxter and Blodgett, 1994) as well as Devonian (Pragian?) corals known elsewhere only from peri-cratonal North America (Road River Formation, Yukon Territory, Canada; W.A. Oliver, 1993, written commun.). Mississippian plants found in chert and sandstone of the De Long Mountains subterrane (Kelly River and Picnic Creek allochthons) have "uniquely Angaran" (Siberian) affinities (Spicer and Thomas, 1987, p. 355).

Deep-water faunas from both siliciclastic and carbonate facies in northern Alaska are mainly cosmopolitan and only rarely provide specific biogeographic information. For example, conodonts from Lower and Middle Ordovician basinal strata in the Brooks Range and the Seward Peninsula are chiefly cosmopolitan deep- and/or cool-water species of the protopanderodid-periodontid biofacies (Dumoulin and Harris, 1994). However, shallower-water Siberian-Alaskan and Laurentian elements such as *Plectodina?* cf. *P.? dolboricus* and *Prattognathus rutriformis* occur as rare postmortem hydraulic additions in this biofacies in the central and western Brooks Range.

Siberian faunal components appear to decrease in number from present-day west to east across northern Alaska. Siberian elements are most abundant and diverse in lower Paleozoic collections from the western Seward Peninsula and western Brooks Range (Table 2), but even these collections contain some Laurentian province forms. Our Ordovician conodont collections show that the ratio of Laurentian province to Siberian-Alaskan species increases across northern Alaska (Table 1), from 1.75 on Seward Peninsula to 4.5 in the western Brooks Range to 5 in the central and eastern Brooks Range. Certain Siberian faunal elements (such as the pentamerid brachiopod *Tcherskidium*) occur throughout northern Alaska and are found as far north and east as the Shublik Mountains (Fig. 2; Blodgett et al., 1992, and this volume; Blodgett, 1998). At least part of the apparent eastward decline in Siberian forms reflects changes in depositional environment. Shallow-water facies of Early and Middle Ordovician age contain notable numbers of Siberian taxa (particularly conodonts), but these facies are rare or absent in the central Brooks Range, where Ordovician basinal strata are best developed. Basinal strata, as noted here, contain largely cosmopolitan forms, and the central Brooks Range intraplatform basin may have blocked the dispersal of Siberian taxa into shallow-water environments of the northeastern Brooks Range.

In summary, both Siberian and Laurentian faunal elements occur at various times and in various fossil groups across northern Alaska. Siberian influences are noted from at least Middle Cambrian through Mississippian time. This pattern of mixed faunal influences is strikingly similar to that seen in the Nixon Fork subterrane in central Alaska (Dumoulin et al., 1998b), which is further discussed in the following.

CENTRAL ALASKA

Unmetamorphosed lower Paleozoic rocks that formed in both shallow- and deep-water environments are widely distributed across central Alaska, and a complex and contentious nomenclature designates these strata. Chiefly shallow-water facies of Late Proterozoic through Devonian age in the Medfra Quadrangle and adjacent areas have been considered part of the Nixon Fork terrane (Patton et al., 1994; Silberling et al., 1994). Coeval deep-water strata in this area have been called the Minchumina terrane to the north and the Dillinger terrane to the south. Shallow- and deep-water facies in both terranes are overlain by heterogeneous, Devonian through Cretaceous lithologies of the Mystic terrane. Decker et al. (1994) established the Farewell terrane, which they interpreted as a coherent, but locally highly deformed, continental margin sequence, to include all of these units. This interpretation has been widely accepted, but which continent the sequence formed along remains the subject of much debate. Some have suggested that the Farewell terrane is a displaced fragment of the North American continental margin (Decker et al., 1994), whereas others believe that it rifted away from the Siberian craton (Blodgett, 1998).

In this chapter we follow the usage of Bundtzen et al. (1997) and Blodgett (1998). We use "Nixon Fork subterrane" for Ordovician through Devonian rocks in the Nixon Fork terrane of Patton et al. (1994), as well as broadly coeval shallow-water strata (Holitna Group) exposed to the south, employ "Dillinger subterrane" for the lower Paleozoic deep-water facies that crop out to the east and southeast of the Nixon Fork rocks, and use "Mystic subterrane" for the Devonian through Cretaceous strata that overlie both older subterranes (Fig. 4). We agree with Decker et al. (1994) that these three subterranes are depositionally related, but retain the subterrane terminology herein because there is no workable stratigraphic nomenclature that encompasses all the rocks under discussion. New and previously published lithologic and faunal data that bear on the origin of the Farewell terrane are summarized in the following.

Lithofacies

Shallow-water rocks of the Nixon Fork subterrane have many similarities with coeval strata in northern Alaska. Lithologic details presented here are from Dutro and Patton (1982), Dumoulin et al. (1999, 2000b), Decker et al. (1994), Measures et al. (1992), and our own observations (Fig. 3, column 7). Rocks considered basement to Farewell terrane Paleozoic strata (Patton et al., 1980, 1994) are exposed in the northeastern Medfra Quadrangle and consist of amphibolite to greenschist facies metasedimentary and metavolcanic rocks. Rhyolitic metatuff and granitic orthogneiss yielded U-Pb zircon ages of 979 +8/−3 Ma and 850 ± 1 Ma, respectively (McClelland et al., 1999). The next-youngest parts of the terrane are exposed in the McGrath and Sleetmute Quadrangles to the south. These rocks are oolitic dolostones and redbeds of probable Late Proterozoic age overlain by mudstones with rare limestone interbeds that contain a rich Middle Cambrian trilobite fauna (Babcock et al., 1993, 1994; St. John and Babcock, 1997).

The Ordovician through Devonian section in the Nixon Fork subterrane is best exposed in the Medfra Quadrangle. Some 5400 m of strata represent a shallow-water platform carbonate succession interrupted by intervals of deeper-water sedimentation in the Silurian and Devonian (Fig. 3) (Dutro and Patton, 1982; Dumoulin et al., 1999, 2000b). Lower Lower Ordovician rocks (Novi Mountain Formation; 900 m thick) weather orange and contain locally abundant argillaceous material, quartz and feldspar silt, ooids, intraclasts, and peloids. Decameter-scale cycles shallow upward from open-marine shales to ooid or intraclast grainstones deposited in nearshore shoals. Upper Lower, Middle, and Upper Ordovician strata (Telsitna Formation; 2000 m thick) are largely micrite and peloidal packstone deposited in middle to inner shelf settings with locally restricted circulation.

Nixon Fork subterrane sections south of the Medfra area contain some similar shallow-water rocks of Early, Middle, and Late Ordovician age, but these facies are intercalated with deeper-water mudstones and limestones (Decker et al., 1994); the percentage of deep-water strata increases to the south. In the Medfra area, deep-water facies first interfingered with shallow-water sediments during the Late Ordovician. Calcareous turbidites, calcitized radiolarite, and lesser shale occur locally in the upper part of the Telsitna Formation and make up all of the succeeding Paradise Fork Formation. The Paradise Fork is ~1000 m thick (Dutro and Patton, 1982) and largely of late Early Silurian (Wenlock) age; the upper part of the unit interfingers with Upper Silurian shallow-water facies. Uppermost beds in the Paradise Fork are extremely condensed and at least as young as Early Devonian (Dumoulin et al., 1999).

Upper Silurian and Devonian shallow-water facies in the Medfra Quadrangle comprise 1000–1500 m of limestone and dolostone (Whirlwind Creek Formation) deposited in a range of subtidal settings with locally restricted circulation. The unit is characterized by decameter-scale cycles of algal laminite that grade up into peloidal and then fossiliferous limestone; notable fossils are ostracodes, gastropods, brachiopods, stromatoporoids, and corals (Dutro and Patton, 1982). Coeval strata to the south include a distinctive algal-sponge mound complex as thick as 500 m that formed a barrier reef along the outer platform margin (Clough and Blodgett, 1985, 1989).

The Dillinger subterrane and related rocks (Minchumina terrane of Patton et al., 1994), exposed southeast of the Nixon Fork subterrane, consist of as much as 1500 m of deep-water strata deposited from Late Cambrian through Early Devonian time (Churkin and Carter, 1996). Hemipelagic shale and fine-grained limestone turbidites predominate in these sections, but coarser-grained turbidites of mixed composition (e.g., the Terra Cotta Mountains Sandstone of late Early–Late Silurian age) occur locally. These coarser turbidites consist chiefly of quartz, feldspar, and calcareous clasts and include notable mica and vol-

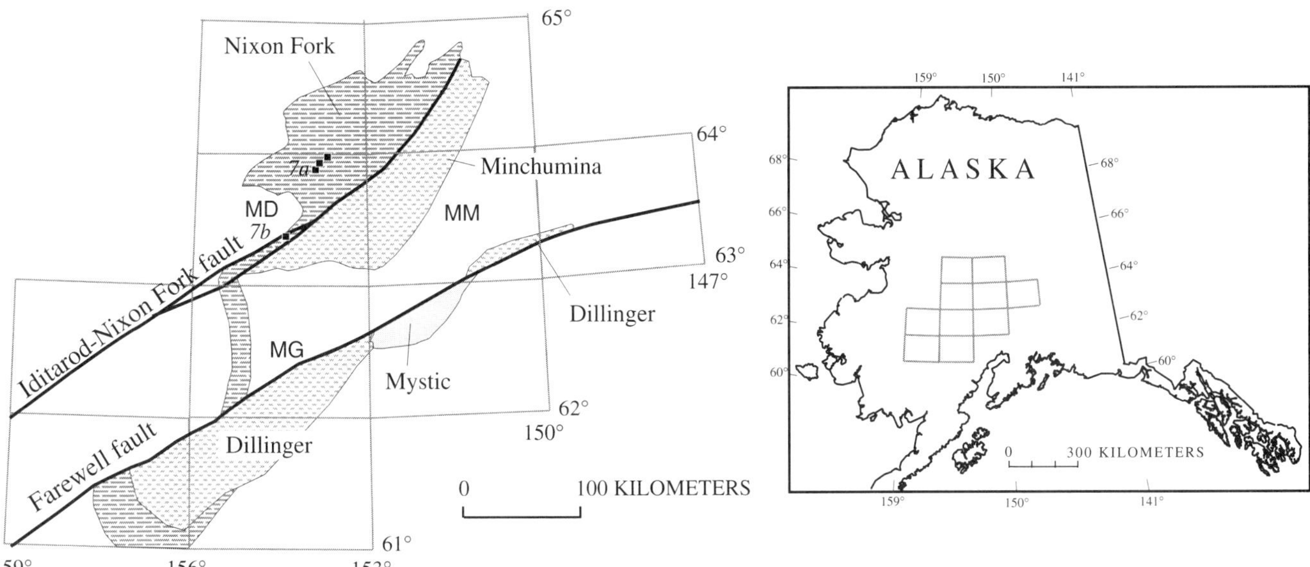

Figure 4. Location of Dillinger, Minchumina, Mystic, and Nixon Fork tectonostratigraphic subterranes in central Alaska. All subterranes shown were included in Farewell terrane of Decker et al. (1994). Dillinger and Nixon Fork south of 63° are modified from Decker et al. (1994) and Silberling et al. (1994); Nixon Fork north of 63° and Minchumina are from Patton et al. (1994); Dillinger north of 63° and Mystic are from Wilson et al. (1998). Selected quadrangles: MD, Medfra; MG, McGrath; MM, Mount McKinley. 7a and 7b are locations of stratigraphic sections shown as column 7 in Figure 3.

canic lithic clasts. Discrete ash layers are intercalated with the Silurian turbidites in both the McGrath and Mount McKinley Quadrangles (Dumoulin et al., 1998a).

Paleobiogeography

Fossil assemblages from the Farewell terrane, like those from northern Alaska, have affinities with both the Siberian and Laurentian paleobiogeographic provinces; ties to faunas from the Urals, Baltica, and terranes of northeastern Russia are also indicated. Early Ordovician conodont collections from the Novi Mountain Formation contain chiefly tropical cosmopolitan forms with lesser Laurentian elements; some Siberian-Alaskan and Laurentian-Siberian components occur in coeval, slightly shallower water strata from the southern part of the Nixon Fork subterrane. Conodont species associations from the Telsitna Formation are also dominated by cosmopolitan forms, but include lesser, roughly equal numbers of tropical cosmopolitan, Laurentian province, and Siberian-Alaskan elements. Siberian-Alaskan elements are most abundant in late Early–Middle Ordovician Telsitna Formation faunas and in Late Ordovician collections from unnamed carbonate rocks at Lone Mountain in the McGrath Quadrangle. Ordovician conodonts from the Farewell terrane have a ratio of Laurentian province to Siberian-Alaskan species of 1.66 (Table 1), which is comparable to, but slightly lower than, the lowest ratio found in northern Alaska (1.75, Seward Peninsula).

Mixed biogeographic affinities are also indicated by a variety of Farewell terrane megafossils. Middle Cambrian trilobites from the southern part of the Nixon Fork subterrane are most like coeval faunas of the East Siberian platform (Palmer et al., 1985; St. John and Babcock, 1997; Kingsbury and Babcock, 1998), and the Mystic subterrane contains Early Permian plants that belong to the Angaran floral province of Siberia (Mamay and Reed, 1984). Other fossils with Siberian or Old World (non-Laurentian) affinities include Middle–Late Ordovician stromatoporoids, gastropods, and brachiopods and late Early Devonian (Emsian) brachiopods, ostracodes, and rugose corals (Dutro and Patton, 1982; Rohr and Gubanov, 1997; Blodgett, 1998; Blodgett et al., this volume; C.W. Stock, 1998, written commun.). Northern Laurentian faunal ties are suggested for other Farewell biota, however, including Middle–Late Ordovician brachiopods (Potter, 1984; Rohr et al., 1992) and early Middle Devonian (Eifelian) trilobites and corals (Blodgett, 1983). Distinctive aphrosalpingid sponges like those in the Silurian reef complexes of the southern Nixon Fork subterrane are known elsewhere only from the Alexander terrane in southeastern Alaska (Fig. 1) and the Ural Mountains (Rigby et al., 1994; Soja, 1994; Soja and Antoshkina, 1997).

COMPARISON OF LITHOFACIES AND PALEOBIOGEOGRAPHY BETWEEN NORTHERN AND CENTRAL ALASKA

Late Proterozoic through Middle Devonian lithofacies and biogeography of the North Alaska carbonate platform correlate remarkably well with those of the Farewell terrane in central Alaska. Sedimentary successions in both areas appear to be built on metamorphosed basement of Late (and locally older?) Proterozoic age, and are characterized by Late Proterozoic ooid-rich dolostones; Middle Cambrian outer shelf deposits; and Ordovician, Silurian, and Devonian shallow-water platform facies. Specific lithologic similarities are numerous and include Lower Ordovician impure limestones rich in detrital material, ooids, and intraclasts; Middle and Upper Ordovician peloidal micrites formed in inner shelf or platform settings with locally restricted circulation; and decameter-scale shallowing-upward cycles in Ordovician, Silurian, and Devonian strata. Deeper-water facies of Middle Ordovician age like those in northern Alaska do not occur in the Medfra Quadrangle but are exposed to the south in the White Mountain area of the McGrath Quadrangle, where they overlie and underlie shallow-water platform strata (Decker et al., 1994). The Silurian deep-water incursion so prominent in the Nixon Fork subterrane has not been seen in platform carbonate rocks of northern Alaska, but isolated sections of Silurian carbonate turbidites crop out adjacent to the platform on Seward Peninsula and in the central Brooks Range.

Biotic similarities between northern and central Alaska are also striking. Ordovician conodont faunas with both Siberian and Laurentian elements are a distinctive feature of both the North Alaska carbonate platform and the Farewell terrane, and Siberian-Alaskan conodonts occur during the same time intervals in both areas and include many of the same species. Middle Cambrian trilobites of Siberian aspect are found in both successions and may include some conspecific forms (Palmer et al., 1985). Ordovician megafossils in the two areas also correlate well (Dutro and Patton, 1981); identical species of gastropod and pentamerid brachiopod (*Tcherskidium*) occur at Lone Mountain (southern Nixon Fork subterrane) and in the York and Shublik Mountains (Blodgett et al., 1992, and this volume; Blodgett, 1998). Middle Ordovician graptolite faunas in the Dillinger subterrane correspond well with those in the Baird Mountains and share strong ties with coeval faunas in Australia and New Zealand (Churkin and Carter, 1996; Carter and Tailleur, 1984). Faunal matches between northern and central Alaska continue into the later Paleozoic: conspecific Eifelian brachiopods are found at White Mountain and in the western Brooks Range (Baxter and Blodgett, 1994), and Angaran (Siberian) plant assemblages occur in upper Paleozoic strata of both regions (Mamay and Reed, 1984; Spicer and Thomas, 1987).

NORTHEASTERN RUSSIA

The Siberian craton, also called the Angara craton (Şengör and Natal'in, 1996) and the North Asia craton (Nokleberg et al., 1994, 1998), is flanked to the north and east by numerous terranes that have been variously defined and named by different authors (Fig. 1). In this chapter, we use the terminology of Nokleberg et al. (1994, 1998) unless otherwise stated. Lower Paleozoic carbonate platform sequences occur both on the Siberian craton and on several of the terranes that surround it. The oc-

currence of Siberian faunal elements in lower Paleozoic strata of northern and central Alaska has led some to propose that these successions are displaced fragments of the Siberian craton (Grantz et al., 1991; Moore et al., 1994; Blodgett, 1998). To test this idea, we compare lithostratigraphy and faunal assemblages of the Siberian craton and various terranes north and east of this craton to those of northern and central Alaska (Fig. 3).

Discussion herein of Russian lithostratigraphy is based largely on relatively detailed Ordovician through Devonian stratigraphic columns compiled by Valentyn Bondarev and colleagues at VNIIOkeangeologiya, Saint Petersburg, in 1992. The columns were prepared as part of the Paleozoic Circumpolar Correlation Project led by Godfrey Nolan, Geological Survey of Canada, and represent several dozen localities across Russia north of lat 65°N. Kanygin et al. (1988) and Oradovskaya (1988) provided excellent summaries (in English) of Ordovician strata of the Siberian platform and northeastern Russia, respectively. Supplementary lithostratigraphic data come from Oradovskaya and Obut (1977), Kos'ko et al. (1990), Nokleberg et al. (1994, 1998), Şengör and Natal'in (1996), and Natal'in et al. (1999). Faunal data include a suite (87 collections) of Ordovician conodonts collected by Gagiev from various parts of the Kolyma-Omolon superterrane and analyzed by Harris and Repetski (see GSA Data Repository, Appendix 1[1]). A single Ordovician conodont collection from the Chukchi Peninsula, collected by Jaime Toro and coworkers, was also examined by Harris. Conodont data from the Ordovician of the Siberian craton were presented by Moskalenko (1970, 1973, 1983) and Abaimova (1975), among others.

Siberian craton

Paleozoic rocks of the Siberian craton overlie unmetamorphosed Middle and Late Proterozoic siliciclastic, carbonate, and volcanic rocks, which in turn overlie an Archean basement of gneiss and schist; two episodes of Proterozoic rifting are documented (Nokleberg et al., 1994). The Bondarev database contains stratigraphic columns for 6 Ordovician, 7 Silurian, and 15 Devonian sections distributed across northern Siberia; a representative composite section for all three systems from the Mojero River area in central Siberia (~lat 67°N, long 105°E) is shown in Figure 3 (column 8). Ordovician strata range from 300 to 1000 m in thickness, formed in shallow-water settings, and are mainly limestone and dolostone with significant siliciclastic interbeds. Gypsum and anhydrite occur locally, particularly in upper Lower Ordovician sections (Kanygin et al., 1988). Shallow-water carbonate rocks predominate in the Silurian (220–750 m thick), but slightly deeper water (outer shelf) graptolitic

[1]GSA Data Repository item 2002078, Appendix 1, is available on request from Documents Secretary, GSA, P.O. Box 9140, Boulder, CO 80301-9140, USA, editing@geosociety.org, or at www.geosociety.org/pubs/ft2002.htm, or on the CD-ROM accompanying this volume.

marl and shaly limestone characterize the early Early Silurian (Llandovery), and evaporitic facies (including gypsum and anhydrite) accumulated during the Late Silurian. Devonian sections (60–1230 m thick; most ≤430 m) contain abundant evaporites and siliciclastic rocks, and subordinate, mainly Givetian, limestone. Alkalic basalt flows accumulated in several major rift basins during Middle Devonian through Mississippian time (Nokleberg et al., 1994).

Several lithologic and faunal features distinguish the Siberian craton succession from the platform successions of northern and central Alaska. The Archean basement, thick sequence of unmetamorphosed Proterozoic strata, and Ordovician shallow-water siliciclastic rocks have no counterparts in the Alaskan successions (Fig. 3). The deeper-water facies of Llandovery age are similar to, but apparently thinner and shorter lived than, the Paradise Fork and coeval strata in the Farewell terrane. The Ordovician and Upper Silurian–Devonian evaporites of Siberia have no analogs in northern or central Alaska. Siberian province conodonts and megafossils link the faunas of Siberia and Alaska, but only a few of the Laurentian conodonts that occur in the Alaskan successions are also found on the Siberian craton (Table 2).

Chukotka

Rocks of the Chukotka Peninsula have long been correlated with coeval strata in northern Alaska, and some have suggested a common origin for both successions. The Arctic Alaska Plate of Halgedahl and Jarrard (1987), for example, is defined as a single crustal block made up of Chukotka, northern Alaska, adjacent continental margins, and a part of the oceanic crust of the Canada basin. The western boundary of this proposed plate is uncertain; the South Anyui suture zone, adjacent to the Kolyma River Delta, is favored by some workers, but others extend the block west to the Laptev Sea (see discussion in Rowley and Lottes, 1988). The Laptev Sea boundary was favored by Şengör and Natal'in (1996) and Natal'in et al. (1999), who included rocks of northeastern Russia (from Chukotka west through Wrangel and Kotelny Islands) and northwestern Alaska (Seward Peninsula and Hammond subterrane of the Arctic Alaska terrane) in their Bennett-Barrovia crustal block and considered all lower Paleozoic carbonate strata in this region to be part of their Novosibirsk carbonate platform.

Few lower Paleozoic rocks crop out on the Chukotka Peninsula and adjacent regions to constrain tectonic reconstructions of this area. The composite lithostratigraphic section from the Chegitun River area (Fig. 3, column 9) is taken from the Bondarev database; additional lithostratigraphic and faunal data are provided by Oradovskaya and Obut (1977) and Natal'in et al. (1999). These strata were considered part of the Seward terrane by Nokleberg et al. (1994, 1998). A Proterozoic metamorphic complex of gneiss, migmatite, amphibolite, marble, and schist (K-Ar and Rb-Sr ages of 1.6 Ga and 1.9 Ga, respectively) is

overlain by schist, carbonate, and quartzite of presumed Late Proterozoic age (Nokleberg et al., 1994); Cretaceous metamorphism has obscured the older history of these rocks (Bering Strait Geologic Field Party, 1997). The lower Paleozoic section is variably metamorphosed and comprises Middle to Upper Ordovician shallow-water carbonate rocks (600–700 m thick) that contain a varied fauna of corals, brachiopods, trilobites, and conodonts; Lower Silurian (upper Llandovery–Wenlockian) deeper water graptolitic black shale and limestone (<100 m), and Upper Silurian–Lower Devonian shallow-water carbonate rocks (450–600 m). The Middle Devonian Tanatap Formation (350–475 m), which consists of polydeformed, greenschist facies phyllites, carbonate turbidites and debris flows, and rare tuffs, structurally overlies the older rocks (Natal'in et al., 1999).

Paleozoic rocks also occur on Wrangel Island, but Paleozoic strata older than Late Silurian have not been documented here. The Paleozoic section overlies Upper Proterozoic clastic and metavolcanic rocks intruded by granite with a U-Pb zircon age of 699 ± 2 Ma (Cecile et al., 1991), which is closely comparable to the U-Pb zircon ages of 676 ± 15 Ma and 681 ± 3 Ma determined for Seward Peninsula orthogneisses (Patrick and McClelland, 1995). Stratigraphic columns in the Bondarev database indicate that Upper Silurian–lower Lower Devonian, shallow-water, carbonate and siliciclastic strata are overlain by deeper water, Lower–Middle Devonian siliciclastic facies, which are in turn overlain by Upper Devonian evaporites and mafic volcanic rocks; the clastic strata in this sequence include both arkosic and quartz-rich layers (Nokleberg et al., 1994).

In general aspect, the lower Paleozoic succession in Chukotka resembles those in northern and central Alaska. Chukotka lacks the Middle Ordovician basinal facies found on Seward Peninsula and in the Brooks Range, but contains deeper-water strata of Silurian age that are precisely coeval with, but apparently thinner than, the Paradise Fork Formation in the Nixon Fork subterrane. A single Ordovician conodont collection from Chukotka (Appendix 1, no. 88; see footnote 1) contains Siberian-Alaskan elements like those found in northern and central Alaska, but includes none of the Laurentian forms that are common in the Alaskan rocks. Megafossils suggest ties between Chukotka and the Alaskan successions; *Tcherskidium* occurs in all three areas (Blodgett et al., 1992) and monorakid trilobites, found on the Siberian craton, Kotelny Island, Omulevka, and Taimyr, are also identified in the York Mountains (Ormiston and Ross, 1976).

Kotelny and Bennett Islands

Kotelny Island (one of the New Siberian Islands; Fig. 1) contains Paleozoic carbonate rocks that have been correlated with those on Chukotka (part of the Novosibirsk carbonate platform of Şengör and Natal'in, 1996) or, alternately, considered part of the Siberian platform (Fujita and Newberry, 1982). Deepwater rocks exposed to the east on Bennett Island are generally interpreted as a lateral facies equivalent of the rocks on Kotelny

Island (e.g., Kos'ko, 1994; Şengör and Natal'in, 1996), although some consider Bennett Island to be a separate terrane (Nokleberg et al., 1994). Stratigraphic columns for Kotelny and Bennett Islands are included in the Bondarev database, described by Kos'ko et al. (1990), and summarized in Figure 3 (column 10).

No Proterozoic rocks are exposed in this area. On Bennett Island, an ~500-m-thick sequence of argillite, lesser siltstone, and limestone that contains Middle Cambrian trilobites with Siberian affinities in its lower part is overlain by 1000–1200 m of Ordovician argillite, siltstone, and quartz sandstone with soft-sediment slump and flow features and Arenig and Llanvirn graptolites (Kos'ko et al., 1990; Şengör and Natal'in, 1996). The lower Paleozoic succession on Kotelny Island begins with a thick (1400–1800 m) Ordovician section of chiefly shallow-water carbonate rocks; an interval of slightly deeper water limestone, marl, and shale accumulated during the Middle Ordovician (early Caradoc). Silurian strata (750–1700 m thick) comprise a thinner sequence of basinal graptolitic mudstone and limestone to the southwest and thicker, shallow-water carbonate rocks to the northeast; the basinal sequence shallows and becomes increasingly calcareous upward. Chiefly shallow-water carbonate facies (1000–1700 m) continue through the Middle Devonian and are succeeded by as much as 8900 m of Upper Devonian siliciclastic rocks (Kos'ko et al., 1990). Paleozoic strata on Kotelny Island contain a diversified fauna of corals, ostracodes, brachiopods, and gastropods similar to faunas of the Siberian platform (Nokleberg et al., 1994). No conodont faunas from Kotelny Island are described in the literature.

Lower Paleozoic lithofacies on Kotelny and Bennett Islands correlate intriguingly well with coeval strata in northern and central Alaska. Trilobite-bearing Middle Cambrian, chiefly siliciclastic rocks on Bennett Island correspond to Middle Cambrian outer shelf deposits in the central Brooks Range (Snowden Mountain area) and southern Farewell terrane. Ordovician, Silurian, and Lower–Middle Devonian shallow-water facies on Kotelny match those in the North Alaska carbonate platform and the Farewell terrane (Nixon Fork subterrane). Upper Arenig–middle Caradoc deep-water strata on Bennett Island are coeval with, but less calcareous than, intraplatform basin facies in northern Alaska and also correlate, in age and lithofacies, with the lower part of the Post River Formation in the Dillinger subterrane (Churkin and Carter, 1996); Bennett Island strata, however, are much thicker than equivalent rocks in Alaska. Deep-water Silurian rocks on southwestern Kotelny Island correspond in age, facies, and thickness to the Paradise Fork Formation in central Alaska. The Late Devonian siliciclastic influx recorded on Kotelny Island roughly correlates with, but appears to be at least twice as thick as, siliciclastic strata of the Endicott Group in northern Alaska.

Kolyma-Omolon superterrane

Paleozoic rocks are widely but discontinuously exposed east of the Siberian platform in an area called the Kolyma-Omolon superterrane, which includes 13 discrete terranes (Nokleberg et al., 1994). Lower Paleozoic platform carbonate rocks occur in the Omolon, Omulevka, and Prikolyma terranes, and deeper-water strata of Ordovician and Devonian age are found in the Rassokha and Argatas terranes (Fig. 1; Nokleberg et al., 1994).

The Bondarev database contains stratigraphic columns from four of the Kolyma-Omolon terranes: Omolon (2 Ordovician and 11 Devonian columns), Omulevka (3 Ordovician, 4 Silurian, and 4 Devonian columns), Prikolyma (2 Silurian and 3 Devonian columns), and Rassokha (2 Ordovician and 2 Devonian columns). Only the Omulevka terrane contains a continuous Ordovician through Devonian succession (Nokleberg et al., 1994); a representative composite section from the Selennyakh Range is shown in Figure 3 (column 11). Some Ordovician conodont faunas from the Omolon, Omulevka, and Prikolyma terranes are analyzed in Appendix 1 (see footnote 1) and summarized in the following.

The Omulevka terrane, westernmost of the five Kolyma-Omolon terranes discussed herein, comprises several discrete fault-bounded blocks separated by belts of Mesozoic sedimentary, volcanic, and plutonic rocks (chiefly the Jurassic-Cretaceous Indigirka-Oloy assemblage) (Nokleberg et al., 1994). The Omulevka succession begins with presumed Upper Proterozoic marble, schist, and metavolcanic rocks unconformably overlain by a thick sequence (1700 m) of boulder conglomerate, marble with Middle and Late Cambrian fossils, schist, metarhyolite, and quartzite (Nokleberg et al., 1994). Ordovician strata (2200–5700 m thick) consist of Lower–Middle Ordovician limestone and shale deposited in a deepening-upward environment: graptolitic shale, limestone, and local tuff of Llanvirn–early Caradoc age, and Upper Ordovician shallow-water limestone and local evaporites. Deep-water limestone and graptolitic shale (300–1450 m thick) accumulated during the Early Silurian and grade upward into shallow-water dolostone, limestone, and locally abundant siliciclastic rocks of Late Silurian (500–1400 m) and Devonian (1200–2500 m) age. Middle Devonian (Givetian) calcareous sandstones are intercalated with trachybasalt flows along the southeastern border of the terrane, and Famennian (and younger) sedimentary strata include subordinate layers of andesite and basalt tuff (Nokleberg et al., 1994).

Deep-water rocks of early Paleozoic age occur in the Rassokha and Argatas terranes, which form discrete, fault-bounded blocks east of the Omulevka terrane and were interpreted by Nokleberg et al. (1994) as fragments of oceanic crust. Ordovician shale, graywacke, and volcanic rocks occur in the Rassokha terrane (Nokleberg et al., 1994); two stratigraphic columns from this terrane in the Bondarev database contain 1000–3700 m of Ordovician strata, including a Llanvirn–early Caradoc interval of graptolitic shale and marl. These rocks are unconformably overlain by 1670–3700 m of Devonian limestone, dolostone, and lesser siliciclastic strata. The Argatas terrane includes Lower Ordovician shale and minor Devonian limestone and sandstone associated with Devonian ophiolitic rocks (Nokleberg et al., 1994).

The Prikolyma terrane is directly east of the Omulevka terrane and includes amphibolite-grade gneiss and schist of Proterozoic and/or early Paleozoic age (Nokleberg et al., 1994). Riphean, Vendian, and Cambrian shallow-marine clastic and carbonate rocks occur in the central part of the terrane and Ordovician carbonate and Lower Silurian clastic strata crop out in the southwestern area (Nokleberg et al., 1994). A thick Silurian-Devonian succession in the Yasachnaya River basin is documented in the Bondarev database. It consists of 450–600 m of Llandovery deep-water graptolitic shale and some coarser-grained clastic rocks, 500 m of Wenlock-Ludlow limestone and dolostone, 200–700 m of Pridoli chiefly clastic rocks, and 1300–2100 m of Devonian siliciclastic and carbonate strata. Mafic volcanic rocks of Eifelian and Frasnian age occur within this succession.

The Omolon terrane adjoins the eastern boundary of the Prikolyma terrane. Archean to Early Proterozoic, amphibolite to granulite facies crystalline basement (U-Pb ages of 2.8–3.4 Ga) is unconformably overlain by lower-grade Proterozoic siliciclastic rocks that grade upward into carbonate strata and are in turn unconformably overlain by rift-related clastic and volcanic rocks of Cambrian age (Nokleberg et al., 1994). Lower and Middle Ordovician units comprise several small tectonic blocks; stratigraphic columns in the Bondarev database for the Molandzha and Kedon zones indicate 1625 and 400–700 m, respectively, of shallow-water carbonate and lesser clastic rocks of Tremadoc, Arenig, and Llanvirn age. Devonian strata are widespread, unconformably overlie older deposits, and consist of mafic to felsic volcanic rocks and nonmarine and lesser shallow-marine siliciclastic and carbonate strata (Nokleberg et al., 1994); 11 stratigraphic columns in the Bondarev database document 600–4600 m of chiefly Middle and Upper Devonian rocks.

Ordovician conodont collections from the Omulevka, Prikolyma, and Omolon terranes of the Kolyma-Omolon superterrane (Appendix 1; see footnote 1) contain abundant (78%) cosmopolitan and tropical cosmopolitan species, an unsurprising result given that deeper-water strata were preferentially sampled (basinal facies worldwide produce largely cosmopolitan forms). Of the collections from the Omulevka and Prikolyma terranes, 10% include or consist exclusively of Siberian province and/or Siberian-Alaskan species (Appendix 1; see footnote 1). Rare to common Laurentian-Siberian faunal elements occur in 15% of the samples, but are much less diverse than those associated with Siberian-Alaskan conodonts in collections from northern and central Alaska. Laurentian-Siberian species include *Oistodus multicorrugatus, Histiodella* n. sp. 2 of Harris et al. (1979), *Scolopodus kelpi, Clavohamulus* n. sp. cf. *C.* n. sp. A of Repetski (1982), and *"Scandodus" robustus* (Table 2). Megafossils from both the North Alaska carbonate platform and the Farewell terrane have been broadly correlated with coeval forms from the Kolyma-Omolon superterrane (e.g., Blodgett et al., 1992, and this volume; Blodgett, 1998).

Lithologic correlations between terranes of the Kolyma-Omolon superterrane and those of northern and central Alaska are generally weak. The Omulevka succession best matches the Alaskan rocks; specific similarities include metamorphosed basement of presumed Late Proterozoic age and Middle Ordovician basinal deposits underlain and overlain by shallow-water carbonate strata. Early Silurian basinal facies correlate well in age, thickness, and lithology with the Paradise Fork Formation and equivalent rocks in the Farewell terrane. Features of the Omulevka succession that have no equivalent in coeval Alaskan rocks include the Cambrian (?) boulder conglomerate, Upper Ordovician evaporites, and locally abundant siliciclastic strata of Late Silurian–Early Devonian age. Ordovician basinal deposits and volcanic rocks of the Rassokha terrane have some similarities with coeval strata of the central Seward Peninsula, but the Seward Peninsula volcanic sequence was not deposited on oceanic crust. Lower Silurian deep-water facies in the Prikolyma terrane correlate broadly with the Paradise Fork, but too few data are available on Cambrian-Ordovician strata in Prikolyma to permit a comparison with Alaskan rocks. The Omolon succession includes Ordovician siliciclastic strata that have no counterpart in Alaska and lacks Silurian rocks. All three terranes of the Kolyma-Omolon superterrane that include lower Paleozoic platform carbonates contain volcanic rocks of Middle and Late Devonian age; broadly coeval volcanic rocks occur locally in northern Alaska but are apparently absent from the Farewell terrane.

DISCUSSION

Lithologic and paleontologic data from lower Paleozoic platform carbonate successions in northern and central Alaska and northeastern Russia provide important limitations on models proposed to explain the tectonic evolution of the Arctic region. In this section we explore the broader implications of our stratigraphic and faunal data. We first consider correlation of lower Paleozoic successions within Alaska, and then consider implications of these correlations for the ultimate origin of Alaskan terranes.

Alaskan successions

Strong lithologic and faunal similarities imply that lower Paleozoic carbonate successions on Seward Peninsula and in the western and central Brooks Range are fragments of a single carbonate platform that formed along a continental margin (Dumoulin and Harris, 1994). Carbonate platform strata in the northeastern Brooks Range (Shublik-Sadlerochit Mountains) may also have been part of this margin, but these rocks differ in several respects from coeval platform rocks to the west.

Lower Paleozoic carbonate strata in the northeastern Brooks Range are the most cratonic sequence known in the range; they overlie a thick (>2500 m) Late Proterozoic passive-margin succession (Clough, 1989; Clough and Goldhammer, 1995) and were deposited almost exclusively during transgressive maxima (Harris et al., 1995). The succession contains numerous disconformities—most strikingly, one spanning the entire Silurian and part of the Early Devonian—and lacks evidence of any deeper-water incursions during the early Paleozoic. Laurentian faunal affinities are strong in Shublik-Sadlerochit strata, and Siberian affinities relatively weak. Upper Cambrian and Ordovician rocks in this area contain mainly tropical cosmopolitan conodont species such as *Gepikodus communis* and *Juanognathus variabilis* and Laurentian forms including *Clavohamulus densus, Phragmodus* n. sp. of Barnes, 1974, belodinids and aphelognathids. *C. densus* is widespread in Laurentia, but has not been found elsewhere in Alaska; *Phragmodus* n. sp. is widespread in northern Canada but is unknown west or south of the central Brooks Range in Alaska. Upper Ordovician strata in the Shublik Mountains contain the pentamerid brachiopod *Tcherskidium,* which is characteristic of peri-Siberian terranes; the species found in the Shublik Mountains is new and also occurs in the Farewell terrane (Blodgett, 1998). We provisionally include the Shublik-Sadlerochit succession in our North Alaska carbonate platform, but acknowledge that the lithologic and faunal evidence that ties these rocks to this platform is less robust than that linking the successions of Seward Peninsula and the western and central Brooks Range.

Similarities between carbonate platform successions in northern and central Alaska (North Alaska carbonate platform and Farewell terrane) suggest that these fragments were in mutual proximity during the early Paleozoic, and may even have been joined. Upper Proterozoic through Lower Devonian lithofacies in the two areas correlate well, as do the conodont and megafossil assemblages. Lithologic and faunal ties between northern Alaska and the Farewell terrane diminish after Middle Devonian time. Minor Upper Devonian siliciclastic rocks are reported from the Farewell terrane (e.g., Blodgett and Gilbert, 1992), but there is no equivalent in this area of the thick, largely fluvial siliciclastic strata of the Upper Devonian–Mississippian Endicott Group. Fluvial siliciclastic strata are only locally abundant in the Farewell terrane and appear to be chiefly Permian (Decker et al., 1994).

Other lower Paleozoic carbonate successions in Alaska cannot be definitively linked to those in northern Alaska and the Farewell terrane using lithologic or faunal evidence. Lower Paleozoic carbonate rocks occur in the Jones Ridge area; the Porcupine, Livengood, and White Mountain terranes of east-central Alaska; and the Alexander terrane of southeastern Alaska (Fig. 1; Nokleberg et al., 1994; Silberling et al., 1994). Strata in the Jones Ridge area (Fig. 1, location 15) contain a typical Laurentian fauna of conodonts and megafossils and are part of the ancestral North American continental margin. Lower Paleozoic successions in the Porcupine, Livengood, White Mountain, and Alexander terranes have some lithologic and megafaunal similarities to those of the Farewell terrane and/or the North Alaska carbonate platform, but do not contain Ordovician conodonts of Siberian aspect. Ties appear strongest between the Alexander and Farewell terranes; both contain distinctive Silurian reef complexes with aphrosalpingid sponges, and volcanic rocks of Silurian age (Soja, 1994; Blodgett, 1998; Blodgett et al., this

volume; Dumoulin et al., 1998a). Conspecific Middle Devonian (Eifelian) gastropods occur in the Livengood and Farewell terranes (Blodgett et al., this volume) and Ordovician volcanic rocks characterize successions of both the central Seward Peninsula and the White Mountain terrane (Harris et al., 1995). A single specimen of *Tcherskidium* sp., an Ordovician pentamerid brachiopod with Siberian affinities, has been found in the Porcupine terrane, but other Ordovician and Devonian megafossils in this terrane are more similar to forms from cratonal North America (Blodgett et al., this volume).

Fossils with Siberian affinities are reported from Alaskan rocks that formed in several tectonic environments. Cambrian trilobites of Siberian aspect occur at Mount Doonerak in rocks interpreted as a subduction-related magmatic arc complex (e.g., Julian and Oldow, 1998). However, megafossils and microfossils characteristic of the Siberian province are also documented in carbonate platform successions in northern and central Alaska (e.g., Dumoulin and Harris, 1994; Blodgett et al., this volume). Platform environments in both northern Alaska and the Farewell terrane persisted from Proterozoic through Devonian time; such long-lived successions generally form along a continental margin, or on fragments derived from such a margin. We next consider the ultimate origin of these Alaskan carbonate platforms.

Arctic correlations

Metamorphic and isotopic data from basement rocks and lithologic and faunal information from lower Paleozoic strata all help to determine the origin of the carbonate platforms in northern and central Alaska. Metamorphosed strata of Proterozoic age appear to underlie both the North Alaska and Nixon Fork platforms, a feature that distinguishes them from coeval carbonate platforms in western Canada (as pointed out by Patton et al., 1994) and the Siberian craton (Nokleberg et al., 1994), which are underlain by thick sequences of unmetamorphosed Proterozoic rocks. Sparse isotopic ages from Proterozoic basement complex rocks—on the Seward Peninsula (Patrick and McClelland, 1995) and in the Sadlerochit Mountains (McClelland, 1997) in northern Alaska, and in the Medfra Quadrangle (McClelland et al., 1999) in central Alaska—are unlike those from the Canadian Cordillera and Arctic Islands, and suggest that the Alaskan rocks were not derived from western Laurentia.

Lower Paleozoic carbonate platform rocks in northern and central Alaska also differ from coeval carbonate platform rocks in both the Canadian Arctic Islands and the Canadian Cordillera. The carbonate platform in the Arctic Islands has many lithologic and faunal mismatches with the age-equivalent platform in North Alaska (Fig. 3, column 12; Dumoulin et al., 1998c, 2000a). Although lower Paleozoic rocks in the Farewell terrane correlate somewhat better lithologically with the Arctic Islands succession—both areas record widespread drowning of platform facies during Silurian time—Ordovician evaporites in the Arctic Islands have no counterparts in central or northern Alaska. Cambrian through Devonian strata in the Canadian

Cordillera have some lithologic similarities to coeval platform facies in northern and central Alaska (cf. Fritz et al., 1991), but faunal evidence distinguishes age-equivalent strata in these areas. Conodonts and megafossils characteristic of the Siberian province that are locally common in lower Paleozoic platform carbonate facies from northern Alaska and the Farewell terrane are absent from coeval facies in the Canadian Arctic Islands and the Canadian Cordillera (Dumoulin et al., 1998c, 2000a). The presence in Alaskan terranes of Siberian forms not seen in well-studied cratonal margin sequences such as the Canadian Rockies and eastern Great Basin suggests that the North Alaska carbonate platform and the Farewell terrane were not attached to Laurentia during the early Paleozoic.

The distribution of Siberian faunal elements in Alaska implies that early Paleozoic carbonate platforms in northern and central Alaska formed in or adjacent to the Siberian faunal province. As noted earlier, however, the Alaskan successions have notable lithologic differences from coeval rocks of the Siberian craton and contain Laurentian faunal elements (both conodonts [Table 1] and megafossils, such as some Ordovician and Devonian corals) that are rare or absent in Siberian craton strata. The lithostratigraphic and megafossil evidence suggests stronger ties between the carbonate platforms in northern and central Alaska and equivalent platforms in terranes north and east of the Siberian craton than between the Alaskan rocks and the Siberian craton. Lower Paleozoic rocks on Kotelny and Bennett Islands, the Chukotka Peninsula, and in the Omulevka terrane have the most compelling lithostratigraphic similarities to the Alaskan successions. Evidence for Early Silurian drowning, like that seen across the Nixon Fork platform and locally, perhaps, along the northern margin of the North Alaska platform, also occurs in the Kotelny, Chukotka, and Omulevka successions; late Early–Middle Ordovician basinal deposits, like those within carbonate platform facies in northern Alaska and the southern part of the Nixon Fork subterrane, are also found in Omulevka and on Bennett Island. Limited Ordovician conodont data from Chukotka and the Kolyma-Omolon superterrane indicate that these "peri-Siberian terranes" contain far fewer Laurentian forms than do coeval Alaskan successions (Tables 1 and 2). Conodont data are not available for Ordovician strata on Kotelny and Bennett Islands.

The faunal data outlined here imply that lower Paleozoic carbonate strata in northern and central Alaska accumulated at or near the juncture of the Siberian and Laurentian faunal provinces. Lithologic and faunal evidence suggests that the Alaskan rocks were not part of the Siberian craton during their deposition, as proposed by Blodgett (1998). The Alaskan strata could have formed on a crustal fragment rifted away from Siberia during the Late Proterozoic (e.g., Şengör and Natal'in, 1996), but more isotopic information is needed from Alaskan, Siberian, and peri-Siberian basement complexes to evaluate this hypothesis. There are intriguing lithologic similarities between the lower Paleozoic successions of northern and central Alaska and those of certain peri-Siberian terranes, most notably Kotelny

and Bennett, Chukotka, and Omulevka; additional detailed conodont, megafossil, and lithologic data are needed to test these correlations. Siberian affinities in northern and central Alaskan biotas are notable from at least Middle Cambrian through Early Devonian (Emsian) time, but the lithologic and faunal ties between northern Alaska, the Farewell terrane, and the peri-Siberian terranes of northeastern Russia diminish after the Middle Devonian. Plant fossils of the Angaran floral realm testify that Siberian biotic affinities persisted (or perhaps recurred) as late as Mississippian time in northern Alaska (Spicer and Thomas, 1987) and Permian time in the Farewell terrane (Mamay and Reed, 1984).

CONCLUSIONS

Strong lithologic and faunal ties link Cambrian through Devonian strata of the North Alaska carbonate platform and the Farewell terrane in central Alaska, and rocks in both areas have some similarities to coeval strata of peri-Siberian terranes such as Chukotka, Kotelny, and Omulevka. Megafossil and microfossil assemblages from northern and central Alaska contain both Siberian and Laurentian province elements; the Siberian elements are absent from coeval sequences in Laurentia, and only a few of the Laurentian forms that occur in the Alaskan successions also reached the Siberian craton. These data suggest that carbonate platform successions in northern and central Alaska were in proximity (possibly joined) during the early Paleozoic, and formed at or near the juncture of the Siberian and Laurentian faunal provinces, possibly on a crustal fragment rifted away from the Siberian craton during Late Proterozoic time.

ACKNOWLEDGMENTS

Mussa Gagiev, our northeast Russian colleague, died shortly after this manuscript was submitted for review. We already miss his geologic and paleontologic knowledge and his joie de vivre. We trust he would approve of our efforts.

REFERENCES CITED

Abaimova, G.P., 1975, Early Ordovician conodonts of the middle fork of the Lena River: Trudy Sibirskogo Nauchno-Issledovatelskogo Instituta, Geologii, Geofiziki i Mineralnogo Sirya (SNIGGIMS), v. 207, 129 p. (in Russian).

Anderson, A.V., Wallace, W.K., and Mull, C.G., 1994, Depositional record of a major tectonic transition in northern Alaska: Middle Devonian to Mississippian rift-basin margin deposits, upper Kongakut River region, eastern Brooks Range, Alaska, *in* Thurston, D.K., and Fujita, K., eds., Proceedings of the 1992 International Conference on Arctic Margins: Anchorage, Alaska, Minerals Management Service, Outer Continental Shelf MMS Report 94–0040, p. 71–76.

Armstrong, A.K., Mamet, B.L., Brosgé, W.P., and Reiser, H.N., 1976, Carboniferous section and unconformity at Mount Doonerak, Brooks Range, northern Alaska: American Association of Petroleum Geology Bulletin, v. 60, p. 962–972.

Babcock, L.E., Blodgett, R.B., and St. John, J., 1993, Proterozoic and Cambrian stratigraphy and paleontology of the Nixon Fork terrane, southwestern Alaska, *in* Ortega-Gutiérrez, F., Coney, P.J., Centeno-García, E., and Gómez-Caballero, A., eds., Proceedings of the First Circum-Pacific and Circum-Atlantic Terrane Conference: Guanajuato, Mexico, Universidad Nacional Autónoma de México, p. 5–7.

Babcock, L.E., Blodgett, R.B., and St. John, J., 1994, New Late (?) Proterozoic-age formations in the vicinity of Lone Mountain, McGrath quadrangle, west-central Alaska, *in* Till, A.B., and Moore, T.E., eds., Geologic studies in Alaska by the U.S. Geological Survey, 1993: U.S. Geological Survey Bulletin 2107, p. 143–155.

Barnes, C.R., 1974, Ordovician conodont biostratigraphy of the Canadian Arctic, *in* Aitken, J.D., and Glass, D.J., eds., Canadian Arctic geology: Calgary, Alberta, Geological Association of Canada and Canadian Society of Petroleum Geologists Special Volume, p. 221–240.

Baxter, M.E., and Blodgett, R.B., 1994, A new species of *Droharhynchia* (Brachiopoda) from the lower Middle Devonian (Eifelian) of west-central Alaska: Journal of Paleontology, v. 68, p. 1235–1240.

Bergström, S.M., 1990a, Biostratigraphic and biogeographic significance of Middle and Upper Ordovician conodonts in the Girvan succession, southwest Scotland, *in* Ziegler, W., ed., First International Senckenberg Conference and Fifth European Conodont Symposium (ECOS V), Contributions 4: Courier Forschungsinstitut Senckenberg, no. 118, p. 1–43.

Bergström, S.M., 1990b, Relations between conodont provincialism and the changing paleogeography during the Early Paleozoic, *in* McKerrow, W.S., and Scotese, C.R., eds., Paleozoic palaeogeography and biogeography: Geological Society [London] Memoir 12, p. 105–121.

Bering Strait Geologic Field Party, 1997, Koolen metamorphic complex, NE Russia: Implications for the tectonic evolution of the Bering Strait region: Tectonics, v. 16, p. 713–729.

Blodgett, R.B., 1983, Paleobiogeographic affinities of Devonian fossils from the Nixon Fork terrane, southwestern Alaska, *in* Stevens, C.H., ed., Pre-Jurassic rocks in western North America suspect terranes: Los Angeles, Pacific Section, Society of Economic Paleontologists and Mineralogists, p. 124–130.

Blodgett, R.B., 1998, Emsian (late Early Devonian) fossils indicate a Siberian origin for the Farewell Terrane, *in* Clough, J.G., and Larson, F., eds., Short notes on Alaska Geology 1997: Alaska Division of Geological and Geophysical Surveys Professional Report 118, p. 53–61.

Blodgett, R.B., and Gilbert, W.G., 1992, Upper Devonian shallow-marine siliciclastic strata and associated fauna and flora, Lime Hills D-4 quadrangle, southwest Alaska, *in* Bradley, D.C., and Dusel-Bacon, C., eds., Geologic studies in Alaska by the U.S. Geological Survey, 1991: U.S. Geological Survey Bulletin 2041, p. 106–115.

Blodgett, R.B., Rohr, D.M., and Clough, J.G., 1992, Late Ordovician brachiopod and gastropod biogeography of Arctic Alaska and Chukotka: International Conference on Arctic Margins Abstracts with Programs, Anchorage, Alaska, p. 11.

Bundtzen, T.K., Harris, E.E., and Gilbert, W.G., 1997, Geology of the eastern half of the McGrath quadrangle, Alaska: Alaska Division of Geological and Geophysical Surveys Report of Investigations 97–14a, scale 1:125 000, 34 p., 1 plate.

Carter, C., and Tailleur, I.L., 1984, Ordovician graptolites from the Baird Mountains, western Brooks Range, Alaska: Journal of Paleontology, v. 58, p. 40–57.

Cecile, M.P., Harrison, J.C., Kos'ko, M.K., and Parrish, R.R., 1991, Precambrian U-Pb ages of igneous rocks, Wrangel Complex, Wrangel Island, USSR: Canadian Journal of Earth Sciences, v. 28, p. 1340–1348.

Churkin, M., Jr., and Carter, C., 1996, Stratigraphy, structure, and graptolites of an Ordovician and Silurian sequence in the Terra Cotta Mountains, Alaska Range, Alaska: U.S. Geological Survey Professional Paper 1555, 84 p.

Clough, J.G., 1989, General stratigraphy of the Katakturuk Dolomite in the Sadlerochit and Shublik Mountains, Arctic National Wildlife Refuge, Alaska: Alaska Division of Geological and Geophysical Surveys Public-Data File 89–4a, 9 p., 1 plate.

Clough, J.G., and Blodgett, R.B., 1985, Comparative study of the sedimentology and paleoecology of middle Paleozoic algal and coral-stromatoporoid reefs in Alaska: Proceedings of the Fifth International Coral Reef Congress, Tahiti, v. 6., p. 593–598.

Clough, J.G., and Blodgett, R.B., 1989, Silurian-Devonian algal reef mound complex of southwest Alaska, *in* Geldsetzer, H.H.J., James, N.P., and Tebbutt, G.E., eds., Reefs: Canada and adjacent areas: Canadian Society of Petroleum Geologists Memoir 13, p. 404–407.

Clough, J.G., and Goldhammer, R.K., 1995, Deposition on a Late Proterozoic carbonate ramp, Katakturuk Dolomite, northeast Brooks Range, Alaska: Geological Society of America Abstracts with Programs, v. 27, p. 10.

Decker, J., Bergman, S.C., Blodgett, R.B., Box, S.E., Bundtzen, T.K., Clough, J.G., Coonrad, W.L., Gilbert, W.G., Miller, M.L., Murphy, J.M., Robinson, M.S., and Wallace, W.K., 1994, Geology of southwestern Alaska, *in* Plafker, G., and Berg, H.C., eds., The geology of Alaska: Boulder, Colorado, Geological Society of America, Geology of North America, v. G-1, p. 285–310.

de Freitas, T.A., Harrison, J.C., and Mayr, U., 1997, Mineral showings and sequence stratigraphic correlation charts of the Canadian Arctic Islands and parts of North Greenland: Geological Survey of Canada Open-File Report 3410, 3 charts.

Dumoulin, J.A., and Harris, A.G., 1988, Off-platform Silurian sequences in the Ambler River quadrangle, *in* Galloway, J.P., and Hamilton, T.D., eds., Geologic studies in Alaska by the U. S. Geological Survey during 1987: U.S. Geological Survey Circular 1016, p. 35–38.

Dumoulin, J.A., and Harris, A.G., 1992, Devonian-Mississippian carbonate sequence in the Maiyumerak Mountains, western Brooks Range, Alaska: U.S. Geological Survey Open-File Report 92–3, 83 p.

Dumoulin, J.A., and Harris, A.G., 1994, Depositional framework and regional correlation of pre-Carboniferous metacarbonate rocks of the Snowden Mountain area, central Brooks Range, northern Alaska: U.S. Geological Survey Professional Paper 1545, 74 p.

Dumoulin, J.A., Bradley, D.C., and Harris, A.G., 1998a, Sedimentology, conodonts, structure, and regional correlation of Silurian and Devonian metasedimentary rocks in Denali National Park, *in* Gray, J.E., and Riehle, J.R., eds., Geologic studies in Alaska by the U.S. Geological Survey, 1996: U.S. Geological Survey Professional Paper 1595, p. 71–98.

Dumoulin, J.A., Bradley, D.C., Harris, A.G., and Repetski, J.E., 1998b, Sedimentology, conodont biogeography, and subsidence history of the Nixon Fork terrane, Medfra quadrangle, Alaska [abs.]: Celle, Germany, Abstracts, Third International Conference on Arctic Margins, p. 49.

Dumoulin, J.A., Harris, A.G., and de Freitas, T.A., 1998c, Facies patterns and conodont biogeography in Arctic Alaska and the Canadian Arctic Islands: Evidence against juxtaposition of these areas during early Paleozoic time [abs.]: Third International Conference on Arctic Margins, Celle, Germany, p. 50.

Dumoulin, J.A., Bradley, D.C., Harris, A.G., and Repetski, J.E., 1999, Lower Paleozoic deep-water facies of the Medfra area, central Alaska, *in* Kelley, K.D., ed., Geologic studies in Alaska by the U.S. Geological Survey, 1997: U.S. Geological Survey Professional Paper 1614, p. 73–103.

Dumoulin, J.A., Harris, A.G., Bradley, D.C., and de Freitas, T.A., 2000a, Facies patterns and conodont biogeography in Arctic Alaska and the Canadian Arctic Islands: Evidence against juxtaposition of these areas during early Paleozoic time: Polarforschung, v. 68, p. 257–266.

Dumoulin, J.A., Bradley, D.C., and Harris, A.G., 2000b, Paleozoic strata of the Dyckman Mountain area, northeastern Medfra quadrangle, Alaska, *in* Kelley, K.D., and Gough, L.P., eds., Geologic studies in Alaska by the U.S. Geological Survey, 1998: U.S. Geological Survey Professional Paper 1615, p. 43–57.

Dutro, J.T., Jr., and Patton, W.W., Jr., 1981, Lower Paleozoic platform carbonate sequence in the Medfra quadrangle, west-central Alaska, *in* Albert, N.R.D., and Hudson, T., eds., The United States Geological Survey in Alaska: Accomplishments during 1979: U. S. Geological Survey Circular 823–B, p. 42–44.

Dutro, J.T., Jr., and Patton, W.W., Jr., 1982, New Paleozoic formations in the northern Kuskokwim Mountains, west-central Alaska, *in* Stratigraphic notes, 1980–1982: U.S. Geological Survey Bulletin 1529–H, p. 13–22.

Ethington, R.L., and Brand, U., 1981, *Oneotodus simplex* (Furnish) and the genus *Oneotodus* (Conodonta): Journal of Paleontology, v. 55, p. 239–247.

Ethington, R.L., and Clark, D.L., 1971, Lower Ordovician conodonts in North America, *in* Sweet, W.C., and Bergström, S.M., eds., Symposium on conodont biostratigraphy: Boulder, Colorado, Geological Society of America Memoir 127, p. 63–82.

Fortey, R.A., Harper, D.A.T., Ingham, J.K., Owen, A.W., and Rushton, A.W.A., 1995, A revision of Ordovician series and stages from the historical type area: Geological Magazine, v. 132, no. 1, p. 15–30.

Fritz, W.H., Cecile, M.P., Norford, B.S., Morrow, D., and Geldsetzer, H.H.J., 1991, Cambrian to Middle Devonian assemblages, *in* Gabrielse, H., and Yorath, C.J., eds., Geology of the Cordilleran Orogen in Canada: Boulder, Colorado, Geological Society of America, Geology of North America, v. G-2, p. 151–218.

Fujita, K., and Newberry, J.T., 1982, Tectonic evolution of northeastern Siberia and adjacent regions: Tectonophysics, v. 89, p. 337–357.

Grantz, A., Moore, T.E., and Roeske, S.M., 1991, Continent-ocean transect A-3: Gulf of Alaska to Arctic Ocean: Geological Society of America, scale 1:500 000, 3 sheets.

Halgedahl, S., and Jarrard, R., 1987, Paleomagnetism of the Kuparuk River Formation from oriented drill core: Evidence for rotation of the Arctic Alaska Plate, *in* Tailleur, I.L., and Weimer, P., eds., Alaskan North Slope geology, Volume 2: Bakersfield, California, Pacific Section, Society of Economic Paleontologists and Mineralogists and the Alaska Geological Society, Book 50, v. 2, p. 581–617.

Harris, A.G., Bergström, S.M., Ethington, R.L., and Ross, R.J., Jr., 1979, Aspects of Middle and Upper Ordovician conodont biostratigraphy of carbonate facies in Nevada and southeast California and comparison with some Appalachian successions: Provo, Utah, Brigham Young University, Geology Studies, v. 26, p. 7–43.

Harris, A.G., Dumoulin, J.A., Repetski, J.E., and Carter, C., 1995, Correlation of Ordovician rocks of northern Alaska, *in* Cooper, J.D., Droser, M.L., and Finney, S.C., eds., Ordovician odyssey: Short papers for the 7th International Symposium on the Ordovician system: Fullerton, California, Pacific Section, SEPM (Society for Sedimentary Geology), Book 77, p. 21–26.

Julian, F.E., and Oldow, J.S., 1998, Structure and lithology of the lower Paleozoic Apoon assemblage, eastern Doonerak window, central Brooks Range, Alaska, *in* Oldow, J.S., and Avé Lallemant, H.G., eds., Architecture of the central Brooks Range fold and thrust belt, Arctic Alaska: Geological Society of America Special Paper 324, p. 65–80.

Kanygin, A.V., Moskalenko, T.A., and Yadrenkina, A.G., 1988, The Siberian platform, *in* Ross, R.J., Jr., and Talent, J.A., eds., The Ordovician System in most of Russian Asia: International Union of Geological Sciences, no. 26, p. 1–27.

Kingsbury, S.A., and Babcock, L.E., 1998, Biogeography and paleogeographic implications of early Middle Cambrian trilobites and enigmatic fossils from the Farewell terrane, southwestern Alaska: Geological Society of America Abstracts with Programs, v. 30, p. 27.

Kos'ko, M.K., 1994, Major tectonic interpretations and constraints for the New Siberian Islands region, Russian Arctic, *in* Thurston, D.K., and Fujita, K., eds., Proceedings of the 1992 International Conference on Arctic Margins: Anchorage, Alaska, Minerals Management Service Outer Continental Shelf MMS Report 94–0040, p. 195–200.

Kos'ko, M.K., Lopatin, B.G., and Ganelin, V.G., 1990, Major geological features of the Islands of the East Siberian and Chukchi Seas and the northern coast of Chukotka: Marine Geology, v. 93, p. 349–367.

Lehnert, O., 1995, Ordovizische Conodonten aus der Präkordillere Westargentiniens: Ihre Bedeutung für Stratigraphie und Paläogeographie: Erlanger Geologische Abhandlungen, v. 125, p. 1–193.

Löfgren, A.M., Repetski, J.E., and Ethington, R.L., 1998, Some trans-Iapetus conodont faunal connections in the Tremadoc, *in* Bagnoli, G., ed., Euro-pean Conodont Symposium 7 Abstracts, Bologna-Modena, 1998: 1–2 Typografica Compositori Bologna, p. 65.

Mamay, S.H., and Reed, B.L., 1984, Permian plant megafossils from the conglomerate of Mount Dall, central Alaska Range, *in* Coonrad, W.L., and Elliot, R.L., eds., The United States Geological Survey in Alaska: Accomplishments during 1981: U.S. Geological Survey Circular 868, p. 98–102.

Mayfield, C.F., Tailleur, I.L., and Ellersieck, I., 1988, Stratigraphy, structure, and palinspastic synthesis of the western Brooks Range, northwestern Alaska, *in* Gryc, G., ed., Geology and exploration of the National Petroleum Reserve in Alaska, 1974–1982: U.S. Geological Survey Professional Paper 1399, p. 143–186.

Mayfield, C.F., Curtis, S.M., Ellersieck, I., and Tailleur, I.L., 1990, Reconnaissance geologic map of the De Long Mountains A-3 and B-3 quadrangles and parts of the A-4 and B-4 quadrangles, Alaska: U.S. Geological Survey Miscellaneous Investigations Series Map I-1929, scale 1:63 360, 2 sheets.

McClelland, W.C., 1997, Detrital zircon studies of the Proterozoic Neruokpuk Formation, Sadlerochit and Franklin Mountains, northeastern Alaska: Geological Society of America Abstracts with Programs, v. 29, no. 5, p. 28.

McClelland, W.C., Kusky, T., Bradley, D., Dumoulin, J.A., and Harris, A.G., 1999, The nature of "Nixon Fork basement," west-central Alaska: Geological Society of America Abstracts with Programs, v. 31, p. A78.

Measures, E.A., Rohr, D.M., and Blodgett, R.B., 1992, Depositional environments and some aspects of the fauna of Middle Ordovician rocks of the Telsitna Formation, northern Kuskokwim Mountains, Alaska, *in* Bradley, D.C., and Dusel-Bacon, C., eds., Geologic studies in Alaska by the U.S. Geological Survey, 1991: U.S. Geological Survey Bulletin 2041, p. 186–201.

Moore, T.E., and Nilsen, T.H., 1984, Regional sedimentological variations in the Upper Devonian and Lower Mississippian (?) Kanayut Conglomerate, Brooks Range, Alaska: Sedimentary Geology, v. 38, p. 465–497.

Moore, T.E., Wallace, W.K., Bird, K.J., Karl, S.M., Mull, C.G., and Dillon, J.T., 1994, Geology of northern Alaska, *in* Plafker, G., and Berg, H.C., eds., The geology of Alaska: Boulder, Colorado, Geological Society of America, Geology of North America, v. G-1, p. 49–140.

Moore, T.E., Wallace, W.K., Mull, C.G., Adams, K.E., Plafker, G., and Nokleberg, W.J., 1997, Crustal implications of bedrock geology along the Trans-Alaska Crustal Transect (TACT) in the Brooks Range, northern Alaska: Journal of Geophysical Research, v. 102, p. 20645–20684.

Moskalenko, T.A., 1970, Conodonts of the Krivoluk Stage (Middle Ordovician) of the Siberian Platform: Transactions of the Institute of Geology and Geophysics, Siberian Branch, USSR Academy of Sciences, v. 61, 116 p. (in Russian).

Moskalenko, T.A., 1973, Conodonts of the Middle and Upper Ordovician of the Siberian Platform: Transactions of the Institute of Geology and Geophysics, Siberian Branch, USSR Academy of Sciences, v. 137, 143 p. (in Russian).

Moskalenko, T.A., 1983, Conodonts and biostratigraphy in the Ordovician of the Siberian Platform: Fossils and Strata, no. 15, p. 87–94.

Natal'in, B.A., Amato, J.M., Toro, J., and Wright, J.E., 1999, Paleozoic rocks of northern Chukotka Peninsula, Russian Far East: Implications for the tectonics of the Arctic region: Tectonics, v. 18, no. 6, p. 977–1003.

Nelson, B.K., Nelson, S.W., and Till, A.B., 1993, Nd- and Sr-isotope evidence for Proterozoic and Paleozoic crustal evolution in the Brooks Range, northern Alaska: Journal of Geology, v. 101, p. 435–450.

Nokleberg, W.J., Moll-Stalcup, E.J., Miller, T.P., Brew, D.A., Grantz, A., Reed, J.C., Jr., Plafker, G., Moore, T.E., Silva, S.R., and Patton, W.W., Jr., 1994, Tectonostratigraphic terrane and overlap assemblage map of Alaska: U.S. Geological Survey Open-File Report 94–194, scale 1:2 500 000, 1 sheet, 54 p.

Nokleberg, W.J., Parfenov, L.M., Monger, J.W.H., Norton, I.O., Khanchuk, A.I., Stone, D.B., Scholl, D.W., and Fujita, K., 1998, Phanerozoic tectonic evolution of the circum-North Pacific: U.S. Geological Survey Open-File Report 98–754, 125 p.

Oldow, J.S., Boler, K.W., Avé Lallemant, H.G., Gottschalk, R.R., Julian, F.E., Seidensticker, C.M., and Phelps, J.C., 1998, Stratigraphy and paleogeographic setting of the eastern Skajit allochthon, central Brooks Range, Arctic Alaska, *in* Oldow, J.S., and Avé Lallemant, H.G., eds., Architecture of the central Brooks Range fold and thrust belt, Arctic Alaska: Boulder, Colorado, Geological Society of America Special Paper 324, p. 109–125.

Oradovskaya, M.M., 1988, Northeastern and Far Eastern USSR, *in* Ross, R.J., Jr., and Talent, J.A., eds., The Ordovician System in most of Russian Asia: International Union of Geological Sciences, no. 26, p. 85–115.

Oradovskaya, M.M., and Obut, A.M., 1977, Stratigrafiya, korrelyatsiya, paleogeografiya ordovikskikh i siluriyskikh otlozheniy na Chukotskom poluostrove (Stratigraphy, correlation, and paleogeography of the Ordovician and Silurian strata in Chukotka Peninsula) *in* Obut, A.M., ed., Stratigrafiya i fauna ordovika i silura Chukotskogo poluostrova: Novosibirsk, Russia, Nauka, p. 4–42 (in Russian).

Ormiston, A.R., and Ross, R.J., Jr., 1976, *Monorakos* in the Ordovician of Alaska and its zoogeographic significance, *in* Gray, J., and Boucot, A.J., eds., Historical biogeography, plate tectonics, and the changing environment: Corvallis, Oregon State University Press, p. 53–59.

Palmer, A.R., Egbert, R.M., Sullivan, R., and Knoth, J.S., 1985, Cambrian trilobites with Siberian affinities, southwestern Alaska [abs.]: American Association of Petroleum Geologists Bulletin, v. 69, no. 2, p. 295.

Patrick, B.E., and McClelland, W.C., 1995, Late Proterozoic granitic magmatism on Seward Peninsula and a Barentian origin for Arctic Alaska–Chukotka: Geology, v. 23, p. 81–84.

Patton, W.W., Jr., Moll, E.J., Dutro, J.T., Jr., Silberman, M.L., and Chapman, R.M., 1980, Preliminary geologic map of the Medfra quadrangle, Alaska: U.S. Geological Survey Open-File Report 80–811A, scale 1:250 000, 1 sheet.

Patton, W.W., Jr., Box, S.E., Moll-Stalcup, E.J., and Miller, T.P., 1994, Geology of west-central Alaska, *in* Plafker, G., and Berg, H.C., eds., The geology of Alaska: Boulder, Colorado, Geological Society of America, Geology of North America, v. G-1, p. 241–269.

Potter, A.W., 1984, Paleobiogeographical relations of Late Ordovician brachiopods from the York and Nixon Fork terranes, Alaska: Geological Society of America Abstracts with Programs, v. 16, p. 626.

Repetski, J.E., 1982, Conodonts from El Paso Group (Lower Ordovician) of westernmost Texas and southern New Mexico: New Mexico Bureau of Mines and Mineral Resources Memoir 40, 121 p.

Rigby, J.K., Nitecki, M.H., Soja, C.M., and Blodgett, R.B., 1994, Silurian aphrosalpingid sphinctozoans from Alaska and Russia: Acta Palaeontologica Polonica, v. 39, no. 4, p. 341–391.

Rohr, D.M., and Gubanov, A.P., 1997, Macluritid opercula (Gastropoda) from the Middle Ordovician of Siberia and Alaska: Journal of Paleontology, v. 71, p. 394–400.

Rohr, D.M., Dutro, J.T., Jr., and Blodgett, R.B., 1992, Gastropods and brachiopods from the Ordovician Telsitna Formation, northern Kuskokwim Mountains, west-central Alaska, *in* Webby, B.D., and Laurie, J.R., eds., Global perspectives on Ordovician geology, Sixth International Symposium on the Ordovician System, Proceedings, University of Sydney, Australia: Rotterdam, A.A. Balkema Press, p. 499–512.

Ross, C.A., and Ross, J.R.P., 1988, Late Paleozoic transgressive-regressive deposition, *in* Wilgus, C.K., Hastings, B.S., Posamentier, H., Van Wagoner, J., Ross, C.A., and Kendall, C.G.St.C., eds., Sea-level changes: An integrated approach: Society of Economic Paleontologists and Mineralogists Special Publication 42, p. 227–247.

Rowley, D.B., and Lottes, A.L., 1988, Plate-kinematic reconstructions of the North Atlantic and Arctic: Late Jurassic to present: Tectonophysics, v. 155, p. 73–120.

Ryherd, T.J., and Paris, C.E., 1987, Ordovician through Silurian carbonate base-of-slope apron sequence, northern Seward Peninsula, Alaska [abs.], *in* Tailleur, I.L., and Weimer, P., eds., Alaskan North Slope geology: Bakersfield, California, Pacific Section, Society of Economic Paleontologists and Mineralogists and the Alaska Geological Society, Book 50, v. 1, p. 347–348.

Ryherd, T.J., Carter, C., and Churkin, M., Jr., 1995, Middle through Upper Ordovician graptolite biostratigraphy of the Deceit Formation, northern Seward Peninsula, Alaska: Geological Society of America Abstracts with Programs, v. 27, no. 5, p. 75.

Scotese, C.R., 1997, Paleogeographic Atlas: PALEOMAP Progress Report 90–0497: Arlington, Texas, University of Texas at Arlington, Department of Geology, 37 p.

Şengör, A.M.C., and Natal'in, B.A., 1996, Paleotectonics of Asia: Fragments of a synthesis, *in* Yin, A., and Harrison, M., eds., The tectonic evolution of Asia: Cambridge, Cambridge University Press, Rubey Colloquium, p. 486–640.

Silberling, N.J., Jones, D.L., Monger, J.W.H., Coney, P.J., Berg, H.C., and Plafker, G., 1994, Lithotectonic terrane map of Alaska and adjacent parts of Canada, *in* Plafker, G., and Berg, H.C., eds., The geology of Alaska: Boulder, Colorado, Geological Society of America, Geology of North America, v. G-1, plate 3, scale 1:2 500 000, 1 sheet.

Soja, C.M., 1994, Significance of Silurian stromatolite-sphinctozoan reefs: Geology, v. 22, p. 355–358.

Soja, C.M., and Antoshkina, A.I., 1997, Coeval development of Silurian stromatolite reefs in Alaska and the Ural Mountains: Implications for paleogeography of the Alexander terrane: Geology, v. 25, p. 539–542.

Spicer, R.A., and Thomas, B.A., 1987, A Mississippian Alaska-Siberia connection: Evidence from plant megafossils, *in* Tailleur, I.L., and Weimer, P., eds., Alaskan North Slope geology: Bakersfield, California, Society of Economic Paleontologists and Mineralogists, Pacific Section, and the Alaska Geological Society, Book 50, v. 1, p. 355–358.

St. John, J.M., and Babcock, L.E., 1997, Late Middle Cambrian trilobites of Siberian aspect from the Farewell terrane, southwestern Alaska, *in* Dumoulin, J.A., and Gray, J.E., eds., Geologic studies in Alaska by the U.S. Geological Survey, 1995: U.S. Geological Survey Professional Paper 1574, p. 269–281.

Sweet, W.C., 1984, Graphic correlation of upper Middle and Upper Ordovician rocks, North American Midcontinent Province, U.S.A., *in* Bruton, D.L., ed., Aspects of the Ordovician System: Palaeontological Contributions from the University of Oslo, no. 295, p. 23–50.

Sweet, W.C., and Bergström, S.M., 1986, Conodonts and biostratigraphic correlation: Annual Review of Earth and Planetary Science, v. 14, p. 85–112.

Sweet, W.C., Ethington, R.L., and Barnes, C.R., 1971, North American Middle and Upper Ordovician conodont faunas, *in* Sweet, W.C., and Bergström, S.M., eds., Symposium on conodont biostratigraphy: Boulder, Colorado, Geological Society of America Memoir 127, p. 163–193.

Tailleur, I.L., 1965, Low-volatile bituminous coal of Mississippian age on the Lisburne Peninsula, northwestern Alaska, *in* Geological Survey Research 1965: U.S. Geological Survey Professional Paper 525–B, p. 34–38.

Till, A.B., and Dumoulin, J.A., 1994, Seward Peninsula: The Seward and York terranes, *in* Plafker, G., and Berg, H.C., eds., The geology of Alaska: Boulder, Colorado, Geological Society of America, Geology of North America, v. G-1, p. 141–152.

Till, A.B., Dumoulin, J.A., Gamble, B.M., Kaufman, D.S., and Carroll, P.I., 1986, Preliminary geologic map and fossil data, Solomon, Bendeleben, and southern Kotzebue quadrangles, Seward Peninsula, Alaska: U.S. Geological Survey Open-File Report 86–276, scale 1:250 000, 3 sheets, 74 p.

Wang, Z.-H., Bergström, S.M., and Lane, H.R., 1996, Conodont provinces and biostratigraphy in Ordovician of China: Acta Palaeontologica Sinica, v. 35, p. 26–59.

Wilson, F.H., Dover, J.H., Bradley, D.C., Weber, F.R., Bundtzen, T.K., and Haeussler, P.J., 1998, Geologic map of central (Interior) Alaska: U.S. Geological Survey Open-File Report 98–133, scale 1:500 000, 3 sheets, 63 p.

Witmer, R.J., Haga, H., and Mickey, M.B., 1981, Biostratigraphic report of thirty-three wells drilled from 1975 to 1981 in National Petroleum Reserve in Alaska: U.S. Geological Survey Open-File Report 81–1166, 47 p.

MANUSCRIPT ACCEPTED BY THE SOCIETY MAY 15, 2001.

Geological Society of America
Special Paper 360
2002

Tectonic setting of Mesozoic magmatism: A comparison between northeastern Russia and the North American Cordillera

Elizabeth L. Miller
Department of Geological and Environmental Sciences, Stanford University, Stanford, California 94305, USA
Mikhail Gelman
Northeast Research Institute, Far East Branch, Russian Academy of Sciences, Magadan, Russia
Leonid Parfenov
Yakutia Institute of Geology, Academy of Sciences, 39 Lenin Prospekt, Yakutsk 677982 Russia
Jeremy Hourigan
Department of Geological and Environmental Sciences, Stanford University, Stanford, California 94305, USA

ABSTRACT

Early to Middle Triassic rift-related basaltic magmatism occurred along the entire length of the northern circum-Pacific margin and was followed by Triassic to Early Jurassic (230–190 Ma) development of marine island arcs and deep-water basins. The inception of deformation in the Middle Jurassic brought an end to this tectonic regime. In northeastern Russia and Alaska, deformation was over by Middle Jurassic (Bathonian), but shortening continued in an Andean setting until ca. 160–150 Ma in the United States and Canada. Marine basins closed, marginal arc systems were accreted, and calc-alkaline batholiths developed in concert with backarc deformation, while in Alaska and Russia, mostly subsidence occurred. The Siberian margin subsequently experienced deformation, terrane accretion, magmatism, and batholith emplacement between ca. 150–125 Ma, while a conspicuous lull in magmatism and tectonism occurred in the Canadian and U.S. Cordillera. During ca. 125–170 Ma, the Okhotsk volcanic belt developed in northeastern Russia during minor extension. Magmatism began again in the Cordillera ca. 130–120 Ma, but instead was accompanied by significant crustal shortening and plutonism, continuing into the Tertiary.

This comparison indicates that striking similarities exist for event boundaries for magmatism and deformation, but fundamental differences are evident in the tectonic setting for magmatism through time, suggesting that plate motions, coupled at least at the scale of the northern Pacific, are responsible for the recorded events and their duration(s). Along-strike changes appear to take place in Alaska, which has a Mesozoic history similar to Russia, and was also affected by Arctic plate motions.

INTRODUCTION

The Mesozoic represents an important time span of subduction-related magmatism along the northern circum-Pacific margin. As a result, volcanic and plutonic belts of Mesozoic age now form fundamental elements of the geology of the Pacific margins of both North America and Northeast Asia (Fig. 1). The emplacement of these belts was followed by continued magmatism into the Cenozoic (Fig. 1). At the scale of the northern Pacific, magmatic belts appear to have remained relatively stationary through time in the western U.S. and Canadian Cordillera (Fig. 1). In contrast, a progressive migration of mag-

Miller, E.L., Gelman, M., Parfenov, L., and Hourigan, J., 2002, Tectonic setting of Mesozoic magmatism: A comparison between northeastern Russia and the North American Cordillera, *in* Miller, E.L., Grantz, A., and Klemperer, S.L., eds., Tectonic Evolution of the Bering Shelf–Chukchi Sea–Arctic Margin and Adjacent Landmasses: Boulder, Colorado, Geological Society of America Special Paper 360, p. 313–332.

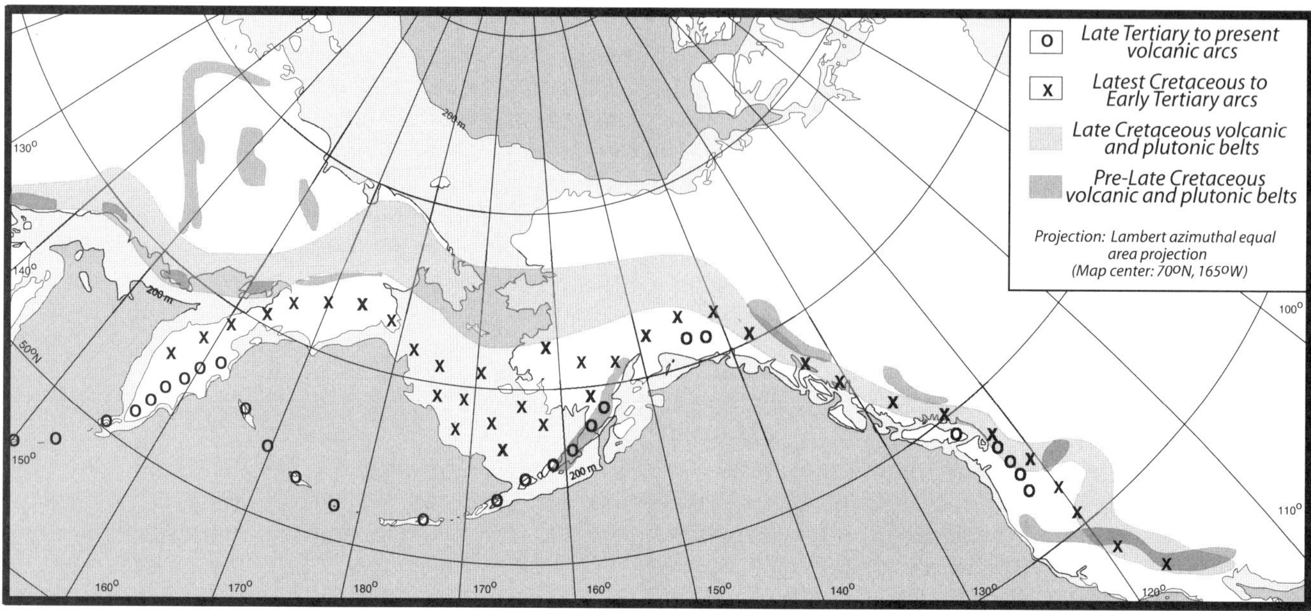

Figure 1. Index map of northern circum-Pacific illustrating approximate age and distribution of magmatic belts discussed in text developed along Pacific continental margin of U.S. and Canadian Cordillera, Alaska, and northeastern Russia. Our map projection (Lambert azimuthal equal-area projection with map center at 70°N, 165°W) and base map (*Geologic Map of Circum-Pacific Region,* Arctic sheet [Moore, 2000]) emphasize along-strike continuity of North Pacific margin.

matism toward the Pacific seems to have occurred in the Alaska and northeastern Russia sectors of the belt. This migration locally included subduction zone jumps with the formation of backarc basins during the Tertiary (Fig. 1). In this chapter we explore some of the basic differences in the tectonic setting of magmatism on either side of the North Pacific by comparing data from northeastern Russia to that from the Cordillera, focusing on the late Mesozoic, a time noted for extensive and simultaneous development of volcanic and plutonic belts along most of the length of the margin.

Compared to the Cordillera, less is known about the timing of magmatic events in northeastern Russia. Our preliminary comparison suggests that tectonic and magmatic events in northeastern Russia are similar to those of the Cordillera but that in several cases nearly opposite tectonic environments characterize these time intervals. For example, a magmatic episode coeval with crustal shortening on one side of the Pacific may coincide with a coeval period of magmatism on the other side of the Pacific that occurred in a neutral to extensional tectonic environment. Similarly, time intervals characterized by magmatic lulls on one side of the Pacific may correspond to periods of increased magmatism on the other side of the Pacific. These systematic variations, and the locations where changes take place along strike, may provide insight into the nature of ancient Pacific margin plate interactions and their controls on continental-margin orogenesis, magmatism, and basin development. Our discussion and comparison mostly emphasize magmatic belts and basins categorized as "overlap assemblages" by Nokleberg

et al. (1994), although earlier allochthonous assemblages are also described and discussed. By the Cretaceous, the displacement of the magmatic belts we focus on is not critical.

Armstrong (1988) was among the first to briefly compare magmatic histories on either side of the Pacific Ocean, pointing out that the ages of magmatic events in Alaska and Japan appeared to be out of phase with those in the Canadian Cordillera, a conclusion further evident from this discussion of data available from all of the North Pacific. Because the geochronologic and geologic database for this region is evolving, our summary represents a beginning framework to build upon or dismantle as new information becomes available.

Methodology

Magmatic belts of the North Pacific provide a long record of the history of subduction along this plate margin. The two main goals of this chapter are a comparison of the timing of magmatism and a comparison of the tectonic setting for this magmatism along the northern circum-Pacific margin in the Mesozoic. Geologic mapping, increased availability and precision of geochronologic techniques, major and trace element analyses, and isotopic studies have led to a tremendously increased database for the magmatic belts along the Cordilleran and Alaskan portion of the North Pacific margin. Several important compilations summarize and interpret this existing database (e.g., Armstrong, 1988; Armstrong and Ward, 1991, 1993; Lipman, 1992; Souther, 1991; Woodsworth et al., 1991; Brew,

1994; Moll-Stalcup, 1994; Barker, 1994; Miller, 1994; Nokleberg et al., 1994). We have relied heavily on these compilations in our analysis. Far fewer data and summaries exist, particularly in English, for the history of magmatism in northeastern Russia. Notable is an ongoing compilation of all radiometric data in northeastern Russia by Akinin and Kotlyar (1997) and an evolving database of $^{40}Ar/^{39}Ar$ dates on the granites and volcanic rocks of northeastern Russia by Layer et al. (2001, personal commun.). We relied mostly on the summary and discussion papers in English by Parfenov (1984, 1991), Parfenov et al. (1993), Nokleberg et al. (1994), Filatova (1995), and Zonenshain et al. (1990), and in Russian by Parfenov (1984), Parfenov et al. (1993), Sokolov et al. (1997), and Filatova (1988).

To interpret ancient tectonic settings for magmatism, we utilized what is known about the contemporary tectonic settings for magmatism around the Pacific today. These contemporary settings range from Andean- to Southwest Pacific–style subducting margins (e.g., Uyeda and Kanamori, 1979; Uyeda, 1987), representing compressional to extensional settings. These differing tectonic settings dictate the general nature of magmatism, the degree of preservation of magmatic belts in the geologic record (and whether mostly volcanic or mostly plutonic rocks are preserved), magma source regions, the isotopic and compositional variation of magmas, the type and distribution of ore deposits, and whether or not marine sedimentary basins form fundamental components of magmatic belts. In most cases, the simplest associations or relations are easy to document. For instance, determining the general tectonic setting for magmatism based on the presence or absence of nonmarine and/or shallow-marine versus deep-marine deposits, the degree of preservation and total thickness of volcanic successions, and whether magmatism was coeval with thrust faulting or normal faulting and marine basin development is a relatively simple task for the Mesozoic of the North Pacific.

Choice of time intervals

Geochronologic, stratigraphic, and paleontologic data provide constraints on the timing of magmatism along the northern circum-Pacific margin. We initially chose our time intervals with Cordilleran events in mind. We subsequently discovered that the nature and timing of magmatic events in northeastern Russia were distinct from Cordilleran-wide events, but that the time intervals in question had common boundaries. The critical time intervals and their boundaries, discussed here, depicted in our map compilations (Figs. 2, 3, 4, 5) and our summary timing diagrams (Figs. 6 and 7), are 250–190 Ma, 190–150 Ma, 150–125 Ma, and 125–70 Ma. The boundaries between these intervals are working boundaries and are clearly approximate. Our discussion of each time interval is preceded by a brief synopsis of the key similarities and differences in the nature of events occurring along the North Pacific margin followed by a discussion of geologic relations in northeastern Russia, Alaska, Canada, and the western U.S. Cordillera. Our compilations include data from all of northeastern Russia to the Bering Strait and data from the Cordillera to the U.S.-Mexican border (Figs. 2, 3, 4, 5). The depiction and outline of magmatic belts shown in the maps are generalized and in some cases exaggerated in size because illustrating individual exposures of rocks of a particular age range was impractical at the scale of our compilations and not essential, given the preliminary nature of our discussion.

250–190 MA TIME INTERVAL: EARLY TRIASSIC TO EARLY JURASSIC

This time interval is further subdivided into two intervals lasting from ca. 250 to 230 Ma and from ca. 230 to 190 Ma, informally designated as the age of rifting and the age of marine island arcs, respectively (Figs. 6 and 7). Early and Middle Triassic rifting, accompanied by basaltic magmatism, appears to have been important along the entire northern circum-Pacific. This rifting occurred locally within continental-shelf regions as well as within offshore oceanic basin settings, as represented by the stratigraphy of many of the accreted terranes of the North Pacific. Triassic rifting and basaltic magmatism were followed by the widespread development and subsequent preservation of volcanic sequences erupted mainly in marine island-arc settings during the Late Triassic and Early Jurassic.

Early to Middle Triassic: The age of rifting

Northeastern Russia. Early Triassic diabase dikes and sills intruded into shelfal sequences are reported from many areas in northeastern Siberia. Probably the most spectacular examples are those of Chukotka (Fig. 2), where thick (to 5.5 km) marine turbidite sequences of Triassic age are intruded by diabase, gabbro-diabase, rare gabbro, quartz diabase, and quartz gabbro sills of tholeiitic composition, to 0.5 km thick (Bychkov and Gorodinsky, 1992; Gelman, 1963; Ivanov and Milov, 1975). Mafic magmatism and associated basin development is thought to be Early to Middle Triassic because the sills occur mainly in Early Triassic (and underlying Permian strata), but not in Late Triassic strata. Whole-rock K-Ar ages vary from 250 to 190 Ma. Younger ages are also reported, but are likely related to Cretaceous thermal overprint (I.A. Zagruzina, 2001, personal commun.). Although no obvious unconformities are present within the Triassic marine section, Middle Triassic fossils have not been reported from the Chukotka Triassic sequences (Bychkov and Gorodinsky, 1992). The Triassic sediments and diabase sills of Chukotka are believed to be related to basin development across an older Paleozoic shallow-water shelf succession, which was deposited on Precambrian continental basement of Chukotka (Bychkov and Gorodinsky, 1992; Gelman, 1963; Ivanov and Milov, 1975).

Basalt and ultramafic to gabbroic plutons are present in various places on the eastern tip of the Chukchi Peninsula, but it is not clear if they are part of the Triassic of Chukotka described here or part of the Paleozoic to early Mesozoic oceanic and arc rocks of the South Anyui zone (shown in Fig. 4) (Lychagin et

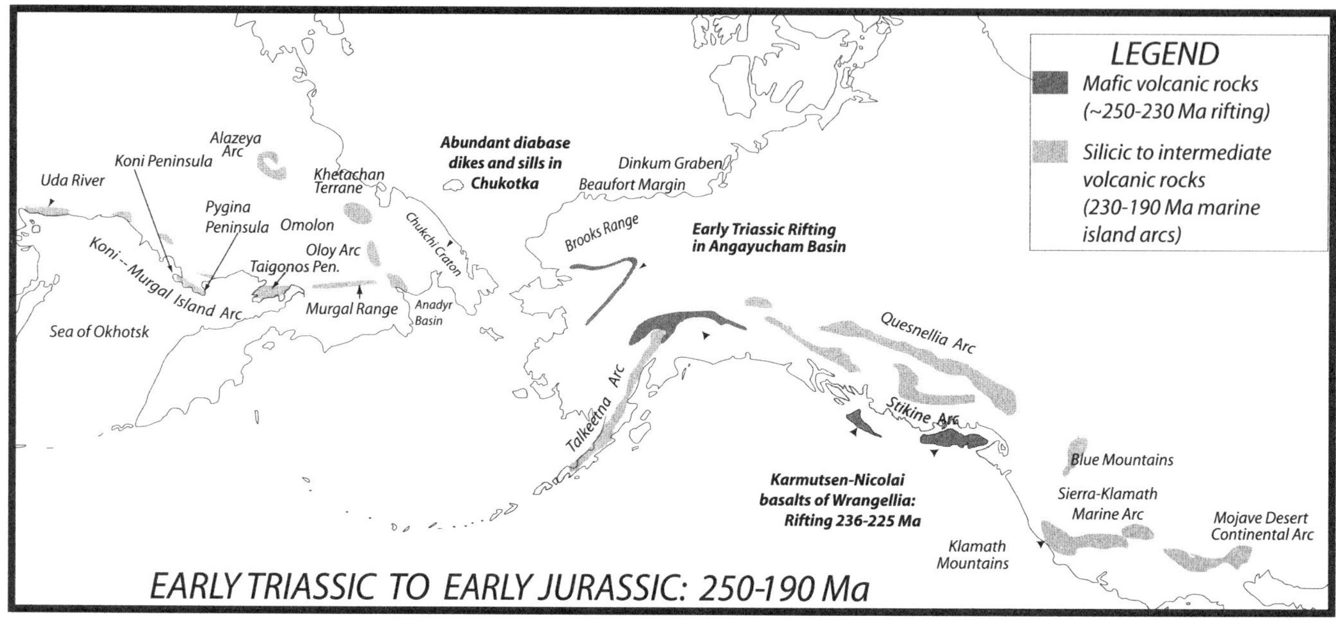

Figure 2. Triassic to Early Jurassic northern circum-Pacific magmatism. Note subdivision of this time interval into two time intervals: (1) 250–230 Ma, time span characterized by rifting and basaltic magmatism and (2) 230–190 Ma, age of development of marine island arcs. Magmatism associated with these two time intervals is combined on this map.

al., 1991a, 1991b; Palymskaya and Palymskiy, 1975; Tynanakergav and Bychkov, 1987).

In the northern Verkhoyansk region (Fig. 2), basalt flows and basaltic tuffs occur within Late Permian and Early Triassic stratigraphic units, and diabase sills and dikes occur in underlying Carboniferous to Permian deposits (Andrianova and Andrianov, 1970; Vikhert, 1957).

The Omolon massif represents a continental block within the interior portion of the Verkhoyansk belt (Fig. 2). Alkalic basalt sills and flows are common within the massif, but these

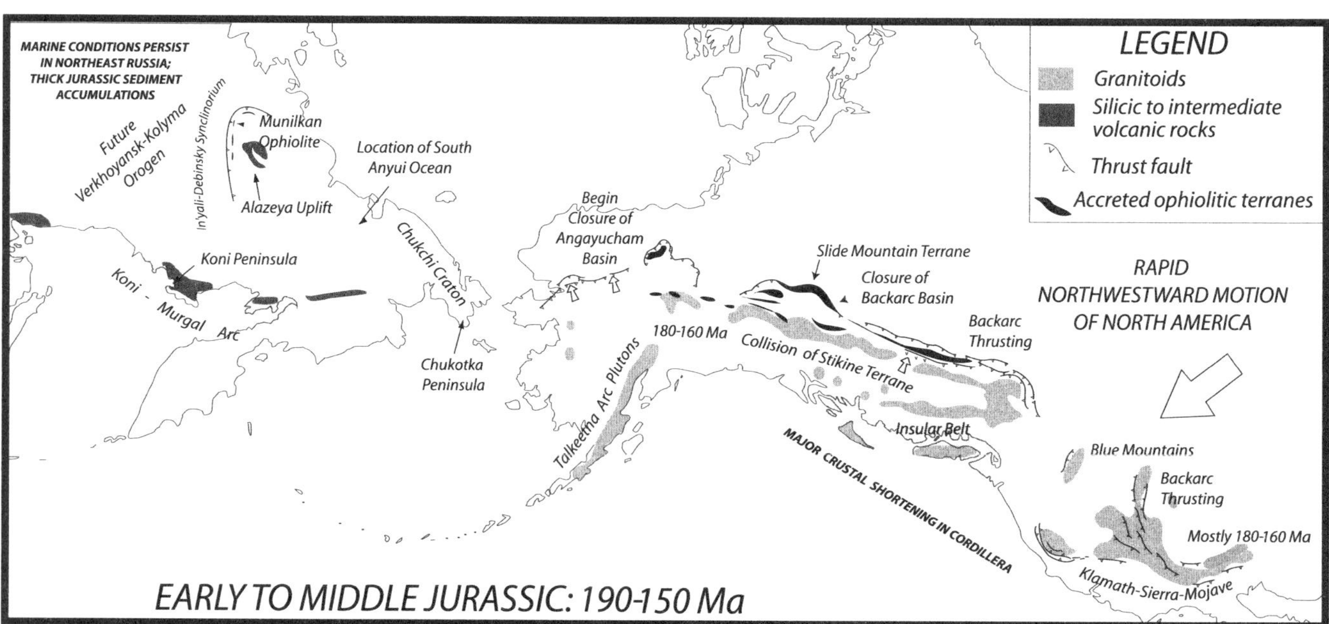

Figure 3. Early to Middle Jurassic magmatic belts of northern circum-Pacific. During this time span, major differences are developed in tectonic setting for magmatism from east to west along margin.

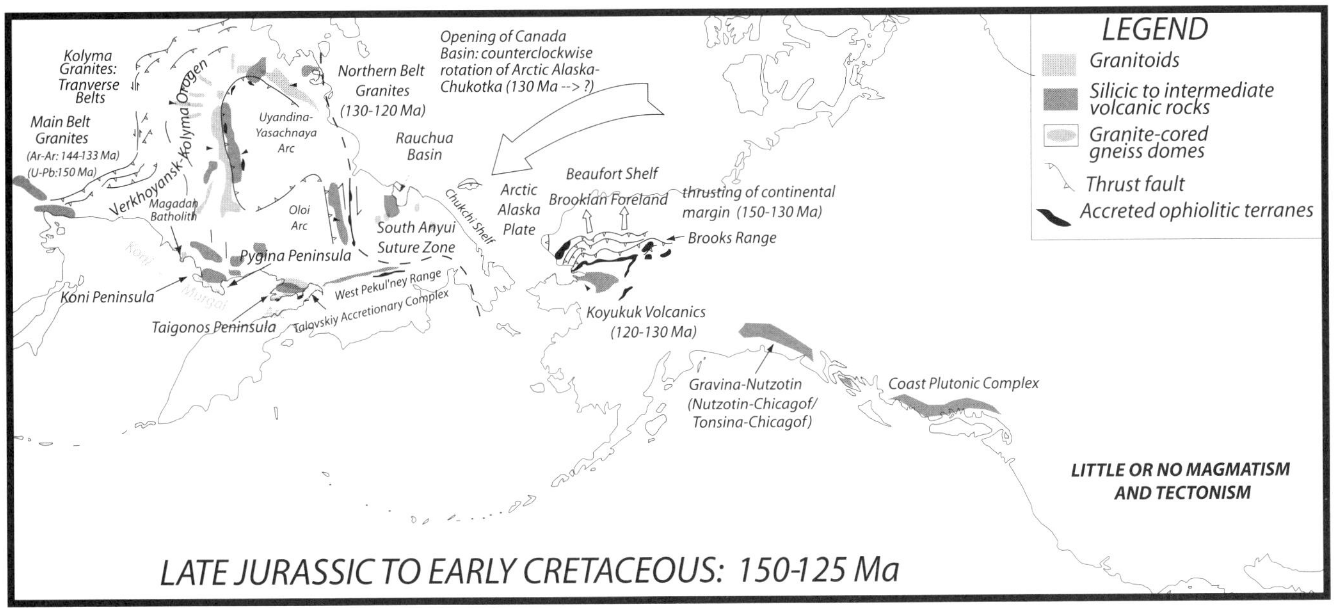

Figure 4. Late Jurassic to Early Cretaceous magmatic belts of northern circum-Pacific. Note shift from west to east in differences developed in Early to Middle Jurassic time span.

are likely younger than the occurrences described here because they are associated with Late Triassic and Early Jurassic sediments (Lychagin, 1975, 1982; Terekhov et al., 1984).

Alaska. It has been widely noted that there are no counterparts to the thick Triassic sequences of Chukotka in northern Alaska. Triassic rifting does not appear to have affected the Paleozoic shelfal sequences that are now part of the Brooks Range and North Slope of Alaska. The subsurface Dinkum

graben on the Beaufort Shelf is an exception, but is thought to have formed during Late (?) Triassic faulting and was subsequently filled with sediments mostly in the Early Jurassic (Grantz et al, 1990).

Basaltic pillow lavas and pelagic sediments of Triassic age are found in the allochthonous Angayucham terrane, a sequence of imbricated oceanic sediments and ophiolitic rocks that form the structurally highest allochthons or thrust plates of the Brooks

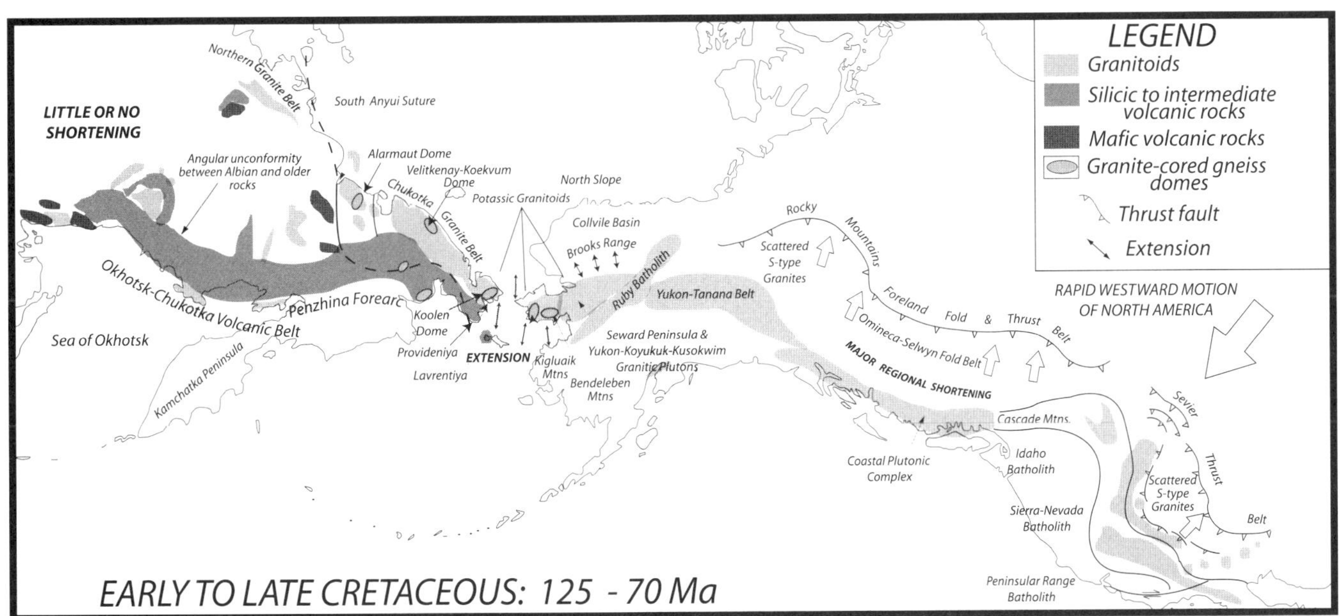

Figure 5. Late Cretaceous magmatic belts of northern circum-Pacific.

Mesozoic Continental Margin Tectonism

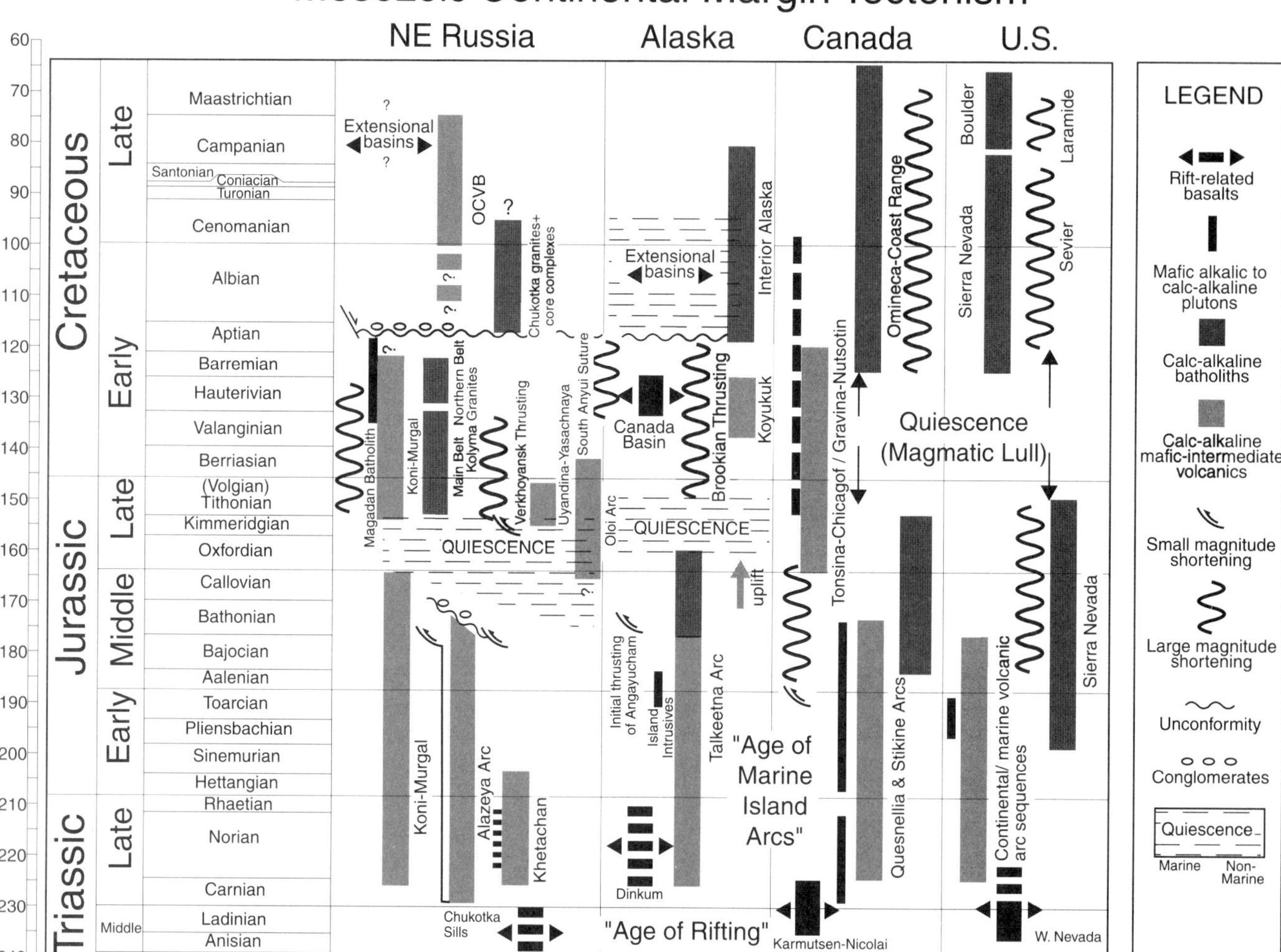

Figure 6. Summary chart showing timing of events in northeastern Russia, Alaska, and Canadian and U.S. Cordillera. OCVB is Okhotsk-Chukotka volcanic belt.

Range (Fig. 2). Triassic basalts of the Angayucham terrane are geochemically similar to within-plate basalts and are analogous to those from modern oceanic plateaus and ocean islands (Pallister et al., 1989).

Canadian Cordillera. Important Triassic rifting events occur in parts of the Cordillera considered to be allochthonous and accreted later during the Mesozoic. Although the proximity of these terranes to the continental margin is a matter of debate, the timing of events is similar to that described herein. Most notable are the thick successions (to 6 km) of tholeiitic lavas composing the allochthonous Triassic Karmutsen and Nicolai basalts of the Wrangellia terrane in the Canadian and Alaskan Cordillera (Fig. 2). These lavas probably erupted in <10 m.y. between the late Landinian and late Carnian (ca. 230 Ma), producing an oceanic plateau built on rifted crust of a late Paleozoic island-arc suc-

cession. This oceanic plateau subsided during the Late Triassic (Muller, 1977; Jones et al., 1977; Barker et al., 1985).

Basinal sequences in Canada correlative to the Angayucham terrane form part of the Slide Mountain terrane (shown in Fig. 3) but are almost entirely late Paleozoic age; Triassic sediments are known only locally. No basalt-bearing successions of Triassic age are reported from the Slide Mountain terrane (Monger, 1991).

Western U.S. Cordillera. Along the western continental margin of the U.S. Cordillera, rift-driven subsidence followed Permian-Triassic deformation and uplift. Subsidence began as early as Ladinian-Carnian time and was ongoing in Norian time, as recorded by several-kilometer-thick sequences of Middle Triassic shelf, shelf-basin, and basinal sequences in Nevada and California (Silberling, 1973; Wyld, 1991; Saleeby and Busby-Spera, 1992).

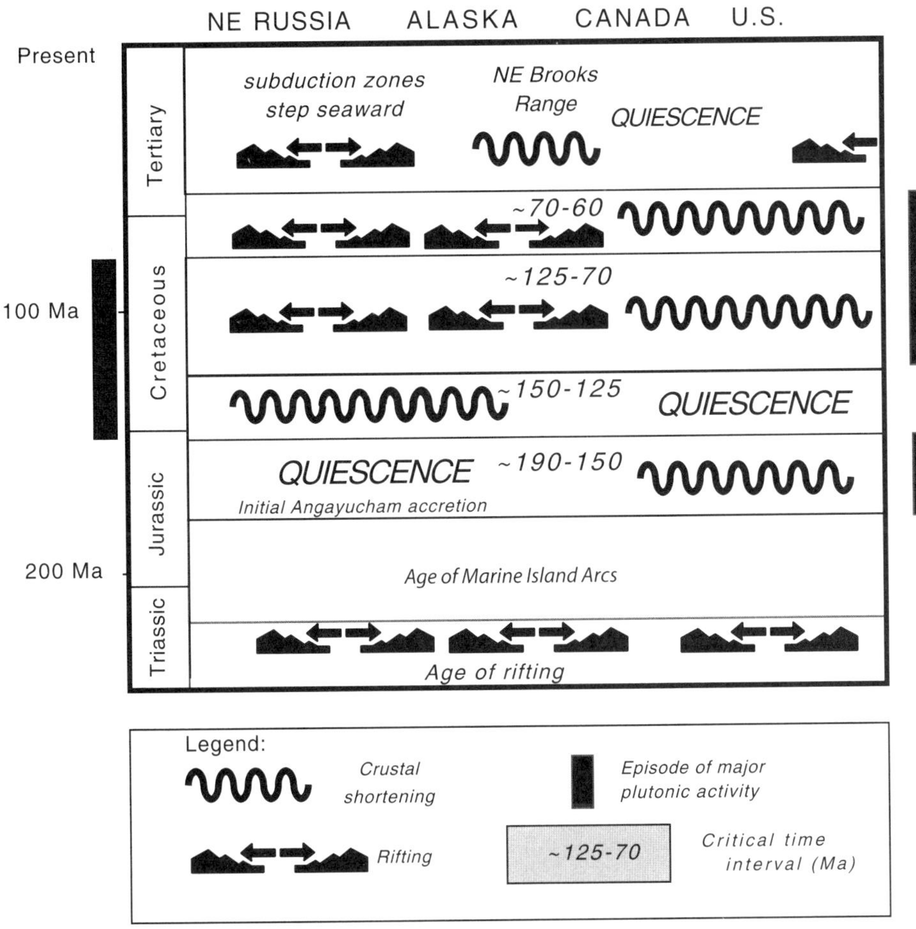

Figure 7. Simplified chart showing generalized tectonic regimes operating along parts of northern circum-Pacific margin through time.

Late Triassic and Early Jurassic: The age of marine island arcs

During the Late Triassic to Early Jurassic, arc volcanism and the development of island-arc sequences was widespread in both northeastern Russia and the Cordillera (Figs. 6 and 7). Not all of the island-arc sequences in question were tied to, or accreted to, their respective continental margins at the time they formed. However, together they provide a widespread record of subduction-related volcanism during the Late Triassic and Early Jurassic. Although disconformities and unconformities are present in these island-arc successions, there is little or no evidence for major deformation during their time of development (Figs. 6 and 7).

Northeastern Russia. In northeastern Russia, Chukotka, and along the northern coast of the Okhotsk Sea, volcaniclastic and pyroclastic strata characterize the Upper Triassic, especially during the Norian-Rhaetian. These strata were derived from volcanic arc systems between the eastern Siberian continental margin and the ancient Pacific Ocean (Bychkov, 1995) and were deposited in shelf basins. Triassic arc-related activity is also

represented by volcanic rocks in the allochthonous Alazeya arc sequence in the interior portion of the Verkhoyansk-Kolyma belt in northeastern Russia (Fig. 2). The Late Triassic to Early Jurassic Alazeya arc sequence consists of volcaniclastic sandstone, grit, conglomerate and tuff with rare basalt, andesite, keratophyre, and dacite flows deposited unconformably above older Permian and Carboniferous arc sequences (Parfenov, 1991; Parfenov et al., 1993). The clastic rocks are derived almost exclusively from andesitic and subordinate andesitic basalt, basalt, and dacitic composition volcanic sources, as evidenced by their petrography (Parfenov, 1991). Unconformities occur at the base of the Upper Triassic and Lower Jurassic. Another exposure of rocks of this age occurs in the Khetachan terrane on the east bank of the Kolyma River (Nokleberg et al., 1994; Parfenov et al., 1993). This sequence consists chiefly of Upper Triassic to Lower Jurassic volcanic and sedimentary units including tuff, volcanogenic and/or volcaniclastic sandstone, grit, and conglomerate with lesser mafic to intermediate tuff and calc-alkaline lava flows and interbedded siltstone and/or argillite and limestone. The section is intruded by Late Triassic syenite and diorite, and is unconformably overlain by Middle Jurassic to Early Cretaceous sedimentary and volcanic sequences of the Oloy arc (cf. Indigirka-Oloy assemblage in Nokleberg et al., 1998).

Late Triassic to Early Jurassic arc volcanic rocks also constitute a 2500-km-long discontinuously exposed belt that is parallel to the present-day northern margin of the Sea of Okhotsk (Fig. 2). The main regions of exposure of this belt are along the mouth of the Uda River, the Koni and the Pygina Peninsulas (e.g., Vorob'yev, 1985), the Taigonos Peninsula (Nekrasov, 1976; Zaborovskaya, 1978), and the Murgal Range (Filatova, 1988) (Fig. 2). These exposures consist of 3–7 km of mostly andesite and basaltic andesite with associated volcanogenic marine sediments, and are referred to as the Koni-Murgal arc. Basement rocks of the Koni-Murgal arc are exposed and well studied in two localities: (1) along the Uda River where Archean gneiss, schist, and granitoids compose the arc basement and (2) on the Taigonos Peninsula, where Mississippian limestone, sandstone, and shale compose its basement rocks. In the arc-sequence exposures of the Koni Peninsula, the first major pulse of volcanism is heralded by the deposition of coarse-grained volcanogenic sediments of Norian age (Vorob'yev, 1985). The lack of coarse volcaniclastic rocks and tuffaceous sediments in Early Jurassic age sediments, however, suggests a period of volcanic quiescence (Vorob'yev, 1985).

Calc-alkaline diorite to monzodiorite suites and dunite-clinopyroxenite and gabbro bodies are present along the southern margin of the Chukchi craton, possibly within the allochthonous terranes of the South Anyui suture zone, as briefly discussed in the previous section. These may be part of Late Triassic to Early Jurassic sequences, but their exact age and tectonic setting are not certain.

Alaska and Canada. Late Triassic to Early Jurassic magmatism in the northern Cordillera is mostly represented by allochthonous island-arc volcanic sequences of mafic to intermediate composition with a few plutons, generally of gabbroic composition. The main regions of exposure of rocks of this age include volcanic and associated sedimentary rocks of the Talkeetna island arc in Alaska (Barker, 1994), arc sequences that conformably overly the Karmutsen basalts of the Wrangellia terrane, and the volcanic rocks of the Quesnellia and Stikine terranes of the Canadian Cordillera (Souther, 1991) (Fig. 2). Magmatism in the Talkeetna arc consists of Upper Triassic (Norian) and Lower Jurassic andesite flows, breccia, and volcaniclastic rocks of the Talkeetna Formation (Barker, 1994). Wrangellia has a varied calc-alkaline assemblage of Early to Middle Jurassic age consisting of intermediate to acid volcanic and volcaniclastic rocks including tholeiitic andesite, dacite, and rhyolite, and calc-alkaline dacite and rhyolite (Barker, 1994). Volcanism was explosive in this arc and was associated with intrusion of plutons (the Early to Middle Jurassic island intrusions) dated as 190–180 Ma.

The Quesnellia and Stikine terranes of the Canadian Cordillera (Fig. 2) are similar in terms of their distribution and Triassic to Jurassic age range; both are characterized by a hiatus in magmatism during the Late Triassic–Early Jurassic (Souther, 1991). Island-arc magmatism in the Quesnellia arc is represented mostly by the Upper Triassic Nicola Group, which consists of intermediate composition feldspar and feldspar-augite porphyry, volcaniclastic rocks, and alkaline augite porphyry pillow lavas. The assemblage is interpreted as a calc-alkaline arc complex developed above an east-dipping subduction zone, flanked on the east by a backarc basin where a large volume of alkalic lavas erupted (Souther, 1991). Associated small and scattered Alaska-type ultramafic and mafic intrusions include dunite, wehrlite, clinopyroxenite, hornblendite, and hornblende gabbro or diorite. Plutons are concentrically zoned, more mafic rocks being in the core (Woodsworth et al., 1991). Orthopyroxene is absent and plagioclase is rare and interstitial. Early Jurassic arc volcanism in the Quesnellia arc (Fig. 2) is represented by the Rossland volcanics of the Nicola Group, and by the Hazelton Group agglomerates, lavas, and volcaniclastic augite-rich ankaremite to plagioclase-phyric andesite (Souther, 1991). Augite-porphyry basalts are characteristic of both groups. Early Jurassic alkaline plutons include syenite, monzonite, and monzodiorite dated as 209–170 Ma (Woodsworth et al., 1991). Upper Triassic rocks of the Stikine arc (Fig. 2) include andesites and basalts, augite, and feldspar porphyry breccias and flows (Souther, 1991). Plutons of the Stikine arc include comagmatic tholeiitic alkalic to subalkalic diorite, quartz monzodiorite, and monzodiorite dated as 215–230 Ma (Woodsworth et al., 1991, Souther, 1991). Early Jurassic intrusion of scattered small plutons, comagmatic with volcanic rocks of the younger Hazelton Group (and equivalents), are commonly highly alkaline in composition and include biotite and hornblende granite to quartz diorite with K-feldspar megacrysts (Woodsworth et al., 1991).

Western U.S. Cordillera. In the western U.S. Cordillera, volcanic sequences of Late Triassic–Early Jurassic age are widespread. A margin-parallel belt of Triassic and Early Jurassic volcanic and lesser plutonic rocks traverses the Mojave Desert, Sierra Nevada, and Klamath Mountains region (Fig. 2). Like their Canadian counterparts, volcanic successions of this age in these locations are marine island-arc sequences. Late Triassic volcanic rocks in the Klamath Mountains occur in the Rattlesnake Creek terrane, and represent basement for younger Jurassic sequences. These rocks and their equivalents in the Foothills belt of the Sierra Nevada and in western Nevada consist of basalt and basaltic andesite and form a high-K calc-alkaline suite (Quinn et al., 1997). In western Nevada, basalts, basaltic andesites, and andesites occur together with late-stage quartz-diorite plutons that are as old as 196–190 Ma (Quinn et al., 1997). In southern California and Arizona, thick sequences of age-equivalent volcanic rocks accumulated in extensional fault-bound grabens within a continental arc setting. Arc sequences in the western U.S. Cordillera thus range from continental to offshore marine, moving northward along the margin from Arizona to northern California (Saleeby and Busby-Spera, 1992; Karish et al., 1987).

Summary

Island-arc volcanic sequences of Triassic to Early Jurassic age are well preserved throughout the northern circum-Pacific in northeastern Russia, Alaska, and the Cordillera (Figs. 2, 6, and 7). These volcanic successions are generally associated with marine sedimentary rocks and volumetrically minor plutonic rocks. Except for volcanic successions of this age in the southwestern United States, and parts of the Koni-Murgal magmatic belt, developed on continental or shelfal sequences, most of the magmatic rocks of this age must have resided an unknown distance offshore during their formation. Although unconformities are present, all of these rock sequences indicate little or no tectonic disruption by compressive deformation. On the contrary, these sequences mostly record subsidence and local extensional faulting during the time span in question (Figs. 6 and 7).

EARLY TO MIDDLE JURASSIC (190–150 MA) TIME INTERVAL

During the time span 190–150 Ma, continental-margin magmatism continued in both northeastern Russia and in the Cordillera, but there are important contrasts across the Pacific in terms of the tectonic setting for this magmatism (Figs. 3, 6, and 7). Initial closure of ocean basins and accretion of arc sequences began along the entire length of the margin in the early part of this time interval, but deformation ceased and deposition continued in Alaska and Russia in the Late Jurassic. In the Canadian and U.S. Cordillera, deformation continued as eastward subduction beneath the continental margin took place, ultimately resulting in a major episode of crustal shortening simultaneous with the initial development of extensive belts of plutons and batholiths (Figs. 3, 6, and 7).

Northeastern Russia. Along the Verkhoyansk-Kolyma orogen of northeastern Russia (Fig. 3), isolated exposures of ophiolitic rocks and oceanic sediments form an inner belt of structurally high allochthons (e.g., Munilkan ophiolite) (Fig. 3) (Parfenov, 1991). On the basis of lithology, geochemical composition, and structural position above Paleozoic platform carbonates, this oceanic assemblage is interpreted as backarc or oceanic crust developed on the Pacific side of the Siberian Shelf (Oxman et al., 1995). Initial formation of this mafic crust occurred during a rifting event that affected the outer edge of the late Precambrian to Paleozoic continental shelf in Late Devonian time. Rifted portions of the margin subsequently evolved as plateaus or highs at some distance away from the margin (Oxman et al., 1995). Initial deformation in the Middle Jurassic occurred when the Alazeya arc collided with the platformal fragments, and ophiolite sequences were initially emplaced. Boulder conglomerates and olistostromes of ophiolitic lithologies occur together with the obducted sequences and yield Bathonian to Callovian fossils, indicating deformation by thrust faulting at this time. Metamorphic white mica ages thought to

be coeval with early deformation of the oceanic terrane are dated as 174 Ma (^{40}Ar/^{39}Ar, Oxman et al., 1995). Regionally, this event is documented by the development of an angular unconformity at the base of Bathonian deposits that is traceable for broad distances across these various paleogeographic elements, including across folded Late Triassic and Lower Jurassic volcanogenic rocks of the Alazeya arc (Fig. 2). Apparent tectonic quiescence characterized this region from the Middle to Late Jurassic. Subsequent deformation of this composite terrane (which included the Alazeya arc) occurred in the latest Jurassic to earliest Cretaceous, when all of these rocks were finally thrust over the eastern edge of the Siberian continental margin. This younger event is discussed in more detail in the next section.

Early to Middle Jurassic volcanic rocks of the Koni Peninsula, part of the Koni-Murgal arc (Figs. 2, 3, 6, and 7) are divided into two series of rocks. The Odyan series consists of basaltic andesites and andesitic lavas and pyroclastic flows. The overlying Labirintov series contains volcanic rocks ranging in composition from basalt to rhyolite. The sequence ranges from 3.5 to 5.5 km in total thickness (Vorob'yev, 1985).

Alaska. In Alaska, the oceanic Angayucham basin developed by rifting in the Devonian (Pallister, 1989), simultaneous with rifting along the Siberian margin. Intermittent rifting persisted into the Triassic, based on the age of sediments associated with mafic lavas in the sequence (Pallister et al., 1989). Deformation of this basin is dated by the initial thrust emplacement of ophiolitic basement rocks of an island-arc complex northward over ocean-floor sediments and basalts. The youngest sediments deposited prior to thrusting are Bathonian-Bajocian or Middle Jurassic in age (Mayfield et al., 1988). Amphibolite facies metamorphism generated along the sole of the thrust fault that emplaces the arc basement rocks is dated as 165–170 Ma by ^{40}Ar/^{39}Ar on hornblende (Wirth et al., 1993; Boak et al., 1987). The Angayucham terrane was not actually thrust onto the continental shelf proper until the latest Jurassic (Tithonian) to earliest Cretaceous (discussed in a subsequent section).

Plutonic rocks of the Talkeetna arc of Alaska were emplaced during the Middle Jurassic and are mostly quartz diorites and tonalites that range in age from 174 to 158 Ma (Miller, 1994) (Fig. 3). The plutonic rocks define a calc-alkaline magmatic suite that varies in composition from hornblende gabbro through hornblende-biotite diorite, hornblende-biotite quartz diorite, tonalite, and rare granodiorite and quartz monzonite. Plutons form a continuous belt 15–35 km wide and 740 km long and intrude the Triassic–Early Jurassic Talkeetna Formation volcanic-arc sequence. During and after this phase of intrusion, the arc underwent uplift and erosion. The Talkeetna arc is generally considered to be allochthonous, and added to Alaska's collage of terranes only in the Late Cretaceous (e.g., Nokleberg et al., 1998). However, arc-plutonic rocks similar in age and composition to the Talkeetna arc are inferred to have been present, and shed debris northward during Brookian thrusting that placed ophiolitic arc basement rocks and basinal sediments onto the

Brookian shelf in the latest Jurassic–Early Cretaceous (Miller and Hudson, 1991).

Canadian Cordillera. The Canadian Cordillera underwent major changes in paleogeography and a major change in tectonic setting during the Middle Jurassic. Marine basins represented by the Slide Mountain terrane and its equivalents closed by thrust faulting, juxtaposing the Quesnellia arc of the Stikine terrane against the continental margin between ca. 190 and 182 Ma (Woodsworth et al., 1991). The continental margin continued to deform, and thrust faulting migrated east toward the foreland or craton. Plutonic rocks of Middle Jurassic age throughout the Canadian Cordillera are distinct in trend in that they cut across all of the various accreted terranes and intrude the continental margin proper. They are more quartzose in composition compared to older volcanic and plutonic rocks (Woodsworth et al., 1991). Middle Jurassic magmatism in Canada records a variation from more alkalic plutons intruded between ca. 180 and 170 Ma, to calc-alkaline plutons intruded between ca. 170 and 160 Ma, to the intrusion of two-mica granites 160–150 Ma (Woodsworth et al., 1991). Presumably, the observed compositional variation through time reflects increasing crustal contribution to the magmas, which in turn most likely reflects ongoing crustal thickening during magmatism. Murphy et al. (1995) reported age data on Jurassic deformation that suggest it occurred mostly between ca. 187 Ma and 174 Ma. Volcanic rocks of Middle Jurassic age are rare to absent in Canada, except for exposures in the Insular belt, which, at the time, is believed to have resided farther outboard of the Canadian continental margin.

Western U.S. Cordillera. Jurassic plutons in the Sierra Nevada batholith of the western United States occur both on the eastern flank of the Sierra and in the western foothills, and are intruded by more voluminous Cretaceous plutons of the batholith. The older plutons of the Sierra Nevada, although now scattered, represent part of a major belt of magmatism that extends from Arizona to the Sierra Nevada and to the Blue Mountains, Oregon, and was once likely continuous with the belt of Jurassic plutons in the southern Canadian Cordillera (Fig. 3). It is a broad belt in which plutons are scattered as far inland as westernmost Utah (Fig. 3): younger Basin and Range extension is partially but not entirely responsible for this. Chen and Moore (1982) defined a main phase of Jurassic plutonism that lasted from ca. 195 to 150 Ma. This interval of magmatic activity matches the age range of peak Jurassic magmatism in the Canadian Cordillera (e.g., Armstrong, 1988).

As in Canada, the Middle to Late Jurassic brought dramatic changes in paleogeography to the western U.S. Cordillera. Increased plutonism (with lesser volcanic rocks of this age preserved) during the interval 180–150 Ma was accompanied by many hundreds of kilometers of crustal shortening perpendicular to the continental margin. This shortening closed intra-arc and backarc basins, terminated marine deposition, and accreted arc complexes to the margin. Middle Jurassic backarc deformation at ~lat 39°N began earlier in the region of elevated heat flow within and behind the arc, and migrated east with time toward the continental interior (Smith et al., 1993). In general, deformation appears to be localized adjacent to Jurassic plutonic complexes. Where well dated, deformation and metamorphism is early to late Middle Jurassic (ca. 181–165 Ma) along the eastern margin of the magmatic arc in western Nevada, Middle to early Late Jurassic (ca. 172–161) in the backarc region of western Nevada, and Late Jurassic (160–144 Ma) in eastern Nevada and western Utah (Smith et al., 1993). A nearly identical pattern in the age of subduction-related plutonism across the transect is also seen, making a strong argument that the two are linked, and that heating by magmatism allowed the upper parts of the crust to deform during intrusion (Smith et al., 1993). The intrusion of the Independence dike swarm along the eastern side of the Sierra Nevada batholith at 148 Ma marks an end to the pulse of Jurassic pluton emplacement and also represents the end of crustal shortening during this time interval (Chen and Moore, 1979).

The timing of Middle Jurassic orogenesis and magmatism in both the Canadian and U.S. Cordillera corresponds very closely with a period of rapid northwestward motion of North America within an absolute reference frame as well as high orthogonal convergence and subduction rates along the western margin of the continent (Gordon et al., 1984; Page and Engebretson, 1984; Engebretson et al., 1985).

LATE JURASSIC TO EARLY CRETACEOUS (150–125 MA) TIME INTERVAL

There were major differences in the nature of magmatism and its tectonic setting along strike of the northern circum-Pacific margin during the latest Jurassic to Early Cretaceous time interval. Deformation and magmatism (with significant plutonism) characterized the northeastern Russian segment of the margin, but a lull in magmatism occurred along the Canadian and western U.S. Cordilleran margin; this lull is thought to be related to development of a transform plate margin during this time span (Fig. 4) (Armstrong and Ward, 1991) (Figs. 6 and 7). The change between the distinct magmatic and tectonic histories represented by these two portions of the northern circum-Pacific margin takes place across Alaska, the western part of which has a tectonic history that is more similar to that of northeastern Russia (Figs. 6 and 7).

Northeastern Russia. From ca. 150 to 125 Ma in northeastern Russia there was abundant volcanism, represented by lavas ranging in composition from basaltic to intermediate to felsic, and voluminous plutonism. The main belts of volcanism include the Uyandina-Yasachnaya and Koni-Murgal arcs (Fig. 4). Along the inner or Pacificward part of the Verkhoyansk-Kolyma orogenic belt, gently dipping Late Jurassic rocks of the Uyandina-Yasachnaya volcanic arc overly complexly folded Lower Jurassic and older arc-related units. The overlying volcanic rocks consist of submarine lavas and volcaniclastic rocks of basaltic, andesitic, dacitic, and rhyolitic composition and associated sedimentary rocks. The volcanic rocks are tholeiitic, calc-alkaline, and subalkaline with high TiO_2 contents (Parfenov, 1991; Natapov and Surmilova, 1986; Ged'ko, 1988). The volcanic rocks are conformably overlain by as much as 6 km of Volgian shallow-water clastic sediments grading upsection into Neocomian coal-bearing strata (Parfenov, 1991). The polarity of the Uyandina-Yasachnaya arc is not clear, but it probably developed above a northeast-dipping subduction zone (Parfenov, 1991). The arc, and the composite terranes it overlies, collided with the Siberian margin in late Kimmeridgian to Volgian time. As collision proceeded, folding migrated cratonward, toward the Siberian platform, where it continued into the Cretaceous (Parfenov, 1991). Within the Alazeya uplift in the interior portion of the Verkhoyansk-Kolyma orogenic belt (Fig. 3), folded and thrust-faulted Lower Jurassic sedimentary and volcanic rocks are unconformably overlain by flat-lying sands, silts, and muds of Middle and Late Jurassic age (Parfenov, 1991), indicating that the interior portion of the orogen had ceased deforming while the exterior portion continued to deform.

As deformation stepped eastward toward the Siberian craton, granites of the Main belt of the Verkhoyansk-Kolyma orogenic belt were emplaced (Fig. 4). Main belt granites developed subparallel to the trend of the Uyandina-Yasachnaya volcanic belt and consist mainly of biotite, two-mica, and amphibole-biotite granites (Parfenov, 1991). They are associated with cassiterite-quartz and gold–rare metal mineralization and occur in laccoliths or sheetlike bodies. Compilation of biotite and hornblende $^{40}Ar/^{39}Ar$ dates from granite bodies of the Main belt by L. Parfenov, P. Layer, and K. Fujita (1999, personal commun.) indicate ages from 144 to 135 Ma. A single U-Pb age of 150 Ma indicates that plutons in this belt may be slightly older than the reported $^{40}Ar/^{39}Ar$ range of ages (Akinin and Kotlyar, 1997). Some ages as old as 160–170 Ma are reported for the belt based on Rb-Sr whole-rock analyses, and these occur in a region where there are no Late Jurassic to Early Cretaceous strata present (in the In'yali-Debinski synclinorium, Fig. 3) (V. Akinin and I. Kotlyar, 1988, personal commun.). Although it seems that the general stratigraphic relations support the proposed Late Jurassic–earliest Cretaceous age of deformation in this belt, some evidence exists for an earlier beginning of deformation, metamorphism, and plutonism in the belt. The unusual aluminous composition and mineralogy of the Main belt plutons in the Verkhoyansk-Kolyma belt has led to the interpretation that the Main belt granites formed by melting of continental-margin sedimentary rocks during and after thrust faulting (Parfenov, 1991). Granites of the Northern belt (Fig. 4) are somewhat different in composition and age and consist of tonalites, granodiorites, and, less frequently, leucogranites. Tin mineralization is common, as are silver and copper. Compilation of biotite and hornblende $^{40}Ar/^{39}Ar$ dates from Northern belt granites (L. Parfenov, P. Layer, and K. Fujita. 1999, personal commun.) indicate an age range from 130 to 120 Ma, slightly younger than the Main belt granites. Transverse belts of granite extend for several hundred kilometers into the Siberian platform, cutting across the general trend of folded structures, and appear to die out toward the craton (Fig. 4). These consist of chains, stocks, dikes, and lenticu-

lar bodies of diorite, granodiorite, and andesine granites, often younger by a few tens of million years than the Main belt granites (Layer et al., 1995; Parfenov, 1991). Cassiterite-silicate-sulfide mineralization is common in these transverse belts of granite.

Late Jurassic volcanic and volcaniclastic rocks crop out in three major regions along the southeastern part of the Pacific margin of northeastern Russia: (1) Taigonos Peninsula, (2) Koni-Pygina Peninsula, and (3) West Pekul'ney Range (Fig. 4). The southern Taigonos Peninsula exposes andesites, andesitic basalt, and basalt, inferred to have developed along the main axis of arc magmatism. The northern portion of the peninsula consists largely of volcanogenic sedimentary rocks, presumably deposited in a backarc setting. Late Jurassic–Early Cretaceous deformation produced elongate southeast-vergent folds with steep axial planes (Nekrasov, 1976; Zaborovskaya, 1978). Upper Jurassic–Hauterivian deposits in the Murgal Range consist of basalt, andesitic basalt, andesite, and their pyroclastic equivalents. The West Pekul'ney Range contains Upper Jurassic–Hauterivian andesites, dacites, and rhyolites of island-arc affinity in association with marine volcanogenic coarse clastics, sandstones, and siltstones.

The polyphase Magadan batholith on the Koni-Pygina Peninsula contains several intrusive suites, including gabbro, gabbrodiorite, diorite, and granodiorite ranging in age from 135 to 115 Ma, corresponding to the late stages of Koni-Murgal arc magmatism (Andreeva, 1998). In the Murgal Range, sporadic plutons of pre-Albian age include granodiorite, quartz diorite, and tonalite. The coastal granite and metamorphic belt on the Taigonos Peninsula also contain intrusive rocks within this age range (Zaboraovskaya, 1978).

The West Pekul'ney and Talovskiy accretionary wedge complexes are along the southeastern flank of the Koni-Murgal island arc (Fig. 4). Upper Jurassic–Lower Cretaceous deposits are characterized by intraformational disconformities and stratigraphic gaps as well as by abundant ophiolitic detritus (Sokolov, 1988). Crosscutting relationships in the Kuyul terrane provide the best insight into the timing of accretionary deformation. In the Kuyul terrane, folded Berriasian and Hauterivian sediments containing ophiolite fragments are unconformably overlain by undeformed late Albian terrigenous deposits, indicating that accretion-related deformation in the late Albian ceased prior to inception of continental-margin arc magmatism in the Okhotsk-Chukotka volcanic belt (Sokolov, 1992; Alekseev, 1982).

Alaska-Chukotka. During the latest Jurassic to Early Cretaceous, Arctic plate motions had important effects on both magmatism and deformation in the Chukotka Peninsula and the North Slope and Brooks Range of Alaska (Fig. 4). These plate motions involved formation of the oceanic Canada Basin, now bordered on its southern side by the Siberian, Chukchi, and Beaufort shelves. Together with the landmasses represented by the North Slope of Alaska, the Brooks Range, and the Chukotka Peninsula of northeastern Russia, these elements collectively define the Arctic Alaska–Chukotka plate (Fig. 4). Grantz et al.

(1990, 1994) suggested that the Canada Basin first began to develop in the Early Jurassic, beginning with formation of a few small rift basins between 190 and 144 Ma; the main phase of rifting was in the Hauterivian, and was mostly over by ca. 120 Ma. Although there is no universally accepted tectonic model for the formation of the Canada Basin, many believe that it opened by rotation about a pole located in northern Canada near the Canada-Alaska border (Grantz et al., 1990, 1994), a hypothesis supported by new magnetic-anomaly maps of the Arctic and the observed anomalies in the Canada Basin (Grantz et al., 1990, 1994) (Fig. 4). The South Anyui zone is believed to be the main suture zone between Siberia and its attached terranes and the Chukotka Peninsula part of the Arctic Alaska–Chukotka terrane in northeastern Russia (Fig. 4). Studies in the South Anyui zone indicate that earliest Cretaceous deep-water sediments are the youngest rocks involved in deformation and that crosscutting plutons of the northern belt that range in age from 130 to 120 Ma postdate deformation (e.g., Nokleberg et al., 1994). Possible complications to these constraints on the timing of events include the report of plutons as old as 150 Ma in northern Chukotka (Gelman, 1995; Akinin and Kotlyar, 1997). If better documented, these ages could be significant in terms of whether deformation is earlier than (and unrelated to), or coeval and related to the opening of the Canada Basin. Stratigraphic relations also indicate that deformation ended prior to Albian time; the suture zone is overlapped by Albian strata that form the basal units of the Okhotsk-Chukotka volcanic belt (Fig. 5). To the north of the suture, deformation may have ended earlier. The deformed Triassic–Early Jurassic turbidite and shale section along the northern Chukotka coastline is straddled by unconformably overlying sediments of the Rauchua basin (Fig. 4), comprising gently dipping Upper Jurassic (Volgian) to Lower Cretaceous sediments with some intercalated tuffaceous and volcanic rocks of intermediate composition (Nokleberg et al., 1994; Paraketsov and Paraketsova, 1989).

In northern Alaska, deformation in the Brooks Range is dated on the basis of the age of the Okpikruak Formation as Late Jurassic to Early Cretaceous. The Okpikruak Formation was deposited synchronously with the development of north-verging thrust faults that involved the Angayucham marginal-basin deposits as well as Arctic Alaska continental-shelf strata (Mayfield et al., 1988) (Fig. 4). A flood of magmatic arc and mafic rock debris similar in age and composition to rocks of the Talkeetna arc (Miller and Hudson, 1991) was shed northward onto Arctic Alaska beginning in the Tithonian, indicating collision of a volcanic-plutonic arc edifice with the Arctic continental margin at this time (Patton and Box, 1989). This was the culminating deformation that placed the oceanic Angayucham terrane onto the continental shelf. Based on the development of regional unconformities, it is believed that this phase of deformation was mostly over by the Albian, between 130 and 113 Ma (see review in Miller and Hudson, 1991) (Figs. 6 and 7). Regional deformation changed quickly in character at that time and a series of Albian and younger extensional basins formed south of the Brooks Range. The faults bounding these basins cut and offset the earlier thrust faults of the Brooks Range (Miller and Hudson, 1991).

The preceding summary suggests that deformation along the southern edge of the Arctic Alaska–Chukotka plate, from the South Anyui zone to the Brooks Range, Alaska, is everywhere similar in age. Deformation is closely bracketed in many different places as latest Jurassic to Early Cretaceous, and is likely mostly Early Cretaceous, having ceased prior to deposition of Albian sediments and intrusion of crosscutting plutonic rocks that are as old as 130–120 Ma. Possible older ages of plutons reported (as old as 150 Ma) from Chukotka (Gelman, 1995; Akinin and Kotlyar, 1997) would compromise this interpretation.

In northern Alaska, magmatic rocks of this time interval are sparse and are represented only by the Koyukuk volcanic succession (Fig. 4). The Koyukuk volcanic succession consists of shallow-marine to nonmarine mafic to intermediate composition volcaniclastic rocks that yield K-Ar hornblende ages of 120–128 Ma, a biotite age of 137 Ma, and Berriasian-Valanginian fossils (Box and Patton, 1989). Volcanic rocks are tholeiitic to shoshonitic in composition (Box and Patton, 1989). Based on regional relations, the sequence is believed to depositionally overlie Middle Jurassic plutonic rocks along a significant angular unconformity (Box and Patton, 1989). The volcanic rocks are conformably overlain by Hauterivian-Barremian volcaniclastic rocks and mafic to intermediate flows deposited in transitional water depths, and are in turn overlain by siliciclastic turbidites, derived from the uplift and erosion of the Brooks Range and surrounding metamorphic highlands. Box and Patton (1989) argued that the volcanic rocks represent an island arc that collided with the continental margin between ca. 145 and 130 Ma. Miller and Hudson (1991) alternatively suggested that the volcanic rocks mostly postdate thrusting and are precursors to regional extension following the major episode of thrust faulting. In the second interpretation, increased water depths indicate the onset of extension-driven subsidence in the Hauterivian-Barremian (Miller and Hudson, 1991).

In southeastern Alaska and continuing into northern Canada, the Gravina-Nutzotin belt (Fig. 4) consists of volcanic, sedimentary, and plutonic rocks deposited on and intruded into older parts of the allochthonous Wrangellia superterrane and interpreted as an island-arc complex. Intercalated volcanic and sedimentary rocks range in age from Late Jurassic (Oxfordian) to Early Cretaceous (Albian), and associated quartz-diorite, tonalite, and granodiorite plutons yield ages ranging from 164 to 94 Ma (Miller, 1994; Nokleberg et al., 1994). These magmatic rocks form part of the extensive Gravina arc of Nokleberg et al. (1998), a marine arc sequence and associated sedimentary basin deposits formed offshore (?) of Alaska (Nokleberg et al., 1998); the development of the arc was approximately coeval with major deformation of the edge of the Siberian margin.

Canadian and Western U.S. Cordillera. In contrast to the fairly complex sequence of events outlined here for Alaska and northeastern Russia, the Canadian and western U.S. Cordillera is characterized mostly by magmatic and tectonic quiescence during the latest Jurassic to Early Cretaceous. Few plutons and volcanic rocks of this age range exist (e.g., Armstrong, 1988; Armstrong and Ward, 1993). In Canada the exceptions are 140–150 Ma plutons in northwestern British Columbia–Alaska–Yukon, possibly some plutons in the Coast Plutonic Complex, and the Francoise Lake Suite (140 Ma) in one small area of central British Columbia (Armstrong and Ward, 1993). The magmatic lull in Canada spans the approximate time interval 140–125 Ma. If the intervening ages of some of the plutons in the Coast belt are taken into account, the magmatic gap is 145–130 Ma (Friedman and Armstrong, 1995; Nokleberg et al., 1994). In the western U.S. Cordillera, the time span of the magmatic lull is tightly bracketed and ranges in age from ca. 148 Ma, when the Independence dike swarm was intruded, to ca. 120 Ma, when plutonic activity related to the Sierran arc began anew in the Cretaceous (Armstrong, 1988; Chen and Moore, 1982; Smith et al., 1993). During the time span of this magmatic lull, no shortening-related deformation or basin-forming events are evident in the Cordillera. If subduction was taking place along this part of the Pacific margin, it would have had to be located some distance offshore.

EARLY TO LATE CRETACEOUS (125–70 MA) TIME INTERVAL

The 125–70 Ma time span was characterized by voluminous magmatism along the entire length of the northern circum-Pacific margin (Fig. 5). Magmatism encroached far into the craton and is characterized mostly by plutons in the Cordillera and Alaska and by thick sequences of volcanic rocks in Russia. This culminating magmatic episode took place in very different tectonic settings on either side of the northern Pacific. In Russia, volcanic rocks of the Ohkotsk-Chukotsk volcanic belt accumulated during an interval of time characterized by tectonic quiescence or perhaps even weak or localized extension. In the western U.S. and Canadian Cordillera, magmatism of this age was accompanied by significant shortening of continental crust in the arc and backarc regions, leading to the development of the Rocky Mountain and Sevier foreland fold and thrust belts (Figs. 5, 6, and 7). The change between the two very different tectonic regimes along this margin seems to take place within the Alaskan sector of the margin (Figs. 5, 6, and 7). Crustal extension took place in the Bering Strait region during this time and continued intermittently into the Tertiary (and to the present day), but in eastern Alaska, thrust faulting began in the latest Cretaceous–early Tertiary and was coeval with Laramide thrusting in the Canadian Rockies (Moore et al., 1994).

Northeastern Russia. The Andean-style Okhotsk-Chukotka volcanic and plutonic belt of Russia (Fig. 5) is as wide as several hundred kilometers and nearly 3000 km long, extending from the Bering Strait to the Uda River region. The interior or seaward zone of the Okhotsk-Chukotka volcanic and plutonic belt is characterized mainly by intrusive complexes. Volcanic sequences to 3 km thick and shallow cogenetic plutonic com-

plexes characterize the exterior zone of the Okhotsk-Chukotka volcanic and plutonic belt (Belyi, 1977a, 1977b, 1978). Where available, pressure-temperature data suggest deeper levels of exposure for the plutonic complexes in the seaward part of the belt (e.g., Andreeva, 1998). Surface exposure of deep-seated rocks, the intrusion of which was approximately coeval with the volcanic accumulations on the continent side of the belt, imply a significant amount of differential uplift of the interior part of the belt with respect to the exterior part of the belt. The tectonic significance of this differential uplift and erosion is not known. The basement of the Okhotsk-Chukotka volcanic and plutonic belt is composed of variable units, but in general is represented by variably deformed and intruded continental-margin strata or accreted island-arc successions. The earliest deposits within the belt are often coal-bearing continental deposits of Early Cretaceous age (Belyi, 1978). Flora within these deposits are Albian and suggest that volcanism began by at least 105 Ma (Belyi, 1994). Calc-alkaline andesite and andesitic basalt make up the earliest volcanic units of the Okhotsk-Chukotka volcanic and plutonic belt. The uppermost portions of the volcanic sequence are characterized by voluminous rhyolites that were associated with large caldera-forming eruptions. Overlying capping plateau basalts are the youngest units of the Okhotsk-Chukotka volcanic and plutonic belt. Belyi (1978) suggested that the association of rhyolite and basalt compose a bimodal volcanic sequence of late Santonian to earliest Campanian age.

The exact duration and timing of magmatic activity represented by the Okhotsk-Chukotka volcanic and plutonic belt are not well defined. Many of the age constraints placed on the Okhotsk-Chukotka volcanic and plutonic belt are based on the floral assemblages present in the associated continental deposits, which are often difficult to correlate with the marine faunal record. Radiometric dating of igneous rocks is limited primarily to Rb-Sr whole-rock isochron analyses and K-Ar dating of whole rock and individual minerals; both methods have low accuracy. There is a notable discrepancy between ages yielded by radiometric methods (which are generally younger) and those inferred from the floral assemblages in associated continental deposits. A study by Kelley et al. (1999) investigating the age of the Chauna Group in the Okhotsk-Chukotka volcanic belt in the region of the Chukotka Peninsula illustrates this problem. The Chauna Group has a characteristic fossil plant assemblage, often referred to as the Chauna floral assemblage, that has no direct equivalents with assemblages elsewhere in Russia and Alaska. This assemblage has been considered Albian in age. Single-crystal high-precision $^{40}Ar/^{39}Ar$ dating of feldspar, biotite, and amphibole from four samples of units within the Chauna Group are all Coniacian and within error of a mean age of 88 ± 1.5 Ma. These dates suggest the eruption of a large volume of material over a short time span in this part of the Okhotsk-Chukotka volcanic and plutonic belt. Thus the age of the main phase of volcanism and age of the Chauna flora in the northern part of the belt need to be revised from late Albian–earliest Cenomanian to Turonian-Campanian (Kelley et al.,

1999). Further high-precision (U-Pb and ^{40}Ar/^{39}Ar) studies will allow researchers to clarify whether the source of error is within the floral stratigraphy or the absolute dating method, and more important, will provide better data for the duration of magmatism along the margin during development of the Okhotsk-Chukotka volcanic and plutonic belt.

On the Chukotka Peninsula, Cretaceous granitic rocks and dome-shaped metamorphic culminations cover a broad region to the northern Arctic coastline (where they presumably continue offshore) and are intrusive into Paleozoic and older strata and their thick cover of Triassic clastic rocks. The ages of the granitic bodies in this belt are not well known, but as discussed in the following section, many of them are unconformably overlain by flat-lying to gently dipping volcanic rocks and associated sediments of the Okhotsk-Chukotka volcanic and plutonic belt. In the Providenya and Lavrentia region of the eastern Chukotka Peninsula, Cretaceous plutons span the interval 118–95 Ma, whereas volcanic rocks of the Okhotsk-Chukotka volcanic and plutonic belt are somewhat younger and range from 94 to 84 Ma (Bering Strait Geologic Field Party, 1997; Rowe, 1998; Akinin and Calvert, this volume, Chapter 8). The Alarmaut dome and Velitkenay uplift (Gelman, 1995) are the largest of these domes along the Arctic coast of Chukotka (Fig. 5), but similar structures are inferred in the subsurface based on geophysical data (Gelman, 1995). Rocks in the centers of these uplifts are characterized by subhorizontal metamorphic foliations and features that indicate large strains, such as boudinage of units and extreme flattening and stretching of fossils. Cretaceous granitoid plutons in the cores of the domes have sill-like geometry. The sill-shaped plutonic complexes are inferred to be more deeply seated plutons than those present in the surrounding higher structural levels of the Chukotka granite belt that are typically crosscutting. The sills in the cores of the domes consist of intermediate and felsic rocks, and shallower plutons on the flanks of the dome are granitic in composition. Deeper-seated intrusions are coarser grained and more equigranular, and do not have chilled margins. Migmatites and amphibolite facies regional metamorphic rocks are associated with the granites in the domes. Maximum metamorphic temperatures represented are ~680 °C, and pressures are ~4–5 kbar (Gelman, 1995). The pressure and temperature values are controversial because they represent greater depths than can be accounted for by the known thickness of overlying Triassic country rocks. It is also known that metamorphic grade along the flanks of the domes decreases to greenschist facies and locally, metasedimentary rocks are andalusite bearing, thus more compatible with the inferred lesser stratigraphic and/or structural overburden for metamorphism in the region. The difference in inferred pressures for regional metamorphism and intrusion of granites in the cores versus the flanks of the domes suggests a vertical and possibly diapirically driven component of uplift to the metamorphic domes. Minor folds, ductile shear zones, and steeper foliation are characteristic of the flanks of the domes and perhaps compatible with this interpretation. Metamorphic rocks of these two structures are not well dated, but reported whole-rock K-Ar ages are as old as 145 Ma (Gelman, 1962). Plutons have yielded K-Ar whole-rock ages ranging from 125 to 70 Ma, but mostly 110–95 Ma, and are unconformably overlain by Cretaceous volcanic rocks of the Okhotsk-Chukotka volcanic and plutonic belt (Gelman, 1962).

In the Koolen dome, near the Bering Strait (Fig. 5), peak metamorphic conditions and the main deformational event leading to the formation of the metamorphic culmination involved granitic rocks dated as 104 and 108 Ma by the U-Pb method on zircon and monazite (Bering Strait Geologic Field Party, 1997). These deformational fabrics are sharply crosscut by undeformed granitic rocks dated as 94 Ma (U-Pb, monazite), indicating peak metamorphism and formation of the gneiss dome in the middle to Late Cretaceous. Peak metamorphic conditions are represented by mineral assemblages that yield pressures to 7 kbar and temperatures in excess of 700 °C (Akinin and Calvert, this volume, Chapter 8). Metamorphic conditions varied with time, however, and together with ^{40}Ar/^{39}Ar data indicate that rocks that initially equilibrated in the mid-crust underwent a rapid component of vertical uplift to near surface conditions in the middle to Late Cretaceous (Akinin and Calvert, this volume, Chapter 8). Regional relations together with seismic reflection data in the Bering Strait discussed by Klemperer et al. (this volume, Chapter 2) and Miller et al. (this volume, Chapter 10) suggest that the granites and gneiss domes of Chukotka and of the Seward Peninsula, western Alaska, are similar in age and origin and may be the result of a major transfer of heat from the mantle by mafic magmatism. This heating resulted in melting and remobilization of the crust during regional deformation associated with north-south stretching in the Bering Strait and surrounding regions.

Alaska. In Alaska, air-fall and water-lain tuff horizons are common within Cretaceous deposits of the North Slope Colville basin. These have been studied and dated by high-precision ^{40}Ar/^{39}Ar analyses of biotite and feldspar and yield ages ranging from 100 to 67 Ma, but most are 93–85 Ma (Bergman et al., 1995; Kelley et al., 1999; S. Bergman, 1999, personal commun.). A likely source region for these tuffs is the similar-age caldera complexes of the Okhotsk-Chukotka volcanic and plutonic belt in northeastern Russia.

Mid-Cretaceous granodiorite and granite bodies present across interior Alaska also represent voluminous intrusion ca. 118–80 Ma. These plutons are commonly associated with a widespread low-pressure-high-temperature metamorphic overprint where andalusite was generally stable (Dusel-Bacon, 1994). Localized basin development occurred during the same time span and led to creation of the Yukon-Koyukuk and Kuskokwim basins. Granitic magmatism involved variable, but often significant, melting and assimilation of continental crust (Hudson, 1994; Arth et al., 1989a, 1989b; Miller, 1994). Miller and Hudson (1991) inferred an extensional tectonic setting for Alaska during 120–70 Ma. Associated with this belt of plutons are lesser-volume mafic-potassic and ultrapotassic plutons described by Miller (1972, 1989, 1994) and Amato and Wright

(1997). These magmatic rocks were grouped as the Omineca-Selwyn plutonic belt by Nokleberg et al. (1994). Amato and Wright (1997) summarized geochronologic, geochemical, and isotopic data from these granitoid magmas in Alaska, and Rowe (1998) summarized similar data from Russia. Isotopic values indicate a mantle component to the magmas, modified by a variable but often significant crustal component. The variation of these values in any given location with time suggests a decreasing crustal component during the time of magmatism, at least for the Alaska plutons. Amato and Wright (1997) interpreted data from Alaskan plutons and suggested that they were intruded within a continental arc undergoing fragmentation during extension, perhaps as a result of retreat or rollback of a north-dipping slab beneath Alaska toward the Pacific in the Late Cretaceous.

Canadian and Western U.S. Cordillera. Magmatic rocks of middle to Late Cretaceous age are voluminous in the Canadian and western U.S. Cordillera (Fig. 5). In contrast to the inferred tectonic setting for magmatism in northeastern Russia and Alaska, plutons in the Cordillera were emplaced during a protracted time span of crustal shortening that resulted in the development of the Rocky Mountains and Sevier foreland fold and thrust belts (Fig. 5). During this time span, plutons of the Coast Plutonic Complex of Canada were emplaced as well as the bulk of the Sierra Nevada and Peninsular Ranges batholiths of the western U.S. Cordillera. The Boulder batholith was emplaced in the very latest Cretaceous (Fig. 5). Only local evidence exists for the large-volume ash flows and caldera complexes that must have once represented the high levels of these batholiths. In general, erosion has stripped these to expose only plutonic roots of these complexes today. In Canada, the most voluminous magmatism in this general period occurred between 90 and 95 Ma and ranges in composition from sodic to normal calc-alkaline, to aluminous, and to exceptionally alkaline across the magmatic belt at its greatest width (Armstrong and Ward, 1993; Woodsworth et al., 1991). Late Cretaceous plutons in the Coast Plutonic Complex and Cascade Mountains of the western United States were emplaced concurrently with structures formed during contraction and dextral transpression and transtension, and were accompanied by regional metamorphism (Rubin et al., 1990; Jorneay and Friedman, 1993; Nokleberg et al., 1994). The extremely radiogenic Sr of the aluminous plutons in the eastern part of the belt establishes that melting and bulk assimilation of old, tectonically thickened and magmatically heated crust was involved in their genesis (Armstrong and Ward, 1993). Across the Sierra Nevada batholith, elements such as K, Rb, U, and Th increase while Ca and Sr and related elements decrease eastward. These trends are likely to be related to a combination of magma source regions, magma flux, basement rock types, crustal thickness, and amount of crustal assimilation (Lipman, 1992). Plutons of the 120–70 Ma age range occur from the Sierra Nevada eastward to the Utah border. They were intruded during a time span of deformation that appears to have begun in the Sierran arc region and then spread rapidly across the backarc region to the Sevier fold and thrust belt, reaching its eastern extent by Albian time (Lawton and Trexler, 1991). Deformation took place simultaneously across the orogen ca. 100 Ma and continued to 70–80 Ma (Smith et al., 1993). The locus of deformation in the western United States then stepped eastward during the Laramide orogeny (not discussed herein), which began in the Maastrichtian (74–65 Ma).

To the east of the Sierra Nevada, scattered two-mica and garnet-bearing granitic plutons of Late Cretaceous age are peraluminous in composition (Miller and Bradfish, 1980; Miller and Barton, 1990). Their peraluminous composition, isotopic signatures, high volatile content, and abundant associated pegmatites indicate that they were derived from partial melting of hydrous aluminous sources, probably upper crustal sources carried to depth during formation of the foreland fold and thrust belt (Miller and Gans, 1989; Miller and Barton, 1990; Wright and Wooden, 1991). As discussed here, similar plutons occur in the eastern portion of the magmatic belt of the Canadian Cordillera and in the Selwyn fold belt of the northern Canadian Rockies (Fig. 5) (Woodsworth et al., 1991).

The Late Cretaceous magmatic event in the Canadian and western U.S. Cordillera correlates closely with a period of inferred rapid northwestward motion of North America within an absolute reference frame as well as with high orthogonal convergence or subduction rates along the western margin of North America (Gordon et al., 1984; Page and Engebretson, 1984; Engebretson et al., 1985).

SUMMARY

Figures 6 and 7 summarize the magmatic and tectonic history of the northern circum-Pacific margin in Mesozoic time. The events discussed in this chapter are shown in a time-line framework in Figure 6 and the interpretation of the tectonic regimes characterizing the Pacific margins of northeastern Russia, Alaska, Canada, and the U.S. Cordillera through time are shown in Figure 7. The commonalities and contrasts discussed in some detail here are readily apparent in these summary figures.

During the late Early, Middle, and early Late Triassic, basaltic magmatism and rifting affected all parts of the northern circum-Pacific margin, except perhaps for Alaska. In the Late Triassic to Early Jurassic (230–190 Ma), mostly andesitic marine island arcs developed along the same length of the circum-Pacific margin. Deep-marine sedimentation prevailed in and around these arc systems and autochthonous continental-shelf regions underwent subsidence and accumulated sediments throughout much of this time span. We thus infer that neutral to extensional tectonics characterized the length of the overriding continental plates and arc systems during this time interval. In the interval 190–155 Ma, deformation began along parts of the northern circum-Pacific. This deformation closed marine basins and accreted offshore arc systems to the continental margin. Initial Jurassic deformation along the margin was apparently arrested in the Bathonian in northeastern Russia and in Alaska, where basinal sedimentation resumed. In Canada and the United

States, however, Jurassic deformation continued into the time interval 160–150 Ma, and continental-crust and overlying continental-margin sediments were overthrust by accreted terranes and thickened by thrust faulting within the arc and backarc regions. Major Jurassic plutonic belts were intruded in the time interval 190–155 Ma along the Canadian and United States continental margins.

During the interval 155–125 Ma, widespread accretion of arc terranes and deformation by thrust faulting occurred in northeastern Russia together with intrusion of the Main belt granites of the Verkhoyansk-Kolyma orogen and the granites of the Koni-Murgal arc trend. Thrust faulting in the Brooks Range, Alaska, emplaced ophiolitic rocks on the continental shelf and imbricated shelfal successions during arc collision. Deformation in northern Alaska and along the Chukotka Peninsula was possibly linked to the rotation and southward motion of the Arctic Alaska–Chukotka plate during the formation of the Canada Basin, but coeval events farther south in Russia must have been related to Pacific plate interactions. In contrast, the same 155–125 Ma time span was largely an interval of tectonic and magmatic quiescence along the Canadian and U.S. Cordilleran margin. The magmatic lull in Canada took place from ca. 140 to 125 Ma and in the western United States, from ca. 150 to 120 Ma (Armstrong, 1988; Armstrong and Ward, 1993). Deformation in Russia and Alaska during this same time interval ceased ca. 120 Ma. This is well documented by the overlap of terrestrial sedimentary and volcanic sequences as well as by cross-cutting granitic rocks. Thereafter, sedimentary basins were developed locally on the Chukotka Peninsula and deeper-water turbidite basins developed across central Alaska.

Magmatism and deformation in Alaska and Russia since ca. 125 Ma are inferred to have occurred in conjunction with localized extensional tectonism, particularly in northern and central Alaska, the Bering Strait region, and parts of the Chukotka Peninsula, where metamorphic gneiss domes developed. Beginning ca. 120 Ma and continuing into the Late Cretaceous, the Okhotsk-Chukotsk volcanic belt and associated plutons developed along the length of the northeastern Russian continental margin. The Okhotsk-Chukotsk volcanic and plutonic belt is an unusually broad belt and is little deformed, and thus probably formed in a neutral to weakly extensional tectonic regime. Magmatism was also voluminous in the Canadian and U.S. Cordillera during the interval 125–70 Ma, but in contrast to Alaska and Russia, continued shortening of continental crust occurred during this magmatism, forming the Rocky Mountain and Sevier foreland fold and thrust belts. Although we do not discuss younger time intervals in this chapter, it appears that the pattern of magmatism and deformation developed in the Late Cretaceous along the northern circum-Pacific margin persisted into the Tertiary with contrasting tectonic regimes on either side of the North Pacific (Figs. 1 and 7). The Laramide orogeny was the most important shortening event to affect the U.S. and Canadian Cordillera and easternmost Alaska during the early Tertiary. North-south shortening in eastern Alaska changed to transtension and then north-south extension moving west toward the Bering Shelf region. This deformation was accompanied by a southward jump in the locus of subduction from the Bering Shelf to the present-day Aleutians in the Eocene, trapping oceanic crust in the backarc region (Scholl et al., 1975; Marlow et al., 1994). In northeastern Russia, development of the accretionary complexes of Kamchatka occurred outboard of the Sea of Okhotsk, which possibly formed by rifting in the early Tertiary (Hourigan et al., 1999; Mann, 1998).

ACKNOWLEDGMENTS

The Continental Dynamics Program, National Science Foundation, award EAR-93-17087 funded part of the research presented in this chapter. We are grateful to the Exxon Exploration Company, Arco Oil and Gas Company, and Placer Dome Exploration, Inc., for support of the Russian-Alaska Tectonics Research Consortium at Stanford University, which provided critical ongoing support for the collaboration and information exchange needed to complete this chapter.

REFERENCES CITED

Akinin, V.V., and Kotlyar, I.N., 1997, GEOCHRON: A computer database of isotopic dating of minerals, rocks and ores from the North-East of Russia, *in* Byalobzheskiy, S.G., and Belyi, V.F., eds., Magmatism and mineralization in north-eastern Russia: Magadan, Russia, Northeast Research Institute, Far East Branch, Russian Academy of Sciences, p. 202–210.

Alekseev, E.S., 1982, Main features of the development and structure of the southern Koryak highland: Geotectonics, v. 1, p. 85–96 (in Russian).

Allmendinger, R.W., 1991, Fold and thrust tectonics of the western United States exclusive of the accreted terranes, *in* Burchfiel, B.C., Lipman, P.W., and Zoback, M.L., eds., The Cordilleran Orogen: Conterminous U.S.: Boulder, Colorado, Geological Society of America, Geology of North America, v. G-3, p. 583–608.

Amato, J.A., and Wright, J.E., 1997, Potassic mafic magmatism in the Kigulaik gneiss dome, northern Alaska: A geochemical study of arc magmatism in an extensional tectonic setting: Journal of Geophysical Research, v. 102, p. 8065–8084.

Andreeva, N.V., 1998, Estimation of the depth of the Magadan pluton by virtue of the hornblende geobarometer, *in* Byalobzheskiy, S.G., and Belyi, V.F., eds., Magmatism and mineralization in north-eastern Russia: Magadan, Russia, Northeast Research Institute, Far East Branch, Russian Academy of Sciences, p. 192–203.

Andreeva, N.V., Davydov, I.A., Lyuskin, A.D., 1998, The main stage of intrusive magmatism in the northern Sea of Okhotsk region according to isotope-based dating, *in* Byalobzheskiy, S.G., and Belyi, V.F., eds., Magmatism and mineralization in north-eastern Russia: Magadan, Russia, Northeast Research Institute, Far East Branch, Russian Academy of Sciences, p. 175–191.

Andrianova, V.A., and Andrianov, V.N., 1970, Nekotorye novye dannye o vulcanisme na rubezhe permskogo i triasovogo periodov v oblasti Verkhoyanskoy geosynklinali [Some new data on volcanism during the latest Permian to earliest Triassic timespan in the Verkhoyansk geosyncline], *in* Materialy po geologii i poleznym iskopaemym Yakutskoy ASSR [Contributions to geology and mineral deposits of the Yakut ASSR], Issue XVI: Yakutsk, Yakutskoe knizhnoe isdatel'stvo [Yakutsk Publishing House], p. 137–145.

Armstrong, R.L., 1988, Mesozoic and early Cenozoic magmatic evolution of the Canadian Cordillera, *in* Clark, S.P., Burchfiel, B.C., and Suppe, J., eds.,

Processes in continental lithospheric deformation: Boulder, Colorado, Geological Society of America Special Paper 218, p. 55–110.

Armstrong, R.L., and Ward, P.L., 1991, Evolving geographic patterns of Cenozoic magmatism in the North American Cordillera: The temporal and spatial association of magmatism and metamorphic core complexes: Journal of Geophysical Research v. 96, p. 13201–13224.

Armstrong, R.L., and Ward, P.L., 1993, Late Triassic to earliest Eocene magmatism in the North American Cordillera: Implications for the Western Interior basin, *in* Caldwell, W.G.E., and Kauffman, E.G., eds., Evolution of the Western Interior basin: Geological Association of Canada Special Paper 39, p. 49–72.

Arth, J.G., Criss, R.E., Zmuda, C.C., Foley, N.K., Patton, W.W. Jr., and Miller, T.P., 1989a, Remarkable isotopic and trace element trends in potassic through sodic Cretaceous plutons of the Yukon-Koyukuk basin, Alaska, and the nature of crustal lithosphere beneath the Koyukuk terranes: Journal of Geophysical Research v. 94, p. 15957–15968.

Arth, J.G., Zmuda, C.C., Foley, N.K., Criss, R.E., Patton, W.W. Jr., and Miller, T.P., 1989b, Isotopic and trace element variations in the Ruby batholith, Alaska, and the nature of the deep crust beneath the Ruby and Angayucham terranes: Journal of Geophysical Research, v. 94, p. 15941–15955.

Barker, F., 1994, Some accreted volcanic rocks of Alaska and their elemental abundances, in Plafker, G., and Berg, H.C., eds., The Geology of Alaska: Boulder, Colorado, Geological Society of America, Geology of North America, v. G-1, p. 555–587.

Barker, F., Plafker, G., and Brown, A.S., 1985, Karmutsen Formation, Queen Charlotte Island and Vancouver Island: An arc-rift ferrotholeiite of Wrangellia: Geological Society of America, Cordilleran Section, Abstracts with Programs v. 17, no. p. 340.

Belyi, V.F., 1977a, Structurno-Formatsionaya Karta Okhotsk-Chukotskovo Volcanogennovo Poyasa: Northeast Research Institute, Far East Branch, Russian Academy of Sciences, scale 1:1 500 000.

Belyi, V.F., 1977b, Stratigraphy and structure of the Okhotsk-Chukotka volcanic belt: Moscow, Nauka, 172 p.

Belyi, V.F., 1978, Formations and tectonics of the Okhotsk-Chukotka volcanic belt: Moscow, Nauka, 216 p.

Belyi, V.F., 1994, Geologiia Okhotsk-Chukotskovo Volcanogennovo Poyasa: Magadan, Russsia, Northeast Research Institute, Far East Branch, Russian Academy of Sciences, 76 p.

Belyi, V.F., 1997, On the problem of mid Cretaceous phytostratigraphy and paleofloristics of Northeastern Asia: Stratigraphy and Geologic Correlation, v. 5, no. 2, p. 51–59 (in Russian).

Bergman, S.C., Decker, J., and Talbot, J., 1995, Upper Cretaceous tephra deposits, Canning River area, North Alaska, Geological Society of America Abstracts with Programs, v. 27, no. 5, p. 5.

Bering Strait Geologic Field Party, 1997, Koolen metamorphic complex, northeastern Russia: Implications for the tectonics evolution of the Bering Strait region: Tectonics, v. 16, p. 713–729.

Boak, J.M., Turner, D.L., Henry, D.J, Moore, T.E., and Wallace, W.K., 1987, Petrology and K-Ar ages of the Misheguk Igneous sequence: An allochthonous mafic and ultramafic complex and its metamorphic aureoloe, western Brooks Range, Alaska, *in* Tailleur, I., and Weimer, P., eds., Alaskan North Slope Geology, v. 2, Pacific Section, Society of Economic Paleontologists and Mineralogists, Bakersfield, California, and Anchorage, Alaska, p. 737–745.

Brew, D.A., 1994, Latest Mesozoic and Cenozoic magmatism in southeastern Alaska, *in* Plafker, G., and Berg, H.C., eds., The Geology of Alaska: Boulder, Colorado, Geological Society of America, Geology of North America, v. G-1, p. 621–656.

Bychkov, Yu.M., 1995, Provinces of the Marine Boreal Upper Triassic, *in* Proceedings of the International Conference on Arctic Margins, Magadan, Russia, 1994, p. 36–42.

Bychkov, Yu.M., and Gorodinsky, M.E., 1992, Comparative geology of northern Chukotka and the northern Canadian Cordillera, Proceedings, International Conference on Arctic Margins, Anchorage, Alaska, 1992, p. 49–53.

Chen, J.H., and Moore, J.G., 1979, Late Jurassic Independence dike swarm in eastern California: Geology, v. 7, p. 129–133.

Chen, J.H., and Moore, J.G., 1982, Uranium-lead isotopic ages from the Sierra Nevada batholith, California: Journal of Geophysical Research, v. 87, p. 4761–4784.

Dudkinskiy, D.V., Efremov, S.V., Kozlov,V.D., et al., 1992, Geokhimicheskie osobennosti i resultaty Rb-Sr dtirovaniya redkometal'nykh granitoidov vostochnogo poberezh'ya Chaunskoy guby [Geochemistry features and Rb-Sr ages of tin-bearing granites on the Chaun Bay east coast]: Doklady Akademii Nauk (Rossiya), v. 325, no. 5, p. 1039–1043.

Dusel-Bacon, C., 1994, Metamorphic history of Alaska, *in* Plafker, G., and Berg, H.C., eds., Geology of Alaska: Boulder, Colorado, Geological Society of America, The Geology of North America, v. G-1, p. 495–533.

Engebretson, D.C., Cox, A., and Gordon, R.G., 1985, Relative motions between oceanic and continental plates in the Pacific Basin: Geological Society of America Special Paper 206, 59 p.

Filatova, N.I., 1974, Formations and tectonics of the Okhotsko-Chukotskiy volcanic belt in the Penzhina River Basin: Geotectonics, no. 2, p. 110–118.

Filatova, N.I., 1988, Circum-oceanic volcanic belts: Moscow, Nedra, 264 p. (in Russian).

Filatova, N.I., 1995, The history of the Cretaceous environments of the northeastern Asian continental margin, Russia, Island-arc:Victoria, Blackwell Scientific Publications, v. 4, no. 2, p. 128–139.

Friedman, R.M., and Armstrong, R.L., 1995, Jurassic and Cretaceous geochronology of the southern Coast Belt, British Columbia, 49°–51°N, *in* Miller, D.M., and Busby, C., eds., Jurassic magmatism and tectonics of the North American Cordillera: Boulder, Colorado, Geological Society of America Special Paper 299, p. 95–139.

Gabrielse, H., and Yorath, C.J., 1991, Tectonic synthesis, *in* Gabrielse, H., and Yorath, C.J., eds., Geology of the Cordilleran Orogen in Canada: Geological Survey of Canada, Geology of Canada, no. 4, p. 677–705.

Ged'ko, M.I., 1988, The Late Jurassic Uyandina-Yasachnaya island-arc (northeast USSR): Geotectonics, v. 22, p. 263–273.

Gelman, M.L.,1961, Amfibolitovaya fatsiya mezozoyskogo metamorfizma v nizov'yakh r. Kolymy [Mesozoic amphibolite facies metamorphism in the Lower Kolyma River], *in* Materialy po geologii i poleznym iskopaemym Severo-Vostoka SSSR [Contributions to geology and mineral deposits of the North-East USSR]: Magadan, Russia, Magadanskoe knizhnoe izdatel'stvo [Magadan publishing house], p.105–129.

Gelman,M.L., 1962, Glubinnaya fatsiya melovykh granitoidov v Anyuyskoy zone [Deep seated granitoides in the Anuy zone], *in* Materialy po geologii i poleznym iskopaemym Severo-Vostoka SSSR [Contributions to geology and mineral deposits of the North East USSR]: Magadan, Russia, Magadanskoe knizhnoe izdatel'stvo [Magadan publishing house], Issue 16, p. 213–233.

Gelman, M.L., 1963, Triasovaya diabasovaya formatisiya Anyuiskoi zony (Chukotka) [Triassic diabase association in the Anuy fold zone (Chukotka)]: Geologiya i Geofizica, no. 2, p. 127–134.

Gelman, M.L., 1995, Phanerozoic granite-cored domes of the Siberian Northeast, *in* Simakov, K.V., and Thurston, D.K., eds., Proceedings of the International Conference on Arctic Margins (Magadan, Russia, September 1994), Magadan: SVNTs DVO RAN, p. 203– 209.

Gordon, R.G., Cox, A., and O'Hare, S., 1984, Paleomagnetic Euler poles and the apparent polar wander and absolute motion of North America since the Carboniferous: Tectonics, v. 3, no. 5, p. 499–537.

Gorodinsky, M.E., 1980, Geological map of northeastern Russia: Ministry of Geology, scale 1:1 500 000.

Grantz, A., May, S.D., Taylor, P.T., and Lawver, L.A., 1990, Canada Basin, *in* Grantz, A., Johnson, L., and Sweeney, J.F., eds., The Arctic Ocean Region: Boulder, Colorado, Geological Society of America, Geology of North America, v. L, p. 379–402.

Grantz, A., May, S.D., and Hart, P.E., 1994, Geology of the Arctic continental margin of Alaska, *in* Plafker, G., and Berg, H.C., eds., The Geology of Alaska: Boulder, Colorado, Geological Society of America, Geology of North America, v. G-1, p. 17–48.

Harper, G.D., and Wright, J.E., 1984, Middle to Late Jurassic tectonic evolution of the Klamath Mountains, California-Oregon: Tectonics, v. 3 p. 759–772.

Hourigan, J., Miller, E.L., and Akinin, V.V., 1999, Land-based constraints on the tectonic setting of the margin of Russia prior to the formation of the Sea of Okhotsk: Eos (Transactions, American geophysical Union), v. 80, no. 46, p. F946.

Hudson, T., 1983, Calc-Alkaline plutonism along the Pacific rim of southern Alaska, *in* Roddick, J.A., ed., Circum-Pacific plutonic terranes: Boulder, Colorado, Geological Society America Memoir 159, p. 159–169.

Hudson, T.L., 1994, Crustal melting events in Alaska, *in* Plafker, G., and Berg, H.C., eds., The Geology of Alaska: Boulder, Colorado, Geological Society of America, Geology of North America, v. G-1, p. 657–670.

Ivanov, O.N., and Milov, A.P., 1975, Diabasovaya formatsiya Chukotskoy skladchatoy sistemy i ee svyaz's basitovym magmatismom severnogo sektora Tichookeanskogo podvizhnogo poyasa [The diabase association in the Chukchi fold system and its relation to the basic magmatism in the northern portion of the Pacific Mobile Belt], *in* Shatalov, E.T., ed., Magmatism Severo-Vostochnoy Azii [Magmatism of North East Asia], part 2, p. 274–278.

Jones, D.L., Silberling, N.J., and Hillhouse, J., 1977, Wrangellia: A displaced terrane in northwestern North America: Canadian Journal of Earth Sciences, v. 14, p. 2565–2577.

Journeay, J.M., and Friedman, R.M., 1993, The Coast Belt thrust system: Evidence of Late Cretaceous shortening in Southwest British Columbia: Tectonics, v. 12, no. 3, p. 756–775.

Karish, C.R., Miller, E.L., and Sutter, J.F., 1987, Mesozoic tectonic and magmatic history of the central Mojave Desert: Arizona Geological Digest, v. 18, p. 15–32.

Kelley, S., Spicer, R.A., and Herman, A.B., 1999, New $^{40}Ar/^{39}Ar$ dates for Cretaceous Chauna Group Tephra, Northeastern Russia, and their implications for the geologic history and floral evolution of the North Pacific region: Cretaceous Research, v. 20, no. 1, p. 97–106.

Lawton, T.F., and Trexler, J.H., Jr., 1991, Piggyback basin in the Sevier orogenic belt, Utah: Implications for development of the thrust wedge: Geology, v. 19, p. 827–830.

Lawver, L.A., and Scotese, C.R., 1990, A review of tectonic models for the evolution of the Canadian basin, *in* Grantz, A., Johnson, G.L., and Sweney, J.F., eds., The Arctic Ocean region: Boulder, Colorado, Geological Society of America, Geology of North America, v. L, p. 593–618.

Layer, P.W., Parfenov, L.M., and Trunilina, U.A., 1995, Age and tectonic significance of granitic belts within the Verkhoyansk fold and thrust belt, Yakutia, Russia, Abstracts with Programs, Geological Society of America Cordilleran Section Meeting, Fairbanks, Alaska, v. 27, no. 5, p. 60.

Lipman, P.W., 1992, Magmatism in the Cordilleran United States: Progress and problems, *in* Burchfiel, B.C., Lipman, P.W., and Zoback, M.L., eds., The Cordilleran Orogen: Conterminous U.S.: Boulder, Colorado, Geological Society of America, Geology of North America, v. G-3, p. 481–514.

Lychagin, P.P., 1975, Ranneyurskie shchelochnye porody na Omolonskom massive [Early Jurassic alkali rocks in the Omolon massif], *in* Materialy po geologii i poleznym iskopaemym Severo-Vostoka SSSR [Contributions to geology and mineral deposits of the North East USSR]: Magadan, Russia, Magadanskoe knizhnoe izdatel'stvo [Magadan publishing house], Issue 25, p. 62–69.

Lychagin, P.P., 1982, Shchelochnye basity Severo-Vostjka SSSR [Alkali basic rocks on the USSR North-East]: Tichookeanskaya geologiya, no. 6, p. 85–93.

Lychagin, P.P., Byalobzheskiy, S.G., Kolyasnikov, Yu.A., Korago, E.A., and Likman, V.B., 1991a, Geologiya i petrologiya Gromadnensko-Vurguveemskogo gabbronoritovogo massiva (Yuzhno-Anyuyskaya skladchataya zona) [Geology and petrology of gabbronorite compose the Gromada-Vurguveem pluton in the South Anuy folded zone]: Magadan, Russia, SVKNII DVO RAN, 47 p.

Lychagin, P.P., Byalobzheskiy, S.G., Kolyasnikov, Yu.A., and Likman, V.B., 1991b, Magmaticheskaya istoriya Yuzhno-Anyuyskoy skladchatoy zony [Magmatic evolution of the South Anuy folded zone], *in* Sidorov, A.A., and Milov, A.P., eds. Geologiya zony perekhoda continent-okean na Severo-Vostoke Azii [Geology of the continent-ocean transition zone in North-East Asia]: Magadan, Russia, SVKNII DVO RAN, p. 140–157.

Mann, P., 1998, Geosat gravity anomalies of the Sea of Okhotsk: Opening by long term backarc extension behind the Kamchatka-Kuril arc system? Proceedings: III International Conference on arctic Margins (ICAM III), Celle, Germany, October 12–16, 1998.

Marlow, M.S., Cooper, A.K., and Fisher, M.A., 1994, Geology of the eastern Bering Sea continental shelf, *in* Plafker, G., and Berg, H.C., eds., The Geology of Alaska: Boulder, Colorado, Geological Society of America, Geology of North America, v. G-1, p. 271–284.

May, S.A., and Butler, R.F., 1986, North American Jurassic apparent polar wander: Implications for plate motion, paleogeography and Cordilleran tectonics: Journal of Geophysical Research, v. 91, p. 11519–11544.

Mayfield, C.F., Taillerur, I.L., and Ellersieck, I., 1988, Stratigraphy, structure and palinspastic synthesis of the western Brooks Range, northwest Alaska, *in* Gryc, O., ed., Geology and exploration of the National Petroleum Reserve in Alaska, 1974–1982: U.S. Geological Survey Professional Paper 1399, p. 143–186.

Miller, C.F., and Barton, M.D., 1990, Phanerozoic plutonism in the Cordilleran Interior, U.S.A., *in* Mahlburg, S.K., and Rapela, C.W., eds., Plutonism from Antarctica to Alaska: Boulder, Colorado, Geological Society of America Special Paper 241, p. 213–231.

Miller, C.F., and Bradfish, L., 1980, An inner Cordilleran belt of muscovite-bearing plutons: Geology, v. 8, p. 412–416.

Miller, E.L., and Gans, P.B., 1989, Cretaceous crustal structure and metamorphism in the hinterland of the Sevier thrust belt, western U.S. Cordillera: Geology, v. 17, p. 59–62.

Miller, E.L., and Hudson, T.L., 1991, Mid-Cretaceous extensional fragmentation of a Jurassic–Early Cretaceous compressional orogen, Alaska: Tectonics, v. 10, p. 781–796.

Miller, T.P., 1972, Potassium-rich alkaline intrusive rocks of western Alaska: Geological Society of America Bulletin, v. 83, p. 2111–2128.

Miller, T.P., 1989, Contrasting plutonic rock suites of the Yukon-Koyukuk basin and the Ruby Geanticline, Alaska: Journal Geophysical Research, v. 94, p. 15969–15987.

Miller, T.P., 1994, Pre-Cenozoic plutonic rocks in mainland Alaska, *in* Plafker, G., and Berg, H.C., eds., The Geology of Alaska: Boulder, Colorado, Geological Society of America, Geology of North America, v. G-1, p. 535–554.

Moll-Stalcup, E.J., 1994, Latest Cretaceous and Cenozoic magmatism in mainland Alaska, *in* Plafker, G., and Berg, H.C., eds., The Geology of Alaska: Boulder, Colorado, Geological Society of America, Geology of North America, v. G-1, p. 589–619.

Monger, J.W.H., 1991, The geology of the Canadian Cordillera, *in* Smith, P.L., ed., Field guide to the paleontology of southwestern Canada, Vancouver, British Columbia: Geological Association of Canada, Cordilleran Section, p. 6–18.

Monger, J.W.H., and Nokleberg, W.J., 1995, Evolution of the northern American Cordilla: Generation, fragmentation, displacement and accretion of successive North American plate-margin arcs, *in* Coyner, A.R., ed., Geology and Ore deposits of the American Cordillera, Symposium Proceedings, Reno, Nevada, Geological Society of Nevada, 1995, p. 1133–1152.

Moore, G.W., 2000, Geologic map of the Circum-Pacific Region, *in* Moore, G.W., Arctic Sheet: U.S. Geological Survey Circum-Pacific Map Series, Map CP-XX, scale 1:10 000 000.

Moore, T.E., Wallace, W.K., Bird, K.J., Karl, S.M., Mull, C.G., and Dillon, J.T., 1994, Geology of Northern Alaska, *in* Plafker, G., and Berg, H.C., eds., The Geology of Alaska: Boulder, Colorado, Geological Society of America, Geology of North America, v. G-1, p. 49–140.

Muller, J.E., 1977, Evolution of the Pacific Margin, Vancouver Island, and adjacent regions: Canadian Earth Sciences, v. 14, no. 9, p. 2062–2085.

Murphy, D. C., van der Heyden, P., Parrish, R.R., Klepacki, D.W., McMillan, W., Struik, L.C., and Gabites, J., 1995, New geochronological constraints on Jurassic deformation of the western edge of North America, southeastern Canadian Cordillera, *in* Miller, D., and Busby, C., eds., Jurassic magmatism and tectonics of the North American Cordillera: Boulder, Colorado, Geological Society of America Special Paper 299, p. 159–171.

Natapov, L.M., and Surmilova, E.P., eds., 1986, Geologic map of the USSR: Leningrad, scale 1:1 000 000.

Nekrasov, G.E., 1976, Tectonics and magmatism of the Taigonos Peninsula and Northwest Kamchatka: Moscow, Nauka, 160 p.

Newberry, R.J., Layer, P.W., Gans, P.B., Goncharov, V.I., Goryachev, N.A., and Voroshin, S.V., 1997, Preliminary revised chronology of magmatism and mineral deposition in northeastern Russia (Magadan region) and the Seward Peninsula, Alaska, based on $^{40}Ar/^{39}Ar$ dating, *in* Gold mineralization and granitoid magmatism of the North Pacific: Magadan, Russia, p. 302–304.

Nokleberg, W.J., Parfenov, L.M., Monger, J.W.H., Baranov, B.V., Byalobzhesky, S.G., Bundtzen, T.K., Feeney, T.D., Fujita, K., Gordey, S.P., Grantz, A., Khanchuk, A.I., Natal'in, B.A., Natapov, L.M., Norton, I.O., Patton, W.W., Jr., Plafker, G., Scholl, D.W., Sokolov, S.D., Sosunov, G.M., Stone, D.B., Tabor, R.W., Tsukanov, N.V., Vallier, T.L., and Wakita, K., 1994, Circum-North Pacific tectono-stratigraphic terrane map: U.S. Geological Survey Open-File Report 94–714, scale 1:5 000 000, 2 sheets, and scale 1:10 000 000, 2 sheets, 211 p.

Nokleberg, W.J., Parfenov, L.M., Monger, J.W.H., Norton, I.O., Khanchuk, A.I., Stone, D.B., Scholl, D.W., and Fujita, K., 1998, Phanerozoic tectonic evolution of the circum-north Pacific: U.S. Geological Survey Open-File Report 98–754, p. 125.

Oxman, V.S., Parfenov, L.M., Prokopiev, A.V., Timofeev, V.F., Tretyakov, F.F., Nedosekin, Y.D., Layer, P.W., and Fujita, K., 1995, The Chersky Range ophiolite belt, Northeast Russia: Journal of Geology, v. 103 p. 539–557.

Page, B.M., and Engebretson, D.C., 1984, Correlation between the geologic record and computed plate motions for central California: Tectonics, v. 3, p. 133–156.

Pallister, J.S., Budhan, J.R., and Murchey, B.L., 1989, Pillow basalts of the Angayucham terrane: Oceanic plateau and island-arc crust accreted to the Brooks Range: Journal of Geophysical Research, v. 94, p. 15901–15924.

Palymskaya, Z.A., and Palymskiy, B.F., 1975, Pozdnepaleozoyskiy intruzivnyy magmatizm vostochnoy chasti Anyuysko-Oloyskogo bloka (Zapadnaya Chukotka) [Late Paleozoic intrusions in the east portion of the Anuy-Oloy terrane, West Chukotka], *in* Shatalov, E.T., ed., Magmatism Severo-Vostochnoy Azii [North-East Asia magmatism], part 2, p. 51–58.

Paraketsov, K.V., and Paraketsova, G.I., 1989, Stratigraphy and fauna of Upper Jurassic and Lower Cretaceous deposits in northeast USSR: Moskow, Nedva, 297 p.

Parfenov, L.M., 1984, Continental margins and island-arcs of the Mesozoids, northeast Asia: Novosibersk, Nauka, 192 p.

Parfenov, L.M., 1991, Tectonics of the Verkhoyansk-Kolyma Mesozoides in the context of plate tectonics: Tectonophysics, v. 199, p. 319–342.

Parfenov, L.M., Nataopov, L.M., Sokolov, S.D., and Tsukanov, N.V., 1993, Terrane analysis and accretion in North-East Asia: The Island-Arc, no. 2, p. 35–53.

Patton, W.W., and Box, S.E., 1989, Tectonic setting of the Yukon-Koyukuk basin and its borderlands, western Alaska: Journal Geophysical Research, v. 94, p. 15807–15820.

Plafker, G., and Berg, H.C., 1994, Overview of the geology and tectonic evolution of Alaska, *in* Plafker, G., and Berg, H.C., eds., The Geology of Alaska: Boulder, Colorado, Geological Society of America, Geology of North America, v. G-1, p. 989–1021.

Price, R.A., 1981, The Cordilleran foreland thrust and fold belt in the southern Canadian Rocky Mountains, *in* McClay, K.R., and Price, N.J., eds, Thrust and nappe tectonics, Geological Society [London] Special Publication 9, p. 427–448.

Quinn, M.J., Wright, James E., and Wyld, S.J., 1997, Happy Creek igneous complex and tectonic evolution of the early Mesozoic arc in the Jackson Mountains, Northwest Nevada: Geological Society of America Bulletin, v. 109 (4), p. 461–482.

Rowe, H., 1998, Petrogenesis of plutons and hypabyssal rocks of the Bering Strait Region, Chukotka, Russia [M.S. thesis]: Rice University, Houston, Texas, 90 p.

Rubin, C.M., Miller, E.L., and Toro, J., Deformation of the northern Circum-Pacific Margin: Variations in tectonic style and plate tectonic implications: Geology, v. 23, p. 897–900.

Saleeby, J.B., and Busby-Spera, C., 1992, Early Mesozoic tectonic evolution of the western U.S. Cordillera, *in* Burchfiel, B.C., Lipman, P.W., and Zoback, M.L., eds., The Cordilleran Orogen: Conterminous U.S.: Boulder, Colorado, Geological Society of America, Geology of North America, v. G-3, p. 107–168.

Scholl, D.W., Buffington, E.C., and Marlow, M.S., 1975, Plate tectonics and the structural evolution of the Aleutian–Bering Sea region, *in* Forbes, F.B., ed., Contributions to the geology of the Bering Sea basin and adjacent regions: Geological Society of America Special Paper 151, p. 1–31.

Silberling, N.J., 1973, Geologic events during Permian-Triassic time along the Pacific margin of the United States, *in* Logan, A., and Hill, L.V., eds., The Permian and Triassic systems and their mutual boundary: Calgary, Canada, Canadian Society of Petroleum Geologists Memoir 2, p. 345–362.

Smith, D.L., Miller, E.L., Wyld, S.J., and Wright, J.E., 1993, Progression and timing of Mesozoic crustal shortening in the northern Great basin, western U.S.A, *in* Dunne, G.C., and McDougall, K.A., Mesozoic paleogeography of the western United States, Field Trip Guidebook-Pacific Section, Society of Economic Paleontologists and Mineralogists, v. 71, p. 389–405.

Sokolov, S.D., 1988, Accretion tectonics of the Koryak-Kamchatka segment of the Pacific Belt: Moscow, Geological Institute, AN SSSR, 444 p. (in Russian).

Sokolov, S.D., 1992, Accretionary tectonics of the Koryak-Chukchi segment of the Pacific Belt, Geologicheskiy Institut (Moscow), v. 479, 182 p.

Sokolov, S.D., Didenko, A.N., Grigoriev, M.V., Akeksyutin, M.V., Bondarenko, G.E., and Kylov, K.A., 1997, Paleotectonic reconstructions for northeast Russia: Problems and uncertainties: Geotectonics, v. 31, no. 6, p. 498–515.

Souther, J.G., 1991, Volcanic regimes, *in* Gabrielse, H., and Yorath, C.J., eds., Geology of the Cordilleran Orogen in Canada: Boulder, Colorado, Geological Society of America, Geology of North America, v. G-2, p. 457–490.

Terekhov, M.I., Lychagin, P.P., and Merzlyakov, V.M., et al., 1984, Ob'yasnitel'naya zapiska k geologicheskoy karte mezhdurech'ya Sugoya, Korkodona, Oloya i Gizhigi masshtaba [Geology of the Sugoy, Korkodon, Omolon, Oloy and Giziga Rivers basins area map]: Magadan, Russia, SVKNII DVNTs AN SSSR 144 p. scale 1: 500 000.

Tynankergav, G.A., and Bychkov, Yu.M., 1987, Kremnisto-vulkanogenno-terrigennye verkhnetriassovye otlozheniya zapada Chukotskogo poluostrova [A Late Triassic sequence consisting of cherts, volcanics, and terrigenous sediments on the West Chukchi Peninsula]: Doklady Akademii Nauk SSSR, v. 296, no. 3, p. 698–700.

Uyeda, S., 1987, Chilean vs. Mariana type subduction zones with remarks on arc volcanism and collision tectonics: Circum-Pacific orogenic belts and the evolution of the Pacific Ocean Basin: Geodynamics Series, American Geophysical Union, v. 18, p. 1–7.

Uyeda, S., and Kanamori, H., 1979, Backarc opening and the mode of subduction: Journal of Geophysical Research, v. 84, p. 1049–1061.

Vikhert, A.V., 1957, Triassovoye diabasy zapadnogo sklona Zapadnogo Verkhoyn'ya [Triassic diabase on the west slope of the west Verchoyan region]: Doklady Akademii Nauk SSSR, v. 114, no. 1.

Vorob'yev, Y.Y., 1985, Triasovo-yurskiy vulkanizm pova Koni (Severo-Vostok SSSR) [Triassic-Jurassic volcanism on the Koni Peninsula]: Tikhookenaskaya Geologiya [Pacific Geology], v. 1985, no. 4, p. 39–44.

Ward, P.L., 1995, Subduction cycles under western North America during the Mesozoic and Cenozoic eras, *in* Miller, D.M., and Busby, C., eds., Jurassic magmatism and tectonics of the North American Cordillera: Boulder, Colorado, Geological Society of America Special Paper 299, p. 1–46.

Wirth, K.R., Bird, J.M., Blythe, A.E., Harding, D.J., and Heizler, M.T., 1993, Age and evolution of western Brooks Range ophiolites, Alaska: Results from $^{40}Ar/^{39}Ar$ thermochronometry: Tectonics, v. 12 (2), p. 410–432.

Woodsworth, G.J., Anderson, R.G., and Armstrong, R.L., 1991, Plutonic regimes, *in* Gabrielse, H., and Yorath, C.J., eds., Geology of the Cordilleran Orogen in Canada: Boulder, Colorado, Geological Society of America, Geology of North America, v. G-2, p. 491–531.

Worral, D.M., 1991, Tectonic history of the Bering Sea and the evolution of Tertiary strike-slip basins of the Bering Shelf: Boulder, Colorado, Geological Society of America Special Paper 257, 120 p.

Wright, J.E., and Wooden, J.L., 1991, New Sr, Nd, and Pb isotopic data from plutons in the northern Great basin: Implications for crustal structure and granite petrogenesis in the hinterland of the Sevier thrust belt: Geology, v. 19, p. 457–460.

Wyld, S.J., 1991, Geology and geochronology of the Pine Forest Range, northwest Nevada: Stratigraphic, structural and magmatic history and regional implications [Ph.D. thesis], Stanford University, Stanford, California, 149 p.

Zaborovskaya, N.B., 1978, The geology of the interior zone of the Okhotsk-Chukotka volcanic belt: Moscow, Nauka, 199 p.

Zhulanova, I.L., 1974, Tectonics and development history in the northern part of the Taigonos Peninsula: Geotectonika, v. 1, no. C, p. 112–118.

Zonenshain, L.P., Kuzmin, M.I., Natapov, L.M., and Page, B.M., 1990, Geology of the USSR: A plate tectonic synthesis: Geodynamics Series, American Geophysical Union, 242 p.

MANUSCRIPT ACCEPTED BY THE SOCIETY MAY 15, 2001.

Geological Society of America
Special Paper 360
2002

Plate kinematic evolution of the present Arctic region since the Ordovician

Lawrence A. Lawver*
Institute for Geophysics, University of Texas at Austin,
4412 Spicewood Springs Road #600, Austin, Texas 78759-8500, USA
Arthur Grantz*
Department of Geological and Environmental Sciences,
Stanford University, Stanford, California 94305, USA
Lisa M. Gahagan*
Institute for Geophysics, University of Texas at Austin,
4412 Spicewood Springs Road #600, Austin, Texas 78759-8500, USA

ABSTRACT

The Laurentian, Baltic, and Siberian blocks of the modern circum-Arctic were scattered across a broad area between 10°S and 30°N during the Late Ordovician. Closure of the Iapetus Ocean during the Ordovician and Early Silurian brought Baltica and the Chukotka block in contact with Laurentia, simultaneously producing the Scandian phase of the Caledonian orogeny and creating the supercontinent of Laurussia. The assembly of Laurussia brought the cores of most of the scattered circum-Arctic landmasses into roughly their present-day relative positions. This collisional event also attached Pearya to northern Ellesmere Island, transpressionally sutured Chukotka to the general region of the present-day Canadian Arctic Islands, and attached the Seward Peninsula to Laurentia. Devonian rifting, perhaps driven by a mantle plume, formed the Vilyui basin of eastern Siberia, opened the Oimyakon oceanic basin offshore, and rotated the Siberian block away from North America and toward Baltica.

The major continental blocks that encircle the modern Arctic Ocean have migrated generally northward since the Carboniferous, although some, such as Siberia, have passed over the pole and are now moving southward. During the Jurassic and into the Early Cretaceous, the Chukotka-Barents shelf region passed over a nearly fixed Siberian Traps–Iceland hotspot. Opening of the Amerasia basin of the Arctic Ocean by seafloor spreading may have been initiated as seafloor spreading associated with continued subduction along the northeastern Pacific Rim, but additional stress produced by a hotspot converted what might have been originally orthogonal opening to rotational opening. Volcanism at the margins of the opposing plates of the newly forming Amerasia basin over the site of the fixed Siberian Traps–Iceland hotspot is thought to have created the major high-standing Alpha and Mendeleev submarine ridge system of the northern Amerasia basin. Beginning in the late Paleocene, North

*E-mails: Lawver, lawver@ig.utexas.edu; Grantz, agrantz@pacbell.net; Gahagan, plates@ig.utexas.edu.
Figures in this chapter appear in color on the CD-ROM accompanying this volume and on Plate 6A. The CD-ROM also includes an animation of the paleolocations of the arctic continental fragments from 450 Ma to present in 3-m.y. steps.

Lawver, L.A., Grantz, A., and Gahagan, L.M., 2002, Plate kinematic evolution of the present Arctic region since the Ordovician, *in* Miller, E.L., Grantz, A., and Klemperer, S.L., eds., Tectonic Evolution of the Bering Shelf–Chukchi Sea–Arctic Margin and Adjacent Landmasses: Boulder, Colorado, Geological Society of America Special Paper 360, p. 333–358.

Atlantic seafloor spreading split Greenland and the Lomonosov Ridge from Eurasia, creating the Eurasia basin of the Arctic Ocean by slow spreading along the Nansen-Gakkel Ridge that continues to the present day, and produced transtension through parts of far-eastern Siberia.

INTRODUCTION

Regional paleoreconstructions are generally of two distinct types. One type uses computer-generated rotation files to reconstruct digitized plate outlines using the relative motion between the plates, based, where possible, on dated marine magnetic anomalies, relational matching of geological features across boundaries, and estimation of paleopositions based on measured and dated paleomagnetic poles with respect to an assumed north pole where direct determination of paleoposition is not possible. The other type uses pictorial representations of changing plate shapes that do not rigorously control the relative positions of the

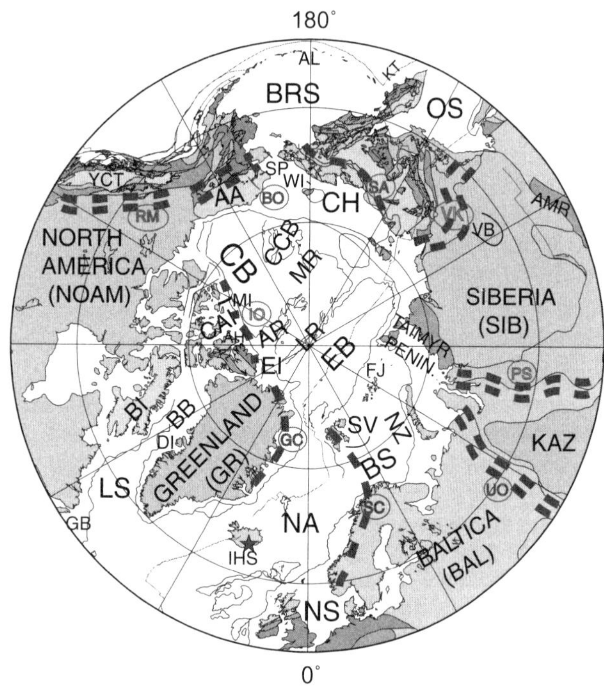

Figure 1. Polar stereographic projection of present Arctic configuration. Each of terrane pieces used in rotational database are shown color coded in Plate 6A-1. As in all figures and plates, major tectonic blocks and terranes are identified using acronyms defined in Table 1. Double-dashed thick black lines are sutures, foreland fold and thrust belts, or zones of major plate collisions. BO represents location of Brookian orogeny. PS represents closure of paleo–Asian Ocean. RM represents Rocky Mountain foreland fold and thrust belt. UO represents closure of Uralian Ocean. Verkhoyansk represents Verkoyansk foreland fold and thrust belt. Single-dashed thick black line represents suture (e.g., SA is South Anyui suture), or represents later-divided parts of Ordovician Scandian orogeny (e.g., GC is Greenland Caledonides and SC is Scandinavian Caledonides).

reconstructed pieces. In this chapter we use the first method and focus on the kinematic evolution of the present Arctic region (Fig. 1). Our database includes the positions of the major continental plates through time, particularly those defined by old cratons, as well as various other plates or blocks that may include, but are not limited to, tectonostratigraphic terranes as defined by Howell et al. (1985). The latter include fault-bounded crustal bodies defined on the basis of stratigraphy and tectonic processes involving dislocations of tens or more kilometers, island arcs, subduction or accretion complexes, and trapped ocean crust. The outlines of the plates are geological boundaries based on published literature supplemented with the inferred position of present-day ocean-continent boundaries. Proxies for the ocean-continent boundaries are determined from satellite-derived gravity signatures, as described in Lawver et al. (1999). For this study we have digitized additional plates, including some of the numerous terranes shown on the terrane map of Northeast Asia (Fig. 2) by Parfenov et al. (1993), and the Lithotectonic Terrane map of Alaska and adjacent parts of Canada (Fig. 3) by Silberling et al. (1994). With the exception of Ellesmere Island, which we have broken into tectonic slivers in order to show deformation during the Eurekan orogeny, the terranes shown in this chapter are rigid representations manipulated as polygons defined by latitude and longitude points rotated through time. In a few cases we show overlapped terranes. Following Silberling et al. (1994), we show the allochthonous Angayucham terrane (see 3a in Fig. 3) of the Ocean composite terrane obducted onto Arctic Alaska during the Jurassic to form the Arctic Alaska composite terrane. In another case, we maintain the outline of the Taimyr block (Fig. 1) through time, but represent temporal deformation and extension of the block by overlapping Taimyr with Baltica or Siberia at different times. In our reconstructions the digitized plates represented by polygons were moved about on the computer screen while maintaining their correct areal extent on an orthographic representation of the globe. When satisfied with the movement of the blocks and terranes through time, we input the poles of rotation determined by the interactive software into a rotational database (from which the figures herein were produced).

Relative plate motions for many of the marine terranes and blocks for the past 160 m.y. are now known, based on marine magnetic anomaly patterns (Cande et al., 1989; Royer et al., 1990) and tectonic lineations derived from satellite-gravity data (Sandwell and Smith, 1997). Difficulties are encountered, however, when an absolute framework is sought to describe the motion of continental blocks. Müller et al. (1993) produced a reasonable absolute framework for 130 Ma to the present for most

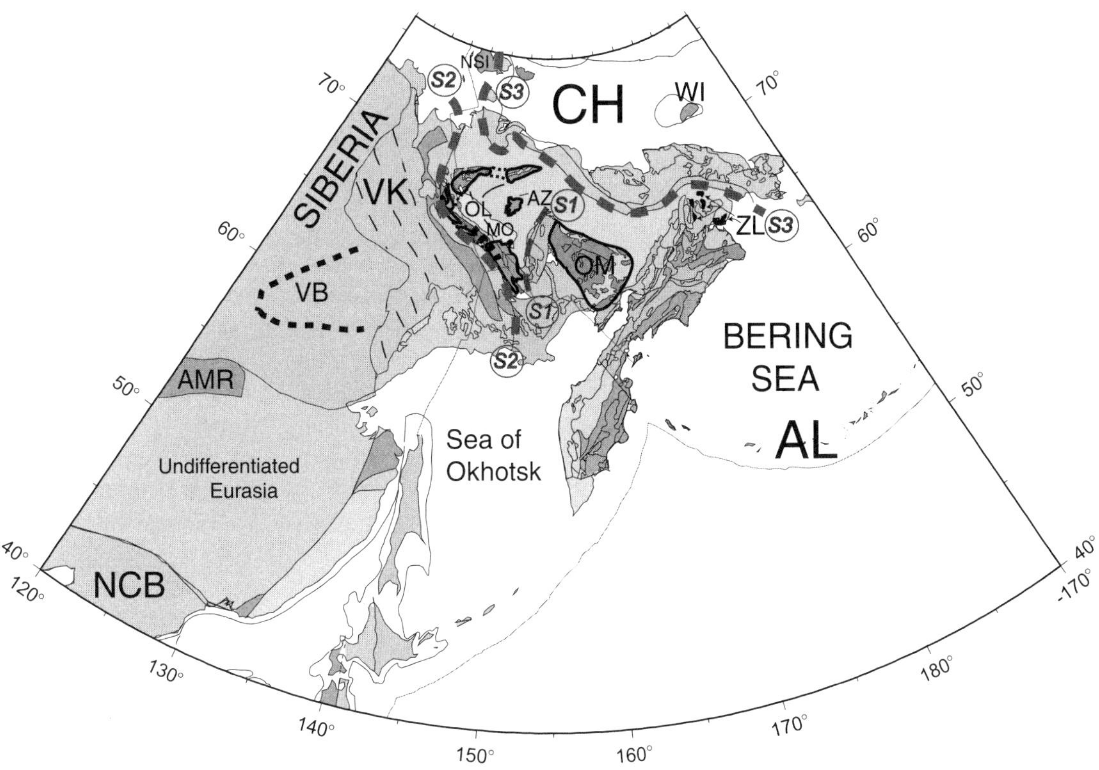

Figure 2. Polar stereographic projection of terranes of northeastern Siberia south to North China block of East Asia. Terranes of Northeast Asia were digitized from map of Parfenov et al. (1993). Abbreviations are as in Table 1. Thick single, dashed lines represent following. S1 is suture between Omolon block with Kolymian loop and Omulevka terranes. S2 is collision of Omulevka terranes and Omolon massif with Siberia resulting in Verkoyansk fold and thrust belt. S3 is South Anyui suture zone, which is one of the collisional zones between the Chukotka block and Siberia. Thick, black dashed line represents outline of Vilyui basin. Thinner dashed lines represent Verkoyansk fold and thrust area.

Indo-Atlantic continental blocks using six hotspots that were assumed to be fixed with respect to each other for the time period considered. The relative motion of the Indo-Atlantic continental blocks as a group was adjusted en masse on the surface of a sphere to best fit the dated tracks of the six hotspots. From this, Müller et al. (1993) produced an absolute framework for a region that extended from North and South America on the west to the eastern margins of Australia and Asia. They were unable, however, to reconcile continental movements with the motion of the Pacific plate because the many subduction zones rimming the Pacific region produce inexact boundary conditions. Combining relative plate motions with paleomagnetic measurements allows some constraints to be imposed on plate motions prior to 130 Ma, but it is impossible to delimit the paleolongitude of continental blocks solely with paleomagnetic data. Consequently, many Paleozoic and Mesozoic plate tectonic reconstructions lack some critical constraints.

Paleomagnetic data are used to approximate the positions of the major continental blocks for times earlier than 130 Ma, and the validity of these positions were corroborated, where possible, with known geologic relations and/or events. The paleo-magnetic poles of Van der Voo (1993) for North America are used to tie our hierarchical plate rotation model to the earth in what we refer to as an absolute plate reference frame. Paleomagnetic poles for other continental blocks, such as those of Smethurst et al. (1998) for Siberia, are rotated into reasonable agreement with the North American poles for specific times where appropriate data exist. The Smethurst et al. (1998) poles for Siberia tend to place it a few degrees north of previously published poles, causing our reconstructions to differ from those of Zonenshain et al. (1990) and Golonka (2000). Paleomagnetic poles for other blocks are primarily from Van der Voo (1993) and are rotated into as close agreement as possible with both the North American and Siberian poles. The paleomagnetic pole of Van der Pluijm et al. (1993) is used for Baltica at 450 Ma, but after Baltica collides with North America during the Scandian phase of the Caledonian orogeny its position and consequently that of Eurasia are dependent on the Van der Voo (1993) poles for North America. Nokleberg et al. (1998) compiled a table of paleomagnetic data for northeastern Russia, Alaska, and the Canadian Cordillera that they used to define many of the small terranes of the Arctic. We have not used these data directly in

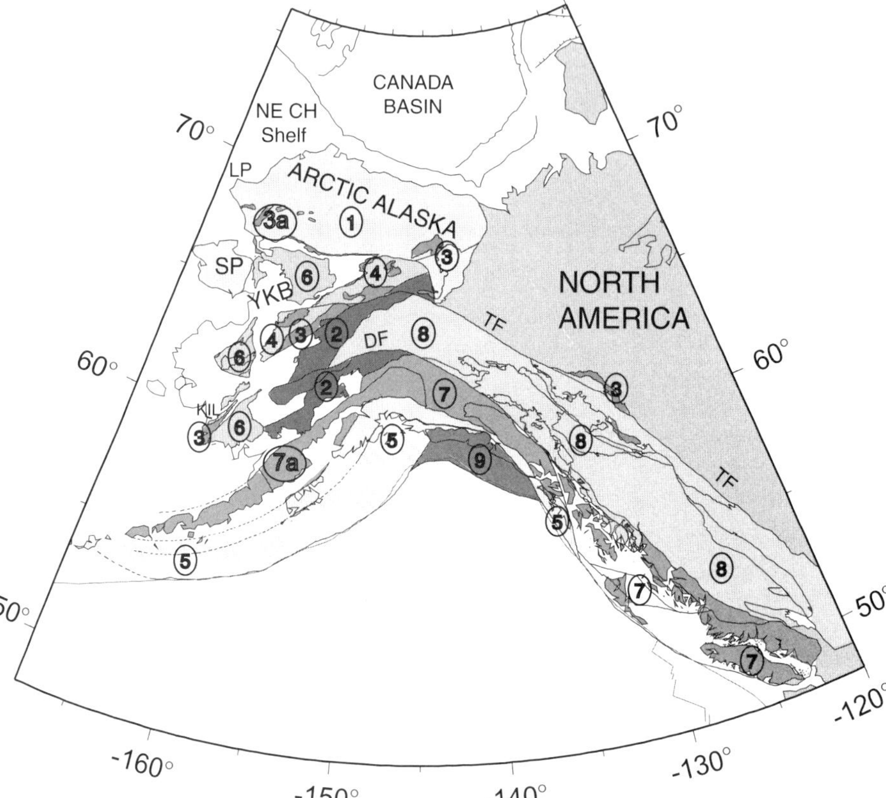

Figure 3. Polar stereographic projection of terranes of Alaska taken from Silberling et al. (1994) lithoterrane map of Alaska and adjacent Canada. Alaskan composite terranes are as identified by number in Table 2. DF is Denali fault; NE CH is Northeast Chukchi; TF is Tintina fault. Stripped part of terrane 5, Southern Margin terrane, represents continental margin where definitive data do not exist. See Table 1 for abbreviations.

this study but tried to make the motion of our small-terrane pieces consistent with the relationships that Nokleberg et al. (1998) showed in their reconstructions.

Relative motion between North America and Eurasia in the Late Cretaceous and Cenozoic is based on identified marine magnetic anomalies of Late Cretaceous to Paleogene age in the Labrador Sea (Srivastava et al., 1992) and of Cenozoic age in the Norwegian-Greenland Sea (Talwani and Eldholm, 1977). The absolute motion of Baltica (as part of Eurasia) is therefore determined directly for the Late Cretaceous and Cenozoic without dependence on paleomagnetic data from Baltica. Avoiding continental overlap is required, but published paleolongitudinal conjectures are not. For times prior to 130 Ma, the positions of tectonic blocks without paleomagnetic data are determined by geologic relations between known blocks, general faunal realms, or other evidence. For times when paleomagnetic data are needed but lacking and other data are ambiguous, motions of the continental blocks are assumed to be simple extrapolations between times for which there are data. If simple extrapolation caused overlap of continental blocks, then the most plausible adjustment was used, and then checked both forward and backward in time.

Our kinematic model for the evolution of the Arctic region relies on a number of recent models as well as the many references mentioned throughout this text. In particular, we use the model of Parfenov (1997) for the motion and relations of many

of the large and small terranes of northeastern Eurasia. We use Golonka (2000) for the late Paleozoic motions of some of the major plates that are not easily defined with the paleomagnetic data mentioned earlier. We do not, however, agree with Golonka's placement of Siberia against Barentsia, which he showed as a separate plate. Instead, guided by the paleomagnetic data of Smethurst et al. (1998), we include Barentsia as part of Baltica in the Middle to Late Ordovician. Nokleberg et al. (1998) provided a comprehensive summary of the northeastern Siberian terranes and many of the Alaskan and northwestern North American terranes, but used the cartoon method to produce the time slices they showed. In general, we adopt the designations of Plafker and Berg (1994a) for terranes and seas and try to incorporate the movement they show for Alaskan terranes. Our model does not show paleoseafloor spreading centers because the available evidence does not provide information as to their actual location or paleo-orientation. It is difficult to determine the temporal extent and paleopositions of subduction zones because subduction of young oceanic crust (younger than 25 Ma) may not leave a magmatic signature (DeLong et al., 1978) and because initial subduction may occur for a considerable period of time before a magmatic arc develops behind it. Consequently, lack of evidence for a paleosubduction zone does not necessarily mean that subduction was not occurring, and furthermore, a magmatic arc may continue to be active after the related subduction has stopped.

Acronyms designate continents and fragments of continents, tectonostratigraphic terranes, and ancient oceans and seas in the figures (listed in Table 1), and we follow the time scale of Gradstein and Ogg (1996) throughout. A synopsis of the composite tectonostratigraphic terranes for northeastern Siberia mentioned in this chapter is presented in Table 2 and shown in Figure 2 and a synopsis for Alaska is presented in Table 3 and shown in Figure 3.

TECTONIC EVENTS IN THE FUTURE ARCTIC BEFORE THE ASSEMBLY OF PANGEA (PRE–LATE PENNSYLVANIAN)

Our reconstructions of the paleogeography of the Arctic begin with the Ordovician locations of the modern circum-Arctic landmasses, which are reconstructed in Plate 6B-1 with respect to the Ordovician paleomagnetic pole for North America from Van der Voo (1993). The landmasses are scattered between 10°S and 30°N, with North America rotated ~90° clockwise from its present orientation and straddling the equator. Siberia, based on paleomagnetic poles at 452 Ma (Rodionov, 1966) and 449 Ma (Torsvik et al., 1995), and Baltica were in the temperate zone. Arctic Alaska, at ~30°N, is attached to North America while other coeval Arctic blocks, such as Chukotka, the Omolon block, the northern Taimyr Peninsula, and fragments of far eastern Siberia such as the Omulevka block are shown in plausible, but not tightly delimited positions based on geological relations. The shape of Baltica, shown with Novaya Zemlya and the northern continental fragment of the Taimyr Severna Zemlya block attached, is taken from Nikishin et al. (1996). The southern margin of the Taimyr Peninsula and the Omulevka block remain in relative proximity to each other from our first reconstruction at

TABLE 1. ACRONYMS USED IN FIGURES AND IN TEXT

Acronyn	Name	Acronym	Name
AA	Arctic Alaska	NA	North Atlantic
AB	Amerasia Basin	NCB	North China Block
ACT	Arctic Composite Terrane	NE	Northeastern Chukchi Sea shelf
AFR	Africa	NG	Nansen-Gakkel Ridge
AH	Axel Heiberg Island	NOAM	North America
AL	Aleutian Island Arc	NS	Nares Strait
AMR	Amuria	NZ	Novaya Zemlyna
AR	Alpha Ridge	OB	Oimyakon Basin
AS	Angayucham Sea	OCT	Oceanic Composite Terrane
AZ	Alezaya Arc	OL	Omulevka Block
BAL	Baltica	OM	Omolon Massif
BB	Baffin Bay	OS	Sea of Okhotsk
BI	Baffin Island	P-AV	Peri-Avalonian Block
BRS	Bering Sea	P-LR	Peri-Laurentian Block
BS	Barents Shelf	PAO	PaleoAsian Ocean
CAI	Canadian Arctic Islands	PB	Porcupine Basin
CB	Canada Basin	PD	Piedmont Terrane (SE NOAM)
CCT	Central Composite Terrane	PY	Pearya
CCB	Chukchi Continental Borderland	RT	Ruby Terrane
CCS	Cache Creek Sea	SAB	South Anyui Basin
CH	Chukotka	SAM	South America
DI	Disko Island, West Greenland	SCB	South China Block
EB	Eurasian Basin	SCT	Southern Composite Terrane
EI	Ellesmere Island	SHS	Siberian Traps hot-spot
EURA	Eurasia	SIB	Siberia
FJ	Franz Josef Islands	SP	Seward Peninsula
GB	Grand Banks	ST	Stikine Terrane
GR	Greenland	SV	Svalbard
IB	Iberia	TP	Taimyr Peninsula
IHS	Iceland hot-spot	UO	Uralian Ocean
KA	Koyukuk arc	UY	Uyandina arc
KAZ	Kazakhstan	VB	Vilyui Basin
KT	Kuril Trench	VK	Verkhoyansk Block
LP	Lisburne Peninsula	WCT	Wrangellia Composite Terrane
LR	Lomonosov Ridge	WI	Wrangell Island
LS	Labrador Sea	YCT	Yukon Composite Terrane
MI	Melville Island	YT	Yakutat Terrane
MOS	Mongol-Okhotsk Sea	ZL	Zolotogorskiy terrane
MR	Mendeleev Ridge		

TABLE 2. SELECTED LITHOTECTONIC BLOCKS AND TERRANES OF EURASIA*

Abbreviation	Name	Explanation
AMR	Amuria block	Includes Argun and Gonzha cratonal terranes of Nokleberg et al. (1998). They describe them as having rifted from the southern margin of Siberia during the Late Proterozoic.
AZ	Alazeya Arc terrane	An intensely deformed tectonic melange with a thick sequence of Carboniferous to Lower Jurassic littoral-marine and shallow-marine tuff. The terrane was the product of an elongate subduction zone where subduction ended in the Early to Middle Jurassic with accretion to the Kolyma-Omolon superterrane (Nokleberg et al., 1998).
CH	Chukotka block	Passive continental-margin terrane of a long-lived Late proterozoic, Paleozoic and early Mesozoic, Atlantic-type (Nokleberg et al., 1998). The Late Proterozoic continental-margin arc activity seen in Arctic Alaska is not seen in Chukotka.
NCB	North China Block	Craton that may have rifted from the Gondwanide or Rodinian margin during Late Proterozoic.
OL	Omulevka Block	Passive continental margin terrane with oldest unit containing presumed Late Precambrian Marble, schist, and metavolcanic rocks overlain by Middle and Upper Cambrian fossils, schist, metarhyolite, and quartzite. Upper sequences contain Ordovician through Triassic fossiliferous carbonate rocks, siltstone, mudstone, marl, and some Devonian volcanic rocks (Nokleberg et al., 1998). Assumed to have derived from the Siberian craton called the North Asian Craton by Nokleberg et al. (1998).
OM	Omolon Massif	According to Nokleberg et al. (1998), it consists of poorly exposed Archean to Early Proterozoic crystalline basement overlain by Proterozic conglomerate, sandstone, and siltstone and Vendian dolomite. These are overlain by rift-related Cambrian units and widespread sills and stocks of Middle Cambrian layered gabbro. Lower and Middle Ordovician shallow-marine units are unconformably overlain by by Middle and Upper Devonian calc-alkalic lava, rhyloite tuff, trachyte, trachyandesite, granitc plutonic rocks, and basalt, the latter possibly related to formation of the Vilyui Basin and opening of the Omyikon Basin.
SIB	Siberia Cratonic Block	Called the North Asian Craton by Nokleberg et al. (1994), it consists of Archean and Early Proterozoic metamorphic basement, and Late Proterozic, Paleozoic, and Mesozoic sedimentary cover. Paleomagnetic data indicate the craton formed from accretion of major Archean and Proterozoic fragments between about 850 and 650 Ma (Gusev et al., 1985).
TP	Taimyr Peninsula	It consists of a collage of Late Proterozoic through Mesozoic terranes and overlap assemblages according to Nokleberg et al. (1998). It is divided into three distinct parts by Vernikovsky (1997): the northern part is considered part of the Kara continental margin as part of Baltica; the Central Taimyr Neoproterozoic accretionary belt including 850 Ma granites to Vendian to Early Carboniferous sedimentary cover; and the South Taimyr Paleozoic-Mesozoic fold belt, part of the passive margin of the Siberian continent.
VB	Vilyui Basin	The eastern Siberian Craton was uplifted in the Early Devonian with extensive basic dike swarms and kimberlite bodies up to 700 km long and 150 km wide emplaced along the north-west and south margins of the Vilyui Basin (Parfenov, 1997). As the Vilyui Basin formed, the Omulevka block and the Omolon Massif were rifted from the eastern edge of the Siberian Craton.
VK	Verkoyansk Block	The Verkhoyansk fold and thrust belt contains a number of passive continental margin terranes enclosed in a collage of accreted oceanic and island arc terranes (Nokleberg et al., 1998). We refer to these terranes collectively as the Verkoyansk Block and show it diagramatically with some pieces being rifted from Siberia including the OL and OM blocks with the opening of the Oimyakon Basin and some remaining along the passive margin of Siberia and then reaccreted as the Oimyakon Basin closed in the Mesozoic.
ZL	Zolotogorskiy terrane	According to Nokleberg et al. (1998) it is divided into three units with the oldest being Devonian(?) schist, sparse marble, and metavolcanic. The middle unit consists of Carboniferous and Permian sandstone and siltstones that contain corals, brachyopods, bryozoa, and crinoids. The third layer consists of Upper Jurassic to Lower Cretaceous conglomerate intruded by Early Cretaceous granite and overlain by the Okhotsk-Chukotka volcanic-plutonic belt.

*From Golonka (2000), Parfenov (1997), and Nokleberg et al. (1998).

450 Ma (Plate 6B-1) until they assume their present respective positions during the Permian (Plate 6B-7), in accordance with Zonenshain et al. (1990).

Arctic Alaska is shown closed against the Canadian Arctic Islands in Plate 6B-1, a position it occupied prior to its counter-clockwise rotation away from the Canadian Arctic Islands in late Mesozoic time (Carey, 1958; Rickwood, 1970; Tailleur, 1969, 1973). We follow Plafker and Berg (1994a) in showing the Yukon composite terrane to be south of its present position, along the western margin of Laurentia from the Ordovician to the Late Devonian. The Late Proterozoic rocks of the Yukon, Arctic Alaska, and Central composite terranes, which include the cores of the Nixon Fork terrane and the Ruby Mountains terrane, were interpreted by Plafker and Berg (1994a) to have been formed mainly in the Cordilleran miogeosyncline, although some of the terranes of the Arctic Alaska and Central compos-

TABLE 3. COMPOSITE LITHOTECTONIC TERRANES OF ALASKA*

Abbreviation	Name	Explanation
ACT (1)	Arctic composite terrane	Arctic Alaska, Porcupine, and Seward Peninsula terranes and probably part of St. Lawrence Island. Arctic Alaska terrane consists of the North Slope, Endicott, Delong Mountains, Coldfoot, and Hammond subterranes. The Arctic composite terrane constitutes the Alaska segment of the Arctic Alaska-Chukotka microplate, which was separated from the North American craton by rotational rifting in the Early Cretaceous.
CCT (2)	Central composite terrane	Several small northeast-southwest–trending terranes of central Alaska consisting of fragments of miogeoclinal sedimentary sequences, and of ultramafic and deep marine bedded rocks of the Livengood terrane. Includes the moderate-sized Nixon Fork, Dillinger, Minchumina, and Mystic terranes and the small Wickersham, Baldry, Minook, and White Mountains terranes.
KIL	Kilbuck terrane and Idono Complex	Small slivers of Early Proterozoic gneiss and schist of upper amphibolite to granulite facies in southwestern Alaska.
NOAM	Authochthonous North America	North American craton and Cordilleran miogeocline.
OCT (3)	Oceanic composite terrane	Devonian to Early Jurassic oceanic basalt, volcanogenic, and oceanic sedimentary rocks and minor ultramafic rocks of the Angayucham, Tozitna, Innoko, Goodnews, (3a) = Misheguk Mountain, and probably Seventymile terranes in north-central Alaska.
RT (4)	Ruby terrane	Structurally complex and metamorphosed continental margin sedimentary rocks and basalt with mid-Paleozoic and mid-Cretaceous plutonic rocks in north-central Alaska. Probably originally continuous with the Arctic Alaska and Yukon composite terranes.
SCT (5)	Southern Margin composite terrane	Complexly deformed accretionary prism of Upper Triassic to Paleogene continental terrace sedimentary and oceanic sedimentary and volcanic rocks of the Chugach, Ghost Rocks, and Prince William terranes. Local Early Jurassic and mid-Cretaceous blueschist metamorphism.
TCT (6)	Togiak-Koyukuk composite terrane	Late Triassic to Early Cretaceous intraoceanic volcanic arc rocks and their intrusive equivalents of the Togiak, Koyukuk, Nyac, and Innoko terranes and older Triassic and Paleozoic oceanic sedimentary and ophiolite assemblages.
WCT (7)	Wrangellia composite terrane	Late Proterozoic (?) to Early Jurassic volcanic arc and associated syn- and post-arc plutonic rocks, oceanic plateau (?) volcanic rocks, and post-arc sedimentary assemblages of the Wrangellia, (7a) = Peninsular, Alexander, and northern part of the Taku terrane; Late Jurassic to mid-Cretaceous magmatic arc, and flysch deposits of the Gravina-Nutzotin belt.
YCT (8)	Yukon composite terrane	Crystalline rocks of the Yukon-Tanana, Tracy Arm, and Taku terranes and the overlying arc-related rocks of the Stikine terrane in east-central and southeastern Alaska. The crystalline rocks were derived from the western margin of the North American craton.
YT (9)	Yakutat terrane	Allochthonous fragment of the continental terrace of the Pacific coast of North America emplaced by 600 km of post-Oligocene right-slip on the Queen Charlotte–Fairweather fault system and associated underthrusting of the Southern margin composite terrane.

*From Plafker and Berg, 1994.

ite terranes may originally have been part of Chukotka or possibly Siberia (Dumoulin et al., this volume; Blodgett et al., this volume).

The southern Wrangellia composite terrane (7 in Fig. 3; WCT in Plate 6B) contains Early Ordovician to Late Devonian magmatic rocks. It is assumed to have been within 15° of latitude of its present position with respect to the North American craton during those times (Plafker and Berg, 1994a). We place the Wrangellia composite terrane near Siberia in Plate 6B-1, whereas Nokleberg et al. (1998) suggested that it might have been derived from the Barents Sea region, which would be difficult given the plate geometries we show in Plate 6B-1. Plafker and Berg (1994a) rejected the Southern Hemisphere location of Wrangellia suggested by Gehrels and Saleeby (1987) because Devonian invertebrate fossils from the Alexander terrane of the Wrangellia composite terrane have North American affinities (Savage, 1988). The weight of the evidence (Blodgett et al.,

this volume) indicates that the Middle Cambrian to Lower Devonian invertebrate faunas of the Alexander terrane of the Wrangellia composite terrane, the Nixon Fork and Livengood terranes of the Central composite terrane, and the York and Coldfoot-Hammond subterranes of the Arctic Alaska composite terrane have faunal affinities with Siberia. Moreover, the Cambrian or Ordovician through Carboniferous invertebrate faunas of these terranes are dissimilar to those of North America, including Arctic Canada. Hahn and Hahn (1993) and John Carter (1999, personal commun.) also indicated that at least some elements of the Carboniferous brachiopod faunas of southeastern Alaska have Eurasian, rather than North American affinities. The Alexander, Nixon Fork, Livengood, Coldfoot-Hammond, and York terranes therefore appear to have been in the Northern Hemisphere, and close to or part of Siberia, from at least the Ordovician to the Carboniferous. Many of the Late Triassic faunas found in the Central and Wrangellia composite terranes, however, have tropical paleogeographic aspects and are unlike coeval faunas in northwestern North America (R. Blodgett, 1999, personal commun.). According to Blodgett, little is known about the travels of these terranes between their probable location on the periphery of Siberia in the Devonian to Carboniferous and their amalgamation with southern and central Alaska during the Late Jurassic and Early Cretaceous. Our placement of the Wrangellia composite terrane in Plate 6B-1 (450 Ma) near Siberia is based solely on the faunal data. There is no control on its subsequent migration to its speculative position in the eastern Pacific off North America during the Middle Devonian (Plate 6B-3). Subcomponents of the Wrangellia composite terrane may have drifted into the tropics or been amalgamated with tropical components, before it joined with North America during the Jurassic.

Closure of the Iapetus Ocean and assembly of Laurussia: The Scandian phase of the Caledonian orogeny

The first major event in the long process by which the scattered Arctic landmasses were assembled into the continental framework of the modern Arctic was the Scandian phase of the Caledonian orogeny, when Baltica collided with North America in the region of Greenland during the late Early Silurian (Wenlock) to late Early Devonian (?Emsian). Baltica, Pearya, and possibly Chukotka were added to Laurentia to form Laurussia (Fig. 4 and Plate 6B-2). Closure of the Iapetus Ocean assembled the principal continental components of the modern Arctic (Siberia, Chukotka, Baltica, Greenland, the Canadian Arctic Islands, and Arctic Alaska) between the equator and lat 30°N.

McKerrow et al. (2000) redefined the classic Caledonian orogeny of northwestern Europe to include all the phases from Late Cambrian to Late Devonian age associated with the closure of those parts of the Iapetus Ocean situated between Laurentia to the northwest and Baltica and Avalonia to the southeast and east. Whereas McKerrow et al. (2000, p. 1151) stated "The term 'Caledonian orogeny' should be restricted to tectonic events within, and on the borders of, the Iapetus Ocean," Figure 4 indicates that the Iapetus Ocean extended as far as the margins of Arctic Alaska. The Scandian phase of the orogeny resulted from the continent-continent collision of Baltica with the eastern Greenland section of Laurentia (Stephens and Gee, 1985) during Wenlock to Emsian (?) time (Silurian to Early Devonian). More recent work seems to place the Scandian phase of the Caledonian orogeny for Svalbard in the Silurian, based on Ar/Ar dates (Gee et al., 1994), and closure of Baltica with Greenland appears to have been completed in the Silurian according to MacNiocall et al. (1997). Ziegler (1988, 1989) proposed that the Scandinavian Caledonides have a northward trend in the southwestern Barents Sea (BS in Fig. 1) and link up with the Innuitian orogen (Fig. 1) through the Svalbard Caledonides. Gudlaugsson et al. (1998) preferred the model of Doré (1991), which has the main arm of the Caledonides extending northeast across the Barents Sea as a direct continuation of the structural axis of the Scandinavian–East Greenland Caledonides. In their model,

Figure 4. Major Late Ordovician to Early Silurian tectonic events that affected structurally imprinted future Arctic plates. Closure of Iapetus Ocean produced Caledonian deformation along Greenland and Scandinavia, as indicated by single thick lines off eastern Greenland, Canadian Arctic Islands, Chukotka, and western Baltica (BAL). Branching dashed thick line is Caledonian deformation traced into Barents Sea region (Gudlaugsson et al., 1998). Based on deformational events on Wrangell Island (Kos'ko et al., 1993), we have placed Chukotka block (CH) on incoming Baltica plate along with Pearya terrane (PY), which accreted to Ellesmere Island (EI) and parts of Seward Peninsula (SP), which accrete to Arctic Alaska (AA). SIB is Siberia. Single-headed arrows indicate general plate motion directions but are not constrained. Double thick lines indicate collision of peri-Avalon block with Laurentia prior to Late Ordovician.

a separate north-oriented arm underlies the northwestern Barents Sea and western Svalbard (SV in Fig. 1). It is this western arm that links the Caledonides of northwestern Europe with the Innuitian orogen of northwestern North America, as Ziegler (1988, 1989) suggested, and associated the Scandian phase of the Caledonian orogeny with the major mid-Paleozoic tectonic events of Laurentia. The geometry of the colliding plates (Figs. 4 and 5) suggests that the earliest part of the Innuitian orogeny of northern Greenland and northern North America and the Scandian phase of the Caledonian orogeny of northwest Europe are the same event. This is expressed in both the early Paleozoic rocks of northwestern Europe (Roberts and Sturt, 1980) and the pre–late Visean (pre–middle Mississippian) rocks of Greenland and northern North America (Trettin, 1991a).

Caledonian deformation of northern Greenland

Higgins et al. (1991) suggested that the most precise time record of the closing of the Iapetus Ocean is provided by a west-

Figure 5. Major Middle to Late Devonian tectonic events that structurally imprinted the future Arctic plates. To southeast of Laurussia, Rheic Ocean was closing during Middle to Late Devonian. NE shows northeast Chukchi Sea shelf region with double line approximating position of the present-day Herald arch. Late Devonian to Carboniferous Ellesmerian orogeny resulted from either collision of some plate, perhaps Siberia (SIB) with northern margin of Laurussia or flat-slab subduction along that margin. Verkoyansk region of present-day eastern Siberian craton was rifted at this time, perhaps by mantle plume in Vilyui basin region. We infer that the mantle plume may have also opened the Oimyakon basin (OB) to west and imparted eastward motion to Siberia. Double-headed arrows indicate active or assumed active seafloor spreading. Single-headed arrows indicate general plate motion directions but are not well defined. See Table 1 for abbreviations.

ward-prograding, chert-pebble conglomerate in the Franklinian basin in northern Greenland. Higgins et al. (1991) thought that these mid-Wenlock (late Early Silurian) deposits had their source in upthrust Ordovician chert sequences exposed to erosion in the rising Caledonian mountain belt to the east. These relationships suggest that the Scandian phase of the Caledonian orogeny began in the Early Silurian prior to mid-Wenlock time, or before 425 Ma. The final Paleozoic orogenic episode in the Franklinian basin according to Surlyk and Hurst (1984) was north-south folding during the Late Devonian–Early Mississippian Ellesmerian orogeny (Trettin, 1991b), which corresponds to the Late Devonian Svalbardian phase of the Caledonian orogeny as defined by Roberts and Sturt (1980).

Emplacement of Pearya

Several tectonic events in northwestern North America can be related to the Scandian phase of the Caledonian orogeny. Perhaps the best documented is the suturing of the Pearya terrane

(PY in Plate 6B-1) to the Clements Markham arc of northern Ellesmere Island (Trettin, 1991a; Klaper, 1992) by sinistral transpression as Baltica collided obliquely with Greenland and North America. According to Trettin (1991a), the Pearya terrane was part of the European Caledonides until at least late Middle Ordovician time. It was then emplaced by sinistral transpression upon the Franklinian mobile belt in northern Ellesmere Island beginning in the late Early Silurian. Emplacement was dated by the age of the first Pearya terrane-derived sediment in the Franklinian mobile belt; Trettin (1991c) tentatively assigned a Late Silurian age to the entire emplacement event, while Klaper and Ohta (1993) concluded that the emplacement event predates the Late Silurian.

Emplacement of Chukotka

A key element in reconstructions of the western Arctic is the origin and accretionary history of the Chukotka block (CH) (Plates 6B-1 and 6B-2), which consists of the western part of the modern Chukchi shelf, Wrangell Island (Kos'ko et al., 1993) (WI, Fig. 1), the East Siberian shelf to the west of Wrangell Island (in modern coordinates), and adjacent northeastern Russia. Basement exposures in the center of Wrangell Island consist of Late Proterozoic felsic to intermediate volcanic and volcaniclastic rocks and phyllitic schist and slate intruded by plutonic rocks (Kos'ko et al., 1993). The volcanic rocks yield a U-Pb crystallization age from zircons of 633 +21/−12 Ma; a porphyritic granite sill yields a zircon age of 699 ± 2 Ma; and a leucogranite had an imprecise age of 0.7 Ga (Cecile et al., 1991). These ages and rock types are distinct from the Late Proterozoic equivalents found on the Canadian Arctic Islands and suggest that Chukotka with Wrangell Island and the Canadian Arctic Islands were not one unit during the Late Proterozoic (Cecile et al., 1991). These dates from Wrangell Island are broadly similar, however, to metamorphic ages from the western part of the southern Brooks Range of northern Alaska and to northern Chukotka (Cecile et al., 1991). The rocks underlying the schist belt of the western southern Brooks Range are metasedimentary and metavolcanic rocks intruded by 705 Ma (Hub Mountain) and 750 Ma (Mount Angayukaqsraq) plutonic rocks (Karl et al., 1989; Karl and Aleinikof, 1990). Two similar dates, 681 ± 3 Ma and 676 ± 15 Ma from granitic orthogneiss bodies on the Seward Peninsula, were reported by Patrick and McClelland (1995). The similarities between the Late Proterozoic intrusions on Wrangell Island, the western southern Brooks Range sequence, and the Seward Peninsula suggest that they may have been part of the same block during the Late Proterozoic. If these correlations are valid, the suture between Chukotka and Arctic Alaska is south and west of the present-day location of the Lisburne Peninsula, but north of the Seward Peninsula and the Hub Mountain and Mount Angayukaqsraq areas of the southwestern Brooks Range, which appear to be parts of the ancestral Chukotka. Seward Peninsula is therefore shown separated from Arctic Alaska in Figure 4 (also on Plate 6B-1), but transferred along with the western part of the southern Brooks Range to Arctic Alaska in Plate 6B-2. From west of the Lisburne Peninsula the suture probably trended north or northwest (in modern coordinates) between the Herald arch and Wrangell Island on the southwest and west and a thick section of gently dipping strata that underlies the northeast Chukchi Sea on the east (see Grantz et al., 1987, and Fig. 11, 1990). These gently dipping strata are inferred to include unmetamorphosed, or only mildly metamorphosed Late Proterozoic rocks at mid-crustal depths, and to be part of the ancestral Arctic Alaska.

Kos'ko et al. (1993) and Natal'in et al. (1999) provided insight into the age of suturing of Chukotka to Arctic Alaska. Kos'ko et al. (1993) concluded that deformation and uplift of the upper Proterozoic rocks of Wrangell Island, some western parts of the southern Brooks Range, and possibly the New Siberian Islands and adjacent areas of northeastern Russia occurred between the latest Proterozoic and the Early Silurian. Although they related this deformation to the Middle Ordovician M'Clintock orogeny of northern Ellesmere Island (Trettin, 1987), two of the three dates that they cited for the Wrangell Island deformation are Early Silurian (430 and 432 Ma, Kos'ko et al., 1993). This suggests that the Wrangell Islands and, by extrapolation, possibly all of Chukotka were emplaced against North America during an Early Silurian collision (cf. Plates 6B-1 and 6B-2). This collision was only slightly earlier than the tentatively dated emplacement of the Pearya terrane onto the Franklinian mobile belt in the late Early Silurian to Early Devonian (Trettin, 1991a, 1991c), and broadly correlative with the Scandian phase of the Caledonian orogeny.

Caledonian deformation in Arctic Alaska

The Scandian phase of the Caledonian orogeny may also be represented by the regional deformation within Arctic Alaska. Here a strongly deformed and indurated sequence of marine argillite with interbeds of fine-grained graywacke, quartzose sandstone, carbonate, and chert containing Early Silurian graptolites and Ordovician and Silurian chitinozoans (Carter and Laufield, 1975) in the subsurface of the North Slope subterrane of Arctic Alaska are overlain by steeply dipping, but unmetamorphosed nonmarine clastic sediment containing Middle (perhaps Lower) Devonian plant remains (Collins, 1958). The Middle (Lower?) Devonian nonmarine redbeds are truncated by the same regional unconformity that overlies the Ordovician and Early Silurian argillite and sandstone sequence of Arctic Alaska, but lack the penetrative deformation of the older beds. This suggests that the Middle (Lower?) Devonian redbeds postdate the main deformation of the argillite and sandstone sequence (Moore et al., 1994) and that an orogenic event of Late Silurian and/or Early Devonian age consolidated and penetratively deformed the Ordovician and Silurian argillite and sandstone sequence and elevated it above sea level. The regional unconformity is between moderately deformed argillite and graywacke containing Middle Ordovician graptolites and Early Silurian conodonts and mildly

deformed Lower Mississippian nonmarine beds of the Kapaloak Formation (Endicott Group) on the Lisburne Peninsula (Grantz et al., 1983; Moore et al., 1994, this volume). Similarly, complexly deformed chert and phyllite of Cambrian to Ordovician age in the eastern Brooks Range are overlain unconformably by mildly deformed shallow-marine beds with Middle Devonian (Eifelian) invertebrates (Anderson et al., 1994). These deformation events are all attributed to the Scandian phase of the Caledonian orogeny, which was defined as late Early Silurian to Early Devonian by McKerrow et al. (2000).

The presence of Cambrian to Devonian marine conodont and megafossil assemblages of Siberian affinity or a mixture of Siberian and North American taxa (see especially Dumoulin et al., this volume; Blodgett et al., this volume) in several terranes in central and northern Alaska appear to be explained by (1) the relative proximity of Chukotka and other continental fragments to Laurentia (Plate 6B-1) during the Ordovician and perhaps the Cambrian, and (2) the suturing of Chukotka and other continental fragments with Laurentia in the Scandian phase of the Caledonian orogeny (Figs. 4 and 5). Our reconstructions suggest that the Late Silurian to possibly Early Devonian regional deformation observed in the North Slope subterrane of Arctic Alaska is a product of the left-lateral transpression that closed this part of the Iapetus Ocean between 450 and 420 Ma (cf. Plates 6B-1 and 6B-2) and should be assigned to the Scandian phase of the Caledonian orogeny. This event antedated the classic Ellesmerian orogeny, dated by Trettin (1991b) as latest Devonian to earliest Mississippian, by ~50 m.y.

Motion of pre-Pangean Siberia

Paleomagnetic data (Smethurst et al, 1998) suggest that during the Late Silurian (420 Ma, Plate 6B-2), Siberia was situated between 25°N and 50°N, <10° north of our reconstructed position of Chukotka. The locations of Chukotka and Verkhoyansk with respect to Siberia are best approximations from Parfenov (1997) and Golonka (2000). Although there are reasonable paleomagnetic data to locate Siberia at 450 Ma (Smethurst et al., 1998), its locations with respect to North America in Plates 6B-2 and 6B-3 are based on extrapolation between paleomagnetic poles at 435 Ma (Smethurst et al., 1998; Rodionov et al., 1982) and 380 Ma (Kamysheva, 1973). The available paleomagnetic data are insufficient to determine how far Siberia was separated from northern Laurussia during the critical time (Early Silurian) when the Pearya terrane and possibly Chukotka were accreted to North America. Kolesov and Stone (this volume) use the apparent polar wander (APW) path for Siberia of Didenko and Pechersky (1993). This APW path, which is undocumented, shows Siberia farther south during the Silurian than shown by Smethurst et al. (1998), and produces an unacceptable overlap between Siberia and the Taimyr Peninsula.

Comparison of the 420 Ma reconstruction (Plate 6B-2) with the 450 Ma reconstruction (Plate 6B-1) shows that Siberia was moving northwestward as the Iapetus Ocean closed. Golonka

(2000) showed a strike-slip fault operative between Siberia and Laurentia as the Iapetus Ocean was undergoing final closure during the Early Silurian (443–428 Ma, ca. 435 Ma). In our reconstructions, this hypothesized left-lateral fault (Fig. 5) is placed between Chukotka and North America from Svalbard to Arctic Alaska, indicating that there was transpression with oblique collision between Baltica and Greenland during the Early Silurian. Because we have no contemporary geological data on the location of Siberia with respect to Chukotka or North America, and the paleomagnetic data do not delimit Siberia for the critical Late Silurian period, we cannot determine whether Siberia was involved in the Scandian phase of the Caledonian orogeny.

Bennett-Barrovia block

Şengör and Natal'in (1996) combined the Bennett massif, defined as the Precambrian basement under the East Siberian and Chukchi Seas, with the idea of Barrovia land (Tailleur, 1973) to describe a Bennett-Barrovia block. Natal'in et al. (1999) showed their Bennett-Barrovia block as roughly equivalent to offshore Chukotka from west of the New Siberian Islands to the northern margin of the Seward Peninsula, the Hammond terrane of the southern Brooks Range, and the northeastern part of the Chukchi shelf. Tailleur (1973) defined Barrovia as the provenance for the old Arctic Alaska sedimentary basin (Tailleur and Brosgé, 1970) and showed it as encompassing the entire present northern Alaska margin, not just the area to the west of Point Barrow, as Natal'in et al. (1999) showed it in their Figure 7. We place the boundary between the Chukotka block and Laurentia during the Scandian phase of the Caledonian orogeny on the west side of the gently deformed lower Paleozoic and Precambrian strata that underlie the northeastern Chukchi shelf (Grantz et al., 1990), which we consider to be part of Arctic Alaska. Figure 5 shows that the area in question (labeled NE) would be difficult to emplace without producing significant deformation in the pre-Silurian strata of the northeast Chukchi shelf. We therefore place the Chukotka–Arctic Alaska boundary along the southwestern edge of the Herald thrust and the western side of the northeast Chukchi shelf, rather than in the eastern Chukchi Sea on a path that is parallel to the northeast-trending coast of Alaska northeast of the Lisburne Peninsula, as shown by Natl'in et al (1999). We thus consider the northeast Chukchi shelf (NE) as shown in Figure 5 to be part of North America in the Late Ordovician, rather than part of the Bennett-Barrovia block of Natal'in et al. (1999).

Natal'in et al. (1999) proposed that their Bennett-Barrovia block collided with North America during the Late Silurian to Early Devonian, but they assigned the collision to the Ellesmerian orogeny. However, Lerand (1973), who defined the orogeny, and Trettin (1991a, 1991b) and Embry (1991), who summarized the early Paleozoic and Devonian geology of the Canadian Arctic Islands, dated the Ellesmerian orogeny as Late Devonian to Early Mississippian. The age of the collision of Bennett-Barrovia with North America proposed by Natal'in et al. (1999) cor-

responds, however, to the Scandian stage of the Caledonian orogeny and the age tentatively proposed by Trettin (1991a, 1991c) for the emplacement of Pearya onto the Franklinian mobile belt. We therefore consider the collision of Chukotka with North America to be more or less coeval with the collision of Baltica with northern North America and that the suture, which is west of both the Lisburne Peninsula and the northeast Chukchi shelf, formed during the Scandian phase (late Early Silurian to Early Devonian) of the Caledonian orogeny.

ARCTIC DURING THE ASSEMBLY OF PANGEA (DEVONIAN AND EARLY MISSISSIPPIAN)

During Middle to Late Devonian time (Fig. 5) the Rheic Ocean closed, culminating in the collision of Gondwana with Laurussia and the formation of Pangea (Fig. 6). Two major events affected the position and configuration of the future Arctic continental landmasses during the Devonian. These events were formation of the open-sided Vilyui basin on the Siberian craton and the Late Devonian to Mississippian Ellesmerian orogeny.

Figure 6. Major Carboniferous tectonic events that structurally imprinted future Arctic plates. The Rheic Ocean had closed by mid-Carboniferous time and the Appalachian orogeny (AP) continued with right-lateral deformation along a wide region. To the northwest, the Cache Creek Sea (CCS) was opening, rifting Yukon composite terrane (YCT) from its position along Laurussia. Paleo-Asian (PAO) and Uralian Oceans (UO) were closing as the Kazahkstan block (KAZ) converged with Siberia (SIB) and Baltica (BAL). Ocean crust was being subducted under Siberia (Vernikovsky, 1997) as it moved toward Taimyr Peninsula (TP). Double-ended arrows indicate assumed seafloor spreading. Single-headed arrows indicate general plate motion directions but are not well defined. See Table 1 for abbreviations.

Motion of Siberia and the position of continental fragments later accreted to Siberia

During the Devonian (Plates 6B-3 and 6B-4) a large embayment in the Pacific margin resulted from the motion of Siberia with respect to Chukotka and Arctic Alaska, and was called the Angayucham Ocean by Nokleberg et al. (1998). The 330 Ma reconstruction (Plate 6B-5) differs from those of Nokleberg et al. (1998) and Parfenov (1997) in that we place Siberia farther east relative to Arctic Alaska than is shown in their reconstructions. Our reconstructions are based on overlapping the paleomagnetic 95% confidence ellipses for Siberia (Smethurst et al., 1998) with those of North America (Van der Voo, 1993), combined with reasonable assumptions concerning placement of Siberia with respect to Baltica. In addition to the Angayucham Sea (Fig. 6) between Siberia and Chukotka, there was a narrow sea (Vernikovsky, 1997) between Siberia and the northern Taimyr Peninsula margin of Laurussia. The eastward motion of Siberia in an absolute framework starting in Late Devonian (cf. Plates 6B-4 and 6B-5) suggests that the opening of the Oimyakon basin and the narrow sea between Siberia and Taimyr Peninsula may have been contemporaneous, and possibly related.

Devonian dike swarms traceable for 700 km along the northwest and south edges of the major Vilyui sedimentary basin (Vilyui basin, Fig. 5 and Plate 6B-5) on Siberia date the age of the initial formation of the Vilyui basin (Parfenov, 1997). The ocean crust of the Oimyakon basin (Plate 6B-5) may also have been initiated by this eruptive episode because Middle to Late Devonian volcanic rocks are found in both the Omulevka and the Omolon blocks (Nokleberg et al., 1998), which were separated from Siberia by opening of the Oimyakon Sea. Khramov (1984, 1986) presented paleomagnetic data for the Omulevka block that suggest it was part of the Siberian platform during the Ordovician and Silurian. A prime candidate for the cause of the Middle Devonian rift event is the impact of a mantle plume under Siberia. We suggest that the Vilyui basin represents the failed arm of one of Burke and Dewey's (1973) plume-generated triple junctions, and that the Oimyakon basin represents the successful arms. Based on the paleomagnetic data of Smethurst et al. (1998), Siberia moved northwestward with respect to Laurussia during Ordovician to Early Devonian time (Plates 6B-1 to 6B-3), but by Late Devonian time (Plate 6B-4) Siberia had apparently changed direction to move northeast while undergoing significant clockwise rotation (Plate 6B-5). Such a reversal of plate motion and rotation is consistent with interaction with a Middle Devonian mantle plume.

Şengör and Natal'in (1996) suggested that the Vilyui rift opened during the Late Devonian (360 Ma). Following Parfenov (1997), we show a widening basin (Fig. 5) between Siberia and the Omulevka and Omolon blocks. There are no geological data for the position of the Omulevka or Omolon blocks during the Carboniferous and Permian, and consequently the width of the intervening sea, the Oimyakon basin, is not known. Although Parfenov (1991) cited paleomagnetic data that place the Omulevka block 20°–30° south of Siberia during the Middle Devonian to Mississippian, close scrutiny of Parfenov's (1991) Figure 14 suggests that the Omulevka block was then 10°–18° north of, rather than south of, Siberia. Kolesov and Stone (this volume) have new paleomagnetic data for the Omulevka block during the Late Devonian that agree with the interpretation shown in Figure 5. They also compiled available data for the Omulevka block with respect to Siberia for younger times through the Late Cretaceous that place the Omulevka block as much as 25° or more south of Siberia during the Triassic. The plate motions implied by the values listed in Kolesov and Stone (this volume) range up to 100 mm/yr, and the locations of Siberia are as much as 15° north of where Smethurst et al. (1998) showed it for the Triassic. We resolve the uncertainty by simply placing the Omulevka block ~1000 km west of Siberia at 300 Ma (Plate 6B-6) until we begin to close the Oimyakon basin in the Middle Jurassic (Plates 6B-11 and 6B-12). Parfenov (1997) and Golonka (2000) both suggested that the Oimyakon basin started to close ca. 250 Ma with the caveat that it may have remained open until the Valanginian or Hauterivian (ca. 130 Ma). Blodgett (1998) suggested a Siberian affinity for late Early Devonian fossils from the Nixon Fork subterrane of central Alaska (see Central composite terrane in Table 2), which indicates that terranes that eventually accreted to North America, as well as terranes like the Omulevka and Omolon blocks that later rejoined Siberia, were rifted from Siberia during the Devonian.

Ellesmerian orogeny

According to Trettin (1991b, 1991c), a tectonic event of uncertain origin produced the Ellesmerian orogeny in the Franklinian mobile belt of the northern and central Canadian Arctic Islands and northern Greenland, and he suggested convergence of an unidentified plate with the northern margin of Laurussia as the most probable cause. Klaper (1992) attributed the extensive compressional deformation found on Ellesmere Island to this orogeny, and the cause to be collision of a continental Siberian plate with the northern margin of Laurussia during the Middle and Late Devonian (Plates 6B-3 to 6B-5). Trettin (1991a, 1991b) tentatively placed the Ellesmerian orogeny at the end of a sequence of orogenic events that range in age from middle Givetian or possibly Eifelian (Middle Devonian) to the Tournaisian (Early Mississippian), and indicated that the age of the orogeny sensu stricto (Trettin, 1991b) is latest Devonian (Famennian) to Early Mississippian (Tournaisian). This event roughly correlates with the Svalbardian or Acadian phase of the Caledonian orogeny.

Our reconstructions for the Middle Devonian to Late Mississippian (Plates 6B-3 to 6B-5) do not make a definitive case for a collision between Siberia and Laurussia that would correlate with the Svalbardian phase of the Caledonian orogeny. Plate 6B-4 shows a possible interaction of Siberia with the Taimyr Peninsula, but it is unlikely that that alone would have led to an Ellesmerian orogeny along the Canadian Arctic Islands as far west as Melville Island (MI on Plate 6B-4). Zonenshain et al. (1990), however, provided support for the origin of the Ellesmerian orogeny in a Svalbardian phase collision between northern Laurussia and a continental mass to the north, but as in the Canadian Arctic Islands, identification of the causative agent remains uncertain. They suggested that the North Taimyr–Severnaya Zemlya continental fragment, which was between Siberia and Baltica, collided with northern Laurussia near the end of the Devonian to the beginning of the Carboniferous (360 Ma, Plate 6B-4), but Nikishin et al. (1996) showed little or no movement between Severnaya Zemlya, Taimyr, and Baltica after the Late Cambrian, and showed Baltica as part of Laurussia during the Late Devonian and Early Mississippian.

The Ellesmerian orogeny consists mainly of faulting in northern Ellesmere Island and thin-skinned folding above the base of Ordovician evaporites from southern Ellesmere Island on the northeast to Melville Island and adjacent islands on the southwest (Trettin, 1991a). The orogeny appears to be absent from the eastern Brooks Range, because Early Devonian (Eifelian) to Cretaceous strata there lack compressive deformation (Anderson et al., 1994). Evidence for Ellesmerian deformation on the North Slope of Alaska may be preserved, however, in the Topagoruk test well, ~90 km south-southeast of Point Barrow. Collins (1958) reported unmetamorphosed medium gray chert conglomerate and dark gray carbonaceous lutite at the bottom of the well that dip 30°–65° and contain Middle (Early?) Devonian plant debris. These beds are overlain unconformably by red to gray lutite and sandstone with some red chert conglomerate that dip 8° or less. The redbeds could not be dated, but are overlain abruptly by light gray sandstone- and siltstone-bearing marine fossils of Permian age. The steep dips in the Middle (Early?) Devonian nonmarine beds may represent the Late Devonian or Early Mississippian Ellesmerian orogeny.

The areas of known or inferred Ellesmerian deformation are 1000 km or more from the presumed suture between Siberia with North Taimyr–Severnaya Zemlya, or with Chukotka as they existed during the Middle to Early Mississippian (Plates 6B-4 to 6B-5). If collision and convergence at this suture generated the stress that created the Ellesmerian orogeny, it was a far-field effect that was transmitted across Chukotka and left little deformation in the early Paleozoic and Precambrian strata that underlie the northeast Chukchi shelf. It would also imply that the early Paleozoic crust beneath the Franklinian mobile belt and the Arctic Alaska sedimentary basin of northern Alaska were mechanically more susceptible to compression than the crust that underlies the northeast Chukchi Sea and Chukotka. Dalziel et al. (2000) explained similar far-field effects in the Gondwanide fold belt, where orogenic deformation occurs 1500 km behind an active subduction zone, by flat-slab subduction. Sanford et al. (1985) suggested that compressional stresses at cratonal margins might reactivate fault-bounded basement segments far into craton interiors. Another explanation for the Ellesmerian orogeny may be reactivation of convergence, and subsequent redevelopment of subduction or compression along

the suture between Chukotka and the northern margin of Laurussia, along which the Iapetus Ocean closed during the earlier Scandian phase of the Caledonian orogeny. The Ellesmerian orogeny, which roughly correlates with the Svalbardian phase (Roberts and Sturt, 1980) or Acadian phase (McKerrow et al., 2000) of the Caledonian orogeny. Deep seismic reflection profiling across the Canadian Arctic Islands may be needed to definitively determine the tectonic associations of the Ellesmerian orogeny.

Subduction and rifting along the western margin of Laurussia (North America)

According to Howell et al. (1985), the western margin of North America underwent a late Precambrian regional rifting event that produced a passive margin that received shelf and slope sediment through much of the Paleozoic. Oceanic island arcs and backarc basins of Cambrian to Late Devonian age that were unrelated to North America, but were eventually accreted to it during the Mesozoic (Howell et al., 1987), were oceanward of this margin. These arcs and backarc basins were accreted to the western and Arctic margins of North America from at least northern California (Rubin et al., 1991) to western Arctic Alaska (Plafker and Berg, 1994a). Plafker and Berg (1994a) noted dates for magmatism related to these arcs that range from 350 to 370 Ma for their Yukon composite terrane to 365 to 410 Ma for the Coldfoot terrane of Arctic Alaska in the southern Brooks Range. There is also evidence for a primitive Devonian arc on the outboard Alexander terrane of the Wrangellia composite terrane (Table 1). During the Late Devonian, an elongate sliver of continental rock with strong lithologic affinities to western North America drifted away from the continent and left an ocean basin between the sliver and North America (Plates 6B-5 to 6B-7). This sliver, consisting mainly of crystalline rocks, was named Stikinia by Tempelman-Kluit (1979) and the Yukon-Tanana terrane by Mortensen (1992). The crystalline rocks are overlain depositionally by Upper Triassic to Middle Jurassic marine clastic and volcanic and carbonate arc rocks and intruded by comagmatic igneous rocks of the Stikine terrane. Together the Yukon-Tanana and Stikine terranes form the Yukon composite terrane of Plafker and Berg (1994a). The ocean basin that formed between the Yukon composite terrane and North America was named the Anvil Ocean by Templemen-Kluit (1979), but was referred to as the Cache Creek Sea by Monger and Berg (1987) and Plafker and Berg (1994a). The latter indicated that the Cache Creek Sea is of at least Early Mississippian to Middle (?) Jurassic age and formed behind a Late Devonian to Early Mississippian magmatic belt. Nokleberg et al. (1998) adopted the interpretation of Mihalynuk et al. (1994) and used the term Cache Creek Ocean for oceanic crust outboard of the Yukon composite terrane, crust which in reality is simply part of the ancestral Pacific Ocean, as shown in Nokleberg et al. (1998, Fig. 8). They also used the term Slide Mountain Ocean for the region that Plafker and Berg (1994a) referred to as the Cache Creek

Sea. We restrict our use of the term "ocean" to its modern usage and use "sea" for marginal oceanic basins that may or may not have been fully open to the Pacific Ocean, much like the present-day Japan Sea or the Sea of Okhotsk. In general, we adopt the place-name usage of Plafker and Berg (1994a, 1994b) herein.

Mortensen (1992) suggested that the magmatic arc along the outboard margin of the Yukon composite terrane was associated with continentward subduction from the Late Devonian to the Early Mississippian, but that its polarity had reversed by mid-Permian time. This agrees well with Tempelman-Kluit's (1979) timing of rifting of his Anvil Ocean, which he believed to have been initiated as a backarc basin in response to Late Devonian continentward subduction outboard of the Yukon composite terrane. The existence of evidence for a mid-Permian reversal of subduction direction suggests that the Cache Creek Sea was sufficiently wide, and subduction sufficiently long lived, to generate a westward-subducted slab that imprinted the geologic record of the Yukon composite terrane. Gordey et al. (1987) suggested that by mid-Mississippian time the Cache Creek Sea had become sufficiently deep and wide that clastic sediment from the rifted sliver (Yukon composite terrane) could no longer reach the margin of the North American craton. While many authors (as summarized in Plafker and Berg, 1994a) indicate the width of the Cache Creek Sea as unknown, it is commonly suggested, on the basis of sedimentation patterns, that it may have been thousands of kilometers wide. We show a width of 1000–1200 km for the Cache Creek Sea based on a rough analogy with the Cenozoic South China Sea, where the Palawan, Reed Bank, and other blocks were rifted from the South China block margin and transported ~1200 km southward from the Eocene until spreading stopped in the Miocene (Lee and Lawver, 1995).

TECTONIC ACTIVITY IN THE ARCTIC DURING THE TIME OF PANGEA (LATE PENNSYLVANIAN TO MIDDLE JURASSIC)

Pennsylvanian

The mid-Paleozoic Rheic Ocean (Plates 6B-2 to 6B-4) was essentially closed by 300 Ma (Fig. 6 and Plate 6B-6) as North America sutured to Africa and the Piedmont block (the present southeastern United States) during the Alleghenian orogen (Hatcher et al., 1990). The resultant supercontinent, Pangea, lasted for at least 100 m.y. By Carboniferous time (Fig. 6) the Cache Creek Sea between the Yukon composite terrane and North America had begun to open. Only the rapidly narrowing paleo–Asian Ocean separated Siberia from the Kazakhstan block (Golonka, 2000). The open sea between the northern Taimyr block and Siberia (Fig. 6) closed when the northern part of the Taimyr Peninsula, as part of the Kara continental block of Pangea, sutured to Siberia (Plate 6B-6) starting ca. 300 Ma, according to Vernikovsky (1997). During the Carboniferous (Fig.

6) Eurasia coalesced with closure of the Uralian Ocean between Baltica and Kazakhstan and the paleo–Asian Ocean between Kazakhstan and Siberia (Golonka, 2000). Vernikovsky (1997) illustrated the collision of the northern Taimyr Peninsula block with the Pangean margin through subduction beneath Siberia. Such Siberian-directed subduction would also explain the westward motion of Kazakhstan and closure of the paleo–Asian Ocean (cf. Figs. 6 and 7). The presence of as much as 4 km of eastward-wedging middle Carboniferous through Triassic sediment in the Vilyui basin of eastern Siberia (Parfenov, 1997) implies that the eastern end of the basin was still open at that time, and that the Omulevka block did not rejoin Siberia until sometime after the Triassic.

Triassic to Early Jurassic

By Triassic time (Fig. 7) the Cache Creek Sea was closing and northern Eurasia was largely assembled by the suturing of Siberia, Kazakhstan, and Baltica following closure of the Ural and paleo–Asian Oceans. Through time the nucleus of the future circum-Arctic landmasses generally moved northward; a major acceleration took place when the Central Atlantic opened dur-

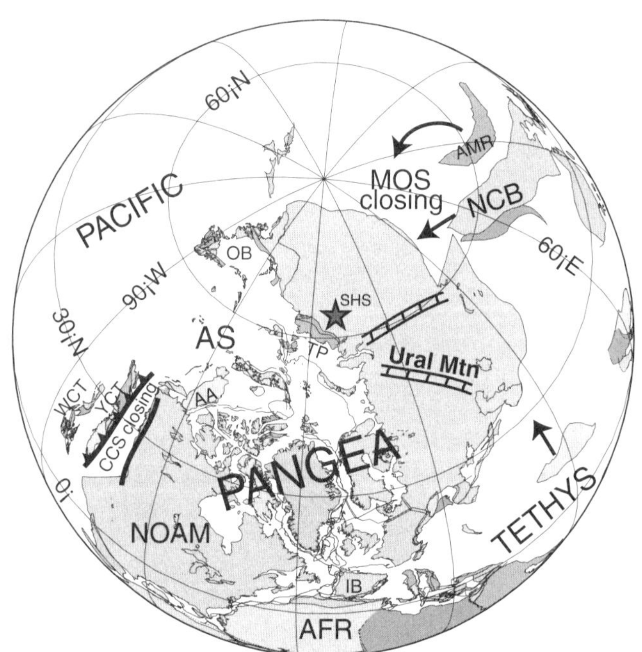

Figure 7. Major Triassic tectonic events that structurally imprinted the Arctic plates. Large star (SHS) indicates location of eruption of Siberian Traps. Yukon composite terrane was undergoing westward-directed subduction as Cache Creek Sea closed. Mongol-Okhotsk Sea was closing as Amuria and North China blocks approached Pangea. Double-ended arrows indicate active or assumed active seafloor spreading. Single-headed arrows indicate general plate motion directions, but are not well defined. Ladder structures indicate recently closed Uralian and paleo–Asian Oceans (see Fig. 6). See Table 1 for abbreviations.

ing the Early to Middle Jurassic (Klitgord and Schouten, 1986). Eruption of the Siberian Traps at the Permian-Triassic boundary ca. 248 Ma (Renne et al., 1995) was the major tectonic event in the future Arctic region between the Early Permian and the Middle Triassic (Plates 6B-7 and 6B-8). The work of Nie et al. (1990), using biomes to delineate the Late Permian paleogeography of the northern continental plates, placed the point of eruption of the 248 Ma Siberian Traps close to the current latitude of the Iceland hotspot. Siberia in a paleomagnetically acceptable latitudinal position has been shifted longitudinally with the rest of Pangea such that a fixed location of the present-day Iceland hotspot (taken to be ~65°N, 17°W; Lawver and Müller, 1994) coincides with the initial eruption point of the Siberian Traps. Although many people do not believe that hotspots are fixed (Norton, 1995, 2000), we believe that the Indo-Atlantic hotspots can be assumed to have been relatively fixed for the past 250 m.y. The voluminous magmatism of the North Atlantic igneous province (ca. 55 Ma, White and McKenzie, 1989) can be explained by the lateral flow model of Sleep (1997), wherein plume material flowed to the thinner part of the lithosphere as the keel of Greenland passed overhead.

Lawver and Müller (1994) traced the track of a fixed Iceland hotspot in the absolute reference frame of Müller et al. (1993) from its present location to northern Ellesmere Island at 130 Ma (see Plate 6B-12 at 120 Ma). As delimited by the paleomagnetic data of Van der Voo (1993) and Smethurst et al. (1998), the extension back in time of the absolute reference frame of Müller et al. (1993) produces a Triassic to Early Jurassic track for a Siberian Traps hotspot that agrees with the early age relationship for the emplacement of the Siberian Traps as determined by Basu et al. (1995). They found slightly younger magma to the east and north of the initial point of eruption (see Plate 6B-9) for a possible 210 Ma position of the hotspot. The location of the fixed Iceland hotspot at 150 Ma (Plate 6B-11) is only a few hundred kilometers from Kong Karls Land, Svalbard, where Bailey and Rasmussen (1997) postulated a Siberian Traps–Iceland hotspot may have produced the Late Jurassic (Early Kimmeridgian) basalts they studied there.

BREAKUP OF PANGEA AND ITS IMPACT ON THE ARCTIC (MIDDLE JURASSIC TO PRESENT)

Rapid northwestward motion of North America (Fig. 8), beginning with the opening of the Central Atlantic in the Jurassic (Klitgord and Schouten, 1986), appears to have strongly affected the western margin of North America, Arctic Alaska, and Chukotka. Considered broadly, Mesozoic and younger Alaska is the product of dextral strike-slip displacement and tectonic convergence between the oceanic plates of the northeastern Pacific Ocean and northwestern North America as North America moved northwestward in an absolute framework. Arctic Alaska became the Arctic Alaska composite terrane when the upper part of the Oceanic composite terrane was obducted onto it from the

south during the Late Jurassic (Plates 6B-9 and 6B-10). The Ocean composite terrane includes pieces of the ultramafic Misheguk Mountain allochthonous terrane (3a in Fig. 3) that were emplaced upon oceanic rocks of the Ocean composite terrane by convergence within the Angayucham Sea during the Middle Jurassic (Wirth, 1991; Wirth and Bird, 1992), and an attenuated degree of convergence between the Pacific and the Arctic Alaska composite terrane continues to the present. A comprehensive summary of the geology and tectonic development of Alaska during this time was given in Plafker and Berg (1994b).

Accretion of Alaska by orthogonal convergence between the Farallon and North American plates (Jurassic to mid–Early Cretaceous)

From Aalenian (ca. 190 Ma) or earlier to the mid-Cretaceous, northeastward convergence between the Pacific Ocean and North America uniformly stressed the western continental

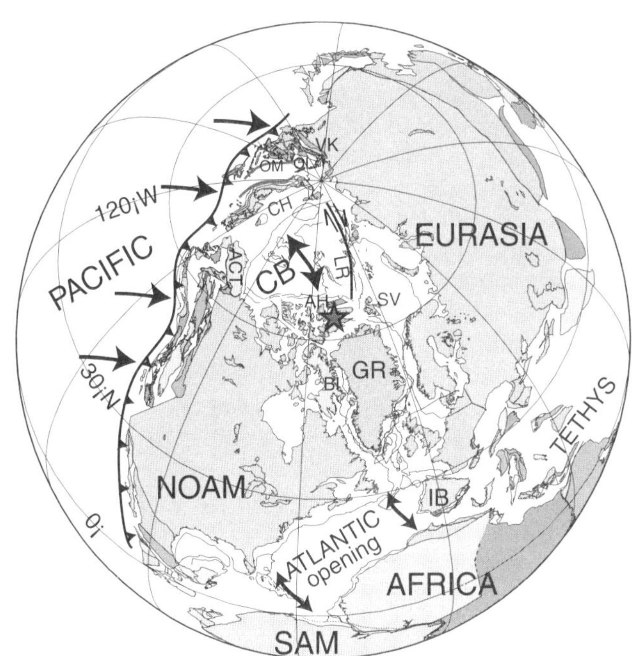

Figure 8. Major mid-Cretaceous tectonic events that structurally imprinted the Arctic plates. Opening of Central Atlantic produced significant northwest motion of North America, Greenland, and Eurasia. Continentward subduction was active along most of Pacific margin of North America and future Arctic region. During Late Jurassic into Early Cretaceous, Canada basin opened about a point south of Mackenzie Delta, rotating Arctic Alaska composite terrane and Chukotka away from Canadian Arctic Islands. A right-lateral strike-slip fault operated along Lomonosov Ridge margin of Eurasia. Siberian Traps–Iceland hotspot is shown under Ellesmere Island and is assumed to have been responsible for Early Cretaceous flood basalts found on Axel Heiberg Island. Double-ended arrows indicate active or assumed active seafloor spreading. Single-headed arrows indicate general plate motion directions but are not well defined. See Table 1 for abbreviations.

margin of North America (Fig. 8) as far north as south-central Alaska, and probably to or beyond the present western end of the Arctic Alaska composite terrane. This convergence closed the Cache Creek Sea off the future coast of British Columbia and southeastern Alaska between 240 and 180 Ma (Late Triassic to earliest Middle Jurassic); sutured the Yukon composite terrane against North America, mainly during the Early Jurassic; and sutured the Wrangell composite terrane against the Yukon composite terrane during the Early Cretaceous (Plates 6B-12 and 6B-13). Closure of the Cache Creek Sea, accommodated by subduction toward the Pacific (McClelland et al., 1992) beneath the Yukon composite terrane (Plates 6B-9 and 6B-10), produced both the Late Triassic to Early Jurassic Quesnellia arc and the Late Triassic to Middle Jurassic Stikine arc (Plafker and Berg, 1994a). Closure of the Cache Creek Sea ca. 175 Ma was roughly coeval with the initiation of seafloor spreading in the Central Atlantic (Klitgord and Schouten, 1986) and the initial breakup of Pangea. Opening of the Central Atlantic and the subsequent northwestward motion of North America may thus have had a substantial far-field effect on the tectonics of the Pacific margin from California to Alaska, and possibly as far as Amuria in northeastern Eurasia (Fig. 8).

Continued north-directed convergence during the Early Cretaceous successively thrust more distal oceanic allochthons upon more proximal continental margin allochthons of the Arctic Alaska composite terrane to largely complete the structural stacking of the Brooks Range orogen by late Early Cretaceous (Albian) time, although Late Cretaceous and/or Paleogene convergence on low-angle thrust faults also took place (Moore et al., 1994). Mid–Early Cretaceous convergence between the Pacific basin and North America also sutured a series of ensimatic islands arcs formed in the Pacific Ocean to western Alaska (Fig. 8 and Plates 6B-11 to 6B-13). The first of these to dock was the Early Cretaceous Koyukuk arc (KA, Plate 6B-10). The Koyukuk arc was sutured to the Angayucham terrane of the Ocean composite terrane, which had been obducted onto the southern margin of the Arctic Alaska composite terrane during the Late Jurassic and early Early Cretaceous (Plates 6B-11 to 6B-14). Later arrivals include the Jurassic and Early Cretaceous Togiak and Nyac arc terranes (Box and Patton, 1989). These terranes are listed separately in Table 1 but are incorporated in the Koyukuk terrane in Plates 6B-10 and 6B-11.

Opening of the Amerasia basin

The Amerasia basin of the Arctic Ocean opened when the Arctic Alaska composite terrane and Chukotka rotated counterclockwise away from the Canadian Arctic Islands in early Early Cretaceous time about a pole of rotation at or south of the present Mackenzie Delta. This rotation was first hypothesized by Carey (1958) and elaborated on by Rickwood (1970), Tailleur (1969, 1973), and many others. Although alternative kinematic schemes have been suggested (see summary by Lawver and Scotese, 1990), congruent lines of evidence now indicate that

the basin formed by some variant of rotational rifting. The evidence includes (1) matching bathymetry across the Canada basin (the southern section of the Amerasia basin) (Carey, 1958); (2) facies trends in the Mississippian to Triassic strata of northern Alaska that indicate that these beds had sources north of Alaska, probably in the Canadian Arctic Islands (Rickwood, 1970, and many later workers); (3) matching stratigraphy and structural features across the Canada basin (Grantz et al., 1979, 1998; Embry, 1990; Embry et al., 1994); (4) a fan of lineated seafloor magnetic anomalies in the Canada basin that converge toward the Mackenzie Delta (Taylor et al., 1981; Brozena et al., 1998); (5) a narrow, linear negative-gravity anomaly in the satellite-derived gravity data of Laxon and McAdoo (1994) that is along the axis of symmetry of the magnetic anomalies and closely resembles the gravity anomalies seen at other abandoned seafloor-spreading centers; and (6) well-developed rift shoulders and rift margin grabens that contain Jurassic and Cretaceous marine sediment parallel to both the Canadian (Miall, 1979; Dixon, 1982; Harrison et al., 1999) and Alaskan (Grantz and May, 1982) margins of the Amerasia basin. A paleomagnetic estimate of the degree of rotation (~66°) was obtained by Halgedahl and Jarrard (1987) on samples from test wells on the rift shoulder of the Canada basin in northern Alaska. The satellite altimetry data of Laxon and McAdoo (1998) and aerogeophysical work (Brozena et al., 1998) suggest that opening of the basin was a multistage process that can be generalized as rotation about a pole south of the Mackenzie Delta. Brozena et al. (1998) speculated that opening started with an orthogonal phase that was followed by rotation, while Lane (1997) argued for strictly orthogonal opening without rotation.

Similar timing of the extensional events that opened the Amerasia basin and the convergence at the North Pacific Rim that created the main features of the Brooks Range orogen suggested to Tailleur (1969) that these processes were tectonically linked. Work by Harris (1989), Wirth (1991), and Wirth and Bird (1992) indicated that the initial convergence that created the Brooks Range allochthons may have been as young as 171–163 Ma (mid–Middle Jurassic) when 187–184 Ma (late Early Jurassic [Toarcian]) ophiolitic ultramafic rocks of the future Misheguk Mountain terrane (Patton et al., 1994) of the southwestern Brooks Range were thrust over an accretionary prism consisting of Upper Devonian to Lower Jurassic oceanic rocks of the future Angayucham terrane, also part of the Oceanic composite terrane. Stacking of the multiple nappes that characterize the Brooks Range orogen was initiated by the north-directed emplacement of the Angayucham terrane upon the continental rocks of the Arctic Alaska composite terrane during the Late Jurassic (?) and early Early Cretaceous. This event is recorded by the abrupt deposition of the south-sourced, synorogenic, flyschoid Okpikruak Formation, of this age, upon north-sourced stable outer shelf sedimentary deposits of the Arctic Alaska composite terrane exposed in Brooks Range allochthons (Crane, 1987; Mayfield et al., 1988; Moore et al., 1994; Mull et al., 1982). A southern source for the Okpikruak Formation is demonstrated by the clasts and olistostromes of Angayucham terrane lithologies that it contains. The oldest strata in the northern foothills of the central and western Brooks Range that have not undergone large-scale tectonic (thrust) transport are those of the Albian age Fortress Mountain Formation (Mayfield et al., 1988), indicating that the convergence that created the Brookian orogeny was largely completed by Albian time. The convergence that created the Brookian orogeny therefore ranges from at least the Middle Jurassic to post-Neocomian but pre-Albian (i.e., Barremian or Aptian) Early Cretaceous.

Initiation of the rift stage of the extension that created the Amerasia basin is recorded by the oldest deposits in rift margin grabens that formed on the rift shoulders of the Amerasia basin. The Dinkum and related grabens on the Alaskan rift shoulder (Grantz and May, 1982; Hubbard et al., 1987) contain clastic deposits that, at their base, are of Hettangian? or Sinemurian age (Mickey et al., this volume), confirming that extension began early in the Early Jurassic. The rift shoulder along the Canadian Arctic Islands margin of the Amerasia basin contains grabens with strata as old as late Toarcian (Harrison et al., 1999), and the initial age of margin-parallel faults in the southwestern Arctic Islands and Mackenzie Delta region indicate that continental rifting there began in early Middle Jurassic (Aalenian) time (Embry and Dixon, 1994). In the Canada basin, as in many areas where seafloor spreading initially occurs within a continent, rapid sedimentation covered the spreading center and greatly subdued or precluded formation of coherent marine magnetic anomalies (Lawver and Hawkins, 1978). Lineated, but currently uncorrelated seafloor magnetic anomalies in the Canada basin extend to the gravity anomaly taken to mark its extinct spreading axis (Brozena et al., 1998; Grantz et al., 1998), indicating that the seafloor spreading in the basin was completed by the onset of the Cretaceous normal magnetic superchron, near the beginning of Aptian time (ca. 120 Ma). The age of the extension that created the basin was thus no earlier than early Early Jurassic, and not later than earliest Aptian. This is similar to the Middle Jurassic (but possibly older) to Barremian or Aptian range of the principal convergence in the Brookian orogeny.

The similarity in age of extension in the Amerasia basin and the main convergence that produced the Brookian orogeny in northern Alaska, and the compatible trends of their extension and convergence axes, suggest a kinematic link between these adjacent tectonic provinces. Both the extension and convergence axes are, moreover, similar in trend to the east- to northeast-trending convergence velocity vectors between the Farallon plate and North America from prior to 175 Ma to after 125 Ma (Engebretson et al., 1985), which encompasses the span of time during which the Amerasia basin opened. It is not clear, however, what specific process may have linked Farallon–North America and Brooks Range convergence with Amerasia basin spreading. Trench rollback related to subduction of the lower crust of the Angayucham Sea and the southern fringes of Arctic Alaska beneath Arctic Alaska (Grantz et al., 1991; Miller and Hudson, 1991), in a manner broadly analogous to that proposed

by Rubin et al. (1995) for crustal extension in north-central Alaska and northeasternmost Russia during the mid-Cretaceous, may provide a viable mechanism, but many uncertainties remain. For example, the main (Middle Jurassic to Early Cretaceous) phase of the Brookian orogeny, which correlates with extension in the Amerasia basin, can only be traced as far west as Wrangell Island (Kos'ko et al., 1993), and its possible continuation northwest beneath the East Siberian Sea to the northwest corner of the Amerasia basin is conjectural. In addition, a large transform fault along the Amerasia side of Lomonosov Ridge, which is required for the rotational opening of the Amerasia basin, has yet to be conclusively demonstrated.

We propose that rifting in the Amerasia basin was triggered when the northern part of a zone of crustal weakness drifted over a hotspot, perhaps the remnant of the mantle plume that created the Siberian Traps at 248 Ma (Renne, 1995). The opening of the Amerasia basin may have been initially orthogonal followed by rotation. Brozena et al. (1998) and Grantz et al. (1998) suggested that the boundary between the irregular magnetic anomalies along the margins of the Amerasia basin and the well-ordered northwest-fanning magnetic lineations in the center of the basin (Taylor et al., 1981) may mark a transition from orthogonal to rotational rifting; however, the age of the transition is not known. We speculate that the rift initially propagated south-southeastward through the early Paleozoic Franklinian deep-water basin, which now underlies the Arctic continental margin of the Canadian Arctic Islands (Trettin et al., 1991) and the Arctic coastal plain of northern Alaska (Moore et al., 1994). It is thought that this deep-water basin contained the zone of crustal weakness along which the axial rift of the Amerasia basin formed. Passage of the rift from the relatively thin Precambrian crust presumed to underlie the Franklinian deep-water basin to the thick Precambrian crust of the cratonic Mackenzie platform is thought to have blocked further propagation to the south-southeast. The marked increase in thickness of Precambrian crust in the vicinity of the Mackenzie Delta coincides with the region where the structural effects associated with the rifting die out toward the pole of rotation in the lower Mackenzie River Valley.

Opening of the Amerasia basin and the kinematically linked closure of the South Anyui basin, the last remnant of the Angayucham Sea, resulted in the collision of Chukotka with some part of Siberia. This collision, which closed the basin and created the South Anyui suture zone, has been dated as pre-Albian (112 Ma) (Parfenov et al., 1995; Parfenov, 1997; Sokolov et al., this volume). This age is compatible with the presence of well-developed lineated magnetic anomalies at the extinct spreading axis of the Canada basin, which had to have formed prior to the start of the Cretaceous normal magnetic superchron (chron 34n; Gradstein et al., 1994) in earliest Aptian time (ca. 120 Ma), because the lineated magnetic anomaly over the abandoned spreading center is reversed. As can be seen in Plate 6B-12, there is a substantial gap between the South Anyui suture zone, which forms the western margin of Chukotka in this figure, and the Omolon block. Whether this gap should be between Chukotka and the Omolon block or between the Omolon block and Siberia is uncertain. Parfenov (1997) has deformation in the Verkoyansk fold and thrust belt continuing until Albian to Campanian time, but he showed the South Anyui basin closed by 110 Ma. The gap shown in our Plates 6B-12 through 6B-15 in this area should not be interpreted to indicate the presence there of an open sea with oceanic crust. Rather, it indicates that a global reconstruction of this region implies that substantial compression occurred in this region into the Cenozoic, particularly as the Eurasian basin opened.

Accretion of Alaska by orthogonal convergence between the Farallon and North American plates: Middle to Late Cretaceous

During the mid-Cretaceous a strong component of dextral strike-slip displacement developed between the Pacific Ocean basin and northwestern North America. This component initiated the northwestward migration of coastal terranes along the western margin of North America and contributed to a 135° clockwise rotation of the Ruby Mountains terrane with respect to the continental margin (Tailleur, 1980; Plafker and Berg, 1994a) (e.g., cf. Plates 6B-11 and 6B-12 to 6B-14). During the Late Cretaceous, northward-dipping subduction continued beneath much of the Pacific margin of Alaska and Siberia (Miller and Hudson, 1991). Between 110 Ma and 80 Ma, a significant magmatic belt and associated thermal metamorphism overprinted a broad east-west–trending belt across north-central Alaska (Miller, 1994), adjacent parts of Canada (Armstrong, 1988), and the Russian far east (Parfenov and Natal'in, 1986). In southern Alaska, the final suturing of the Wrangell composite terrane to North America occurred after the Cenomanian (91 Ma) with closure of the last remnant of the intervening ocean basin (Plates 6B-11 to 6B-13). Cole et al. (1999) also attributed the north-directed shortening in the Cantwell basin between 70 and 60 Ma to final suturing of the Wrangellia composite terrane to southern Alaska.

Accretion and formation of northeastern Eurasia by convergence and collision during the Mesozoic

Through the early Middle Jurassic (Plate 6B-10) the Siberian craton was the northeastern margin of Eurasia. Beginning ca. 150 Ma, as recorded by deformation in the Verkhoyansk fold and thrust belts (Parfenov, 1997), the Omulevka and Omolon blocks began to collide with Siberia (Fig. 8 and Plate 6B-11), initiating closure of the Oimyakon basin and moving the continental margin outboard of the Omulevka and Omolon blocks. Deformation along the collision zone at the platform margin continued, however, until 80 Ma (early Late Cretaceous). Both Parfenov (1997) and Nokleberg et al. (1998) showed the development of a seafloor-spreading center in a "South Anyui basin" during the Early and Middle Jurassic. Parfenov (1997) placed the spreading center between the Koyukuk arc and Eurasia and

connected it to a spreading center between the Alazeya arc and North America, whereas Nokleberg et al. (1998) placed it behind opposite outward-facing arcs, the Koyukuk arc off North America and the Uyandina arc off Eurasia. This Mesozoic seafloor-spreading center may be the equivalent of backarc basin extension behind a major subduction zone along the Pacific margin (Bonderenko and Didenko, 1997) outboard of both the arc shown in Plate 6B-11 and the Omolon block. The Verkoyansk orogen (Parfenov et al., 1995) records many complex tectonic events involving closure of a number of marginal basins (Plates 6B-10 to 6B-12). During the Late Jurassic, the Mongol-Okhotsk Sea closed and major blocks accreted to the southern margin of Siberia, including Amuria and the North China and South China blocks (Parfenov, 1997; Nokleberg et al., 1998), completing the assembly of eastern Eurasia (Plates 6B-10 to 6B-11).

Rotational opening of the Amerasia basin was probably stopped by collision of Chukotka with Eurasia at the South Anyui suture zone (Parfenov et al., 1995; Parfenov, 1997; Sokolov et al., this volume). Parfenov (1997) showed a complete assembly of far eastern Siberia as early as the Albian (ca. 110 Ma). Our rigid plate model, however, suggests that closure between Chukotka and the remainder of far eastern Siberia may not have been completed until sometime in the Cenozoic (cf. Plates 6B-14 and 6B-15). Our model requires some deformation in far eastern Siberia after closure of the South Anyui basin, but whether the deformation was elimination of oceanic crust or convergence and thickening of continental crust is not known. Studies of the present plate boundary between Eurasia and North America (Fujita and Cook, 1990) indicate a series of transtensional structures from the Laptev Sea margin near the New Siberian Islands to the Mona rift at ~65°N, 146°E (Fig. 2). While our model is not definitive, it suggests there may have been as much as 300 km of closure somewhere within far eastern Siberia between the time of collision at the South Anyui suture zone ca. 120 Ma and the present. The only other possibility would be an as-yet unrecognized, but significant plate boundary between Eurasia and North America where the required Late Cretaceous and Cenozoic motion could have been accommodated.

Accretion of Alaska by oblique and normal convergence between the Kula plate and North America during the Late Cretaceous and Paleogene

Formation of the Kula-Pacific spreading center in the northeastern Pacific Ocean ca. 85 Ma (Woods and Davies, 1982) rotated the convergence velocity vectors between the Kula plate and northwestern North America from northeast to north and increased their velocity from 80–100 to 150–200 km/m.y. These changes strengthened the dextral strike-slip component of the motion between the Kula and Farallon plates and the western margin of North America. They also created a complex system of dextral strike-slip faults of large horizontal displacement within and along the continental margin of western North America as far north as south-central Alaska (St. Amand, 1957; Grantz, 1966; Plafker and Berg, 1994a). These faults include the Queen Charlotte, Fairweather, and Denali fault systems of coastal British Columbia; southeast Alaska and interior central Alaska; and the Tintina fault within the Cordillera of western Canada and central Alaska. Displacement on the Tintina fault system was mainly during the Late Cretaceous and early Tertiary, but there may have been Late Jurassic or older offsets (Gabrielse, 1985). The Denali system has been active from the Late (mainly latest) Cretaceous to the Holocene, and the Queen Charlotte–Fairweather system has had mainly late Cenozoic displacement. In south-central and western Alaska, the Kula–North America convergence was in general normal to the continental margin and produced the accretionary prisms of the Chugach and Prince William terranes of the Southern Margin composite terrane, and parts of the Yakutat terrane (see Table 3; Fig. 8 and Plate 6B-14). Far-field effects of the Kula–North American convergence appear to have created duplexes within the allochthons of the central Brooks Range and low-angle thrust faults and foreland folds within the northern foothills of the Brooks Range and the continental shelf and slope of the eastern Beaufort Sea (Grantz et al., 1991; Moore et al., 1994). Geologic and paleomagnetic data (Grantz, 1966; Globerman, 1985; Plafker and Berg, 1994a) support a counterclockwise bending of southwestern Alaska of 45°–60° between ca. 65 and 50 Ma.

The predominant building blocks of which southeast, central, and southwestern Alaska and adjacent parts of British Columbia were assembled are slivers sliced off coastal western North America by dextral strike slip along the Tintina, Denali, Fairweather, and Queen Charlotte fault systems and their predecessors. These fault-bounded slivers became tectonostratigraphic terranes as a result of the tens or hundreds of kilometers of northwest and westward tectonic transport that carried them to their present locations in central and southwestern Alaska (Plates 6B-12 to 6B-16). These predominantly continental terranes are generally outboard (south) of the ensimatic island-arc terranes (Koyukuk, Togiak, and Nyac; see Togiak composite terrane in Table 2) that were accreted to western Alaska between the Jurassic and late Early Cretaceous.

A major jump in the location of the Kula–North American plate convergence from the foot of the continental slope in the Bering Sea (the Beringian margin) to the modern Aleutian arc occurred during the early Eocene. Scholl et al. (1986, 1987) reported that the Aleutian arc formed ca. 50–55 Ma, which is also about the time that the Mid-Atlantic Ridge split the outer Barents Sea shelf from Eurasia to create the Lomonosov Ridge and the Eurasia basin of the Arctic Ocean (Karasik, 1973; Vogt et al., 1979). Subduction at the Aleutian arc consumed the Kula-Pacific spreading center ca. 42 Ma (middle Eocene) (Lonsdale, 1988), ending the strong north-directed convergence (150–200 km/m.y. according to Engebretson et al., 1985) between the Pacific basin and southern Alaska. Thereafter, convergence vectors between the Pacific plate and North America had northwesterly azimuths and velocities of ~60–80 km/m.y. (Engebret-

son et al., 1985, Fig. 8). These vectors were parallel to the western margin of North America as far west as 140°W, and the motion continued to drive displacement along the major, seismogenic Queen Charlotte, Denali, and Fairweather dextral strike-slip fault systems of coastal northwestern North America and interior Alaska. West of 145°W, where the continental margin changes trend from northwest to southwest, the same convergence vectors are orthogonal to the continental margin and drove subduction at the Aleutian Trench. West of 180°W, the arcuate trend of the Aleutian arc and trench progressively bring the convergence velocity vectors and the trench into near parallelism, producing a transition from a fully convergent margin off south-central Alaska to a predominantly dextral strike-slip boundary along the south margin of the western Aleutian Islands. West of the Aleutian Trench, the Pacific plate converges orthogonally with eastern Eurasia along the generally northeast-trending Kuril Trench (Fig. 9) outboard of the Kamchatka Peninsula and the Sea of Okhotsk. Predecessors of this conver-

Figure 9. Major Cenozoic tectonic events that affected the Arctic region shown on present-day configuration. Late Cretaceous seafloor spreading in Labrador Sea and Baffin Bay split Greenland from North America. Cenozoic seafloor spreading began east of Greenland between Rockall Plateau–Hatton Bank and eastern Greenland. This seafloor spreading rifted Lomonosov Ridge from Barents Sea shelf margin. Double-ended arrows indicate active or assumed active seafloor spreading. Bering Sea region behind Aleutian Trench underwent "extrusion" as result of collision and oroclinal bending of Alaska. Right-lateral strike-slip arrows indicate direction of motion along faults along western margin of North America as well as along western end of Aleutian Island arc. Single-headed arrows indicate general plate motion directions but are not well defined. Subduction continues to the west of the Aleutian Trench at the Kurile Trench.

gence may have active when the South Anyui basin closed between 150 and 120 Ma (cf. Plates 6B-11 and 6B-12) and later produced a series of volcanic arcs that accreted to Siberia, starting with the Olyutorka-Kamchatka arc during the Maastrichtian (ca. 70 Ma) (Parfenov, 1994). The lesser Kuril and Kronotsky arcs appear to have been accreted in the Early Eocene (Parfenov, 1994), although Nokleberg et al. (1998) show them as having accreted during the Miocene.

Track of the Iceland hotspot from the Arctic to the Atlantic

According to Lawver and Müller (1994), a fixed present-day Iceland hotspot would have underlain northern Ellesmere Island ca. 130 Ma. The Alpha Ridge (AR in Fig. 1), which Vogt et al. (1979) suggested was a hotspot track, intersects the North American continental margin in the area where Lawver and Müller (1994) showed the fixed Iceland hotspot at the time of the opening of the Amerasia basin. A series of flood basalts recorded in the Canadian Arctic Islands include massive outpourings on Axel Heiberg and Ellesmere Islands during the late Albian (Embry, 1991). Embry (1991) reported early Aptian volcanic flows occur on northern Ellesmere Island, and Tarduno et al. (1998) dated part of the basalts of the Strand Fjord formation on Axel Heiberg Island as 95.3 ± 0.2 Ma (Cenomanian). As the North Atlantic and the Labrador Sea opened, the northeastern margin of the North American continent and Greenland drifted over the fixed hotspot. Uplift of the eastern margin of Baffin Island and flood basalts at Cape Dyer (Clarke and Upton, 1971) on the southern end of Baffin Island are also attributed to passage over the Siberian Traps–Iceland hotspot (Lawver and Müller, 1994). The Disko Island flood basalt along western Greenland (Gill et al., 1992) traces the track of the hotspot during the Paleocene. The hotspot uplifted the eastern margin of central Greenland ca. 40 Ma (Larson, 1990) and generated the alkalic basalts of Iceland and its surrounding platform from 25 Ma to the present.

Propagation of Central Atlantic seafloor spreading into the Arctic region and creation (opening) of the Eurasia basin

During the Late Cretaceous, Central Atlantic seafloor spreading propagated into the Labrador Sea and Baffin Bay (Roest and Srivastava, 1989; Srivastava and Roest, 1999). The pole of motion between North America and Eurasia for the period between the late Albian, by which time most authors would agree that the Amerasia basin had opened, and ca. 58 Ma in the late Paleocene, when the Eurasia basin (Fig. 9) began to open (Kristoffersen, 1990), was located near 68°N, 150°E (Srivastava and Roest, 1989). Such a pole, originally located south of far eastern Siberia, implies continued extension, albeit at a fairly slow rate, from the pole of rotation in far eastern Siberia toward the Baffin Bay region from the late Albian to the late Paleocene. We propose that some of this extension may be represented by the dispersed Late Cretaceous to early Tertiary extension seen in seismic reflection profiles from the Chukchi borderland of the Amerasia basin (Plate 6B-13) (Grantz et al., 1993) and locally within the broad Siberian margin (Drachev et al., 1998).

The major Cenozoic event in the Arctic was the initiation of opening of the Eurasia basin by extension of North Atlantic seafloor spreading (Talwani and Eldholm, 1977) into the Arctic between Greenland and Svalbard (site of the future Fram Strait) by splitting the 50–100-km-wide Lomonosov Ridge from the outer Barents Sea shelf ca. 58 Ma. We suggest that the initial position of the rift ~50–100 km inboard of the ocean-continental boundary of the early Paleocene Barents shelf was a consequence of two factors: the general parallelism of the North Atlantic spreading axis with the shelf margin, and the relative strength of the adjacent oceanic and continental lithospheres (Brace and Kohlsted, 1980). On reaching the Arctic near the continental margin, the Mid-Atlantic Ridge propagated through the weaker continental lithosphere of the outer Barents Sea shelf in preference to the more competent oceanic lithosphere beneath the adjacent Amerasia basin (Fig. 9 and Plates 6B-15 and 6B-16). Opening of the Eurasia basin accelerated the southward motion of Alaska and the westward translation of Arctic Canada while western Eurasia continues to move slightly northward.

SUMMARY AND CONCLUSIONS

The geologic framework of the present-day Arctic region is the product of a relatively small number of first-order plate tectonic events that operated through most of the Phanerozoic.

Eurasia, Pangea, and the major landmasses of the present-day Arctic region were largely assembled by plate migration and collision beginning with closure of the eastern part of the Iapetus Ocean during the Silurian to create Laurussia. Subsequent closure of the Rheic and paleo–Tethys Oceans in the Late Pennsylvanian united Laurussia with Gondwana and created Pangea, bringing most of the present Arctic landmasses into proximity.

Closure of the Iapetus Ocean during the Middle to Late Silurian involved the collision of Baltica, Pearya, and Chukotka with Laurentia and brought Caledonian deformation to Arctic North America. The Scandian phase of the Caledonian orogeny included left-lateral transpression between the Canadian Arctic Islands region and Chukotka, now part of northeastern Siberia.

A Devonian rifting event was possibly caused by a hotspot that produced the Vilyui basin on the Siberian craton. This event rifted a number of blocks from the Siberian margin, opened the Oimyakon Sea, and changed the absolute motion of Siberia from northwestward to northeastward.

We lack definitive evidence for the root cause of the later and distinct Ellesmerian orogeny of northwestern North America. It may have been a far-field effect of a collision of an unknown continental fragment with northern Laurussia during the Late Devonian to Early Mississippian; flat-slab subduction beneath the northern margin of Laurussia, similar to the subduction that caused the Mesozoic Gondwanide fold belt; or thin-skinned convergence generated by reactivation of the suture between Chukotka and Laurussia at which the Iapetus Ocean was closed during the Scandian phase of the Caledonian orogeny.

Emplacement of the Siberian Traps in northern Russia southeast of the modern Kara Sea ca. 250 Ma (Late Permian) was possibly the initial manifestation of the present-day Iceland hotspot, now located at ~65°N, 17°W. Migration of the newly amalgamated Arctic landmasses across this hotspot left a broken trail of igneous provinces and extensional features that include the extensive Siberian Traps, the Alpha and possibly the Mendeleev Ridges of the Amerasia basin, the Kap Washington volcanics of northern Greenland, flood basalts on Axel Heiberg and Ellesmere Islands in the Canadian Arctic Islands, Disko Island on West Greenland, and the volcanic accumulations on the present Iceland platform.

Tensional stresses created in the Arctic landmasses as they drifted across the Siberian Traps–Iceland hotspot, in conjunction with the distal effects of subduction of the Pacific plate beneath Arctic Alaska at the Pacific Rim, are thought to have triggered the rotational rifting that opened the Amerasia basin of the Arctic Ocean during early Early Cretaceous. The rift apparently propagated south-southeast along a zone of weakness within the early Paleozoic Franklinian deep-water basin until it encountered the Mackenzie platform of the Canadian shield in the vicinity of the Mackenzie Delta, which localized the pole of rotation in the lower Mackenzie River Valley.

Convergence of the oceanic plates of the Pacific Ocean with Alaska and northeastern Russia from the Jurassic onward amalgamated the allochthonous lithotectonic terrane fields of western Canada, southern and central Alaska, and Russia west of the Siberian shield and created the broad array of continental fragments, accretionary prisms of both continental margin and oceanic sediments, volcanic arcs, and ophiolites with sporadic blueschist metamorphism that characterize the continents that border the northern rim of the Pacific Ocean.

Slow but progressive migration of the future Arctic landmasses carried them from low latitudes in the Late Silurian to their present polar position by Late Cretaceous time.

Extension of seafloor spreading from the North Atlantic split the Lomonosov Ridge from the outer Barents Shelf beginning ca. 58 Ma, and created the slow-spreading Eurasia basin of the Arctic Ocean as well as transtensional rifts in far eastern Siberia and possibly in the Chukchi continental borderland.

ACKNOWLEDGMENTS

This chapter is the result of numerous discussions with untold numbers of colleagues. Much of our model has resulted from extensive interactions with Leonard Parfenov when he visited the University of Texas Institute for Geophysics in 1995 and with Jan Golonka, who has visited many times between 1995 and 2000. We also benefited from communications concerning the origin of the Amerasia basin with Michael B. Mickey and Alan P. Byrnes. We thank Ian Dalziel, Warren Nokleberg, Ian Norton, George Plafker, David Scholl, David Stone, and Simon Klemperer for critical reviews. Many of the reviewers suggested additions to the paper, including cross sections, tectonic boundaries (e.g., subduction zones, spreading centers), and more detailed comparisons with other models, but space and time did not permit this. Nokleberg et al. (1998) is a U.S. Geological Survey Open-File report that was superseded by a U.S. Geological Survey Professional Paper (Nokleberg et al., 2000).

This work was supported by the Plates Project, an industry-sponsored plate tectonic research program at the Institute for Geophysics, the University of Texas at Austin (Lawver and Gahagan) and National Science Foundation grants OPP95–31249 (Lawver) and OPP99–05790 (Grantz). We also gratefully acknowledge Exxon Exploration Co. for partial support of this study (Grantz). The U.S. Geological Survey provided support to Grantz through an emeritus appointment. This is Institute for Geophysics contribution 1513.

REFERENCES CITED

Aitken, J.D., 1993, Proterozoic sedimentary rocks, *in* Stott, D.F., and Aitken, J.D., eds., Sedimentary cover of the craton in Canada: Boulder, Colorado, Geological Society of America, Geology of North America, v. D-1, p. 81–95.

Anderson, A.V., Wallace, W.K., and Mull, C.G., 1994, Depositional record of a major tectonic transition in northern Alaska: Middle Devonian to Mississippian rift-basin margin deposits, upper Kongakut River region, eastern Brooks Range, Alaska, *in* Thurston, D.K., and Fujita, K., eds., Proceedings of the 1992 International Conference on Arctic Margins: Anchorage, Alaska, Minerals Management Service Outer Continental Shelf Report MMS 94–0040, p. 71–76.

Armstrong, R.L., 1988, Mesozoic and early Cenozoic magmatic evolution of the Canadian Cordillera, *in* Clark, S.P., Burchfiel, B.C., and Suppe, J., eds., Processes in continental lithospheric deformation: Boulder, Colorado, Geological Society of America Special Paper 218, p. 55–91.

Bailey, J.C., and Rasmussen, M.H., 1997, Petrochemistry of Jurassic and Cretaceous tholeiites from Kong Karls Land, Svalbard, and their relation to Mesozoic magmatism in the Arctic: Polar Research, v. 16, p. 37–62.

Basu, A.R., Poreda, R.J., Renne, P.R., Teichmann, F., Vasiliev, Y.R., Sobolev, N.V., and Turrin, B.D., 1995, High-^{3}He plume origin and temporal-spatial evolution of the Siberian flood basalts: Science, v. 269, p. 822–825.

Blodgett, R.B., 1998, Emsian (Late Early Devonian) fossils indicate a Siberian origin for the Farewell terrane, *in* Clough, J.G., and Larson, F., eds., Short notes on Alaskan geology 1997: Alaska Division of Geological and Geophysical Surveys Professional Report 118, p. 53–61.

Bondarenko, G.E., and Didenko, A.N., 1997, New geological and paleomagnetic data on the Jurassic-Cretaceous history of the Omolon Massif: Geotectonics, v. 31, p. 99–111.

Box, S.E., and Patton, W.W., Jr., 1989, Igneous history of the Koyukuk Terrane, Western Alaska: Constraints on the origin, evolution, and ultimate collision of an accreted island arc terrane: Journal of Geophysical Research, v. 94, p. 15843–15868.

Brace, W.F., and Kohlstedt, D., 1980, Limits on lithospheric stress imposed by laboratory experiments: Journal of Geophysical Research, v. 85, p. 6248–6252.

Brozena, J.M., Kovacs, L.C., and Lawver, L.A., 1998, An aerogeophysical study of the southern Canada Basin [abs.]: Third International Conference on Arctic Margins, Celle, Germany, p. 36–37.

Burke, K., and Dewey, J.F., 1973, Plume-generated triple junctions: Key indicators in applying plate tectonics to old rocks: Journal of Geology, v. 81, p. 46–433.

Cande, S.C., LaBrecque, J.L., Larson, R.L., Pittman, W.C., III, Golovchenko, X., and Haxby, W.F., 1989, Magnetic lineations of the world's ocean basins: Tulsa, Oklahoma, American Association of Petroleum Geologists, scale 1:27 400 000, 1 sheet, 13 p. text.

Carey, S.W., 1958, The tectonic approach to continental drift, *in* Carey, S.W., ed., Continental drift: A symposium: Hobart, Australia, Tasmania University, p. 177–355.

Carter, C., and Laufield, S., 1975, Ordovician and Silurian fossils in well cores from North Slope of Alaska: American Association of Petroleum Geologists Bulletin, v. 59, p. 457–464.

Cecile, M.P., Harrison, J.C., Kos'ko, M.K., and Parrish, R., 1991, Precambrian U-Pb ages of igneous rocks, Wrangel Island Complex, Wrangel Island, U.S.S.R.: Canadian Journal of Earth Sciences, v. 28, p. 1340–1348.

Clarke, D.B., and Upton, B.G.J., 1971, Tertiary basalts of Baffin Island: Field relations and tectonic setting: Canadian Journal of Earth Sciences, v. 8, p. 248–258.

Cole, R.B., Ridgway, K.D., Layer, P.W., and Drake, J., 1999, Kinematics of basin development during the transition from terrane accretion to strike-slip tectonics, Late Cretaceous–early Tertiary Cantwell Formation, south central Alaska: Tectonics, v. 18, p. 1224–1244.

Collins, F.R., 1958, Test wells, Topagoruk area, Alaska. 5. Subsurface geology and engineering data: Exploration of Naval Petroleum Reserve No. 4 and adjacent areas, northern Alaska, 1944–1953: U.S. Geological Survey Professional Paper 305–D, p. 265–316.

Crane, R.C., 1987, Cretaceous olistostrome model, Brooks Range, Alaska *in* Tailleur, I.L., and Weimar, P., eds., Alaskan North Slope geology, Volume 1: Bakersfield, California, Pacific Section, Society of Economic Paleontologists and Mineralogists and the Alaska Geological Society, Book 50, p. 433–440.

Dalziel, I.W.D., Lawver, L.A., and Murphy, J.B., 2000, Plumes, orogenesis, and supercontinental fragmentation: Earth and Planetary Sciences Letters, v. 178, p. 1–11.

DeLong, S.E., Fox, P., and McDowell, F.W., 1978, Subduction of the Kula Ridge at the Aleutian Trench: Geological Society of America Bulletin, v. 89, p. 83–95.

Didenko, A.N., and Pechersky, D., 1993, Revised paleozoic apparent polar wander paths for East Europe, Siberia, North China and Tarim plates: Program with Abstracts for L.P. Zonenshain Memorial Conference on Plate Tectonics, Moscow, p. 47.

Dixon, J., 1982, Jurassic and Lower Cretaceous subsurface stratigraphy of the Mackenzie Delta–Tuktoyaktuk Peninsula, N.W.T.: Geological Survey of Canada Bulletin, v. 349, 52 p.

Doré A.G., 1991, The structural foundation and evolution of Mesozoic seaways between Europe and the Arctic: Palaeogeography, Palaeoclimatology, Palaeoecology, v. 87, p. 441–492.

Drachev, S.S., Savostin, L.A., Groshev, V.G., and Bruni, I.E., 1998, Structure and geology of the continental shelf of the Laptev Sea, eastern Russian Arctic: Tectonophysics, v. 298, p. 357–393.

Embry, A.F., 1990, Geological and geophysical evidence in support of the hypothesis of anticlockwise rotation of northern Alaska: Marine Geology, v. 93, p. 317–329.

Embry, A.F., 1991, Mesozoic history of the Arctic Islands, *in* Trettin, H.P., ed., Geology of the Innuitian Orogen and Arctic Platform of Canada and Greenland: Boulder, Colorado, Geological Society of America, Geology of North America, v. E, p. 369–433.

Embry, A.F., and Dixon, J., 1994, The age of the Amerasia Basin, *in* Thurston, D.K., and Fujita, K., eds., Proceedings of the 1992 International Conference on Arctic Margins: Anchorage, Alaska, Minerals Management Service Outer Continental Shelf Report 94–0040, p. 289–294.

Embry, A.F., Mickey, M.B., Haga, H., and Wall, J.H., 1994, Correlation of the Pennsylvanian–Lower Cretaceous succession between northwest Alaska and southwest Sverdrup basin: Implications for Hanna trough stratigraphy, *in* Thurston, D.K., and Fujita, K., eds., Proceedings of the 1992 International Conference on Arctic Margins: Anchorage, Alaska, Minerals Management Service Outer Continental Shelf MMS Report 94–0040, p. 105–110.

Engebretson, D.C., Cox, A., and Gordon, R.G., 1985, Relative motions between oceanic and continental plates in the Pacific basin: Boulder, Colorado, Geological Society of America Special Paper 206, 59 p.

Fujita, K., and Cook, D.B., 1990, The Arctic continental margin of eastern Siberia, *in* Grantz, A., Johnson, G.L., and Sweeney, J.F., eds., The Arctic Ocean region: Boulder, Colorado, Geological Society of America, Geology of North America, v. L, p. 289–304.

Gabrielse, H., 1985, Major dextral transcurrent displacements along the Northern Rocky Mountain Trench and related lineaments in north-central British Columbia: Geological Society of America Bulletin, v. 96, p. 1–14.

Gee, D.G., Björklund, L., Stølen, L.-K., 1994, Early Proterozoic basement in Ny Friesland: Implications for the Caledonian tectonics of Svalbard: Tectonophysics, v. 231, p. 171–182.

Gehrels, G.E., and Saleeby, J.B., 1987, Geologic framework, tectonic evolution, and displacement history of the Alexander terrane: Tectonics, v. 6, p. 151–174.

Gill, R.C.O., Pedersen, A.K., Larsen, J.G., 1992, Tertiary picrites in West Greenland: Melting at the periphery of a plume? *in* Storey, B.C., Alabaster, T., and Pankhurst, R.J., eds., Magmatism and the causes of continental break-up: Geological Society [London] Special Publication 68, p. 335–348.

Globerman, B.D., 1985, A paleomagnetic and geochemical study of Upper Cretaceous to Lower Tertiary volcanic rocks from the Bristol Bay region, southwestern Alaska [Ph.D. thesis]: Santa Cruz, University of California, 292 p.

Golonka, J., 2000, Cambrian-Neogene plate tectonic maps: Rozprawy Habilitacyjne Nr. 350, Krakow, Poland, Wydawnictwa Uniwersytetn Jagielloñskiego (Jagiellonian University Publishing House), 125 p.

Gordey, S.P., Abbott, J.G., Tempelman-Kluit, D.J., and Garielse, H., 1987, "Antler" clastics in the Canadian Cordillera: Geology, v. 15, p. 103–107.

Gradstein, F.M., and Ogg, J.G., 1996, A phanerozoic time scale: Episodes, v. 19, p. 3–5.

Gradstein, F.M., Agterberg, F.P., Ogg, J.G., Hardenbol, J., van Veen, P., Thierry, J., and Huang, Z., 1994, A Mesozoic time scale: Journal of Geophysical Research, v. 99, p. 24051–24074.

Grantz, A., 1966, Strike-slip faults in Alaska: U.S. Geological Survey Open-File Report 267, 82 p.

Grantz, A., and May, S.D., 1982, Rifting history and structural development of the continental margin north of Alaska, *in* Watkins, J.S., and Drake, C.L., eds., Studies in continental margin geology: American Association of Petroleum Geologist Memoir 34, p. 77–100.

Grantz, A., Eittreim, S., and Dinter, D.A., 1979, Geology and tectonic development of the continental margin north of Alaska, *in* Keen, C.E., and Keen, M.J., eds., Crustal properties across passive continental margins: Tectonophysics, v. 59, p. 263–291.

Grantz, A., Tailleur, I.L., and Carter, C., 1983, Tectonic significance of Silurian and Ordovician graptolites, Lisburne Hills, northwest Alaska: Geological Society of America Abstracts with Programs, v. 15, no. 5, p. 274.

Grantz, A., May, S.D., and Dinter, D.A., 1987, Regional geology and petroleum potential of the United States Chukchi shelf north of Point Hope, *in* Scholl, D.W., Grantz, A., and Vedder, J.G., eds., Geology and resource potential of the continental margin of western North America and adjacent ocean basins: Beaufort Sea to Baja California: Houston, Texas, Circum-Pacific Council for Energy and Mineral Resources, Earth Science Series, v. 6, p. 37–58.

Grantz, A., May, S.D., and Hart, P.E., 1990, Geology of the Arctic continental margin of Alaska, *in* Grantz, A., Johnson, G.L., and Sweeney, J.F, eds., The Arctic Ocean region: Boulder, Colorado, Geological Society of America, Geology of North America, v. L, p. 257–288.

Grantz, A., Moore, T.E., and Roeske, S.M., 1991, Gulf of Alaska to Arctic Ocean: Boulder, Colorado, Geological Society of America, Centennial Continent/Ocean Transect # 15, 3 sheets, 72 p.

Grantz, A., May, S.D., Mullen, M.W., Gray, L.B., Lull, J.S., Clark, D.L., and Stevens, C.H., 1993, Northwind Ridge: A continental fragment isolated by Tertiary rifting in the Amerasia Basin, Arctic Ocean [abs.]: Eos (Transactions, American Geophysical Union), Fall Meeting Supplement, p. 614.

Grantz, A., Clark, D.L., Phillips, R.L., Srivastava, S.P., and 14 others, 1998, Phanerozoic stratigraphy of Northwind Ridge, magnetic anomalies in the Canada Basin, and the geometry and timing of rifting in the Amerasia Basin, Arctic Ocean: Geological Society of America Bulletin, v. 110, p. 801–820.

Gudlaugsson, S.T., Faleide, J.I., Johansen, S.E., and Breivik, A.J., 1998, Late Palaeozoic structural development of the southwestern Barents Sea: Marine and Petroleum Geology, v. 15, p. 73–102.

Gusev, G.S., Petrov, A.R., Fradkin, G.S., and others, 1985, Structure and evolution of the earth's crust of Yakutia: Moscow, Nauka, p. 248 (in Russian).

Hahn, G., and Hahn, R., 1993, Neue Trilobiten-Funde aus dem Karbon und Perm Alaskas (New trilobites from the Carboniferous and Permian of Alaska): Geologica et Palaeontologica, v. 27, p. 141–163.

Halgedahl, S., and Jarrard, R., 1987, Paleomagnetism of the Kuparuk River Formation from oriented drill core: Evidence for rotation of the Arctic Alaska plate, *in* Tailleur, I.L., and Weimer, P., eds., Alaska North Slope geology, Volume 2: Bakersfield, California, Pacific Section, Society of Economic Paleontologists and Mineralogists and the Alaska Geological Society, Book 50, p. 581–620.

Harris, R.A., 1989, Processes of allochthon emplacement with special reference to the Brooks Range ophiolite, Alaska and Timor, Indonesia [Ph.D. thesis]: London, University of London, 480 p.

Harrison, J.C., Wall, J.H., Brent, T.A., Poulton, T.P., and Davies, E.H., 1999, Rift-related structures in Jurassic and Lower Cretaceous strata near the Canadian polar margin, Yukon Territory, Northwest Territories, and Nunavut: Geological Survey of Canada, Current Research 1999–E, p. 47–58.

Hatcher, R.D., Jr., Thomas, W.A., Geiser, P.A., Snoke, A.W., Mosher, S., and Wiltschko, D.V., 1990, Alleghenian orogen, *in* Hatcher, R.D., Jr., Thomas, W.A., and Viele, G.W., eds., The Appalachian-Ouachita Orogen in the United States: Boulder, Colorado, Geological Society of America, Geology of North America, v. F, p. 233–318.

Higgins, A.K., Ineson, J.R., Peel, J.S., Surlyk, F., and Sonderholm, M., 1991, Cambrian to Silurian basin development and sedimentation, North Greenland, *in* Trettin, H.P., ed., Geology of the Innuitian Orogen and Arctic Platform of Canada and Greenland: Boulder, Colorado, Geological Society of America, Geology of North America, v. E., p. 111–161.

Howell, D.G., Jones, D.L., and Schermer, E.R., 1985, Tectonostratigraphic terranes of the circum-Pacific region: Principles of terrane analysis, *in* Howell, D.G., ed., Tectonostratigraphic terranes of the circum-Pacific region: Houston, Texas, Circum-Pacific Council for Energy and Mineral Resources, Earth Science Series, v. 1, p. 3–31.

Howell, D.G., Moore, G.W., and Wiley, T.J., 1987, Tectonics and basin evolution of Western North America: An overview, *in* Scholl, D.W., Grantz, A., and Vedder, J.G., eds., Geology and resource potential of the continental margin of western North America and adjacent ocean basins: Beaufort Sea to Baja California: Houston, Texas, Circum-Pacific Council for Energy and Mineral Resources, Earth Science Series, v. 6, p. 17–36.

Hubbard, R.J., Edrich, S.P., and Rattey, R.P., 1987, Geologic evolution and hydrocarbon habitat of the "Arctic Alaska Microplate": Marine and Petroleum Geology, v. 4, p. 2–34.

Kamysheva, G.G., 1973, Pole number 08038, *in* Khramov, A.N., ed., Paleomagnetic directions and paleomagnetic poles: Data for the USSR: Moscow, Materials of the WDC-B, Issue 2, p. 21, 63–64 (in Russian).

Karasik, A.M., 1973, Anomalous magnetic field of Eurasia Basin of the Arctic Ocean: Transactions (Doklady) of the Russian Academy of Sciences, Earth Science Sections, v. 211, p. 1–4 (in Russian).

Karl, S.M., and Aleinikof, J.N., 1990, Proterozoic U-Pb zircon age of granite in the Kallarichuk Hills, western Brooks Range, Alaska: Evidence for Precambrian basement in the schist belt: U.S. Geological Survey Bulletin, Report B-1946, p. 95–100.

Karl, S.M., Aleinikof, J.N., Dickey, C.F., and Dillon, J.T., 1989, Age and chemical composition of the Proterozoic intrusive complex at Mount Angayukaqsraq, western Brooks Range, Alaska, *in* Dover, J.H., and Galloway, J.P., eds., Geological studies in Alaska by the U.S. Geological Survey, 1988: U.S. Geological Survey Bulletin 1903, p. 10–19.

Khramov, A.N., 1984, Paleomagnetic directions and paleomagnetic poles: Data for USSR: Moscow, Summary Catalogue 1, Materials of the WDC-B, p. 94 (in Russian).

Khramov, A.N., 1986, Paleomagnetic directions and paleomagnetic poles: Data for USSR: Moscow, Materials of the WDC-B, Issue 6, 39 p. (in Russian).

Klaper, E.M., 1992, The Paleozoic tectonic evolution of the northern edge of North America: A structural study of northern Ellesmere Island, Canadian Arctic Archipelago: Tectonics, v. 11, p. 854–870.

Klaper, E.M., and Ohta, Y., 1993, Paleozoic metamorphism across the boundary between the Clements Markham foldbelt and the Pearya Terrane in northern Ellesmere Island, Canadian Arctic Archipelago: Canadian Journal of Earth Scineces, v. 30, p. 867–880.

Klitgord, K.D., and Schouten, H., 1986, Plate kinematics of the central Atlantic, *in* Vogt, P.R., and Tucholke, B.E., eds., The western North Atlantic region: Boulder, Colorado, Geological Society of America, Geology of North America, vol. M, p. 351–378.

Kos'ko, M.K., Cecile, M.P., Harrison, J.C., Ganelin, V.G., Khandoshko, N.V., and Lopatin, B.G., 1993, Geology of Wrangel Island, between Chukchi and East Siberian seas, northeastern Russia: Geological Survey of Canada Bulletin, v. 461, 101 p.

Kristoffersen, Y., 1990, Eurasia Basin, *in* Grantz, A., Johnson, G.L., and Sweeney, J.F., eds., The Arctic Ocean region: Boulder, Colorado, Geological Society of America, Geology of North America, v. L., p 365–378.

Lane, L., 1997, Canada Basin, Arctic Ocean: Evidence against a rotational origin: Tectonics, v. 16, p. 363–387.

Larson, H.-C., 1990, The East Greenland Shelf, *in* Grantz, A., Johnson, G.L., and Sweeney, J.F, eds., The Arctic Ocean region: Boulder, Colorado, Geological Society of America, Geology of North America, v. L, p. 185–210.

Lawver, L.A., and Hawkins, J.W., 1978, Diffuse magnetic anomalies in marginal basins: Their possible tectonic and petrologic significance: Tectonophysics, v. 45, p. 323–338.

Lawver, L.A., and Müller, R.D., 1994, The Iceland hotspot track: Geology, v. 22, p. 311–314.

Lawver, L.A., and Scotese, C.R., 1990, A review of tectonic models for the evolution of the Canada Basin, *in* Grantz, A., Johnson, G.L., and Sweeney, J.F, eds., The Arctic Ocean region: Boulder, Colorado, Geological Society of America, Geology of North America, v. L, p. 593–618.

Lawver, L.A., Brozena, J.M., Kovacs, L.C., and Childers, V.A., 1998, Plate kinematics in the Arctic: Mesozoic to recent evolution [abs.]: Eos (Transactions, American Geophysical Union), Fall Meeting Supplement, v. 79, p. F876.

Lawver, L.A., Gahagan, L.M., and Dalziel, I.W.D, 1999, A tight fit-Early Mesozoic Gondwana, a plate reconstruction perspective: Memoirs of the National Institute of Polar Research, International Symposium on the Origin and Evolution of Continents, Tokyo, Japan, no. 53, p. 214–229.

Laxon, S., and McAdoo, D., 1994, Arctic ocean gravity field derived from ERS-1 satellite altimetry: Science, v. 265, p. 621–624.

Laxon, S., and McAdoo, D., 1998, Satellites provide new insights into polar geophysics [abs.]: Eos (Transactions, American Geophysical Union), v. 79, p. 69, 72–73.

Lee, T-y., and Lawver, L.A., 1995, Cenozoic plate reconstruction of the Southeast Asia region: Tectonophysics, v. 251, p. 85–138.

Lerand, M., 1973, Beaufort Sea, *in* McCrossan, R.G., ed., The future petroleum provinces of Canada: Their geology and potential: Canadian Society of Petroleum Geologists Memoir 1, p. 315–386.

Lonsdale, P.F., 1988, Paleogene history of the Kula Plate: Offshore evidence and onshore implications: Geological Society of America Bulletin, v. 100, p. 733–754.

MacNiocaill, C., van der Pluijm, B.A., and Van der Voo, R., 1997, Ordovician paleogeography and the evolution of the Iapetus Ocean: Geology, v. 25, p. 159–162.

Mayfield, C.F., Tailleur, I.L., and Ellersieck, I., 1988, Stratigraphy, structure, and palinspastic synthesis of the western Brooks Range, northwestern Alaska, *in* Gryc, G., ed., Geology and exploration of the National Petroleum Reserve in Alaska, 1974 to 1982: U.S. Geological Survey Professional Paper 1399, p. 143–186.

McClelland, W.C., Gehrels, G.E., and Saleeby, J.B., 1992, Upper Jurassic–Lower Cretaceous basinal strata along the cordilleran margin: Implications for the accretionary history of the Alexander-Wrangellia-Peninsular Terrane: Tectonics, v. 11, p. 823–835.

McKerrow, W.S., MacNiocaill, C., and Dewey, J.F., 2000, The Caledonian Orogeny redefined: Journal of the Geological Society of London, v. 157, p. 1149–1154.

Miall, A.D., 1979, Mesozoic and Cenozoic geology of Banks Island, Arctic Canada: Geological Survey of Canada Memoir 387, 235 p.

Mihalynuk, M.G., Nelson, J., and Diakow, L.J., 1994, Cache Creek terrane entrapment: Oroclinal paradox within the Canadian Cordillera: Tectonics, v. 13, p. 575–595.

Miller, E.L., and Hudson, T., 1991, Mid-Cretaceous extensional fragmentation of a Jurassic–Early Cretaceous compressional orogen, Alaska: Tectonics, v. 10, p. 781–796.

Miller, T.P., 1994, Pre-Cenozoic plutonic rocks in mainland Alaska, *in* Plafker, G., and Berg, H.C., eds., The geology of Alaska: Boulder, Colorado, Geological Society of America, Geology of North America, v. G-1, p. 535–554.

Monger, J.W.H., and Berg, H.C., 1987, Lithotectonic terrane map of western Canada and southeastern Alaska: U.S. Geological Survey Miscellaneous Field Studies Map MF-1874–B, scale 1:2 500 000, 1 sheet, 12 p.

Moore, T.E., Wallace, W.K., Bird, K.J., Karl, S.M., Mull, C.G., and Dillon, J.T., 1994, Geology of northern Alaska, *in* Plafker, G., and Berg, H.C., eds., The geology of Alaska: Boulder, Colorado, Geological Society of America, Geology of North America, v. G-1, p. 535–554.

Mortensen, J.K., 1992, Pre–mid Mesozoic tectonic evolution of the Yukon-Tanana terrane, Yukon and Alaska: Tectonics, v. 11, p. 836–853.

Mull, C.G., Tailleur, I.L., Mayfield, C.F., Ellersieck, I.F., and Curtis, S., 1982, New upper Paleozoic and lower Mesozoic stratigraphic units, central and western Brooks Range, Alaska: American Association of Petroleum Geologists Bulletin, v. 66, p. 348–362.

Müller, R.D., Royer, J.-Y., and Lawver, L.A., 1993, Revised plate motions relative to the hotspots from combined Atlantic and Indian Ocean hotspot tracks: Geology, v. 21, p. 275–278.

Natal'in, B.A., Amato, J.M., Toro, J., and Wright, J.E., 1999, Paleozoic rocks of northern Chukotka Peninsula, Russian Far East: Implications for the tectonics of the Arctic region: Tectonics, v. 18, p. 977–1003.

Nie, S., Rowley, D.B., and Ziegler, A.M., 1990, Constraints on the locations of Asian microcontinents in Palaeo-Tethys during the Late Palaeozoic, *in* McKerrow, W.S., and Scotese, C.R., eds., Palaeozoic palaeogeography and biogeography: Geological Society of London Memoir 12, p. 397–409.

Nikishin, A.M., Ziegler, P.A., Stephenson, R.A., Cloetingh, S.A., Furne, A.V., Fokin, P.A., Ershov, A.V., Bolotov, S.N., Korotoaev, M.V., Alekseev, A.S., Gorbachev, V.I., Shipilov, E.V., Lankreijer, A., Bembinova, E.Yu., and Shalimov, I.V., 1996, Late Precambrian to Triassic history of the East European Craton: Dynamics of sedimentary basin evolution: Tectonophysics, v. 268, p. 23–63.

Nokleberg, W.J., Parfenov, L.M., Monger, J.W.H., Norton, I.O., Khanchuk, A.I., Stone, D.B., Scholl, D.W., and Fujita, K., 1998, Phanerozoic tectonic evolution of the circum–North Pacific, U.S. Geological Survey Open File Report 98–754, 125 p.

Nokleberg, W.J., Parfenov, L.M., Monger, J.W.H., Norton, I.O., Khanchuk, A.I., Stone, D.B., Scotese, C.R., Scholl, D.W., and Fujita, K., 2000, Phanerozoic tectonic evolution of the circum–North Pacific: U.S. Geological Survey Professional Paper 1626, 136 p.

Norton, I.O., 1995, Plate motions in the north Pacific: The 43 Ma non-event: Tectonics, v. 14, p. 1080–1094.

Norton, I.O., 2000, Global hotspot reference frames and plate motions, *in* Richards, M.A., Gordon, R.G., and Van der Hilst, R.D., eds., The history and dynamics of global plate motions: American Geophysical Union Monograph 121, p. 339–357.

Parfenov, L.M., 1991, Tectonics of the Verkhoyansk-Kolyma Mesozoides in the context of plate tectonics: Tectonophysics, v. 199, p. 319–342.

Parfenov, L.M., 1994, Accretionary history of northeast Asia, *in* Thurston, D.K., and Fujita, K., eds., Proceedings of the 1992 International Conference on Arctic Margins: Anchorage, Alaska, Minerals Management Service Outer Continental Shelf Report MMS 94–0040, p. 183–188.

Parfenov, L.M., 1997, Geological structure and geological history of Yakutia, *in* Parfenov, L.M., and Spektor, V.B., eds., Geological monuments of the Sakha Republic (Yakutia): Novosibirsk, Russia, Studio Design, p. 60–77.

Parfenov, L.M., and Natal'in, B.A., 1986, Mesozoic tectonic evolution of northeastern Asia: Tectonophysics, v. 127, p. 291–304.

Parfenov, L.M., Natapov, L.N., Sokolov, S.D., and Tsukanov, N.V., 1993, Terranes and accretionary tectonics of northeastern Asia: Geotectonics, v. 27, p. 62–72.

Parfenov, L.M., Prokopiev, A.V., and Gaiduk, V.V., 1995, Cretaceous frontal thrusts of the Verkhoyansk fold belt, eastern Siberia: Tectonics, v. 14, p. 342–358.

Patrick, B.E., and McClelland, W.C., 1995, Late Proterozoic granitic magmatism on Seward Peninsula and a Barentian origin for Arctic Alaska–Chukotka: Geology, v. 23, p. 81–84.

Patton, W.W., Jr., Box, S.E., Moll-Stalcup, E.J., and Miller, T.P., 1994, Geology of west-central Alaska, *in* Plafker, G., and Berg, H.C., eds., The geology of Alaska: Boulder, Colorado, Geological Society of America, Geology of North America, v. G-1, p. 241–270.

Plafker, G., and Berg, H.C., 1994a, Overview of the geology and tectonic evolution of Alaska, *in* Plafker, G., and Berg, H.C., eds., The geology of Alaska: Boulder, Colorado, Geological Society of America, Geology of North America, v. G-1, p. 989–1021.

Plafker, G., and Berg, H.C., editors, 1994b, The geology of Alaska: Boulder, Colorado, Geological Society of America, Geology of North America, v. G-1, 1055 p.

Renne, P.R., 1995, Excess ^{40}Ar in biotite and hornblende from the Noril'sk 1 Intrusion, Siberia: Implications for the age of the Siberian Traps: Earth and Planetary Science Letters, v. 131, p. 165–176.

Renne, P.R., Zichao, Z., Richards, M.A., Black, M.T., and Basu, A.R., 1995, Synchrony and causal relations between Permian-Triassic boundary crises and Siberian flood volcanism: Science, v. 269, p. 1413–1416.

Rickwood, F.K., 1970, The Prudhoe Bay field, *in* Adkison, W.L., and Brosgé, W.P., eds., Proceedings of the Geological Seminar on the North Slope of Alaska: American Association of Petroleum Geologists, Pacific Section, Los Angeles, p. L1–L11.

Roberts, D., and Sturt, B.A., 1980, Caledonian deformation in Norway: Journal of the Geological Society of London, v. 137, p. 241–250.

Rodionov, V.P., 1966, On the dipole nature of the geomagnetic field in the Late Cambrian and Ordovician in the southern part of the Siberian Platform: Academy of Science of the USSR, Siberian Branch, Geology and Geophysics, v. 1, p. 94–101 (in Russian).

Rodionov, V.P., Osipova, E.P., Pogarskaya, I.A., 1982, Pole number 10023, *in* Khramov, A.N., ed., Paleomagnetic directions and paleomagnetic poles:

Data for the USSR: Moscow, Materials of the WDC-B, Issue 5, p. 14, 38 (in Russian).

Roest, W.R., and Srivastava, S.P., 1989, Sea-floor spreading in the Labrador Sea: A new reconstruction: Geology, v. 17, p. 1000–1003.

Royer, J.-Y., Gahagan, L.M., Lawver, L.A., Mayes, C.L., and three others, 1990, A tectonic chart for the southern ocean derived from Geosat altimetry data: American Association of Petroleum Geologists Studies in Geology, no. 31, p. 89–100.

Rubin, C.M., Miller, M.M., and Smith, G.M., 1991, Tectonic development of Cordilleran mid-Paleozoic volcano-plutonic complexes: Evidence for convergent margin tectonism, *in* Harwood, D.S., and Miller, M.M., eds., Paleozoic and early Mesozoic paleogeographic relations of the Sierra Nevada, Klamath Mountains, and related terranes: Boulder, Colorado, Geological Society of America Special Paper 255, p. 1–16.

Rubin, C.M., Miller, E.L., and Toro, J., 1995, Deformation of the northern circum-Pacific margin: Variations in tectonic style and plate-tectonic implications: Geology, v. 23, p. 897–900.

St. Amand, P., 1957, Geological and geophysical synthesis of the tectonics of portions of British Columbia, the Yukon Territory and Alaska: Geological Society of America Bulletin, v. 68, p. 1343–1370.

Sandwell, D.T., and Smith, W.H.F., 1997, Marine gravity anomaly from Geosat and ERS-1 Satellite Altimetry: Journal of Geophysical Research, v. 102, p. 10039–10054.

Sanford, B.V., Thomson, F.J., and McFall, G.H., 1985, Plate tectonics: A possible controlling mechanism in the development of hydrocarbon traps in southwestern Ontario: Bulletin of Canadian Petroleum Geology, v. 33, p. 52–71.

Savage, N.M., 1988, Devonian faunas and major depositional events in the southern Alexander terrane, southeastern Alaska, *in* McMillan, N.J., Embry, A.F., and Glass, D.J., eds., Devonian of the world, Volume 1, Regional syntheses: Canadian Society of Petroleum Geologists Memoir 14, p. 257–264.

Scholl, D.W., Vallier, T.L., Stevenson, A.J., 1986, Terrane accretion, production, and continental growth: A perspective based on the origin and tectonic fate of the Aleutian–Bering Sea region: Geology, v. 14, no. 1, p. 43–47.

Scholl, D.W., Vallier, T.L., and Stevenson, A.J., 1987, Geologic evolution and petroleum geology of the Aleutian Ridge, *in* Scholl, D.W., Grantz, A., and Vedder, J.G., eds., Geology and resource potential of the continental margin of western North America and adjacent ocean basins—Beaufort Sea to Baja California: Houston, Texas, Circum-Pacific Council for Energy and Mineral Resources, Earth Science Series, v. 6, p. 123–155.

Şengör, C., and Natal'in, B., 1996, Paleotectonics of Asia: Fragments of a synthesis, *in* Yin, An, and Harrision, M., eds., The tectonic evolution of Asia: Cambridge, Cambridge University Press, p. 486–640.

Silberling, N.J., Jones, D.L., Monger, J.W.H., Coney, P.J., Berg, H.C., and Plafker, G., 1994, Lithotectonic terrane map of Alaska and adjacent parts of Canada, *in* Plafker, G., and Berg, H.C., eds., The geology of Alaska: Boulder, Colorado, Geological Society of America, Geology of North America, v. G-1, Plate 3, scale 1:2 500 000, 1 sheet.

Sleep, N.H., 1997, Lateral flow and ponding of starting plume material: Journal of Geophysical Research, v. 102, p. 10001–10012.

Smethurst, M.A., Khramov, A.N., and Torsvik, T.H., 1998, The Neoproterozoic and Paleozoic palaeomagnetic data for the Siberian Platform: From Rodinia to Pangea: Earth-Science Reviews, v. 43, p. 1–24.

Srivastava, S.P., and Roest, W.R., 1989, Sea floor spreading history II–VI Labrador Sea, *in* Bell, J.S. (coordinator), East Coast Basin Atlas Series: Labrador Sea: Dartmouth, Nova Scotia, Geological Survey of Canada, Atlantic Geoscience Centre, p. 100–109.

Srivastava, S.P., and Roest, W.R., 1999, Extent of oceanic crust in the Labrador Sea: Marine and Petroleum Geology, v. 16, p. 65–84.

Srivastava, S.P., Kovacs, L.C., Verba, V.V., Maschenkov, S., Pogrebitsky, Y., Levesque, S., Stark, A., Oakey, G., Verhoef, J., and Macnab, R., 1992, Opening of the Canada Basin and its relation to the opening of the rest of the Arctic Basin and the North Atlantic: International Conference on Arctic Margins Abstracts with Programs, Anchorage, Alaska, p. 57.

Stephens, M.B., and Gee, D.G., 1985, A tectonic model for the evolution of the eugeoclinal terranes in the central Scandinavian Caledonides, *in* Gee, D.G., and Sturt, B.A., eds., The Caledonide orogen-Scandinavia and related areas: Chichester, UK, Wiley, p. 953–978.

Surlyk, F., and Hurst, J.M., 1984, The evolution of the early Paleozoic deep-water basin of North Greenland: Geological Society of America Bulletin, v. 95, p. 131–154.

Tailleur, I.L., 1969, Speculations on North Slope geology: Oil and Gas Journal, v. 67, p. 128–130.

Tailleur, I.L., 1973, Probable rift origin of Canada Basin, Arctic Ocean: American Association of Petroleum Geologists Memoir 19, p. 526–535.

Tailleur, I.L., 1980, Rationalization of Koyukuk "crunch," northern and central Alaska: American Association of Petroleum Geologists Bulletin, v. 64, p. 792.

Tailleur, I.L., and Brosgé, W.P., 1970, Tectonic history of northern Alaska, *in* Adkison, W.L., and Brosgé, W.P., eds., Proceedings of the Geological Seminar on the North Slope of Alaska: American Association of Petroleum Geologists, Pacific Section, Los Angeles, p. E1–E19.

Talwani, M., and Eldholm, O., 1977, Evolution of the Norwegian-Greenland Sea: Geological Society of America Bulletin, v. 88, p. 969–999.

Tarduno, J.A., Brinkman, D.B., Renne, P.R., Cottrell, R.D., Scher, H., and Castillo, P., 1998, Evidence for extreme climatic warmth from Late Cretaceous Arctic vertebrates: Science, v. 282, p. 2241–2244.

Taylor, P.T., Kovacs, L.C., Vogt, P.R., and Johnson, G.L., 1981, Detailed aeromagnetic investigation of the Arctic Basin, 2: Journal of Geophysical Research, v. 86, p. 6323–6333.

Tempelman-Kluit, D.J., 1979, Transported cataclasite, ophiolite and granodiorite in the Yukon: Evidence of arc-continent collision: Geological Survey of Canada Paper 79–14, 27 p.

Torsvik, T.H., Tait, J., Moralev, V.M., McKerrow, W.S., Sturt, B.A., and Roberts, D., 1995, Ordovician paleogeography of Siberia and adjacent continents: Journal of the Geological Society of London, v. 152, p. 279–287.

Trettin, H.P., 1987, Pre-Carboniferous geology, M'Clintock Inlet map-area, northern Ellesmere Island, interim report and map (340E, H): Geological Survey of Canada Open-File 1652, 210 p.

Trettin, H.P., 1991a, Late Silurian–Early Devonian deformation, metamorphism and granitic plutonism, Northern Ellesmere and Axel Heiberg Islands, *in* Trettin, H.P., ed., Geology of the Innuitian Orogen and Arctic Platform of Canada and Greenland: Boulder, Colorado, Geological Society of America, Geology of North America, v. E, p. 295–301.

Trettin, H.P., 1991b, Introduction (Silurian–Early Carboniferous deformational phases and associated metamorphism and plutonism, Arctic Islands), *in* Trettin, H.P., ed., Geology of the Innuitian Orogen and Arctic Platform of Canada and Greenland: Boulder, Colorado, Geological Society of America, Geology of North America, v. E, p. 295.

Trettin, H.P., 1991c, Summary (Silurian–Early Carboniferous deformational phases and associated metamorphism and plutonism, Arctic Islands), Chapter 12, *in* Trettin, H.P., ed., Geology of the Innuitian Orogen and Arctic Platform of Canada and Greenland: Boulder, Colorado, Geological Society of America, Geology of North America, v. E, p. 337–341.

Trettin, H.P., Mayr, U., Long, G.D.F., and Packard, J.J., 1991, Cambrian to Early Devonian basin development, sedimentation, and volcanism, Arctic Islands, *in* Trettin, H.P., ed., Geology of the Innuitian Orogen and Arctic Platform of Canada and Greenland: Boulder, Colorado, Geological Society of America, Geology of North America, v. E, p. 165–238.

Van der Pluijm, B.A., Van der Voo, R., Potts, S.S., and Stamatakos, J., 1993, Early Silurian paleolatitude for central Newfoundland from paleomagnetism of the Wigwam Formation: Discussion: Canadian Journal of Earth Sciences, v. 30, p. 644–645.

Van der Voo, R., 1993, Paleomagnetics of the Atlantic, Tethys and Iapetus Oceans: Cambridge, Cambridge University Press, 411 p.

Vernikovsky, V.A., 1997, Neoproterozoic and Late Paleozoic Taimyr orogenic and ophiolitic belts, north Asia: A review and models for their formation, *in* Xu, Z., Ren, Y., and Qiu, X., eds., Proceedings, 30th International Geological Congress, Beijing, v. 7, p. 121–138.

Vogt, P.R., Taylor, P.T., Kovacs, L.C., and Johnson, G.L., 1979, Detailed aeromagnetic investigations on the Arctic Basin: Journal of Geophysical Research, v. 84, p. 1071–1089.

White, R.S., and McKenzie, D.P., 1989, Magmatism at rift zones: The generation of volcanic continental margins and flood basalts: Journal of Geophysical Research, v. 94, p. 7685–7729.

Wirth, K.R., 1991, Processes of lithosphere evolution: Geochemistry and tectonics of mafic rocks in the Brooks Range and Yukon-Tanana region, Alaska [Ph.D. thesis]: Ithaca, New York, Cornell University, 367 p.

Wirth, K.R., and Bird, J.M., 1992, Chronology of ophiolite crystallization, detachment, and emplacement: Evidence from the Brooks Range, Alaska: Geology, v. 20, p. 75–78.

Woods, M.T., and Davies, G.F., 1982, Late Cretaceous genesis of the Kula plate: Earth and Planetary Science Letters, v. 10, p. 376–381.

Ziegler, P.A., 1988, Evolution of the Arctic–North Atlantic and the Western Tethys: American Association of Petroleum Geologists Memoir 43, 198 p.

Ziegler, P.A., 1989, Evolution of Laurussia: Dordrecht, Netherlands, Kluwer Academic Publishers, 102 p.

Zonenshain, L.P., Kuzmin, M.I., and Natapov, L.M., 1990, Geology of the USSR, *in* Page, B.M., ed., A plate-tectonic synthesis: American Geophysical Union, Geodynamics Series, v. 21, 242 p.

MANUSCRIPT ACCEPTED BY THE SOCIETY MAY 15, 2001.

Geological Society of America
Special Paper 360
2002

Geographic information systems compilation of geophysical, geologic, and tectonic data for the Bering Shelf, Chukchi Sea, Arctic margin, and adjacent landmasses

Simon L. Klemperer*
Department of Geophysics, Stanford University, Stanford, California 94305-2215, USA
Mark L. Greninger*
Earth Systems Program, Stanford University, Stanford, California 94305-2210, USA
Warren J. Nokleberg
U.S. Geological Survey, Menlo Park, California 94025, USA

ABSTRACT

The accompanying CD-ROM contains a compilation of geophysical, geologic, and tectonic data for the Bering Shelf, the Chukchi Sea, the Arctic margin, and adjacent landmasses. These data sets focus on Alaska, the Russian Far East, and the continental shelves that link these two landmasses. For compatibility with other available geographic information system (GIS) products, our GIS compilation extends from 120°E to 115°W, and from 40°N to 80°N. This area encompasses the region from the modern Pacific plate boundary of the Japan, Kurile, and Aleutian subduction zones, the Queen Charlotte transform fault, and the Cascadia subduction zone (in the south) to the continent-ocean transition from the Eurasian and North American continents to the Arctic Ocean (in the north); and from the diffuse Eurasian–North American plate boundary, including the probable Okhotsk plate (in the west) to the Alaskan-Canadian Cordilleran fold belt (in the east). The CD-ROM comprises thematic layers of spatial data sets for topography, gravity field, magnetic field, earthquakes, volcanoes, geology, tectonostratigraphic terranes, and cultural reference features, and also includes metadata (data about the data) for all these data sets. The spatial data sets can be viewed, analyzed, and plotted with commercial GIS software (ArcView and ARC/Info) or through a freeware program (ArcExplorer) that is included on this CD-ROM. This GIS compilation provides data for studies of the Mesozoic and Cenozoic collisional and accretionary tectonics that assembled this continental crust and of the neotectonics of active and passive plate margins in this region, and for constructing and interpreting geophysical, geologic, and tectonic models of the region.

*E-mail, Klemperer: sklemp@stanford.edu. Current address, Greninger: Rydek Computer Associates, Los Angeles County Department of Urban Research, 222 S. Hill Street, 5th Floor, Los Angeles, California 90012, USA.

Klemperer, S.L., Greninger, M.L., and Nokleberg, W.J., 2002, Geographic information systems compilation of geophysical, geologic, and tectonic data for the Bering Shelf, Chukchi Sea, Arctic margin, and adjacent landmasses, *in* Miller, E.L., Grantz, A., and Klemperer, S.L., eds., Tectonic Evolution of the Bering Shelf–Chukchi Sea–Arctic Margin and Adjacent Landmasses: Boulder, Colorado, Geological Society of America Special Paper 360, p. 359–374.

INTRODUCTION

In order to facilitate understanding of the tectonics of the geologically complex region between North America and Asia that includes Alaska and northeast Russia across the Bering Strait, we have assembled a geographic information systems (GIS) compilation of geological and geophysical data (Fig. 1). This region also presents difficulties for conventional and digital geographic mapping and for geologic mapping, in that the area includes Arctic regions, crosses longitude 180°, is partially marine, and straddles the United States/Russia political boundary. In Arctic latitudes the most common map projection (Mercator) introduces large distortions, so Polar and Lambert projections are often also used, creating the necessary but difficult ask of transforming data between projections. Many digital data sets are stored from 180°W to 180°E and some querying routines do not allow access of data sets straddling longitude 180°. Traditional geological tools allow us to map land areas, not marine areas, so that the geology of the submerged Bering and Chukchi continental shelves, extending more than 500 km offshore from mainland Alaska and Chukotka, is even less well known than that of Siberia, even though over most of this region the coastline is only an accidental manifestation of modern-day sea level rather than a fundamental geological boundary. The politicization of geologic maps leads us to commonly ignore the large part of the North American plate that is in Russia (the best geologic map of Alaska shows Russia as a white outline; the converse is true for the best geologic map of Russia). These four difficulties, i.e., variable projections, arbitrary longitude reference system, coastlines, and political boundaries, led us to create a GIS (Greninger et al., 1996) to facilitate our crustal studies of the Bering and Chukchi Seas, and adjacent onshore Alaska and Chukotka. That original GIS has been extended in the present compilation to the Circum–North Pacific area from 120°E to 115°W and from 40°N to 80°N to coincide with an earlier compilation, the Circum–North Pacific terrane map (Nokleberg et al., 1994a). The bulk of our digital data compilation has been made available as a U.S. Geological Survey (USGS) Open-File Report (Greninger et al., 1999), and is reprinted here with additions and corrections.

The map-making software most commonly used by academic geophysicists is probably Generic Mapping Tools (GMT) (Wessel and Smith, 1995) and the newer Interactive Generic Mapping Tools (iGMT) (Becker and Braun, 1998), both freeware packages that run on any Unix computer. GMT has been widely used to create images of superimposed point data sets (e.g., hypocenters) and raster data sets (e.g., topography), and many of the data sets it supports are common to this database (ETOPO5, GTOPO30, satellite-derived free-air gravity). GMT and iGMT are mapping, not GIS, software, so despite their widespread use we have chosen to develop a GIS compilation based on the Environmental Systems Research Institute (ESRI) ARC/Info and ArcView programs, for the following three reasons. (1) These programs permit database queries that are not available in GMT and iGMT. (2) We hope that use of ARC/Info and ArcView will encourage new, innovative ways of working with multiple geophysical and geological data sets such as geologic maps of the Russian Far East, Alaska, and the Canadian Cordillera. (3) Because ESRI software is available for both Windows and Unix computers, a greater opportunity exists for scientists in the former Soviet Union to have access to this GIS compilation.

We included data sets that we believe may assist geological and geophysical research, and that we are able to disseminate as digital data: topography, gravity field, magnetic field, earthquakes, volcanoes, geology, and tectonostratigraphic terranes, and sociopolitical fiducial information, though we have limited ourselves to data sets that can be fitted onto a single CD-ROM (640 Mb). The accompanying CD-ROM contains two separate products: (1) a compilation of data sets in ARC/Info format; and

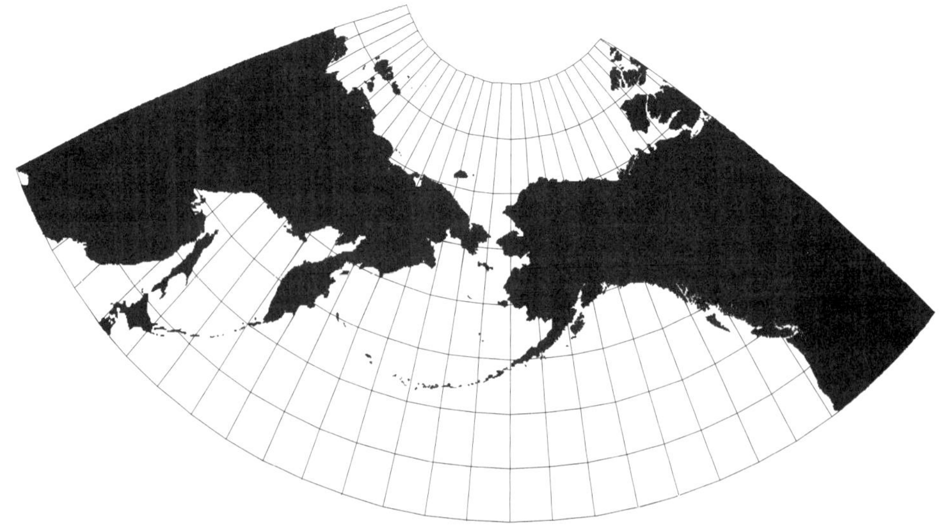

Figure 1. Location map showing full extent of geographic information system database. Few data sets cover entire region; extent of each data set is listed in metadata for that data set on the accompanying CD-ROM.

(2) a collection of displays (views) of combinations of these data sets in ArcView and ArcExplorer projects. The data sets in ARC/Info format are available to any GIS user for further manipulation and analysis, but it is beyond the scope of this chapter to provide a GIS tutorial: training is available online, e.g., through ESRI at http://campus.esri.com/campus/catalog/courses.cfm, and in readily available textbooks (e.g., ESRI, 1999; Hutchinson and Daniel, 2000). The ArcView and ArcExplorer projects can be used as displays for interpretation without any additional GIS manipulation. These projects include views to illustrate the type and quality of data available, and to illustrate some different ways in which data may be combined by utilizing the ability of GIS software to accept spatial queries.

The digital data consist of vector, raster, and point data. Vector data (e.g., the geographic base map) have a location precision that depends upon the scale of the original map and the density of the digitization points. Because of the large area covered, we have included no data sets or maps originally compiled at scales greater than 1:250 000, and some of our source data sets were compiled at scales as small as 1:10 000 000. Raster data (gridded data) have a resolution that depends upon the grid spacing and cell size. Our raster data sets range from <1 km to >8 km spacing. Point data (e.g., trackline shot locations or earthquake epicenter locations) are only limited by the precision of their locations.

All data sets are recorded in a Lambert azimuthal projection with a center of projection at 165°W, 70°N as originally utilized for the Circum–North Pacific terrane map (Nokleberg et al., 1994a) that forms the principal geologic base for our GIS compilation. Although it is common for digital data releases to be prepared as unprojected data, i.e., as data referenced to a geographic (latitude-longitude) coordinate system with no assigned map projection, all data sets in this GIS compilation are projected to a single chosen map projection in order to avoid problems associated with data sets that straddle longitude 180°. We do this so that this GIS compilation is accessible to scientists who lack proprietary GIS software and who use the freeware ArcExplorer provided on this CD-ROM. ArcExplorer displays our unprojected spatial data sets centered on the Greenwich, or zero meridian, split into separate far western and far eastern sections. Advanced users, i.e., those using programs such as ArcView or ARC/Info, have the option and ability to unproject and reproject the spatial data to a chosen projection. This procedure may lead to some loss of precision (due to a data conversion within ESRI software from double-precision to single-precision coordinates), but this loss of precision is not significant for data sets at scales of 1×10^6 and smaller. The geological and geophysical potential-field data sets on this CD-ROM are not intended for use at scales larger than 1:1 000 000.

All cartographic data sets for this GIS compilation are in ARC/Info coverage or GRID format. Various text files, including an explanatory README file, a detailed listing of all metadata (DOCUMENT), and detailed explanations to the Circum–North Pacific terrane map, are given in text (*.txt), Word 6 (*.doc), and Adobe Acrobat (*.pdf) formats. In addition, a suite of 178 stratigraphic columns for the Circum–North Pacific terrane map is provided in Adobe Acrobat format. The digital cartographic data for this compilation were compiled with ARC/Info 7.0.3 on a Sun Microsystems SPARC Station 20 computer running Solaris OpenWindows version 5.4 (Sun OS 2.4).

SYSTEM REQUIREMENTS

The data and text on this CD-ROM require a computer and software able to read ESRI data formats. Appropriate software packages include ARC/Info version 7.1.2 or higher, ArcView 3.1 or higher, and ArcExplorer. Full system requirements for each software package can be found at the Internet (Worldwide Web) homepage for ESRI, http://www.esri.com. In order to run ArcView 3.1, a Windows computer is required with a Pentium processor with 24 Mb of RAM (32 Mb recommended). A Pentium-II processor with a speed ≥200 MHz is recommended for handling the large data files. All systems require a color monitor that can display 256 colors. This CD-ROM was produced in accordance with the ISO 9660 and Macintosh HFS standards. All ASCII text on this CD-ROM file can be accessed from DOS, Windows, Macintosh, and Unix computers.

Instructions for installing the software on the disk, and complete details of the data contained on the disk, are found on the CD-ROM in the file /readme/document.***, in multiple file formats, text (*.txt), Word (*.doc), and portable document format (*.pdf).

GIS PROGRAMS FOR VIEWING OR MANIPULATING THESE DATA: ARC/INFO

ARC/Info is a full-featured GIS software program sold by Environmental Systems Research Institute, Inc. (ESRI), and is generally used on Unix-based workstations although it is also available for WindowsNT. (The separate software package PC ARC/Info is available for Windows (3.x, 95, 98, and NT), as well as DOS.) For the GIS compilation on this CD-ROM, the data are saved in ARC/Info 7.1.2, utilizing both coverage and GRID formats. ARC/Info can be utilized to view and manipulate the data sets on this CD-ROM exactly as compiled.

ArcView

ArcView (version 3.1) is a desktop GIS software program sold by Environmental Systems Research Institute, Inc. Although ArcView does not have all the features of ARC/Info, it is easier to learn; some recent ArcView extensions (e.g., 3D Analyst) are not available in ARC/Info 7.1.2. ArcView 3.1 runs on Unix computers and on Windows computers (3.x, 95, 98, or NT).

ArcExplorer

ArcExplorer is a freeware GIS-viewing program created by and available from Environmental Systems Research Institute,

Inc. (ESRI) by download from http://www.esri.com. A copy is provided on this CD-ROM. The directory /norpac/setup/arcexplo contains installers for ArcExplorer 1.1 for Windows 95/98/NT. NT machines must have Service Pack 3 installed. In addition to viewing GIS data in ARC/Info, ArcView, and other digital formats, ArcExplorer permits GIS data queries. ArcExplorer is currently only available for Windows computers (95, 98, or NT 4.0). ArcExplorer can access the GIS data contained in ARC/Info 7.1.2 coverages on this CD-ROM. Because ArcExplorer cannot display gridded data sets, images of these data sets are created as ***.bil files, which are displayed for the following data sets: (1) two raster topographic data sets (/norpac/data/topogrfy/raster/etopo5 and /gtopo2); (2) three gravity data sets (/norpac/data/gravity/dgrav, /geosat, and /sea-surf); and (3) five magnetic data sets (/norpac/data/magnetic/ak_mag, /arct_mag, /asia_mag, /dnag_mag, and /russ_mag).

USE OF GIS COMPILATION ON UNIX, MACINTOSH, AND DOS/WINDOWS COMPUTERS

Depending on system configuration and user needs, the data can be viewed directly from the CD-ROM, or the data can be downloaded onto a computer hard drive for viewing, manipulation, and plotting.

Unix

To mount this CD-ROM on a Unix-based computer, become the root user, then: % su (not necessary for a Silicon Graphics workstation).

If a /cdrom directory does not exist, create one:
cd/
#mkdir cdrom

Use the command appropriate to the UnixTM host:

DG AViiON:	mount -o noversion,ro -t cdrom/<dev> /cdrom
DEC ALPHA:	mount -t cdfs -r -o nodefperm,noversion /<dev> /cdrom
DECstation:	mount -t cdfs -r -o nodefperfm /<dev> /cdrom
HP 700/8x7:	mount -rt cdfs /<dev> /cdrom (or use sam)
IBM RS/6000:	mount -v `cdrfs´ -p˝ -r˝ /<dev> /cdrom (or use smit)
Silicon Graphics:	mount -o setx -t iso9660 /dev/ssi/<dev> /cdrom
Sun Solaris 1.x:	mount -rt hsfs /<dev> /cdrom
Sun Solaris 2.3:	Use volume management software to mount and access the CD-ROM. Sun Workstations running the Common Desktop Environment will auto-mount the CD-ROM.

Macintosh

ArcView 3.1 and ArcExplorer are not available for the Macintosh computer. The document, Adobe Acrobat PDF, and readme files can be viewed, manipulated, and printed with a Macintosh computer. ArcView 3.0 for Macintosh can be used to create new ArcView project files.

DOS/Windows

To mount this CD-ROM on a DOS/Windows-based computer, insert the CD-ROM into a drive, and open the CD-ROM window. Follow the instructions in the next section.

OPENING THE DATA SETS AND EXAMPLES OF GIS COMPILATION

In the directory /norpac/setup, two ArcView projects (norpac1.apr, norpac2.apr) and one ArcExplorer project (norpac.aep) are provided to give users easy access to data on the CD-ROM. Note that norpac1.apr requires that Spatial Analyst, an ArcView Extension that allows advanced manipulation and analysis of raster or point data (such as aeromagnetic, gravity, and earthquake data) compiled in ARC/Info GRID format, be installed as well as ArcView 3.1. Use of norpac2.apr is designed for ArcView users who do not also have Spatial Analyst. Use of the ArcExplorer 1.1 file, norpac.aep, requires installation of the ArcExplorer program that is provided on the CD-ROM.

Use of GIS Compilation with ArcView 3.1

If ArcView 3.1 or higher is installed on your computer, do the following.

1. Place the CD-ROM in the CD Drive. Open the CD-ROM window.

2. Find the ArcView project setup.apr located in the directory /norpac/setup.

3. Open the ArcView project by clicking on setup.apr, either from a file manager or from within ArcView.

4. ArcView should start with a small window with the title, "Cannot find the NorPac Data" followed by the sentence, "Enter the location of the NorPac Data." In the white box, replace "Drive Name" with the drive name and directory, "N:/norpac/" where "N" is the letter or name of the CD or hard drive. Next click "OK." A new window will appear with the title, "Found Spatial Analyst!" followed by the sentence, "Select a Project to load." In the white box, select either "Norpac1.apr (Spatial Analyst)," if your computer has both ArcView 3.1 and Spatial Analyst installed, or "Norpac2.apr (non-Spatial Analyst)," if your computer has only ArcView 3.1 installed. The selected ArcView project should load automatically. (Note that if Spatial Analyst is not installed, Norpac2.apr will automatically load.) A window will appear with the title norpac1.apr or norpac2.apr, depending on the preceding selection. In the white box on the left side of the window will be a list of views that can be opened, viewed, manipulated, or printed within ArcView. Each view is a digital map of part of the GIS compilation. Each view has one or more themes (layers) that can be selected (made visible) or deselected (made invisible).

The views are: Active Earth Example; Magnetic—Lithologic Correlation Example; Topographic Example; Cultural Features; Geology—Alaska; Geology—Bering-Chukchi shelves; Geology—Russia; Gravity—DGRAV; Gravity—Geosat; Gravity—Seasurface; Magnetics—Alaska; Magnetics—Arctic; Magnetics—DNAG; Magnetics—East Asia; Magnetics—Russia; Reflection Profile Tracklines; Seismicity; Terranes—Alaska; Terranes—Canadian Cordillera; Terranes—Circum–North Pacific; Topography—Raster; Topography—Vector; and Volcanoes—Active.

A more detailed description of these views, including the theme name of each layer shown on screen, the relevant data set location, and legend location, is contained on the CD-ROM in the file /readme/document.***, in the section on Description of Views. Note that if new themes and views are added to either norpac1.apr or to norpac2.apr and the ArcView project is saved, the ArcView project file needs to be edited before subsequent use of setup.apr. Using a text editor such as WordPad, the path name N:/norpac/ (where N is the designator of the CD drive) in either norpac1.apr or norpac2.apr (as appropriate) will need to be replaced by $norpac/.

Use of GIS Compilation with ArcExplorer 1.1

If ArcExplorer 1.1 is installed on your computer:
1. Place the CD-ROM in the CD Drive. Open the CD-ROM window.

2. Find the ArcView project "norpac.aep" located in the directory /norpac/setup.

3. Copy the ArcExplorer project, "norpac.aep" into a directory on your hard drive. Deselect the Read-only box for the properties of this file. In order to accomplish this on a Windows computer, right-click on the file name and select "Properties."

4. Open the norpac.aep file in a word processor program, such as WordPad. Substitute the text string, D:\norpac\ with N:\norpac\ where "N" is the letter or name of the CD drive. Save the file as a text file with a new file name. Note that a back slash (\) is used in the substitution.

5. Start the ArcExplorer project by clicking on the new file name. The ArcExplorer project will start. In the gray box on the left side will be a list of preselected themes that can be opened (by selecting), viewed, and printed. Each theme is one part of the GIS compilation. A full list of the theme names and their "translated" names is on the CD-ROM in the file /readme/document.***.

DIRECTORY ORGANIZATION

The /norpac directory of this CD-ROM contains four directories, /data, /readme, and /setup (Table 1), plus the /index directory used for Acrobat Search. All directory and subdirectory names are restricted to eight lowercase characters for the benefit of DOS users. In all tables the directories and subdirectories are

TABLE 1. DIRECTORY ORGANIZATION WITHIN /norpac

Directory	Description of files
data	GIS data in ARC/Info 7.1.2 and GRID formats.
	Data directories are organized by data type in the following subdirectories: cultural, geology, gravity, magnetic, seismcty, shiptrax, terranes, topogrfy, and volcano. Each of these contains a subdirectory for each data set, which in turn contains all necessary files for use of the data with ESRI Arc software. All the vector data sets are presented as ARC/Info coverages accessible by ArcView and ARC/Info. Because Arc/Info GRID format files are not accessible by ArcExplorer, these data sets are represented as ***.bil images that can be opened in ArcExplorer. Subdirectories within geology, seismcty, and terranes also contain explanatory texts and figures, in ***.txt, ***.doc, and ***.pdf formats, within sub-subdirectories called /explanat. All data directories act as ARC/Info workspaces and include an ARC/Info file.
	Two additional subdirectories exist. The directory /norpac/data/examples contains shape files for new coverages created from this CD-ROM that are used to produce example views of data compilations. The directory /norpac/data/legends contains suggested color tables and legends for different data sets. The directories /examples and /legends are not ARC/Info workspaces and have no ARC/Info files.
readme	Information for this CD-ROM in multiple file formats, text (***.txt), Word (***.doc), and portable document format (***.pdf). ***pdf files can be read with Adobe Acrobat Reader 4.0, freeware contained in /norpac/setup/Acrobat.
	readme.*** contains brief ASCII information about the CD-ROM.
	document.*** (this file) contains full documentation for all data sets on this CD-ROM.
	Subdirectory cpyright contains copyright notices associated with multiple-generation data.
setup	Installation files for ArcExplorer, the freeware software for users of this CD-ROM who lack access to the proprietary ArcView and ARC/Info GIS software. The files are contained in the subdirectory /norpac/setup/explorer. Files for Acrobat, freeware text-viewing software for PDF files, are contained in the subdirectory /norpac/setup/Acrobat.
	Project files for use with ArcView (norpac1.apr for users with the extension Spatial Analyst installed, norpac2.apr for those without) and for use with ArcExplorer (norpac.aep).

listed alphabetically. However, the ArcView windows with the available views for the projects norpac#.apr are listed alphabetically by "theme name," as listed in the contents of /data.

In the following sections we discuss the directories data, readme, and setup, and then the subdirectory /data/examples. Complete data about the data sets (metadata files) are on the CD-ROM in the files /readme/document.***, in multiple file formats, text (*.txt), Word (*.doc), and portable document format (*.pdf). These metadata listings contain the following information for each data set, to the extent available: the data type; the data set geographic extent, projection, source scale, or source grid interval; the data range (maximum and minimum values), resolution, and/or error estimates; a data dictionary of arc, polygon, and point attributes; data documentation files and data legends available on the CD-ROM; and the primary reference to, and any further information available about, the data set.

CONTENTS AND DESCRIPTION
OF /norpac/setup

The directory /norpac/setup contains subdirectories /explorer and /Acrobat, which hold the freeware programs ArcExplorer and Adobe Acrobat, respectively.

The directory /norpac/setup/Acrobat contains installers for Adobe Acrobat Reader 3.01 and 4.0 for both Windows 95/98/NT and Macintosh. The latest version of Adobe Acrobat Reader can also be downloaded free via the Internet from the Adobe homepage on the Worldwide Web at http://www.adobe.com.

The directory /norpac/setup also contains the files norpac .aep, norpac1.apr, and norpac2.apr. The ***.aep and ***.apr files are project files for ArcExplorer and ArcView, respectively. If either ArcView and/or ArcExplorer are installed, the ***.apr and ***.aep files can be loaded from within these programs, using the instructions given in the preceding section on opening the data sets, to display preset examples of the data. In general, users need not modify these files, but will wish to create new project files for individual needs.

CONTENTS AND DESCRIPTION
OF /norpac/data

For a complete list of the available data sets, see Table 2.

CONTENTS AND DESCRIPTION
OF /data/examples

Example compilations (coverages) of individual data sets (themes) from this GIS

In order to display some features (and limitations) of the available data sets, and to illustrate some ways in which our GIS compilation might stimulate data analysis, we have created three example "views" of various combinations of data sets using ArcView or ArcExplorer projects. Our "Topography" example (Fig. 2) shows the different resolution raster and vector data available for different subareas, and highlights both the range of data availability and also data problems for a single data type. Our "Active Earth" example (Fig. 3) combines tectonic information (faults, volcanoes, earthquake locations) with topographic and gravity data to show the relationship between these features and to synthesize the neotectonic activity of our region. Our "Magnetic-Lithologic Correlation" example (Fig. 4) overlays lithologic data on aeromagnetic data onshore, suggesting a correlation between high-frequency magnetic signature and igneous rocks that may then allow recognition of different geologic provinces offshore. These three views are only a few examples of many possible visualizations of combinations of the spatial data in this GIS compilation. For all three examples, the CD-ROM contains specific views and color scales that highlight features of interest. Users are encouraged to experiment with their own views, to zoom into different areas, and to select different display styles, in order to emphasize different features and to portray different and new geological relationships.

"Topography" example, /data/examples/topog_ex (Figure 2)

Table 3 lists the data sets that are superimposed in the "Topography" example view (Fig. 2); the higher-numbered layers are above and partly concealing the lower-numbered layers. Topography, i.e., the departure of the land and seafloor elevation from sea level, is the most basic geophysical data set, yet data availability and data quality vary widely. The coverages in this example (Fig. 2) have been chosen to show the range of different topographic data sets publicly available, both raster and vector, and land and marine, and to focus on typical data quality and reliability issues.

In the "Topography" example, the base layer is low-resolution (1:10 M) vector bathymetry with contours at 1000 m intervals, plus the −200 m isobath. Its land areas are obscured by layer 2, the best raster topography available globally, GTOPO30. This nominal 30 arc-second topography has points at ca. 900 m spacing, corresponding to a vector map on a scale of ~1:1 M (on which this raster data set would provide a data point every 0.25 mm). GTOPO30 has superimposed on it low-resolution drainages for geographic reference (layer 3), and more important, the contour lines (layer 4) from the low-resolution basal layer. Layers 1 and 4 are the same data set displayed with different legends, the former with color-shaded bathymetry (using the same legend as layers 5, 6, and 7), the latter with line contours only and no shading. Comparison of these contour lines with the raster GTOPO30 shows how the low-resolution contour data generalize the more detailed raster data. In general the contours match well with, but are obviously a generalization of, the raster topography; e.g., the narrow drainages dissecting the south flank of the Brooks Range are visible in the raster data (the chosen color scale emphasizes the change from the 200 m contour by a change from green to very pale green), but not in the 200 m contour line.

TABLE 2. DATA SETS CONTAINED WITHIN /norpac/data

Data directory	Subdirectory	File name (data set name) in GIS	Theme name used in views presented in projects norpac1.apr and norpac2.apr	Brief description of data and references
cultural		boundary	International/state boundaries	International, provincial, and state boundaries (ESRI's ArcWorld database, http://www.esri.com)
		cities	Cities	Significant population centers (ESRI's ArcWorld database, http://www.esri.com)
		features	Manmade features	Other sociocultural features, such as airports and golf courses (ESRI's ArcWorld database, http://www.esri.com)
		latlong5	Latitude/longitude grid (5°)	Latitude and longitude grid at 5° spacing (ESRI's ArcWorld database, http://www.esri.com)
		latlon12	Latitude/longitude grid (12°)	Latitude and longitude grid at 12° spacing (ESRI's ArcWorld database, http://www.esri.com)
examples				Shape files used in constructing the example views. "Topography," "Active Earth," and "Magnetic-Lithologic correlation" examples are discussed in text
geology	onshore	ak_geol	Geologic map of Alaska	Geologic map of Alaska (Beikman, 1980)
		rus_flts	Fault map of Russia	Fault map of Russia (Nalivkin, 1994; GlavNIVC, 1998)
		rus_geol	Geologic map of Russia	Geologic map of Russia (Nalivkin, 1994; GlavNIVC 1998)
	offshore	isopachs	Sedimentary basin isopachs, Bering-Chukchi shelves	1-km isopachs in selected Mesozoic and Cenozoic basins (Plate 1; Miller et al., this volume; Preface)
		basnflts	Intrabasin faults, Bering-Chukchi shelves	Faults within selected Mesozoic and Cenozoic basins (Plate 1; Miller et al., this volume; Preface)
		dredge	Dredge samples from the Beringian margin	Lithology of dredge samples from the Beringian margin (Plate 1; Miller et al., this volume; Preface)
		wells	Exploration wells, Bering-Chukchi shelves	Summary information for selected exploration wells (Plate 1; Miller et al., this volume; Preface)
gravity		dgrav	Onshore Bouguer and offshore free-air gravity (mGal)	Onshore Bouguer gravity anomalies and offshore free-air gravity anomalies; dgrav.bil has been created[†] (Hittelman, 1994; Committee for the Gravity Anomaly Map of North America, 1987)
		geosat	Satellite free-air gravity (mGal)	Satellite-derived free-air gravity, offshore areas only; geosat.bil has been created[†] (Smith and Sandwell, 1997)
		sea_surf	Satellite sea-surface height (m)	Satellite-derived sea-surface heights; sea_surf.bil has been created[†] (Hittelman, 1994)
legends				Contains suggested color-bars and contour information for each data set as ***.avl files; automatically accessed by ArcView as needed; contains subdirectories for each of the other subdirectories to /norpac/data (i.e., cultural, geology, gravity, etc.)

(continued)

TABLE 2. DATA SETS CONTAINED WITHIN /norpac/data (continued)

Data directory	Subdirectory	File name (data set name) in GIS	Theme name used in views presented in projects norpac1.apr and norpac2.apr	Brief description of data and references
magnetic		ak_mag	Magnetic map of Alaska (nT)	Magnetic map of Alaska (mainly onshore); ak_mag.bil has been created[†] (Saltus and Simmons, 1997)
		arct_mag	Magnetic map of the Arctic (nT)	Magnetic map of the Arctic (largely offshore); arct_mag.bil has been created[†] (Verhoef et al., 1996; Macnab et al., 1995)
		asia_mag	Magnetic map of Far East Asia (nT)	Magnetic map of Far East Asia; asia_mag.bil has been created[†] (Geological Survey of Japan and CCOP, 1996)
		dnag_mag	Magnetic map of North America (nT)	Magnetic map of North America, onshore and offshore; dnag_mag.bil has been created[†] (Committee for the Magnetic Anomaly Map of North America, 1987; Hittelman et al., 1989)
		russ_mag	Magnetic map of Russia (nT)	Magnetic map of Russia; russ_mag.bil has been created[†] (Anonymous, 1995)
seismcty		isc_cat	Global seismicity, 1964–1991	Catalog of global seismicity, 1964–1991 (Whiteside et al., 1996)
		ak_seis	Alaskan seismicity, 1898–1998	Alaska State seismicity, 1898–1998 (Hansen et al., 1999)
		chukseis	Chukotka and western Bering Sea seismicity	Catalog of Chukotka and western Bering Sea seismicity (Fujita et al., this volume and 2001)
shiptrax		ew94_09	Ship trackline, EW94-09	Ship trackline for geophysical cruise EW94-09 (Fliedner and Klemperer, 1999; Holbrook et al., 1999)
		ew94_10	Ship trackline, EW94-10	Ship trackline for geophysical cruise EW94-10 (Brocher et al., 1995; Klemperer et al., this volume, Chapter 1)
terranes	alaska	explanat		ak_terr.*** (map explanations)
		terr_lnd	Alaska terrane map, 1:2.5M	Tectonostratigraphic terrane and overlap map of Alaska (scale, 1:2.5M) (Nokleberg et al., 1994b)
	canada	assemblg	Canadian Cordillera—terranes/tectonic assemblages	Terrane and tectonic-assemblage maps of the Canadian Cordillera; hotlinks have been created[††] (Journeay and Williams, 1995)
	nor_pac		Circum–North Pacific Terrane Map	Circum-North Pacific Tectonostratigraphic Terrane Map; hotlinks have been created[††] (Nokleberg et al., 1994a)
		basins	Circum–North Pacific basins	Outlines of Mesozoic and Cenozoic basins
		coast	Circum–North Pacific coastline	Coastline; same as topogrfy/vector/contours, but presented as sea-level contour only
		explanat		assemb (text files for assemblage descriptions; columns (stratigraphic columns); terranes (text files for terrane descriptions); terrane.*** (map explanation)
		flts_lnd	Circum–North Pacific onshore faults	Onshore faults
		flts_ocn	Circum–North Pacific offshore faults	Offshore faults
		flts_pos	Circum–North Pacific post-accretionary faults	Post-accretionary faults, onshore and offshore
		mag_lins	Circum–North Pacific magnetic lineaments	Magnetic lineaments

		ocn_geol	Circum–North Pacific oceanic geology	Oceanic geology
		sea_mnts	Circum–North Pacific seamounts	Seamounts
		terr_lnd	Circum–North Pacific onshore terranes—terranes/overlap abbreviations/overlap assemblages	Onshore terranes
		terr_ocn	Circum–North Pacific offshore terranes	Offshore terranes (Nokleberg et al., 1994a)
topogrfy	raster			raster contains gridded topographic data
		etopo5	5′ topography/bathymetry (m)	5′-sampled land topography and marine bathymetry; etopo5.bil has been created[†] (Hittelman et al., 1994; NOAA, 1988)
		gtopo2	2′ topography/bathymetry (m)	2′-sampled land topography and marine bathymetry (only available to 70°N); gtopo2.bil has been created[†] (Smith and Sandwell, 1997)
		gtopo30	30″ land topography (m)	30″-sampled topography (land only) (U.S. Geological Survey, 1997)
		hillshad	hillshade of 30″ land topography	30″-sampled topography illuminated by sun from 315° at 45° elevation
	vector			vector contains contours (elevation or bathymetry) and hence polygons lying between fixed contour values (Alaska Biological Science Center, 1998)
		ak_shelf	U.S. shelf bathymetry (m) (1:0.25M)	Bathymetry for U.S. waters shallower than 200 m, Beaufort Sea to the Aleutians (most detailed, 1:0.25M)
		ber_chuk	Bering/Chukchi Seas bathymetry (m) (1:2.5M)	Bathymetry for Bering and Chukchi Seas, U.S. and Russian waters (intermediate detail, 1:2.5M)
		chukchi	Chukchi Sea/Bering Straits bathymetry (m) (1:1M)	Bathymetry for Chukchi Sea and Bering Straits, U.S. and Russian waters (more detailed, 1:1M)
		nor_pac	1:10M topography /bathymetry (m)	Circum–North Pacific topography and bathymetry (least detailed, 1:10M)
		rivers	Rivers	Major drainages, circum–North Pacific (Moore, 1990; Nokleberg et al., 1994a)
		shorline	World Vector Shoreline (1:0.25M)	World Vector Shoreline, designed for use at scales of up to 1:250 000 (http://web.ngdc.noaa.gov/mgg/fliers/93mgg01.html)
volcano		volcano	Volcanoes	Historically active volcanoes (Simkin & Siebert, 1994; Simkin et al., 1994)

[†] For some gridded data sets, an image file, /data/data-directory/data-subdirectory/filename.bil, is created to allow viewing of these data sets in the freeware ArcExplorer program, which otherwise cannot display these data sets.

[††] For some data sets, hotlinks are created to link features in views to other data sources, typically text descriptions of geologic units. When the linked feature is clicked with the Hot Link tool, ArcView opens the linked data source. Hotlinks are created for the map explanations for the Circum-North Pacific tectonostratigraphic terrane map and for the terrane and tectonic assemblage maps of the Canadian Cordilleran. Refer to ArcView manuals for use of the Hot Link tool.

Layers 5, 6, and 7 of the "Topography" example are successively more detailed bathymetric coverages (1:2.5 M; 1:1 M; 1:0.25 M, respectively) (cf. layers 1 and 4, 1:10 M), each data set encompassing a successively smaller area (Bering and Chukchi Seas; Chukchi Sea and Bering Straits; U.S. continental shelf [waters shallower than 200 m]) (cf. layer 1, with full coverage). Depth resolution increases correspondingly, from 1000 m throughout (layer 1), to 10 m in shallow water to 1000 m in deep water (layer 5), to 10 m throughout (layer 7). An identical color bar has been chosen to shade layers 1, 5, 6, and 7, so that the boundaries between the data sets are only visible where dense contours give way to sparse contours, and due to a thin outline drawn for each layer. The nesting of these four layers shows clearly how topographic resolution increases in countries rich enough to be rich in data (U.S. waters) and in shallow, nearshore waters. The uppermost of these bathymetric layers is used to illustrate a typical problem with digital data sets used uncritically. There are some coastal regions of shallow tidal waters for which topogrfy/vector/ak_shelf lacks data (so-called "holiday areas" in which no data were collected). The polygons are attributed with the elevation value +999, in contrast to the true but unknown negative value. When shaded uncritically, these areas may appear as land in a display (e.g., see the proximal regions of the Yukon Delta entering into Norton Sound which in reality have depths of –9 to –1 m, but are coded as positive values and shaded light green by the legend legends/topog_ex/ak_shelf.avl).

The relative agreement of different data sets is tested by comparing their coastlines with generally excellent results. With reference to the World Vector Shoreline, the 0 m (sea-level) contour derived from GTOPO30 is consistent to within 1 pixel (900 m) at Saint Lawrence Island in the approximate center of our coverages (Fig. 2). Note that although GTOPO30 is sampled at a 30 arc-second interval, in some areas this represents an oversampling of originally coarser data. This oversampling is interpreted as the cause of some artifacts, e.g., sharp boundaries in the elevation samples, both north-south and east-west, bounding the northern promontory of the Seward Peninsula. Athough barely visible in this "Topography" example, these artifacts are clearly seen in the "Active Earth" example. A complete 1-km-sampled digital elevation model of land areas of the Earth (NOAA's Global Land One-kilometer Base Elevation [GLOBE] Project) is under development at http://web.ngdc.noaa.gov/seg/topo/globe.html.

With respect to the World Vector Shoreline, the vector coverage topogrfy/vector/nor_pac (layers 1 and 4 in "Topography" example) is offset by 0–4 km around Saint Lawrence Island, appears consistent around Kamchatka, is displaced ~8 km north near Barrow, and is displaced ~4 km southwest near Anchorage (Fig. 2). We believe that these errors arose during the process of hand-tracing the coastline from the 1:10 M paper map (Moore, 1990) onto mylar, followed by scanning: a 1 mm error at the 1:10 M scale results in 10 km of relative displacement. Displacements of these magnitudes should also be anticipated in the

other coverages of the Circum–North Pacific terrane map, and represent the limits of useful accuracy in our database. Other errors are present in topogrfy/vector/nor_pac: several islands are missing (including both the Diomede Islands and even Nunivak, ~50 000 km^2!; Fig. 2). We do not know if similar errors are present in the Russian waters that are not known to us personally. At least a few polygons are miscoded (see the closed contour and apparent landmass east of the Shumagin Islands, highlighted in bright yellow in Fig. 2 by the legend legends/topog_ex/ n_p_eror.avl designed for this purpose). The higher-resolution vector coverages (topogrfy/vector/ak_shelf, topogrfy/vector/ chukchi, and topogrfy/vector/ber_chuk) all appear to be consistent with the World Vector Shoreline. Although not shown here, the Canadian terrane map terranes/canada/assemblg appears to be very well correlated with the World Vector Shoreline, and the Alaska terrane map terranes/alaska/terr_lnd seems to be offset by only ~3 km to the southwest, an error probably acquired during hand-digitization from a Decade of North American Geology (DNAG) map with its center of projection at ~100°W, well outside the present coverage.

Layers 8 and 9, ship tracks, are included in the "Topography" example as a visual centerpiece and because an original in-

Figure 2. Topography of Bering Shelf–Chukchi Sea and adjacent landmasses superimposed with tracklines of geophysical cruises discussed by Klemperer et al. (Chapter 1, this volume). Depths and elevations not specified in legend are shown by intermediate shades of greens and blues. See text for description of data sets shown, and discussion of potential pitfalls with compiled data sets, including holiday areas and miscoded polygons. A is Anchorage; B is Barrow; D is Diomede Islands; Nv is Nunivak Island; S is Shumagin Islands; St.L is Saint Lawrence Island. Image is on CD-ROM at /norpac/data/examples/topog_ex.

Figure 3. Free-air gravity of Bering Shelf–Chukchi Sea and relief of adjacent landmasses (sun shaded with sun angle of 45° elevation from northwest) superimposed with earthquake locations (color coded by depth; size of symbol corresponds to magnitude), Holocene volcanoes, and faults. See text for description of data sets shown. H is Hagemeister Island; K is Kaltag fault; St.M is Saint Matthew Island; T is Togiak-Tikchik strand of Denali fault and Togiak Lake; Z is Zemchug Canyon. Image is on CD-ROM at /norpac/data/examples/act_erth.

Figure 4. Magnetic anomalies of Bering Shelf–Chukchi Sea and adjacent landmasses superimposed with outline geological maps of Russia and Alaska. Because the four different magnetic data sets shown have not been merged, four different legends have been defined in which identical colors correspond to identical ranges of standard deviations away from the specific data mean, in order to minimize offscts in the dichromatic red-blue color scale where different data sets abut. Igneous (volcanic, plutonic, ultramafic) units are tinted yellow. See text for description of data sets shown. An is Anadyr basin; G is Goodnews arch; M-H is Matthew-Hall basin; Nav is Navarin basin; Nv is Nunivak Island; Pb is Pribilof Islands; St.G is Saint George basin; St.M is Saint Matthew Island. Corresponding image on CD-ROM at /norpac/data/examples/ mag_lith has the additional feature that the gray and yellow shading, distinguishing igneous and nonigneous rocks, is transparent, allowing correlation of magnetic features from onshore to offshore. This transparency is not preserved in this figure printed from a JPEG file.

75°N
70°N
65°N
60°N
55°N
50°N
180°
170°W
160°W
Chukotka
Brooks
Range
An
Nav
St. M
M-H
Nv
G
G
G
Pb
St.G
Bristol
Bay
Geologic compilations
Ak_geol
Russ_geol
Not igneous
Igneous
Magnetic compilations
Russ_mag
Dnag_mag
Ak_mag
Arct_mag
≤ 3 σ
≤ 2 σ
≤ 1 σ
≤ 0.5 σ
≤ .25 σ
≥ 3 σ
≥ 2 σ
≥ 1 σ
≥ 0.5 σ
≥ .25 σ
mean
no data

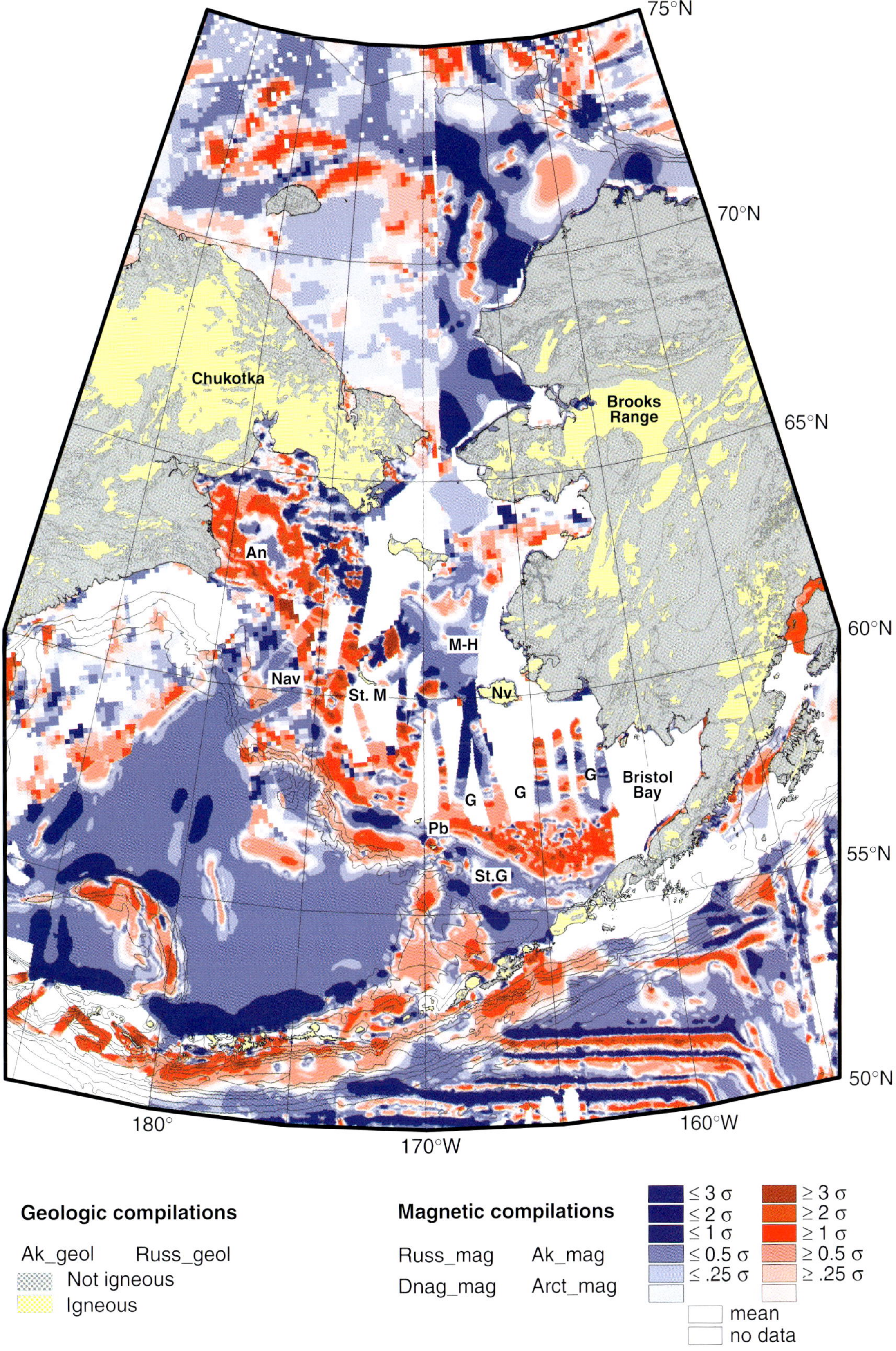

Figure 4

70°N
65°N
60°N
55°N
50°N
180°
170°W
160°W
Seward
Norton
Sound
K
St.M
T
H
Bristol
Bay
Alaska Peninsula
Z
Aleutian Islands
0 to 15 km
16 to 25 km
26 to 35 km
36 to 45 km
46 to 60 km
61 to 80 km
81 to 100 km
101 to 150 km
151 to 250 km
251 to 400 km
401 to 632 km
Earthquakes,
ISC 1964-1991
3.5 ≤ mag < 4.0
4.0 ≤ mag < 4.5
4.5 ≤ mag < 5.0
5.0 ≤ mag < 5.5
5.5 ≤ mag < 6.0
6.0 ≤ mag < 6.5
magnitude 6.5
Free-air gravity (Geosat)
≤ -120 mGal ≅ 3σ
≤ -80 mGal ≅ 2σ
≤ -50 mGal ≅ 1.25σ
≤ -30 mGal ≅ 0.75σ
≤ -15 mGal ≅ 0.3σ
-15 to -5 mGal
≥ +120 mGal ≅ 3σ
≥ +80 mGal ≅ 2σ
≥ +50 mGal ≅ 1.25σ
≥ +30 mGal ≅ 0.75σ
≥ +15 mGal ≅ 0.3σ
+15 to +5 mGal
-5 to +5 mGal
fault
Holocene volcano

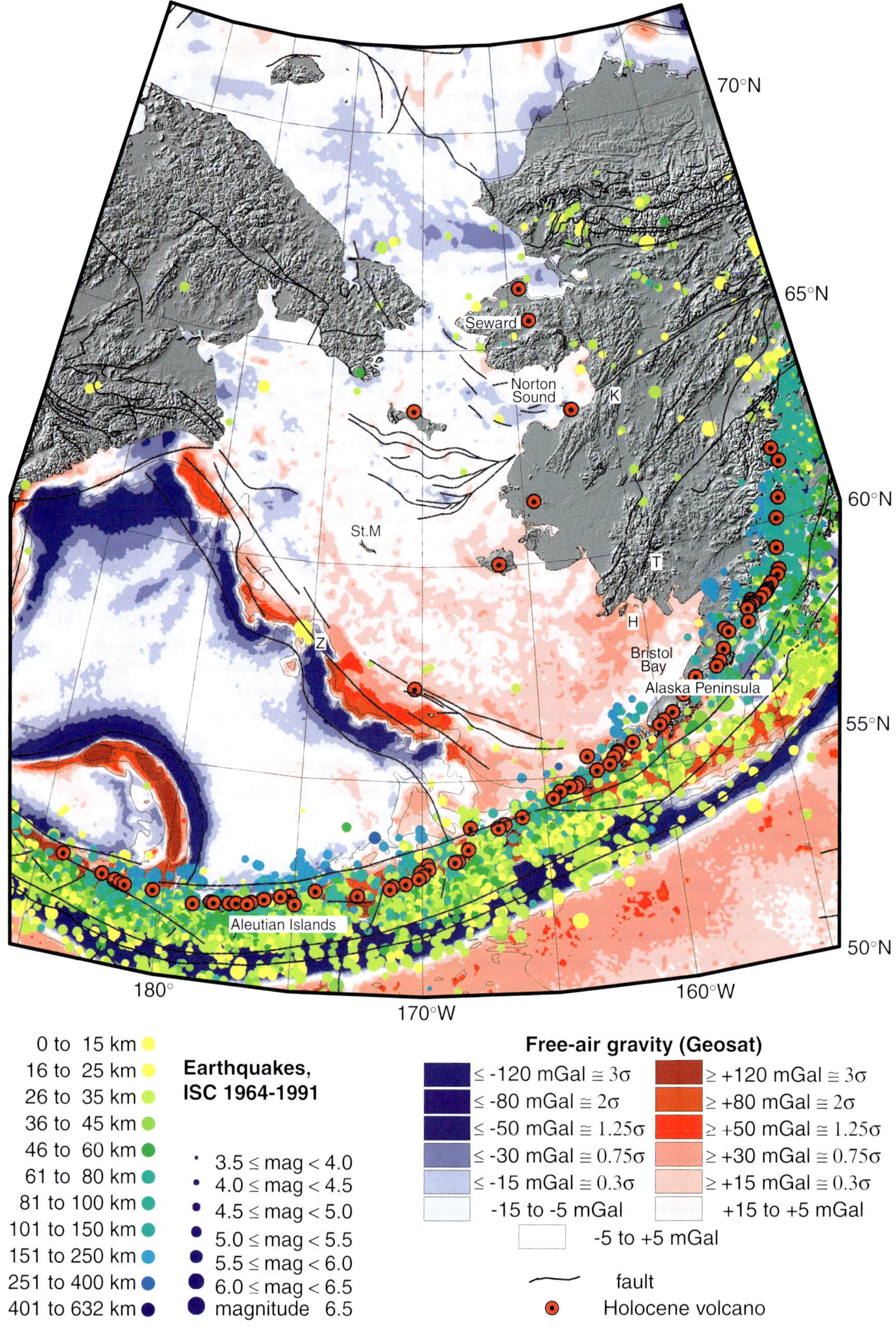

Figure 3

75°N
70°N
65°N
60°N
55°N
50°N
180°
170°W
160°W
Brooks Range
Seward
Norton Sound
St.L
Nv
D
B
A
S
raster topography
4000 to 6000 m
3000 to 3999 m
2000 to 2999 m
1000 to 1999 m
750 to 999 m
500 to 749 m
200 to 249 m
150 to 199 m
0 to 24 m
vector bathymetry
1 to 9 m
10 to 29 m
30 to 59 m
60 to 99 m
100 to 199 m
200 to 999 m
2000 to 5999 m
6000 to 9999 m
mis-coded polygons
miscoded areas of nor_pac
holiday areas of ak_shelf
marine seismic cruises
EW94-09 tracklines
EW94-10 tracklines

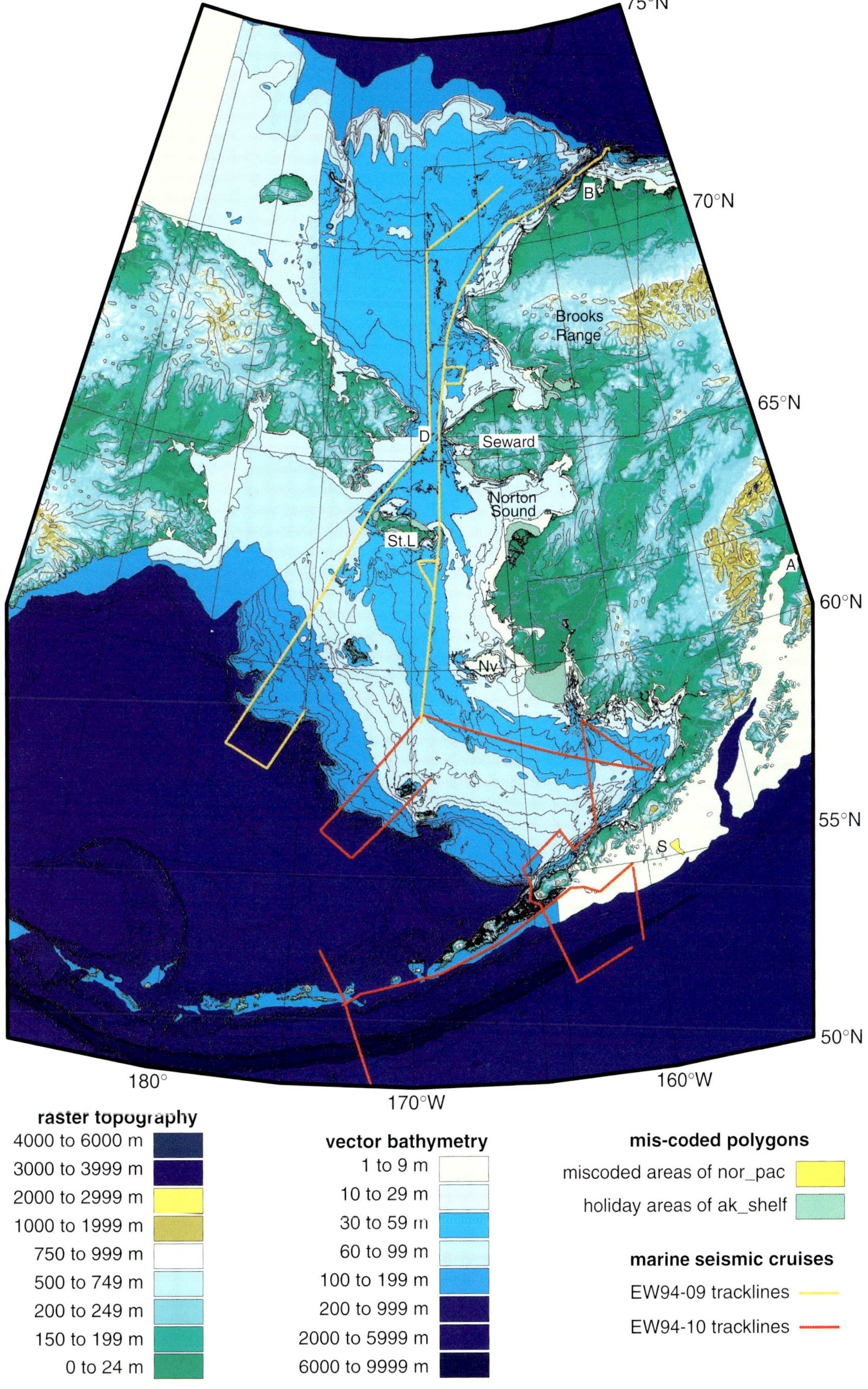

Figure 2

TABLE 3. DATA SETS DISPLAYED IN "TOPOGRAPHY" EXAMPLE

	Data set location	Legend location in /norpac/data/legends/topog_ex (locations of the form ***2.avl are for use with norpac2.apr)	Data description
10	cultural/latlong5	latlong5.avl	Latitude and longitude grid at 5° spacing
9	shiptrax/ew94_09	ew94_09.avl	Ship trackline for geophysical cruise EW94-09
8	shiptrax/ew94_10	ew94_10.avl	Ship trackline for geophysical cruise EW94-10
7	topogrfy/vector/ak_shelf	ak_shelf.avl (ak_shelf2.avl)	Bathymetry for U.S. waters shallower than 200 m, Beaufort Sea to the Aleutians (1:250 000)
6	topogrfy/vector/chukchi	topovect.avl (topovect2.avl)	Bathymetry for Chukchi Sea and Bering Straits, U.S. and Russian waters (1:1 000 000)
5	topogrfy/vector/ber_chuk	topovect.avl (topovect2.avl)	Bathymetry for Bering and Chukchi Seas, U.S. and Russian waters (1:2 500 000)
4	topogrfy/vector/nor_pac	n_p_eror.avl (n_p_eror2.avl)	Circum–North Pacific region (1:10 000 000), as topographic contours and a single polygon coded in error (see text)
3	topogrfy/vector/rivers	rivers.avl	Major drainages, circum–North Pacific region
2	topogrfy/raster/gtopo30	gtopo30.avl	30″-sampled land topography, color-shaded topography
1	topogrfy/vector/nor_pac	topovect.avl (topovect2.avl)	Circum–North Pacific region (1:10 000 000), color-shaded bathymetry

tent of this GIS database was to support interpretations of seismic profiles acquired along these tracklines.

"Active Earth" example, /data/examples/act_erth (Figure 3)

Table 4 lists the data sets that are superimposed in this view; the higher-numbered layers are above and partly concealing the lower-numbered layers. These data sets were chosen to show how a map can be created from our GIS database to illustrate the relations between active seismicity, mapped faults, volcanoes, and topography. This view illustrates the neotectonic setting of the study area.

Although topography is the most basic data set for studying tectonics, because marine bathymetry is shown in the "Topography" example, for the base layer in this example we display the satellite-derived free-air gravity, /gravity/geosat, using a standard blue-to-red dichromatic color bar (blue represents gravity lows; red represents gravity highs). These data constitute a topographic proxy in marine areas. On land the GTOPO30 data set is used, as in the "Topography" example, and because the marine gravity data set had its land areas masked using the nor_pac coastline, the offset between nor_pac and GTOPO30 appears as an offset, or doubled, coastline wherever the nor_pac coastline is outside (seaward of) the GTOPO30 coastline. This double coastline, e.g., adjacent to northern Alaska, represents the discrepancy between the nor_pac coastline and the GTOPO30 coastline, rather than a discrepancy between nor_pac and /gravity/geosat.

Subsurface density variations are important to the gravity field, but the most significant features are the major changes in water depth at the Aleutian trench, and across the Beringian margin and the Arctic shelf edge, as made clear by the superimposition of the second layer, bathymetric contours from /topogrfy/vector/nor_pac. Abrupt bathymetric features correspond well with corresponding features of the gravity field, e.g., the topographic promontory of the Bering Shelf edge, just west of Zemchug Canyon (Fig. 3).

For layer 3, rather than simply repeat the color-scale bar used for the GTOPO30 layer in the "Topography" example, a hillshade is applied to provide a real sense of the topography. The hillshade was created using the ESRI Spatial Analyst (ARC GRID has an equivalent program) with a sun angle of 45° elevation from the northwest. This sun angle highlights the data errors in GTOPO30 on the northern Seward Peninsula (Fig. 3) (described in "Topography" example), and highlights (as dark lineaments) the topographic lineaments and valleys corresponding to the southwest-northeast family of faults that transect southern and western Alaska.

Earthquake epicenters are overlaid as seven different themes from the International Seismological Centre (ISC) catalog, /seismcty/isc_cat. In order to capture both the depth and magnitude information on a single view, seven subsets of the catalog are derived corresponding to magnitude ranges of one-half unit, from $3.5 \leq$ magnitude < 4.0 to magnitude ≥ 6.5. The earthquakes of each magnitude range are displayed with a different-size circle, and these symbols are color coded by hypocentral depth (dichromatic yellow to blue representing shallow to deep sources). This color variation clearly shows the northward dip of the Benioff zone beneath the Alaskan Peninsula and the Aleutian Islands (Fig. 3). There is no easy way to use the focal mechanism of individual earthquakes as an individual symbol within ArcView, so focal-plane solutions are not included in the database. However, solutions for individual earthquakes are readily available, e.g., from the Harvard centroid-moment tensor (CMT) catalog at http://www.seismology.harvard.edu/CMTsearch.html.

Coverages of volcano and fault locations complete this example. The volcanoes (black spots on red circles, Fig. 3) show

TABLE 4. DATA SETS DISPLAYED IN "ACTIVE EARTH" EXAMPLE

	Data set location	Legend location in /norpac/data/legends/act_erth	Data description
14	terranes/nor_pac/flts_pos	flts.avl	Postaccretionary faults, onshore and offshore onshore faults
13	terranes/nor_pac/flts_lnd	flts.avl	Onshore faults
12	terranes/nor_pac/flts_ocn	flts.avl	Offshore faults
11	volcano	volcano.avl	Volcanoes active during the Holocene
10	examples/mag35 derived from seismcty/isc_cat	mag35.avl	Earthquakes of 3.5 ≤ magnitude < 4.0, extracted from Catalog of Global Seismicity, 1964–1991
9	examples/mag40 derived from seismcty/isc_cat	mag40.avl	Earthquakes of 4.0 ≤ magnitude < 4.5, extracted from Catalog of Global Seismicity, 1964–1991
8	examples/mag45 derived from seismcty/isc_cat	mag45.avl	Earthquakes of 4.5 ≤ magnitude < 5.0, extracted from Catalog of Global Seismicity, 1964–1991
7	examples/mag50 derived from seismcty/isc_cat	mag50.avl	Earthquakes of 5.0 ≤ magnitude < 5.5, extracted from Catalog of Global Seismicity, 1964–1991
6	examples/mag55 derived from seismcty/isc_cat	mag55.avl	Earthquakes of 5.5 ≤ magnitude < 6.0, extracted from Catalog of Global Seismicity, 1964–1991
5	examples/mag60 derived from seismcty/isc_cat	mag60.avl	Earthquakes of 6.0 ≤ magnitude < 6.5, extracted from Catalog of Global Seismicity, 1964–1991
4	examples/mag65 derived from seismcty/isc_cat	mag65.avl	Earthquakes of magnitude ≥ 6.5, extracted from Catalog of Global Seismicity, 1964–1991
3	topogrfy/raster/hillshad	hillshad.avl	Hillshade derived from 30″-sampled land topography illuminated by sun from 315° at 45° elevation
2	topogrfy/vector/nor_pac	nor_pac.avl	Circum–North Pacific region (1:10 000 000)
1	gravity/geosat	geosat.avl	Satellite-derived free-air gravity, offshore areas only

not only the Pacific Ring of Fire, but also the young centers of the Bering Strait basaltic province that are associated with sparse crustal seismicity in the Seward Peninsula (Fig. 3). In many cases, the locations of earthquakes should match modern topographic lineaments; however, the discrepancies between the earthquake, topography, and fault data sets, consistent with the previously discussed errors in the digitized /topogrfy/vector/nor_pac contours, can cause apparent offsets. For example, the Kaltag fault (K in Fig. 3) plots ~5 km southeast of the Unalakleet River and associated topographic lineament where they and the fault enter Norton Sound, and the Togiak-Tikchik strand of the Denali fault (T in Fig. 3) plots ~5 km southeast of its associated topographic lineament and Togiak Lake, close to its entry into Bristol Bay near Hagemeister Island.

"Magnetic-Lithologic Correlation" example, /data/examples/mag_lith (Figure 4)

Table 5 lists the data sets that are superimposed in this view; the higher numbered layers above and partly concealing the lower numbered layers (Fig. 4). These coverages are chosen to exemplify the possible use of GIS databases to correlate different data sets, and then to use these correlations to interpret other areas. The correlation we seek to explore here is that of igneous rocks with magnetic anomalies. Although the relation is not ubiquitous, igneous rocks, particularly mafic and ultramafic varieties, generally have more magnetite and other magnetic minerals than sedimentary rocks, and so are associated with high magnetic anomalies. Although more-magnetic lithologies may be present

at depth beneath sedimentary basins, their associated anomalies are dampened by the presence of the sedimentary rocks above them; below the mid-crust most rocks lose their magnetic signature due to temperatures being above the Curie limit of magnetization.

Layers 1, 2, 4, and 5 in this view are all magnetic coverages, displayed with data of higher quality and higher resolution toward the top of the stack of layers. The shallowest magnetic layers, 4 and 5, are the land coverages for Alaska and Russia, typically derived from dense flight-line coverage. The deeper layers, 1 and 2, have their lower-resolution land parts obscured by the upper layers, but provide coverage offshore that is typically less well sampled by sparse marine geophysical tracklines. Placing topographic contours (/topogrfy/vector/nor_pac) above the marine magnetic coverages, but below the land magnetic coverages, provides bathymetric contours that identify the location of the shelf edge, without obscuring or confusing the land areas on which the geology is overlain. Because both russ_mag and ak_mag obscure the coastlines and some nearshore bathymetry, the World Vector Shoreline is displayed as layer 8. Note that where offshore data were sampled along a single ship track, even though the information is strictly correct only along that line, the grids typically show data several grid cells wide, and so appear on our image as narrow bands of color-coded information rather than as lines (outer Bristol Bay and the northeast Pacific Ocean are areas where the magnetic information comes from sparse ship tracks). For all four magnetic coverages a dichromatic red-blue color scale is used (as for the gravity coverage in the "Active Earth" example). In order to visually match

TABLE 5. DATA SETS DISPLAYED IN "MAGNETIC-LITHOLOGIC CORRELATION" EXAMPLE

	Data set location	Legend location in /norpac/data/legends/mag_lith	Data description
9	cultural/latlong5	shorline.avl	Latitude and longitude grid
8	topogrfy/vector/shorline	shorline.avl	World Vector Shoreline
7	examples/rus_geol.shp	igneous.avl	Geologic map of Russia with igneous lithologies selected
6	examples/ak_geol.shp	igneous.avl	Geologic map of Alaska with igneous lithologies selected
5	magnetic/russ_mag	russ_mag.avl	Magnetic map of Russia
4	magnetic/ak_mag	ak_mag.avl	Magnetic map of Alaska (largely onshore)
3	topogrfy/vector/nor_pac	nor_pac.avl	Circum–North Pacific region (1:10 000 000)
2	magnetic/dnag_mag	dnag_mag.avl	Magnetic map of North America, onshore and offshore
1	magnetic/arct_mag	arct_mag.avl	Magnetic map of the Arctic (largely offshore)

the different coverages where they abut, because each magnetic coverage has a slightly different mean and range we have defined four different legends in which identical colors correspond to identical ranges of standard deviations away from the specific data mean.

In order to compare geology with magnetic data we need to display two areal coverages so that both are visible simultaneously. Hence the onland part of the geological map of Russia (polygons of /geology/onshore/rus_geol with attribute agecod ≤ 256) and the entire geological map of Alaska (/geology/onshore/ak_geol) are shown as transparent overlays over the magnetic grids. Because it is not possible to label individual geological units without further obscuring the magnetic display, we chose instead to highlight all igneous lithologies in yellow, and to shade gray all other lithologies. This example does not present well in norpac2.apr, for which Spatial Analyst is not available to provide the dichromatic red-blue color scale for the magnetic field. The Russian geology and Alaskan geology data sets are exported as shape files, then igneous units (intrusive, volcanic, and ophiolitic) were selected by Boolean query of the legends, and given the value "True" for the new polygon attribute (Igneous) in attribute files ak_geol.dbf and rus_geol.dbf. These igneous units are, on the Alaska map, any unit with its identifying attribute (mapunit_ab) containing the letter "i," except units identified as "Ice" (i.e., all intrusives), and any identifier containing the letter "v" (i.e., volcanics), or ending in the letter "u" (i.e., ultramafics, but avoiding abbreviations for time periods, "Upper ***") On the Russia map, these are lithologic units that did not contain the label "acoustic" (corresponding to undifferentiated offshore acoustic basement) or the label "continential" (*sic*) (corresponding to platform deposits); all other units are labeled "False."

When the areas of igneous outcrop in the southern Brooks Ranges or in Chukotka are examined, this "Magnetic-Lithologic Correlation" example shows that igneous rocks in these regions correlate with high-spatial-frequency, high-intensity magnetic anomalies. In contrast, areas underlain by sedimentary rocks, such as the North Slope of Alaska, have quieter magnetic signatures. Can this correlation be taken offshore, into regions that are poorly mapped geologically? On the Bering Shelf the style of magnetic anomalies varies widely, and has been described both as showing a fairly simple pattern (low-high-low from north to south; Marlow et al., 1976), the high-anomaly zone marking a submarine, Late Cretaceous–early Tertiary volcanic belt (Worrall, 1991) and as a more complex pattern of magnetic domains interpretable as a pattern of accreted terranes (McGeary and Ben-Avraham, 1981), although the degree of sedimentary burial of the basement also contributes significantly to the observed magnetic signature. Thus Worrall (1991) identified as volcanics the zone from the Anadyr basin, including the Saint Matthew–Nunivak arch, that broadens eastward to include the Goodnews arch and much of Bristol Bay. McGeary and Ben-Avraham (1981) interpreted the same data in more detail, e.g., describing a small region of lower anomalies within the high-anomaly zone midway between Nunivak Island and the Pribilof Islands. Marlow et al. (1976) noted that the northern boundary of the region of high-amplitude, high-frequency magnetic anomalies is close to the southern boundary of the Saint Matthew basin–Hall basin, whereas the southern limit of the high-intensity magnetic belt generally is coincident with the northern boundaries of the large outer shelf basins such as Saint George and Navarin, raising the possibility that any magnetic-lithologic correlation in this region is inappropriate. A GIS database and display such as ours cannot resolve such uncertainties, but they can make it easier to understand the limits of the available data and to test the possibilities.

ASSOCIATED STUDIES

This GIS compilation on this CD-ROM is part of a project on the major mineral deposits, metallogenesis, and tectonics of the Russian Far East, Alaska, and the Canadian Cordillera. The project aims to provide critical information on bedrock geology and geophysics, tectonics, major metalliferous mineral resources, metallogenic patterns, and crustal origin and evolution of mineralizing systems for these regions. Published major companion studies for the project are: (1) a report on the metallogenesis of mainland Alaska and northeastern Russia (Nokleberg et al., 1993); (2) a tectonostratigraphic terrane map of the Circum-North Pacific at 1:5 000 000 scale with a detailed explanation of map units and stratigraphic columns (Nokleberg et al., 1994a); (3) a tectonostratigraphic terrane map of Alaska at 1:2 500 000 scale (Nokleberg et al., 1994b); (4) a summary terrane map of the Circum–North Pacific at 1:10 000 000 scale (Nokleberg et al., 1997a); (5) detailed tables of mineral deposits and placer districts for the Russian Far East, Alaska, and the Canadian Cordillera in paper format (Nokleberg et al., 1996) and in CD-ROM format (Nokleberg et al., 1997b); (6) a GIS presentation of a summary terrane map, mineral deposit maps, and metallogenic belt maps of the Russian Far East, Alaska, and the Canadian Cordillera (Nokleberg et al., 1998); and (7) a study of the Phanerozoic tectonic evolution of the Circum–North Pacific (Nokleberg et al., 2000).

ACKNOWLEDGMENTS

We thank the many organizations that contributed data, directly or indirectly, to this compilation, including the following: Alaska Biological Science Center, Alaska Division of Oil and Gas, Defense Mapping Agency, Eros Data Center, European Space Agency, Exxon Production Research Company, Geological Society of America (GSA), Geological Survey of Canada (GSC), Geological Survey of Japan, International Seismological Centre (ISC), Japan Oceanographic Data Center, Lamont-Doherty Earth Observatory, Michigan State University, Ministry of Geology (former U.S.S.R.), National Aeronautics and Space Administration (NASA), National Earthquake Information Center (NEIC), National Geophysical Data Center (NGDC), National Ocean Service, National Oceanic and Atmospheric Administration (NOAA), Naval Research Laboratory (NRL), Research Information Center (GlavNIVC, Russian Ministry of Natural Resources [RMNR]), Russian Academy of Sciences, Scripps Institution of Oceanography, Smithsonian Institution Global Volcanism Program, Society of Exploration Geophysicists, Stanford University, United Nations Environment Programme/Global Resource Information Database (UNEP/ GRID), U.S. Agency for International Development (USAID), U.S. Geological Survey (USGS), and the University of Alaska, Fairbanks.

We thank the following individuals (listed here alphabetically) who contributed data to this compilation. Douglas S. Aitken, Brian S. Bennett, U.S. Geological Survey, Menlo Park, California, USA; Keri L. Brennan, Chapleau, Ontario, Canada; Ronald W. Buhmann, National Geophysical Data Center, National Oceanic and Atmospheric Administration (NOAA), Boulder, Colorado, USA; Stanislav G. Byalobzhesky, Russian Academy of Sciences, Magadan, Russia; Kazuya Fujita, Michigan State University, East Lansing, Michigan, USA; Steven P. Gordey, Geological Survey of Canada, Vancouver, Canada; Arthur Grantz, Department of Geological and Environmental Sciences, Stanford University, Stanford, California, USA; Roger A. Hansen, University of Alaska, Fairbanks, Alaska, USA; Paul P. Hearn Jr., U.S. Geological Survey, Reston, Virginia, USA; Thomas G. Hildenbrand, U.S. Geological Survey, Menlo Park, California, USA; Allen M. Hittelman, National Geophysical Data Center, NOAA, Boulder, Colorado, USA; J. Murray Journeay, Geological Survey of Canada, Vancouver, Canada; Alexander I. Khanchuk, Russian Academy of Sciences, Vladivostok, Russia; Boris Khlebnikov, Research Information Center, GlavNIC, Russian Ministry of Natural Resources, Moscow, Russia; Kevin Mackey, Michigan State University, East Lansing, Michigan, USA; D. Paul Mathieux, U.S. Geological Survey, Reston, Virginia, USA; Dean L. Miller, U.S. Geological Survey, Menlo Park, California, USA; James W.H. Monger, Geological Survey of Canada, Vancouver, Canada; Boris A. Natal'in, Russian Academy of Sciences, Khabarovsk, Russia; Leonid M. Parfenov, Russian Academy of Sciences, Yakutsk, Russia; Richard W. Saltus, U.S. Geological Survey, Denver, Colorado, USA; David T. Sandwell, University of California San Diego, San Diego, California, USA; Lee Siebert, Tom Simkin, Smithsonian Institution, Washington, D.C., USA; Gregory Ulmishek, U.S. Geological Survey, Reston, Virginia, USA; and Frederic H. Wilson, U.S. Geological Survey, Anchorage, Alaska, USA.

We particularly thank the contributors for providing their data, assistance in compiling, and reviewing of the GIS compilation. We thank V.J.S. Grauch (U.S. Geological Survey, Denver, Colorado) for assistance in obtaining the aeromagnetic data set of the former Soviet Union from the National Geophysical Data Center. Compilation of the digital data sets was undertaken at the Stanford University School of Earth Sciences Geographic Information Systems Laboratory, http://pangea.Stanford.EDU/ gis/, with "Research Experience for Undergraduate" funds from National Science Foundation grant EAR-93-17087 from the Continental Dynamics Program. We thank Jeffrey C. Wynn, Frances R. Mills, Dogan Seber, and an anonymous reviewer for their constructive input.

REFERENCES CITED

Alaska Biological Science Center, 1998, Bering and Chukchi Sea Ecosystem Database, Bathymetry Coverages: http://www.absc.usgs.gov/research/ bering/bathy/index.htm.

Anonymous, 1995, Magnetic anomalies and tectonic elements of Northeast Eurasia, 1:10 000 000: Geological Survey of Canada Open-File 2574, and http://agcwww.bio.ns.ca/pubprod/eurasia.gif.

Becker, T.W., and Braun, A., 1998, New program maps geoscience datasets interactively: Eos (Transactions, American Geophysical Union), v. 79, p. 505, 508, and http://www.seismology.harvard.edu/~becker/igmt/.

Beikman, H.M., 1980, Geologic map of Alaska: U.S. Geological Survey, scale 1:2 500 000, 2 sheets.

Brocher, T.M., Allen, R.M., Stone, D.B., Wolf, L.W., and Galloway, B.K., 1995, Data report for onshore-offshore wide-angle seismic recordings in the Bering-Chukchi Sea, western Alaska and eastern Siberia: U.S. Geological Survey Open-File Report 95–0650, 57 p., and http://pangea.stanford .edu/~sklemp/bering_chukchi/alaska.overview.html.

Committee for the Gravity Anomaly Map of North America, 1987, Gravity anomaly map of North America: Boulder, Colorado, Geological Society of America, scale 1:5 000 000, 5 sheets.

Committee for the Magnetic Anomaly Map of North America, 1987, Magnetic anomaly map of North America: Boulder, Colorado, Geological Society of America, scale 1:5 000 000, 5 sheets.

ESRI, 1999, Getting to know ArcView GIS: The geographic information system (GIS) for everyone: Redlands, California, Environmental System Research Institute (ESRI), 3rd ed., 538 p. and CD-ROM.

Fliedner, M.M., and Klemperer, S.L., 1999, Composition of an island-arc: Wide-angle studies in the eastern Aleutian islands, Alaska: Journal of Geophysical Research, v. 104, p. 10667–10694.

Fujita, K., Mackey, K., Gunbina, L., and Gordeev, E., 2001, Seismicity catalog for Chukotka and the western Bering Sea: Manuscript, 3 p., contained on CD-ROM included with this volume in files /norpac/data/seismcty/explanat/chukseis.***.

Geological Survey of Japan and Coordinating Committee for Coastal and Offshore Geoscience Programmes in East and Southeast Asia (CCOP), 1996, Magnetic anomaly map of East Asia: Geological Survey of Japan Digital Geoscience Map 2 (P-1), scale 1:4 000 000, CD-ROM.

GlavNIVC, 1998, Natural resources GIS of Russia: Washington, D.C., American Geological Institute, CD-ROM.

Greninger, M.L., Klemperer, S.L., and Nokleberg, W.J., 1996, Geographic Information System (GIS) database of the geology, geophysics, deep-crustal structure and tectonics of the Russian Far East, Alaska, Canadian Cordillera and adjacent offshore regions [abs.]: Eos (Transactions, American Geophysical Union), v. 77, no. 46, p. F669, and http://geo.stanford.edu/~wabs/index.html.

Greninger, M.L., Klemperer, S.L., and Nokleberg, W.J., 1999, Geographic Information Systems (GIS) compilation of geologic and geophysical data for the circum–North Pacific: U.S. Geological Society Open-File Report 99–422, version 1.0.

Hansen, R.A., Rowe, C.R., Fogelman, K., and staff of the Alaska Earthquake Information Center, 1999, Catalog for seismic events in Alaska: 1898 through 1998, compiled from State Seismologist's Report, Earthquakes in Alaska: Fairbanks, University of Alaska, Earthquake Information Center, Geophysical Institute, http://www.aeic.alaska.edu/.

Hittelman, A.M., Kinsfather, J.O., and Meyers, H., 1989, Geophysics of North America CD-ROM and User's Manual: Boulder, Colorado, National Oceanic and Atmospheric Administration, National Geophysical Data Center.

Hittelman, A.M., Dater, D.T., Buhmann, R.W., and Racey, S.D., 1994, Gravity CD-ROM and User's Manual: Boulder, Colorado, National Oceanic and Atmospheric Administration, National Geophysical Data Center, p. 39.

Holbrook, W.S., Lizarralde, D., McGeary, S., and Bangs, N., 1999, Structure and composition of the Aleutian island arc and implications for continental crustal growth: Geology, v. 27, p. 31–34, and http://www.gg.uwyo.edu/faculty/holbrook/aleut/aleut.html.

Hutchinson, S. and Daniel, L., 2000, Inside ArcView GIS: Albany, New York, OneWord Press, 3rd ed., 488 p. and CD-ROM.

Journeay, J.M., and Williams, S.P., 1995, GIS map library, a window on Cordilleran geology: Computer data and programs: Canada Geological Survey Open-File 2948, CD-ROM.

Macnab, R., Verhoef, J., Roest, W., and Arkani-Hamed, J., 1995, New database documents the magnetic character of the Arctic and North Atlantic: Eos (Transactions, American Geophysical Union), v. 76, p. 449, 458.

Marlow, M.S., Scholl, D.W., Cooper, A.K., and Buffington, E.C., 1976, Structure and evolution of Bering Sea shelf south of St. Lawrence Island: American Association of Petroleum Geologists Bulletin, v. 60, p. 161–183.

McGeary, S.E., and Ben-Avraham, Z., 1981, Allochthonous terranes in Alaska: Implications for the structure and evolution of the Bering Sea shelf: Geology, v. 9, p. 608–614.

Moore, G.W., 1990, Geographic map of the circum–Pacific region, Arctic sheet: Circum-Pacific Council for Energy and Mineral Resources, scale 1:10 000 000, 1 sheet.

Nalivkin, D.V., editor, 1994, Geologic map of Russia: USSR Research Geological Institute (VSEGEI) in collaboration with institutes, research institutes and manufacturing associates of the USSR Ministry of Geology, scale 1:2 500 000, 16 sheets.

NOAA, 1998, Data Announcement 88–MGG-02, Digital relief of the surface of the earth: Boulder, Colorado, National Oceanic and Atmospheric Administration, National Geophysical Data Center, and http://web.ngdc.noaa.gov/mgg/global/etopo5.HTML.

Nokleberg, W.J., Bundtzen, T.K., Grybeck, D., Koch, R.D., Eremin, R.A., Rozenblum, I.S., Sidorov, A.A., Byalobzhesky, S.G., Sosunov, G.M., Shpikerman, V.I., and Gorodinsky, M.E., 1993, Metallogenesis of mainland Alaska and the Russian Northeast: Mineral deposit maps, models, and tables, metallogenic belt maps and interpretation, and references cited: U.S. Geological Survey Open-File Report 93–339, 222 p., 1 map, scale 1:4 000 000, and 5 maps, scale 1:10 000 000.

Nokleberg, W.J., Parfenov, L.M., and Monger, J.W.H., and Baranov, B.V., Byalobzhesky, S.G., Bundtzen, T.K., Feeney, T.D., Fujita, K., Gordey, S.P., Grantz, A., Khanchuk, A.I., Natal'in, B.A., Natapov, L.M., Norton, I.O., Patton, W.W., Jr., Plafker, G., Scholl, D.W., Sokolov, S.D., Sosunov, G.M., Stone, D.B., Tabor, R.W., Tsukanov, N.V., Vallier, T.L., and Wakita, K., 1994a, Circum–North Pacific tectono-stratigraphic terrane map: U.S. Geological Survey Open-File Report 94–714, scale 1:5 000 000, 2 sheets, and scale 1:10 000 000, 2 sheets, 211 p.

Nokleberg, W.J., Moll-Stalcup, E.J., Miller, T.P., Brew, D.A., Grantz, A., Reed, J.C., Jr., Plafker, G., Moore, T.E., Silva, S.R., and Patton, W.R., Jr., 1994b, Tectonostratigraphic terrane and overlap assemblage of Alaska: U.S. Geological Survey Open-File Report 94–194, scale 1:2 500 000, 1 sheet, 26 p.

Nokleberg, W.J., Bundtzen, T.K., Dawson, K.M., Eremin, R.A., Goryachev, N.A., Koch, R.D., Ratkin, V.V., Rozenblum, I.S., Shpikerman, V.I., Frolov, Y.F., Gorodinsky, M.E., Melnikov, V.D., Ognyanov, N.V., Petrachenko, E.D., Petrachenko, R.I., Pozdeev, A.I., Ross, K.V., Wood, D.H., Grybeck, D., Khanchuk, A.I., Kovbas, L.I., Nekrasov, I.Ya., and Sidorov, A.A., 1996, Significant metalliferous lode deposits and placer districts for the Russian Far East, Alaska, and the Canadian Cordillera: U.S. Geological Survey Open-File Report 96–513–A (paper format), 385 p.

Nokleberg, W.J., Parfenov, L.M., Monger, J.W.H., Baranov, B.V., Byalobzhesky, S.G., Bundtzen, T.K., Feeney, T.D., Fujita, K., Gordey, S.P., Grantz, A., Khanchuk, A.I., Natal'in, B.A., Natapov, L.M., Norton, I.O., Patton, W.W., Jr., Plafker, G., Scholl, D.W., Sokolov, S.D., Sosunov, G.M., Stone, D.B., Tabor, R.W., Tsukanov, N.V., and Vallier, T.L., 1997a, Summary circum–North Pacific tectono-stratigraphic terrane map: Geological Survey of Canada Open-File Report 3428, scale 1:10 000 000, 1 sheet, and U.S. Geological Survey Open-File Report 96–727, scale 1:10 000 000, 1 sheet.

Nokleberg, W.J., Bundtzen, T.K., Dawson, K.M., Eremin, R.A., Goryachev, N.A., Koch, R.D., Ratkin, V.V., Rozenblum, I.S., Shpikerman, V.I., Frolov, Y.F., Gorodinsky, M.E., Melnikov, V.D., Ognyanov, N.V., Petrachenko, E.D., Petrachenko, R.I., Pozdeev, A.I., Ross, K.V., Wood, D.H., Grybeck, D., Khanchuk, A.I., Kovbas, L.I., Nekrasov, I.Ya., and Sidorov, A.A., 1997b, Significant metalliferous lode deposits and placer districts for the Russian Far East, Alaska, and the Canadian Cordillera: U.S. Geological Survey Open-File Report 96–513–B, CD-ROM.

Nokleberg, W.J., West, T.D., Dawson, K.M., Shpikerman, V.I., Bundtzen, T.K., Parfenov, L.M., Monger, J.W.H., Ratkin, V.V., Baranov, B.V., Byalobzhesky, S.G., Diggles, M.F., Eremin, R.A., Fujita, K., Gordey, S.P., Gorodinskiy, M.E., Goryachev, N.A., Frolov, Y.F., Grantz, A., Khanchuck, A.I., Koch, R.D., Natal'in, B.A., Natapov, L.M., Norton, I.O., Patton, W.W., Jr., Plafker, G., Pozdeev, A.I., Rozenblum, I.S., Scholl, D.W., Sokolov, S.D., Sosunov, G.M., Stone, D.B., Tabor, R.W., Tsukanov, N.V., and Vallier, T.L., 1998, Summary terrane map, mineral deposit maps, and metallogenic belt maps of the Russian Far East, Alaska, and the Canadian Cordillera: U.S. Geological Survey Open-File Report 98–136, CD-ROM.

Nokleberg, W.J., Parfenov, L.M., Monger, J.W.H., Norton, I.O., Khanchuk, A.I., Stone, D.B., Scotese, C.R., Scholl, D.W., and Fujita, K., 2000, Phanerozoic tectonic evolution of the circum–North Pacific: U.S. Geological Survey Professional paper 1626, 122 p.

Saltus, R.W., and Simmons, G.C., 1997, Composite and merged aeromagnetic data for Alaska: A web site for distribution of gridded data and plot files:

U.S. Geological Survey Open-File Report 97–520, p. 14, and http://minerals.cr.usgs.gov/publications/ofr/97–520/alaskamag.html.

Simkin, T., and Siebert, L., 1994, Volcanoes of the world: A regional directory, gazetteer, and chronology of volcanism during the last 10,000 years: Tucson, Arizona, Geoscience Press, 349 p.

Simkin, T., Unger, J.D., Tilling, R.I., Vogt, P.R., and Spall, H., 1994, This dynamic planet: World map of volcanoes, earthquakes, impact craters, and plate tectonics: Denver, Colorado, U.S. Geological Survey, scale 1:30 000 000, 1 sheet, and http://www.volcano.si.edu/gvp/.

Smith, W.H.F., and Sandwell, D.T., 1997, Global seafloor topography from satellite altimetry and ship depth soundings: Science, v. 277, p. 1956–1962, and http://topex.ucsd.edu/marine_grav/mar_grav.html.

Ulmishek, G., 1998, Natural resources GIS of Russia: Washington, D.C., American Geological Institute, CD-ROM.

U.S. Geological Survey, 1997, GTOPO30 Global 30 Arc Second Elevation Data Set, at http://edcwww.cr.usgs.gov/landdaac/gtopo30/gtopo30.html.

Verhoef, J., Roest, W.R., Macnab, R., Arkani-Hamed, J., and members of the Project Team, 1996, Magnetic anomalies of the Arctic and North Atlantic Oceans and adjacent land areas: Geological Survey of Canada Open-File 3125a, CD-ROM, and at http://agcwww.bio.ns.ca/pubprod/of3125etc.html.

Wessel, P., and Smith, W.H.F., 1995, New version of the Generic Mapping Tools released: Eos (Transactions, American Geophysical Union), v. 76, p. 329.

Whiteside, L.S., Dater, D.T., Dunbar, P.K., Racey, S.D., Buhmann, R.W., and Hittelman, A.M., 1996, Earthquake seismicity catalog volumes 1 and 2, Global and regional, 2150 B.C.–1996 A.D., Volume 2: Boulder, Colorado, National Geophysical Data Center, 8 p. and 2 CD-ROMs.

Worrall, D.M., 1991, Tectonic history of the Bering Sea and the evolution of Tertiary strike-slip basins of the Bering Shelf: Geological Society of America Special Paper 257, 120 p.

MANUSCRIPT ACCEPTED BY THE SOCIETY MAY 15, 2001.

Index